The Third Hutton Symposium
on the Origin of Granites and Related Rocks

Proceedings of a symposium held at the
University of Maryland at College Park
27 August to 2 September 1995

Edited by

M. Brown and P. A. Candela
Department of Geology
University of Maryland at College Park
College Park, Maryland 20742

D. L. Peck
U.S. Geological Survey
National Center, MS 959
Reston, Virginia 22092

W. E. Stephens
Department of Geology
University of St. Andrews
Purdie Building
St. Andrews, Fife KY16 9ST, U.K.

R. J. Walker and E-an Zen
Department of Geology
University of Maryland at College Park
College Park, Maryland 20742

Copublished in volume format by arrangement with,
and with the permission of, the Royal Society of Edinburgh

SPECIAL PAPER

315

1996

The papers in this volume were originally published together as *Transactions of the Royal Society of Edinburgh: Earth Sciences,* Volume 87, Parts 1 and 2 (ISSN 0263-5933). Copies of that issue are available from the Royal Society of Edinburgh, 22 George Street, Edinburgh EH2 2PQ, Scotland.

Copublished as a limited edition in volume format in the United States by the Geological Society of America, Inc., 3300 Penrose Place, P.O. Box 9140, Boulder, Colorado 80301

Printed in Huddersfield, U.K., and bound in the U.S.A.

Library of Congress Cataloging-in-Publication Data
Hutton Symposium on the Origin of Granites and Related Rocks (3rd :
   1995 : College Park, Md.)
      The Third Hutton Symposium on the Origin of Granites and Related
   Rocks : proceedings of a symposium held at the University of
   Maryland at College Park, 27 August to 2 September 1995 / edited by
   M. Brown ... [et al.].
         p.   cm. -- (Special paper ; 315)
      "Papers in this volume were originally published together as
   Transactions of the Royal Society of Edinburgh: Earth sciences,
   volume 87, parts 1 and 2" -- T.p. verso.
      Includes bibliographical references and index.
      ISBN 0-8137-2315-9
      1. Granite--Congresses. I. Brown, Michael, 1947 Mar. 19-
   II. Title.   III. Series: Special papers (Geological Society of
   America) ; 315.
   QE462.G7H88   1995b
   552'.3--dc20   96-30882
                                                              CIP

Cover Photo: Granite as far as the eye can see. This view is toward the northeast, up Tenaya Canyon, from the top of Half Dome (elevation 2,698 m), Yosemite National Park, California. Most of the rocks in view are part of the 1,200 km^2 Tuolumne Intrusive Suite (ca. 87 Ma), a classic example of a normally zoned group of five plutons in the Sierra Nevada batholith. The vertical relief in the field of view includes more than 1,000 m of nearly continuous exposure. Photograph courtesy of Phil Piccoli, Department of Geology, University of Maryland at College Park.

10 9 8 7 6 5 4 3 2 1

Third Hutton Symposium
The Origin of Granites and Related Rocks

Proceedings of a Symposium held at the
University of Maryland at College Park
27 August to 2 September 1995

Edited by M. Brown, P. A. Candela, D. L. Peck, W. E. Stephens,
R. J. Walker and E-an Zen

CONTENTS

Transactions of the Royal Society of Edinburgh: Earth Sciences, **87**, v. 1996

Preface

Michael Brown

Towards the end of his life, James Hutton (1726–97) set forth his 'Theory of the Earth' in which he recognised the length of geological time, the uniformity of geological processes through time, and the role of igneous activity, sedimentation and erosion in forming the Earth as we know it. The four-part two-volume book was preceded by a public presentation of many of these ideas in 1785, in a paper on 'Examination of the System of the Habitable Earth with Regard to its Duration and Stability' read to Fellows of the Royal Society of Edinburgh on March 7 and April 4. Subsequently, the paper was published in Volume I of the *Transactions of the Royal Society of Edinburgh* in 1788. In the ten years between presentation of the paper to the Royal Society of Edinburgh and publication of his book, Hutton occupied himself in accumulating the 'proofs and illustrations' of his *Theory of the Earth* that enhance the fuller exposition of the *Theory* as presented in the book. For the three years that followed his 1785 presentations to the Royal Society of Edinburgh, Hutton examined several Scottish granites seeking key field relationships to solve questions in his mind concerning the origin of granite. The results were presented to Fellows of the Royal Society of Edinburgh on 4 January 1790, and subsequently published in Volume III of the *Transactions* in 1794. Hutton found unequivocal evidence to show that molten granite had invaded the surrounding stratified rock. This represents the first demonstration of the class of rocks that we now call *intrusive igneous rocks*. The subject of the origin of granites and related rocks has remained of intense topical interest since the time of Hutton. Implicit in Hutton's conclusions concerning granite are the ideas of crustal anatexis, melt segregation, magma transfer and granite emplacement into lower-grade upper-crustal anatexis, melt segregation, magma transfer and granite emplacement into lower-grade upper-crustal rocks. The *Theory of the Earth* was published in 1795, so it was appropriate last summer to assess how far we had progressed in our understanding of these processes in 200 years.

An outgrowth of the successful symposium on 'The Origin of Granites' organised by the Royal Society of Edinburgh and the Royal Society of London in 1987 to celebrate the bicentenary of the work of James Hutton, the third meeting in the Hutton Symposium Series was held in College Park, Maryland, from the 27 August to 2 September 1995. An Organising Committee of Michael Brown (Chair), P.A. Candela, J. Ehlen, J. Hammarstrom, E-an Zen, E.J. Krogstad, E.L. McLellan, D. Peck, P.M. Piccoli, K. Ratajeski, S. Shirey, P. Tomascak and R.J. Walker indentified several themes as foci for the Symposium. Oral presentations were invited to contribute to these themes, and these are the papers presented in this volume. Most contributed papers were presented in poster form during informal afternoon sessions, and two evenings were devoted to discussions on *current debates and future research directions*. Partly to identify themes for these two evening discussion sessions, but also to promote an international exchange of ideas, D.B. Clarke established the E-mail Granite Research Group, one of the most active discussion networks on the internet and a direct outgrowth of the Hutton Symposium Series.

Detailed field studies, such as those completed by Hutton 200 years ago, in combination with the results of several decades of experimental investigation have produced a firm foundation on which our understanding of crustal melting is built. During the decade of the Hutton Symposium Series, improvements in analytical and experimental techniques, coupled with rigorous theoretical underpinning, have led to significant advances in our understanding of the physics and chemistry of melts, and of the processes that are responsible for the chemical variation within and among granites. Papers that address the generation and modification of granitic magmas include those by Thompson, Patino Douce, Barboza & Bergantz, Bea, Watson, Holtz *et al.*, Dingwell *et al.*, Nekvasil & Carroll, Hogan, Chappell, Collins, Nakajima, Rapela & Pankhurst, Drummond *et al.*, Flinders & Clemens, Poli *et al.*, Wiebe, Baker & Rutherford, Pichavant *et al.*, Krogstad & Walker and Johnson *et al.*

There has been considerable debate concerning melt segregation and the mechanisms of ascent of granitic magma through the continental crust. The role of deformation in melt segregation is addressed in papers by Rushmer and Sawyer. During the past decade support for diapirism as the ascent mechanism has been waning as emphasis has been increasingly placed on the relationship between plutons and shear zones. More recently, magma transfer through the crust in dykes has become popular. The controversy is apparent in the papers of Weinberg and Petford. During magma emplacement crust must be displaced at a rate equal to the growth of the pluton. Further, the evidence cited by Hutton and many others subsequently shows a variety of small-scale mechanisms that may act in combination to enhance the space available for magma accumulation at the emplacement site. Multiple horizontal and vertical material transfer processes acting in different proportions from pluton to pluton may be more realistic than the extremes implied by a single model either of ascent or of emplacement, as argued by Paterson *et al.* Mineral equilibrium methods to estimate depth of emplacement are reviewed by Anderson.

Our knowledge of the internal structure of granite-related pegmatites and controls on processes such as fractional crystallisation is good in some systems, but our knowledge of various pegmatite groups, and the ultimate origin of granite-related pegmatite melts is less clear. Granitic pegmaties are reviewed in the paper by London. The role of fluids has been realised for some time. Granite-related hydrothermal systems and mineral deposits (granite porphyries and granite skarns) develop from the synergistic effects of a multitude of processes and conditions. All mineral deposits have a common requirement—a process by which an element is brought from a dispersed state to a state of higher concentration. Many of these issues are addressed in the papers by Hanson, Barton, Blevin *et al.*, Lowenstern & Sinclair and Pichavant *et al.* Understanding those processes that produce economically important mineral deposits is fundamental if geology is to continue to support the economic development of humankind. Finally, D.B. Clarke assesses the status of granite science two centuries after publication of *Theory of the Earth*.

For the 1995 meeting, we gratefully acknowledge the following sponsors: Department of Geology, University of Maryland at College Park; U.S. Geological Survey, Reston; and Department of Terrestrial Magnetism and Geophysical Laboratory, Carnegie Institution of Washington. The Organising Committee acknowledges a grant from the National Science Foundation to subsidise the cost of travel to the meeting by U.S.-based graduate students, and the assistance of staff in the Department of Geology at UMCP. The Fourth Hutton Symposium on the Origin of Granites and Related Rocks will be held in France in 1999, *plus ça change plus c'est la même chose*!

Transactions of the Royal Society of Edinburgh: Earth Sciences, **87**, 1–10, 1996

Fertility of crustal rocks during anatexis

Alan Bruce Thompson

ABSTRACT: After many years of systematic experimental investigations, it is now possible to quantify the conditions for optimum fertility to melt production of most common crustal rock types as functions of temperature and a_{H_2O}, to about 30 kbar pressure. Quartzo-feldspathic melting produces steady increases in melt proportion with increasing temperature. The exact melt fraction depends on the mineral mode relative to quartz–feldspar eutectics and the temperatures of mica dehydration melting reactions. Mica melting consumes SiO_2 from residual quartz during the formation of refractory Al_2SiO_5, orthopyroxene, garnet or cordierite.

A simple graphical interpretation of experimental results allows a deduction of the proportions of mica and feldspar leading to optimum fertility. In effect, the mica dehydration melting reactions, at specific pressure and a_{H_2O}, are superimposed on quartz–feldspar melting relations projected onto Ab–An–Or. Fertility to melt production varies with the mica to feldspar ratio and pressure. Pelites are more fertile than psammites at low pressures (e.g. 5 kbar), especially if they contain An_{40} to An_{50} plagioclase. At higher pressure (e.g. 10–20 kbar) and for rocks containing albitic plagioclase, psammites are more fertile than pelites. For a typical pelite (e.g. with An_{25} at 20 kbar), the cotectic with muscovite lies at higher a_{H_2O} (≈ 0.6) and X_{Ab} (≈ 0.42) than with biotite ($a_{H_2O} \approx 0.35$; $X_{Ab} \approx 0.32$), thus dehydration melting of muscovite requires 10% more plagioclase for fertility than does biotite.

The first melts from dehydration melting of muscovite (with $Plg + Qtz$) are more sodic and form at lower temperatures than the first melts from $Bio + Plg + Qtz$. With increasing pressure, to at least 30 kbar, granite minimum and mica dehydration melts become more sodic. This indicates that a_{H_2O} of such melts is greater than 0.3.

KEY WORDS: optimum melt production, crustal anatexis, dehydration melting, metasediments, mica–feldspar ratio, pressure, a_{H_2O}.

Calculated pressure–temperature–time (P–T–t) paths for continental collision (e.g. Thompson & Connolly 1995) have shown that the attainment of temperatures in excess of 800°C within the continental crust requires unusual tectonic circumstances. Several experimental investigations have determined that at 10 kbar (≈ 35 km) the dehydration melting solidus (with albite and quartz) for muscovite is near 725°C, and that for biotite is above 780°C. Thus in continental collision orogens only small melt fractions are likely even in quartzo-feldspathic rocks, which represent the lowest temperature crustal anatexites. It is important to note here that studies using trace elements in mica and feldspar (e.g. Harris & Inger 1992) to evaluate the degree of melting also suggest anatectic melt proportions of 10–15% to be responsible for some leucogranites.

Experiments on pelite (Vielzeuf & Holloway 1988; Patiño Douce & Johnston 1991) and psammite compositions (Vielzeuf & Montel 1994; Patiño Douce & Beard 1995) have provided modal and chemical information on minerals and quenched melts, when the latter exceeded 20%. A few workers (LeBreton & Thompson 1988; Skjerlie & Johnston 1992, 1993; Gardien et al., 1995) have studied anatexis at low melt fractions. The principal experimental difficulty in obtaining compositions from small melt fractions arises because it is no trivial matter to analyse small regions of glass with an electron microprobe beam. Glass is easily partly vapourised and the chemistry from nearby or subsurface crystals is often included in the analysis. For future applications of petrological results to problems of crustal anatectic melt generation, it is important to obtain information on the distribution and ranges of composition of these small melt fractions.

The purpose of this paper is to obtain such details from the available experimental studies at various degrees of melting and to suggest ways in which the interpretation of rocks and experimental charges may be developed to fully quantify the thermal and chemical evolution of anatectic melts.

1. The problem of fertility (of crustal rocks to melt production)

Melt production depends on the solidus temperature for a particular mineral assemblage and the modal proportions compared with the nearest eutectic. The first melts generated from a wide range of rock types [granitoid intrusive and felsic extrusive rocks, pelitic and psammite (including greywacke) metasedimentary rocks] have compositions close to the minimum/eutectic of quartz–feldspar (Qtz–Fsp) assemblages of the 'granite' system (Ab + Or + Qtz; Tuttle & Bowen 1958; Luth et al. 1964; Huang & Wyllie 1975; Winkler, summary from 1979).

Water greatly lowers the melting temperature of these minerals because of its high solubility in silicate melts at high pressure. The compositions of the granite minimum move away from Qtz towards Fsp with increasing P_{H_2O}. Therefore the fertility to melt production (where large melt fractions are generated over the smallest temperature intervals) in quartzo-feldspathic rocks of different modes varies with depth in the continental crust.

The porosity of crystalline rocks in the middle and lower crust is so low that the amount of free H_2O expected to be available for H_2O-saturated melting is very small—insufficient to produce more than a few per cent of minimum melt (Thompson & Connolly 1995). The principal source for H_2O

is that stored in the crystal structures of micas and amphiboles. For the rock compositions of interest here we will concentrate on the fluid-absent (dehydration) melting of the two micas: muscovite and biotite. Even though Burnham (1967, 1979) has suggested that at the dehydration melting solidus for Mus + Plag + Qtz ($\approx 750°C$, 10 kbar) the X_w^m is about 0·6, and at the Bio + Plag + Qtz solidus (≈ 800–$850°C$ at 10 kbar) the X_w^m is about 0·35, we do not know how X_w^m relates to melt fraction. In any case, the fertility of rocks to crustal anatexis seems to be controlled very much by the proportion and composition of the micas and plagioclase undergoing melting with quartz. To simplify discussion, the symbol X_w^m is used to denote the mole fraction of H_2O in the melt and is considered equal to activity of H_2O in the melt (Burnham 1979). The symbol $a_{H_2O}^{fluid}$ is used to denote water activity in the fluid and can be changed due to dilution by another component. The symbol a_{H_2O} refers to water activity in the environment.]

Optimum fertility refers to a mineral mode (micas and feldspars present in eutectic proportion) that will produce the maximum amount of melt at, or just above, the solidus for the nearest eutectic. A 'fertile' composition which has the minerals of the eutectic reaction, but not in the eutectic proportion, will produce some melt at the dehydration melting solidus. Although precise modes can be deduced appropriate to rock fertility, for the purpose of the discussion in this paper pelites contain more mica than feldspar and quartz, whereas psammites contain more feldspar and quartz than mica. Much information about fertility to crustal anatexis is contained in experimental data on the change in composition and temperature of the 'granite' minimum, with increasing pressure and decreasing a_{H_2O}.

2. Change in composition of the haplogranite minimum and eutectic melt composition with pressure and a_{H_2O}

In their pioneering experimental work, Tuttle and Bowen (1958) showed that the minimum in Ab + Or + Qtz at H_2O saturation ($P_{total} = P_{H_2O}$) migrated with increasing pressure towards Ab from 1 to 3 kbar. Subsequent high-pressure experiments demonstrated the dramatic increase in the feldspar to quartz ratio with increasing pressure (from about Fsp/ Qtz $\approx 58/42$ at 1 atm to about 86/14 at 30 kbar, wt% units shown in Fig. 1). The experiments of Luth et al. (1964) from 4 to 10 kbar (see also Luth 1976) and the estimates of Huang

and Wyllie (1975) at 20–30 kbar indicated that the migration of the minimum from 5 to 30 kbar becomes more parallel to the Ab + Qtz sideline. These results are shown in Figure 1, together with various more recent studies on the system Ab + Or + Qtz + H_2O + CO_2 (Holtz & Johannes 1991; Ebadi & Johannes 1991).

Some simplified statements about the migration of the minimum/eutectic in the Ab + Or + Qtz projection at various a_{H_2O} may be made with these data. The H_2O saturated minimum/eutectic ($a_{H_2O} = 1$) migrates, with increasing pressure, to much lower temperatures and towards Ab_{25}, almost parallel to the Ab + Qtz sideline, i.e. from $T = 770°C$, $Ab_{30} Or_{31} Qtz_{39}$ wt% (500 bar P_{H_2O}, Tuttle & Bowen 1958: 34) to $T = 650°C$, $Ab_{67} Or_{13} Qtz_{20}$ wt% (30 kbar, Huang & Wyllie 1975). The anhydrous minimum ($a_{H_2O} = 0$), migrates with increasing pressure to higher temperatures and towards $Ab_{40}Or_{60}$, almost parallel to the Or + Qtz sideline, i.e. from $T = 960°C$, $Ab_{27}Or_{32}Qtz_{41}$ wt% (1 atm Schairer 1950; Bowen & Tuttle 1958) to $T = 1060°C$, $Ab_{31}Or_{48}Qtz_{21}$ wt% (10 kbar, Ebadi & Johannes 1991). For conditions of $a_{H_2O} < 1$, the migration of the minimum/eutectic composition in the Ab + Or + Qtz projection is almost radial from the low pressure minima towards the Ab + Or sideline.

An extremely important additional observation from Figure 1 is that irrespective of the a_{H_2O} the minimum/eutectic occurs at approximately the same Fsp/Qtz ratio at any particular pressure. The projected X_{Or} values corresponding to these shifts are shown in Figure 2a, where it can be seen that dP/dX_{Or} changes from negative to positive between $a_{H_2O} = 0·5$ and 0·3. The changes in the (Ab + Or)/Qtz ratio of the minimum/eutectic composition in the Ab + Or + Qtz projection is shown as a function of temperature in Figure 2b. Here the compositional shifts with changing pressure and a_{H_2O} can be compared with the compositions defining the 'thermal valley' at $P_{H_2O} = 1$ kbar (Bowen & Tuttle 1958, fig. 39). This simplified behaviour allows a straightforward thermodynamic treatment of the migration of the minimum temperature and composition with both pressure and a_{H_2O}, within a relatively small triangular volume of Ab + Or + Qtz + a_{H_2O} composition space.

3. Micas and haplogranites (Ab + Or + Qtz + H_2O) at various a_{H_2O}

There have been several experimental studies during the last decade in the H_2O undersaturated region of the system $NaAlSi_3O_8(Ab) + KAlSi_3O_8(Or) + SiO_2(Qtz) + H_2O$ + diluting components. The compositions of coexisting phases as functions of P, T, and a_{H_2O} are vital in understanding the dehydration melting reactions of micas.

3.1. Distinct P–T regions for minima and eutectica in Ab + Or + Qtz + H_2O at various values of a_{H_2O}

Luth et al. (1964) concluded that at H_2O-saturation the four-component system Ab + Or + Qtz + H_2O changed from minimum to eutectic melting at pressures above about 3.6 kbar. By including information on the pressure dependence of the alkali feldspar critical temperature, the interpretation of Stewart & Roseboom (1962) would displace this intersection to lower than 3 kbar P_{H_2O}. The P–T of the critical curve, presented by Waldbaum and Thompson (1969), allowed revision of this intersection point to about 2.5 kbar. The projection of the P–T of the alkali feldspar critical curve (AFC) into Ab + Or + Qtz phase relations, under anhydrous and H_2O saturation conditions, was shown by Huang and Wyllie (1975).

It is now possible to separate in P–T space a region of Ab + Or + Qtz + H_2O minima at higher temperatures than the critical curve from eutectic behaviour at lower temperatures

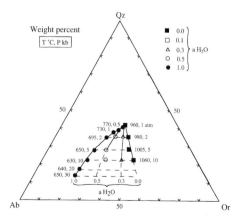

Figure 1 Minimum and eutectic compositions (wt%) determined at various water activities in the system Ab + Or + Qtz + H_2O (+CO_2). Modified from data summarised by Ebadi and Johannes (1991; figure 13: 293) with additional data for H_2O-saturated eutectics at 20 and 30 kbar from Huang and Wyllie (1975). It is easily seen that the ratio (Ab + Or)/Qtz increases with pressure, apparently independently of a_{H_2O}.

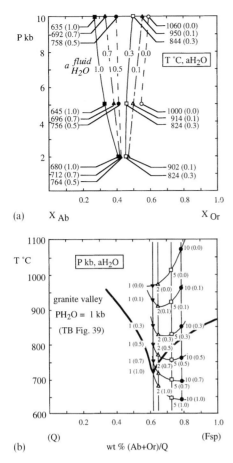

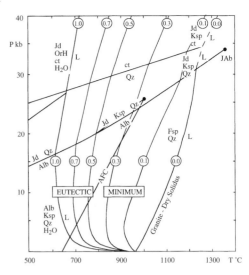

Figure 3 Minima and eutectic regions in $Ab+Or+Qtz$ at various $a_{H_2O}^{fluid}$ values as functions of pressure and temperature. The alkali feldspar ($Ab+Or$) critical curve (AFC) from Waldbaum and Thompson (1969) lies between anhydrous (dry granite) and H_2O saturated solidi (Luth *et al.* 1964, Huang & Wyllie 1975). The extrapolation into the H_2O undersaturated region is possible using the experimental results on $Ab+Or+Qtz+H_2O+CO_2$ (to 10 kbar, Ebadi & Johannes 1991) and 15 kbar (Keppler 1989). The AFC separates higher temperature–lower $a_{H_2O}^{fluid}$ ($Ab+Or+Qtz$) minima from lower temperature–higher $a_{H_2O}^{fluid}$ eutectica. The broken lines show extrapolations to higher pressures within the stability of feldspar, within the stability field of jadeite plus alkali feldspar (or orthoclase hydrate, Or H, $KAlSi_3O_8 \cdot H_2O$; Huang & Wyllie 1975), plus coesite.

Figure 2 Projection of $Ab+Or+Qtz+H_2O$ minimum/eutectic as a function of pressure (kbar), temperature (°C) and a_{H_2O} in H_2O+CO_2 fluids. (a) Projected onto $Ab+Or$ join (X_{Or}, mol%) plotted from data in table 2 of Ebadi and Johannes (1991: 289). The change in slope of dP/dX_{Or} occurs between $a_{H_2O}=0.3–0.5$. (b) Projected as the ratio ($Ab+Or$)/Q (wt%) using data from Ebadi and Johannes (1991; table 2; fig. 13), Luth *et al.* (1964) and Tuttle and Bowen (1958). The liquidus surface for $Ab_{50}Or_{50}$ for $P_{H_2O}=1$ kbar (granite valley) is shown from Tuttle and Bowen (1958, fig. 39).

(also at quartz/saturation). The ternary minimum in $Ab+Or+Qtz$ lies at a lower temperature than the binary $Ab+Or$ minimum at any particular $a_{H_2O}^{fluid}$. Thus the critical curve at quartz saturation, separating minimum from eutectic conditions in haplogranites, will lie some degrees lower than illustrated for the AFC here. In Figure 3, the position of the AFC in $P–T$ space is compared with the $Ab+Or+Qtz+H_2O$ solidus at various $a_{H_2O}^{fluid}$ using the experimental results of Keppler (1989) and Ebadi & Johannes (1991). As can be seen in Figure 3, the $P–T$ region inferred for the formation of many granulites and for crustal anatexis (e.g. at 10 kbar at temperatures greater than 800°C) lie in the minimum region, not the eutectic region.

The important geological consequence of melting at a haplogranite minimum is that the melting temperature and anatectic melt composition, for any particular pressure and a_{H_2O}, are determined by the bulk composition (particularly the Ab/Or ratio, Fig. 2). Conversely, at eutectica the initial melt composition and temperature are independent of the phase proportions. It remains to be investigated whether natural rock compositions will exhibit Ab/Or minima as opposed to eutectica, when the potassic feldspar in any rock is completely substituted by muscovite or biotite.

3.2. Dehydration melting of mica relative to the haplogranite system

Micas in lower crustal rocks are important sources of H_2O for anatexis. In rocks that contain mica, feldspar and quartz,

but without excess H_2O, fluid-absent (dehydration) melting occurs at temperatures higher than appropriate H_2O-saturated solidi. In the $P–T$ diagram of Figure 4, the dehydration melting of muscovite granite (Huang & Wyllie 1981), of $Mus+Ab+Qtz$ (Petö & Thompson 1974; Petö 1976) and the dehydration melting interval for $Bio+Plag+Qtz$ (LeBreton & Thompson 1988; Vielzeuf & Holloway 1988) are shown relative to the haplogranite melting results of Figure 3 and the alkali feldspar critical curve (AFC).

Important for our purposes from Figure 4 is that muscovite dehydration melting (with albite and quartz) occurs at lower temperatures than the AFC, whereas biotite dehydration melting (with plagioclase and quartz) occurs at higher temperatures than the AFC. We will next consider melting reactions with binary K–Na feldspars, i.e. mica instead of K-feldspar and albite as plagioclase.

3.3. Differences in dehydration melting of muscovite-bearing compared with biotite-bearing assemblages

In mica dehydration melting the amount of melt produced is determined by the proportions of mica to feldspar, relative to the composition of the minimum or eutectic between them, as well as the width of the melting interval caused by crystalline solutions (e.g. see Burnham 1967, 1979; Thompson 1982, 1988). Muscovite and biotite require excess Al_2O_3, MgO (and FeO) above the feldspar composition as well as H_2O to form from their equivalent anhydrous assemblages. It was suggested by Burnham (1979) that initial muscovite dehydration melting corresponds to an X_w^m of about 0.6 and initial biotite dehydration melting about 0.35. The observations of LeBreton and Thompson (1988, fig. 4) suggest that, depending on the biotite composition, X_w^m may vary. For example, more Fe-rich biotite may begin dehydration melting at 780°C, $X_w^m=0.5$ and extend to 800°C, $X_w^m=0.4$ for more Mg, F or Ti richer compositions, and thus further into the projected 'minimum' regions.

Because muscovite dehydration melting for pelites and

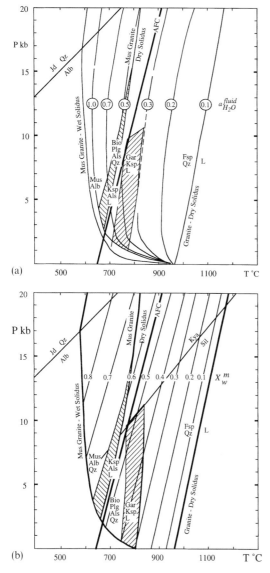

Figure 4 *P–T* diagram showing the H_2O saturated (wet) and dehydration melting (dry) solidi for a muscovite granite (Huang & Wyllie 1981) with dehydration melting of muscovite- and biotite-bearing quartzo-feldspathic assemblages relative to haplogranite melting and the AFC (from Fig. 3). The reaction Mus + Alb + Qtz → Ksp + Als + L (broken line, Petö & Thompson 1974; Petö 1976) is taken to mark the lower *T* limit of plagioclase reactions (see Thompson & Tracy 1979). The interval for the reaction Bio + Plag + Als + Qtz → Gar + Ksp + L begins at about 790°C at 10 kbar (dotted line, LeBreton & Thompson 1988) and ends at about 860°C (Vielzeuf & Holloway 1986). The dehydration melting of muscovite lies at lower temperatures than AFC—that is, within the region of Ab + Or + Qtz eutectica, whereas the dehydration melting of biotite lies within the region of Ab + Or + Qtz minima. (a) Contoured for $a_{H_2O}^{fluid}$ from Fig. 3; (b) contoured for X_w^m obtained by Thompson (1988) using the methods of Burnham (1979).

granites lies on the low-temperature (two-phase) side of the AFC, the melting of Mus + Alb + Qtz will be eutectic. The eutectic will occur over a narrow temperature interval (< 20°C) as muscovite does not show extensive solid solution (compared with biotite). Thus the melting of common natural proportions of muscovite and albite will generate quartz-saturated solidi very close in temperature to one another and obviously melts of the same composition.

The dehydration melting interval of Bio + Plag + Qtz is spread out by more than 100°C because of biotite crystalline solution (mainly Fe–Mg, but involving Al_2O_3, TiO_2 and F in many instances). Furthermore, the extent of biotite dehydration melting appears to be strongly controlled by the coexisting

plagioclase composition as well as by its amount. According to the picture revealed in Figure 4, dehydration melting of Bio + Plag + Qtz will be related to minima and not the eutectic region of Ab + Or + Qtz + H_2O (projected to lowered a_{H_2O}). In this instance, dehydration melting of biotite in specific assemblages will begin at very different temperatures depending on X_{Or} in albite, Fe/Mg in biotite and the proportions of these minerals to one another. This explains why there is considerably more variability in the solidus temperatures and melt compositions for biotite dehydration melting (LeBreton & Thompson 1988; Vielzeuf & Holloway 1988; Patiño Douce & Johnston, 1991; Vielzeuf & Montel 1994; Gardien *et al.* 1994) compared with muscovite dehydration melting. These features also partly explain why there are also greater differences in melt amount from biotite dehydration melting experiments compared with muscovite dehydration melting experiments (Skjerle & Johnston 1992, 1993; Gardien *et al.* 1995). These aspects will be considered further in the discussion of the effect of natural rock compositions on fertility.

Until now we have mainly been concerned with the significance of feldspar composition, or projected X_{Or} ratio at SiO_2 saturation, for the phase relations relevant to understanding crustal anatexis. A further fact to be considered is that the amount of quartz relative to mica exerts strong controls on the types of melting reactions and the width of melting temperature intervals because of the nature of the residual minerals.

4. Dehydration melting reactions and residual mineralogy

Both micas melt incongruently to give H_2O undersaturated liquids that are richer in H_2O than the micas themselves. The dehydration melting of muscovite releases $KAlSi_3O_8$ and H_2O to the melt, but consumes one mole of SiO_2 from the rock to form residual Al_2SiO_5 (often sillimanite), according to the simplified reaction

$$KAl_3Si_3O_{10}(OH)_2 + SiO_2 = KAlSi_3O_8 + Al_2SiO_5 + H_2O \tag{1}$$

The dehydration melting behaviour of phlogopite is analogous, but consumes three moles of SiO_2 during the formation of residual enstatite according to

$$KMg_3AlSi_3O_{10}(OH)_2 + 3SiO_2 = KAlSi_3O_8 + 3MgSiO_3 + H_2O \tag{2}$$

Because both micas exhibit the Tschermak's exchange $(Al_2Mg_{-1}Si_{-1})$, garnet or cordierite or other $MgO + Al_2O_3$ minerals are also commonly produced as residual phases.

An important question is by how much is the Fsp/Qtz ratio decreased to generate the minimum/eutectic composition when coexisting with SiO_2-bearing residual minerals, for rocks that contain mica instead of $KAlSi_3O_8$? Calculations were performed relative to the estimated Ab + Or + Qtz minimum compositions (Fig. 1) for $a_{H_2O} = 0.6$ and 0.35, corresponding to the approximate respective conditions for muscovite and biotite dehydration melting. The recalculated minimum compositions in Table 1 represent the proportions of plagioclase and mica required to produce a sillimanite-, or enstatite-bearing residue coexisting with the granite minimum melt compositions. These displacements of the Ab + Or + Qtz minimum compositions, presented in Figure 5 for pressures to 30 kbar and $a_{H_2O} = 0.6$ and 0.35, are clear indicators that different fertilities are expected when mica dehydration melting produces refractory residual minerals. Very different chemical trends in granite evolution are to be expected because of the

Table 1 Calculation of the amount of quartz required relative to the granite minimum compositions when muscovite and biotite consume quartz during the formation of the residual minerals Al_2SiO_5 and $MgSiO_3$ (Fig. 5, wt%). Reaction (1): Mus (398) + Qtz (60) = Ksp (278) + Al_2SiO_5 (162) + H_2O (18). Reaction (2): Phl (417) + 3 Qtz (180) = Ksp (278) + 3$MgSiO_3$ (301) + H_2O (18).

	Pressure (kbar) for $a_{H_2O} = 0.6$					Pressure (kbar) for $a_{H_2O} = 0.35$				
	2	5	10	20	30	2	5	10	20	30
(a) Granite minimum composition (wt%)										
Qtz	35	28	23	18	13	35	28	23	18	13
Ab	29	28	28	29	30	33	36	38	40	42
Or	36	44	49	53	57	32	36	39	42	45
(b) Equivalent modes (wt%)										
Ab	35*	44*	49*	53*	57*	32†	36†	39†	42†	45†
Mus	41*	40*	40*	41*	43*	—	—	—	—	—
Qtz'	41*	34*	29*	24*	19.5*	56.4†	51.3†	47.6†	44†	40.2†
H_2O	1.87*	1.81*	1.81*	1.88*	1.94*	2.13†	2.33†	2.4†	2.6†	2.7†
Sil	16.9*	16.3*	16.3*	16.9*	17.5*	—	—	—	—	—
Bio	—	—	—	—	—	49†	54†	57†	60†	63†
Ens	—	—	—	—	—	36†	39†	41†	43†	45.5†
Equivalent granite minimum compositions [Qtz' from (b); Ab + Or from (a) wt%]										
Qtz'	39	32	28	23	19	46.5	42	38	35	32
Ab	27	41	46	50	53	26.3	29	31	33	35
Or	34	27	26	27	28	27.2	29	31	32	33

*Using reaction (1).
†Using reaction (2).

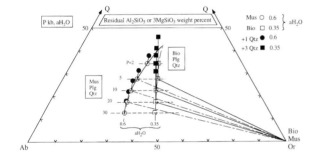

Figure 5 Compositional displacements (wt%) of Ab + Or + Qtz minima due to the consumption of quartz through the formation of residual minerals (one Al_2SiO_5 and three $MgSiO_3$). The closed symbols show the displacements of the modal mineralogy (mica, plagioclase, quartz) that correspond to the eutectica/minima in Ab + Or + Qtz (open symbols) for values of $a_{H_2O} = 0.6$ and 0.35, appropriate to muscovite (circles) and biotite (squares) dehydration melting, respectively from Table 1.

subsequent mineralogical behaviour resulting from the separation of melt from its silica-enriched refractory residue.

5. SiO₂ content of anatectic melts and residual mineralogy

Two apparently opposing effects which change the SiO₂ content are at work during crustal anatexis. The composition of the granite minimum in Ab + Or + Qtz migrates uniformly away from quartz and towards the feldspar sideline with increasing pressure (Fig. 1). Thus quartz-poor compositions, e.g. syenites, would be fertile to anatexis at high pressures, whereas granites with lower Fsp/Qtz ratios would be fertile at lower pressures (Huang & Wyllie 1975). The preliminary experimental data suggest that the value of the ratio (Ab + Or)/Q is independent of a_{H_2O} at any particular pressure (Fig. 2), but is a strong function of temperature (Fig. 1). Thus the modal mineralogy (i.e. the Fsp/Qtz ratio) of felsic intrusives is a crude barometer that applies regardless of differences in a_{H_2O} or cooling histories.

On the other hand, the formation of residual minerals (e.g. sillimanite, enstatite, garnet or cordierite) during mica dehydration melting depletes the bulk composition in SiO₂. Thus

the overall result of underestimating the initial quartz content after considering mica dehydration melting, gives higher interpreted modal Fsp/Qtz ratios, which in turn would lead to overestimates of intrusion depth (pressure). The graph in Figure 5 can be used to assess the quartz contents necessary for optimum fertility in mica-bearing rocks compared with the effect of pressure on the Fsp/Qtz ratio of the haplogranite minimum.

6. Melt evolution of ternary feldspars with quartz at various H₂O activities

The crystal liquid equilibrium of crustal magmas is controlled by a delicate interplay of mafic and felsic components. However, the ternary feldspar system An ($CaAl_2Si_2O_8$) + Ab ($NaAlSi_3O_8$) + Or ($KAlSi_3O_8$) at quartz saturation provides a subset of constraints on the evolution of crustal melts. Several sets of experimental data on the phase relations of coexisting plagioclase, alkali feldspar and melt coexisting with quartz at H₂O saturation will be examined first to give important information on the temperature, pressure and a_{H_2O} conditions for the anatexis of quartzo-feldspathic rocks and the fractional crystallisation of more mafic magmas.

6.1. The five-phase (Plag, Ksp, Qtz, L, H₂O) equilibrium at $a_{H_2O} = 1$ with increasing pressure

The H₂O and quartz saturated eutectics on the binary $CaAl_2Si_2O_8$(An)–$KAlSi_3O_8$(Or) in Figure 6a (closed circles) are from Winkler and Ghose (1973). The equivalent minima and eutectics on the H₂O and quartz saturated $NaAlSi_3O_8$(Ab)–$KAlSi_3O_8$(Or) binary (closed circles) are from Tuttle and Bowen (1958) and Luth et al. (1964).

Piercing points for the five-phase equilibrium Plag, Ksp, Qtz, L, H₂O, investigated at $P_{H_2O} = 1$ kbar by James and Hamilton (1969), are shown by open circles. Their deduced cotectic is shown by the broken line passing through these points. The smoothed cotectic curve across to the An–Or sideline is constructed to be consistent with the Winkler–Ghose (1973) data.

Winkler and co-workers have presented data on the five-phase equilibrium at $P_{H_2O} = 5$ and 7 kbar (see summary by Winkler 1979: 303–4). It is not known whether the unusual

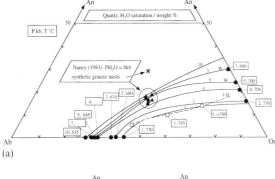

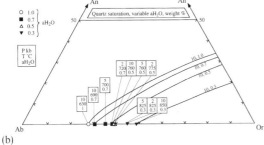

Figure 6 (a) Projection of ternary feldspar cotectics onto the plane An + Ab + Or at quartz and H₂O saturation. The phase relations in Ab + Or + Qtz + H₂O are from Figures 1 and 2 (with P in kbar, T in °C; Tuttle & Bowen 1958; Luth *et al.* 1964); those for An + Or + Qtz + H₂O (closed circles) are from Winkler and Ghose (1973). The open circles (labelled 1, T °C) show piercing points deduced by James and Hamilton (1969) from experiments at 1 kbar. Their cotectic curve (broken line labelled J–H) has been smoothed to be consistent with the data of Winkler and Ghose (1973) for An + Or + Qtz + H₂O. The $P_{H_2O} = 5$ and 7 kbar cotectics are taken from Winkler (1979: 303–4). It is not known whether the reported curvature around An₅–An₁₀ is related to ternary feldspar critical phenomena or a ternary liquid minima. The cotectics at 2, 4 and 10 kbar are constructed from Ab + Or + Qtz + H₂O end-points. Several glass analyses (closed triangles) were reported for experiments by Naney (1983) for a synthetic granite (bulk composition indicated by asterisk) at $P_{H_2O} = 8$ kbar. (b) Migration of ternary feldspar cotectics at quartz saturation at 10 kbar due to lowering a_{H_2O}. These are constructed with data from Ebadi and Johannes (1991) for Ab + Or + Qtz at the indicated a_{H_2O} pressure (kbar) and temperature (°C) and are drawn subparallel to the 10 kbar curve for $a_{H_2O} = 1$ from (a).

curvature of the projected five-phase equilibria at low An contents is an artifact of data interpretation or whether it reflects migration of the neutral point (see Abbott 1978: 250), separating cotectic crystallisation at low An content. The cotectics for 2, 4 and 10 kbar are drawn simply to be consistent with the other cotectics determined experimentally. Deductions about the behaviour of the cotectics at lower a_{H_2O} are based on these smoothed cotectics for $a_{H_2O} = 1$.

6.2. Five-phase cotectics at $a_{H_2O} < 1$

To understand the compositions of initial 'granitic' melts produced by dehydration melting of mica we need to be able to locate the temperatures and compositions of the five-phase equilibria as functions of pressure and a_{H_2O}. However, the only data available are for the Ab + Or sideline of the quartz- and fluid-bearing ternary feldspar space (Keppler 1989; Ebadi & Johannes 1991). In the absence of experimental data and using $P_{fluid} = 10$ kbar as an example, I have extrapolated (in Fig. 6b) the X_{Or} compositional and a_{H_2O} data for the Ab + Or + Qtz sideline (Fig. 2) into the ternary feldspar diagram parallel to the illustrated projected cotectic for the five-phase equilibrium for $a_{H_2O} = 1$ (from Fig. 6a). The errors in this extrapolation will become apparent from future experimental studies. The suggested ranges in composition, a_{H_2O} and temperature for

such experiments are indicated in Figure 6b. The simplest experimental strategy would be first to perform experiments on the end-member system An + Or + Qtz at various a_{H_2O}. It should be noted that Ai and Green (1989) give a projected eutectic composition of Or₇₀ for the anhydrous An + Or eutectic at 10 kbar, but without quartz. The present interpretation requires that the An + Or eutectic at quartz saturation is displaced towards Or₈₅.

The fact that experimentally produced melts plot close to these cotectics [e.g. the results of Naney (1983) for melting of a synthetic granite at $P_{H_2O} = 8$ kbar in Fig. 6a] provides some reassurance that the extrapolated isobaric five-phase cotectics are reasonably well located. However, the melts reported by Naney (1983) do contain some dissolved MgO + FeO and some have lost quartz at high temperatures. This example illustrates well that because of the projected strong curvature of the five-phase cotectics they will not provide suitable barometers for use with modally determined proportions of feldspars from intrusive rocks.

With these provisionally located projected quartzo-feldspathic cotectics at 10 kbar for various a_{H_2O}, it is intended to use dehydration melting experiments on muscovite- and biotite-bearing quartzo-feldspathic rocks (pelites, gneisses, greywackes) to attempt to locate the temperatures and compositions of initial melts at various pressures and a_{H_2O}.

6.3. Composition of initial melts obtained from dehydration melting experiments on mica-bearing quartzo-feldspathic rocks

The fact that several studies have used plagioclase (An₂₃ to An₃₀) instead of albite in dehydration melting studies of micaceous quartzo-feldspathic rocks means that it is clearly necessary to consider melting in projected ternary feldspar space and not just the projected Ab–Or binary.

Although beginning with Mica + Plag + Qtz (+ Sill), the melt composition for dehydration melting at 10 kbar reported by Patiño Douce and Johnston (1991) from temperatures of 825°C and above did not coexist with plagioclase (open circles in Fig. 7). Likewise the melt compositions reported in a similar study by Vielzeuf and Holloway (1988) from temperatures of 875°C and above did not coexist with biotite (open squares in Fig. 7). The compositional trends projected backwards meet around An₆Ab₃₇Or₅₇. This composition is close to a dehydration melt composition (triangles in Fig. 7) reported by LeBreton and Thompson (1988, analyses no. 1, 850°C/10 kbar,

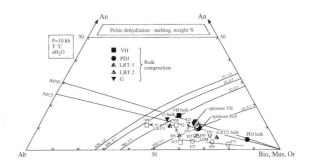

Figure 7 Results of pelitic dehydration melting experiments at 10 kbar (≈ 35 km depth) projected onto Figure 6b. The glass compositions at experimental temperatures (°C) are indicated by open symbols; bulk compositions are indicated by closed symbols (circle, PDJ—Patiño Douce & Johnston 1991; square, VH—Vielzeuf & Holloway 1988; triangle, LBT mixtures 1 and 2—Le Breton & Thompson 1988; inverted triangle, G—Gardien *et al.* 1994, 1995). The optimum bulk compositions for Bio + Plag + Qtz mixtures calculated by Patiño Douce and Johnston (1991: 214) shown by the half-shaded hexagons, lie between the tie lines for plagioclase An₂₅ and An₃₀ with biotite and quartz. The recalculation of analyses is given in Table 2.

Table 2 Calculations of An + Ab + Or ratios from glass compositions reported from various dehydration melting studies at 10 kbar.

Study	Ratio (wt%)		
	An	Ab	Or
Patiño Douce & Johnston (1991)			
Plag	26·1	73·2	0·7
Bulk	4·6	10·6	84·8
825°C	3·6	32·6	63·8
850°C	2·4	26·4	71·2
875°C	1·7	19·7	78·4
Optimum protolith			
PDJ	8·9	28·7	62·4
VH	11	28·7	60·3
Vielzeuf & Holloway (1988)			
Plag	30	70	—
Bulk	15·4	32	32·6
875%	2·6	38·8	58·6
900°C	10·2	36·1	53·7
950°C	11·6	34·1	34·3
1000°C	10·8	42	47·2
Le Breton & Thompson (1988) (all at 850°C)			
Bulk mix 1	10·2	38·1	51·6
Bulk mix 2	5·2	22·5	72·3
No. 1	6·2	37	56·8
No. 85	4·4	22·3	73·2
No. 92	5·4	29·9	64·6
No. 33	6·5	42·4	51·07
No. 54	12	51·2	37
Gardien et al. (1994,1995)			
Bulk	14	37·2	48·8
BPQ, 825°C	11·6	30·6	37·8
BPQM, 825°C	5·8	30·8	63·4
BPQ, 850°C	10·5	31·6	37·8
BPQM, 850°C	5·4	25·4	69·2

coexisting with biotite, plagioclase, Al_2SiO_5, quartz and K-feldspar). Initial melts obtained by Gardien et al. (1994, 1995) from dehydration melting experiments on Bio + Plag + Qtz ± Musc are shown for 825 and 850°C (inverted open triangles in Fig. 7). The data presented in Table 2 were used for constructing Figure 7—for each set the bulk compositions are represented by the closed symbols and the melt composition by open symbols.

Although showing diversity in melt composition (not surprising considering the difficulty of obtaining electron microprobe analyses of small amounts of glass), it is reassuring that relatively small compositional regions around the projected five-phase cotectics are located. The disposition of the various bulk compositions from these studies (closed symbols in Fig. 7) indeed suggest an approach to cotectics from both the plagioclase and biotite sides.

It is also important to note that if the ratio X_{Or} was taken as the single characteristic of a natural initial melt composition (as for the binary Ab–Or system projected from quartz and a_{H_2O} in Fig. 2a), it would lead to a misleading interpretation of melting temperature because in the projected ternary feldspar system cotectic the compositions clearly project far to the X_{Or} side of the Ab–Or minimum.

6.4. Eutectica or minima in ternary feldspar anatexis?
Until now we have not been concerned with whether the division into minima or eutectica obtained in the projected binary system extends into the projected ternary feldspar system. One possibility is that the extension of the AFC into the ternary feldspar system (the composition line joining the binary Ab + Or to the binary An + Or critical points) coincides with a particular five-phase cotectic valley for a given pressure and a_{H_2O}. In this instance the eutectic region would be located

on the An-rich side and the minimum region on the Or (mica)-rich side of the particular cotectic. The consequence of this is that minimum melting would apply for mica to plagioclase ratios greater than the projected ternary cotectica and that eutectic melting will occur for plagioclase to mica ratios greater than the projected ternary cotectica.

The other possibility is that the extension of the AFC into the ternary feldspar system (the composition line joining the projected binary Ab + Or to the projected binary An + Or critical points) lies at a higher temperature than the trace of the melt compositions along the five-phase cotectic valleys. This would mean that the projected ternary feldspar system is everywhere eutectic. Thus solidi in the five-component system (Ab, An, Or, Qtz, H_2O) at a_{H_2O} appropriate to mica dehydration melting all lie in the eutectic region within the projected ternary feldspar system.

Most bulk compositions used to experimentally determine fertility to crustal anatexis (large melt fractions generated over the smallest temperature intervals) were chosen because they contained subequal amounts of quartz and feldspar to provide the low temperature melting fraction and mica to provide H_2O (we might call this 'potential fertility'). All investigated compositions project close to the five-phase cotectic valleys at the appropriate pressures and a_{H_2O}, with the exception of the metapelite (HQ-36) used by Patiño Douce and Johnston (1991). However, even for this mica-rich composition these workers suggest a near-eutectic melting behaviour (A. Patiño Douce 1995, pers. comm.).

There are distinct advantages to the interpretation of fertility to crustal anatexis by assuming that Mica + Plag + Qtz melting is always eutectic. Thus the observed variation in the amount of melt as a function of temperature would be related to the chemistry of coexisting phases more than their modal amounts.

7. aH_2O for common rocks and optimum mineral proportions for fertility

Although not experimentally investigated in detail, it was suggested by Burnham (1979) that initial muscovite dehydration melting corresponds to an X_w^m of about 0·6 and initial biotite dehydration melting to an X_w^m of about 0·35. These values were supported by calculations by Thompson (1988) and LeBreton and Thompson (1988), who suggested that biotite dehydration melting can begin around 790°C where X_w^m is between 0·5 and 0·45. Whatever the true value of X_w^m for a particular dehydration melting reaction, this quantity will not vary much during a particular anatectic episode in the absence of a fluid phase. It is useful to examine the deduced initial melt compositions and the projected five-phase equilibria at these suggested values for X_w^m.

The cotectics illustrated in Figure 8 for 5, 10 and 20 kbar pressure at $X_w^m = 0·6$ and 0·35 were obtained in a similar way to that discussed for Figure 6b. The cotectics show the lowest temperature melts at quartz saturation and are related to the rock compositions for optimum fertility by tie lines to the equilibrium feldspar compositions. These plagioclase and alkali feldspar compositions generate ternary melts that lie at lower temperature, smaller X_{An}, X_{Or}, but larger X_{Ab}, down the appropriate cotectics. The projected An + Ab + Or diagram is a useful tool with which to deduce rock fertility because both muscovite and biotite have very high K/Na ratios and plot virtually at $KAlSi_3O_8$. For plagioclase, the weight per cent and mole per cent are very close and Or components have been ignored.

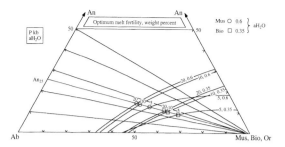

Figure 8 Predicted compositions of optimum fertility for mixtures of muscovite or biotite with plagioclase of indicated composition at various pressures and a_{H_2O}. An example for An_{25} is shown by open circles (muscovite) and open squares (biotite). The projected cotectics onto $An + Ab + Or$ at quartz saturation are shown for $P = 5$, 10, 20 kbar, at $a_{H_2O} = 0.6$ and 0.35 (Figures 5 and 6b).

7.1. Effect of plagioclase composition and proportion on optimum melt fertility

As can be seen in Figure 8, because of the high angles between the projected ternary cotectics and the mica–plagioclase tie lines, very different proportions of plagioclase and mica are required for optimum fertility depending on the plagioclase composition. For albite-rich plagioclase the optimum proportion of plagioclase is greater than that of mica. For anorthite-rich plagioclase (e.g. An_{50}) the optimum proportion of mica is greater than that of feldspar.

For a particular plagioclase composition (e.g. An_{25} in Fig. 8) it can be seen that the optimum fertility composition migrates with increasing pressure towards greater amounts of feldspar. It is also seen that biotite rocks (open squares) require more mica relative to feldspar than muscovite rocks (open circles) or, conversely, muscovite dehydration melts are more sodic than biotite dehydration melts. This is a reflection of the value of X_w^m at the different temperatures of dehydration melting of the two micas.

It is of great interest to compare the optimum fertility compositions deduced here with those calculated by Patiño Douce and Johnston (1991: 214). Their optimum compositions were obtained by comparing their values and Vielzeuf and Holloway's (1988) experimentally determined melt compositions in the projection $Na_2O \cdot Al_2O_3 - K_2O \cdot Al_2O_3 - (MgO + FeO)$. Patiño Douce and Johnston's optimum protoliths for melt fertility were obtained by mass balance calculations and have the following modes (in wt%)

Patiño Douce and Johnston:

41% Bio, 23% Plag, 31% Qtz, 6% Al_2SiO_5

Vielzeuf and Holloway:

36% Bio, 21% Plag, 34% Qtz, 9% Al_2SiO_5

It is important, as these workers noted, that the optimum protolith is closer to a metagreywacke than a pelitic schist. These optimum protolith compositions are shown by the half-closed hexagons in Figure 7. It can be seen that these compositions compare remarkably well with those deduced here (Fig. 8), using different reasoning.

7.2. Effect of pressure on the relative fertilities of psammites (granites, greywackes and gneisses) compared with metapelites

A further implication of the results of Figure 8 is that pelites are more fertile than psammites at low pressures (e.g. 5 kbar) and when the plagioclase is anorthite-rich (e.g. $An_{40}-An_{50}$). Psammites are more fertile than pelites at higher pressures (e.g. 10–20 kbar) and with more albitic plagioclase. For a typical pelite (e.g. with An_{25} in Fig. 8) the cotectic with

muscovite lies at higher a_{H_2O} (≈ 0.6) and X_{Ab} (≈ 0.42) than with biotite ($a_{H_2O} \approx 0.35$; $X_{Ab} \approx 0.32$), thus dehydration melting of muscovite requires 10% more plagioclase for fertility than biotite.

With increasing pressure, to at least 30 kbar, granite minimum and mica dehydration melts become more sodic. The first melts from $Mus + Plag + Qtz$ melting are more sodic than, and form by dehydration melting before (at lower temperatures than), the first melts from $Bio + Plag + Qtz$. This indicates that the a_{H_2O} of these melts is greater than 0.3 (Figs 1, 8). This simple deduction is of great help when evaluating a_{H_2O} during the dehydration melting of biotite.

8. Further refinements on fertility

Some very simple and necessary observations and experiments will enable great refinements to our understanding of the controls of rock fertility. For example, no attempt has been made here to include plagioclase immiscibility gaps, the cases arising from a possible ternary feldspar minimum or the occurrence of H_2O-bearing residual minerals (e.g. cordierite, Stevens et al. 1995). Additional investigations of the compositions and stability of the ternary feldspar minimum, at high pressure and at quartz saturation, are clearly desirable.

8.1. Ternary minimum melting

A quartz-saturated ternary feldspar cotectic melt could form if some ternary composition had a solidus temperature lower than minimum/eutectic melting on the $Ab + Or + Qtz$ sideline (see, for example, Stewart & Roseboom 1962; Carmichael 1963; Abbott 1978). The existence of such a quartz-saturated ternary feldspar cotectic melt would mean that a wide range of bulk compositions would produce the same initial melt composition. For such a case it would be far from easy to distinguish melt evolution at low melt fractions in different lithologies and many difficult to perform experiments would be needed. Until there is a reason to think otherwise, it is certainly more practical to consider projected ternary feldspar liquids to be always less anorthitic than plagioclase.

8.2. Curvature along the cotectic valleys

In the present figures the cotectic valleys have been drawn without inflections and with very little curvature. This means that the tangents at specific compositions along the cotectic are not changing. For the present example then it is sufficient to define optimum fertility by the intersection of a particular tie line with a specific cotectic.

It is suspected, however, that as the five-phase cotectic at 1 kbar P_{H_2O} approaches the $Ab + Or (+ Qtz + H_2O)$ sideline it changes from even to odd through a reaction relationship (i.e. becomes peritectic; see James & Hamilton 1969). Abbott (1978) noted that the curvature may be enough for low An melts to cause K-feldspar to move from an even (cotectic) to odd (peritectic) relationship with the melt. In the present example it has been assumed that all melts are cotectic (or minimum) between mica and plagioclase so that any peritectic reactions that are relevant must involve K-feldspar in addition.

8.3. P–T–X along the cotectic valleys

The present interpretation (Fig. 6a) suggests that the $An + Or + Qtz$ cotectic temperature is about 50°C higher than the $Ab + Or + Qtz$ cotectic for a given pressure and a_{H_2O}. Such small temperature differences along the cotectics mean that the details of continuous reactions would be smeared out in natural migmatites.

Tuttle and Bowen (1958: 136) have shown the strong concentration of granitic rock analyses in the low An region of $An + Ab + Or$. The analyses include pyroclastic volcanic rocks which have equilibrated at low pressure. Such rocks are frequently rich in MgO + FeO and contain biotite, especially when metamorphosed. It is suggested in Figure 8 that the 'biotite' cotectics with plagioclase and quartz do not vary in composition with temperature and pressure from 5 to 20 kbar (this is equivalent to saying that dP/dT for the biotite melting reaction is nearly infinite—vertical in a P–T diagram such as Fig. 4). Thus low pressure pyroclastic volcanic rocks are potentially fertile compositions in terms of An + Ab + Or for anatexis at any pressure in the lower crust. However, volcanic rocks contain only small amounts of micas and like 'granites and granodiorites are therefore too anhydrous to be fertile protoliths for granitoid magmas' (Patiño Douce & Johnston 1991: 214).

8.4. Superimposed mode composition diagrams on cotectic valleys

The position of a cotectic valley is determined by near-eutectic phase relations between plagioclase and K-feldspar (or mica proxy) at quartz saturation. Thus at a given pressure and a_{H_2O}, solidus temperature and composition are not dependent on the proportions of phases, whereas the width of the melting interval is. The depth of the cotectic valley (width of the melting interval) is determined mainly by the X_w^m and the proportion of plagioclase to mica.

The degree of rock fertility depends on the melting reaction temperature of particular mica and plagioclase compositions with quartz. The situation may be visualised with an analogy of erecting a cable across a valley—the feldspar–quartz phase relations describe the valley, the mica compositions locate the cable and the temperature is given by the height. For suitably Fe-enriched mica compositions lower temperature melting reactions can occur near the bottom of the valley. A more Mg-enriched biotite will only react with plagioclase at higher temperatures—near to the top of the valley (e.g. LeBreton et al. 1995; Singh 1995). Continuous reaction paths involving Fe, Mg, Ti and F in biotite and X_{An} in plagioclase will take complex routes up and across the simple cotectic valleys. Another major experimental problem is to work out the systematics of anatexis in such chemically varied cotectic valleys.

9. Concluding remarks

A large amount of experimental data has been examined to obtain bounds on the optimum fertility to melt production of crustal rocks. It can be concluded that the most fertile assemblages, defined by specific modal amounts of common minerals, vary significantly with increasing pressure and decreasing a_{H_2O} (a conclusion also reached by Patiño-Douce, this issue). Even though great extrapolations of experimental results have been made, the conclusions reached in this paper are considered to be valid.

It is recognised that many of the extrapolations herein are made well outside any controlling experimental data. For each example the extrapolations can be used to greatly limit the ranges of rock composition, $a_{H_2O}^{fluid}$, T and P that need to be considered for future experiments. These results are being used to improve existing models for optimum melt production (e.g. Clemens & Vielzeuf 1987) across the whole spectrum of common metasedimentary and magmatic quartzo-feldspathic rock compositions.

Acknowledgements

It is a great pleasure to acknowledge discussions on rock fertility over the years with John Clemens, Véronique Gardien, Nicole LeBreton, Bob Tracy, Peter Ulmer, Daniel Vielzeuf and Steve Wickham. I thank Dave Hamilton for communication about his work with Dick James. The manuscript was carefully reviewed by Rick Abbott, Alberto Patiño Douce and Jim Beard. I also thank Mike Brown for very useful comments. Ursula Stidwill and Claudia Büchel typed the manuscript and Guy Simpson drafted the final figures. The Schweizerische National Fonds is gratefully acknowledged for financial support.

References

Abbott, R. N. 1978. Peritectic reactions in the system. An–Ab–Or–Qz–H₂O. CAN MINERAL **16**, 245–56.

Ai, Y. & Green, D. H. 1989. Phase relations in the system anorthite–potassium–feldspar at 10kbar with emphasis on their solid solutions. MINERAL MAG **53**, 337–45.

Burnham, C. W. 1967. Hydrothermal fluids at the magmatic stage. *In:* Barnes, H. L. (ed.) *Geochemistry of hydrothermal ore deposits*, 34–67. New York: Holt, Rinehart and Winston.

Burnham, C. W. 1979. Magmas and hydrothermal fluids. *In:* Barnes, H. L. (ed.) *Geochemistry of hydrothermal ore deposits*, 2nd edn, 71–136. New York: Wiley-Interscience.

Carmichael, I. S. E. 1963. The crystallization of feldspar in volcanic acid liquids. Q J GEOL SOC LONDON **119**, 95–131.

Clemens, J. D. & Vielzeuf, D. 1987. Constraints on melting and magma production in the crust. EARTH PLANET SCI LETT **86**, 287–306.

Ebadi, A. & Johannes, W. 1991. Beginning of melting and composition of first melts in the system Qz–Ab–Or–H₂O–CO₂. CONTRIB MINERAL PETROL **106**, 286–95.

Gardien, V., Thompson, A. B., Grujic, D. & Ulmer, P. 1994. The role of the source composition on melt fractions generation during crustal anatexis. EOS, TRANS AM GEOPHYS UNION, **75**, 359–60.

Gardien, V., Thompson, A. B., Grujic, D. & Ulmer, P. 1995. Melt fractions during crustal anatexis. Experimental melting of biotite + plagioclase + quartz ± muscovite assemblages and implications for crustal melting. J GEOPHYS RES **100**, 15581–91.

Harris, N. B. W. & Inger, S. 1992. Trace element modelling of pelite-derived granites. CONTRIB MINERAL PETROL **110**, 46–56.

Holtz, F. & Johannes, W. 1991. Experimental investigation of H₂O-saturated and H₂O-undersaturated liquidus phase relations in the system NaAlSi₃O₈–KAlSi₃O₈–SiO₂–H₂O–CO₂ at 2 and 5kbar. J PETROL **32**, 935–58.

Huang, W. L. & Wyllie, P. J. 1975. Melting reactions in the system NaAlSi₃O₈–KAlSi₃O₈–SiO₂ to 35 kilobars, dry and with excess water. J GEOL **83**, 737–48.

Huang, W. L. & Wyllie, P. J. 1981. Phase relationship of S-type granite with H₂O to 35 kbar: Muscov granite from Harney Peak, South Dakota. J GEOPHYS RES **86**, 1015–29.

James, R. S. & Hamilton, D. L. 1969. Phase relations in the system NaAlSi₃O₈–KAlSi₃O₈–CaAl₂Si₂O₈–SiO₂ at 1 kilobar water vapour pressure. CONTRIB MINERAL PETROL **21**, 111–41.

Keppler, H. 1989. The influence of the fluid phase composition on the solidus temperatures in the haplogranite system. NaAlSi₃O₈–KAlSi₃O₈–SiO₂–H₂O–CO₂. CONTRIB MINERAL PETROL **102**, 321–7.

Le Breton, N. & Thompson, A. B. 1988. Fluid-absent (dehydration) melting of biotite in metapelites in the early stage of crustal anatexis. CONTRIB MINERAL PETROL **99**, 226–37.

Le Breton, N., Scaillet, B. & Pons, J. 1995. Amphibole stability in granitoid melts. Experimental constraints. TERRA ABSTR **7**, 297.

Luth, W. C. 1969. The systems NaAlSi₃O₈–SiO₂ and KAlSi₃O₈–SiO₂ to 20 kbar and the relationship between H₂O content, P_{H₂O}, and P_{total} in granitic magmas. AM J SCI **267-A**, 325–41.

Luth, W. C., Jahns, R. H. & Tuttle, O. F. 1964. The granite system at pressures of 4 to 10 kilobars. J GEOPHYS RES **69**, 759–73.

Naney, M. T. 1983. Phase equilibria of rock forming ferromagnesian silicates in granitic systems. AM J SCI **283**, 993–1033.

Patiño Douce A. E. 1996. Effects of pressure and H₂O content on the compositions of primary crustal melts. TRANS R SOC EDINBURGH **87**, 000–000.

Patiño Douce, A. E. & Beard, J. S. 1995. Dehydration-melting of biotite gneiss and quartz amphibolite from 3 to 15kbar. J PETROL **36**, 707–38.

Patiño Douce, A. E. & Johnston, A. D. 1991. Phase equilibria and melt productivity in the pelitic system: implications for the origin of peraluminous granitoids and aluminous granulites. CONTRIB MINERAL PETROL **107**, 202–18.

Petö, P. 1976. An experimental investigation of melting relations involving muscovite and paragonite in the silica-saturated portion of the system K_2O–Na_2O–SiO_2–H_2O to 15kbar total pressure. PROGR EXP PETROL NERC LONDON **3**, 41–5.

Petö, P. & Thompson, A. B. 1974. Wet and dry melting of white mica–alkali feldspar assemblages. TRANS AM GEOPHYS UNION **55**, 479.

Schairer, J. F. 1950. The alkali–feldspar join in the system $NaAlSi_3O_8$–$KAlSi_3O_8$–SiO_2. J GEOL **58**, 512–8.

Singh, J. 1995. Dehydration melting of tonalites—implications for the origin of continental crust. TERRA ABSTR **7**, 297.

Skjerlie, K. P. & Johnston, A. D. 1992. Vapor-absent melting at 10 kbar of biotite- and amphibole-bearing tonalitic gneiss: implications for the generation of A-type granites. GEOLOGY **20**, 263–6.

Skjerlie, K. P. & Johnston, A. D. 1993. Fluid-absent melting behaviour of an F-rich tonalitic gneiss at mid-crustal pressures: implications for the generation of anorogenic granites. J PETROL **34**, 785–815.

Stevens, G., Clemens, J. D. & Droop, G. T. R. 1995. Hydrous cordierite in granulites and crustal magma production. GEOLOGY **23**, 925–8.

Stewart, D. B. & Roseboom, E. H. 1962. Lower temperature terminations of the three-phase region plagioclase–alkali feldspar–liquid. J PETROL **3**, 280–315.

Thompson, A. B. 1982. Dehydration melting of pelitic rocks and the generation of H_2O–undersaturated granitic liquids. AM J SCI **282**, 1567–95.

Thompson, A. B. 1988. Dehydration melting of crustal rocks. REND SOC ITAL MINERAL PETROL **43**, 41–60.

Thompson, A. B. & Connolly, J. A. D. 1995. Melting of the continental crust: some thermal and petrological constraints on anatexis in continental collision zones and other tectonic settings. J GEOPHYS RES **100**, 15565–79.

Thompson, A. B. & Tracy, R. J. 1979. Model systems for anatexis of pelitic rocks. CONTRIB MINERAL PETROL **70**, 429–38.

Thompson, J. B. & Thompson, A. B. 1976. A model system for mineral facies in pelitic schists. CONTRIB MINERAL PETROL **58**, 243–77.

Tuttle, O. F. & Bowen, N. L. 1958. Origin of granite in the light of experimental studies in the system $NaAlSi_3O_8$–$KAlSi_3O_8$–SiO_2–H_2O. GEOL SOC AM MEM **74**, 153 pp.

Vielzeuf, D. & Holloway, K. R. 1988. Experimental determination of the fluid-absent melting relations the pelitic system. Consequences for crustal differentiation. CONTRIB MINERAL PETROL **9**, 257–76.

Vielzeuf, D. & Montel, J. M. 1994. Partial melting of metagrey wackes. Part I. Fluid-absent experiments and phase relationships. CONTRIB MINERAL PETROL **117**, 375–93.

Waldbaum, D. R. & Thompson, J. B. 1969. Mixing properties of sanidine crystalline solutions: IV. Phase diagrams from equation of state. AM MINERAL **54**, 1274–98.

Winkler, H. G. F. 1979. Anatexis, formation of migmatites, and origin of granitic magmas. *In Petrogenesis of metamorphic rocks*, 5th edn. New York: Springer.

Winkler, H. G. F. & Ghose, N. C. 1973. Further data on the eutectics in the system Qz–Or–An–H_2O. N JAHRB MINERAL MONATSH **31**, 481–4.

ALAN BRUCE THOMPSON, Erdwissenschaften, ETH Zurich, CH-8092, Switzerland.

Transactions of the Royal Society of Edinburgh: Earth Sciences, **87**, 11–21, 1996

Effects of pressure and H_2O content on the compositions of primary crustal melts

Alberto E. Patiño Douce

ABSTRACT: Melting experiments with and without added H_2O on a model metagreywacke and a natural metapelite demonstrate how pressure and H_2O content control the compositions of melts and residual assemblages. Several effects are observed under isothermal conditions. Firstly, the stability field of biotite shrinks with decreasing pressure and with increasing H_2O content, whereas that of plagioclase shrinks with increasing pressure and H_2O content. Secondly, the ferromagnesian content of melts at the source (i.e. coexisting with their residual assemblages) decreases with decreasing H_2O activity. Thirdly, with increasing pressure the Ca/Mg and Ca/Fe ratios of melts decrease relative to those of coexisting garnet. As a consequence, a wide spectrum of melts and crystalline residues can be generated from the same source material. For example, H_2O-starved dehydration melting of metagreywacke at low pressure ($\leqslant 10$ kbar) generates K-rich (granitic) melts that coexist with pyroxene- and plagioclase-rich residues, whereas melting of the same material at high pressure (≈ 15 kbar) and with minor H_2O infiltration can generate leucocratic Na-rich and Ca-poor (trondhjemitic) melts that coexist with biotite- and garnet-rich residues. An increased H_2O content stabilises orthopyroxene at the expense of garnet + biotite + plagioclase, causing melts to shift towards granodioritic or perhaps tonalitic compositions.

KEY WORDS: crustal anatexis, granites, trondhjemites, biotite, plagioclase, garnet, partition coefficients.

Granitoid igneous rocks of all geological ages are an essential component of the continental crust. In a broad sense, this observation must reflect two facts. Firstly, that the processes that give rise to granitoid magmas have operated in the earth since the early Archean. Secondly, that the generation of granitoid magmas is a straightforward, and perhaps even unavoidable, process. This is so both because granitoid melts are the lowest temperature melts that are produced from common crustal rocks when the geotherm is perturbed, and because granitoid melts are at the end of the liquid line of descent of many primary basaltic magmas (e.g. Tuttle & Bowen 1958: 78–9). These observations are almost universally accepted, but they do not provide an insight into the next level of analysis. That is the issue of why there is such a remarkable diversity among granitoid rocks, which encompass such diverse types as quartz diorites, tonalites, trondhjemites, granodiorites and true granites, and, within these types, such varieties as peraluminous, metaluminous and peralkaline.

It has been surmised for a long time that this diversity may arise at least in part from the melting of different crustal source materials, such as basaltic amphibolites, tonalitic gneisses, metagreywackes and metapelites. Melting experiments on different metamorphic rocks have confirmed that diversity of source composition is one possible explanation for diversity among granitoid magmas (Green 1976; Helz 1976; Spulber & Rutherford 1983; Conrad et al. 1988; Rutter & Wyllie 1988; Vielzeuf & Holloway 1988; Le Breton & Thompson 1988; Beard & Lofgren 1989, 1991; Wyllie et al. 1989; Patiño Douce & Johnston 1991; Rapp et al. 1991; Rushmer 1991; Winther & Newton 1991; Skjerlie & Johnston 1992, 1993; Skjerlie et al. 1993; Wolf & Wyllie 1991, 1994; Patiño Douce & Beard 1995, in press).

Experimental studies have also shown that changing the conditions under which melting takes place can have effects on melt composition that are comparable in magnitude with the effects of changing source composition. For example, the experiments of Conrad et al. (1988) and Beard and Lofgren (1991) demonstrated some of the important effects of varying H_2O contents on melt compositions, and those of Patiño Douce (1995) and Patiño Douce and Beard (1995) revealed some of the effects of pressure on melt compositions.

In this paper, results are presented of melting experiments on two biotite-bearing starting compositions, a model metagreywacke (biotite–plagioclase–quartz–ilmenite assemblage) and a natural metapelite (biotite–muscovite–quartz–sillimanite–plagioclase–garnet–ilmenite assemblage), with varying amounts of added H_2O. These new data, in conjunction with results from earlier dehydration melting experiments on these and other starting materials (Patiño Douce & Johnston 1991; Patiño Douce & Beard 1995, in press), help to elucidate the effects of H_2O content and pressure on the major element compositions of anatectic melts in their source region (here called primary crustal melts). Two complementary aspects of this issue are discussed. The first part of the paper discusses the isothermal effects of pressure and H_2O content on the compositions of melts and coexisting crystalline residues generated by partial melting of metagreywackes and metapelites. An important outcome of this analysis is to emphasise that, because in many common crustal rocks Na and K reside in minerals with very different behaviours (feldspar and mica, respectively), the melting of feldspar + quartz assemblages is not a good model of crustal anatexis. The second part of the paper examines biotite–melt, garnet–melt and orthopyroxene–melt exchange equilibria.

Although this study focuses on the effects of pressure and a_{H_2O}, these are not the only intensive variables that affect crustal anatexis. Temperature and f_{O_2} are also important, so that the interaction among all four of these variables must be considered to understand how the continental crust melts (e.g. Patiño Douce & Beard, in press).

1. Sources of data and starting materials

One goal of this paper is to synthesise some of the experimental data that have been presented elsewhere into more generalised models for melting of common crustal rocks. In this synthesis are included melt and mineral composition data from dehydration melting experiments (with no added H_2O) on four different starting materials. Patiño Douce and Johnston (1991) studied the dehydration melting of a feldspar-poor metapelite that produced peraluminous granitic melts in equilibrium with garnet- and sillimanite-rich residues (mineral compositions were reported in Patiño Douce et al. 1993). Patiño Douce and Beard (1995) reported melt and mineral compositions for dehydration melting of a biotite–plagioclase–quartz–ilmenite assemblage with Mg#=55 (synthetic biotite gneiss, SBG) and a hornblende–plagioclase–quartz–ilmenite assemblage with Mg#=60 (synthetic quartz amphibolite, SQA). Dehydration melting of the biotite-bearing source (SBG) yielded granitic melts that coexist with residues rich in orthopyroxene + plagioclase at $P \leqslant 10$ kbar. At higher pressures, the residual assemblages also contained garnet and clinopyroxene. Dehydration melting of the amphibole-bearing source (SQA) generated melts of granodiorite composition that coexisted with residual assemblages dominated by clinopyroxene + orthopyroxene + plagioclase (+ garnet at $P \geqslant 10$ kbar). Patiño Douce and Beard (in press) carried out dehydration melting experiments on two Fe-rich biotite–plagioclase–quartz–ilmenite assemblages, one containing biotite of Mg#=23 (synthetic Mg-annite gneiss) and another containing F-rich annite with Mg#=0.4 (synthetic F-annite gneiss). Both of these starting materials yielded granitic melts that coexist with residual assemblages dominated by either garnet or magnetite, depending on the pressure and f_{O_2}.

New data are presented here for melting experiments with varying amounts of added H_2O on two of these starting materials, the metapelite (HQ-36) studied by Patiño Douce and Johnston (1991) and the biotite–plagioclase–quartz–ilmenite assemblage (model metagreywacke) with Mg#=55 (SBG) studied by Patiño Douce and Beard (1995). Full details of the modes and mineral compositions of these starting materials can be found in these earlier publications.

2. Experimental and analytical procedures

2.1. Experimental apparatus and sample containment

All the new experiments were carried out in solid-media piston-cylinder apparatus (PC) at the University of Georgia, with 12.7 mm (0.5 in) diameter NaCl–graphite cell assemblies. All experiments were run at 925°C and pressures ranging from 7 to 15 kbar (Table 1). The apparatus and experimental techniques are described in detail in Patiño Douce and Beard (1994, 1995) and Patiño Douce (1995). Samples were contained in welded Au capsules, 2.4 mm inner diameter with 0.3 mm wall, containing 10 mg of dry sample and the appropriate amount of H_2O (see Table 1), added with a microsyringe. Weight loss was monitored during welding and the capsules were also checked for leaks before the experiments by verifying that no weight loss (± 0.1 mg) occurred after ≈ 2 h in an oven at 130°C. The average duration of the experiments was kept short (≈ 4 days, see Table 1) to minimise H_2O loss by diffusion of either H_2 (Chou 1986) or H_2O (Patiño Douce & Beard 1994) through the capsule material. Capsules were examined

for tears and weighed after the experiments. No weight loss of $\geqslant 0.1$ mg was detected in any of the experiments reported in this paper. The regular variations in melt compositions (including estimated H_2O contents), phase assemblages, modal abundances and mineral compositions suggest that, if H_2O losses occurred within the weighing uncertainties, they were most probably systematic and do not affect the conclusions.

2.2. Oxygen fugacity

No attempt was made to buffer f_{O_2} in the experiments. We have shown before (Patiño Douce & Beard 1994, 1995, in press) that the graphite-based cell assemblies used in these experiments restrict f_{O_2} in the samples to a well-defined interval below the QFM buffer, and that the stabilities and compositions of ferromagnesian phases are not affected by f_{O_2} variations within the range imposed by these cell assemblies. Oxygen fugacity was estimated for some of the experimental products reported here from the equilibrium

$$FeTiO_3 + \tfrac{3}{4}O_2 \rightleftharpoons [TiFe_{-2}]_{bio} + \tfrac{3}{2}Fe_2O_3 \qquad (1)$$

using the empirical calibration of Patiño Douce (1993). The estimated f_{O_2} values (listed in Table 1) are between 0.9 and 1.8 log units more reducing than QFM, in agreement with previous estimates from dehydration melting experiments in piston-cylinder apparatus (Patiño Douce & Beard 1994, 1995, in press).

2.3. Analytical procedures

Electron beam analyses were performed in the JEOL JXA 8600 electron probe at the University of Georgia using an accelerating voltage of 15 kV and sample current of 5 nA. To minimise alkali migration and to ensure accurate targeting, glass analyses were performed in the scanning mode at 50 000× magnification, with the beam rastered over an area approximately 4 μm on a side. All other phases were also analysed in the scanning mode, but with magnifications typically > 100 000×. Na and K were counted first and Na was counted for 10 s. All other elements were counted for 40 s. Variations of K and Na count rates with time at the analytical conditions for glass were investigated in several synthetic hydrous alkali-aluminosilicate glasses. Extrapolation of the Na count rate decay resulted in a correction factor of 20%, which was applied to all Na_2O values in glass analyses (Table 2). No count rate decay was observed for K.

The oxygen contents of the experimental glasses (quenched melt) were measured with the electron probe and were used to estimate the melt H_2O contents (Nash 1992; Patiño Douce & Beard 1994). Synthetic Al_2O_3 was used as a standard for oxygen and the effect of variable carbon-coating thickness of the samples was evaluated by analysing quartz or sillimanite crystals in the experimental products (see Patiño Douce & Beard 1994 for further details). Melt H_2O contents calculated from measured O_2 contents are shown as H_2O^* in Table 2. The analytical totals for glass obtained when H_2O^* values and Na_2O correction factors are included range from 99.1 to 101.4 wt%.

Modal abundances in the experimental products (Tables 3, 4) were calculated by simultaneous mass balance (by least sum of squares minimisation) of K_2O, TiO_2, Al_2O_3, SiO_2, MgO, CaO and FeO*, against the analysed phase compositions. Estimates of modal abundances of melt and total mafic phases, obtained from backscattered electron images, were included as additional linear equations in the least-squares minimisation and were given the same weight as the chemical mass balance equations (the method is described in Patiño Douce & Johnston 1991). Analytical errors and the presence

Table 1 Experimental conditions and phase assemblages. All experiments were performed at 925°C.

Run No.	Duration (h)	P (kbar)	H_2O^* (wt%)	$f_{O_2}^{\dagger}$ ΔQFM	Assemblage
Synthetic biotite gneiss (SBG)					
APD-607	101	7	1	−0·92	Qtz, Plg, Opx, Bio, Ilm, Mlt
APD-606	99	10	1	−1·71	Qtz, Plg, Opx, Bio, Gar, Ilm, Mlt
APD-603	96	10	2	—	Opx, Ilm, Mlt
APD-621	100	15	1	—	Qtz, Plg, Bio, Gar, Amp, Rut, Mlt
APD-608	96	15	6	−1·47	Opx, Bio, Gar, Ilm, Rut, Mlt
APD-615	102	15	8	−1·81	Opx, Bio, Ilm, Mlt
Metapelite (HQ-36)					
APD-609	102	7	2	—	Gar, Spi, Als, Ilm, Mlt
APD-617	62	7	6	—	Gar, Spi, Als, Ilm, Mlt
APD-613	120	10	2	—	Bio, Gar, Als, Rut, Mlt
APD-604	96	10	4	—	Bio, Gar, Als, Rut, Mlt

Notes: *Added H_2O. The uncertainty in the volume of H_2O added is $\leqslant 0·02$ μl. Thus the relative uncertainties in H_2O contents vary from $\pm 20\%$ (for runs with 1 wt% added H_2O) to $\pm 2·5\%$ (for runs with 8 wt% added H_2O). $\dagger f_{O_2}$ calculated from mineral compositions in experimental products (see text).

Table 2 Glass compositions (normalised to 100 wt% anhydrous). Values are averages of six to eight different glass pools. Typical relative uncertainties (two standard deviations of the mean values) are: SiO_2, 1%; TiO_2, 15%; Al_2O_3, 2%; FeO*, 5%; MnO, 100%; MgO, 8%; CaO, 7%; Na_2O, 8%; and K_2O, 4%.

P (kbar)	H_2O^* (wt%)	SiO_2	TiO_2	Al_2O_3	$FeO^{*\dagger}$	MnO	MgO	CaO	$Na_2O^{\ddagger}$	K_2O	Probe§ total	$H_2O^{*¶}$	Total‖
Synthetic biotite gneiss (SBG)													
7	1	74·1	0·28	14·3	1·63	0·05	0·36	1·20	2·17	5·74	97·1	2·8	99·9
10	1	72·3	0·31	14·7	1·99	0·03	0·44	2·12	3·32	4·65	97·1	3·2	100·2
10	2	72·4	0·38	15·2	2·01	0·06	0·57	2·76	2·50	4·26	96·4	4·5	100·8
15	1	72·0	0·28	15·0	1·11	0·02	0·28	0·98	5·83	4·56	95·2	5·3	100·5
15	6	71·3	0·37	15·4	2·50	0·11	0·87	2·92	3·06	3·40	90·2	10·1	100·3
15	8	70·4	0·51	15·7	2·82	0·06	1·03	2·88	2·85	3·75	89·1	11·3	100·4
Bulk**		64·6	2·6	12·5	7·8	0·1	4·7	2·1	1·9	3·7			
Metapelite (HQ-36)													
7	2	73·0	0·41	15·0	3·61	0·02	1·41	0·25	0·74	5·58	94·2	6·7	100·9
7	6	69·0	0·55	16·8	5·66	0·04	2·14	0·20	0·66	5·01	90·8	8·9	99·7
10	2	75·6	0·30	13·8	2·57	0·02	1·03	0·22	0·84	5·55	93·3	8·0	101·4
10	4	75·4	0·30	14·3	2·70	0·05	1·04	0·25	0·81	5·19	92·2	8·8	100·9
Bulk**		58·6	1·3	23·8	8·8	0·2	2·8	0·4	0·5	3·7	—	—	—

Notes: *Bulk added H_2O. †Total Fe as FeO. ‡Reported values include 20% correction factor. §Includes corrected Na_2O values. ¶H_2O estimated from oxygen analyses (see text). ‖Probe total plus H_2O^*. **Anhydrous bulk compositions of starting materials (from Patiño Douce & Johnston 1991; Patiño Douce & Beard 1995).

Table 3 Modal compositions of experimental products: SBG (wt%).

P (kbar)	H_2O^* (wt%)	Qtz	Plg	Bio	Opx	Gar	Amp	Ilm	Rut	Mlt
7	1	5	14	7	19	0	0	1	0	54
10	1	14	10	17	12	1	0	3	0	43
10	2	0	0	0	19	0	0	1	0	80
15	1	24	2	21	0	17	4	0	2	30
15	6	0	0	14	11	2	0	1	1	71
15	8	0	0	10	13	0	0	2	0	75
Bulk†	—	34	27	37	0	0	0	2	0	0

Notes: *Bulk added H_2O. †Starting material (from Patiño Douce & Beard 1995).

Table 4 Modal compositions of experimental products: HQ-36 (wt%).

P (kbar)	H_2O^* (wt%)	Qtz	Plg	Bio	Gar	Spi	Als	Ilm	Rut	Mlt
7	2	0	0	0	12	6	12	2	0	68
7	6	0	0	0	1	10	7	2	0	80
10	2	0	0	5	15	0	18	0	1	61
10	4	0	0	4	16	0	15	0	1	64
Bulk†	—	31	4	30	5	0	19	1	0	0

Notes: *Bulk added H_2O. †Starting material, also contains 10 wt% muscovite (from Patiño Douce & Johnston 1991).

of zoned phases (see later) generate uncertainties in the calculated modes of the order of ± 2 wt% (Patiño Douce 1995). The trends in modal abundances with changes in pressure and H_2O content are strong enough (see Figs 1, 2) that are not likely to be affected by uncertainties of this magnitude.

3. Description of experimental products with added H₂O

Biotite is present in experiments with SBG at 7, 10 and 15 kbar, and in experiments with the metapelite at 10 kbar (Tables 1, 5). Biotite crystals are subhedral to euhedral, up to 30 μm long by 10 μm wide. The Ti contents in biotite are markedly lower in the metapelite experiments than in the SBG experiments, and in both compositions TiO_2 tends to decrease with increasing H_2O contents at constant temperature and pressure (Table 5). At constant H_2O content the modal abundance of biotite increases with pressure (Tables 3, 4).

Garnet is present in all experiments with the metapelite and in some experiments with SBG at 10 and 15 kbar, in which its modal abundance is a strong function of H_2O content (Tables 3, 4). It is always present in euhedral crystals 5–30 μm in diameter. Garnet crystals in SBG experiments are homogeneous, but in experiments with the metapelite they are zoned, containing cores of garnet present in the starting material. The garnet compositions reported in Table 5 correspond to neoblastic rim analyses.

Orthopyroxene is present in experiments with SBG, in which its modal abundance increases with H_2O content and with decreasing pressure (Table 3). Orthopyroxene crystals are prismatic to acicular, commonly 10 μm long and 2–5 μm wide. The Al_2O_3 content in orthopyroxene decreases with increasing H_2O content at constant pressure and temperature (Table 5).

Amphibole of edenitic composition is only present, in small euhedral crystals 5–10 μm long, in one experiment with SBG at 15 kbar (1 wt% added H_2O).

Spinel is present in euhedral grains 2–15 μm in diameter in experiments with the metapelite at 7 kbar. Spinel compositions (Table 5) correspond to a nearly binary hercynite$_{60}$–spinel$_{40}$ solid solution. Dehydration melting experiments with this starting material did not produce spinel at T ≤ 950°C (Patiño Douce & Johnston 1991). The modal abundance and crystal size of spinel in experiments with added H_2O increase with H_2O content, at the expense of garnet and aluminosilicate (Table 4).

Fe–Ti oxides are present in all experiments. Ilmenite (with 2–4 mol% Fe_2O_3) is the only oxide phase in SBG at 7 and 10 kbar, and in the metapelite at 7 kbar. It is accompanied or replaced by euhedral acicular crystals of rutile in higher pressure experiments (Tables 3, 4).

Quartz is only present, in globular crystals 5–20 μm in diameter, in SBG experiments with 1 wt% added H_2O at 7, 10 and 15 kbar. Its modal abundance increases with pressure (Table 3).

Plagioclase is only present in SBG experiments with 1 wt% added H_2O at 7, 10 and 15 kbar, and its modal abundance decreases with increasing pressure (Table 3). Plagioclase

Table 5 Mineral compositions (wt%). Values are averages of four to six different grains for biotite, garnet, orthopyroxene and amphibole: three to four different grains for plagioclase and spinel and two grains for ilmenite. Typical relative uncertainties (two standard deviations of the mean values) are: SiO_2, 2%; TiO_2, 4% (biotite) and 0·5% (ilmenite); Al_2O_3, 3% (silicate minerals) and <0·5% (spinel); FeO*, 2% (silicate minerals and spinel) and <0·5% (ilmenite); MnO, 50%; MgO, 3%; CaO, 5%; Na_2O, 8%; and K_2O, 3%.

P (kbar)	H₂O* (wt%)	Phase	SiO₂	TiO₂	Al₂O₃	FeO*†	MnO	MgO	CaO	Na₂O	K₂O	F	Total‡
Synthetic biotite gneiss (SBG)													
7	1	Bio	37·9	6·1	15·6	15·8	0·11	11·6	0·07	0·42	8·78	0·59	96·7
7	1	Opx	51·0	0·4	5·2	23·3	0·45	19·5	0·57	0·07	—	—	100·4
7	1	Ilm	0·1	50·0	0·3	43·4	0·70	2·8	0·06	—	—	—	97·3
7	1	Plg	56·8	—	27·2	0·4	—	—	9·70	5·09	0·96	—	100·1
10	1	Bio	37·5	4·9	16·0	15·1	0·09	12·5	0·06	0·33	9·43	0·51	96·1
10	1	Opx	50·0	0·4	5·0	24·4	0·54	18·8	0·84	0·06	—	—	100·1
10	1	Gar	38·2	1·2	20·7	23·5	1·35	7·5	5·61	—	—	—	98·0
10	1	Ilm	0·2	50·5	0·3	42·2	0·66	2·9	0·08	—	—	—	96·7
10	1	Plg	55·5	—	27·8	0·4	—	—	10·43	5·03	0·67	—	99·8
10	2	Opx	52·7	0·4	3·6	22·4	0·47	19·9	0·87	0·04	—	—	100·3
10	2	Ilm	0·3	50·4	0·3	41·2	0·60	3·4	0·11	—	—	—	96·4
15	1	Bio	38·2	5·5	16·3	14·2	0·04	11·9	0·05	0·41	9·47	0·42	96·3
15	1	Amp	45·9	1·5	9·9	13·9	0·26	11·3	11·37	1·99	1·16	0·21	97·5
15	1	Gar	38·1	1·3	20·8	24·7	1·24	6·2	7·14	—	—	—	99·6
15	1	Plg	58·3	—	26·4	0·2	—	—	8·08	6·84	0·25	—	100·1
15	6	Bio	38·1	4·1	15·4	14·9	0·09	14·1	0·09	0·16	9·63	0·43	96·8
15	6	Opx	50·6	0·3	4·4	24·7	0·53	19·1	0·67	0·08	—	—	100·2
15	6	Gar	39·3	0·8	21·3	23·2	1·29	8·3	5·00	—	—	—	99·1
15	6	Ilm	0·1	50·7	0·3	42·9	0·45	3·1	0·10	—	—	—	97·6
15	8	Bio	38·1	4·1	15·6	14·0	0·11	14·7	0·06	0·33	9·36	0·59	96·8
15	8	Opx	50·9	0·3	3·8	22·8	0·50	20·1	0·68	0·03	—	—	99·2
15	8	Ilm	0·0	51·7	0·3	42·2	0·61	3·3	0·06	—	—	—	98·1
Metapelite (HQ-36)													
7	2	Gar	38·4	0·9	22·5	27·8	0·26	9·8	0·16	—	—	—	99·9
7	2	Spi	0·4	0·3	60·7	27·5	0·03	10·3	0·02	—	—	—	99·3
7	2	Ilm	0·1	50·4	0·4	42·3	0·20	3·4	0·00	—	—	—	96·8
7	6	Gar	38·6	0·8	22·1	28·7	0·23	9·8	0·14	—	—	—	100·4
7	6	Spi	0·2	0·3	61·4	27·7	0·01	9·7	0·02	—	—	—	99·4
7	6	Ilm	0·1	53·4	0·3	41·2	0·23	3·3	0·00	—	—	—	98·5
10	2	Bio	36·9	3·8	18·1	12·8	0·04	13·1	0·00	0·09	9·35	0·99	94·9
10	2	Gar	38·5	0·7	21·6	29·2	0·23	9·7	0·18	—	—	—	100·2
10	4	Bio	37·3	3·6	18·6	13·6	0·03	12·5	0·01	0·04	9·42	1·02	95·7
10	4	Gar	40·2	0·5	21·7	27·9	0·24	9·3	0·14	—	—	—	100·0

Notes: *Bulk added H_2O. †Total Fe as FeO. ‡Totals for biotite and amphibole are probe totals minus F equivalent.

crystals are equidimensional and ≈ 10 μm in diameter, with Ca-rich rims that become more prominent at low pressure (the compositions reported in Table 5 were all measured in these rims).

Aluminosilicates are present in all experiments with the metapelite, in subhedral prismatic crystals 2–15 μm long. Sillimanite is the stable Al_2SiO_5 phase at 925°C, 7 and 10 kbar.

Glass (quenched melt) is abundant in all experimental products, always forming fully interconnected networks. Glass compositions are uniform throughout the experimental charges, which is reflected in the low standard deviations of their mean analytical values (Table 2). The estimated H_2O contents of the glasses are always lower than measured or calculated H_2O solubilities in silicate melts at the same pressure (e.g. Burnham 1979; Silver et al. 1990).

4. Melting of biotite–plagioclase–quartz–ilmenite assemblage as a function of pressure and H_2O content

Isothermal modal abundance contours for plagioclase and biotite in SBG melting experiments at 925°C are shown in Figure 1 as a function of pressure and bulk H_2O content. These diagrams show that plagioclase and biotite respond differently to changes in pressure and H_2O content. In general, the modal abundances of the two phases decrease with increasing H_2O content, but plagioclase abundance also decreases with increasing pressure, whereas biotite follows the opposite trend with pressure. As a consequence, the two phases are consumed along paths (arrows in Fig. 1) that are almost orthogonal to each other. The divergence between the two paths is important because Na_2O and K_2O, which are essential components of granitoid melts, reside in most amphibolite-grade metagreywackes and orthogneisses in plagioclase and biotite, respectively. The relative proportions in which the two phases are consumed during melting will thus determine the relative proportions of the two alkalis in the melt. In addition, the stability of residual garnet, which has a strong effect on melt CaO content, is also a function of pressure and H_2O content (Fig. 1).

Figure 2 shows changes in modal compositions along two perpendicular isothermal sections, one at constant H_2O content (2·5 wt% bulk H_2O, corresponding to 1 wt% added H_2O) and the other at constant pressure (15 kbar). Figure 3 shows the normative albite–orthoclase–anorthite (Ab–Or–An) contents of the melts along these isothermal sections, plotted against the classification scheme proposed by Barker (1979).

Increasing pressure at constant H_2O content enhances biotite stability (Figs 1, 2) and causes replacement of the assemblage orthopyroxene + plagioclase by the assemblage garnet + quartz (Fig. 2; see also Patiño Douce & Beard 1995). The effect on melt compositions is most clear in the experiments with 1 wt% added H_2O (Fig. 3, middle section), but is also observed in dehydration melting experiments (Fig. 3, bottom section). The Ab/Or ratio of the melts increases continuously with pressure. The An content of the melt increases from 7 to 10 kbar, and then decreases at $P > 10$ kbar (see also Table 2). The reversal in An content reflects the change from the congruent dissolution of plagioclase at $P \leqslant 10$ kbar to the incongruent breakdown of anorthite accompanied by profuse crystallisation of garnet at higher pressure (Fig. 2). Melt compositions in dehydration melting experiments (bottom section of Fig. 3) follow the same trend as in experiments with 1 wt% added H_2O, but they are consistently more potassic than in the experiments with added H_2O. This reflects the preferential destabilisation of biotite relative to plagioclase

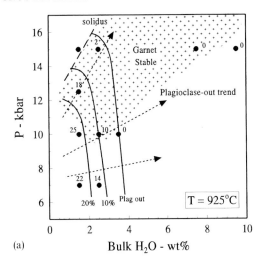

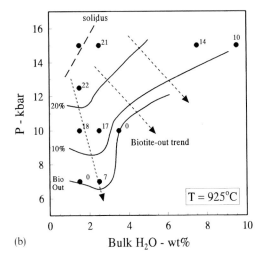

Figure 1 Modal abundances of (a) plagioclase and (b) biotite in experiments with biotite–plagioclase–quartz–ilmenite assemblage (synthetic biotite gneiss, SBG) at 925°C as a function of pressure and bulk H_2O content. Numbers next to data points are wt% plagioclase or biotite from Patiño Douce and Beard (1995) for dehydration melting experiments (1·5 wt% bulk H_2O) and from Table 3 for experiments with added H_2O (1 wt% added $H_2O = 2·5$ wt% bulk H_2O; 2 wt% added $H_2O = 3·5$ wt% bulk H_2O; 6 wt% added $H_2O = 7·5$ wt% bulk H_2O; 8 wt% added $H_2O = 9·5$ wt% bulk H_2O). Solidus approximated from dehydration melting experiment at 15 kbar and 925°C (Patiño Douce & Beard 1995). Stippled area in (a) shows field of garnet stability. Arrows indicate approximate trends of plagioclase and biotite consumption.

with decreasing H_2O activity (e.g. Maaløe & Wyllie 1975; Clemens & Wall 1981; Naney 1983; Conrad et al. 1988; Scaillet et al. 1995).

At constant $P = 15$ kbar, increasing H_2O content causes the assemblage garnet + quartz + biotite ($\pm$ hornblende) to melt incongruently with the crystallisation of orthopyroxene (Fig. 2). The effect on melt composition (Fig. 3, top section, see also Table 2) is to lower the Ab/Or ratio and raise the An content.

Pressure and H_2O content, via the relative stabilities of plagioclase, biotite and garnet, strongly affect the normative Ab–Or–An contents of anatectic melts derived from biotite- and plagioclase-rich sources (e.g. metagreywackes) and can give rise to distinctly different melt compositions from the same source composition. These two intensive variables also control the concentrations of mafic components in granitoid melts at their source region, as discussed in the following section.

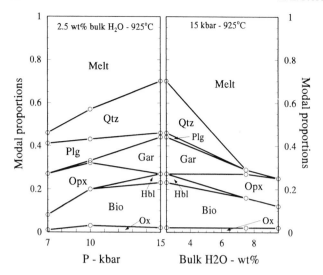

Figure 2 Modal compositions in SBG melting experiments at 925°C at constant bulk H₂O content (1 wt% added H₂O = 2·5 wt% bulk H₂O) and at constant pressure (15 kbar). Modal abundances from Table 3.

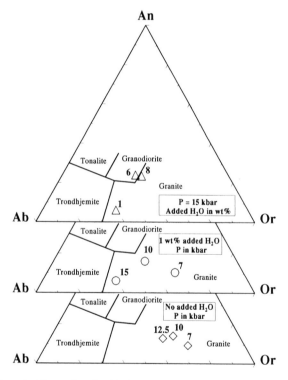

Figure 3 Normative Or–Ab–An contents in SBG melts at 925°C (fields after Barker 1979). Lower panel: dehydration melting experiments (from Patiño Douce & Beard 1995); numbers next to data points are pressures in kbar. Center panel: experiments with 1 wt% added H₂O (from Table 2); numbers next to data points are pressures in kbar. Upper panel: experiments at 15 kbar (from Table 2); numbers next to data points are amounts of added H₂O in wt%.

5. Effects of pressure and H₂O content on ferromagnesian components in primary crustal melts

Although ferromagnesian components make up only a minor proportion of granitoid rocks, their concentrations, both in absolute terms and relative to the concentrations of other components, are often distinctive of magmas formed by specific processes or under specific *P–T* conditions (e.g. White & Chappell 1983; Wyllie *et al.* 1989; Carroll & Wyllie 1990; Patiño Douce 1995). The solubilities of FeO and MgO in

felsic melts increase with temperature (e.g. Naney 1983; Naney & Swanson 1980; Puziewicz & Johannes 1990; Patiño Douce & Johnston 1991). Numerous experimental studies have shown, however, that in strongly H₂O-undersaturated felsic melts coexisting with refractory mafic phases, such as those produced during the dehydration melting of quartzo-feldspathic protoliths, the concentrations of ferromagnesian components remain notably low (≈1–5 wt%) even at temperatures of 1000°C or more (Stern & Wyllie 1978; Puziewicz & Johannes 1988, 1990; Patiño Douce & Johnston 1991; Rapp *et al.* 1991; Skjerlie & Johnston 1993; Patiño Douce & Beard 1995, in press; Patiño Douce 1995; Dooley & Patiño Douce 1996). Given that, at constant temperature and pressure, MgO and FeO concentrations in granitoid melts vary directly with H₂O content (e.g. Naney 1983; Conrad *et al.* 1988; Puziewicz & Johannes 1990; Holtz & Johannes 1991), the leucocratic nature of the melts produced by dehydration melting reflects their low H₂O contents.

The individual effects of pressure and H₂O content are shown in Figure 4. Despite the fairly large uncertainties in H₂O contents (boxes in Fig. 4) and the fact that the 10 kbar data for SBG are not well behaved (for which there is no obvious explanation), two trends emerge from these diagrams. Firstly, at constant temperature and pressure the melts become more mafic with increasing H₂O content (as many previous studies have shown; see earlier). Secondly, at constant temperature and H₂O content the melts become more felsic with increasing pressure (see also Patiño Douce & Beard 1995; Patiño Douce 1995). Water activity increases with H₂O content at constant pressure, and with decreasing pressure at constant H₂O content. Therefore the trends in Fig. 4 show either that melt fraction increases with H₂O activity, causing the melt compositions to shift towards the (more mafic) protolith compositions, or that decreasing H₂O activity causes refractory mafic components to crystallise and to be replaced in the melt by fusible quartzo-feldspathic components, while

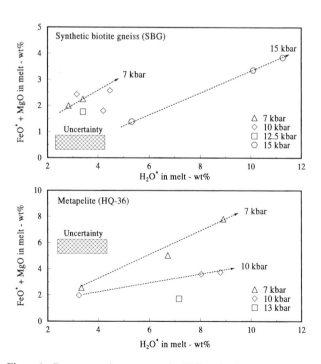

Figure 4 Ferromagnesian contents in SBG melts (upper panel) and metapelite melts (lower panel) at 925°C as a function of estimated melt H₂O contents (H₂O*, see text). Dehydration melting data (lowest H₂O contents) are from Patiño Douce and Johnston (1991) and Patiño Douce and Beard (1995). All other data are from this paper (Table 2). Stippled boxes show uncertainties in H₂O* and FeO* + MgO contents.

preserving the melt fraction (Patiño Douce & Beard 1995). In any case, melts formed under a fixed H_2O budget (e.g. by dehydration melting) become more leucocratic with increasing depth of melting.

6. Exchange reactions between granitoid melts, biotite, garnet and orthopyroxene

This section examines Fe–Mg exchange reactions between granitoid melts, biotite, garnet and orthopyroxene, as well as Ca–Mg and Ca–Fe exchange reactions between garnet and melt. Data for this analysis were obtained by Patiño Douce and Johnston (1991), Patiño Douce et al. (1993) and Patiño Douce and Beard (1995, in press). All the data included in the analysis were obtained from piston-cylinder experiments at $f_{O_2} \leqslant QFM$. Throughout this discussion I assume that all Fe is present as Fe^{2+}.

The equilibrium constants describing the mineral–melt exchange reactions are of the general form

$$K_{i/j}^{x/melt} = \left(\frac{iO}{jO} \right)_x \cdot \left(\frac{jO}{iO} \right)_{melt} \qquad (2)$$

where x can be biotite, orthopyroxene or garnet; i is either Fe or Ca; and j is either Mg or Fe (see Table 6 and Figs 5, 6). These equilibrium constants can be approximated by functions of the following form

$$\ln K = \frac{a}{T} + b\,\frac{P}{T} + c \qquad (3)$$

where a, b and c are constants over restricted P–T intervals. The analysed oxide wt% values for the exchange reactions were fitted to equations of this form by least-squares regression weighed on the basis of the least-median of squares, using the PROGRESS algorithm of Rousseeuw and Leroy (1987; this weighing procedure identifies outliers and dampens their effect on the calculated regression parameters). Regression results are given in Table 6. Additionally, Figure 5 shows all the values of $\ln K_{Fe/Mg}$ for biotite–melt, garnet–melt and orthopyroxene–melt, plotted against the inverse of temperature, and Figure 6 shows all the values of $\ln K_{Ca/Mg}$ and $\ln K_{Ca/Fe}$ for garnet/melt, plotted against the ratio P/T [see Equation (3)].

Fe–Mg exchange between garnet and melt and between biotite and melt is sensitive to temperature and pressure, as shown by the positive correlation between $\ln K_{Fe/Mg}$ and $1/T$ (Fig. 5) and by the regression coefficients summarised in Table 6. The regressed values for garnet–melt Fe–Mg exchange are in good agreement with the previous results of Ellis (1986) and Green (1977). With increasing pressure, the Mg# of granitoid melts increases relative to the Mg# of coexisting garnet or biotite (Table 6). The effect of pressure on garnet–melt Fe–Mg exchange was first reported by Green (1977,

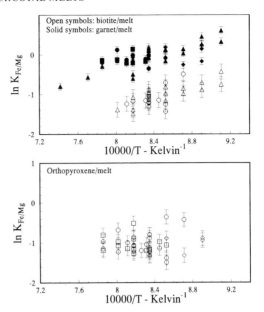

Figure 5 ln K values for biotite–melt, garnet–melt and orthopyroxene–melt Fe–Mg exchange reactions as a function of the inverse of temperature. Uncertainties estimated from analytical uncertainties. Symbols indicate different bulk compositions as follows: circles, SBG (from Patiño Douce & Beard 1995 and this paper); triangles, metapelite HQ-36 (from Patiño Douce & Johnston 1991 and this paper); squares, synthetic quartz amphibolite (from Patiño Douce & Beard 1995); diamonds, synthetic Mg-annite gneiss (from Patiño Douce & Beard, in press); inverted triangles, synthetic F-annite gneiss (from Patiño Douce & Beard, in press).

his fig. 4) and these results confirm the direction and approximate magnitude of Green's experimental observations. The results presented here also show that the effect of pressure on biotite–melt Fe–Mg exchange is approximately of the same magnitude as that on garnet–melt Fe–Mg exchange (Table 6).

In contrast with the results for garnet–melt and biotite–melt, Fe–Mg exchange between orthopyroxene and melt appears to be independent of temperature, and the values show considerably more scatter than those of the equilibrium constants for the other two Fe–Mg exchange reactions (Fig. 5). An attempt to fit Equation (3) to the orthopyroxene–melt data showed that none of the regression coefficients was statistically significant. A possible explanation for this is that $\ln K_{Fe/Mg}$ for orthopyroxene–melt has a relatively small temperature dependency and that the temperature range of the experiments used to attempt the calibration is rather narrow ($T = 900$–$1000^\circ C$). The average of the measured values of $K_{Fe/Mg}$ for orthopyroxene–melt is ≈ 0.36, which is comparable with the range of $K_{Fe/Mg}$ values (0.4–0.5) predicted by Ellis (1986) for this temperature range.

Several experimental studies have shown that garnet

Table 6 Regression parameters for mineral/melt exchange reactions. Calculated with PROGRESS regression algorithm (Rousseeuw & Leroy 1987). Data from Patiño Douce and Johnston (1991), Patiño Douce et al. (1993), Patiño Douce and Beard (1995, in press) and this paper.

Phases x–melt	Exchange i–j	Coefficients*			
		a (Δa)	b (Δb)	c (Δc)	r^2
Bio–melt	Fe–Mg	7444 (837)	39·4 (11·1)	−7·697 (0·743)	0·85
Gar–melt	Fe–Mg	5939 (697)	40·4 (11·0)	−5·371 (0·628)	0·75
	Ca–Mg	4504 (956)	173·6 (14·9)	−6·985 (0·865)	0·80
	Ca–Fe[†]	−1435	133·2	−1·614	—
Gar–melt (Ellis 1986)	Fe–Mg[‡]	4975	16·6	−4·270	—

Notes: *a, b and c as given in Equation (3) (see text), in K kbar. Δa, Δb and Δc are the standard errors of the regression coefficients calculated by PROGRESS. [†]Coefficients for Ca–Fe exchange are not regression coefficients, but were calculated from coefficients for Fe–Mg and Ca–Mg exchange reactions. [‡]Ellis (1986) did not report standard errors of regression coefficients.

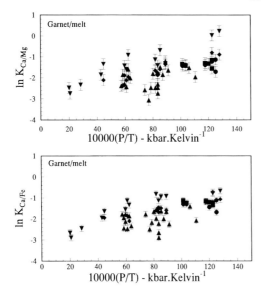

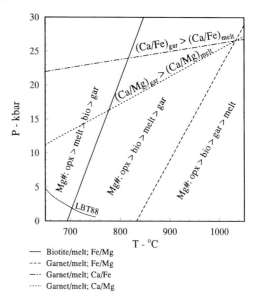

Figure 6 ln K values for garnet–melt Ca–Mg and Ca–Fe exchange reactions as a function of the ratio P/T [see Equation (3)]. Symbols as in Figure 5.

Figure 7 ln $K = 0$ isopleths for biotite–melt and garnet–melt exchange reactions, indicating the P–T conditions at which each of the element partitionings inverts. The curve labelled LBT88 is the H_2O-saturated metapelite solidus proposed by Le Breton and Thompson (1988) (see text).

coexisting with rhyolitic to andesitic liquids becomes enriched in the grossular component with increasing pressure (Green & Ringwood 1972; Green 1977; Green 1992). Figure 6 shows the values of ln $K_{Ca/Mg}$ and ln $K_{Ca/Fe}$ for all the garnet–melt pairs in the database, plotted against the ratio P/T [see Equation (3)]. These diagrams show that the Ca:Mg and Ca:Fe ratios of garnet increase with increasing pressure, relative to the corresponding ratios in coexisting melt. This agrees with Green's experimental results. Regression of the data yielded a good fit to the Ca–Mg exchange reaction, which confirms the strong pressure effect (Table 6). The Ca–Fe exchange data yielded rather poor fit statistics, particularly for the temperature coefficient. Of the three garnet–liquid exchange equilibria, however, only two are linearly independent. The coefficients for the Ca–Fe exchange reaction are thus a linear combination of the Fe–Mg and Ca–Mg coefficients, and the Ca–Fe coefficients shown in Table 6 were calculated in this manner. The calculated temperature coefficient for the Ca–Fe exchange is smaller than the temperature coefficients for the other garnet–melt exchange equilibria, which may account for the poor statistical fit to the Ca–Fe exchange data.

The simple functional relationships of garnet–melt and biotite–melt exchange reactions with pressure and temperature suggest that these equilibria can be used as thermometers and barometers for volcanic rocks containing biotite or garnet phenocrysts. In particular, the fact that Ca–Fe exchange between garnet and melt is a strong function of pressure, whereas Fe–Mg exchange is a strong function of temperature (Table 6, Fig. 7), supports the suggestion of Green (1977) that a combination of these equilibria can be used to provide accurate pressure and temperature estimates for garnet-bearing volcanic rocks. However, the plots of the equilibrium constants versus temperature and pressure (Figs 5, 6) show some scatter, which is reflected in correlation coefficients that range from 0·75 to 0·85 (Table 6). This scatter suggests that there are other factors (perhaps compositional effects) that must be resolved before a quantitative application to the thermobarometry of volcanic rocks is attempted.

The data used in this regression analysis were obtained at pressures between 5 and 15 kbar. If these data can be extrapolated to pressures of 20–25 kbar, then biotite–melt and garnet–melt exchange parameters predict reversals in the

relative values of the Fe–Mg, Ca–Mg and Ca–Fe ratios at P–T conditions that overlap with possible magmatic conditions in the crust and upper mantle. Figure 7 shows ln $K = 0$ isopleths for the exchange reactions given in Table 6. At pressures higher than those of the Ca–Fe and Ca–Mg isopleths, melt has lower Ca–Fe and Ca–Mg ratios, respectively, than coexisting garnet. At temperatures lower than those of the biotite–melt and garnet–melt Fe–Mg isopleths, the Mg# of melt is higher than the Mg# of coexisting biotite and garnet, respectively. My results for the garnet–melt Fe–Mg exchange equilibrium agree with those of Green (1977), Ellis (1986) and Le Breton and Thompson (1988), all of whom suggested that, at crustal pressures, Fe–Mg exchange between garnet and melt reverses at temperatures of the order of 900°C.

Because the reversal of biotite–melt Fe–Mg partitioning occurs at temperatures that are attainable during the crystallisation of H_2O-rich granitic magmas, there should be an upper bound to the Fe–Mg enrichment of H_2O-rich granitic liquids that can be caused by the fractional crystallisation of biotite. Maximum Fe–Mg enrichment of the liquid should occur at the temperature at which partitioning reverses, because below this temperature biotite fractionation lowers the Fe–Mg ratio of the liquid.

Ellis (1986) and Le Breton and Thompson (1988) suggested that, owing to the reversal of garnet–melt Fe–Mg partitioning, H_2O-saturated melting of natural metapelites could be eutectic, of the general form: garnet + biotite + H_2O → melt (for simplicity, only Fe–Mg bearing phases are shown here). These workers also suggested that a consequence of the eutectic behaviour would be a minimum temperature melting point for rocks of intermediate Fe–Mg ratio. The postulated eutectic behaviour, however, is unlikely. This is so because the H_2O-saturated solidus of quartzo-feldspathic rocks lies almost entirely within a P–T region in which both biotite and garnet are more ferroan than melt (i.e. Mg#$_{melt}$ > Mg#$_{biotite}$ > Mg#$_{garnet}$; Fig. 7). The H_2O-saturated melting reaction is thus likely to be peritectic, of the general form: biotite + H_2O → garnet + melt (see also Grant 1985). In any event, garnet does have a higher Fe–Mg ratio than melt over a wide range of pressures and temperatures within which crustal anatexis is possible, under both fluid-absent and fluid-present conditions (Fig. 7), suggesting that the generation and segregation of granitoid melts from metapelites and metagreywackes may lead to Fe enrichment of the residual mid- to lower continental crust.

7. Conclusions and geological implications

One of the goals of this study is to emphasise that the compositions of primary crustal melts are controlled both by source composition and by intensive variables during melting. The effect of source composition on melt composition is certainly important, as demonstrated by a comparison of melts obtained by the dehydration melting of metagreywackes and metapelites, which are generally of granitic composition, with melts obtained by dehydration melting of basaltic to andesitic amphibolites, which encompass granodiorites, tonalites and trondhjemites (see references cited in the introductory section). However, the experimental results suggest that some of the rich diversity among silicic to intermediate magmas may also be caused by different conditions during melting, in particular by differences in pressure and H_2O content. For example, a wide spectrum of melt compositions can be produced from a fixed biotite- and plagioclase-rich crustal source (e.g. a metagreywacke). This is a consequence of two facts.

Firstly, in biotite–plagioclase gneisses, K_2O resides in biotite, whereas Na_2O and CaO reside in plagioclase, and the two phases respond differently to changes in pressure and the availability of externally derived H_2O (Fig. 1). Low-pressure ($\leqslant 10$ kbar), H_2O-starved dehydration melting of biotite–plagioclase gneisses is governed by the breakdown of biotite and thus yields granitic melts that coexist with pyroxene + plagioclase residues. If melting of the same source takes place at $P > 10$ kbar and with a small influx of H_2O (≈ 1 wt% H_2O), then garnet + biotite are stabilised at the expense of orthopyroxene + plagioclase and the melts shift towards trondhjemitic compositions. In other words, a small amount of externally derived H_2O depresses the plagioclase + quartz solidus more than the biotite breakdown reaction, and if this happens at a pressure that is high enough to stabilise garnet, then the melt will be enriched in the Ab component and depleted in the An component. Greater H_2O influx stabilises orthopyroxene at the expense of garnet ± biotite ± amphibole ± plagioclase and the result is to produce melts of granodiorite (or perhaps even tonalite) composition, that coexist with a pyroxene-rich residue.

Secondly, granitoid melts that coexist with their refractory residues in the source region become more mafic with increasing H_2O activity (Fig. 4; see also Naney 1983; Conrad et al. 1988; Puziewicz & Johannes 1990; Holtz & Johannes 1991). On the one hand, this means that, under a fixed H_2O budget, relatively sodic melts generated deep within thickened continental crust ($P \approx 15$ kbar) would also be more leucocratic than K-rich melts generated from similar sources at shallower levels. On the other hand, the effect of H_2O activity also means that primary granitoid melts with notably different concentrations of ferromagnesian components (e.g. tonalites versus trondhjemites, dacites versus rhyolites) may form at similar temperatures and pressures, but under different degrees of H_2O undersaturation, perhaps caused by different H_2O influxes into the zone of magma generation. In particular, melting of biotite–plagioclase gneisses with a significant influx of externally derived H_2O can generate melts that are granodioritic to tonalitic, not only in terms of their Ab–Or–An contents (see earlier), but also in terms of their FeO + MgO contents (higher than those of granites and trondhjemites).

The possible derivation of trondhjemites from K-rich rocks, and the conditions under which this can happen, merit further discussion. Trondhjemitic and tonalitic rocks often have strongly fractionated rare earth element (REE) patterns and are enriched in both Sr and Eu. This is the case for many Archean tonalite–trondhjemite suites (e.g. Tarney et al. 1979; Jahn et al. 1984; Martin 1987) as well as for many younger low-K igneous rocks (e.g. Defant & Drummond 1990; Drummond & Defant 1990). The combined major and trace element characteristics of trondhjemites have therefore been interpreted as resulting from high–pressure partial melting of K-poor source rocks, at P–T conditions at which garnet predominates over plagioclase in the residuum. Experiments have shown that melts of trondhjemitic composition can be generated at high pressure ($\geqslant 15$ kbar) by the melting of silica-rich tonalites in the presence of externally derived H_2O (e.g. Johnston & Wyllie 1988) and by the dehydration melting of basaltic amphibolites (e.g. Rapp et al. 1991; Winther & Newton 1991; see also review by Wyllie et al. 1996). Many trondhjemite–tonalite suites may indeed have originated from the partial melting of low-K metamorphosed igneous sources, but at least some trondhjemite–tonalite magmas may have derived from the melting of relatively K-rich metagreywackes at high pressure and under a small H_2O influx. The high Na–K ratio, low Ca, high Sr and Eu and fractionated REE patterns of the melts would arise from the incongruent breakdown of plagioclase to garnet (plus albite component in the melt), coupled with the stabilisation of residual biotite (Figs 1, 2). The biotite-rich residues that are generated in this fashion (Figs 1, 2) would give rise to the accumulation of heat-producing elements in the deep crust and could hence facilitate the inception of subsequent magmatic pulses. The partial melting of metagreywackes under small H_2O influx can also explain the origin of migmatites with trondhjemitic leucosomes and biotite + garnet-rich melanosomes (e.g. Whitney & Irving 1994).

On the basis of the strong effect of pressure on garnet–melt Ca–Mg and Ca–Fe partitioning (Table 6, Fig. 7), we can also predict that melts generated in equilibrium with garnet deep within thickened continental crust, or in subducted oceanic crust, must be strongly depleted in Ca and have higher $Al_2O_3/(CaO + Na_2O + K_2O)$ ratios than melts coexisting with garnet at lower pressure. This is a potentially important phase equilibrium constraint on the origin of intermediate to silicic calc-alkaline magmas. In effect, it complements the previous experimental findings of Carroll and Wyllie (1990) and Patiño Douce (1995) in the sense that calc-alkaline magmas are more likely to be produced within the top 30–40 km of the continental lithosphere than at greater depths.

Experiments such as those discussed here and in earlier contributions (see, in particular, Conrad et al. 1988) show that changing pressure and a_{H_2O} can affect the major element compositions of primary crustal melts to the point of concealing the nature of the source material. This should not be seen as an obstacle to understanding the origin of natural granitoid magmas. Rather, the contribution of experimental petrology lies in the fact that, if the nature of the source material can be reasonably constrained (e.g. on the basis of field or tectonic associations and trace element compositions), then the conditions of magma generation (e.g. pressure and the availability of externally derived H_2O) can also be reliably estimated.

8. Acknowledgements

This work was supported by NSF grants EAR-9118418 and EAR-9316304 to Patiño Douce. The microprobe at the University of Georgia was acquired with NSF grant EAR-8816748 and a matching grant from the University of Georgia Research Foundation. Many discussions with Jim

Beard have helped to develop the material presented in this paper. Very constructive reviews by Trevor Green, François Holtz and Bob Luth helped to substantially improve the manuscript.

9. References

Barker, F. 1979. Trondhjemite: definition, environment and hypotheses of origin. In: Barker, F. (ed.) Trondhjemites, dacites, and related rocks. DEV PETROL 6, 1–12.

Beard, J. S. & Lofgren, G. E. 1989. Effect of water on the compositions of partial melts of greenstone and amphibolite. SCIENCE 244, 195–7.

Beard, J. S. & Lofgren, G. E. 1991. Dehydration melting and water-saturated melting of basaltic and andesitic greenstones and amphibolites at 1, 3, and 7 kb. J PETROL 32, 365–401.

Burnham, C. W. 1979. The importance of volatile constituents. In Yoder, H. S. (ed.) The evolution of the igneous rocks, 439–482. Princeton: Princeton University Press.

Carroll, M. R. & Wyllie, P. J. 1990. The system tonalite–H_2O at 15 kbar and the genesis of calc–alkaline magmas. AM MINERAL 75, 345–57.

Chou, I.-Ming 1986. Permeability of precious metals to hydrogen at 2 kb total pressure and elevated temperature. AM J SCI 286, 638–58.

Clemens, J. D. & Wall, V. J. 1981. Origin and crystallization of some peraluminous (S-type) granitic magmas. CAN MINERAL 19, 111–31.

Conrad, W. K., Nicholls, I. A. & Wall, V. J. 1988. Water-saturated and -undersaturated melting of metaluminous and peraluminous crustal compositions at 10 kb: evidence for the origin of silicic magmas in the Taupo Volcanic Zone, New Zealand, and other occurrences. J PETROL 29, 765–803.

Defant, M. J. & Drummond, M. S. 1990. Derivation of some modern arc magmas by melting of young subducted lithosphere. NATURE 347, 662–5.

Dooley, D. F. & Patiño Douce, A. E. 1996. Vapor-absent melting of F- and Ti-rich phlogopite + quartz; effects on phlogopite stability and melt compositions. AM MINERAL 81, 202–12.

Drummond, M. S. & Defant, M. J. 1990. A model for trondhjemite–tonalite–dacite genesis and crustal growth via slab melting: Archean to modern comparisons. J GEOPHYS RES 95, 21 503–21.

Ellis, D. J. 1986. Garnet–liquid Fe^{2+}–Mg equilibria and implications for the beginning of melting in the crust and subduction zones. AM J SCI 286, 765–91.

Grant, J. A. 1985. Phase equilibria in partial melting of pelitic rocks. In Ashworth, J. R. (ed.) Migmatites, 86–144. Glasgow: Blackie.

Green, T. H. 1976. Experimental generation of cordierite- or garnet-bearing granitic liquids from a pelitic composition. GEOLOGY 4, 85–8.

Green, T. H. 1977. Garnet in silicic liquids and its possible use as a P–T indicator. CONTRIB MINERAL PETROL 65, 59–67.

Green, T. H. 1992. Experimental phase equilibrium studies of garnet-bearing I-type volcanics and high-level intrusives from Northland, New Zealand. TRANS R SOC EDINBURGH EARTH SCI 83, 429–38.

Green, T. H. & Ringwood, A. E. 1972. Crystallization of garnet-bearing rhyodacite under high–pressure hydrous conditions. J GEOL SOC AUST 19, 203–12.

Helz, R. T. 1976. Phase relations of basalts in their melting ranges at $P(H_2O) = 5$ kb. Part II. Melt compositions. J PETROL 17, 139–93.

Holtz, F. & Johannes, W. 1991. Genesis of peraluminous granites I. Experimental investigation of melt compositions at 3 and 5 kb and various H_2O activities. J PETROL 32, 935–58.

Jahn, B. M., Vidal, P. & Kroner, A. 1984. Multichronometric ages and origin of Archean tonalitic gneiss in Finnish Lapland: a case for long crustal residence time. CONTRIB MINERAL PETROL 86, 398–408.

Johnston, A. D. & Wyllie, P. J. 1988. Constraints on the origin of Archean trondhjemites based on phase relations of Nûk gneiss with H_2O at 15 kbar. CONTRIB MINERAL PETROL 100, 35–46.

Le Breton, N. & Thompson, A. B. 1988. Fluid-absent (dehydration) melting of biotite in metapelites in the early stages of crustal anatexis. CONTRIB MINERAL PETROL 99, 226–37.

Maaløe, S. & Wyllie, P. J. 1975. Water content of a granite magma deduced from the sequence of crystallization determined exper-imentally with water-undersaturated conditions. CONTRIB MINERAL PETROL 52, 175–91.

Martin, H. 1987. Petrogenesis of Archean trondhjemites, tonalites, and granodiorites from eastern Finland: major and trace element geochemistry. J PETROL 28, 921–53.

Naney, M. T. 1983. Phase equilibria of rock-forming ferromagnesian silicates in granitic systems. AM J SCI 283, 993–1033.

Naney, M. T. & Swanson, S. E. 1980. The effect of Fe and Mg on crystallization in granitic systems. AM MINERAL 65, 639–53.

Nash, W. P. 1992. Analysis of oxygen with the electron microprobe: applications to hydrated glasses and minerals. AM MINERAL 77, 453–6.

Patiño Douce, A. E. 1993. Titanium substitution in biotite: an empirical model with applications to thermometry, O_2 and H_2O barometries, and consequences for biotite stability. CHEM GEOL 108, 133–62.

Patiño Douce, A. E. 1995. Experimental generation of hybrid silicic melts by reaction of high-Al basalt with metamorphic rocks. J GEOPHYS RES 100, 15 623–39.

Patiño Douce, A. E. & Beard, J. S. 1994. Water loss from hydrous melts during fluid-absent piston–cylinder experiments. AM MINERAL 79, 585–8.

Patiño Douce, A. E. & Beard, J. S. 1995. Dehydration-melting of biotite gneiss and quartz amphibolite from 3 to 15 kbar. J PETROL 36, 707–38.

Patiño Douce, A. E. & Beard, J. S. Effects of P, $F(O_2)$ and Mg/Fe ratio on dehydration melting of model metagreywackes. J PETROL, in press.

Patiño Douce, A. E. & Johnston, A. D. 1991. Phase equilibria and melt productivity in the pelitic system: implications for the origin of peraluminous granitoids and aluminous granulites. CONTRIB MINERAL PETROL 107, 202–18.

Patiño Douce, A. E. Johnston, A. D. & Rice, J. M. 1993. Octahedral excess mixing properties in biotite: a working model with applications to geobarometry and geothermometry. AM MINERAL 78, 113–31.

Puziewicz, J. & Johannes, W. 1988. Phase equilibria and compositions of Fe–Mg–Al minerals and melts in water-saturated peraluminous granitic systems. CONTRIB MINERAL PETROL 100, 156–68.

Puziewicz, J. & Johannes, W. 1990. Experimental study of a biotite-bearing granitic system under water-saturated and water-undersaturated conditions. CONTRIB MINERAL PETROL 104, 397–406.

Rapp, R. P., Watson, E. B. & Miller, C. F. 1991. Partial melting of amphibolite/eclogite and the origin of Archean trondhjemites and tonalites. PRECAMBRIAN RES 51, 1–25.

Rushmer, T. 1991. Partial melting of two amphibolites: contrasting experimental results under fluid absent conditions. CONTRIB MINERAL PETROL 107, 41–59.

Rutter, M. J. & Wyllie, P. J. 1988. Melting of vapour-absent tonalite at 10 kbar to simulate dehydration-melting in the deep crust. NATURE 331, 159–60.

Rousseeuw, P. J. & Leroy, A. N. 1987. Robust regression and outlier detection, 75–245. New York: Wiley.

Scaillet, B., Pichavant, M. & Roux, J. 1995. Experimental crystallization of leucogranitic magmas. J PETROL 36, 663–705.

Silver, L. A., Ihinger, P. D. & Stolper, E. 1990. The influence of bulk composition on the speciation of water in silicate glasses. CONTRIB MINERAL PETROL 104, 142–62.

Skjerlie, K. P. & Johnston, A. D. 1992. Vapor-absent melting at 10 kbar of a biotite- and amphibole–bearing tonalitic gneiss: implications for the generation of A-type granites. GEOLOGY 20, 263–6.

Skjerlie, K. P. & Johnston, A. D. 1993. Fluid-absent melting behavior of an F-rich tonalitic gneiss at mid-crustal pressures: implications for the generation of anorogenic granites. J PETROL 34, 785–815.

Skjerlie, K. P., Patiño Douce, A. E. & Johnston, A. D. 1993. Fluid-absent melting of a layered crustal protolith: implications for the generation of anatectic granites. CONTRIB MINERAL PETROL 114, 365–78.

Spulber, S. D. & Rutherford, M. J. 1983. The origin of rhyolite and plagiogranite in oceanic crust: an experimental study. J PETROL 24, 1–25.

Stern, C. R. & Wyllie, P. J. 1978. Phase compositions through crystallization intervals in basalt–andesite–H_2O at 30 kbar with implications for subduction zone magmas. AM MINERAL 63, 641–63.

Tarney, J., Weaver, B. & Drury, S. A. 1979. Geochemistry of Archaean trondhjemitic and tonalitic gneisses from Scotland and East Greenland. In: Barker, F. (ed.) Trondhjemites, dacites, and related rocks. DEV PETROL 6, 275–99.

Tuttle, O. F. & Bowen, N. L. 1958. Origin of granite in the light of experimental studies in the system $KAlSi_3O_8$–$NaAlSi_3O_8$–SiO_2–H_2O. GEOL SOC AM MEM **74.**

Vielzeuf, D. & Holloway, J. R. 1988. Experimental determination of the fluid-absent melting relations in the pelitic system. Consequences for crustal differentiation. CONTRIB MINERAL PETROL **98,** 257–76.

White, A. J. R. & Chappell, B. W. 1983. Granitoid types and their distribution in the Lachlan Fold Belt, Southeastern Australia. *In* Roddick, J. A. (ed.) *Circum-Pacific plutonic terranes.* GEOL SOC AM MEM **159,** 21–34.

Whitney, D. L. & Irving, J. A. 1994. Origin of K-poor leucosomes in a metasedimentary migmatite complex by ultrametamorphism, syn-metamorphic magmatism, and subsolidus processes. LITHOS **32,** 173–92.

Winther, K. T. & Newton, R. C. 1991. Experimental melting of hydrous low-K tholeiite: evidence on the origin of Archean cratons. BULL GEOL SOC DENMARK **39,** 213–28.

Wolf, M. B. & Wyllie, P. J. 1991. Dehydration-melting of solid amphibolite at 10 kbar: textural development, liquid interconnectivity and applications to the segregation of magmas. MINERAL PETROL **44,** 151–79.

Wolf, M. B. & Wyllie, P. J. 1994. Dehydration-melting of amphibolite at 10 kbar: the effects of temperature and time. CONTRIB MINERAL PETROL **115,** 369–83.

Wyllie, P. J., Carroll, M. R., Johnston, A. D., Rutter, M. J., Sekine, T. & van der Laan, S. R. 1989. Interactions among magmas and rocks in subduction zone regions: experimental studies from slab to mantle to crust. EUR J MINERAL **1,** 165–79.

Wyllie, P. J., Wolf, M. B. & van der Laan, S. R. 1996. Conditions for formation of tonalites and trondhjemites: magmatic sources and products. *In* de Wit, M. J. & Ashwal, L. D. (eds) *Tectonic evolution of greenstone belts,* 258–67. Oxford: Oxford University Press.

ALBERTO E. PATIÑO DOUCE, Department of Geology, University of Georgia, Athens, GA 30602, U.S.A.

Transactions of the Royal Society of Edinburgh: Earth Sciences, **87**, 23–31, 1996

Dynamic model of dehydration melting motivated by a natural analogue: applications to the Ivrea–Verbano zone, northern Italy

Scott A. Barboza and George W. Bergantz

ABSTRACT: Dehydration melting of crustal rocks may commonly occur in response to the intrusion of mafic magma in the mid- or lower crust. However, the relative importance of melt buoyancy, shear or dyking in melt generation and extraction under geologically relevant conditions is not well understood. A numerical model of the partial melting of a metapelite is presented and the model results are compared with the Ivrea–Verbano Zone in northern Italy. The numerical model uses the mixture theory approach to modelling simultaneous convection and phase change and includes special ramping and switching functions to accommodate the rheology of crystal–melt mixtures in accordance with the results of deformation experiments. The model explicitly includes both porous media flow and thermally and compositionally driven bulk convection of a restite-charged melt mass. A range of melt viscosity and critical melt fraction models is considered. General agreement was found between predicted positions of isopleths and those from the Ivrea–Verbano Zone. Maximum melt velocities in the region of porous flow are found to be 1×10^{-7} and 1×10^{-1} m per year in the region of viscous flow. The results indicate that melt buoyancy alone may not be a sufficient agent for melt extraction and that extensive, vigorous convection of partially molten rocks above mafic bodies is unlikely, in accord with direct geological examples.

KEY WORDS: melt segregation, partial melting, convection, mixture theory, underplating, phase change, critical melt fraction, lower crust, Ivrea Zone.

Many continental granitoids contain at least some component of crustally derived material (Wyllie 1977; Pitcher 1993; Brown 1994; Harris *et al.* 1995). One mechanism for generating crustal melts is dehydration melting following the ponding of mantle-derived magma within or at the base of the crust (Huppert & Sparks 1988; Bergantz & Dawes 1994). Of particular interest is the melting of pelitic rocks, as there is ample field evidence for this process during contact metamorphism (Grant & Frost 1990; Harris *et al.* 1995; Symmes & Ferry 1995). However, little is known about the rates of melt generation and mechanisms of melt transport. Some studies have indicated that melt may be removed from the source region through an efficient mechanism. For example, in the Ballachulish aureole, granitic melt appears to have segregated into leucosomes in less than 5×10^4 years (Buntebarth 1991; Sawyer 1994). Sawyer (1991) proposed that tonalitic melt in the Grenville front probably separated from its source in less than 100 years.

Clemens and Mawer (1992) have argued that the mechanism for melt removal may be fracturing resulting from the positive ΔV of hydrate breakdown during dehydration melting, whereas other studies have emphasised the role of deformation and shear (Sawyer 1994; Brown *et al.* 1995; Davison *et al.* 1995; Rushmer 1995). Melt segregation in the mantle, on the other hand, is thought to result primarily from buoyancy forces which initiate the porous flow of melt (Nicolas 1989). It is likely that each of these mechanisms (fracturing, shear and buoyancy), to some extent, contributes to the extraction of crustal melts from their source. However, the relative importance of buoyancy or shear in the melt transport process is poorly understood.

The thermal evolution of underplating regions has also been

a subject of much debate. Analogue experiments and mathematical models have led to the proposal that underplating of basaltic magmas will lead to bulk melting of the country rock and the thermal evolution of underplating regions will be dominated by convective processes (Huppert & Sparks 1988). However, there is little direct geological or geochemical evidence to support this view. As a result, the style and vigour of convection in both the basaltic sill and the overlying country rock have been subject to much debate (Huppert & Sparks 1991; Marsh 1991) and numerous alternative models have been advanced (Bergantz & Dawes 1994).

One of the central themes of this study is that rheological variations associated with the increase in melt fraction during melting provides the dominant control on the dynamics of the growing region of partial melt. The notion of a distinct critical melt fraction (CMF), at which the skeleton of restitic crystals breaks down and stresses are supported by the fluid phase, requires substantial revision for geological systems undergoing partial melting. The rheological transitions in partially molten rocks can be complex and difficult to generalise (Rushmer 1995; Rutter & Neumann 1995). The rheological model invoked in this study is an extension of mixture theory for two-phase systems (Agarwal & O'Neill 1988; Ni & Beckermann 1991; Oldenburg & Spera 1992). The CMF can still play a part in the rheological characterisation of partially molten material, but it does not represent an abrupt transition from a rigid skeleton to a crystal-laden fluid, but rather a gradual transition between the two.

The intent of this study is to explore the relationship between melt generation, the dynamics of melt movement and the thermal evolution of an underplated region adhering to documented geological conditions. To this end a thermomechanical model has been developed: (1) to explore the time and length scales of dehydration melting and the dynamics of melt movement following underplating; and (2) to evaluate the influence of thermally and compositionally induced buoyancy as a mechanism for the extraction of felsic melt. A numerical study is particularly well suited to address such questions as experiments carried out with laboratory analogues cannot be used to assess the relative importance of such processes in a physically realistic manner (Bergantz & Dawes 1994; Bergantz 1995; Jaupart & Tait 1995). A similar modelling approach has been used by Irvine (1970), Bergantz (1989) and Bowers et al. (1990) for evaluating partial melting during contact metamorphism. However, these studies allowed no convection in the region of partial melting and cannot resolve issues related to the dynamics of melt movement.

The geometrical and thermal relationships used in this study are based on current models for the emplacement dynamics of the underplated igneous complex at the Ivrea–Verbano Zone (IVZ) (Quick et al. 1994). A simplified version of this process is summarised in Figure 1. Mantle-derived magma near its liquidus temperature (1200°C) is emplaced in a small, spatially stable sill at the base of an initially unmelted sequence of pelitic rocks in the lower crust. The growth of a solid, conductive quenched margin causes the contact to stabilise at a temperature intermediate between that of the country rock and the initial temperature of the intruding basalt. Melting in the country rock proceeds as the intrusion cools and buoyant instabilities develop in the growing lens of felsic melt. The movement of melt that results from these instabilities advects heat away from the contact with the intrusion and yields a region of partial melting whose geometry and rate of growth differs significantly from both conductive models and those that assume convection is turbulent.

1. The Ivrea–Verbano Zone: a field example

The results of this study will be compared with the IVZ in northern Italy, arguably the premier locality for the study of dehydration melting subsequent to underplating. The IVZ (Fig. 2) is a lower crustal section consisting of a 140 km long sequence of steeply dipping pelitic and mafic rocks metamorphosed under amphibolite to granulite facies conditions (Zingg 1980). To the NW, the IVZ is separated from rocks of the

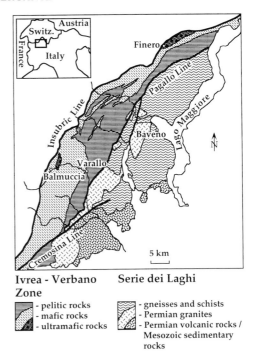

Figure 2 Geological sketch map of the southern Alps west of the Lago Maggiore (modified from Zingg 1980). Inset shows location of area in northern Italy.

Austroalpine Domain by the Insubric Line—a major shear zone that marks the southern extent of Alpine deformation (Schmid et al. 1987). To the SE, the IVZ is separated from rocks of intermediate crustal depth within the Serie dei Laghi unit by the Cremosina and Pogallo Lines (Boriani et al. 1977, 1988; Zingg 1983).

The IVZ is composed of two major lithologic divisions—the Kinzigite Formation and the Mafic Complex. The Mafic Complex is primarily composed of rocks of gabbroic composition, but also includes large peridotite bodies near the Insubric Line and dioritic rocks near the contact with the Kinzigite Formation (Rivalenti et al. 1975, 1980; Sinigoi et al. 1994). The Kinzigite Formation is composed predominantly of sillimanite-bearing paragneisses termed 'kinzigites' and 'stronalites' in the amphibolite and granulite grade, respectively (Zingg et al. 1980). The rocks of the Kinzigite Formation were exposed to increasingly higher degrees of metamorphism towards the contact with the Mafic Complex and, importantly, are increasingly depleted in the granitophile elements with increasing degree of metamorphism (Schmid & Wood 1976; Schmid 1978/79). Together, the IVZ and the Serie dei Laghi are thought to represent an uplifted cross-section through the continental crust (Mehnert 1975; Fountain 1976).

2. Dehydration melting

The melting relationships used in this model are that of the fluid-absent (dehydration) melting of metapelitic rock. Dehydration melting is defined as the melt reaction that occurs when all of the water in the system is initially contained within the hydrous mineral assemblage and is transferred directly to the melt during the melting reaction without ever appearing as a product or reactant (Rushmer 1991). There are numerous studies which emphasise the importance of this process for crustal growth and evolution (Brown & Fyfe 1970; Thompson 1982; Yardley 1986; Whitney 1988; Beard & Lofgren 1989). We choose to focus on dehydration melting in this study because it has been suggested as: (1) a major source of contamination in intruding mafic magmas (e.g. Sinigoi et al.

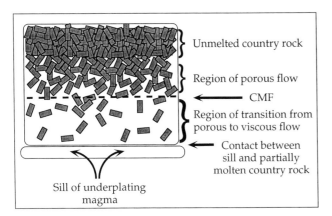

Figure 1 Schematic diagram for melting model. A sill of mantle-derived magma is intruded at the base of a sequence of pelitic rocks. Heat transfer to the sequence causes partial melting of the pelitic rocks and results in buoyant instabilities in the pelite–melt mixture. These instabilities cause the upwards flow of material from the region near the contact with the sill, further enhancing the heating and melting of the pelitic sequence above. When the local volume fraction of melt exceeds the critical melt fraction (CMF), the rheological transition between the porous and viscous flow end-member rheological regimes is initiated.

1994); (2) a source of migmatites and other water-undersaturated granitic melts (Rushmer 1991); (3) a cause of some granulite formations (Mehnert 1968; Fyfe 1973; Wickham 1987); and (4) a source of melts of tonalitic to trondhjemitic composition (Rushmer 1991).

3. Methodology

3.1. General nature of melt movement

A region undergoing melting is essentially a solid–liquid mixture with multiple phase changes. Unlike a pure substance, there is no distinct crystallisation front; rather the solid melts over a temperature range (Fig. 1). Mixture theory provides an approach for the development of the equations describing the transport of energy, mass, momentum and chemical species. A comprehensive discussion of the theory and derivation of the full set of continuum governing equations is discussed in detail elsewhere and will not be revisited here (Bennon & Incropera 1987; Prakash & Voller 1989; Oldenburg & Spera 1991). A detailed discussion with regards to the derivation and specific application is given in Barboza (1995). The key feature of the mixture theory approach is that it explicitly allows for the full range of dynamic conditions, from porous media flow at low melt fraction to fully liquid behaviour at high melt fraction.

The melt fraction at which the solid skeleton breaks down and the flow regime changes from one of porous flow to that of viscous flow with suspended solids is the CMF. The fraction of melt is less than the CMF in the region of porous flow where the solid skeleton is considered to be fixed relative to the percolating interstitial melt. A wide range of estimates for the CMF of a number of compositions are available. Experimental deformation of partially molten granite yields estimates of CMF $\approx$ 25–30% (Arzi 1978; van der Molen & Paterson 1979). Estimates based on the distribution of the phenocryst content of erupted lava and theoretical arguments based on contiguity limits yield more conservative estimates of CMF $\approx$ 50% (Marsh 1981; Miller et al. 1988). However, the experiments of Rutter and Neumann (1995) produced no sharp discontinuity in strength with increasing melt fraction. In light of this result, we have used a ramping, switching function to model the transition in drag from Darcy-dominated drag to suspension-dominated drag (Figure 3). Hence the transition from Darcy-dominated porous to suspension-dominated viscous flow did not occur at a distinct value of the CMF. References to the CMF in this study should be understood to be a reference melt fraction at which the transition to suspension-dominated flow is initiated, not as indicating an abrupt transition in rheology. Other rheological models could have been chosen; our choice reflects a best estimate taken from current experimental results. The permeability in the region of porous flow was calculated using the Blake–Carmen–Kozeny relation.

Melt movement in both the porous flow and viscous flow regime was initiated by buoyant instabilities that arise from variations in the density of the melting pelite. These density changes are strongly dependent on the distribution of suspended solids, the melt composition and the temperature. The magnitude of the buoyancy force resulting from these density changes is incorporated in the thermal and solutal expansion coefficients in the buoyancy source term for the momentum equation. The expansion coefficients were derived from the experimental data with calculations using the MELTS algorithm (Ghiorso & Sack 1995).

The melt viscosity was calculated from the experimental data by the Shaw model (Shaw 1972) and the Krieger and Dougherty relation was used to account for the influence of

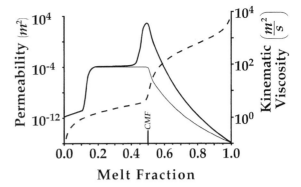

Figure 3 Plot of the kinematic viscosity (solid lines) and permeability (broken line) as a function of the melt fraction. The thin solid line gives the variation of the viscosity of the melt only and the thick solid line is the model mixture viscosity, which includes the influence of suspended solids. The low viscosity at low melt fractions arises because of the high water content of the experimental melts (Vielzeuf & Holloway 1988). The critical melt fraction (CMF) is the melt fraction at which the transition from porous to viscous flow is initiated. The CMF, in this instance, is 0·5. Simulations using different values for CMF had qualitatively similar relationships.

suspended solids (Wildemuth & Williams 1984; Oldenburg & Spera 1992; Bergantz & Dawes 1994). As with permeability, a switching function has been incorporated in the viscosity relation to ensure that the influence of suspensions does not affect the effective viscosity in the region of porous flow (Oldenburg & Spera 1992). Figure 3 depicts the kinematic viscosity as a function of melt fraction for the melt and for the magma (melt + solid suspensions). The temperature and composition dependence of the viscosity of the melt was calculated from the experimental melt compositions and temperatures as a function of the observed melt fractions and, therefore, assumes equilibrium melting (Vielzeuf & Holloway 1988). The melt viscosity is constant between melt fractions of 0·15 and 0·5 because the melt generated within this range takes place at the biotite dehydration invariant point, so temperature and composition do not change. The large increase in effective viscosity just above the CMF approximates the influence of suspended solids on the viscosity of the magma during viscous flow. Although the Krieger and Dougherty relation yields excellent agreement with experimental data (Wildemuth & Williams 1984), some work has demonstrated that, under certain conditions, the Shaw model may overestimate the viscosity of a silicate melt by up to two orders of magnitude (Schulze et al. 1994). Thus the model simulations were conducted using the entire range of proposed felsic melt viscosities.

3.2. Thermodynamic model

The modelling strategy adopted in this study is that of mixture theory where the multi-phase governing equations are averaged, yielding a single set of governing equations for the melt–restite mixture. Closure of these governing equations for the solid–liquid mixture requires that additional thermodynamic functions be developed to link the enthalpy and composition to the melt fraction. By assuming local equilibrium, the phase diagram provides the necessary relationships.

The dehydration melting of both muscovite and biotite (neglecting the influence of biotite solid solution) takes place at invariant points. The composition of the melt produced at these invariant points will remain constant as long as the restitic phases are in equilibrium with the invariant point melt composition. In addition, excess alumina and mafic oxides have very low solubilities in melts dominated by the orthoclase and H_2O components. The particular mica composition undergoing melting will therefore influence the composition of the restitic phases, but have only a limited influence on the composition of the melt (Patiño Douce & Johnston 1990). As the composition of the melt produced during the dehydration melting of either mica is approximately the same, the melt generated at either invariant point and the suite of resulting restite compositions define a line within the compositional volume. Removal of this melt will cause the composition of the remaining unmelted restite to evolve linearly (Patiño Douce, pers. comm.) and a simple pseudo-binary phase diagram may capture much of the melting systematics of some pelitic rocks. Errors in melt fraction estimates resulting from this approximation will be minimal at the relatively low temperatures of interest where the refractory phases are not involved in the melting reaction. In addition, smaller errors are expected for pelite compositions in which melting occurs dominantly at one or the other of the invariant points.

The pelite composition used in the experiments of Vielzeuf and Holloway (1988), in which melting takes place predominately by biotite dehydration melting, was selected as our model composition. Three benefits are obtained by choosing this particular pelite. Firstly, as discussed earlier, errors in the predicted melt composition will be minimised when most of the melting occurs at an invariant point. Secondly, this composition is optimum for melt production. By 'optimum', we mean that this composition produces more melt at lower temperatures than most naturally occurring crustal rocks (Bergantz & Dawes 1994). This composition provides a useful end-member for evaluating the general dynamics of melt production and extraction in the lower crust. Finally, the composition corresponds reasonably well to the average metagreywacke composition found in the Kinzigite Formation of the IVZ (Schnetger 1994). The major difference in mineralogy is that the Kinzigite Formation metagreywacke contains 6% K-feldspar and 3% muscovite, whereas the experimental pelite composition contains 0 and 9%, respectively, of these minerals.

3.2.1. Building the binary. The first step in the thermodynamic parameterisation is to build a pseudo-binary from which to generate liquid fraction and composition data. The 10 kbar NKM ($Na_2O \cdot Al_2O_3$–$K_2O \cdot Al_2O_3$–$MgO + FeO$)-, silica-, alumina-, titania-saturated pseudo-ternary (Fig. 4) was chosen on which to plot the experimental melt compositions. The phase relations, invariant points and cotectics were derived by Patiño Douce and Johnston (1990) using their own data with those of Vielzeuf and Holloway (1988) and Le Breton and Thompson (1988). We observe that the suite of melt compositions along with the invariant points form an approximately linear array across this pseudo-ternary. Our methodology was to extend a binary from the M apex of the pseudo-ternary through the array of compositions and invariant points to the intersection with the N–K join (M'). The experimental melt compositions and derived invariant points were then projected onto the binary join and composition was parameterised as a fractional distance along the pseudo-binary. The liquidus line was constrained by the projected experimental melt compositions and the temperatures at which they were generated. The solidus line was derived using the lever rule by fitting the variation of liquid fraction with temperature observed in the experiments and calculated from mass balance with that predicted by the model (Fig. 5).

3.3. Model description

We consider melting of a sequence of pelitic rocks in the two-dimensional, 5 km × 5 km computational domain overlying the contact with a sill of basaltic magma in the lower crust. The geometry of the sill underlying the computational domain

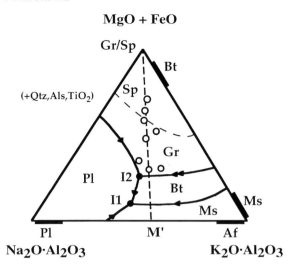

Figure 4 10 kbar NKM-, silica-, alumina- and titania-saturated pseudo-ternary liquidus phase relations projected from the KNMH volume (Patiño Douce & Johnston 1990). Closed circles are the biotite dehydration invariant point (I2) and the muscovite dehydration invariant point (I1). Open circles are the projections of the Vielzeuf and Holloway (1988) experimental liquid compositions. Thick broken line between the MgO + FeO apex and M' is the pseudo-binary used in the model. Mineral abbreviations: Sp = spinel; Gr = garnet; Pl = plagioclase feldspar; Af = alkali feldspar; Ms = muscovite; Qtz = quartz; Als = aluminosilicate; Bt = biotite.

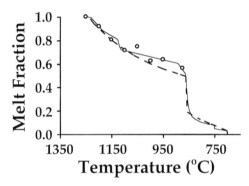

Figure 5 Predicted variation of melt fraction as a function of temperature using a lever rule formulation and the phase relationships depicted in Figure 4. Broken line is the model melt fraction curve; the solid line is that calculated by Vielzeuf and Holloway (1988) using inferred melting relationships. The open circles are the melt fractions observed in the melting experiments (Vielzeuf & Holloway 1988).

was based on current estimates of the sill width during emplacement of the Mafic Complex at the IVZ (Quick et al. 1994). The sill was stipulated to be 2 km in width and long enough that the important features of the convection could be captured in two dimensions (Fig. 1). No assumption was made about the sill thickness except that it contained enough basaltic magma to maintain the contact temperature throughout the duration of the simulations. This will be addressed in more detail in the following.

The computational domain was oriented such that the right wall was parallel to the vertical plane of symmetry of the sill (Fig. 1). The results thus depict melting in a semi-infinite half-space over one limb of the intruding sill. A 65 × 65 variable spaced grid was used and the domain was separated into regions for the purposes of grid point distribution. The highest density of nodes was placed in a 2 km × 2 km region surrounding the simulated contact with the sill. All simulations were halted before the melting front left the region of higher resolution grid spacing. A grid refinement study was also undertaken to ensure that the solutions were independent of spatial and temporal grid spacing.

At time $t=0$, the temperature along the base of the computational domain up to 1 km from the right-hand wall was elevated to a value of $T_{contact}$, simulating the intrusion of a 2 km wide sill of underplating mafic magma. The contact temperature was maintained through the duration of the simulations. The other walls of the domain were insulated, but all simulations were halted before an appreciable increase in temperature (1–5°C) was observed at the roof or left wall. This ensured that the thermal boundary conditions did not significantly influence the solution. No-slip boundary conditions were used on the left wall, roof and floor. Free-slip conditions were allowed on the right wall as it was the axis of symmetry of melt movement.

The material in the domain was initially the unmelted fertile pelite composition of Vielzeuf and Holloway (1988). The initial pressure and temperature in the domain were 10 kbar and 600°C—about 150°C below the solidus of the model composition pelite and corresponding to the approximate depth at which the Mafic Complex was probably intruded (Zingg 1983). These conditions correspond to a relatively high continental geotherm (17°C/km), but are similar to predicted geotherms for modern underplating regions such as in NE Japan and the Rio Grande Rift (Kay & Kay 1980; Hyndman 1981). Two sets of simulations (case 1 and case 2) were performed using different contact temperatures ($T_{contact}$) over the range of estimates for these rheological parameters. A summary of the different combinations of $T_{contact}$ and rheological parameters used in the simulations is given in Table 1, along with the other model thermo-physical properties.

4. Results and discussion

4.1. Thermal evolution and the distribution of melt above the underplating sill

The vigour of convection and the amount of melt produced in the pelite was found to be strongly dependent on $T_{contact}$ and the combinations of values for the rheological parameters of the system (CMF and μ^l/μ^l_{Shaw}). Note that μ^l/μ^l_{Shaw} is the viscosity of the melt used by a particular simulation divided by the viscosity predicted by the Shaw model. A simple way of evaluating the convective vigour is by monitoring the evolution of the melt distribution in the partially molten pelite above the sill and by comparing the melt distribution for simulations with different values of $T_{contact}$ and the rheological

Table 1 Data for the model.

Property (units)	Numerical value
Property data	
Specific heat (J/kg K)	1.04×10^3
Thermal conductivity (W/m K)	1.9
Density (kg/m³)	2.507×10^3
Kinematic viscosity (m²/s)	See text
Schmidt number (μ^l/D)	1.0×10^6
Latent heat (J/kg)	1.0×10^5
Permeability coefficient (m²)	5.56×10^{-10}
Thermal expansion coefficient (K⁻¹)	1.05×10^{-4}
Solutal expansion coefficient (C⁻¹)	1.55×10^{-1}
Initial conditions	
Pelite starting composition	0.325
Initial temperature (°C)	600.0

Contact temperatures and rheological parameters used in simulations

Contact temperature (°C)	Case 1 = 1000.0
	Case 2 = 900.0
Range of CMF	0.35, 0.4, 0.45, 0.5, 0.55
Range of μ^l/μ^l_{Shaw}*	1.0, 0.5, 0.1, 0.01

* μ^l/μ^l_{Shaw} is the ratio of the melt viscosity used in the simulations to that predicted by the Shaw model using MELTS.

parameters. We use profiles of the melt distribution rather than the total mass of melt to emphasise the field comparison because the model pelite composition exhibits large jumps in melt fraction which correspond with the disappearance of easily recognisable minerals from the assemblage. The profiles are monitored along the right wall of the computational domain and three cases are presented for comparison (Fig. 6). Plot A is a case 1 ($T_{contact} = 1000°C$) simulation in which the convection was observed to be vigorous relative to the other simulations. Plot B (case 1) and plot C (case 2) are both simulations in which convection was suppressed. The sharp increase in melt fraction furthest from the contact with the sill is caused by muscovite dehydration melting. The second large increase in melt fraction closer to the contact is biotite dehydration. The spatial position of the muscovite-out (I1) and biotite-out (I2) transitions are labelled on plot B.

Note that in the case 1 simulations the distance the biotite-out transition had propagated from the contact was strongly dependent on the presence of convection. The distance the muscovite-out transition had propagated, on the other hand,

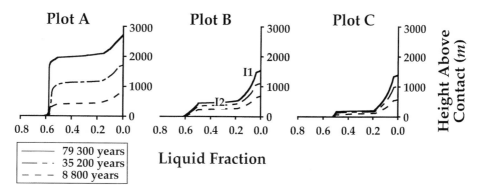

Figure 6 Evolution of melt distribution. Plot A is a plot of the distribution of melt within the pelite–melt mixture with height above the apex of the underplating sill during a convection-dominated simulation. Plot A is a case 1 simulation (see text) in which $\mu^l/\mu^l_{Shaw} = 0.1$ and CMF = 0.45. Three curves are shown in plot A depicting the melt fraction at different times during the simulation. Plots B and C are conduction-dominated simulations from cases 1 and 2, respectively. The sharp increase in melt fraction furthest from the contact is the location of the muscovite melting reaction (I1). The second large increase in liquid fraction closer to the contact is that of the biotite melting reaction (I2). The position of the invariant points is labelled on one curve in plot B or reference. CMF = critical melt fraction; μ^l/μ^l_{Shaw} is the viscosity used in a particular simulation divided by the calculated viscosity using the Shaw model.

was relatively unaffected by advective transport of heat through most the duration of the simulations. For instance, in the presence of convection after 3.5×10^4 years (plot A in Fig. 6), muscovite was no longer stable in the assemblage at a height of about 1·6 km above the contact with the sill. The position of the muscovite-out isograd is about 1·1 km in the conduction case—only 0·5 km closer to the contact. The biotite-out transition, on the other hand, had propagated 1·2 km from the contact with the intrusion in the convective case (plot A), but only 0·4 km in the conduction case (plot B). We conclude that, in the field, the muscovite-out transition will probably lie relatively close to a conductive isotherm whether or not buoyancy-driven melt movement was an important component of the thermal evolution of the region.

In the IVZ, the muscovite-out isograd lies approximately 1 km stratigraphically above the contact with the Mafic Complex (Zingg 1980) in the vicinity of what is believed to be the locus of the intrusion (Quick *et al.* 1994). The simulations demonstrate that, even with the lowest rates of heat flux to the country rock overlying the sill (plot C), the muscovite-out transition will propogate 1 km after only 3.5×10^4 years. The exposed portion of the Mafic Complex south of Val Strona di Omegna covers an area of over 350 km^2 (Fig. 2) and geophysical studies indicate that the complex may project many tens of kilometres into the crust (Giese 1968; Kissling 1980; Zingg 1983). These observations appear to require the unlikely combination of extremely high magma supply rates to the mafic complex (higher than 0·4 km^3 per year), coupled with limited heating of the overlying country rock, as indicated by the relatively short distance the muscovite-out transition propagated from the contact. For comparison, estimates of the average effusion rates for Hawaii and the Columbia River Basalt Group are of the order of 0·1 km^3 per year (Swanson *et al.* 1989). We believe that it is more likely that the temperature of the overlying country rock reached a steady-state temperature lower than might be expected given the volume of magma that made up the Mafic Complex. This indicates that magmatism during the emplacement of the Mafic Complex was probably episodic, punctuated by long periods of stagnation.

4.2. Convective velocities and the rheological and thermal envelope of convection

A more rigorous evaluation of convective vigour was undertaken by monitoring the time evolution of the heat absorption ratio (H_{abs}). H_{abs} is defined as the total amount of latent and sensible heat absorbed by the system (H_{conv}) in the presence of convection relative to that calculated by a simulation performed under the same conditions in which the heat transfer was forced to occur by conduction only (H_{cond}).

$$H_{abs} = \frac{H_{conv}}{H_{cond}}$$

For the purposes of discussion, simulations in which $H_{abs} = 1$ were termed 'conduction-dominated' as conduction was the dominant mode of heat transfer in the system. Simulations in which $1 < H_{abs} < 2$ were termed 'convection-influenced' and those in which $H_{abs} \geq 2$ were termed 'convection-dominated' as the heat transfer in the presence of convection was at least twice that of conduction alone.

Convective velocities in all simulations in cases 1 and 2 were variable, but generally small. Maximum convective velocities in the region of porous flow were of the order of 1×10^{-7} m per year and porous flow transported an insignificant amount of heat. Maximum convective velocities in the region of viscous flow, on the other hand, were of the order of 1×10^{-1} m per year. Viscous flow was thus sufficient, in some simulations, to influence the thermal evolution. H_{abs} values larger than 5 were

observed for simulations in which the rheological parameters were optimum for convection ($\mu^l/\mu^l_{shaw} = 0.01$, CMF = 0·35) in case 1. Convective vigour, however, was dramatically less in case 2 simulations where the contact temperature was lower (900°C). In addition to the contact temperature, the influence of convection was observed to be a strong function of time, the CMF and the viscosity.

4.2.1. Influence of the rheological parameters on convective vigour. The influence of the contact temperature and rheological parameters on convective vigour can be illustrated by plotting H_{abs} over the range of estimates of the viscosity and the CMF 1×10^4 a after the initiation of the simulations (Fig. 7A).

Convective velocities high enough for the contact metamorphism to be convection-influenced are restricted to a limited range of estimates of the melt viscosity and CMF. More conservative estimates of these parameters tended to yield conduction-dominated solutions for all simulations in either case. Maximum magnitudes of convective velocities observed were approximately 1×10^{-1} m per year, but, more commonly, maximum magnitudes were between 1×10^{-5} and 1×10^{-3} m per year. Note, in particular, the dramatic decrease in convective vigour when the contact temperature was 900°C. Convection influenced the thermal evolution of only those case 2 simulations with the most optimum estimates of the rheological parameters ($\mu^l/\mu^l_{shaw} = 0.01$ and CMF = 0·35). The contact temperatures used in the simulations (1000 and 900°C) are probably higher than in natural systems, so the weak convection observed in these simulations is probably more vigorous than to be expected in natural underplating regions. In general, buoyancy-driven convective stirring of a restite-charged, partially molten pelitic sequence is unlikely to occur. Such convection appears to require both extreme conditions of sustained, high-temperature magmatism and more optimum estimates of the rheological parameters.

4.2.2. Evolution of convective vigour with time. It is also informative to measure the time that elapsed before H_{abs} exceeded 1·0. We termed that period the 'convective rise time' and it was found to be strongly dependent on both the rheological parameters (viscosity and CMF) and the contact temperature ($T_{contact}$). Convective rise times were typically on the order of 1×10^3 a for case 1 and substantially longer ($2-6 \times 10^4$ a) for case 2 (Fig. 7B).

These estimates should be taken as the minimum amount of time the contact temperature must be maintained to establish convection with sufficient vigour to influence the regional thermal evolution. Thus not only does vigorous convective stirring of the partially molten pelite require high contact temperatures and optimum estimates of the rheology, but it also requires a contact metamorphic event of significant duration. Buoyancy-driven viscous flow of partially molten country rock is likely to be isolated to regions experiencing long periods of sustained magmatism.

The minimum mass of basalt required to maintain the contact temperature through the duration of the convective rise time may be estimated by assuming that the composition of the basalt was that of a typical MORB and was intruded at a temperature of 1200°C and cooled to the contact temperature (1000 or 900°C). Latent heat released during crystallisation through this temperature range was accounted for by estimating the percentage crystallisation of the basalt with the MELTS algorithm (Ghiorso & Sack 1995). The mass of basalt was calculated by equating the amount of heat absorbed by the country rock with that released by the simulated intrusion (Fig. 7C). These values should be regarded as minimum estimates of the mass of basalt. In evaluating Fig. 7C, we might consider a sill about the dimensions of estimates of the size of mid-ocean ridge magma chambers

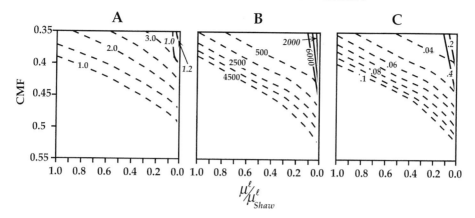

Figure 7 Contours of three parameters for case 1 ($T_{contact} = 1000°C$) and case 2 ($T_{contact} = 900°C$) simulations through the range of rheological parameters. Case 1 values are contoured by broken lines and case 2 values are plotted with solid lines with italicised contour labels. (A) Contours of the range of heat absorption ratios (H_{abs}) after 1.0×10^4 years. Simulations in which $H_{abs} = 1.0$ were termed 'conduction-dominated', those with $1.0 < H_{abs} \{2.0$ were termed 'convection-influenced', and those with $2.0 < H_{abs}$ were termed 'convection-dominated'. (B) Contours of the time (years) the contact temperature must be maintained for convection to become sufficiently vigorous to influence the regional thermal evolution. Weak convection was present, but was insufficiently vigorous to increase the amount of heat transferred to the unmelted pelite in conduction dominated simulations. (C) Contours of the minimum mass (in kg/10^{10}) of basalt per unit length of sill required to be intruded for convection in the pelite–melt mixture to become sufficiently vigorous to influence the thermal evolution of the domain. We assume the underplating cooled from the initial temperature of the intrusion (1200°C) to $T_{contact}$. Abbreviations as in Figure 6.

(Sinton & Detrick 1992). If we assume a sill 2 km in width and with a thickness of 100 m, it contains approximately 3×10^8 kg of basalt per metre of sill width in each limb. Cooling this amount of basalt would not yield the minimum amount of energy required to initiate the viscous flow of the partially molten sequence of pelitic rocks during the course of any simulation.

4.2.3. Application to the Ivrea-Verbano Zone. Temperatures of between 750 and 800°C have been estimated for the kinzigites and stronalites in the IVZ (Zingg 1983; Sills & Tarney 1984). It should be noted that these estimates place the temperature of the kinzigites close to that of the biotite dehydration reaction (12). Although the kinzigites near the contact with the Mafic Complex in Val Sesia have probably undergone a large degree of partial melting (Zingg 1980), they commonly retain a significant amount of biotite. It is thus unlikely that contact temperatures between the Kinzigite Formation and the under-lying Mafic Complex were as high as 1000 or 900°C for significant periods of time. If the melting systematics of the model pelite are representative of the melting of a typical IVZ metagreywacke, the simulations indicate that it is likely that heat transport in both the kinzigites and stronalites was dominated by conduction and convection played a limited part in the transport of heat. In addition, given the relatively low temperatures indicated by the thermometry and the extremely limited convection observed in case 2 ($T_{contact} = 900°C$) simu-lations, conduction is probable whether or not enough melt was produced to exceed the CMF.

4.3. Melt extraction

Convective velocities in the region of porous flow were insufficient to drive significant melt extraction during the course of any simulation. Maximum convective velocities observed in the region of viscous flow were about 1×10^{-7} m per year. The simulations were conducted over a period of 10^5 years so the maximum distance melt migrated from the source during the course of the simulations was of the order of centimetres. Given that the duration of a typical crustal melting event is of the order of $10^5–10^7$ years (Sawyer 1994), the *maximum* distance that melt could be expected move

within a partially molten pelite overlying a sill is probably less than a metre. These velocities are a maximum estimate because the contact temperature was held constant throughout the simulations. This cannot be the case in a natural underplating sill because, among other reasons, basalt would be nearly solidified and cooling conductively at 900°C and 10 kbar. The observed convective velocities are thus an end-member estimate of the natural system. In conclusion, the simulations indicate that the buoyancy of the melt alone is not sufficient to drive a significant extraction of melt from the porous skeleton. Deformation, fracturing or both may be the dominate mechanism for melt extraction in the continental crust and buoyancy-driven porous flow is, at best, a second-order process.

4.3.1. Existence of a low melt fraction window of extractable melt. In cases where melt extraction at low melt fraction (e.g. Sawyer 1994) is determined to be significant, it is possible to estimate from which region melt is most likely to be extracted. Given the non-linear variations in the viscosity (Fig. 3) of the melt residing within the porous skeleton, however, the propensity for melt extraction is unclear. Darcy's law implies that the propensity for melt extraction under a given pressure gradient can be estimated from the ratio of the permeability of the porous medium to the viscosity of the melt (K/μ^l) if melt movement is pervasive. Higher values of K/μ^l would indicate a higher propensity for melt extraction and K/μ^l should be directly proportional to the differential velocity between the melt and porous skeleton. A two order of magnitude increase in K/μ^l thus indicates a two order of magnitude increase in average differential velocities. A plot of K/μ^l with liquid fraction using the model permeability and viscosity relationships is shown in Fig. 8.

Melt existing in regions of a partially molten pelite that possess a melt fraction between 0.05 and 0.15 appears to have a relatively high propensity to undergo melt movement. In fact, the K/μ^l value predicted to exist within this melt fraction range is not exceeded until the system reaches a melt fraction of nearly 0.35. This melt fraction range was also correlated with a peak in the magnitudes of convective velocities in all simulations. Although the observed velocities were extremely small (1×10^{-7} m per year), the pressure gradient that

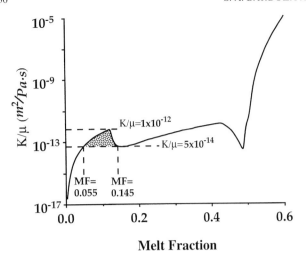

Figure 8 Plot depicting the relative propensity for melt movement within the porous skeleton as a function of the melt fraction. Higher values indicate a higher propensity for melt movement. Note that the values reached between melt fractions of 0·05 and 0·15 are not exceeded until after nearly 40% of the pelites has been melted. Stippled region is the 'low melt fraction window of extractable melt' discussed in the text. Broken lines are reference values.

generated the melt movement was due solely to the buoyancy of the melt and neglected other possible modes of transport. If, for example, dyking were to occur in the region, the velocity of the melt in the porous skeleton might be much higher. In such a circumstance we postulate that the melt existing in the range of melt fraction between 0·05 and 0·15 is the most likely to be extracted, barring the accumulation of a substantial percentage of melt ($>40\%$) in the region.

5. Conclusions and summary

Phase relations for the dehydration melting of pelitic rocks have been parameterised for use in a thermo-mechanical model. Numerical simulations of the dehydration melting and subsequent buoyant melt migration of a sequence of pelitic rocks following the underplating of mafic magma in the lower crust have been performed. The total amount of melt produced is critically dependent on the dynamics of melt movement in the pelite–melt mixture.

Important factors influencing the vigour of melt and melt plus restite movement are the value of the CMF, viscosity and the temperature of the igneous contact. Recall that the CMF, as used in this study, does not provide an abrupt transition from porous-dominated to suspension-dominated drag. Instead, the CMF is a reference value for the initiation of this transition. Simulations in which conservative estimates of the CMF ≈ 0.5 and the melt viscosity calculated by the Shaw model were used tended to yield conduction-dominated solutions. Other estimates of the melt viscosity yielded weak convection that influenced the thermal evolution of crustal rocks to a varying degree.

The simulations demonstrate that vigorous convective stirring in the region of viscous flow and a significant amount of melt extraction within the porous skeleton will probably not occur except under extreme conditions (i.e. long, sustained periods of magmatism; large amounts of very high temperature magma. Kilometre-scale buoyancy-driven convective stirring is, in general, unlikely to occur given current rheological models and heating rates. In addition, buoyancy-driven extraction of melt at low melt fraction appears to be a second-order process. Shear, extensional fracturing or some other mechanism appears to be required to achieve significant melt extraction.

6. Acknowledgements

We gratefully acknowledge Jim Quick, Alberto Patiño Douce, and Silvano Sinigoi for many fruitful discussions and assistance with the ongoing field studies. The editorial assistance of E-An Zen and comments from two reviewers is also gratefully acknowledged. Partial funding for this work was provided by NSF grant 9508291, a Royalty Research Fund grant from the University of Washington, and a AWU Faculty Fellowship to Battelle, Pacific Northwest Laboratory.

7. References

Agarwal, P. K. & O'Neill, B. K. 1988. Transport phenomena in multi-particle system—I. Pressure drop and friction factors: unifying the hydraulic-radius and submerged-object approaches. CHEM ENG SCI **43**, 2487–99.

Arzi, A. A. 1978. Critical phenomena in the rheology of partially melted rocks. TECTONOPHYSICS **44**, 173–84.

Barboza, S. A. 1995. *The dynamics of dehydration melting and implications for melt extraction in the lower crust following underplating: an example from the Ivrea-Verbano Zone, northern Italy.* M.S. Thesis, University of Washington.

Beard, J. S. & Lofgren, G. E. 1989. Effects of water on the composition of partial melts of greenstone and amphibolite. SCIENCE **244**, 195–7.

Beckermann, C. & Viskanta, R. 1988. Double-diffusive convection during dendritic solidification of a binary mixture. PHYSIOCHEM HYDRODYN **10**, 195–213.

Bennon, W. D. & Incropera, F. P. 1987. A continuum model for momentum, heat and species transport in binary solid–liquid phase change systems-I. Model formulation. INT J HEAT MASS TRANSFER **30**, 2161–70.

Bergantz, G. W. 1989. Underplating and partial melting: implications for melt generation and extraction. SCIENCE **245**, 1093–5.

Bergantz, G. W. 1995. Changing techniques and paradigms for the evaluation of magmatic processes. J GEOPHYS RES **100**, 17, 603–13.

Bergantz, G. W. & Dawes, R. 1994. Aspects of magma generation and ascent in continental lithosphere. *In* Ryan, M.P. (ed.) *Magmatic systems*, 291–317. San Diego: Academic Press.

Boriani, A., Bigioggero, B. & Origoni Giobbi, E. 1977. Metamorphism, tectonic evolution, and tentative stratigraphy of the 'Serie dei Laghi'—geological map of the Verbania Area (Northern Italy). MEM IST GEOL MINERAL UNIV PADOVA **32**, 1–25.

Boriani, A., Burlini, L., Caironi, V., Origoni, E. G., Sassi, A. & Sesana, E. 1988. Geological and petrological studies on the Hercynian plutonism of Serie dei Laghi—geological map of its occurrence between Valsesia and Lago Maggiore (N-Italy). REND SOC ITAL MINERAL PETROL **43–2**, 367–84.

Bowers, J. R., Kerrick, D. M. & Furlong, K. P. 1990. Conduction model for the thermal evolution of the Cupsuptic aureole, Maine. AM J SCI **290**, 644–65.

Brown, G. C. & Fyfe, W. S. 1970. The production of granitic melts during ultrametamorphism. CONTRIB MINERAL PETROL **28**, 310–8.

Brown, M. 1994. The generation, segregation, ascent and emplacement of granite magma: the migmatite-to-crustally-derived granite connection in thickened orogens. EARTH SCI REV **36**, 83–130.

Brown, M., Averkin, Y. A., McLellan, E. L. & Sawyer, E. W. 1995. Melt segregation in migmatites. J GEOPHY RES **100**, 15, 655–79.

Buntebarth, G. 1991. Thermal models of cooling. *In* Voll, G., Topel, J., Pattison, D.R.M. and Seifert, F. (eds.) *Equilibrium and kinetics in contact metmorphism: the Ballachulish Igneous Complex and its aureole*, 379–404, Heidelberg: Springer-Verlag.

Clemens, J. D. & Mawer, C. K. 1992. Granitic magma transport by fracture propagation. TECTONOPHYSICS **204**, 339–60.

Davison, I., McCarthy, M., Powell, D., Torres, H. H. F. & Santos, C. A. 1995. Laminar fow in shear zones: the Pernambuco Shear Zone, NE Brazil. J STRUCT GEOL **17**, 149–61.

Fountain, D. M. 1976. The Ivrea Verbano and Strona Ceneri zones, northern Italy, a cross-section of the continental crust: new evidence from seismic velocities of rock samples. TECTONOPHYSICS **33**, 145–65.

Fyfe, W. S. 1973. The granulite facies, partial melting and the Archean crust. PHIL TRANS R SOC LONDON SER A **273**, 457–61.

Giese, P. 1968. Die Struktur der Erdkruste im Bereich der Ivrea-Zone. Ein Vergleich verschiedener, seismischer Interpretationen und der Versuch einer petrographisch-geologischen Deutung. SCHWEIZ MINERAL PETROGR MITT **48**, 261–84.

Ghiorso, M. S. & Sack, R. O. 1995. Chemical mass transfer in magmatic processes IV. A revised and internally consistent thermodynamic model for the interpolation and extrapolation of liquid–solid equilibria in magmatic systems at elevated temperatures and pressures. CONTRIB MINERAL PETROL 119, 197–212.

Grant, J. A. & Frost, B. R. 1990. Contact metamorphism and partial melting of pelitic rocks in the aureole of the Laramie anorthosite complex, Morton Pass, Wyoming. AM J SCI 290, 425–72.

Harris, N., Ayres, M. & Massey, J. 1995. Geochemistry of granitic melts produced during the incongruent melting of muscovite: implications for the extraction of Himalayan leucogranite magmas. J GEOPHYS RES 100, 15,767–77.

Huppert, H. E. & Sparks, R. S. J. 1988. The generation of granitic magmas by intrusion of basalt into continental crust. J PETROL 29, 599–624.

Huppert, H. E. & Sparks, R. S. J. 1991. Comments on 'On convective style and vigor in sheetlike magma chambers' by Bruce D. Marsh. J PETROL 32, 851–4.

Hyndman, D. W. 1981. Controls on source and depth of emplacement of granitic magma. GEOLOGY 9, 244–9.

Irvine, T. N. 1970. Heat transfer during solidification of layered intrusions. I. Sheets and Sills. CAN J EARTH SCI 7, 1031–61.

Jaupart, C. & Tait, S. 1995. Dynamics of differentiation in magma reservoirs. J GEOPHYS RES 100, 17, 615–36.

Kay, R. W. & Kay, S. M. 1980. Chemistry of the lower crust: inferences from magmas and xenoliths. In National Research Council. Geophysics Study Committee (eds) Continental Tectonics, 139–50 Washington DC: National Academy of Sciences.

Kissling E. 1980. Krustenaufbau und Isostasie in der Schweiz. Ph.D. Thesis, ETH, Zurich.

Le Breton, N. & Thompson, A. B. 1988. Fluid-absent (dehydration) melting of biotite in metapelites in the early stages of anatexis. CONTRIB MINERAL PETROL 99, 226–37.

Litvinovsky, B. A. & Podladchikov, Y. Y. 1993. Crustal anatexis during the influx of mantle volatiles. LITHOS 30, 93–107.

Marsh, B. D. 1981. On the crystallininty, probability of occurrence and rheology of lava and magma. CONTRIB MINERAL PETROL 78, 85–98.

Marsh, B. D. 1991. Reply to comments of Huppert and Sparks. J PETROL 32, 855–60.

Mehnert, K. R. 1968. Migmatites and the origin of granitic rocks, 335–42. Amsterdam: Elsevier.

Mehnert, K. R. 1975. The Ivrea zone, a model of the deep crust. N JAHRB MINERAL ABH 125, 156–99.

Miller, C. F., Watson, E. B. & Harrison, T. M. 1988. Perspectives on the source, segregation and transport of granitoid magmas. TRANS R SOC EDINBURGH: EARTH SCI 79, 135–56.

Ni, J. & Beckermann, C. 1991. A volume-averaged two-phase model for transport phenomena during solidification. MET TRANS B 22, 349–61.

Nicolas, A. 1989. Structures of ophiolites and dynamics of ocean lithosphere, 367. Dordrecht: Kluwer Academic.

Oldenburg, C. M. & Spera, F. J. 1991. Numerical modeling of solidification and convection in a viscous pure binary eutectic system. INT J HEAT MASS TRANSFER 34, 2107–21.

Oldenburg, C. M. & Spera, F. J. 1992. Hybrid model for solidification and convection. NUMER HEAT TRANSFER B 21, 217–29.

Patiño Douce, A. E. & Johnston, A. D. 1990. Phase equilibria and melt productivity in the pelitic system: implications for the origin of peraluminous granitoids and aluminous granulites. CONTRIB MINERAL PETROL 107, 202–18.

Pitcher, W. S. 1993. The Origin of Granite, 321. Glasgow: Blackie Academic.

Prakash, C. & Voller, V. R. 1989. On the numerical solution of continuum mixture model equations describing binary solid-liquid phase change. NUMER HEAT TRANSFER B 15, 171–89.

Quick, J. E., Sinigoi, S. & Mayer, A. 1994. Emplacement dynamics of a large mafic intrusion in the lower crust, Ivrea-Verbano Zone, northern Italy. J GEOPHYS RES 99, 21,559–73.

Rivalenti, G., Garuti, G. & Rossi, A. 1975. The origin of the Ivrea-Verbano basic formation (Western Italian Alps)—whole rock geochemistry. BOLL SOC GEOL ITAL 94, 1149–86.

Rivalenti, G., Garuti, G., Rossi, A., Siena, F. & Sinigoi, S. 1980. Existence of different peridotite types and of a layered igneous complex in the Ivrea-zone of the Western Alps. J PETROL 22, 127–53.

Rushmer, T. 1991. Partial melting of two amphibolites: contrasting experimental results under fluid-absent conditions. CONTRIB MINERAL PETROL 107, 41–59.

Rushmer, T. 1995. An experimental deformation study of partially molten amphibolite: application to low-melt fraction segregation. J GEOPHYS RES 100, 15,681–95.

Rutter, E. H. & Neumann, D. H. K. 1995. Experimental deformation of parially moten Westerly granite under fluid-absent conditions, with implications for the extraction of granitic magmas. J GEOPHYS RES 100, 15,697–715.

Sawyer, E. W. 1991. Disequilibrium melting and the rate of melt-residuum separation during migmatization of mafic rocks from the Grenville Front, Quebec. J PETROL 32, 701–38.

Sawyer, E. W. 1994. Melt segregation in the continental crust. GEOLOGY 22, 1019–22.

Schnetger, B. 1994. Partial melting during the evolution of the amphibolite-to granulite-facies gneisses of the Ivrea Zone, northern Italy. CHEM GEOL 113, 71–101.

Schmid, R. 1978/79. Are the metapelites of the Ivrea-Verbano Zone restites? MEM IST GEOL MINERAL UNIV PADOVA 33, 67–9.

Schmid, S. M. & Wood, B. J. 1976. Phase relationships in granulitic metapelites from the Ivrea-Verbano zone. CONTRIB MINERAL PETROL 54, 255–79.

Schmid, S. M., Zingg, A. & Handy, M. 1987. The kinematics of movements along the Insubric Line and the emplacement of the Ivrea Zone. TECTONOPHYSICS 135, 47–66.

Schulze, F., Behrens, H. & Holtz, F. 1994. Effect of water on the viscosity of haplogranitic melts. Experimental investigation using the falling sphere method. EOS, TRANS AM GEOPHYS UNION 75 (44), 724.

Shaw, H. R. 1972. Viscosities of magmatic liquids: an empirical method of prediction. AM J SCI 272, 870–93.

Sills, J. D. & Tarney, J. 1984. Petrogenesis and tectonic significance of amphibolites interlayered with meta-sedimentary gneisses in the Ivrea Zone, Southern Alps, NW Italy. TECTONOPHYSICS 107, 187–206.

Sinigoi, S., Quick, J. E., Clemens-Knott, D., Mayer, A., Dimarchi, G., Mazzucchelli, M., Negrini, L. & Rivalenti, G. 1994. Chemical evolution of a large mafic intrusion in the lower crust, Ivrea-Verbano Zone, northern Italy. J GEOPHYS RES, 99, 21,575–90.

Sinton, J. M. & Detrick, R. S. 1992. Mid-ocean ridge magma chambers. J GEOPHYS RES 97, 197–216.

Swanson, D. A., Cameron, K. A., Evarts, R. C., Pringle, P. T. & Vance, J. A. 1989. Cenozoic volcanism in the Cascade Range and Columbia Plateau, southern Washington and northermost Oregon. NEW MEXICO BUR MINES MINERAL RES MEM 47, 1–50.

Symmes, G. H. & Ferry, J. M. 1995. Metamorphism, fluid flow and partial melting in pelitic rocks from the Onawa contact aureole, central Maine, USA. J PETROL 36, 587–612.

Thompson, A. B. 1982. Dehydration melting of pelitic rocks and the generation of H_2O–undersaturated granitic liquids. AM J SCI 282, 1567–95.

van der Molen, I. & Paterson, M. S. 1979. Experimental deformation of partially-melted granite. CONTRIB MINERAL PETROL 70, 299–318.

Vielzeuf, D. & Holloway, J. R. 1988. Experimental determination of the fluid-absent melting relations in the pelitic system. CONTRIB MINERAL PETROL 98, 257–76.

Whitney, J. A. 1988. The origin of granite: the role and source of water in the evolution of granitic magmas. GEOL SOC AM BULL 100, 1886–97.

Wickham, S. M. 1987. The segregation and emplacement of granitic magmas. J GEOL SOC LONDON 144, 281–97.

Wildemuth, C. R. & Williams, M. C. 1984. Viscosity of suspensions modeled with a shear dependent maximum packing fraction. RHEOL ACTA 23, 627–35.

Wyllie, P. J. 1977. Crustal anatexis: an experimental review. TECTONOPHYSICS 43, 41–71.

Yardley, B. W. D. 1986. Is there water in the deep continental crust? NATURE 323, 111.

Zingg, A. 1980. Regional metamorphism in the Ivrea Zone (Southern Alps, N-Italy): field and microscopic investigations. SCHWEIZ MINERAL PETROGR MITT 60, 153–79.

Zingg, A. 1983. The Ivrea and Strona-Ceneri Zones (Southern Alps, Ticino and North Italy): a review. SCHWEIZ MINERAL PETROGR MITT 63, 361–92.

SCOTT A. BARBOZA and GEORGE W. BERGANTZ, Department of Geological Sciences and Volcano Systems Center, Box 351310, University of Washington, Seattle, WA 98195, U.S.A.
E-mail: barboza@u.washington.edu

Transactions of the Royal Society of Edinburgh: Earth Sciences, **87**, 33–41, 1996

Controls on the trace element composition of crustal melts

F. Bea

ABSTRACT: The behaviour of trace elements during partial melting depends primarily on their mode of occurrence. For elements occurring as trace constituents of major phases (e.g. Li, Rb, Cs, Eu, Sr, Ba, Ga, etc.), slow intracrystalline diffusion ($D \approx 10^{-16}$ cm^2 s^{-1}) at the temperature range of crustal anatexis causes all effective crystal–melt partition coefficients to have a value close to unity and impedes further melt–restite re-equilibration. Usually, therefore, the trace element composition of crustal melts simply depends on the mass balance between the proportion and composition of phases that melt and the proportion and composition of newly formed phases. The behaviour of trace elements occurring as essential structural components in accessory phases (e.g. P, La–Sm, Gd–Lu, Y, Th, U, Zr, Hf, etc.) depends on the solubility, solution kinetics, grain size and the textural position of accessory phases. In common crustal protoliths a significant mass fraction of monazite, zircon, xenotime, Th-orthosilicates, uraninite, etc.—but not apatite—is included within other major and accessory phases. During low melt fraction anatexis the amount of accessory phases available for the melt is not sufficient for saturation, thus producing leucosomes with concentrations of La–Sm, Gd–Lu, Y, Th, U and Zr lower than expected from solubility equations. Low concentrations of these elements may also occur if the melt is prevented from reaching equilibrium with the accessories due to fast segregation. However, the first mechanism seems more feasible as leucosomes that are undersaturated with respect to monazite and zircon are frequently saturated, even oversaturated, with respect to apatite.

KEY WORDS: crustal anatexis, partition coefficients, melting rate, melt segregation rate, fractionation equations, disequilibrium melting, accessory minerals.

The trace element composition of partial melts may initially be considered to be the result of equilibrium or quasi-equilibrium crystal–melt partitioning controlled by (1) the chemical and mineralogical composition of the source rock, (2) the proportions of phases that melt, (3) the degree of partial melting and (4) trace element crystal–melt partition coefficients (Shaw 1970, 1978, 1979; Henderson 1982: 222–7). Depending on the melt segregation rate, two extreme situations may occur. If the melt remains in contact with restitic solids long enough to re-equilibrate completely after the last incremental melting step, the process is called batch melting. If, on the other hand, the melt is extracted from its source as soon as it is generated, thus impeding further melt–solid equilibration, the process is termed fractional or Rayleigh-type melting. As both models assume that the effective (K_{eff}) and equilibrium (K_{eq}) partition coefficients are the same, they are hereafter referred to as equilibrium melting models (note that Henderson uses this term only for batch melting).

Equilibrium melting models are applicable insofar as the trace elements under consideration obey Henry's law during melting, all source minerals are equally accessible to the melt and kinetic effects do not perturb solid–melt partitioning and re-equilibration to a great extent (Henderson & Williams 1979; Hart & Allègre 1980). As a result of the ubiquitous presence of accessory phases saturated with trace elements (Bea 1996) and the low temperature of melting, crustal melts hardly satisfy any of these requirements. It is therefore not surprising that the trace element composition of migmatite leucosomes and 'pure melt' anatectic leucogranites is rarely in accordance with equilibrium melting models (Dougan 1981;

Weber *et al.* 1985; Bea 1991; Sawyer 1991; Watt & Harley 1993; Carrington & Watt 1995). The presence of accessory phases saturated with trace elements can be included in equilibrium models through suitable experimentally determined solubility equations and by supposing that crustal protoliths contain enough accessory phases in relation to their solubility to be considered a infinite reservoirs (Watson & Harrison 1983, 1984; Montel 1986; Rapp & Watson 1986; Pichavant *et al.* 1992; Wolf & London 1994, 1995). Nevertheless, discrepancies between equilibrium and the actual composition of anatectic segregates are still the rule rather than the exception.

The aim of this paper is to discuss which factors other than the efficiency of melt–restite segregation (see Chappell, this issue) may cause the composition of crustal anatectic melts to depart from equilibrium. These factors are evaluated separately according to the residence of trace elements. For elements residing as traces within major phases (e.g. Li, Rb, Cs, Sr, Ba, Eu, Ga, Tl, Pb, V, Zn, etc.), the factors to be evaluated, directly related to the kinetics of intracrystalline diffusion, are: (1) the dependence of effective partition coefficients on the melting rate; and (2) the relation between the homogenisation rate of restitic crystals and the melt segregation rate. For elements that are essential structural components of accessory phases (e.g. La–Sm, Gd–Lu, Y, Th, U, Zr, Hf, etc.), the factors to be considered, chiefly related to accessory grain size and textural position, are: (1) the availability for reaction with the melt; (2) the effects of selective inclusion by major phases; and (3) the relations between the melt segregation rate and accessory solution kinetics. Finally, the possible influence of pressure on the composition of some restitic minerals and hence on the chemistry of coexisting melts is also briefly considered.

1. Controls on trace elements residing within major minerals

1.1. Intracrystalline diffusion

The diffusion coefficient of trace elements in common minerals at the temperature range of crustal anatexis may be estimated from Arrhenius relations calculated by several workers from experimental data (summarised in Brady 1995). Figure 1 shows Arrhenius plots from Sr and Sm in synthetic diopside (Sneeringer *et al.* 1984) and titanite (Cherniak 1995), Rb in alkali feldspar (Giletti 1991), Pb in apatite (Cherniak *et al.* 1991), Sr in K-feldspar (Cherniak & Watson 1992), apatite (Watson *et al.* 1985; Cherniak & Ryerson 1993), plagioclase (Cherniak & Watson 1994) and amphibole (Brabander & Giletti 1995). At 850 and 750°C, diffusion coefficients for these elements in feldspars cluster around $D = 10^{-15}$ cm^2 s^{-1} and $D = 10^{-16}$ cm^2 s^{-1}, respectively, whereas in mafic minerals these values are smaller by between one order of magnitude (synthetic diopside) to three orders of magnitude (titanite). It therefore seems reasonable to assume that a value of $D \approx 10^{-16}$ cm^2 s^{-1} may well represent the situation for most instances of crustal anatexis (but see later discussion).

1.2. Effective partition coefficients and the melting rate

During either crystallisation or partial melting, the effective values of trace element crystal–melt partition coefficients may considerably deviate from equilibrium values due to the sluggishness of intracrystalline diffusion. Henderson & Williams (1979) and Hart & Allègre (1980), using the expressions derived by Burton *et al.* (1953) and Smith *et al.* (1955), respectively, studied the variations of effective partition coefficients for a growing crystal as a function of the diffusion coefficient and the growth rate. These equations may be easily modified to describe the situation of a melting crystal (Bea

1991). Smith's equation (1955) is

$$K_{eff} = \tfrac{1}{2}\{1 + erf[\sqrt{(M/D)x}/2] + (2K_{eq} - 1)$$
$$\times \exp[-K_{eq}(1 - K_{eq})(M/D)x]$$
$$\times erf[(2K_{eq} - 1)/2\sqrt{(M/D)x}]\} \qquad (1)$$

where K_{eff} and K_{eq} are the effective and the equilibrium crystal–melt partition coefficients, respectively, M is the linear dissolution rate, D is the diffusion coefficient, and x is the linear dimension of the crystal.

Solving this equation for a crystal with a diameter of 2 mm, trace elements with a diffusion coefficient of 10^{-16} cm^2 s^{-1} and equilibrium crystal–melt partition coefficients of 10, 5, 0·2, and 0·1 (Fig. 2), reveals that the effective and equilibrium partition coefficients are equal only when melt rates are slower than $M = 10^{-18}$ cm s^{-1} ($3·3 \times 10^9$ a to completely melt a 2 mm crystal) and all effective partition coefficients become equal to unity when melting rates are faster than $M = 10^{-15}$ cm s^{-1} ($3·3 \times 10^6$ a to completely melt a 2 mm crystal). The equation of Burton *et al.* (1953) produces practically the same results.

These melting rates are apparently too slow compared with reasonable expectation in normal geological situations. Therefore, melts produced with all effective partition coefficients equal to unity are probably the rule rather than the exception. In this instance, the concentration of a given trace element in the melt would depend only on the proportion and composition of phases that contribute to the melt, and the proportion and composition of newly formed phases (Dougan 1981; Bea 1991)

$$C_{melt} = \sum_{i=1}^{n} C_m Z_m - \sum_{i=0}^{n} C_n P_n \qquad (2)$$

where C_{melt} is the concentration of the trace element in the melt, C_m is the concentration of the trace element in the source minerals, C_n is the concentration of the trace element in the new solid phases, Z_m is the weight fraction of source minerals that melt and P_n is the weight fraction of new phases.

1.3. Influence of the melt segregation rate

If all $K_{eff} = 1$, the melt segregation rate does not affect the melt composition of trace elements residing as traces within minerals. However, if $K_{eff} = K_{eq}$, then the melt segregation rate determines whether restites have enough time after each incremental melting step to re-equilibrate with the whole melt, thus causing either batch or fractional melting. Once again, the critical factor is the rate of intracrystalline diffusion within

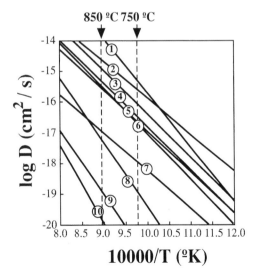

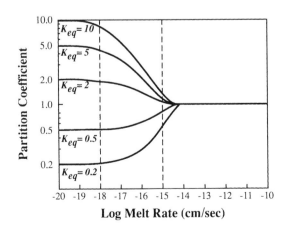

Figure 1 Arrhenius relations for the diffusion of certain trace elements in common silicates at the temperature range of crustal anatexis. 1, Rb in K-feldspar (Giletti 1991); 2, Pb in apatite (Cherniak *et al.* 1991); 3, Sr in oligoclase (Cherniak & Watson 1994); 4, Sr in apatite (Cherniak & Ryerson 1993); 5, Sr in K-feldspar (Cherniak & Watson 1992); 6, Sr in amphibole (Brabander & Giletti 1995); 7, Sm in synthetic diopside (Sneeringer *et al.* 1984); 8, Sr in synthetic diopside (Sneeringer *et al.* 1984); 9, Sr in titanite (Cherniak 1995); 10, Sm in titanite (Cherniak 1995). Note that a value of $D = 10^{-16}$ cm^2 s^{-1} might well represent the situation for most trace elements in major minerals.

Figure 2 Variation of effective partition coefficients with the melting rate. Calculated from the equation of Smith (1959) for trace elements, assuming $D = 10^{-16}$ cm^{-2} s^{-1} and a crystal radius of 0·1 cm. If the crystal melts in less than 3·3 Ma, all $K_{eff} \approx 1$. Only when melting takes more than 3·3 Ma are K_{eff} and K_{eq} equal.

restitic crystals, simply as the slowest of all the processes involved. To estimate how long melt may reside within its source before attaining equilibrium, we can calculate the time required for restitic crystals of appropriate size, surrounded by a melt of different composition, to reach complete homogeneity. This problem may be approached numerically by assuming that the crystal is spherical and the melt behaves as an infinite reservoir of constant concentration. In this instance, the appropriate equation is

$$\frac{C-C_1}{C_0-C_1}=1+2\sum_{n=1}^{\infty}(-1)^n\exp\left(-Dn^2\pi^2t/a^2\right)\qquad(3)$$

where C is the concentration at the centre of the crystal, C_1 is the initial concentration of the crystal, C_0 is the concentration of the melt, D is the diffusion coefficient, t is the time and a is the crystal radius (Crank 1975, equation 6.19).

Solving this equation for a $0{\cdot}1$ cm radius crystal (Fig. 3) results in homogenisation times for elements with $D=10^{-16}$ cm^2 s^{-1} and $D=10^{-17}$ cm^2 s^{-1} of about 2 and 20 Ma, respectively. As melt segregation may take place on a time scale of about $0{\cdot}1$–1 Ma (Brown *et al.* 1995), it seems that melt-restite re-equilibration under crustal conditions is normally likely to be very imperfect, and therefore the conditions for Rayleigh-type partial melting, at least with respect to the melt segregation rate, are probably fairly common. However, as far as the author is aware, no migmatite leucosome has been proved to have a composition compatible with Rayleigh-type melting, which at low melt fractions produces melts very enriched in elements with $K_{eq}^{\text{crystal melt}}\ll 1$. This seems to support the proposed idea that crustal melts are usually generated by disequilibrium melting (i.e. with $K_{eff}\neq K_{eq}$) due to fast melting rates.

1.4. Discussion: geochemical and isotopic evidence

Hofman & Hart (1978), assuming diffusion coefficients greater than $D=10^{-13}$ cm^2 s^{-1}, suggested that a fluid-free upper mantle may maintain a state of local disequilibrium for 10^8–10^9 a, whereas a partially molten mantle will tend to equilibrate locally in less than 10^5–10^6 a. As diffusion coefficients at the temperature range of crustal anatexis are about three orders of magnitude smaller than this value, we would expect equilibration times in a partially molten crust to increase with respect to the partially molten mantle, if not by the same factor then at least significantly, as suggested by the calculations earlier in the paper. The results of these calculations, however, may change considerably if diffusion in restitic crystals is enhanced, or if mechanisms faster than

lattice diffusion operate in the melt–solid equilibration. Diffusion may be enhanced by the effect of dislocations, especially during deformation (Yund *et al.* 1989), or along discrete zones of high crystal defect density (Hodges & Bowring 1995) (but note that the assumed value of $D=10^{-16}$ cm^2 s^{-1} is probably too high for diffusion in mafic minerals; see Section 1.1). Melt–crystal equilibration may occur by dissolution/re-precipitation. During re-precipitation (growth), the value of K_{eff} depends on diffusion in the melt, not in the crystal. As diffusion in the melt is always faster than in crystals (see Brady 1995), effective partition coefficients—and hence the whole crystal–melt partitioning—should be closer to equilibrium than during dissolution. Intracrystalline diffusion would certainly limit this effect to those parts of the crystal involved in the phenomenon, but it may be of considerable importance for newly formed or extensively recrystallised phases.

Finding natural examples to test whether crustal anatexis occurred as equilibrium ($K_{eff}=K_{eq}$) or disequilibrium ($K_{eff}\neq K_{eq}$) melting is very difficult, given the inherent chemical and isotopic heterogeneity of the continental crust as well as the long post-anatexis history of most crustal melts. Nevertheless an attempt is made with the following three examples.

Evidence based on a partially molten xenolith. In the central part of the Avila batholith, composed of peraluminous Hercynian granites and granodiorites, there is a N–S swarm of Triassic camptonite dykes which occasionally contain partially molten xenoliths of country rocks (Bea & Corretgé 1986). A detailed study was made of a xenolith of exceptionally fresh migmatite with a glass fraction of about 10–15 vol% distributed irregularly but showing some preference for accumulation along grain boundaries, especially when one of the phases is garnet or biotite. Analyses with an ultraviolet laser ablation inductively coupled plasma mass spectrometry (UV-LA-ICP-MS) probe in one of the glass pockets located along a garnet–K-feldspar boundary revealed that the glass composition is heterogeneous (Fig. 4). Near the garnet, the glass has a composition very close to that of garnet, with high V, Co, Ni, Y and Sc contents, low Rb, Ba, Sr and Pb contents, rare earth element (REE) chondritic patterns enriched in heavy REEs (HREEs), and a negative Eu anomaly. Near the K-feldspar, the glass has a composition very similar to that of feldspar, with high Rb, Ba, Sr and Pb contents, low V, Co, Ni, Y and Sc contents, REE patterns enriched in light REEs (LREEs) and a positive Eu anomaly. It seems, therefore, that the trace element composition of glass is almost the same as that of the mineral from which it originated, indicating that effective crystal–melt partition coefficients had a value equal to or near unity, at least in this instance where melting was probably very rapid.

Chemical and isotopic evidence based on the Hoyazo de Níjar peraluminous volcanic rocks, SE Iberia. The present day outcrop of the upper Miocene Hoyazo de Níjar volcano is a volcanic vent ($\approx 0{\cdot}25$ km^2) mainly composed of a garnet- and cordierite-bearing dacite, with many inclusions of sillimanite $\pm$ garnet $\pm$ biotite $\pm$ spinel rocks with occasional interstitial glass, that have been interpreted as restites (Zeck 1970). Garnet within dacites appears as isolated crystals with diameters ranging from ≈ 2 to 20 mm, with exactly the same major and trace element composition and zoning patterns as garnet crystals from restitic inclusions. We therefore assume that they are also restitic. Depending on the local composition of the surrounding glass, garnet crystals may be either xenomorphic, with corroded faces suggesting dissolution into the melt, or idiomorphic, with stepped faces suggesting growth from the melt. This indicates that garnet was almost in

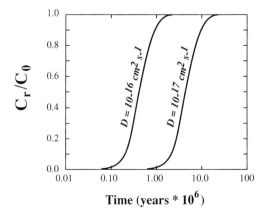

Figure 3 Homogenisation curves of a spherical crystal (Crank, 1975, equation 6·19). For a diffusion coefficient of $D=10^{-16}$ cm^2 s^{-1}, a 2 mm diameter crystal needs about 3 Ma to homogenise, this time increasing to 30 Ma for $D=10^{-17}$ cm^2 s^{-1}.

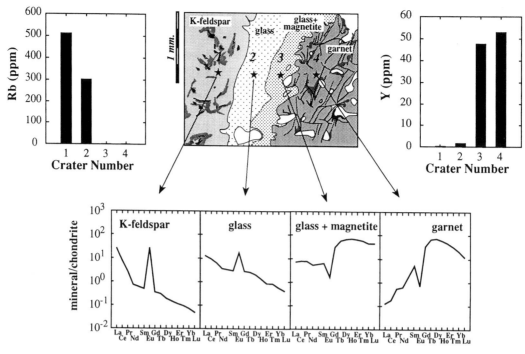

Figure 4 Chondrite-normalised REE patterns and Rb and Y contents of glass and neighbouring crystals in a partially molten felsic xenolith from a camptonite dyke (Avila batholith, central Iberia). Note that near garnet the glass has a composition close to that of garnet, whereas near feldspar the composition of glass is similar to that of feldspar. Analyses performed by UV LA-ICP-MS.

equilibrium with the melt, and small fluctuations caused it either to grow or to dissolve. Glass analysis with the UV-LA-ICP-MS probe at increasing distances from a corroded garnet crystal (Fig. 5) revealed that exactly at the garnet–glass interface, the glass has a REE chondritic pattern which is very similar to that of garnet for HREEs; however, it also has moderate LREF contents and a negative slope from La to Sm. As the distance from garnet increases, the garnet-like HREE pattern is progressively attenuated and a moderate

negative Eu anomaly becomes evident, until at a distance of about 500 μm from the garnet the REE pattern of glass is practically the same as that of the whole rock. As the HREE garnet–melt equilibrium partition coefficients are much higher than unity (Nicholls & Harris 1980; Sisson & Bacon 1992), the observed HREE diffusion field in the melt is not consistent with an equilibrium melting model, but rather suggests that $K_{\mathrm{eff}}^{\mathrm{garnet/melt}}$ was close to unity.

Rb–Sr studies on Hoyazo de Níjar rocks also reveal much

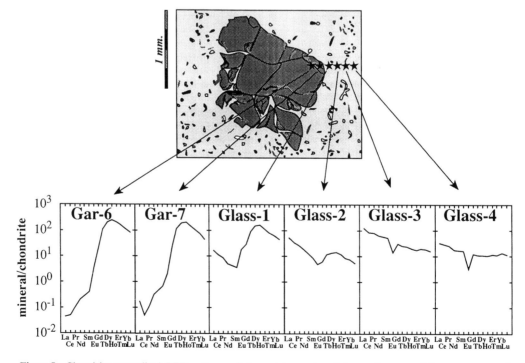

Figure 5 Chondrite-normalised REE patterns of glass at increasing distances from a melting garnet crystal (Hoyazo de Níjar dacite, SE Iberia). Note that the garnet-like HREE pattern is progressively attenuated as the distance from garnet increases, until at a distance of about 500 μm the REE pattern of glass is practically the same as that of whole rock. Analyses performed by UV-LA-ICP-MS.

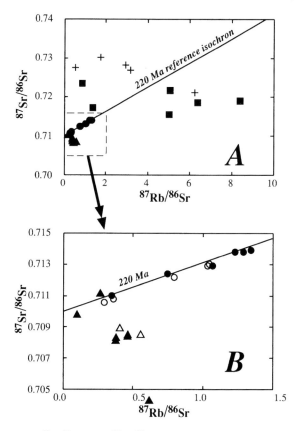

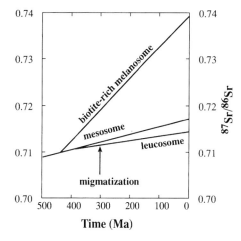

Figure 7 Strontium development diagram for coexisting mesosome, melanosome and leucosome of Peña Negra migmatites, central Iberia. Leucosome and melanosome lines intercept the mesosome line at 360 and 450 Ma, respectively, despite migmatisation occurring at 310–300 Ma. This discrepancy is easily understood if restitic biotite retained most of its pre-anatexis isotopic composition, thus indicating disequilibrium melting.

Figure 6 ^{87}Sr/^{86}Sr versus ^{87}Rb/^{87}Sr plot for samples from the Hoyazo de Níjar volcano, SE Iberia. Circles, dacites; crosses, biotite–almandine–sillimanite restites; squares, spinel–cordierite restites. Closed symbols, data from (Munksgaard 1984); open symbols, data from Bea *et al.* (unpublished data). Note that, in spite of the volcano being younger than 10 Ma (Dabrio *et al.* 1981), the dacites fit a correlation line with a slope corresponding to 220 Ma.

evidence of isotopic disequilibrium during partial melting. Remarkably, despite geological evidence indicating the volcano is no older than 10 Ma (Dabrio *et al.* 1981), dacites closely fit a ^{87}Sr/^{86}Sr versus ^{87}Rb/^{87}Sr correlation line with a slope corresponding to a 220 Ma isochron (Munksgaard 1984) (Fig. 6). Restite inclusions are completely disequilibrated with the melt: spinel–cordierite–sillimanite inclusions have fairly constant ^{87}Sr/^{86}Sr ratios which for ^{87}Rb/^{86}Sr > 1 are lower than coexisting dacites. Biotite-rich restites, on the other hand, have much higher ^{87}Sr/^{86}Sr ratios than dacites, indicating that restitic biotite retained an elevated fraction of Sr with its pre-melting isotopic composition. Munksgaard (1984) suggested that the ^{87}Sr/^{86}Sr versus ^{87}Rb/^{87}Sr correlation line observed in dacites might represent a pre-eruptive isochron, which would mean that the melt itself was isotopically heterogeneous. However, this is not clear as the effect could also be attributed to variable modal proportions of restitic biotite crystals within dacites (Bea *et al.*, unpublished data). In any case, the data indicates that isotopic equilibrium was far from being reached during melting.

Evidence from Peña Negra regional migmatites, central Iberia. Although mass balance studies of leucosome composition with different models reveal that those which assume that $K_{eff} = 1$ usually fit analytical data much more closely than models assuming $K_{eff} = K_{eq}$ (e.g. Dougan 1981; Bea 1991), the protracted post-anatectic history and extensive recrystallisation of regional migmatites makes it difficult to find direct evidence of melting with $K_{eff} \approx 1$. Some information, however, may be derived from Sr isotopes. In our experience, ^{87}Sr/^{86}Sr versus ^{87}Rb/^{86}Sr plots of coexisting mesosome, leucosome and melanosome rarely fit a straight line, as must happen if

isotopic equilibrium was attained during melting. Figure 7 shows the ^{87}Sr/^{86}Sr development diagram for coexisting mesosome, melanosome and leucosome in the low-pressure Peña Negra migmatites, where anatexis occurred within the stability field of biotite (Pereira & Bea 1994). The leucosome appears as small veins composed of quartz, K-feldspar, acid oligoclase and cordierite. The restitic melanosome is composed mainly of biotite and sillimanite and appears as a selvage between the leucosome and mesosome. The mesosome is composed of quartz, K-feldspar, oligoclase, biotite, cordierite and minor sillimanite. Despite the fact that the metamorphic peak occurred at 310–300 Ma (Pereira *et al.* 1992), Sr development lines for the leucosome and melanosome intercept the mesosome development line at 360 and 450 Ma, respectively. These discrepancies are difficult to explain by the post-migmatisation redistribution of Rb and Sr, but are easily understood if we suppose that at the time of melt segregation ^{87}Sr/^{86}Sr ratios in the leucosome and melanosome were, respectively, lower and higher than equilibrium values, simply because restitic biotite retained its pre-melting isotopic composition.

We therefore suggest that crustal anatexis frequently occurs as disequilibrium melting, as the balance between the melting rate and the intracrystalline diffusion causes effective crystal–melt partition coefficients to be equal to or near unity.

2. Controls on trace elements residing within accessory minerals as essential structural components

In common crustal protoliths and granite rocks, a high proportion (≈ 85 wt%) of elements such as P, Zr, Hf, La–Sm, Gd–Lu, Y, Th and U reside within accessory minerals, usually as essential structural components (Bea 1996). As a consequence, they do not generally obey Henry's law during melt–solid partitioning and their concentrations in partial melts are not ruled by crystal–melt distribution coefficients, but by solubility relations and solution kinetics, which have been determined experimentally for some common accessory phases (summarised in Watson 1988; see also Pichavant *et al.* 1992; Wolf & London 1994, 1995). Assuming that solubility equations give accurate results, melts with a higher than expected concentration of P, Zr, Hf, La–Sm, Gd–Lu, Y, Th and U may be caused by the entrapment of restitic accessories. Melts with lower than expected concentrations of such elements may result either because the amount of available accessory phases was not sufficient to saturate the melt, or because melt segregation was fast with respect to the kinetics of solution of accessory phases so that it impeded the attainment of equilibrium. Migmatite leucosomes with a distinctive chemistry characterised by low REE and Zr contents and, in many instances, positive Eu anomalies, have been interpreted as being due to one of these mechanisms (Sawyer 1991; Watt & Harley 1993; Bea *et al.* 1994; Carrington & Watt 1995).

2.1. Availability of accessory phases at crystal–melt interfaces

The assumption that source rocks contain sufficient accessory phases to saturate the melt cannot be validated solely by modal proportions. The textural position of accessory phases, depending largely on grain size, should also be considered (Watson *et al.* 1989). At low melt fractions, accessory phases located at major phase grain boundaries will generally be available for the melt and so may react with it. In contrast, a large fraction of accessory phases included within major minerals will remain physically isolated from the melt, thus preventing any reaction, or may be entrained as inclusions if major minerals are incorporated into the melt as restitic crystals, thus producing segregates with trace element contents higher than expected from solubility relations. To understand the relationship between the grain size distribution of the accessory phase and the mass fraction included in the major

phases of a common crustal protolith, we carried out a systematic scanning electron microscopy study of Peña Negra migmatites (Bea & Pereira 1990; Pereira & Bea 1994), which may well represent a common source rock for Hercynian granites in Western Europe. Figure 8 shows that about 50 wt% of monazite occurs as grains with an apparent diameter of less than 50 μm, whereas about 50 wt% zircon is found as grains smaller than 80 μm. Apatite has a larger grain size, with only 5 wt% occurring as grains smaller than 100 μm. The weight fraction of these minerals situated at major phase grain boundaries increases with grain size, in agreement with the model of Watson *et al.* (1989), and is estimated in this particular instances of nearly 11 wt% for monazite, 19 wt% for zircon and 76 wt% for apatite (Fig. 9). These data indicate that a substantial fraction of monazite and zircon—but not apatite—may be shielded by major phases at the beginning of melting and therefore the source rock may not behave as an infinite reservoir for these two minerals, especially at very low melt fractions. The situation for accessory phases such as xenotime, uraninite and Th-orthosilicates (huttonite and thorite) is still more complicated, as a significant fraction of these may appear included in other accessory phases (zircon and, to a lesser extent, apatite) which are themselves included in larger phases (Bea 1996).

The fact that accessory phase inclusions are not uniformly distributed among the host phases, but show a strong preference for particular phases (Watson *et al.* 1989; Bea 1996) causes further complications in the melt–solid partitioning of trace elements residing in accessory phases, the behaviour of which during anatexis may thus be physically controlled by that of the host phases.

2.2. Effects of melt segregation rate

Migmatite leucosomes with La–Sm, Gd–Lu, Zr, Y, Th and U contents lower than expected from the experimentally determined solubility of monazite and zircon may also be produced if segregation is fast enough to prevent the melt from becoming saturated in monazite and zircon. Certainly, the possibility of deriving information about the melt segregation rate from the composition of migmatite leucosomes is very attractive, but this idea, apart from failing to explain how small, easily entrained accessory phases can rapidly and efficiently be separated from a viscous melt, is also inconsistent with the fact that most of these leucosomes undersaturated in monazite and zircon have high P contents, of the same order or higher than expected for melts saturated in apatite (Fig. 10). As apatite grains usually have a larger diameter than monazite or zircon, we would expect (1) better efficiency at retaining apatite within restites and (2) more severe limitations on

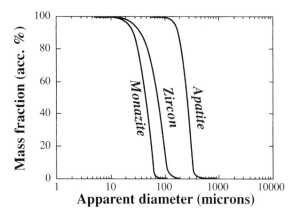

Figure 8 Grain size distribution of key accessories in Peña Negra migmatites. Note that about 50 wt% of monazite is found as grains with a diameter less than 50 μm and about 40 wt% zircon occurs as grains with a diameter less than 80 μm.

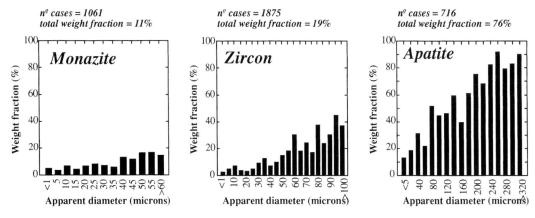

Figure 9 Percentage of key accessory phases in Peña Negra migmatites placed at grain boundaries. Note increase with grain size in accordance with the model of Watson *et al.* (1989).

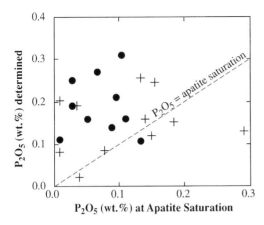

minerals from crustal protoliths reveal some systematic differences in the trace element composition of amphibolite grade with respect to granulite grade minerals, which is probably related to increased pressure. Granulite grade garnets have higher Nd and Sm contents than amphibolite grade garnets, and also show a precipitous negative Eu anomaly (Fig. 11). Likewise, granulite grade feldspars appear to be richer in LREEs than amphibolite grade feldspars.

At present data are lacking to generalise this observation, but if it proves to be the case, it may be a perceptible influence on the chemistry of crustal melts generated under amphibolite with respect to those generated under granulite grade conditions. This still has to be investigated by general comparative studies.

Figure 10 Determined P_2O_5 with respect to calculated P_2O_5 at apatite saturation for low REE and low Zr leucosomes from Iberia and the kinzigite formation of Ivrea-Verbano. Symbols are amphibolite grade (closed circles) and granulite grade (crosses) migmatite leucosomes. Apatite saturation calculated from the model of Harrison & Watson (1984) and empirically corrected for high peraluminousity with the expression of Bea *et al.* (1992).

apatite solubility due to solution kinetics during fast melt extraction. Both factors would contribute to producing melts relatively more depleted (or at least equally depleted) in P than in La–Sm, Gd–Lu, Zr, Y, Th and U, exactly the opposite to what is observed. In contrast, low concentrations of these elements together with high concentrations of P are compatible with the greater availability of apatite for the melt than monazite or zircon, due to their respective textural positions.

Melts with positive Eu anomalies and very low La–Sm, Gd–Lu, Zr, Y, Th and U contents appear to be limited to small leucosome veins (Bea *et al.* 1994) and, at least as far as the author is aware, no granite body has been described with a similar composition. In terms of the large-scale production of granites, both mechanisms—shielding by major phases during melting and fast extraction of melt—seem to have little, if any, importance.

3. Influence of pressure on the trace element composition of major minerals

Ion probe (e.g. Reid 1990; Harris *et al.* 1992; Watt & Harley 1993) and LA-ICP-MS data (e.g. Bea *et al.* 1994; Bea 1996) of

4. Summary and conclusions

The behaviour of trace elements during partial melting depends primarily on their mode of occurrence. For elements occurring as trace constituents of major phases (e.g. Li, Rb, Cs, Eu, Sr, Ba, Ga, etc.), the melting rate determines whether it takes the form of equilibrium melting. Equilibrium melting occurs when effective crystal–melt partition coefficients have equilibrium values, which requires a melting rate slower than $\approx 10^{-18}$ cm s^{-1} if equilibrium is achieved solely by lattice diffusion. If partition coefficients have equilibrium values, the melt segregation rate determines whether melting is of the batch or fractional type. Batch-type melting requires a residence time of the melt within its source at least of the same magnitude as the time required to homogenise restitic crystals by intracrystalline diffusion. For trace elements with diffusion coefficients of 10^{-16} and 10^{-17} cm^2 s^{-1}, a 2 mm sized crystal requires ≈ 2 and ≈ 20 Ma, respectively, to become homogenised. Equilibrium melting with rapid melt segregation rates produces Rayleigh-type melting. At melting rates faster than 10^{-15} cm s^{-1}, all effective crystal–melt partition coefficients become equal to unity, producing disequilibrium melts in which the concentration of a given trace element simply depends on the mass balance between the proportion and composition of phases that melt and the proportion and composition of newly formed phases. Geological evidence indicates that disequilibrium melting (i.e. with $K_{eff} \neq K_{eq}$) may be the rule rather than the exception.

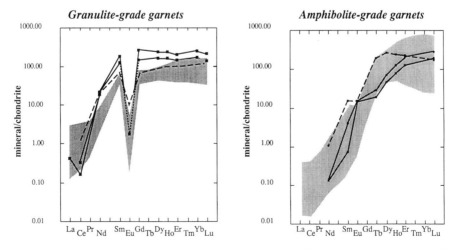

Figure 11 Chondrite normalised REE patterns of garnets from granulite grade migmatites and amphibolite grade migmatites. Shaded areas are LA-ICP-MS data from Bea *et al.* (1994) and Bea (unpublished data). Granulite grade garnets: solid lines, ion probe data from Watt & Harley (1993); broken line, ion probe data from Reid (1990). Amphibolite grade garnets: solid lines, sillimanite gneiss (Harris *et al.* 1992); broken line, kyanite schist (Harris *et al.* 1992). Note the higher Sm–Nd contents and precipitous negative Eu anomaly in granulite grade garnets. Note also that a garnet from a kyanite schist is also enriched in Nd–Sm with respect to a garnet from a sillimanite gneiss and has a more granulite-like REE pattern.

The behaviour of trace elements occurring as essential structural components in accessory phases (e.g. P, La–Sm, Gd–Lu, Y, Th, U, Zr, Hf, etc.) largely depends on the grain size and textural position of the accessory phases containing them. In common crustal protoliths, monazite and zircon usually have a smaller grain size than apatite, which therefore tends to be preferentially placed along grain boundaries. A substantial mass fraction of monazite and zircon, however, appears included within major minerals, principally biotite and garnet, which thus indirectly, but strongly, control the geochemistry of these elements. Migmatite leucosomes with lower than expected concentrations of La–Sm, Gd–Lu, Y, Th, U and Zr commonly occur because there is not enough available mass of monazite and zircon to saturate the melt, due to the effect of shielding by major phases. The same effect might, in principle, be caused by melt segregation fast enough to prevent the melt from reaching equilibrium with the accessory phases. However, the first mechanism is favoured, as melts with lower than expected concentrations of the above elements are usually saturated or oversaturated with respect to apatite.

Acknowledgements

I give many thanks for help from P. G. Montero and L. P. Fernández for their assistance with figures and slides, and from A. Martínez-Barredo and C. Laurin for their help in improving the original English manuscript. Revisions made by E. B. Watson and G. R. Watt, as well as editorial comments by W. E. Stephens greatly helped to improve the manuscript. This work has been financially supported by the Spanish Interministry Commission for Science and Technology (CICYT), projects AMB93-0535 and AMB94-1420.

References

Bea, F. 1991. Geochemical modelling of low melt-fraction anatexis in a peraluminous system: the Peña Negra Complex (central Spain). GEOCHIM COSMOCHIM ACTA 55, 1859–74.

Bea, F. 1996. Residence of REE, Y, Th and W in granites and crustal protoliths; implications for the chemistry of crustal melts. J PETROL 37(3), in press.

Bea, F. & Corretgé, L. G. 1986. Petrography, geochemistry, and differentiation models of lamprophyres from Sierra de Gredos, central Spain. HERCYNICA 2, 1–15.

Bea, F. & Pereira, M. D. 1990. Estudio petrológico del Complejo Anatéctico de la Peña Negra, Batolito de Avila. REV SOC GEOL ESPANA 3, 87–104.

Bea, F., Pereira, M. D. & Stroh, A. 1994. Mineral/leucosome trace-element partitioning in a peraluminous migmatite (a laser ablation-ICP-MS study). CHEM GEOL 117, 291–312.

Brabander, D. J. & Giletti, B. J. 1995. Strontium diffusion kinetics in amphiboles and significance to thermal history determinations. GEOCHIM COSMOCHIM ACTA 59, 2223–38.

Brady, J. B. 1995. Diffusion data for silicate minerals, glasses, and liquids. In Ahrens, T. J. (ed.) Mineral physics & crystallography. A handbook of physical constants, 269–90. Washington: AGU.

Brown, M. Y., Averkin, A., McLellan, E. L. & Sawyer, E. W. 1995. Melt segregation in migmatites. J GEOPHYS RES SOLID EARTH 100, 15 655–79.

Burton, J. A., Prim, R. C. & Slichter, W. P. 1953. The distribution of solute from crystals grown from the melt. Part 1: theoretical. J CHEM PHYS 21, 1987–99.

Carrington, D. P. & Watt, G. R. 1995. A geochemical and experimental study of the role of K-feldspar during water-undersaturated melting of metapelites. CHEM GEOL 122, 59–76.

Cherniak, D. J. 1995. Sr and Sm diffusion in titanite. CHEM GEOL 125, 219–32.

Cherniak, D. J. & Ryerson, F. J. 1993. A study of strontium diffusion in apatite using Rutherford backscattering spectroscopy and ion implantation. GEOCHIM COSMOCHIM ACTA 57, 4653–62.

Cherniak, D. J. & Watson, E. B. 1992. A study of strontium diffusion in K-feldspar using Rutherford backscattering spectroscopy. EARTH PLANET SCI LETT 113, 411–25.

Cherniak, D. J. & Watson, E. B. 1994. A study of strontium diffusion in plagioclase using Rutherford backscattering spectroscopy. GEOCHIM COSMOCHIM ACTA 58, 5179–90.

Cherniak, D. J., Lanford, W. A. & Ryerson, F. J. 1991. Lead diffusion in apatite and zircon using ion implantation and Rutherford backscattering techniques. GEOCHIM COSMOCHIM ACTA 55, 1663–73.

Crank, J. 1975. The mathematics of diffusion, 2th edn. Oxford: Oxford Science Publications.

Dabrio, C. J., Esteban, M. & Martín J. M. 1981. The coral reef of Nijar, Messinian (Uppermost Miocene), Almeria Province, S.E. Spain. J SEDIMENT PETROL 51, 521–39.

Dougan, T. W. 1981. Melting reactions and trace element relationships in selected specimens of migmatitic pelites from New Hampshire and Maine. CONTRIB MINER PETROL 78, 337–44.

Giletti, B. J. 1991. Rb and Sr diffusion in alkali feldspars, with implications for cooling histories of rocks. GEOCHIM COSMOCHIM ACTA 55, 1331–43.

Harris, N. B., Gravestock, W. P. & Inger, S. 1992. Ion-microprobe determinations of trace-element concentration in garnets from anatectic assemblages. CHEM GEOL 100, 41–9.

Hart, S. R. & Allègre, C. J. 1980. Trace-element constraints of magma genesis. In Hargraves, R. B. (ed.) Physics of magmatic processes, 121–59. Princeton: Princeton University Press.

Henderson, P. 1982. Inorganic geochemistry: Oxford: Pergamon Press.

Henderson, P. & Williams, C. T. 1979. Variations in trace element partition (crystal/magma) as a function of crystal growth rate. In Ahrens L. H. (ed.) Origin and distribution of the elements, 191–8. Oxford: Pergamon Press.

Hodges, K. V. & Bowring, S. A. 1995. $^{40}Ar/^{39}Ar$ thermochronology of isotopycally zoned micas: insights from the southwestern USA Proterozoic orogen. GEOCHIM COSMOCHIM ACTA 59, 3205–20.

Hofman, A. W. & Hart, S. R. 1978. An assessment of local and regional isotopic equilibrium in the mantle. EARTH PLANET SCI LETT 38, 44–62.

Montel, J. 1986. Experimental determination of the solubility of Ce-monazite in $SiO_2–Al_2O_3–K_2O–Na_2O$ melts at 800≪C, 2 kb, under H_2O-saturated conditions. GEOLOGY 14, 659–62.

Munksgaard, N. C. 1984. High $\Delta^{18}O$ and possible pre-eruptional Rb–Sr isochrons in cordierite-bearing Neogene volcanics from SE Spain. CONTRIB MINER PETROL 87, 351–8.

Nicholls, I. A. & Harris, K. L. 1980. Experimental rare earth element partition coefficients for garnet, clinopyroxene and amphibole coexisting with andesitic and basaltic liquids. GEOCHIM COSMOCHIM ACTA 44, 287–308.

Pereira, M. D. & Bea, F. 1994. Cordierite-producing reactions at the Peña Negra complex, Avila batholith, central Spain: the key role of cordierite in low-pressure anatexis. CAN MINERAL 32, 763–80.

Pereira, M. D., Ronkin, Y. & Bea, F. 1992. Dataciones Rb/Sr en el Complejo Anatéctico de la Peña Negra (Batolito de Avila, España Central): evidencias de magmatismo pre-hercínico. REV SOC GEOL ESPANA 5, 129–34.

Pichavant, M., Montel, J. M. & Richard, L. R. 1992. Apatite solubility in peraluminous liquids: experimental data and an extension of the Harrison–Watson model. GEOCHIM COSMOCHIM ACTA 56, 3855–61.

Rapp, R. P. & Watson, E. B. 1986. Monazite solubility and dissolution kinetics: implications for the thorium and light rare earth chemistry of felsic magmas. CONTRIB MINER PETROL 94, 304–16.

Reid, M. R. 1990. Ionprobe investigation of rare earth elements distribution and partial melting of metasedimentary granulites. In Vielzeuf, D. & Vidal, P. (eds) Granulites and crustal evolution, 506–22. Amsterdam: Kluwer Academic.

Sawyer, E. W. 1991. Disequilibrium melting and the rate of melt–residuum separation during migmatization of mafic rocks from the Grenville Front, Quebec. J PETROL 32, 701–38.

Shaw, D. M. 1970. Trace element fractionation during anatexis. GEOCHIM COSMOCHIM ACTA 34, 237–43.

Shaw, D. M. 1978. Trace element behaviour during anatexis in the presence of a fluid phase. GEOCHIM COSMOCHIM ACTA 42, 933–43.

Shaw, D. M. 1979. Trace element melting models. In Ahrens, L. H. (ed.) Origin and distribution of elements, 577–86. Oxford: Pergamon Press.

Sisson, T. W. & Bacon, C. R. 1992. Garnet/high silica rhyolite trace element partition coefficients measured by ion microprobe. GEOCHIM COSMOCHIM ACTA 56, 2133–6.

Smith, V. G., Tiller, W. A. & Rutter, J. W. 1955. A mathematical analysis of solute redistribution during solidification. CAN J PHYS **33**, 723–45.

Sneeringer, M., Hart, S. R. & Shimizu, N. 1984. Strontium and samarium diffusion in diopside. GEOCHIM COSMOCHIM ACTA **48**, 1589–608.

Watson, E. B. & Harrison, T. M. 1983. Zircon saturation revisited: temperature and composition effects in a variety of crustal magma types. EARTH PLANET SCI LETT **64**, 295–304.

Watson, E. B. & Harrison, T. M. 1984. Accessory minerals and the geochemical evolution of crustal magmatic systems: a summary and prospectus of experimental approaches. PHYS EARTH PLANET INTER **35**, 19–30.

Watson, E. B., Harrison, T. M. & Ryerson, F. J. 1985. Diffusion of Sm, Sr, and Pb in fluorapatite. GEOCHIM COSMOCHIM ACTA **49**, 1813–23.

Watson, E. B., Vicenzi, E. P. & Rapp, R. P. 1989. Inclusion/host relations involving accessory minerals in high-grade metamorphic and anatectic rocks. CONTRIB MINERAL PETROL **101**, 220–31.

Watt, G. R. & Harley, S. L. 1993. Accessory phase controls on the geochemistry of crustal melts and restites produced during water–undersaturated partial melting. CONTRIB MINERAL PETROL **114**, 550–6.

Weber, C., Barbey, P., Cuney, M. & Martin, H. 1985. Trace element behavior during migmatization. Evidence for a complex melt–residuum–fuid interaction in the St. Malo migmatitic dome (France). CONTRIB MINERAL PETROL **90**, 52–62.

Wolf, M. B. & London, D. 1994. Apatite dissolution into peraluminous haplogranitic melts: an experimental study of solubilities and mechanisms. GEOCHIM COSMOCHIM ACTA **58**, 4127–46.

Wolf, M. B. & London, D. 1995. Incongruent dissolution of REE- and Sr-rich apatite in peraluminous granitic liquids: differential apatite, monazite, and xenotime solubilities during anatexis. AM MINERAL **80**, 765–75.

Yund, R. A., Quigley, J. & Tullis, J. 1989. The effect of dislocations on bulk diffusion in feldspars during metamorphism. J METAMORPH GEOL **7**, 337–41.

Zeck, H. P. 1970. An erupted migmatite from Cerro del Hoyazo, SE Spain. CONTRIB MINERAL PETROL **26**, 225–46.

F. BEA, Department of Mineralogy and Petrology, Campus Fuentenueva, University of Granada, 18002 Granada, Spain.

Transactions of the Royal Society of Edinburgh: Earth Sciences, **87**, 43–56, 1996

Dissolution, growth and survival of zircons during crustal fusion: kinetic principles, geological models and implications for isotopic inheritance

E. Bruce Watson

ABSTRACT: Finite difference numerical simulations were used to characterise the rates of diffusion-controlled dissolution and growth of zircon in melts of granitic composition under geologically realistic conditions. The simulations incorporated known solubility and Zr diffusivity relationships for melts containing 3 wt% dissolved H_2O and were carried out in both one and three dimensions under conditions of constant temperature, linearly time-dependent temperature and for a variety of host system thermal histories. The rate of zircon dissolution at constant temperature depends systematically on time ($t^{-\frac{1}{2}}$), temperature ($\exp T^{-1}$) and degree of undersaturation of the melt with respect to zircon (in ppm Zr). Linear dissolution and growth rates fall in the range 10^{-19}–10^{-15} cm s^{-1} at temperatures of 650–850°C. Radial rates are strongly dependent on crystal size (varying in inverse proportion to the radius, r): for $r > 30$ μm, dissolution and growth rates fall between 10^{-17} and 10^{-13} cm s^{-1}. During crustal magmatism, the chances of survival for relict cores of protolith zircons depend on several factors, the most important of which are: the initial radius of the zircon; the intensity and duration of the magmatic event; and the volume of the local melt reservoir with which the zircon interacts. In general, only the largest protolith zircons (> 120 μm radius) are likely to survive magmatic events exceeding 850°C. Conversely, only the smallest zircons (< 50 μm radius) are likely to be completely consumed during low-temperature anatexis (i.e. not exceeding ≈ 700°C).

The effects of stirring the zircon-melt system are unimportant to dissolution and growth behaviour; except under circumstances of extreme shearing (e.g. filter pressing?), zircon dissolution is controlled by diffusion of Zr in the melt.

KEY WORDS: Zircon dissolution, zircon growth, zircon survival, crustal fusion, isotopic inheritance.

Zircon is indisputably the most important mineral in the geochronology of old crustal rocks. Indeed, recent major advances in our understanding of crustal evolution have followed from isotopic analyses of zircons separated from igneous and metamorphic rocks using both conventional mass spectrometry and the SHRIMP ion microprobe (see Heaman & Parrish 1991 and references cited therein). The enormous value of zircon in geochronology derives from a fortuitous combination of factors, including its tendency to incorporate natural radionuclides (of U, Th, Lu and Sm) and reject Pb, its generally low solubility in crustal melts and fluids (Watson & Harrison 1983; Ayers & Watson 1991), its resistance to chemical and physical breakdown in most geological environments and the remarkably sluggish diffusion of constituent ions (e.g. Cherniak *et al.* 1993, 1995).

In some instances, the properties that make zircons so valuable in geochronology can also greatly complicate the internal chemical and isotopic 'structure' of individual crystals, leading to difficulties in deciphering apparent age information. For example, it has been recognised for some time through U–Pb isotopic studies that the survival of a pre-existing igneous or metamorphic zircon through a melting event commonly results in an old, partially resorbed core mantled by younger igneous growth (e.g. Schärer & Allègre 1982; Harrison *et al.* 1987; Chen & Williams 1990; Williams 1992; Roddick & Bevier 1995). Recent 'chemical imaging' studies of crustal zircons using backscattered electrons and cathodolumi-

nescence (CL) reveal numerous complexities in their chemical structure, including sector- and igneous-growth (oscillatory) zoning, as well as ubiquitous overgrowth rims on older cores (e.g. van Breemen & Hanmer 1986; Miller *et al.* 1992; Paterson *et al.* 1992; Hanchar & Miller 1993; Hanchar & Rudnick 1995). These cores show rounding and other resorption features apparently formed during exposure to melt or fluid before the rim-forming event. One of the most interesting characteristics of overgrowth rims is that their thickness can vary markedly among zircons separated from a single rock (e.g. Paterson *et al.* 1992; see Fig. 1).

This paper focuses on the factors that affect the development and extent of resorption and overgrowth features during the igneous 'processing' of crustal zircons. The overall goal is to shed light on questions that are central to the continued refinement of zircon geochronology as the principal tool for deciphering crustal evolution. These questions include the following. What determines the 'survivability' of individual zircons during crustal melting? What factors produce the variability of core/rim ratios (and consequent magnitude of Pb isotopic inheritance) in zircons that have experienced identical time–temperature histories in the same host rock? This goal is pursued through a quantitative treatment of the dissolution (and growth) kinetics of zircons in granitic melts using a finite difference numerical modelling approach. We try to provide two things: (1) a sufficiently general set of equations and the necessary physical insight to assess zircon behaviour

Figure 1 Backscattered electron image of zircons separated from the Cowra granodiorite, a Devonian S-type granitoid of the Lachlan Belt thought to have been derived from the melting of sediments (I.S. Williams, pers. comm., 1995; see also Wyborn *et al.* 1991). The light-coloured rims of these zircons have been dated at ≈400 Ma. The generally darker cores are relicts of older zircons that survived the 400 Ma event; these contain a number of inherited components ranging from 1·0 to 2·6 Ga. For the purposes of the present study, the most interesting aspect of these zircons is the wide variability in core/rim ratio; note also the off-centre character of the core in the grain at the lower right.

for any crustal melting conditions; and (2) some detailed models of zircon dissolution and growth for several assumed time–temperature paths appropriate to crustal melting. The latter serve as examples to illustrate the importance of various system parameters to the survival of, and overgrowth on, individual zircons.

Although we consider both the dissolution and growth of zircon, by far the greater emphasis is placed on dissolution, because a knowledge of zircon survival during crustal melting is the key to understanding isotopic inheritance. The results confirm the general conclusions of Harrison and Watson (1983) for the constant temperature dissolution of individual zircons. However, the present contribution also sets the stage for a much more refined consideration of zircon behaviour— one that incorporates the effects of heating rate, saturation state of the melt with respect to zircon and volume of the melt reservoir with which dissolving zircons interact.

1. Modelling approach

1.1. General assumptions and constraints

Quantitative treatment of zircon dissolution and growth is made possible by previous experimental investigations that provide data on zircon solubility and Zr diffusivity in granitic melts (Harrison & Watson 1983; Watson & Harrison 1983). If we assume that zircon dissolution and growth rates are limited by Zr diffusion in the contacting melt (i.e. not by interface kinetics), then a knowledge of the saturation level (C_{sat}), the diffusivity (D) of dissolved Zr and the dependencies of these two variables on temperature is sufficient to model zircon behaviour fairly rigorously. Diffusion (as opposed to interface) control is taken as given in this paper. Several arguments can be used to justify this assumption, but the most convincing is simply that zircons can be grown in the laboratory (e.g. Watson 1979, 1980; Watson & Harrison 1983) at rates vastly greater than those computed in the present models, even at comparable temperatures. It thus seems clear

that interface kinetics do not limit dissolution and growth in nature. The assumption of purely diffusion control implicitly ignores effects due to mechanical stirring of the contacting melt. The consequences of possible magma convection and/or crystal settling for the results of our pure diffusion models are discussed in Section 3.

The fundamental factors affecting the rate of diffusive dissolution or growth of a crystal are: (1) its absolute solubility in the melt; (2) the diffusivity(ies) of stability controlling component(s) in the melt; and (3) the extent to which the system deviates from saturation equilibrium. These factors are important in that they determine the magnitude of the flux of components toward or away from a growing or dissolving crystal. This flux, in turn, limits the rate of growth or dissolution. Because zircon has been the subject of previous experimental studies of saturation systematics and dissolution kinetics (Harrison & Watson 1983; Watson & Harrison 1983) and also because its major element composition is simple, these three factors can be readily parameterised for modelling purposes.

Zircon solubility in a silicic melt can be described in terms of the concentration of dissolved Zr required to stabilise $ZrSiO_4$ (Watson 1979; Watson & Harrison 1983); this has been shown to depend systematically on the temperature and melt composition, the latter represented by the parameter $M = (2Ca + Na + K)/(Si \cdot Al)$, where the element symbols represent cation fractions. For typical granitic compositions, which are of particular interest in this study, $M \cong 1·3$; accordingly, zircon solubilities incorporated into the present models pertain to this specific value of M. The temperature dependence of the solubility is given by

$$C_{sat} = (4·414 \times 10^7)/\exp(13\,352/T) \qquad (1)$$

where the solubility (C_{sat}) is expressed as parts per million (ppm) (weight) dissolved Zr and T is absolute temperature (see Harrison & Watson 1983; Watson & Harrison 1983). This equation specifies remarkably low solubilities of zircon, ranging from 23 ppm dissolved Zr at 650°C to 303 ppm at 850°C.

The low solubility of zircon in granitic melts makes it realistic to model the diffusive transport aspects of the dissolution/growth process in terms of a simple, composition-independent diffusivity of Zr in the melt. Silica is also 'absorbed' or 'released' by the growth or dissolution of zircon, but this component diffuses significantly faster than ZrO_2 (e.g. Baker 1991; Chekhmir & Epel'baum 1991), so it is ZrO_2 transport that limits the rate of the overall process. Preliminary computer simulations, using the constant growth rate equations of Smith *et al.* (1955), were run to establish that no significant change in the concentrations of major components develops against a growing or dissolving zircon. The development of a boundary layer enriched or depleted in Si and/or Al could affect the local solubility of zircon by changing the value of M. However, no hint of such a layer was revealed by the preliminary models.

Diffusion purists will recognise that the transport of any species in a complex melt—even one at the dilute concentrations appropriate to Zr in molten granite—is complicated by interactions with other components, including the necessary counter-diffusion of charge-balancing species. Complexities of this sort certainly do enter in as Zr diffuses towards or away from a growing or dissolving zircon, but because the Zr concentrations under consideration are so low, any effects of Zr diffusion on other components must be commensurately small. Moreover, because the most appropriate Zr diffusion data for the present purposes were obtained in zircon dissolution experiments, multicomponent diffusion effects are implicitly taken into account.

Having ruled out possible complications due to major element boundary layers and/or multicomponent diffusion, we adopted the description of Zr diffusion in silicic melts reported previously by Harrison and Watson (1983). Zirconium diffusivity (D) is sensitive to variations not only in temperature, but also in the water content of the melt. To keep this study manageable, we elected to consider melts of granitic composition *containing 3 wt% H_2O only*, for which Zr diffusion is given by

$$D = 0 \cdot 1 \exp(-235\,980/RT) \qquad (2)$$

where D is in cm^2 s^{-1} and R is the gas constant in J k^{-1} mol^{-1}. This Arrhenius relation was obtained by interpolation of data reported in Harrison and Watson (1983), who investigated Zr diffusion in melts containing 0–6 wt% H_2O, but did not perform experiments specifically at 3%. The 3% H_2O value was chosen mainly because some authorities (e.g. Clemens 1984) consider this to be a typical H_2O content for crustal granitic melts (see also the summary by Johnson *et al.* 1994). It is also a 'safe' choice for modelling H_2O-bearing granitic melts in general: most of the change in D with increasing water content occurs over the first 2–3% dissolved H_2O (see fig. 4 of Harrison & Watson 1983), so higher assumed water contents would not greatly affect the conclusions of the present study. Water contents significantly *lower* than 3% would dramatically affect the model results (see fig. 5 of Harrison & Watson 1983), but such dry melts are probably uncommon.

1.2. Modelling details: one-dimensional diffusion

This study incorporates treatments of zircon dissolution (and/or growth) in both one and three dimensions. For reasons that will become clear later, the one-dimensional models are not very relevant to processes occurring on the time-scale of crustal melting events. However, the one-dimensional case is important to understanding the fundamental controls on crystal dissolution and growth and also has applications to more short-term processes such as inclusion formation and the development of fine-scale zoning. The basic techniques and principles applied in this study are more easily conveyed with reference to a one-dimensional example, from which it is a small conceptual leap to three dimensions. Because the greater emphasis of this paper is on dissolution 'regimes', we use a crystal surface dissolving into a contacting, undersaturated melt to illustrate the approach. Figure 2 is a schematic representation of the isothermal, one-dimensional dissolution process. The linear dissolution rate of the planar zircon interface (at $x=0$) is determined by the flux of atoms away from it. This flux, J_{Zr}, is proportional, through D, to the limiting concentration gradient (dc/dx) in the melt as the interface is approached

$$J_{Zr}|_{x=0} = -D \left.\frac{dc}{dx}\right|_{x=0} \qquad (3)$$

Shortly after dissolution begins, $dc/dx|_{x=0}$ is steep, so dissolution is relatively fast. As diffusion in the melt transports Zr away from the interface, the limiting gradient becomes smaller and dissolution slows (see middle and lower panels of Fig. 2). In general, the dissolution rate—expressed as the interface velocity, V—at any point in time is given by

$$V_t = -\frac{J_{Zr}|_{x=0}}{(C_L|_{x=0} - C_S)} \qquad (4)$$

where $C_L|_{x=0}$ is the concentration of Zr in the melt at the interface and C_S is the concentration in the crystal. For mathematical simplicity, changes in the system over time (t) are generally referenced to a stationary crystal/melt interface

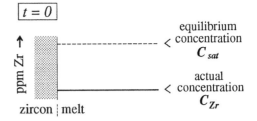

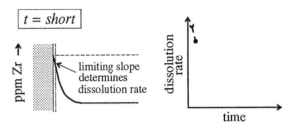

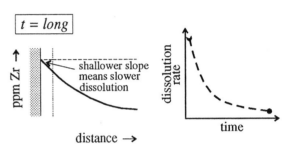

Figure 2 Schematic diagram illustrating diffusion control in the one-dimensional dissolution of a zircon surface. At $t=0$ (top panel), the crystal has just been exposed to a melt that is undersaturated in zircon—i.e. the Zr concentration in the melt (C_{Zr}) is below the level required to stabilize zircon (C_{sat}). After a short time, a steep Zr concentration profile develops in the melt as zircon dissolves and Zr diffuses away; the limiting slope of this profile (approaching the zircon) is proportional to the diffusive flux of Zr and hence to the dissolution rate. With increased time (bottom panel), the gradient shallows and dissolution slows. For the dissolution of a spherical surface the principle is the same as illustrated here, but the nature of the diffusion profile is very different because of the radial dilution effect (see text and Fig. 13).

(at $x=0$), so in effect the concentration profile against the dissolving crystal moves through the interface at rate V. In this case

$$\frac{\partial C_L}{\partial t} = D \frac{\partial^2 C_L}{\partial x^2} + V \frac{\partial C_L}{\partial x}, \qquad x > 0 \qquad (5)$$

$$C_L(x=0, t) = C_{sat}, \qquad x = 0 \qquad (6)$$

$$C_L(x=\infty, t) = C_{sat} - U = \text{const}, \qquad x = \infty \qquad (7)$$

$$C_L(x, 0) = C_{sat} - U, \qquad x > 0 \qquad (8)$$

where U is the degree to which the melt is initially undersaturated in zircon, expressed as ppm Zr below C_{sat} (i.e. $U = C_{sat} - C_{Zr}$ as depicted in Fig. 2). The convention in this paper is to consider U as positive if the melt is undersaturated in zircon; a negative value would thus indicate a zircon-oversaturated melt and the zircon would grow rather than dissolve. The domain $x < 0$ lies within the zircon crystal, where the Zr concentration is fixed by stoichiometry. The boundary condition given by Equation (6)—i.e. that $C_L(x=0)$ is constant over time at C_{sat}—holds because the diffusive flux of Zr away from the interface is balanced by Zr entering the melt as the zircon dissolves. Zhang *et al.* (1989) discussed an analytical solution to Equation (5), for appropriate boundary conditions, that describes C_L as a function of x and t (the original solution is by Neumann, as discussed in Carslaw & Jaeger 1959: 285). The dissolution rate can be deduced from this solution (by differentiation at $x \approx 0$), which results in a predicted inverse dependence of V on the square root of time.

The specific needs and goals of our study necessitated an approach to the solution of Equation (5) rather different from that taken by previous workers. We set out to describe zircon dissolution and/or growth not only for isothermal conditions, but also for various heating (or cooling) trajectories that might apply to crustal magmatic processes. This ambition precluded any hope of a universal analytical solution to the mass balance equations, so we started out directly with a finite difference approach, which makes up in versatility what it lacks in elegance. The existing exact solutions—and specifically the predicted $t^{-\frac{1}{2}}$ dependence of V—served as a good test of our computer code for cases of isothermal dissolution before progressing to consider variable temperature and 'three-dimensional' processes.

Numerical simulation by the explicit finite difference method (e.g. Crank 1975) is a standard technique that does not require detailed explanation. In brief, we used an interface-fixed reference frame and a position-dependent node spacing in which smaller distance elements (δx) were assigned to the region in the melt near the dissolving or growing zircon. The quantity of greatest interest—the dissolution (or growth) rate, V—was computed at each time step (δt) from the finite difference analogue of Equations (3) and (4) combined, i.e.

$$V_{\delta t} = \frac{D(C_2 - C_1)}{\delta x (C_1 - C_S)} \qquad (9)$$

where the interger subscripts are node numbers increasing with distance into the melt away from the interface ($C_1 = C_{sat}$; C_S is the concentration of Zr in zircon). The time steps were chosen to stay well within the stability conditions discussed by Crank (1975: 143). Because of the markedly different densities of zircon (4·65 g cm^{-3}) and hydrous granitic melt ($\approx 2\cdot 4$ g cm^{-3}), all concentrations were expressed for computational purposes as moles of Zr per cubic centimetre; these were later converted to ppm (weight) for convenience of application.

As noted earlier, the performance of the computer code was ultimately tested against the known analytical result, demonstrating that V is linear in $t^{-\frac{1}{2}}$. Preliminary tests were also performed by comparing diffusion profiles generated in isothermal numerical simulations with the corresponding profiles resulting from the analytical solution. In all instances the agreement was excellent. As a routine check for mass conservation in the finite difference simulations, we numerically integrated our final Zr diffusion profiles and compared the result with the amount of Zr lost from the dissolving zircon; these quantities were equal to within 0.5% in every instance.

1.3. Modelling spherical geometry

Some readers will immediately appreciate the limitations of treating zircon dissolution or growth as a one-dimensional problem. A one-dimensional model is appropriate only as long as the diffusion distance in the melt is small in relation to the crystal size. Even for a species as slow diffusing as Zr, however, transport distances exceeding the 100–200 µm size of a typical zircon are achieved in only 10 years or so at 700°C. Geological applications of zircon dissolution and growth models may involve time scales of 10^6 years or more, so it is clearly important to consider the three-dimensional case (as recognised by Harrison & Watson 1983). As in our previous paper, we

used the spherical geometry here for its mathematical simplicity. Although zircons dissolving in magma are not strictly spherical, the shape of the crystal becomes completely unimportant to the behaviour of the system for long dissolution or growth times, because the diffusion field in the melt is vastly larger than in the crystal.

Conversion of the computer code to model the dissolution and growth of spherical zircons involves a conceptual switch from finite difference elements of distance (δx) to concentric volume elements or 'shells' of thickness δr. Mathematically and computationally this change is straightforward, as discussed by Crank (1975: 89): the appropriate equation for diffusion in the melt near a dissolving or growing sphere is

$$\frac{\partial C_L}{\partial t} = D \left[\frac{\partial^2 C_L}{\partial r^2} + \frac{2}{r} \frac{\partial C_L}{\partial r} \right] + V \frac{\partial C_L}{\partial r} \qquad (10)$$

but on making the substitution

$$W = C_L \times r \qquad (11)$$

Equation (10) is transformed to

$$\frac{\partial W}{\partial t} = D \frac{\partial^2 W}{\partial r^2} + V \frac{\partial W}{\partial r} \qquad (12)$$

which is identical to the one-dimensional Equation (5). Thus the computational scheme used for the one-dimensional case can also be used for the spherical geometry simply by replacing δC_L by δW. The physical meaning of this substitution is to convert from an array of elements having the same surface area and volume per unit length (i.e. the one-dimensional case) to concentric shells whose surface areas and volumes depend on their distance from the origin of the spherical coordinate system. The effect of this change on the actual concentration gradient is profound for diffusion length scales much greater than the size of the crystal of interest, as discussed in Section 3.

The power of the finite difference approach described here lies in its ability to simulate the behaviour of zircons in crustal melts even in the face of changes in almost any system parameters. Most importantly, variations in temperature (hence in D and C_{sat}) with time can be handled simply by recalculating D and C_{sat} at each time step from their known dependence on T [see Equations (1) and (2)]. In principle, changes in bulk composition and H$_2$O content of the melt could also be accommodated. A less positive aspect of the finite difference approach is that the results of any one simulation only apply to the specific system conditions and changes assumed, so generalisation may be difficult. We have tried to overcome this limitation by running a large number of simulations that incorporate geologically plausible conditions and include systematic explorations of the effects of various system parameters. On the basis of the results, we provide, where possible, general relationships that should enable the reader to make useful calculations without writing complicated computer programs.

2. Specific models and results

2.1. One-dimensional dissolution and growth

Constant temperature behaviour. The one-dimensional code was used for two basic types of simulation, the first of which involved dissolution at a constant temperature into a granitic melt initially zircon-undersaturated to a degree specified in terms of a Zr deficit, U, in the melt—i.e. ppm Zr below C_{sat} [see Equation (1)]. This kind of model would apply, for example, to a natural situation in which a crustal zircon became entrained in a granitic melt containing insufficient Zr for saturation. A total of 30 simulations was run for various degrees of undersaturation ($U = 20$, 40, 80, 160 and 300 ppm) in the temperature range 650–850°C.

Multiple regression analysis of the correlations resulting from the models yields the following simple result for the isothermal dissolution rate

$$V = -2 \cdot 1 \times 10^{-7} U\, t^{\frac{1}{2}} \exp(-14\,190/T) \qquad (13)$$

where t is expressed in seconds. As expected, V is linear in $t^{-\frac{1}{2}}$. The exponential dependence on $1/T$ is intuitively reasonable for a diffusion-controlled process and the linear dependence on U also makes sense: doubling the value of U, for example, steepens the concentration gradient in the melt at the zircon interface by a factor of two, thereby increasing the flux away from the dissolving crystal by the same amount. The constant within the exponential term (14 190) is equivalent to $\frac{1}{2}(E_a/R)$, where E_a is the activation energy for Zr diffusion in the melt [see Equation (2)] and R is the gas constant. Our analysis returned a halved activation energy because diffusion-controlled interface velocities generally depend on $D^{\frac{1}{2}}$ (see chapter 13 of Crank 1975); in other words, the general form of equation 13 is $V = AU(D/t)^{\frac{1}{2}}$, where A is a constant.

Equation (13) can be used to model one-dimensional zircon dissolution at constant temperature for any plausible set of conditions of T and U. Furthermore, because all correlations in the regression analysis were virtually perfect, the relationship probably holds well beyond the ranges of T, U and t actually examined in the simulations [an equation of the same form would hold for different choices of melt composition (M) and water content, but the constants would differ]. Equation (13) is readily integrated to give the total dissolution distance in a given time

$$\Delta X = -4 \cdot 2 \times 10^{-7} U\, t^{\frac{1}{2}} \exp(-14\,190/T) \qquad (14)$$

This equation has the form $\Delta X = 2AU(Dt)^{\frac{1}{2}}$, which is predicted from analytical solutions (e.g. Carslaw & Jaeger 1959: 285); it yields net dissolution values within 1% of those resulting directly from the numerical simulations. Although strictly applicable to dissolution only, Equations (13) and (14) can be used as good approximations for linear *growth* simply by reversing the sign of U and considering it to represent the degree of *over*saturation of the melt with respect to zircon.

Temperature increasing with time. The geologically more interesting case of zircon dissolution under conditions of progressive heating was also explored using the one-dimensional computer code. The simulations were initiated at an assumed equilibrium temperature of 650°C, which is a reasonable wet solidus for crustal rocks. The temperature was increased to 800°C at constant rates ranging from 0·0001 to 1·0°C year. Although neither as intuitively predictable nor as 'clean' as the constant-temperature results, the results of these progressive heating simulations were readily generalised, through multiple non-linear regression analysis, to the following equations

$$\ln(V \times 10^{18}) = \ln \dot{T}\,[0 \cdot 3837 + 0 \cdot 0684 \ln T^* \qquad (15)$$
$$- 0 \cdot 0135 (\ln T^*)^2 + 0 \cdot 0009 (\ln T^*)^3]$$
$$+ \exp[-0 \cdot 3582 + 0 \cdot 9699 \ln T^*$$
$$- 0 \cdot 1994 (\ln T^*)^2 + 0 \cdot 0204 (\ln T^*)^3]$$

and

$$\ln(\Delta x \times 10^4) = \ln \dot{T}\,[-0 \cdot 6253 + 0 \cdot 0516 \ln T^* \qquad (16)$$
$$- 0 \cdot 0054 (\ln T^*)^2]$$
$$+ [-15 \cdot 749 + 3 \cdot 294 \ln T^*$$
$$- 0 \cdot 6182 (\ln T^*)^2 + 0 \cdot 089 (\ln T^*)^3]$$

where $\dot{T}$ is the heating rate (°C a^{-1}), T^* is the temperature

expressed as a difference from the initial temperature of 650°C (i.e. $T^* = T_{actual} - 650°C$), V is in cm s^{-1} and Δx in cm. In these equations, time is an implicit variable through the heating rate, $\dot{T}$; the simulation times ranged from 150 a ($\dot{T} = 1°C\,a^{-1}$) to 1·5 Ma ($\dot{T} = 0 \cdot 0001°C\,a^{-1}$).

Equations (15) and (16) reproduce the numerical results exactly, but their usefulness is nevertheless limited by the fact that they apply to a specific initial temperature (650°C). Similar relationships could be developed for other starting temperatures, but in view of the limited applicability of the results to geological problems (see later), such an exercise is beyond the scope of this paper. One simulation was initiated at 680°C for comparison with the results for $T_o = 650°$ (Fig. 3), which reveals that the dissolution rate for a starting temperature of 680° quickly rises—after 20–30°C of heating—to approach the value for the same temperature arrived at by starting from 650°C.

Temperature decreasing with time. Additional one-dimensional simulations were run in a *cooling* mode to examine zircon growth. These computations are straightforward and the results are as equally systematic as for the dissolution simulations, but the choice of a starting temperature is much more arbitrary: whereas the wet solidus is a reasonable temperature for initiating dissolution as temperature rises, no generally relevant anchor point exists at which to initiate zircon growth with decreasing temperature. For illustrative purposes we ran a series of growth models starting with a zircon-saturated melt at 800°C. The results are shown in Figure 4, from which two key points are evident: (1) even for the linear cooling paths modelled (0·0001–1·0°C a^{-1}), growth rates are highly non-linear in time, increasing initially to a peak rate at $\approx 780°C$ and decreasing thereafter at an accelerating rate; and (2) when expressed as a function of temperature or normalised time (t/t_{max}, where t_{max} is the total growth time), the growth rates for all assumed values of dT/dt follow identical trajectories that are displaced from each other by an order of magnitude for every two orders of magnitude difference in cooling rate (see Fig. 4). Normalising all growth rates to the initial or maximum value for a given cooling path would collapse all curves in Figure 4 into a single curve.

General conclusions from one-dimensional models. Perhaps the most significant results of the one-dimensional models are the remarkably low dissolution and growth rates that are

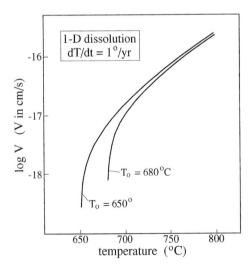

Figure 3 Dependence on temperature of the dissolution rate, V, for one-dimensional dissolution of zircon into granitic melt at a heating rate of 1°C a^{-1}. The curve for an initial temperature (T_o) of 650°C can be calculated from Equation (15), which applies only to this particular value of T_o. The path for $T_o = 680°C$ is shown here for comparison. See text for details.

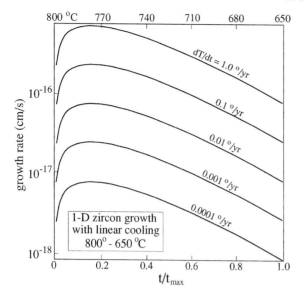

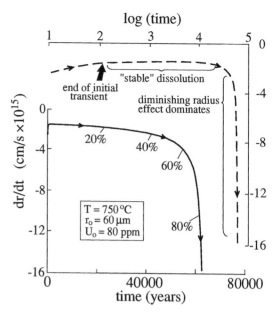

Figure 4 One-dimensional growth rate of zircon from granitic melt as the system cools from 800 to 650°C at various rates. The horizontal axis is expressed in terms of both temperature (top) and time t (normalised to the total cooling interval, t_{max}). The melt is assumed to be in saturation equilibrium with zircon at the outset.

Figure 5 Dissolution rate versus time [and log(time)] for a spherical zircon at constant temperature. The curves apply to the specific circumstances of T, r_o and U_o shown on the figure, but the principal features (labelled on the log version) are typical of all cases. The '%' labels on the lower curve denote reduction in radius. See text for details and Figure 6 for a more general representation of constant temperature, three-dimensional dissolution.

implied for realistic geological conditions. These rates fall in the general range $\approx 10^{-19}$–10^{-15} cm s^{-1} over the temperature interval 650–850°C; this translates into only 0·03 to 30 µm of dissolution or growth per Ma. These extraordinarily low rates result from the fortuitous combination of three key properties of the system: (1) the very low solubility of zircon in melts of granitic composition; (2) the high concentration of Zr in zircon (which must be dispersed during dissolution or localised during growth); and (3) the low diffusivity of Zr in the melts of interest. The case of zircon growth can be thought of in terms of supply and demand: the demand for Zr to make ZrSiO$_4$ is very large ($\approx 500\,000$ ppm), but the supply is sharply limited by the low concentration in the melt (tens to hundreds of ppm) combined with sluggish Zr diffusion.

2.2. Three-dimensional (spherical) models
The treatment of zircons as dissolving or growing spheres was undertaken to simulate more accurately the geological circumstances of zircon interaction with crustal melts. Because of the versatility of finite difference modelling, the choices of what to model were essentially infinite; our challenge was to narrow the possibilities down to a set of models from which useful generalisations (in the form of manageable equations or graphs) could be developed. As for the one-dimensional cases discussed previously, we decided to examine both constant and variable temperature behaviour. In the three-dimensional models we go one step further and discuss dissolution and growth over both 'short' and 'long' times because one or the other approach might be more useful depending on the application. The final models of this section illustrate zircon behaviour under the geologically plausible circumstance of interaction with a melt reservoir of limited volume.

General characteristics of dissolution at constant temperature. To introduce the general systematics of the diffusive dissolution of a sphere, we call attention to Figure 5, which shows numerical results for the specific case of a crystal of 60 µm initial radius (r_o) dissolving at 750°C into a melt that is initially undersaturated in Zr by 80 ppm ($U_o = 80$ ppm). Three general characteristics of this figure are broadly typical of all cases: (1) there exists an initial, brief 'transient' of elevated dissolution rate (before the arrow on the figure) which dissipates as the initially steep concentration gradient

flattens; (2) this transient is followed by a long interval over which the dissolution rate (dr/dt) is relatively constant; and (3) as the radius approaches zero, the dissolution rate accelerates sharply. The initial transient is of relatively little geological interest, being only a few tens to hundreds of years in duration (depending on temperature). The interval of relatively constant dr/dt represents a significant portion of the dissolution history and so takes on greater geological importance. During this period, the 'flattening' of the diffusion gradient (which slows dissolution) is counteracted to a large extent by the effect of diminishing radius (which accelerates dissolution).

Short-term dissolution behaviour and a practical approach for the long term. The words 'short-term' are used here in reference to the point on the dissolution rate versus time curve (Fig. 5) following the initial transient. This point represents the onset of an interval of relatively stable dissolution behaviour and so is worth characterising for a variety of conditions. We wished, in the process, to arrive at a general result that could be implemented with a hand-held calculator. This goal can be achieved without detailed numerical simulations because the crystal radius can be assumed to be constant for the brief time interval of interest. The concentration gradient against the dissolving sphere is given by the solution to Equation (10) (with $V = 0$) for diffusion in a region bounded internally by a spherical surface at which the concentration is held constant [see Crank 1975: 102; this is the solution used by Harrison & Watson (1983) to calculate approximate zircon dissolution times]. The expression given by Crank (1975) describes concentration as a function of radial distance from the surface for a given diffusion time. Because the limiting concentration gradient in the melt at the crystal surface ($dc/dx|_{x=0}$) determines the dissolution rate (see Section 1.2), we can calculate the instantaneous value of dr/dt by taking the derivative of the c versus x curve at $x \approx 0$ for any diffusion time of interest. This procedure was carried out for the 'dissolution' of spherical crystals of 30, 45, 60, 90 and 120 µm radius at temperatures of 650–850°C (50°C intervals). The resulting instantaneous dissolution rates are accurately described (to within $\approx 2\%$) by

$$\frac{dr}{dt} \times 10^{17} = -U\left[\frac{1\cdot25 \times 10^{10}}{r}\right]\exp(-28\,380/T) \quad (17)$$
$$+ 7\cdot24 \times 10^8 \exp(-23\,280/T)\Bigg]$$

where the crystal radius, r, is in cm, dr/dt is in cm s^{-1} and T is the absolute temperature. [Note that the first exponential term contains the constant E_a/R; the significance of the constant in the second exponential term is unclear.] It is important to emphasise that the instantaneous dissolution rates given by Equation (17) strictly pertain only to the immediate 'post-transient' phase of dissolution as indicated on Figure 5 (this condition is reached after dissolution times ranging from ≈ 50 a at 850°C to ≈ 1000 a at 650°C). At this stage, the radius of the dissolving crystal will have changed insignificantly from the initial value, so the stationary interface assumption made in arriving at Equation (17) is valid. Despite the limited range of strict applicability, dr/dt values given by Equation (17) are broadly representative of the dissolution process over a fairly large time interval (see Figure 5; this interval corresponds to the initial time period over which the curves in Fig. 6 can be approximated as linear). Equation (17) is thus a useful relationship for addressing constant temperature dissolution in general. It can also be readily coded for use in a stepwise, iterative mode in which dissolution is allowed to proceed for a short time at some initial rate, the radius then adjusted by the amount dissolved and a new dr/dt calculated for use in the next time step. Use of Equation (17) in this manner is a practical way to simulate, with reasonable accuracy, not only the complete dissolution history of a zircon at constant temperature, but also variable temperature processes such as those described in later sections of this paper. Thermal trajectories can be modelled simply by adjusting the temperature before each new value of dr/dt is calculated. The results of constant temperature 'stepwise' simulations using Equation (17) are included in Figure 6 for comparison with the more rigorous, moving boundary, finite difference method generally used in this paper. The total dissolution time for a zircon of 90 μm radius is reproduced to within about 6%, and deviations from the 'true' curve never exceed about 10% over the entire dissolution interval. For a 30 μm zircon, the outcome is not as good: the total dissolution time is overestimated by about 16% using Equation (17). For most imaginable geological applications, however, Equation (17) can be implemented as an adequate description of the dissolution history of a crustal zircon. For the sake of thoroughness and accuracy, and to provide detailed insight into the importance of various system parameters, we implemented the three-dimensional (spherical) finite difference code (see Section 1.3) for a set of widely varying initial conditions. The results complement and extend those obtainable from Equation (17).

Long-term dissolution behaviour in detail. The modelling efforts described to this point have revealed a systematic dependence of the dissolution rate on temperature (T), undersaturation (U), crystal radius (r) and heating rate ($\dot{T}$). Accordingly, all these parameters were known to be important as we approached the detailed numerical simulations of zircons interacting with melt. In all, about 300 models were run in which the four key parameters were systematically varied. The models are logically divided, as previously, into those involving constant temperature and those in which the temperature varies with time.

Results for constant temperature. The general character of dissolution rate versus time curves for all constant temperature models resembles the example shown in Figure 5—i.e. there is a brief initial transient followed by an extended period of nearly constant (slightly decreasing) dissolution rate and a final interval of accelerating dissolution. The overall results are most easily presented in terms of dimensionless quantities. This is shown graphically in Figure 6, where the dimensionless radius r/r_o (where r_o is the initial radius of the dissolving zircon) is plotted against the product of dimensionless time (Dt/r_o^2) and the undersaturation, U. Use of these variables collapses the 'dimensional' radius versus time dissolution curves for any values of T and U into a single curve pertaining to a specific value of r_o. [Temperature is incorporated into the figure through its effect on D, as expressed in Equation (2).] Figure 6 conveys the same 'time to disappearance' information provided by Harrison and Watson (1983; see their figure 5), but also shows the time evolution of the dissolution process and its dependence on the initial degree of undersaturation of the melt.

Dissolution with progressive heating. The model results for situations involving progressive heating are more difficult to present concisely because the variation of D with T (hence with t) precludes the use of the dimensionless time variable Dt/r_o^2. For this reason we resort to a series of graphs illustrating the effects on the dissolution of changes in the system parameters. For reasons of practicality, this treatment is not comprehensive, but illustrative. As for the one-dimensional models, we focus on an assumed initial temperature of 650°C and we consider heating rates of 0·1, 0·01, 0·001 and 0·0001°C a^{-1}. The results for an initial undersaturation of 20 ppm (equivalent to 3 ppm Zr initially present in the melt) are shown in Figure 7, which summarises dissolution information in two ways: r/r_o versus time for initial radii of 30, 45, 60 and 90 μm, and heating rate versus time for 10, 50 and 100% dissolution of zircons of 30 and 90 μm initial radius. Figure 7 shows that the rate of heating is an important consideration in dissolution—a fact that will also become apparent in the geological simulations discussed in Section 2.3. The effect of dT/dt can be stated in the following way. If the system is approaching a particular peak temperature, faster approach rates (i.e. dT/dt values) will result in less dissolution. The initial temperature (T_o) obviously affects the extent of dissolution for a given heating rate as well. The information in Figure 7 pertains specifically to $T_o = 650$°C, but in Figure 8 we

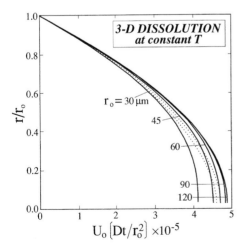

Figure 6 Constant temperature dissolution of zircon spheres expressed in terms of dimensionless radius (r/r_o) and dimensionless time (Dt/r_o^2). The solid curves were obtained from 'full-scale' finite difference simulations as described in Section 1.3 of the text and as used to generate Figures 5 and 7–13. The dotted curves show, for comparison, the approximate results obtained by 'stepwise' application of Equation (17) to zircons of 90 μm (left-hand curve) and 30 μm (right-hand curve) initial radius. For the purposes of this figure, the units of distance and time are centimetres and seconds. See text for discussion.

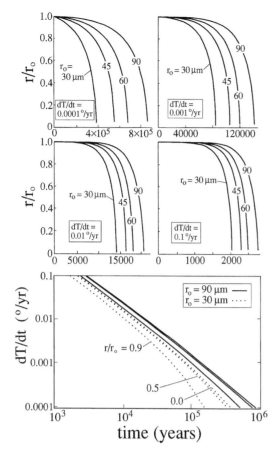

Figure 7 Effect of heating rate (dT/dt) on dissolution of zircon spheres of various initial radii. The clustered diagrams at the top are simple (dimensionless) radius versus time plots. The diagram at the bottom contains the same information plotted as heating rate versus time required for dissolution to $r/r_o = 0.9$, 0.5 and 0.0. The initial temperature, T_o, is 650°C for all cases, and U_o is 20 ppm.

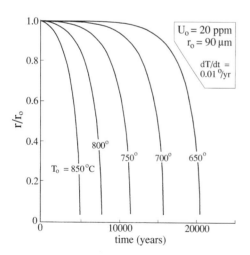

Figure 8 Zircon radius (r/r_o) versus time for heating at a rate of 0.01°C a^{-1} from various initial temperatures (T_o). The initial radius is 90 µm.

illustrate the consequences of changes in T_o for an intermediate heating rate of 0.01°C a^{-1}. Starting at 750°C rather than 650°C reduces the dissolution time of a 90 µm radius zircon by about a factor of two.

Effect of the degree of undersaturation. We examined the effects on dissolution of changing the initial degree of undersaturation (U_o) during heating at various rates from 650°C. For this particular initial temperature, the range of plausible values for U_o is small (see Section 4) because the zircon solubility is so low (≈ 23 ppm); we considered near-

extremes in U_o of 0 and 20 ppm. Figure 9 compares the extent of dissolution (r/r_o) of 60 and 180 µm zircons as a function of time and heating rate for these two U_o values. The effect of changes in U_o is generally small, becoming significant only for low heating rates and initially small zircons. It should be borne in mind, however, that higher initial temperatures would allow a wider range in reasonable values of U_o, so under some circumstances the initial saturation state of the melt could be more important than Figure 9 implies. Recall that the dissolution rate is directly proportional to U during dissolution at constant temperature [see Equation (17)].

2.3. Geological time–temperature paths

Rationale and model types. We completed our three-dimensional modelling efforts by considering zircon behaviour over some hypothetical thermal histories for crustal magmatic events. Difficulties in this endeavour arose not with the modelling itself, but with the choice of time–temperature paths that might be widely accepted as reasonable and realistic. After discussion with several Third Hutton Symposium participants, we settled on three basic thermal path types for which to run illustrative models.

Type 1 (suggested by J.D. Clemens). Intense thermal events—reaching 150–200°C above the solidus—of intermediate duration (10^5 a), in which the temperature increases relatively slowly and drops precipitously after the thermal peak. Such events are imagined to follow from the intrusion of large volumes of mafic magma into the lower crust. The peak temperature could be regarded as coincident with melt segregation from the source, which would be followed by rapid ascent and cooling. Two 'type 1' T–t paths were considered; these are shown in Figure 10 (paths 1 and 2).

Type 2 (suggested by C.F. Miller). Intense thermal events similar in magnitude to type 1, but of generally shorter duration (5000–20 000 a) and involving an abrupt increase in temperature followed in most instances by relatively slow cooling. This kind of T–t path might apply to near-solidus granitic magma or rock into which mafic dykes are injected. Three such paths are modelled (paths 3–5 in Fig. 11).

Type 3. Protracted, 'low intensity' thermal events in which melting is initiated at 650°C, the temperature rises to 700°C in 0.5 Ma, and decreases symmetrically back to 650°C along an overall T–t path that is parabolic in nature (see top right-hand panel of Fig. 12). Such a path might be appropriate to migmatisation in the mid- to lower crust.

For all three thermal event types, we examined the fate of zircons ranging in initial radius from 30 to 200 µm. For type 1 and 2 events, we considered initial undersaturation values (U_o) of 0 and 40 ppm; modelling higher values of U_o is

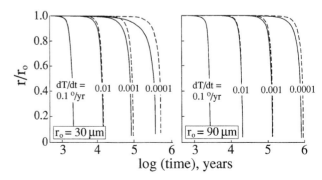

Figure 9 Comparative dissolution curves for zircons of $r_o = 30$ and 90 µm heated at various rates from an initial temperature of 650°C. The solid curves show behaviour for $U_o = 20$ ppm, the broken curves for $U_o = 0$ ppm. See text for discussion.

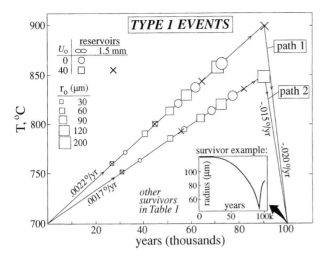

Figure 10 Type 1 time–temperature paths illustrating the fate of zircons of various initial radii (r_o) for two values of U_o (0 and 40 ppm) and two melt reservoir sizes ($r_{res} = 1.5$ mm and ∞). The symbols located on the T–t paths indicate the disappearance points of zircons of particular r_o values. The size of the symbols is scaled to r_o, representing, from the largest to the smallest, $r_o = 200$, 120, 90, 60 and 30 μm. A few of the largest zircons survive the entire melting event; vital statistics on 'survivors' are summarised in Table 1. A radius versus time graph (inset) is shown for the path 1 survivor with the following initial characteristics: $r_o = 120$ μm; $U_o = 0$; $r_{res} = 1.5$ mm (this is the first entry in Table 1). See text for discussion.

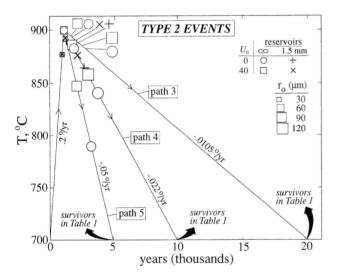

Figure 11 Type 2 thermal events illustrating the fate of zircons of various initial sizes. The symbols and their locations have the same significance as in Figure 10 (however, the four sizes of symbols on the T–t paths represent $r_o = 30$–120 μm only, because the 200 μm crystals survive in all instances). Vital statistics on survivors are summarised in Table 1. See text for discussion.

unrealistic because the solubility of zircon at the starting temperature of 700°C is only 48 ppm. The type 3 event was modelled for $U_o = 20$ ppm only because the solubility of zircon at the initial temperature of 650°C is only 23 ppm.

For physically static melting conditions, it is generally believed that melt originates and persists as small pockets and channels distributed among the residual minerals (e.g. Bulau *et al.* 1979; Jurewicz & Watson 1985; Laporte & Watson 1995). Thus it is reasonable to suppose that any zircons present in the source rock interact, not with an infinite 'ocean' of melt (as we have implicitly assumed to this point), but with melt reservoirs decidedly finite in volume relative to the diffusive transport distances involved. To simulate this added degree of geological realism, we ran some simulations of

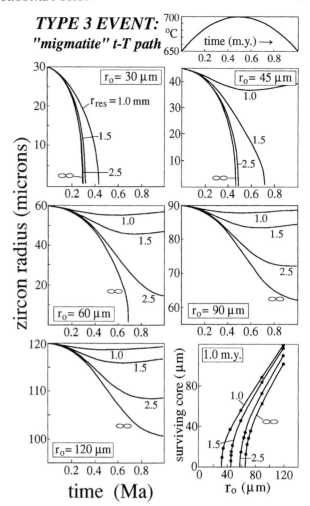

Figure 12 Diagrams depicting the fate of zircons subjected to a million year type 3 (migmatite) event, which is shown as a T–t path at the extreme upper right. The square panels are simple radius versus time graphs for zircons having five different initial radii—i.e. $r_o = 30$, 45, 60, 90 and 120 μm. The four curves on each graph are identified with four different melt reservoir radii ($r_{res} = 1.0$, 1.5 and 2.5 mm and ∞). Note that the size of the melt reservoir can determine not only whether a zircon of a particular initial size will survive the event, but also whether an overgrowth rim will form and how thick it will be (a positive slope on one of these curves indicates regrowth on the old core). The panel at the lower right shows the surviving core radius as a function of r_o for various melt reservoir radii (the dots represent the results of actual numerical simulations; the curves are fitted to the dots). See text for discussion.

zircons interacting with melt reservoirs bounded by a spherical surface at some distance r_{res} from the original zircon–melt interface (usually 1.5 mm, but in some instances 1.0 and 2.5 mm were also considered). These melt pocket sizes were chosen because experimentally produced melting textures suggest melt pocket dimensions of generally the same order or smaller than the average grain size of the restitic material (e.g. Jurewicz & Watson 1985). The limited reservoir can result in zircon dissolution trajectories very different from the 'infinite melt ocean' cases. Interestingly, the limited melt reservoir models also serve as a good representation of zircon dissolution and/or growth in a melt containing a sufficiently high density of individual zircons that their diffusion fields intersect. In other words, the limited reservoir results also apply to the situation in which the inter-zircon distance is equivalent to the reservoir diameter (in this instance, 2–5 mm).

Overview of results. The type 1 thermal events shown in Figure 10 include two T–t paths: No. 1, which reaches a maximum of 900°C and No. 2, which peaks at 850°C. The fates of zircons of various initial radii (r_o) are represented either as symbols (size scaled to r_o) marking the disappearance point on the T–t paths, or—in for surviving crystals—as entries in the inset table. Most zircons dissolving into an infinite melt reservoir do not survive these intense thermal events, disappearing during the heating interval of the T–t cycle. The degree of initial undersaturation ($U_o = 0$ or 40 ppm Zr) makes relatively little difference to the ultimate lifetime or survivability of zircons of all sizes. The volume of the melt reservoir, however, makes a significant difference: restricting the size to a radius of 1.5 mm results in the preservation of 'cores' of those zircons that start out as large crystals (120 and 200 µm in radius). The computer code calculates the amount of new growth on the surviving cores that occurs during the later stages of cooling. The values are reported in Table 1, which reveals, once again, an insignificant effect of the initial undersaturation value, U_o. For instances in which the original zircons do not survive, the Zr put into solution earlier in the T–t history would nucleate as new crystals at some point during cooling. However, because no 'inherited' core survives and because there is no simple way to incorporate nucleation behaviour into our computations, we do not model zircon growth that requires nucleation.

Type 2 thermal events (see Fig. 11) lead to different zircon survival systematics. As in the type 1 cases, it is only the largest crystals that escape complete dissolution. Because heating is so fast in the type 2 paths, however, even small zircons survive up to $\approx 875°C$ (these disappear at only ≈ 750–$760°C$ during the more leisurely heating paths shown in Fig. 10). In the type 2 events, all but the smallest zircons ($r_o = 30$ µm) survive to the peak temperature of 900°C. Nevertheless, because dissolution continues after the thermal maximum, many crystals that survive the peak later disappear on the cooling limb of the path. The main difference between the dissolution behaviour in the two types of events considered thus far is that zircons are generally consumed before and after the peak temperature in the type 1 and type 2 events, respectively. This difference results from the fact that the extent of dissolution depends on the time spent at the temperature (i.e. the integral of the T–t curve), so slow progress towards a high temperature will result in more extensive dissolution before reaching that temperature. A familiar and analogous effect in isotope geochemistry is that of diffusive closure to radiogenic isotope loss from a crystal: the closure temperature (analogous here to the temperature of zircon disappearance) is not a single-valued number, but depends on the cooling rate (analogous to our heating rate) of the system.

The results of the type 3 simulations are presented in a different manner from that used up to this point. To portray clearly the behaviour for several melt reservoir sizes ($r_{res} = \infty$, 2.5, 1.5 and 1.0 mm), we use simple zircon radius versus time plots (Fig. 12). The diagrams are self-explanatory, but some general conclusions are worth noting. Firstly, for the million year time frame considered, there is little qualitative difference in zircon dissolution behaviour for $r_{res} = 2.5$ mm and $r_{res} = \infty$. This similarity occurs because the 2.5 mm radius is some significant fraction of the characteristic (spherical) transport distance for an infinite reservoir. A second point of interest is that, for a given melt reservoir size, the existence and thickness of new overgrowth is related to the initial zircon radius. Also, the size of the relict core from a 'parent' zircon of a particular radius depends on the size of the melt reservoir with which that zircon interacts (see lower right-hand panel of Fig. 12).

3. Consequences of a dynamic system

It was made clear at the beginning of this paper that the system under consideration—zircon surrounded by silicic melt—is assumed to be physically static and that the processes of zircon dissolution and growth are thus influenced solely by diffusion in the melt. It is also clear that this assumption is not strictly accurate for most natural cases. Circumstances can be envisioned under which zircons have some velocity relative to their surroundings and it is important to evaluate the consequences of this motion to conclusions reached by assuming diffusion control.

Some insight into the possible consequences of stirring the system can be gained by inspection of the Zr concentration gradient in the spherical diffusion field surrounding a dissolving zircon. Examples are shown in Figure 13 for dissolution of a zircon of 90 µm initial radius at 750°C for times ranging from 10^3 to 10^5 a (the zircon would be consumed in $\approx 3 \times 10^5$ a). The evolution of these gradients with time differs markedly from the more familiar one-dimensional diffusion case in that a very steep gradient persists near the dissolving crystal throughout the time interval considered. Growth of the overall diffusion profile is characterised by slight broadening of this steep gradient, combined with extension of the 'tail' reaching into the melt reservoir. This behaviour is due to the radial dilution effect in outward diffusion from a spherical surface (see inset of Fig. 13).

To evaluate quantitatively the effect of moving the zircon (or stirring the melt) on the diffusive dissolution rate, we would have to compare the width of the *mechanical* boundary

Table 1 Summary on information on zircons that survive the magmatic events represented by paths 1–5 in Figures 10 and 11. The specific paths are identified in the first column; reservoirs (Res.) are classified either as infinite (I) or limited (L = 1.5 mm). U_o is the initial undersaturation in ppm, r_o the initial radius (µm) and r_c the surviving core radius; 'Rim' refers to the thickness of the overgrowth rim (µm). See text and Figures 10 and 11.

Path	Res.	U_o	r_o	r_c	Rim	Path	Res.	U_o	r_o	r_c	Rim
1	L	0	120	45	38	4	I	40	200	145	—
1	L	0	200	181	10	4	L	0	120	86	8
1	L	40	200	179	11	4	L	0	200	186	8
2	I	0	200	63	—	4	L	40	120	81	8
2	L	0	120	92	7	4	L	40	200	184	6
2	L	0	200	190	4	5	I	0	120	79	—
2	L	40	120	86	8	5	I	0	200	178	—
2	L	40	200	188	4	5	I	40	120	72	—
3	I	0	200	96	—	5	I	40	200	175	—
3	I	40	200	72	—	5	L	0	90	50	2
3	L	0	120	76	23	5	L	0	120	96	2
3	L	0	200	184	10	5	L	0	200	189	2
3	L	40	120	69	23	5	L	40	90	42	2
3	L	40	200	182	10	5	L	40	120	93	2
4	I	0	200	154	—	5	L	40	200	187	2

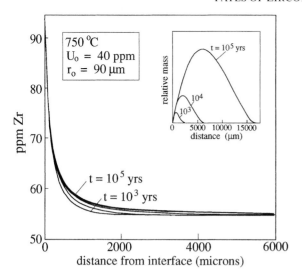

Figure 13 Zirconium concentration profiles against a spherical zircon dissolving into an infinite melt reservoir after 10^3, 3×10^3, 10^4, 3×10^4 and 10^5 years. Note the steep portion of these concentration profiles— the diffusive boundary layer—which remains confined to the region very close to the zircon. As an interesting contrast, the inset diagram depicts the relative mass of Zr from the dissolving zircon as a function of radial distance from the zircon. Although Zr concentration is highest near the dissolving zircon, most of the actual mass put into solution is dispersed to a considerable distance away.

layer that clings to the moving zircon with the width of the steep portion of the diffusion profile for a stagnant system (which can be regarded as the *diffusive* boundary layer). If the mechanical boundary layer is as wide or wider than the diffusive boundary layer, then the motion of the zircon will have little effect on the dissolution process. In other words, if stirring the system erodes only the tail of the diffusion profile, then diffusion control will dominate dissolution and growth. Unfortunately, it is not a simple matter to estimate the width of the mechanical boundary layer around a moving zircon. Although the needed characteristics of the system—i.e. the viscosity of the melt and the relative velocity of the zircon— are known or assumable, an exact solution for the boundary layer width does not exist. The situation is complicated by the fact that the width of the boundary layer is a function of the position around the moving crystal.

Extensive, multidisciplinary interest in the behaviour of dissolving spheres has led to some practical ways of evaluating the relative importance of diffusion and mechanical stirring (e.g. Cussler 1984). One formulation is

$$\frac{kd}{D} = 2 \cdot 0 + 0 \cdot 6 \left(\frac{dv^\circ}{v}\right)^{\frac{1}{2}} \left(\frac{v}{D}\right)^{\frac{1}{3}} \qquad (18)$$

where k is a mass transfer coefficient (units = distance/time) that describes, without specifying the mechanism, the dissolution rate of the sphere into the contacting fluid, d is the sphere diameter, v° is the velocity of the sphere relative to the fluid medium and v is the kinematic viscosity of the fluid (see Cussler 1984, table 9.3-2 and p. 234). As the value of kd/D approaches $2 \cdot 0$ (i.e. as the product on the right-hand side of Equation (18) goes to zero), diffusion control of the dissolution process becomes dominant. If, on the other hand, the right-hand term is of the same order as 2, then advection is of comparable importance to diffusion in controlling dissolution.

For zircon dissolution into granitic melt containing 3% H_2O, D is known [Equation (2)] and v can be estimated from the data of Shaw (1965). The relative velocity of the dissolving

crystal is more open to supposition, but the Stokes' settling velocity is a reasonable maximum. Plugging appropriate values for D, v and v° into the right-hand side of Equation (18) leads to values for kd/D ranging from $2 \cdot 0007$ ($d = 100$ μm; $T = 650°C$) to $2 \cdot 015$ ($d = 300$ μm; $T = 850°C$). These results indicate that diffusion control of zircon dissolution is absolute for the range of conditions considered in this paper. This calculation demonstrates an intuitively reasonable conclusion—i.e. that small crystals interacting with melt as viscous as molten granite carry their diffusion fields around with them.

This result notwithstanding, it is likely that under some circumstances movement or stirring of the melt does affect zircon dissolution and growth. For example, shearing of the melt during segregation or extraction is likely to erode or otherwise modify at least the tails of concentration profiles like those shown in Figure 13. Such modification could be especially important during zircon growth, perhaps leading to the ubiquitous growth zoning generally referred to as oscillatory.

There are probably also circumstances under which mechanical disturbances of the zircon–melt system are minor or non-existent. Some migmatites, for example, are thought to represent *in situ* melt production with little or no loss of melt (e.g. Dougan 1979; Johannes & Gupta 1982; Olsen 1982; Sawyer 1991). In such instances zircons might interact with relatively small pockets of generally sedentary melt, so the only complexities not accounted for in the limited reservoir numerical simulations are that (1) no zircon crystal in a real migmatite would reside at the centre of a melt pocket and (2) the size of the melt pocket would be a function of time. It is reasonable to suppose that zircons interacting with melt would remain attached to a major phase or settle to the bottom of a melt pocket. This physical reality of the system might explain the off-centre character of some inherited zircon cores (see Fig. 1).

We conclude that, in general, diffusion control dominates zircon dissolution and growth behaviour in granitic magmas, at least to the extent that the equations and simulations presented in this paper are reasonably representative of most of the geological processes we seek to understand. Still, it would be naive to argue that fluid mechanical effects never enter into zircon dissolution. It is likely, for example, that melt flow would have some effect during a very dynamic process such as filter pressing. Even then, however, understanding the diffusion-controlled end-member behaviour is crucial.

4. Applications, implications and limitations

The modelling results presented in this paper serve two purposes: they provide not only generalised relationships [e.g. Equations (13) and (17); Fig. 6] that can be used to address a variety of problems, but also some highly specific but (it is hoped) informative simulations of hypothetical geological melting events. The latter serve to illustrate the importance of various system parameters to the overall systematics of zircon dissolution, survival and overgrowth.

Areas of direct and specific application of the results include diffusive boundary layer related problems in zircon growth and dissolution, such as zoning and inclusion formation. Equation (13) is a fundamental and general result that can be applied to any problem for which a knowledge of a linear dissolution or growth rate is needed. Combined with (as yet generally unavailable) data on the diffusivities of key trace elements in the melt (e.g. U, Hf, REEs), growth and dissolution rates of zircon provide a basis for addressing questions related to kinetic disequilibrium in trace element partitioning and consequent intracrystalline zoning. Watson and Liang (1995) have already applied the growth rate results to the problem of sector zoning in zircon.

The 'bigger picture' area of application—to zircon geochronology and the problem of Pb isotopic inheritance—has been emphasised throughout this paper. Figure 6 and Equation (17) are probably the most useful sources of information for readers who wish to model geological circumstances of their own choosing. Strictly speaking, Figure 6 and Equation (17) apply to zircon dissolution at constant temperature; however, as noted in Section 2.2.2, Equation (17) can be implemented in a stepwise mode to obtain approximate results for variable temperature processes. Use of a time-integrated average temperature for a T–t path of interest would also yield a meaningful result. The most difficult value to constrain for any attempted model might well be U_o. We have generally assumed that the level of dissolved Zr in a melt produced in the crust must be close to zero initially, because the melt is produced from major minerals containing very little Zr (there is an implicit assumption here that major minerals dissolve into the melt faster than zircon does). The upper bound for permissible dissolved Zr is given by the solubility of zircon—which is only 23 and 48 ppm at 650 and 700°C, respectively [see Equation (1)]—so U_o is constrained within a fairly narrow range for anatectic situations. Because of this constraint, the dissolution behaviour of zircons participating in crustal fusion is not greatly affected by the allowable variation in reasonable choices of U_o (see Fig. 9). Dissolving xenocrystic zircon in a pre-existing melt is, however, another matter. In this situation, there is a considerably greater range in plausible values of U_o. The effects of changes in U_o are readily evaluated for constant temperature dissolution using Equation (17) or Figure 6. For variable temperature situations, the importance of the assumed value of U_o depends on the heating rate and the initial size of the zircon (see Fig. 9).

Whatever the modelling application, the limitations of our results must be borne in mind. For example, although Equation (13) is a completely general result for the specific one-dimensional system considered, it is nevertheless limited in its range of applicability by the initial assumptions made in its derivation. None of the equations or models applies to melts deviating significantly from a typical granitic composition ($M \approx 1.3$; see Watson & Harrison 1983), nor do they apply quantitatively to melt with H_2O contents other than 3 wt% (as noted earlier, higher H_2O contents would make little difference to the overall conclusions; H_2O contents less than about 2 wt% would produce markedly different results). It is worth reiterating that the general form of the equations and summary curves presented [e.g. Equations (13) and (17) and Fig. 6] would be the same for melt compositions and H_2O contents other than those considered, but the various constants would differ. In Equation (13), for example, the constants both within and outside the exponential would be affected by changes in melt composition and/or the governing Arrhenius relation for diffusion. On the other hand, the linear dependence of the one-dimensional dissolution rate on U_o and $t^{-\frac{1}{2}}$, and the log-linear dependence on T^{-1}, are universal characteristics of diffusion-controlled dissolution.

The illustrative models of hypothetical geological melting events have their own set of built-in assumptions and limitations. A key premise in any model of anatectic zircon dissolution is that zircons present in the source rock are not occluded within major mineral phases during melting—i.e. that most or all zircons actually 'see' the melt. Divergent views on the validity of this assumption have been expressed. Watson et al. (1989) argued on theoretical grounds, with supporting experiments and measurements on natural rocks, that most of the mass of zircon (and other accessory minerals) in high-grade metamorphic rocks is present along grain boundaries and so would probably be exposed to an anatectic melt. However, the principle invoked in their arguments—interfacial energy minimisation—is probably relevant mainly to instances of protracted, high-temperature metamorphism. The conclusions of Watson et al. (1989) do not apply to inclusion–host relations developed during igneous crystallisation (see Green & Watson 1982; Bacon 1989). Bea (1996; this issue) provides convincing examples supporting the view that the main mass of accessory minerals is included in major phases, and so would be generally isolated from a partial melt. This debate will probably not be resolved soon, if indeed any resolution is needed. The probable reality is that zircons are distributed differently in different rocks, reflecting in some way their unique crystallisation and annealing histories. Several additional points are worth noting, however: (1) imaging studies of zircons separated from granites (e.g. Hanchar & Miller 1993; Paterson et al. 1992) reveal the nearly universal presence of igneous overgrowth on older cores (see Fig. 1), indicating exposure to the melt; (2) major mineral phases do not necessarily simply dissolve from the grain margin inward during melting—extensive recrystallisation is entirely possible, so zircons that are occluded at the start of melting may not stay that way throughout the process; (3) biotite is a common host for zircon and biotite breakdown is thought to be a common trigger for melting in the crust; and (4) zircon thermometry based on the assumed exposure of residual zircons to melt seems to work in at least some instances (e.g. Copeland et al. 1988; Brouand et al. 1990; Barrie 1995).

Other limitations of the geological simulations presented in this paper relate to our choices of T_o, T–t paths and the assumed physical circumstances of melting. Our hope is that we have provided enough examples and made explicit evaluations of enough of the key variables to satisfy the needs of most readers.

Despite the highly specific nature of the thermal event simulations (types 1–3), some broad conclusions can be drawn on the basis of the collective results. Progressing from the obvious to the speculative, these include the following:

1. Regardless of the nature of the magmatic event, large zircons are more likely to survive crustal melting (as relict cores) than small ones. This is not a new conclusion; it was reached previously by Harrison and Watson (1983) and follows from common sense.

2. Zircon survival through protracted, low-temperature events is more likely than survival through brief, hot events (compare the type 1 model with types 2 and 3).

3. The probability of core survival depends not only on the initial size of the zircon and its thermal history, but also on the volume of the melt reservoir with which it interacts. In other words, the spatial distribution of melt can be important to zircon survival and radiogenic Pb inheritance: the smaller the reservoir of melt in contact with an individual zircon, the higher the probability that a relict core will survive. Interestingly, this statement can be carried to the extreme of $r_{res} = 0$, where survival of the zircon is guaranteed (this is the situation advocated by Bea 1996, this issue).

4. In cases where a zircon interacts with a limited melt volume and a core survives the melting event, an overgrowth rim will develop in the later stages of cooling. The radial thickness of that rim is inversely proportional to the original size of the zircon (for a given value of r_{res}) and inversely proportional to r_{res} (for a given initial zircon radius).

5. In light of the preceding observations, we can summarise the factors that favour the presence of isotopic inheritance in crustal zircons. They are: (a) the presence in the source rock of large zircons (or general occlusion of zircons within major phases); (b) either a very brief or a rather cool magmatic event; and (c) failure of the melt to coalesce into a large body. None of these factors is a prerequisite for inheritance, but the combination of two or more of them would almost guarantee its occurrence, given a sufficiently low diffusivity for Pb in zircon.

6. Acknowledgements

The author is indebted to Yan Liang, John Hanchar, Dave Wark, Calvin Miller and Daniele Cherniak, as well as to many of the Third Hutton Symposium participants, for sharing their ideas about zircons, granitoid magmatism and kinetics. The manuscript benefited from the official reviews of Peter Michael, Bruce Paterson and Mark Harrison, and from the informal comments of John Hanchar, Calvin Miller and Ian Williams. The research leading to this paper was supported by the National Science Foundation under grant no. EAR-9205793.

References

Ayers, J. C. & Watson, E. B. 1991. Solubility of apatite, monazite, zircon and rutile in supercritical aqueous fluids with implications for subduction zone geochemistry. PHIL TRANS R SOC LONDON A35, 365–75.

Bacon, C. R. 1989. Crystallization of accessory phases in magmas by local saturation adjacent to phenocrysts. GEOCHIM COSMOCHIM ACTA 53, 1055–66.

Baker, D. R. 1991. Interdiffusion of hydrous dacitic and rhyolitic melts and the efficacy of rhyolite contamination of dacitic enclaves. CONTRIB MINERAL PETROL 106, 462–73.

Barrie, C. T. 1995. Zircon thermometry of high-temperature rhyolites near volcanic-associated massive sulfide deposits, Abitibi subprovince, Canada. GEOLOGY 23, 169–72.

Bea, F. 1996. Controls on the trace element composition of crustal melts. TRANS R SOC EDINBURGH EARTH SCI 87, 000–000.

Brouand, M., Banzet, G. & Barbey, P. 1990. Zircon behavior during crustal anatexis. Evidence from the Tibetan slab migmatites (Nepal). J VOLCANOL GEOTHERM RES 44, 143–61.

Bulau, J. R., Waff, H. S. & Tyburczy, J. A. 1979. Mechanical and thermodynamic constraints on fluid distribution in partial melts. J GEOPHYS RES 84, 6102–8.

Carslaw, H. S. & Jaeger, J. C. 1959. Conduction of heat in solids, 2nd edn. New York: Oxford University Press.

Chekhmir, A. S. & Epel'baum, M. B. 1991. Diffusion in magmatic melts: new study. In Perchuk, L. L. & Kushiro, I. (eds) Physical chemistry of magmas. ADV PHYS GEOCHEM 9.

Chen, Y. D. & Williams, I. S. 1990. Zircon inheritance in mafic inclusions from Bega Batholith granites, southeastern Australia; an ion microprobe study. J GEOPHYS RES 95, 17,787–96.

Cherniak, D. J., Hanchar, J. M. & Watson, E. B. 1993. Rare earth diffusion in zircon. EOS, TRANS AM GEOPHYS UNION 74, 651.

Cherniak, D. J., Hanchar, J. M. & Watson, E. B. 1995. Hf and rare earth diffusion in zircon. EOS, TRANS AM GEOPHYS UNION 76, 704.

Clemens, J. D. 1984. Water contents of silicic to intermediate magmas. LITHOS 17, 272–87.

Copeland, P., Parrish, R. R. & Harrison, T. M. 1988. Identification of inherited Pb in monazite and its implication for U–Pb systematics. NATURE 333, 760–3.

Crank, J. 1975. The mathematics of diffusion, 2nd edn. London: Oxford University Press.

Cussler, E. L. 1984. Diffusion: mass transfer in fluid systems. Cambridge: Cambridge University Press.

Dougan, T. W. 1979. Compositional and modal relationships and melting reactions in some migmatitic metapelites form New Hampshire and Maine. AM J SCI 279, 897–935.

Green, T. H. & Watson, E. B. 1982. Crystallization of apatite in natural magmas under high-pressure, hydrous conditions, with

particular reference to 'orogenic' rock series. CONTRIB MINERAL PETROL 79, 96–105.

Hanchar, J. M. & Miller, C. F. 1993. Zircon zonation patterns as revealed by cathodoluminescence and backscattered-electron images. CHEM GEOL 110, 1–13.

Hanchar, J. M. & Rudnick, R. L. Revealing hidden structures: the application of cathodoluminescence and backscattered-electron imaging to dating zircons from lower crustal xenoliths. LITHOS 36, 289–303.

Harrison, T. M. & Watson, E. B. 1983. Kinetics of zircon dissolution and zirconium diffusion in granitic melts of variable water content. CONTRIB MINERAL PETROL 84, 67–72.

Harrison, T. M., Aleinikoff, J. N. & Compston, W. 1987. Observations and controls on the occurrence of inherited zircon in Concord-type granitoids, New Hampshire. GEOCHIM COSMOCHIM ACTA 51, 2549–58.

Heaman, L. M. & Parrish, R. R. 1991. U–Pb geochronology of accessory minerals. In Heaman, L. M. & Ludden, J. N. (eds) Applications of radiogenic isotope systems to problems in geology. MINERAL ASSOC CAN 19, 59–102.

Johannes, W. & Gupta, L. N. 1982. Origin and evolution of a migmatite. CONTRIB MINERAL PETROL 79, 114–23.

Johnson, M. C., Anderson, A. T. & Rutherford, M. J. 1994. Pre-eruptive volatile contents of magmas. In Carroll, M. R. and Holloway, J. R. (eds) Volatiles in magmas. REV MINERAL 30, 281–330.

Jurewicz, S. R. & Watson, E. B. 1985. The distribution of partial melt in a granitic system: the application of liquid-phase sintering theory. GEOCHIM COSMOCHIM ACTA 49, 1109–22.

Laporte, D. & Watson, E. B. 1995. Experimental and theoretical constraints on melt distribution in crustal sources: the effect of crystalline anisotropy on melt interconnectivity. CHEM GEOL 124, 161–84.

Miller, C. F., Hanchar, J. M., Wooden, J. L., Bennett, V. C., Harrison, T. M., Wark, D. A. & Foster, D. A. 1992. Source region of a granitic batholith: evidence from lower crustal xenoliths and inherited accessory minerals. TRANS R SOC EDINBURGH EARTH SCI 83, 49–62.

Olsen, S. N. 1982. Open and closed-system migmatites in the Front Range, Colorado. AM J SCI 282, 1596–622.

Paterson, B. A., Stephens, W. E., Rogers, G., Williams, I. S., Hinton, R. W. & Herd, D. A. 1992. The nature of zircon inheritance in two granite plutons. TRANS R SOC EDINBURGH EARTH SCI 83, 459–71.

Roddick, J. C. & Bevier, M. L. 1995. U–Pb dating of granites with Paleozoic plutons, Canadian Appalachians. CHEM GEOL 119, 307–29.

Sawyer, E. W. 1991. Disequilibrium melting and the rate of melt-residuum separation during migmatization of mafic rocks from the Grenville front, Québec. J PETROL 32, 701–38.

Schärer, U. & Allègre, C. J. 1982. Uranium–lead system in fragments of a single zircon grain. NATURE 295, 585–7.

Shaw, H. 1965. Comments on viscosity, crystal settling, and convection in granitic magmas. AM J SCI 263, 120–52.

Smith, V. G., Tiller, W. A. & Rutter, J. W. 1955. A mathematical analysis of solute redistribution during solidification. CAN J PHYS 33, 724–45.

van Breemen, O. & Hanmer, S. 1986. Zircon morphology and U–Pb geochronology in active shear zones: studies on syntectonic intrusions along the northwest boundary of the Central metasedimentary belt, Grenville Province, Ontario. CURR RES PART B GEOL SURV CAN PAP 86-1B, 775–84.

Watson, E. B. 1979. Zircon saturation in felsic liquids: experimental data and applications to trace element geochemistry. CONTRIB MINERAL PETROL 70, 407–19.

Watson, E. B. 1980. Some experimentally-determined zircon/liquid partition coefficients for the rare earth elements. GEOCHIM COSMOCHIM ACTA 44, 895–7.

Watson, E. B. & Harrison, T. M. 1983. Zircon saturation revisited: temperature and composition effects in a variety of crustal magma types. EARTH PLANET SCI LETT 64, 295–304.

Watson, E. B. & Liang, Y. 1995. A simple model for sector zoning in slowly-grown crystals: implications for growth rate and lattice diffusion, with emphasis on accessory minerals in crustal rocks. AM MINERAL 80, 1179–87.

Watson, E. B., Vicenzi, E. P. & Rapp, R. P. 1989. Inclusion/host relations involving accessory minerals in high-grade metamorphic and anatectic rocks. CONTRIB MINERAL PETROL 101, 220–31.

Williams, I. S. 1992. Some observations on the use of zircon U–Pb

geochronology in the study of granitic rocks. TRANS R SOC
EDINBURGH EARTH SCI **83,** 447–58.

Wyborn, D., White, A. J. R. & Chappell, B. W. 1991. Enclaves in the
S-type Cowra granodiorite. *In Excursion guide for the Second*

Hutton Symposium on Granites and Related Rocks. AUST BUR
MINERAL RESOUR GEOL GEOPHYS REC **1991/24,** 12–29.

Zhang, Y., Walker, D. & Lesher, C. E. 1989. Diffusive crystal
dissolution. CONTRIB MINERAL PETROL **102,** 492–513.

E. BRUCE WATSON, Department of Earth & Environmental Sciences, Rensselaer Polytechnic Institute, Troy,
NY 12180, U.S.A.

Transactions of the Royal Society of Edinburgh: Earth Sciences, **87**, 57–64, 1996

Water contents of felsic melts: application to the rheological properties of granitic magmas

F. Holtz, B. Scaillet, H. Behrens, F. Schulze and M. Pichavant

ABSTRACT: New experimental determinations of water solubility in haplogranitic melts (anhydrous compositions in the system Qz–Ab–Or and binary joins) and of the viscosity of hydrous $Qz_{28}Ab_{38}Or_{34}$ melts (normative proportions) and natural peraluminous leucogranitic melt (Gangotri, High Himalaya) are used to constrain the evolution of viscosity of ascending magmas, depending on their P–T paths.

At constant pressure, in the case of fluid-absent melting conditions, with water as the main volatile dissolved in the melts, the viscosity of melts generated from quartzo-feldspathic protoliths is lower at low temperature than at high temperature (difference of 1–2 log units between 700 and 900°C). This is due to the higher water contents of the melts at low temperature than at high temperature and to the fact that decreasing temperature does not counterbalance the effect of increasing melt water content. In ascending magmas generated from crustal material the magma viscosity does not change significantly whatever the P–T path followed (i.e. path with cooling and crystallisation; adiabatic path with decompression melting) as long as the crystal fraction is low enough to assume a Newtonian behaviour (30–50% crystals, depending on size and shape). Comparison of the properties of natural and synthetic systems suggests that both water solubility and the viscosity of multicomponent natural felsic melts (with less than 30–35% normative Qz) can be extrapolated from those of the equivalent synthetic feldspar melts.

KEY WORDS: water solubility, viscosity, aluminosilicate melts, granites, crystallisation paths.

Water is the most important magmatic volatile, both in terms of its abundance and its influence on melt properties and crystallisation pathways. The rheological properties of melts and magmas are known to be dramatically dependent on dissolved water and are of particular importance in quantifying convection mechanisms and crystal and xenolith settling rates in magmas, atomic diffusivities, chemical equilibrium/disequilibrium processes, the transport of magmas in the crust (dykes or diapirs), volatile exsolution and bubble growth rates in volcanic processes (e.g. Marsh 1981; Clemens & Mawer 1992; Pinkerton & Stevenson 1992; Dingwell *et al.* 1993; Petford *et al.* 1993). Thus an accurate modelling of most magmatic processes requires rheological data (mainly viscosity) for hydrous melts. To avoid empirical models (which are often of limited applicability), predictions of hydrous melt properties (phase relations, viscosity, density) require in turn the dissolution mechanisms of water in melts (water species) to be known.

Although considerable progress has been made in the last two decades, the mechanism(s) of incorporation of hydrous species in felsic melts is still poorly understood and under debate (e.g. Burnham 1975; Silver & Stolper 1985; Kohn *et al.* 1989; Dingwell & Webb 1990; Silver *et al.* 1990; Pichavant *et al.* 1992; Nowak & Behrens 1995; Romano *et al.* 1995). The problems arise mainly (1) from the difficulty of determining the speciation of water in melts (and not in glasses), (2) from a lack of fundamental experimental data on the properties of melts such as viscosity, density and the partial molar volume of water and (3) from a lack of models to relate the water speciation and the properties of hydrous melts. Water solubility data are available, but the analytical and experimental methods are not always appropriate to the determination of water solubility in melts, especially at high pressure (Behrens 1995; Holtz *et al.* 1995).

In this study some recent experimental data on water solubility and viscosity determinations in natural and synthetic felsic systems are analysed. Discrepancies with common assumptions or previous data sets are outlined to avoid the misuse of experimental data when applied to granite petrogenesis. The combined effects of changing temperature, pressure, liquid composition and water content of the melt on the rheological properties of melts and ascending magmas are discussed in the light of the experimental data. Examples of applications to theoretical and natural case studies are given.

1. Water solubility

1.1. Effect of pressure

With the exception of some studies on natural systems (e.g. Goranson 1931; Jahns & Burnham 1958; Friedman *et al.* 1963; Hamilton *et al.* 1964; Holtz *et al.* 1993), most of the experimental determinations of water solubility in felsic melts have been performed on albite (see review in Behrens 1995) or quartzo-feldspathic compositions (Qz–Ab and Qz–Or, Oxtoby & Hamilton 1978a, b). To date, most of the data used to constrain water solubility in felsic melts were taken from the data set of Oxtoby and Hamilton (1978a, b) or from results obtained for ternary Qz–Ab–Or eutectic or minimum compositions by Tuttle and Bowen (1958). Oxtoby and Hamilton (1978a, b) noted the important water solubility increase with pressure at $P \leqslant 3$ kbar, but determined water solubility to be only slightly or not dependent on pressure at $P \geqslant 3$–4 kbar (Fig. 1). The investigations of haplogranitic compositions (normative Qz contents between 25 and 45%) by Holtz *et al.* (1995) and of pure Ab and Or compositions by Behrens (1995) show that the water solubility is significantly underestimated at $P \geqslant 3$ kbar by Oxtoby and Hamilton (1978a,

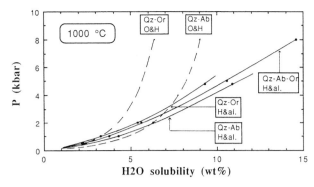

Figure 1 Pressure dependence of H₂O solubility at 1000°C for Qz₃₂Ab₆₈, Qz₃₀Or₇₀, Qz₂₈Ab₃₈Or₃₄ anhydrous melt compositions (proportions are given in wt% normative). Broken curves (labelled O & H) are the results obtained by Oxtoby and Hamilton (1978a, b); solid curves (labeled H & al.) and dots are from results of Holtz et al. (1995). Note the discrepancy between the two data sets at $P \geqslant 3$ kbar (for explanation, see text).

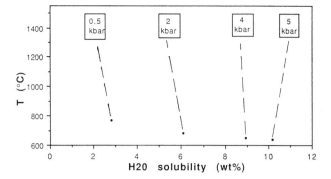

Figure 2 Temperature dependence of H₂O solubility at 0·5, 2, 4 and 5 kbar for a haplogranitic melt composition (Qz₂₈Ab₃₈Or₃₄). The closed dots represent the solidus temperatures at water-saturated conditions. Source of data: Holtz et al. (1995).

Table 2 Temperature dependence of water solubility for compositions close to thermal minima and eutectics in the system Qz–Ab–Or–H₂O up to 1200°C.

Pressure (kbar)	Temperature dependence (wt% H_2O/°C)
0·5	$-1·6 \times 10^{-3}$
1	$-1·6 \times 10^{-3}$
2	$-1·6 \times 10^{-3}$
3	$-0·4 \times 10^{-3}$
4	$-0·4 \times 10^{-3}$
5	$+2·6 \times 10^{-3}$

b) (Fig. 1) as a result of use of an inappropriate analytical technique for the determination of water contents of glasses (for details see Behrens 1995; Holtz et al. 1995). Holtz et al. (1995) and Behrens (1995) observed an almost linear increase in water solubility with pressure at $P > 2$–3 kbar (Fig. 1, Table 1). The slope of the water solubility curve is dependent on temperature. At $P \geqslant 2$ kbar, typical values for the increase in water solubility as a function of pressure are $+1·3$–$1·4$ wt% H_2O/kbar at 900°C and $+1·8$ wt% H_2O/kbar at 1200°C for the haplogranitic compositions investigated by Holtz et al. (1995).

1.2. Effect of temperature

The effect of temperature on water solubility in felsic haplogranitic melts was investigated by Hamilton and Oxtoby (1986) and Holtz et al. (1992, 1995) at several pressures. Both sets of data are in good agreement. Holtz et al. (1995) showed that, for a composition close to that of the thermal minimum in the system Qz–Ab–Or–H₂O (2 kbar), water solubility ranges from retrograde (with increasing temperature) at $P \leqslant 4$ kbar to prograde at 5 kbar (Fig. 2) and higher pressures. The effect of temperature on the water solubility of compositions close to the eutectics and minima can be predicted using the linear regressions given in Table 2. The pressure at which the water solubility is constant at any temperature depends on the melt composition. For Ab melts, this pressure is near 5 kbar (Hamilton & Oxtoby 1986; Paillat et al. 1992), whereas it is close to 2 kbar for Or melts (Behrens et al., in prep.) and between 1·5 and 2·5 kbar for pure silica melts (Holtz et al., in prep.).

1.3. Effect of composition

The effect of the anhydrous melt composition on water solubility has been investigated in felsic compositions mainly by Oxtoby and Hamilton (1978a, b), Voigt et al. (1981), Holtz et al. (1992, 1995), Behrens (1995) and Romano et al. (1996). In haplogranitic melts (all Al is charge-balanced by one alkali cation) with a given Qz content, the main feature is that the water solubility increases with increasing Ab component of the melt at $P \geqslant 1$ kbar (see Fig. 3 for 5 kbar data). At 0·5 kbar, only minor changes have been observed with variations in the Ab and Or content. The effect of Qz content on water solubility in haplogranitic melts is not linear along silica–feldspar joins, as shown in Figure 4. At low Qz contents (approximately $<30\%$ normative Qz) the water solubility is approximately constant with changing Qz content, whereas it decreases with a further increase in the Qz content. This compositional dependence has been determined experimentally for ternary compositions only (Holtz et al. 1995, unpublished data). Extrapolations of the results of Holtz et al. (1995) to Qz–Ab and Qz–Or compositions imply that the results of Oxtoby and Hamilton (1978a, b), suggesting a maximum

Table 1 Water solubility and viscosity of 'minimum' or eutectic compositions in the system Qz–Ab–Or–H₂O.

Pressure (kbar)	Composition (Qz/Ab/Or in normative %)	Water-saturated solidus temperature (°C)	Water solubility at solidus temperature (wt% H_2O)	Viscosity at solidus temperature (poises) (5)	Water solubility at 800°C (4)	Viscosity of water-saturated melt at 800°C (5)	Water content at the liquidus at 800°C (6)	Viscosity of melt at the liquidus at 800°C (5)
0·5	40/30/30 (1)	770 (1)	2·7	$10^{6·75}$	2·65	$10^{6·55}$	2·1	$10^{6·89}$
1	38/33/29 (1)	720 (1)	4·0	$10^{6·51}$	3·8	$10^{5·99}$	2·3	$10^{6·76}$
2	35/40/25 (1, 3)	685 (1)	6·1	$10^{6·01}$	5·9	$10^{5·28}$	2·8	$10^{6·46}$
5	31/47/22 (3)	645 (3)	9·7	$10^{5·40}$	10·1	$10^{4·37}$	3·6	$10^{6·08}$
10	23/56/21 (2)	620 (2)	$\approx 16·0$	—	$\approx 17·0$	—	$\approx 5·0$	—

Note: Sources of data: (1) Tuttle & Bowen (1958); (2) Luth et al. (1964); (3) Holtz et al. (1992); (4) Holtz et al. (1995), water solubility obtained by extrapolation of the experimental data; (5) Schulze et al. (in press), calculated on the basis of composition Qz₂₈Ab₃₈Or₃₄; and, (6) Holtz & Johannes (1994).

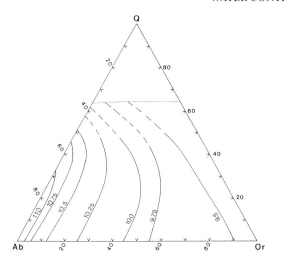

Figure 3 Superliquidus H_2O solubility, expressed as wt% H_2O, in the system Qz–Ab–Or at 5 kbar and 900°C (from Holtz *et al.* 1995; Behrens & Holtz, unpublished data). Each curve is a water solubility isopleth (water solubility values in wt% H_2O are reported on the curves). The dotted curve represents the intersection of this isothermal surface with the quartz liquidus surface (at water-saturated conditions at 900°C).

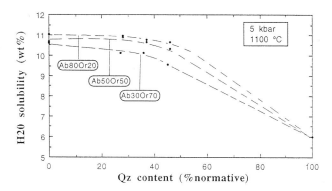

Figure 4 H_2O solubility of haplogranitic melts along feldspar–silica joins at 5 kbar and 1100°C (the normative compositions are given for each curve). Dots represent compositions for which experimental data are available. Source of data: Holtz *et al.* (1995) and Behrens and Holtz (unpublished data).

water solubility along the Qz–Or join for intermediate compositions, have to be reconsidered (see analytical problems described earlier).

The water solubility is also shown to change when the anhydrous melt composition departs from the haplogranite Qz–Ab–Or system. Water solubility increases only slightly with increasing non-charge-balanced Al content (peraluminous compositions), but increases more significantly with increasing alkali content (peralkaline compositions, e.g. Dingwell *et al.* 1984; Holtz *et al.* 1994; Behrens 1995; Linnen *et al.*, in press). At pressures above 1 kbar, the minimum water solubility is observed for haplogranitic compositions.

1.4. Geological implications

The water solubility is attained at the equilibrium liquid ⇌ liquid + vapour and this reaction controls the onset of degassing in magmas, provided that they do not contain other volatiles with low solubilities such as CO_2. In addition to the crystallisation of anhydrous phases (leading to an increase in the water content of residual melts), decompression is the most important parameter controlling the onset of degassing reactions in a magma. Peralkaline melts may contain significantly higher water contents than subaluminous compositions

at pressures corresponding to conditions close to the surface.

The water solubility of melts close to the *P–T* conditions of the water-saturated solidus is also of importance. Once water-saturated conditions are reached (after subsequent crystallisation), the crystal fraction of a magma is strongly dependent on the water solubility at the given *P–T* conditions. For example, assuming the crystallisation of anhydrous phases from a pure melt containing 4 wt% H_2O in its initial stage (water is considered to be the only volatile in the system), it can be shown from the available phase diagrams and mass balance calculations that the crystal fraction (F_{Cr}) a few degrees above the solidus must be at least 59 wt% at 5 kbar, but only 34 wt% at 2 kbar, assuming water solubilities of 10 and 6 wt% H_2O, respectively ($F_{Cr} = 1 - F_L$ and $F_L = X_{wo}/X_w$, where F_L is the fraction of liquid, X_{wo} and X_w correspond to the bulk water content and the melt water content, respectively).

2. Viscosity

2.1. Effect of the water content of the melt

The viscosity of melts is known to decrease dramatically with increasing water content (Figs 5 and 6). However, few systematic studies have been performed to determine experimentally the viscosity of hydrous felsic melts. Until recently, experimental data were available for Ab and Or melts only (Kushiro 1978; Dingwell 1987; Persikov *et al.* 1990; White & Montana 1990) and for some natural obsidian compositions (e.g. Friedman *et al.* 1963; Shaw 1963; Burnham 1964; Baker & Vaillancourt 1995).

Two models have been proposed to predict the viscosities of hydrous melts (Shaw 1972; Persikov *et al.* 1990). The model proposed by Persikov (1991) and Persikov *et al.* (1990) needs the proportions of molecular water and hydroxyl groups in the melt to be known accurately, a question still under debate (e.g. Nowak & Behrens 1995). At water contents higher than 2–3 wt% H_2O, the viscosity differs by almost two log units if the proportions of the two water species are those determined from glasses (e.g. Silver *et al.* 1990) or from melts at 800°C (proportions after Nowak & Behrens 1995). This makes an accurate prediction of the viscosity using the model of Persikov (1990) difficult.

The model of Shaw (1972) is based on experimental data obtained for melts with 4·3 and 6·2 wt% H_2O (Shaw 1963)

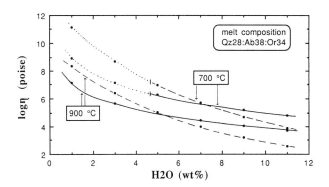

Figure 5 Viscosity of a haplogranitic melt composition ($Qz_{28}Ab_{38}Or_{34}$) at 700 and 900°C as a function of the melt water content. Solid curves are calculated from experimental results of Schulze *et al.* (in press). Broken curves are calculated after the model of Shaw (1972). Dotted curves represent conditions below the liquidus temperature (at these conditions, the water content of the melt is too low to obtain 100% melting; see also Fig. 7). Note the discrepancy between the two calculated curves (for explanation, see text). Dots do not represent experimental data, but results from calculations.

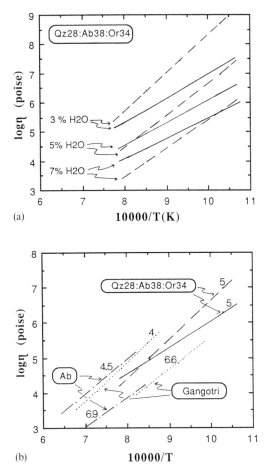

Considering that the possible viscosities of felsic melts are mainly within $10^{4.5}$ and 10^7 poises at temperatures lower than $800°C$ (see Table 1 and further sections), Arrhenian equations derived from experimental data obtained using the falling sphere technique (e.g. Schulze *et al.* in press) can be considered to describe adequately the viscosity of hydrous melts at geologically relevant conditions.

The effect of water on the viscosity of the composition $Qz_{28}Ab_{38}Or_{34}$, calculated using the equation given by Schulze *et al.* (in press; equation retrieving the experimental data for this composition) is compared with that calculated after the model of Shaw (1972) in Figure 5. Between 700 and $900°C$, the model of Shaw (1972) overestimates the viscosity at low water content of the melt and underestimates the viscosity at high water content of the melt. The activation energy of viscous flow (corresponding to the slope of the curves shown on Fig. 6a) for the composition investigated by Schulze *et al.* (in press) differs significantly from that predicted by Shaw (1972). In other words, in the haplogranite composition investigated by Schulze *et al.*, the effect of the first 2 wt% water on melt depolymerisation is larger than that predicted by the model of Shaw (1972).

2.2. Effect of pressure

Pressure is known to affect only slightly the viscosity of hydrous granitic melts. The effect of pressure on the viscosity of hydrous quartzo-feldspathic $Qz_{28}Ab_{38}Or_{34}$ (5·9 wt% H_2O), Ab (2·8 wt% H_2O) and Or (0·2 wt% H_2O) melts has been determined experimentally in the pressure range 3–10 kbar, 2·5–15 and 15–25 kbar, respectively (Schulze *et al.* in press; Dingwell 1987; White & Montana 1990, respectively). No significant viscosity variation was observed for $Qz_{28}Ab_{38}Or_{34}$ and Or melts (variation of less than 0·15 log units over the pressure range investigated) and the variations observed for Ab melts are within the uncertainty range. Considering that pressure variations of more than 5–10 kbar are not expected to occur between the melting area and the emplacement and crystallisation of granitic rocks, the effect of pressure can be neglected in the determination of viscosities of hydrous felsic melts.

Figure 6 Arrhenius plot showing the effect of temperature (in K) on the viscosity of hydrous melts. (a) viscosity of $Qz_{28}Ab_{38}Or_{34}$ melt containing 3, 5 and 7 wt% water calculated based on experimental results of Schulze *et al.* (in press, solid curves) and after the model of Shaw (1972, broken curves). Note that the activation energy of viscous flow (given by the slope of the curve) is different for the two data sets. (b) viscosity of $Qz_{28}Ab_{38}Or_{34}$ melt (solid curve after Schulze *et al.*, in press, broken curve after Shaw 1972), albite melt (dash-dot curves labelled Ab after Persikov *et al.* 1990) and a natural leucogranitic High Himalayan melt composition (dotted curves labelled Gangotri after Scaillet *et al.*, in press). The melt water content expressed as wt% H_2O is given for each curve.

2.3. Effect of anhydrous melt composition

The activation energy of viscous flow is dependent on the water content of the melt, but also on its anhydrous composition. At a given water content, the activation energy of viscous flow can change significantly as a function of the melt composition (Fig. 6b). Although available empirical models fail in reproducing the experimental results obtained in some synthetic systems (haplogranite, for example, Schulze *et al.* in press), it is observed that complex multicomponent granitic or rhyolitic liquids often have viscosities which are well accounted for by the model of Shaw (1972), at least in the range 4–7 wt% water. For instance, the results of experimental viscosity determinations of a leucogranitic composition (Gangotri Himalayan leucogranite; see composition GB4 in Scaillet *et al.* 1995b) containing 4·0 and 6·6 wt% H_2O are plotted in Figure 6b. The maximum difference of viscosity between the calculated (after Shaw 1972) and experimental values is less than 0·2 log units, suggesting that the activation energy of viscous flow for a given melt water content can be reasonably predicted by the model of Shaw (1972) for natural granites of eutectic-like compositions with 4–7 wt% water. This is also confirmed by experimental data on a natural obsidian (Baker & Vaillancourt 1995).

and is mostly valid for high water contents (4–7 wt%), but does not reproduce the experimental determinations of Schulze *et al.* (in press) obtained for a haplogranitic composition close to that of the thermal minimum at 2 kbar ($Qz_{28}Ab_{38}Or_{34}$). The data of Schulze *et al.* (in press) have been obtained over a wide range of melt water contents (1·0–8·2 wt% H_2O) and allow the calculation of the viscosity for any water content of the melt of this composition, assuming an Arrhenian behaviour of the viscosity (linear correlation between the logarithm of viscosity and $1/T$). Although it has been demonstrated that viscosity is in general non-Arrhenian over a large temperature range (e.g. Urbain *et al.* 1982) and that the deviation of viscosity from Arrhenian laws also increases with increasing water content (Richet *et al.* in press), deviations from such a behaviour are small for the composition and the experimental temperature range investigated by Schulze *et al.* (in press). The extrapolation of the data of Schulze *et al.* (in press; viscosity determined in the viscosity range $10^{3.5}$–$10^{5.2}$) to higher viscosities would lead to an underestimation of the viscosity of less than 0·2 and of approximately 0·5 orders of magnitude for viscosities around 10^7 and 10^{10} poises, respectively, and melt water content around 4 wt% H_2O (determinations from an andesite melt, Richet *et al.* in press).

3. Viscosity variations in ascending hydrous magmas

The experimental results just detailed allow the estimation of the viscosity of melts at their source and the viscosity variations of melts and magmas during ascent and cooling. In the following discussion, it is assumed that water is the only volatile dissolved in the melt and that the system is not saturated with respect to water. Such conditions should be close to those produced by dehydration melting because the magmas generated in this way are H_2O-undersaturated. A theoretical approach is discussed first based on results obtained for the synthetic system Qz–Ab–Or. The viscosity variation in natural systems is then discussed.

3.1. Viscosity range of melts at equilibrium with quartz and feldspars

The water content of a melt has to be known (in addition to temperature) to place constraints on the viscosity of melts. Therefore, an estimation of the possible range of water contents of melts at any given pressure and temperature is a fundamental problem. Phase relationships, as shown in Figure 7 (temperature as a function of the bulk water content of the system), allow the determination of the maximum and minimum water content of haplogranitic melts (see also Holtz & Johannes 1994) as a function of temperature at given pressure. Strictly, this type of phase relationship is only valid for eutectic or minimum compositions (in the binary Qz–Ab, Qz–Or and ternary Qz–Ab–Or systems). Small deviations from a eutectic composition (5–10% normative) will significantly affect the liquidus temperatures (see, for example, phase relations at a given pressure, Tuttle & Bowen 1958) but do not significantly affect the maximum and minimum water contents of melts. At the solidus (or eutectic) temperature, the water content of the melt is fixed (Fig. 7) for a given pressure. At a given hyperliquidus temperature (liquid field in Fig. 7), the possible variation of the water content of the melt corresponds to the interval between the liquidus curve and the water solubility curve. The range of possible water contents of the melt increases with temperature. In the liquid + crystal field, the melt water content is fixed for a given temperature and the crystal and melt proportions can be calculated using the level rule. As an example, the possible variation of water content of the melt at 800°C is given for the minima and eutectic compositions at various pressures in Table 1. At a given temperature, the variations are more important at high pressure than at low pressure (Fig. 7) and this also affects the possible viscosity variations (Table 1).

Considering that only small amounts of water are available for melting in the source regions of granites and that melting occurs at relatively high pressure, it can be assumed that the melting conditions of quartzo-feldspathic rocks (greywackes, orthogneisses) occur in the field liquid + crystals (at the solidus temperature or at higher temperatures) in the phase equilibrium diagram of Figure 7. In this instance (quartz and feldspars are present in the residue), the water content of the melt is fixed by the position of the liquidus curve and this allows the viscosity of such melts to be determined at a given pressure and temperature [using the equation given by Schulze et al. (in press) for haplogranite compositions]. The results are shown in Figure 8 as a function of temperature. Clearly, at a given pressure, melts generated at low temperatures are less viscous than melts generated at high temperatures, because low temperature melts necessarily contain more water than high temperature melts along the liquidus (the water content of the melt changes along the curves shown in Fig. 8 and can be determined from Fig. 7 for the 1 and 5 kbar curves).

The evolution of viscosity shown in Figure 8 has important implications for the melt extraction processes and the ascent mechanisms and rates of magmas (dykes versus diapirs, e.g. Clemens & Mawer 1992; Petford et al. 1993). Melts generated at high pressure and low temperature are the best candidates for melt segregation from the source because they have the lowest viscosities. The viscosity curves shown in Figure 8 can be considered as the highest possible viscosities at the given pressure and temperature.

3.2. Viscosity range of ascending haplogranite magmas

In addition to the determination of melt viscosities at the source, it is of importance to evaluate the variation of melt viscosity during magma ascent. Magmas can ascend along different P–T paths, depending on the cooling history and the ascent velocity through the crust. The limiting cases are isobaric cooling (no ascent) and adiabatic ascent (no heat exchange). The corresponding P–T paths are shown on Figure 9. Magmas are usually composed of both crystals and melts and it is emphasised that crystallisation will occur in magmas if ascent is accompanied by significant cooling (path BC in Fig. 9), but that decompression melting will occur in magmas which have experienced only slight heat exchange with the surrounding rocks (path A in Fig. 9).

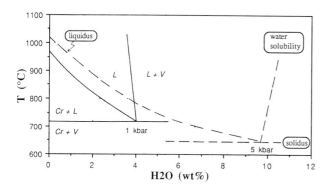

Figure 7 Phase relations for minimum or eutectic compositions of the system Qz–Ab–Or–H_2O at 1 and 5 kbar (solid curves and broken curves, respectively), showing the solidus, liquidus and water solubility curves. The amount of water (H_2O wt%) corresponds to the water present in the system. L, Aluminosilicate liquid; Cr, crystals (quartz and alkali feldspar); and V, supercritical vapour composed of H_2O mainly. Sources of data: Tuttle and Bowen (1958) for solidus temperatures; Holtz and Johannes (1994) for liquidus curves; and Holtz et al. (1995) for water solubility.

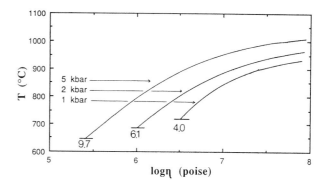

Figure 8 Viscosity of melts of minimum or eutectic composition at the liquidus temperature, coexisting with quartz and alkali feldspar. The water content of the melt (for a given pressure) is not constant along each curve and corresponds to that at the liquidus temperature (water contents for 1 and 5 kbar can be determined from the liquidus curves in Fig. 7). The water content at the solidus (horizontal bar) is given in wt% H_2O. The viscosities are calculated after the experimental data of Schulze et al. (in press).

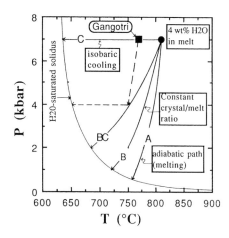

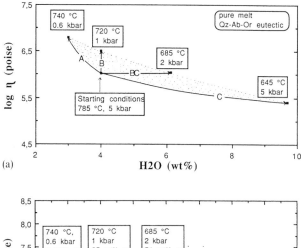

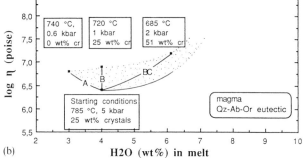

Figure 9 Possible *P–T* paths for ascending magmas with bulk composition corresponding to that of the minimum or eutectic composition in the system Qz–Ab–Or (solid curves, starting conditions represented by the black dot). Water is assumed to be the only volatile in the system. Path A represents an adiabatic ascent (and hence with decompression melting of quartzo-feldspathic minerals and decreasing melt water content). Path B represents an ascent at constant crystal to melt ratio (and hence no change in the water content of the melt; 4 wt% H_2O). Path C represents an isobaric cooling (and hence crystallisation and increasing melt water content). Path BC represents an ascent accompanied by crystallisation (due to heat loss to the wallrocks). The broken curve and black square represent the *P–T* path of the Gangotri leucogranitic magma (constraints from Scaillet *et al.* 1995a, b).

In the case of a pure quartzo-feldspathic system with an eutectic-like composition, the *P–T* paths for a constant crystal to melt ratio can be determined easily and correspond to liquidus curves for a given amount of water (Holtz & Johannes 1994). As an example, a *P–T* path for a constant crystal to melt ratio following the liquidus curve of 4 wt% water is shown on Figure 9. Depending on the bulk water content of the system, and using liquidus curves for given amounts of water (Holtz & Johannes 1994), the melt fraction can be directly calculated using the lever rule.

The possible evolution of viscosities of pure melts along different *P–T* paths has been calculated for synthetic quartzo-feldspathic systems with eutectic-like compositions. If the starting conditions for the magma are considered to be 5 kbar and 785°C (a point along the path at constant crystal to melt ratio shown on Fig. 9), and if quartz and feldspar minerals always remain present in the magma, the melt viscosity increases along the adiabatic path (melt water content decreases with increasing melt fraction) and decreases along paths with important heat loss (path C in Fig. 10a; melt water content increases with crystallisation). However, in both instances the viscosity variation is relatively low. The maximum viscosity variation between the starting conditions and the solidus is $\pm 0{\cdot}6$ log units (Fig. 10a). This small variation is the result of the counterbalancing effects of water content and temperature on viscosity along the *P–T* paths (the melt water content increases with decreasing temperature because crystallisation of the anhydrous phases occurs). It can be observed that no variation of melt viscosity is anticipated in magmas starting at 5 kbar and reaching the solidus at 2 kbar. If the starting conditions for magmas are at lower temperature or at lower pressure, the viscosity variation between the two boundary conditions would be lower than that shown on Figure 10a.

The determination of magma (crystal + liquid) viscosity requires that the crystal abundances and morphologies are accounted for. As shown by several studies (e.g. Marsh 1981; Pinkerton & Stevenson 1992; Dingwell *et al.* 1993; Lejeune &

Figure 10 Evolution of the viscosity of ascending and/or cooling haplogranitic melts (a) and magmas (b) with an anhydrous composition close to the minimum and eutectic compositions. Water is assumed to be the only volatile in the system. Formation conditions for the melt are 785°C and 5 kbar; the melt contains 4 wt% H_2O (water content at the liquidus temperature); and the bulk water content is taken to be 3 wt% H_2O, implying a crystal fraction of 25 wt%. The different paths correspond to those described in Figure 9: A = adiabatic ascent, B = constant melt to crystal ratio, C = isobaric cooling and BC = path with solidus temperature reached at 2 kbar. The dotted area represents the possible viscosity range of melts and magmas for this example (see above conditions). The *P–T* conditions at the solidus and crystal fraction at a temperature just above the solidus are given for each path. The melt viscosity is calculated after Schulze *et al.* (in press). The magma viscosity is calculated using the Einstein–Roscoe equation. In (b) the viscosity for magmas with isobaric cooling is not given (equations not valid for low melt fraction; see text).

Richet 1995), the magma viscosity can be estimated relatively accurately if the crystal fraction remains low (less than 30–40%, depending on the crystal shape) by using the Einstein–Roscoe equation or its derivatives (see discussions in Marsh 1981; Pinkerton & Stevenson 1992). At higher crystallinities (> 50%), very large viscosity changes are expected with only small variations of crystal fraction (see, for example, experimental determinations by Lejeune & Richet 1995) and any viscosity prediction becomes difficult because the magma can no longer be considered as a crystal mush with homogeneous physical properties. To illustrate the effect of crystals on magma viscosity, the Einstein–Roscoe equation can be used for the example described in Figure 9 (crystals are assumed to be homogeneous in size and spherical). If the amount of crystals in the magma is assumed to be 25 wt% at the starting conditions, it can be shown that the viscosity at the solidus would be approximately similar for any *P–T* path reaching the solidus at $P < 2$ kbar (Fig. 10b). A maximum viscosity increase of approximately $0{\cdot}5$ log units would be observed along these *P–T* paths. The viscosity is expected to increase more significantly for *P–T* paths close to isobaric cooling (solidus reached at 2–5 kbar), but can not be calculated in this particular example (see later). One important conclusion from these calculations is that, for a given bulk water content, the viscosity behaviour of magmas is not expected to change significantly in ascending haplogranitic magmas, irrespective of the P–T path followed, as long as the crystal fraction remains lower than 30–40 vol% (Fig. 10b).

The previous discussion shows that several important parameters have to be known to determine accurately the melt and magma viscosity of natural systems, in particular the temperature and water content of melts (or magma). In the case of dehydration melting, if the prevailing pressure and temperature are known, a_{H_2O} can be constrained and the water content of the melt can be estimated. The melt composition should also be constrained to calculate the viscosity. The most appropriate natural examples to determine melt and magma viscosities are compositions which have been investigated experimentally to determine their phase relationships.

For example, the experimental results obtained for the Gangotri High Himalayan tourmaline–muscovite peraluminous leucogranite (Scaillet et al. 1995b) indicate a water content of the melt close to 7 wt% H_2O. The starting P–T conditions were close to 750–770°C and 7 kbar. The magma ascended rapidly along dykes as an almost crystal-free liquid and cooling occurred isobarically at 4 kbar (Scaillet et al. 1995a). In addition, for this particular composition, melt viscosities have been determined experimentally and the experimental phase relationships allow the melt fraction to be constrained along crystallisation paths (Scaillet et al. 1996).

The melt and magma viscosities calculated for the Gangotri leucogranite during its ascent and crystallisation (see path in Fig. 9) are shown on Figure 11. Because the starting magma is considered to be pure melt, no significant change in viscosity is expected to occur during ascent (if cooling is excluded) and the variations are due to isobaric crystallisation only. Phase equilibria show that the magma is still 70 wt% liquid at 660°C. The magma viscosity remains approximately constant as long as the crystal fraction is lower than 40–50 vol%, in agreement with calculations from the synthetic system (see earlier). It is interesting to note that approximately the same viscosity evolution of the magma is observed if the P–T path followed by the Gangotri granite is applied to the synthetic haplogranite system (eutectic compositions). The slightly higher viscosities calculated for the haplogranite system (Schulze et al. in press) (0·5 log units) are due to differences in melt compositions and to the fact that water solubility is

9·5 wt% H_2O at the solidus for the Gangotri granite (instead of 8·5 wt% H_2O for the haplogranite composition).

4. Extrapolation of water solubility and viscosity of natural felsic melts from experimental data

One important question in predicting the physical and chemical properties of natural melts is whether experimental results from synthetic systems (e.g. Qz–Ab–Or system for felsic melts) can be extrapolated to multicomponent systems. Experimental viscosity and water solubility results in synthetic systems show that the water incorporation mechanisms are probably significantly different in Ab-rich melts and in more Qz-rich melts (Figs 3, 4). The approximately constant water solubility at a given pressure, temperature and Ab/Or ratio for compositions with normative Qz contents between 0 and ≈ 35 wt% can be interpreted to reflect a 'feldspar-type' water incorporation mechanism whereas an interplay of a 'feldspar-type' and a 'silica-type' mechanism may occur at higher Qz contents. There is a lack of experimental data to confirm this observation for viscosity. In particular, viscosity determinations in Ab melts with low water contents are necessary to test this hypothesis.

Although the uncertainty on the available experimental data for albite melts is high, it is emphasised that the behaviour of the viscosity of Ab melts (Dingwell 1987; Persikov et al. 1990) is close to that of natural leucogranitic compositions (Fig. 6b) and that it is also well accounted for by the model of Shaw (1972), at least in terms of the activation energy of viscous flow and for water contents between 3 and 7 wt% water. The experimental results suggest that both the water solubility and viscosity of natural leucogranitic compositions can be extrapolated from investigations on alkali feldspar melts. In case of viscosity, experimental data at water contents between 2 and 8 wt% H_2O are only available for Ab melts (Dingwell 1987; Persikov 1990), but differences between Ab and Or melts are not expected to be large (White & Montana 1990). The determination of the water solubility of natural compositions requires their Qz:Ab:Or normative proportions are calculated. The water solubility should be close to that given by the composition with the same proportions (see Fig. 3 and Holtz et al. 1992, 1995 for various pressures). It is emphasised that the water solubility should also be determined accurately from synthetic alkali feldspar compositions if only the Ab:Or proportions of the natural composition are taken into account, and if the normative Qz content of the rock is lower than 30–35% (water solubility is constant at a given Ab:Or ratio, irrespective of Qz content between Qz_0 and $Qz_{30–35}$, Fig. 4). In peraluminous compositions, the viscosity as well as the water solubility seem to be only little affected by excess alumina (Fig. 6b, Dingwell et al. 1984; Holtz et al. 1994; Behrens 1995; Linnen et al. 1996).

Acknowledgements

This work was partly supported by Procope (German–French co-operation programme N° 93018). We appreciated the comments of S. Ohlhorst and the reviews of D. B. Dingwell and H. Nekvasil.

References

Baker, D. R. & Vaillancourt, J. 1995. The low viscosities of F + H_2O-bearing granitic melts and implications for melt extraction and transport. EARTH PLANET SCI LETT **132**, 199–211.

Behrens, H. 1995. Measurements of solubilities of water in melts of albitic and orthoclasic compositions. EUR J MINERAL **7**, 905–20.

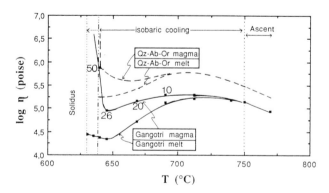

Figure 11 Evolution of the viscosity of the Gangotri leucogranitic melt and magma (solid curves) with cooling in the case of a P–T path similar to that represented in Figure 9 (path with isobaric cooling at 4 kbar, 7 wt% water in the melt; data from Scaillet et al. in press). For comparison, the melt and magma viscosities for a similar P–T path and melt water content in the case of a haplogranitic minimum or eutectic composition have been calculated (broken curves). The vertical dot and dash-dot curves at 630 and 640°C, respectively, represent the solidus temperatures for the Gangotri and synthetic compositions, respectively. The melt water content and the crystal to melt ratio increase with decreasing temperature along the curves. The values reported along the Gangotri magma viscosity curve correspond to the volume fraction of crystals in the magma.

Burnham, C. W. 1964. Viscosity of a water-rich pegmatite. GEOL SOC AM SPEC PAP **76**, 26.

Burnham, C. W. 1975. Water and magmas: a mixing model. GEOCHIM COSMOCHIM ACTA **39**, 1077–84.

Clemens, J. D. & Mawer, C. K. 1992. Granitic transport by fracture propagation. TECTONOPHYSICS **204**, 339–60.

Dingwell, D. B. 1987. Melt viscosities in the system $NaAlSi_3O_8$–H_2O–F_2O_{-1}. *In* Mysen, B.O. (ed.) *Magmatic processes: physicochemical principles.* GEOCHEM SOC SPEC PUBL **1**, 423–33.

Dingwell, D. B. & Webb, S. L. 1990. Relaxation in silicate melts. EUR J MINERAL **2**, 427–49.

Dingwell, D. B., Harris, D. M. & Scarfe, C. M. 1984. The solubility of H_2O in melts in the system SiO_2–Al_2O_3–Na_2O–K_2O at 1 to 2 kbars. J GEOL **92**, 387–95.

Dingwell, D. B., Bagdassarov, N. S., Bussod, G. Y. & Webb, S. L. 1993. Magma rheology. *In* Luth, R.W. (ed.) *Experiments at high pressure and applications to the Earth's mantle.* MINERAL ASSOC CAN SHORT COURSE **21**, 131–96.

Friedman, I., Long, W. & Smith, R. L. 1963. Viscosity and water content of rhyolite glass. J GEOPHYS RES **68**, 6523–35.

Goranson, R. W. 1931. The solubility of water in granitic magmas. AM J SCI **22**, 481–502.

Hamilton, D. L. & Oxtoby, S. 1986. Solubility of water in albite melt determined by the weight-loss method. J GEOL **94**, 626–30.

Hamilton, D. L., Burnham, C. W. & Osborn, E. F. 1964. The solubility of water and effects of oxygen fugacity and water content on crystallization in mafic magmas. J PETROL **5**, 21–39.

Holtz, F. & Johannes, W. 1994. Maximum and minimum water contents of granitic melts: implications for chemical and physical properties of ascending magmas. LITHOS **32**, 149–59.

Holtz, F., Behrens, H., Dingwell, D. B. & Taylor, R. P. 1992. Water solubility in aluminosilicate melts of haplogranitic compositions at 2 kbar. CHEM GEOL **96**, 289–302.

Holtz, F., Dingwell, D. B. & Behrens, H. 1993. Effects of F, B_2O_3, and P_2O_5 on the solubility of water in haplogranite melts compared to natural silicate melts. CONTRIB MINERAL PETROL **113**, 492–501.

Holtz, F., Behrens, H., Dingwell, D. B. & Scaillet, B. 1994. Influence of aluminum on water solubility and structure of granitic melts. TERRA ABSTR 1 (6), 24.

Holtz, F., Behrens, H., Dingwell, D. B. & Johannes, W. 1995. Water solubility in haplogranitic melts. Compositional, pressure and temperature dependence. AM MINERAL **80**, 94–108.

Jahns, R. H. & Burnham, C. W. 1958. Experimental studies of pegmatite genesis: the solubility of water in granitic melts. GEOL SOC AM BULL **69**, 1544–55.

Kohn, S. C., Dupree, R. & Smith, M. E. 1989. A multinuclear magnetic resonance study of the structure of hydrous albite glasses. GEOCHIM COSMOCHIM ACTA **53**, 2925–35.

Kushiro, I. 1978. Viscosity and structural changes of albite ($NaAlSi_3O_8$) melt at high pressures. EARTH PLANET SCI LETT **41**, 87–90.

Lejeune, A. M. & Richet, P. 1995. Rheology of crystal-bearing silicate melts: an experimental study at high viscosities. J GEOPHYS RES **100**, 4215–29.

Linnen, R. L., Pichavant, M. & Holtz, F. The combined effects of f_{O_2} and melt composition on SnO_2 solubility and tin diffusivity in haplogranitic melts. GEOCHIM COSMOCHIM ACTA, in press.

Marsh, B. D. 1981. On the cristallinity, probability of occurrence, and rheology of lava and magma. CONTRIB MINERAL PETROL **78**, 85–98.

Nowak, M. & Behrens, H. 1995. The speciation of water in granitic glasses and melts determined by in situ near-infrared spectroscopy. GEOCHIM COSMOCHIM ACTA **59**, 3445–50.

Oxtoby, S. & Hamilton, D. L. 1978a. The discrete association of water with Na_2O and SiO_2 in NaAl silicate melts. CONTRIB MINERAL PETROL **66**, 185–8.

Oxtoby, S. & Hamilton, D. L. 1978b. Calculation of the solubility of water in granitic melts. *In*: McKenzie, W.S. (ed.) *Progress in experimental petrology.* NAT ENVIRON RES COUN PUB SER D **11**, 37–40.

Paillat, O., Elphick, S. C. & Brown, W. L. 1992. The solubility of water in $NaAlSi_3O_8$ melts: a re-examination of Ab–H_2O phase relationships and critical behaviour at high pressures. CONTRIB MINERAL PETROL **112**, 490–500.

Persikov, E. S. 1991. The viscosity of magmatic liquids: experiment, generalized patterns. A model for calculation and prediction. Applications. *In* Perchuk, L.L. & Kushiro, I. (eds.) ADV PHYS CHEM **9**, 1–40.

Persikov, E. S., Zharikov, V. A., Bukhtiyarov, P. G. & Polskoy, S. F. 1990. The effects of volatiles on the properties of magmatic melts. EUR J MINERAL **2**, 621–42.

Petford, N., Kerr, R. C. & Lister, J. R. 1993. Dike transport of granitoid magmas. GEOLOGY **21**, 845–8.

Pichavant, M., Holtz, F. & McMillan, P. 1992. Phase relations and compositional dependence of water solubility in quartz–feldspar melts. CHEM GEOL **96**, 303–19.

Pinkerton, H. & Stevenson, R. J. 1992. Methods of determining the rheological properties of magmas at sub-liquidus temperatures. J VOLCANOL GEOTHERM RES **53**, 47–66.

Richet, P., Lejeune, M., Holtz, F. & Roux, J. Water and the viscosity of andesite melts. CHEM GEOL, in press.

Romano, C., Dingwell, D. B. & Behrens, H. 1995. The temperature dependence of the speciation of water in $NaAlSi_3O_8$–$KAlSi_3O_8$ melts: an application of fictive temperatures derived from synthetic fluid-inclusions. CONTRIB MINERAL PETROL **122**, 1–10.

Romano, C., Dingwell, D. B., Behrens, H. & Dolfi, D. 1996. Solubility of water along the joins $NaAlSi_3O_8$–$KAlSi_3O_8$, $NaAlSi_3O_8$–$LiAlSi_3O_8$, $LiAlSi_3O_8$–$KAlSi_3O_8$. AM MINERAL **81**, 452–61.

Scaillet, B., Pêcher, A., Rochette, P. & Champenois, M. 1995a. The Gangotri granite (Garhwal Himalaya): laccolithic emplacement in an extending collisional belt. J GEOPHYS RES **100**, 585–607.

Scaillet, B., Pichavant, M. & Roux, J. 1995b. Experimental crystallization of leucogranite magmas. J PETROL **36**, 663–705.

Scaillet, B., Holtz, F. & Pichavant, M. 1996. Rheological properties of granitic magmas in their crystallization range. *In* Bouchez, J.L., Hutton, D.W.H. & Stephens, W.S. (eds) *Granite: from segregation of melt to emplacement fabrics.* Kluwer, Dordrecht, 1–22.

Scaillet, B., Holtz, F., Pichavant, M. & Schmidt, M. O. E. The viscosity of Himalayan leucogranites: implications for mechanisms of granitic magma ascent. J GEOPHYS RES, in press.

Schulze, F., Behrens, H., Holtz, F., Roux, J. & Johannes, W. The influence of water on the viscosity of a haplogranitic liquid. AM MINERAL, in press.

Shaw, H. R. 1963. Obsidian–H_2O viscosities at 1000 and 2000 bars in the temperature range 700° to 900°C. J GEOPHYS RES **68**, 6337–42.

Shaw, H. R. 1972. Viscosities of magmatic liquids: an empirical method of prediction. AM J SCI **272**, 870–93.

Silver, L. A. & Stolper, E. M. 1985. A thermodynamic model for hydrous silicate melts. J GEOL **93**, 161–78.

Silver, L. A., Ihinger, P. D. & Stolper, E. M. 1990. The influence of bulk composition on the speciation of water in silicate glasses. CONTRIB MINERAL PETROL **104**, 142–62.

Tuttle, O. F. & Bowen, N. L. 1958. Origin of granite in the light of experimental studies in the system $NaAlSi_3O_8$–$KAlSi_3O_8$–SiO_2–H_2O. GEOL SOC AM MEM **74**.

Urbain, G., Bottinga, Y. & Richet, P. 1982. Viscosity of liquid silica, silicates and aluminosilicates. GEOCHIM COSMOCHIM ACTA **46**, 1061–72.

Voigt, D. E., Bodnar, R. J. & Blencoe, J. G. 1981. Water solubility in melts of alkali feldspar composition at 5 kbar, 950°C. EOS, TRANS AM GEOPHYS UNION **62**, 428.

White, B. S. & Montana, A. 1990. The effect of H_2O and CO_2 on the viscosity of sanidine liquid at high pressures. J GEOPHYS RES **95**, 15683–93.

FRANCOIS HOLTZ, BRUNO SCAILLET and MICHEL PICHAVANT, Centre de Recherches sur la Synthèse et la Chimie des Minéraux, CRSCM-CNRS, 1A, rue de la Férollerie, 45071 Orleans, France.
HARALD BEHRENS and FRANK SCHULZE, Institut für Mineralogie, Universität Hannover, Welfengarten 1, 30167 Hanover, Germany.

Transactions of the Royal Society of Edinburgh: Earth Sciences, **87**, 65–72, 1996

Granite and granitic pegmatite melts: volumes and viscosities

D. B. Dingwell, K.-U. Hess and R. Knoche

ABSTRACT: Progress in the understanding of the volumes and viscosities of granitic and related pegmatitic melts generated by experimental studies are reviewed. The results of a series of investigations of the volumes and viscosities of melts derived from a haplogranitic base composition, HPG8, located near the 2 kbar water-saturated minimum melt composition in the albite–orthoclase–silica system are discussed. Melt volumes, obtained using a combination of dilatometric and calorimetric methods at 1 atm and relatively low temperatures yield an internally consistent set of partial molar volumes for 18 components in granitic melts. These partial molar volumes, combined with an estimate for water, allow the estimation of melt densities for granitic and related pegmatitic magmas.

Melt viscosities, obtained using a combination of high and low range viscometry techniques, provide a template for the estimation of melt viscosities in more complex natural systems. The parameterisation of the non-Arrhenian temperature-dependence of the viscosity of such melts is presented, together with some structural implications of the variation of melt viscosity with temperature and composition. Outstanding questions related to the PVT equation of state of granitic melts and to the mechanical response to shear stresses are discussed, with an outlook for the experimental solutions to those questions in the next few years.

KEY WORDS: granitic melt, viscosity, density, volume, metastable, glass.

The experimental confirmation of Hutton's (1785) proposed magmatic origin of granitic rocks by Tuttle and Bowen (1958) made clear that a sensible description of the intensive and extensive parameters operating at the magmatic to magmato-hydrothermal stages of granite and pegmatite petrogenesis needed quantification. A few years later, experimental investigations of the volumes and viscosities of granitic melts were underway which mapped out the fundamental variations to be expected in these properties under crustal conditions (Burnham 1963; Shaw 1963; Burnham & Davis 1971; Shaw 1972), resulting in the discussion by Jahns and Burnham (1969) of the role of water-saturated magma in the internal structure of zoned pegmatites. Over the following 20 years these properties of silicic melts received intermittent attention (e.g. Kushiro 1978a, b; Dingwell 1987; Persikov *et al.* 1990; Schulze *et al.* in press), but most of the attention of experimental petrologists was focused on documenting the stability field of specialised granitic melts (e.g. Manning 1981; Pichavant 1981; London 1992), the kinetics of crystallisation (e.g. Naney & Swanson 1980), the solubility of water in such melts (e.g. Holtz *et al.* 1992, 1995) and chemical partitioning between melts and fluids. Increasingly, interest in the magma physics of such systems has turned to the nature of the brittle–ductile transition, degassing kinetics (Sakuyama & Kushiro 1978; Sparks 1978; Bagdassarov *et al.* 1994) and mechanical behaviour of silicic magmas (e.g. Dingwell & Webb 1989; Bagdassarov & Dingwell 1992; 1993a, b; Hurwitz & Navon 1994; Sparks *et al.* 1994; Alidibirov & Dingwell 1996).

The earlier work on melt volumes and viscosities was performed on single selected compositions (synthetic or natural) which were intended to serve as good bases for the bulk of granitic magmas and were demonstrably applicable to the petrogenesis of large-volume calcalkaline intermediate to acidic intrusive rocks. In recent years the investigation of some

more extreme rock compositions, thought by many to represent highly fractionated magmatic to magmato-hydrothermal derivatives of granitic magmas, have been documented in great detail (Cerny *et al.* 1985; London 1992). Additionally, mounting evidence from rare, well-documented eruptive volcanic glasses (e.g. Burt *et al.* 1982, Pichavant *et al.* 1987) and melt inclusions in hypabyssal rocks (e.g. Thomas 1995) has demonstrated that the range of composition of natural silicic melts is much broader than the haplogranitic system albite–orthoclase–quartz. The feasibility of forming such extreme differentiates existing to very low temperatures at high crustal levels has been experimentally demonstrated by several phase equilibrium studies (Manning 1981; Pichavant 1981; London 1992).

Recognising a need for the physical description of the melts involved in such systems, a few years ago we embarked on a series of studies aimed at the experimental determination of the physical and chemical properties of these melts. To date we have studied the viscosity (Dingwell *et al.* 1992, 1993; Hess *et al.* 1995, 1996, in press), the density (Knoche *et al.* 1992, 1995; Dingwell *et al.* 1993) and the surface tension (Bagdassarov & Dingwell 1994) of such melts as well as the chemical diffusivities of components in such melts (Chakraborty *et al.* 1993; Mungall *et al.*, unpublished data). The present paper focuses on our progress in the description of the volume and viscosity of granitic and related pegmatitic melts.

1. Strategy

For the investigation of the effects of a wide range of components on the physical properties of granitic melts a base composition near the 2 kbar $p\mathrm{H_2O}$ minimum melt composition in the albite–orthoclase–quartz (ab–or–qz) system was chosen. It is one of a series of 16 haplogranitic glasses which were synthesised by the direct fusion of Al_2O_3, SiO_2, K_2CO_3 and Na_2CO_3 at high temperatures ($\geqslant 1600°C$) and whose compositions were first reported in Holtz *et al.* (1992). These glasses were chosen to cover the compositional range of late stage granitic melts in ab–or–qz space.

To this base composition were added various amounts of the individual components whose influence on volume and viscosity was to be investigated. Most of these were added as oxides or carbonates, with the exceptions of F, P and B. The components chosen included those deemed relevant to the chemical enrichments (large ion lithophile elements, high field strength elements) occurring in specialised granites and granitic apophyses (aplites, pegmatites) such as Cs_2O, Rb_2O, Li_2O, Ta_2O_5, Nb_2O_5, WO_3 as well as components whose chemical properties were thought to be useful in helping complete the picture of the structural basis of melt property variations, e.g. excess Al_2O_3, K_2O and Na_2O as well as BaO and SrO. CaO and MgO were added to evaluate the trends of granitic magmas (*sensu lato*) towards more intermediate compositions. TiO_2 was added due to the peculiarly composition-dependent part it appears to play in the structure of silicate melts.

The preparation of granitic glasses in the large quantities required for physical property determinations is difficult due to the high viscosity of these liquids at temperatures easily accessible in the laboratory. The high viscosity prevents the complete reaction of the starting reagents, generating instead local, corundum-saturated regions of surrounded by inhomogeneous melt. Classically, this problem has been overcome by repeated fusion and crushing cycles (e.g. Schairer & Bowen 1956). The homogeneity of samples down to the length scale of chemical microanalysis (e.g. electron microprobe) can be ensured with this method. Repeated handling of the samples combined with the repeated fusion of powders carries with it, however, the dangers of a significant cumulative contamination effect and measurable volatilisation of alkalis. Fusing such powders under conditions sufficient for complete fining of the melt does not ensure that crystallites of unreacted materials do not persist in the sample. These unreacted grains can significantly destabilise the melts with respect to crystallisation when they are subsequently investigated in the metastable, subliquidus regime (see later). Heating of the melts beyond 1800°C does reduce the viscosity to levels where homogenisation and fining proceed rapidly, but also has the consequence of rapid alkali loss (Dingwell, unpublished data). Therefore, to generate bubble- and crystal-free homogeneous glasses which would be durable in subsequent investigations conducted in the metastable liquid state, a method of high temperature fusion, combined with mechanically assisted homogenisation of the initially partially molten mixes, was chosen. The homogenisation was accomplished by the rotation of a platinum/rhodium spindle within the samples at temperatures above 1600°C for periods of up to two weeks. The samples so treated were removed from the furnace and allowed to cool in their Pt crucibles in air. This relatively slow cooling tempers the 100 g glass batch sufficiently that it can withstand the stresses of subsequent mechanical treatment such as drilling and cutting.

The experimental strategy used in the determination of the volumes and viscosities of haplogranitic melts includes the investigation of such melts under metastable conditions. This approach has been outlined previously and is briefly described here. For clarity, the isobaric case of a congruently melting composition is presented in Figure 1. Three essential features are displayed: (1) the glass transition curve; (2) the envelope of detectable crystallisation (analogous to an isothermal time–temperature–transformation diagram); and (3) the melting temperature. The determination of a reliable set of volumes

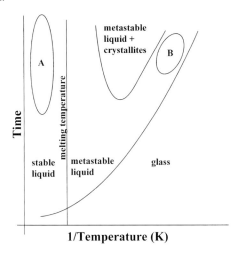

Figure 1 Illustration in time–temperature space of the strategy involved in obtaining the most complete possible description of melt properties by combining superliquidus and subliquidus investigations of melt properties. The metastable region of supercooled liquid behaviour between the zone of significant crystallisation and the glass transition (B) provides a second operating window (in addition to the superliquidus region, (A) where careful measurement of the temperature-dependent properties of granitic liquids is possible. It is the glass-forming ability of many granitic and derivative melts which allows a low temperature window for investigation. Essential to obtaining starting material for such low temperature investigations is that the starting melt can be quenched quickly enough to the glassy state to avoid significant crystallisation or unmixing during cooling. The combination, where possible, of high and low temperature data for melt properties yields a significant improvement in the description of the temperature dependence of melt properties. Fortuitously, for the case of granitic melts, the temperatures corresponding to the crystallisation interval of common water-rich systems are very similar to the temperature range of the glass transition for the equivalent water-free compositions. Thus studies of the water-free system just above the glass transition temperature can be used as constraints on the properties of such melts at the temperatures of interest. Redrawn from Dingwell (1993).

and viscosities for granitic melts in the temperature and pressure range pertinent to the crystallisation of specialised granites and their derivative pegmatites requires a different approach, one that has more in common with solid state methods of materials characterisation than with traditional liquid state methods. The possibility of using the latter approach rests on one central and important aspect of granitic liquids—their slow crystallisation kinetics. The 'reluctance' of silicic supercooled liquids to crystallise in laboratory experiments has been a perennial source of concern and challenge to experimental petrologists attempting to define the phase relations of granitic systems (Tuttle & Bowen 1958). This behaviour is the same as that which would be described by a glass chemist as good glass-forming behaviour. Granitic melts are good glass-formers. Thus the metastable extension of liquid properties below the melting point (or liquidus) into the supercooled liquid regime can be studied extensively without the disturbance of incipient crystallisation (Dingwell 1993).

Practically speaking, this entails the reheating of granitic glasses to temperatures above the glass transition, but not into the region of significant crystallisation. If the time–temperature path or 'excursion' above the glass transition is chosen carefully enough (using, in certain cases, simple trial and error) to avoid any volatilisation, foaming, crystallisation, liquid–liquid unmixing, then an 'operational window' may be defined. This window may be specific to a particular experimental protocol. The operational window can be confirmed by a combination of chemical, structural and textural observations on the samples before and after property measurements. We have used a variety of tools, including infra-red spectroscopy, ^{57}Fe Mössbauer spectroscopy, electron microprobe and inductively-coupled plasma atomic emission spectrometry (ICP–AES) analysis, scanning electron microscopy (SEM) and transmission electron microscopy (TEM) imaging.

2. Volume studies

2.1. Methods

In the determination of the volume of granitic liquids, the experimental challenge provided by their very high viscosities in the temperature range of interest must be overcome. Methods based on immersion fail. The glass-forming tendency that such silicic melts offer, however, enables the solution to this problem by using scanning methods for the determination of derivative thermodynamic properties that are normally reserved for solid-state studies. The use of a combination of heat capacities of granitic melts derived from scanning calorimetry and thermal expansivity values of such melts derived from scanning dilatometry to constrain the volume–temperature relationship of supercooled granitic liquids was introduced by Webb *et al.* (1992). It is based on the principle of the equivalence of enthalpy and volume relaxation in silicate melts (Dingwell 1995a). The method involves (1) the determination of the heat capacity of the sample from room temperature through the glass transition temperature and to a few tens of degrees beyond; (2) determination of the density of the glass at room temperature using immersion techniques and of the expansivity of the glass up to the glass transition temperature (recorded as a peak expansivity value); (3) the normalisation of the calorimetric heat capacity curve and the dilatometric thermal expansivity curve to common constraints (see later) and the derivation of the liquid thermal expansivity from a comparison of the normalised curves; and (4) the compilation of integrated volume data (using the liquid expansivities and room temperature volumes) to obtain a multi-linear regression of partial molar volumes.

2.2. Results

The experimental details of the calorimetry and dilatometry have been presented in full previously (e.g. Webb *et al.* 1992). Examples of calorimetric and dilatometric data are illustrated in Figure 2a and 2b. The normalisation of the calorimetric and dilatometric data, illustrated in Figure 2c, is accomplished on the basis of the temperature-dependent glassy heat capacity and expansivity set equal to zero and the transient peak heat capacity and expansivity values set equal to unity (Fig. 2c). The normalised liquid extension of the expansivity curve is simply set equal to the liquid limb of the heat capacity curve. This is the critical step which relies on the assumption that the kinetic factors controlling the shape and height of the peak are identical for volume and enthalpy relaxation. These factors include the temperature dependence of the viscosity, the thermal history of the sample and the parameters controlling the non-exponential and non-linear aspects of relaxation in silicate melts (see references cited in Dingwell 1995a). All evidence accumulated to date for these silicic melts supports this assumption (Dingwell 1995a; Stevenson *et al.* 1995).

The scheme of partial molar volumes presented in the following is based on a comparison of integrated volume and expansivity data for each sample calculated near the glass transition temperature of that sample. To produce an isothermal set of molar volumes for the samples, a common temperature must be chosen. A value of 750°C was chosen because it represents a moderate extrapolation from the actual

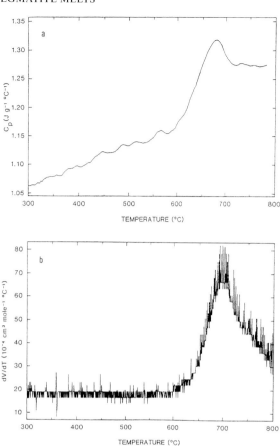

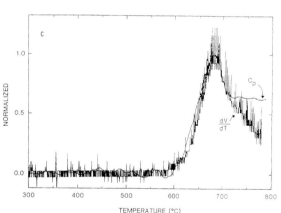

Figure 2 Results of scanning calorimetric and dilatometric determinations of the heat capacity and thermal expansivity of a granitic melt, respectively. The scanning traces reveal three domains of derivative thermodynamic property behaviour, namely, the low-temperature glassy region, the high-temperature liquid region and the glass transition region at intermediate temperatures marked by a peak in the derivative property and of a shape whose details are controlled by the viscosity of the melt and temperature dependence of the melt structure, the heating rate of the measurement and the thermal history of the glass. The expansivity curve, although highly accurate, has the flaw of collapsing above the glass transition temperature due to viscous deformation of the sample. The normalisation of these curves on the basis of the (temperature-dependent) glassy property value = 0 and the peak value = 1 allows a quantitative comparison to be made of the shapes the glass transition peaks. Using this normalisation the liquid expansivity is estimated from the liquid heat capacity using the principle of equivalence of enthalpy and volume relaxation. Integrated volumes up to the glass transition temperature, together with the estimated expansivity data, are used to construct an isothermal comparison of melt volumes from a sample set whose variations in composition lead to variations in the glass transition temperature itself. Redrawn from Knoche *et al.* (1992).

glass transition temperatures and because it lies within the crystallisation temperature range for many granitic and derivative systems. An isothermal comparison of supercooled liquid densities for the measured samples, corrected to 750°C using the liquid expansivity data obtained from the calorimetric/dilatometric comparison, yields the data presented in Figure 3.

The range in molar volumes corresponding to the density variations in Figure 3 yields, through the multilinear regression of Knoche *et al.* (1995), a set of partial molar volumes (in cm³/mol) that range from 10·53 ($\pm$0·29) for MgO to 69·09 ($\pm$1·82) for P_2O_5. On the basis of one oxygen (or fluorine for F_2O_{-1}) the range is from 10·53 ($\pm$0·29) for MgO to 55·38 ($\pm$1·69) for Cs_2O. For a complete listing of the partial molar volumes and expansivities, refer to Knoche *et al.* (1995). To within the precision of the present determinations of molar volumes, the multilinear regression successfully reproduces the molar volume data for the investigated melts. Whether further interaction terms between the melt components may appear in more complex compositions cannot be decided on the basis of the data of Figure 3 and should be investigated in future. Future studies of this type involving hydrous melts await, in part, improvements in the design of high pressure push-rod dilatometry.

3. Viscosity studies

3.1. Methods

The determination of the Newtonian viscosity of these silicic melts has been accomplished using two separate types of experiments: the concentric cylinder and micropenetration viscometry methods. The combination of concentric cylinder determinations obtained at superliquidus conditions and micropenetration determinations at subliquidus, supercooled conditions was necessary to provide a sufficiently wide range of viscosity data to enable accurate generalisation of the temperature dependence of the viscosity. This problem is especially acute for silicic melts because, although the degree to which they depart from a simple Arrhenian temperature dependence of viscosity is smaller than for many basic melts, nevertheless the range of reciprocal temperature space over which we must be concerned with the properties of silicic magmatic systems is much greater than that for basic magmas. The viscosity methods have been described in detail previously (concentric cylinder viscometry, Dingwell 1989, 1990; micropenetration viscometry; Dingwell *et al.* 1992, 1993; Hess *et al.*

1995, 1996) and therefore only the essential points are reviewed here.

The concentric cylinder viscometry is performed in the viscosity range $1–10^4$ Pa s. The strain rate is set in such experiments and the resultant torque exerted on a spindle by the liquid during deformation is recorded. The strain rate can be varied over a small range at each temperature of measurement. Sample stability is checked by repeated measurements. This method is calibrated against an international viscosity standard glass and yields a precision (at 1σ) of $\pm0\cdot05$ $\log_{10}$ viscosity units.

The micropenetration method is performed in a viscosity range of $10^{8\cdot5}–10^{11\cdot5}$ Pa s. The stress is set in such experiments and the resulting penetration depth is determined. The penetration of a hemispherical indenter into the free polished surface of the supercooled liquid is monitored. This method is relative and has been confirmed to an accuracy (at 1σ) of $\pm0\cdot06$ $\log_{10}$ viscosity units. It has also been applied to sufficiently stable supercooled water-bearing melts (Dingwell *et al.*, in press; see later).

3.2. Results

3.2.1. Composition dependence. A comparison of the influence of various components on the viscosity of a single granitic base composition (HPG8) is presented in Figure 4 at the 800°C isotherm. The individual oxides added are expressed as weight per cent (Fig. 4a) and as mole fraction on the basis of one anion (Fig. 4b). The excess alkalis express, together with water, a relatively tight single trend of decreasing viscosity

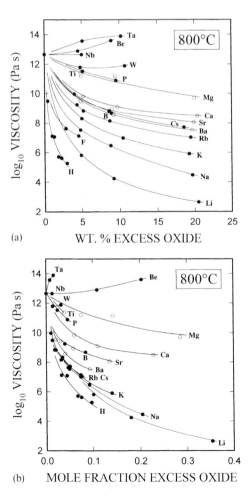

(a)

(b)

Figure 4 Effects of various added components on the viscosity of a haplogranitic melt compared at 800°C and 1 bar. Data from Hess *et al.* (1995a, In press and unpublished data) and Dingwell *et al.* (1995).

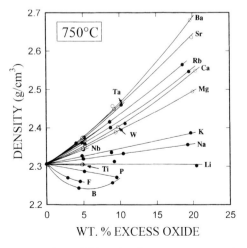

Figure 3 Effect of the addition of various components on the density of a granitic base composition melt at 750°C and 1 bar pressure. Data summarised in Knoche *et al.* (1995).

with added oxide. Other geochemical groups are less tightly constrained. In general, the viscosity increases between groups in the order water, alkalis, alkaline earths and high field strength elements when compared on a mole fraction basis. Within groups clear trends are also visible, with the viscosity increasing from Li to Cs and from Ba to Be. The opposing trends of the relative viscosities versus cation field strength in the alkalis versus the alkaline earths is a perplexing problem in structural models for viscosity.

3.2.2. Temperature dependence. The combined results of viscosity determinations from superliquidus and subliquidus temperatures are plotted together in Figure 5 versus the reciprocal temperature for the example of the addition of Li_2O to the haplogranitic base composition. The contrasting effects of Li_2O addition on the temperature dependence of the melt viscosity in the high and low viscosity regimes can be reconciled with an increasingly non-Arrhenian variation of the viscosity with temperature as the Li_2O content is increased (Fig. 5). It is this non-Arrhenian nature of silicate melts in general which makes the extrapolation of high temperature viscosity data to the low temperatures attendant on granitic and derivative pegmatitic crystallisation from the melt unreliable. Data obtained with a combination of viscometry techniques are the solution to this problem.

3.2.3. The special case of water. The viscosity of hydrous granitic or rhyolitic melts has been investigated several times in the superliquidus region using variations on the falling sphere method (Shaw 1963; Burnham 1963; Schulze et al., in press). Those studies have invariably indicated the classical effects of the addition of a 'depolymerising' component to a tectosilicate melt composition on melt viscosity, namely strong non-linear decreases in melt viscosity and its temperature dependence. Falling sphere viscometry is well suited to the determination of melt viscosity at high pressure because the encapsulation of the material presents no direct impediment to the strain marker, a falling sphere. Nevertheless, such methods have been perennially restricted to operate within a relatively small window of viscosity (cf. Dorfman et al., in press). Thus the determination of non-Arrhenian temperature dependence for hydrous melts also requires the incorporation of complementary viscosity data at higher viscosity values and

lower temperatures obtained using alternative techniques. Such techniques are exemplified by the applications of parallel-plate (Richet et al., in press) and micropenetration (Dingwell et al., in press) dilatometric viscometry techniques to supercooled hydrous silicate melts at 1 bar pressure on melts which have been quenched from higher pressure. The experiments of Dingwell et al. (in press) have been performed on the same haplogranitic melt discussed earlier (HPG8) at 1 bar pressure and temperatures just above the glass transition. Sample stability was carefully controlled by infra-red spectroscopic determinations of the water content before and after viscometry, as well as checks for drift in viscosity data over time. The data obtained in that study are thus directly comparable with the data obtained for anhydrous oxide additions to the same melt (Hess et al., 1995, 1996). Comparison can also be made with the results of the falling sphere investigation of an almost identical base composition AOQ with added water by Schulze et al. (in press). The former comparison has been included in Figure 4. The latter comparison yields a description of the non-Arrhenian temperature dependence of the viscosity of hydrous granitic melts, presented here as Figure 6. On the basis of Figure 6 petrologists are for the first time in a position to accurately estimate the viscosity of hydrous granitic melts, including the effect of a non-Arrhenian temperature dependence. Equations quantifying the non-Arrhenian temperature dependence of the viscosity of hydrous granitic melts have been developed by Dingwell (unpublished data). The degree of non-Arrhenian behaviour decreases strongly in the order $CaO > Na_2O > H_2O$ compared at equimolal additions, with the result that the latest data for the temperature dependence of the viscosity of hydrous granitic melts exhibit unexpectedly small departures from Arrhenian behaviour (Dingwell, in press).

3.2.4. Pressure dependence. Figure 6 is a polybaric comparison of viscosity data. A significant pressure dependence of the viscosity of hydrous granitic melts over a few kilobars pressure

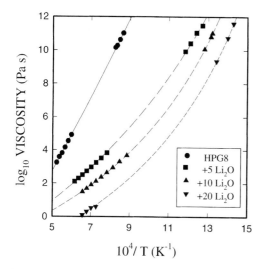

Figure 5 Effect of Li_2O on the shear viscosity of a model granitic melt composition. Note the departure of the viscosity temperature relationship from Arrhenian behaviour with increasing Li_2O component. This non-Arrhenian temperature dependence of melt viscosity remains perhaps the most important challenge to the construction of a reliable scheme for the prediction of melt viscosities at low temperatures in multicomponent silicate melts. Data from Hess et al. (1995).

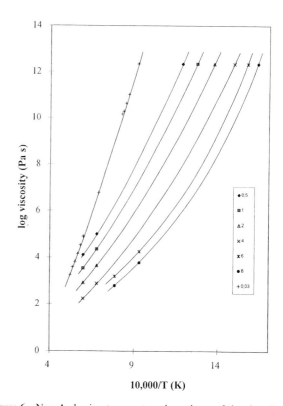

Figure 6 Non-Arrhenian temperature dependence of the viscosity of hydrous granitic melts. This diagram was constructed using a variety of data sources listed by Dingwell (in press). Water contents are listed in the inset in wt%. Redrawn from Dingwell (in press).

would invalidate the comparison. The information available on the pressure dependence of the viscosity of hydrous silicic melts in general does not indicate a strong pressure dependence (Kushiro 1978a; Dingwell 1987; Schulze *et al.*, in press). Confirmation of this comes from the calorimetric glass transition study of Rosenhauer *et al.* (1979), where only a very slight pressure dependence of the melt viscosity isokom (the glass transition, see later) was observed. These observations can be taken to concur with the fact that volume-based models of viscous flow appear to be inferior in predictive power to those based on the configurational entropy of viscous flow (e.g. Richet 1984).

3.2.5. The glass transition or brittle–ductile transition. One important aspect of the mechanical behaviour of viscous silicate melts is the quantification of the glass transition temperature for magmatic and volcanic time-scales. Phenomenologically, the glass transition is merely the transition from the liquid-like metastable equilibrium behaviour of a silicate melt to the solid-like (for the amorphous state, disequilibrium) behaviour or 'glassy' behaviour of the melt (Wong & Angell 1976). The rheological consequences of the glass transition should be obvious. Stresses which are applied to a liquid are able to be dissipated in a steady state such that all strain energy is consumed in creating the entropy associated with viscous flow. There is no memory effect in such flow and the strain rate generated by a given stress is simply described by a coefficient, the Newtonian viscosity. All of the viscosity data discussed here belong to this scenario. Should the strain rate overcome a critical level during melt flow, due to a combination of high stress and high viscosity, then the melt will be driven into a non-steady-state deformation regime which can ultimately lead to the elastic storage (build-up) of stresses to a point where the stored stress may exceed the mechanical strength of the melt. The result is brittle failure and possibly fragmentation of the melt.

There are several sources of data on the glass transition of silicic melts. The glass transition can be directly determined in scanning calorimetry or dilatometry studies as described earlier. Additionally, major configurational elements of the melt structure are effectively frozen at the glass transition so that spectroscopic observations of such features as a function of temperature also provide glass transition data (Nowak & Behrens 1995; Shen & Keppler, 1995). Also, the glass transition in silicic melts approximates an isokom of viscosity (Dingwell 1995a). Thus if we set a value of viscosity we can compare the relative temperatures of the glass transition along with the brittle–ductile transition of the melt for differing compositions from viscosity data just above the glass transition.

Figure 7 compares the glass transition temperatures of

granitic melts with the addition of various components based on dilatometric and calorimetric determinations. The cooling rate, the viscosity at the glass transition and the relaxation time are all simply linked (Dingwell 1995a; Stevenson *et al.* 1995). If a granitic melt was cooled with the rate corresponding to a relaxation time for shear flow defined by a viscosity of $10^{12.38}$ Pa s, then the glass transition will be encountered at the temperature given in Figure 7. It is of the utmost importance to appreciate the kinetic nature of the transition. The activation energy of viscous flow near the glass transition temperatures commonly experienced in nature is such that an order of magnitude increase in the cooling rate of a melt or in the viscosity will lead to a 15–30°C increase in the glass transition temperature.

The vast majority of magmatic processes can be adequately modelled using the assumption of a Newtonian melt viscosity (Dingwell & Webb 1989) because the strain rates required to exceed the steady-state conditions require very large stresses to be operative on highly viscous melts. Experimentally, it has been observed for a wide variety of melts and temperatures that the onset of non-Newtonian viscous flow, a precursor of the brittle failure of melts, occurs 3 $\log_{10}$ units of strain rate slower than the 'relaxation' strain rate which is obtained from the ratio of the shear modulus of the melt to the shear viscosity (i.e. the Maxwell relation; Webb & Dingwell 1990). This simple rule allows the prediction of the kinetic conditions required for departure from the steady-state Newtonian condition.

In the event of an overstepping of the critical conditions, the mechanical strength of the melt comes into play as a controlling parameter in the subsequent deformation history of the magma. Determinations of the breaking strength of melt fibres yield values in the range 10^8 Pa, a full two orders of magnitude below the theoretical strength (elastic constants) of the fibres. This value is almost certainly further reduced by the presence of bubbles, crystals and diffusive gradients within the melt phase (Mungall *et al.*, in press; Romano *et al.*, in press).

4. Outlook

The data describing the granitic and pegmatitic melt properties presented here have been accumulated largely in the past five years. This progress has been possible due to a combination of factors, including a clear recognition of the need for property data in the *P–T–X* range relevant to the magmatic stage of the petrogenesis of specialised granite and granitic pegmatites, theoretical insights into the nature of supercooled liquids and relaxation in such liquids and the development of novel experimental approaches to the determination of melt properties. Nevertheless, the present state of knowledge is far from adequate for petrogenetic modelling. Volume and viscosity data for these melt compositions at pH_2O values of hundreds to thousands of bars are still lacking. *In situ* hydrothermal dilatometry, volumometry and viscometry must be developed to fill in the picture of volumes and viscosities in melts whose compositions approach the natural systems in complexity. We will need all the experimental tools in our arsenal and, especially, complementary approaches and apparatus to achieve this goal.

5. Acknowledgements

I thank the organisers for the invitation to participate in the Hutton symposium and the editors and reviewers handling this paper for their assistance, skill and patience. Most of the work described herein has been funded by the Deutsche

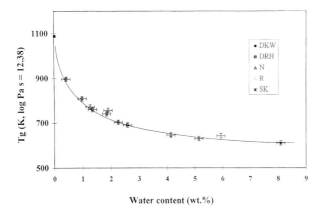

Figure 7 Variation of the glass transition temperature of granitic melts with water content. The data sources are listed by Dingwell (in press). Redrawn from Dingwell (in press).

Forschungsgemeinschaft and the European Commission. This is gratefully appreciated. Reviews by A. Dana Johnston and an anonymous reviewer improved the paper. We thank E-an Zen for his editorial handling of the manuscript.

6. References

Alidibirov, M. & Dingwell, D. B. 1996. Magmafragmentation by rapid decompression. NATURE 380, 146–9.

Bagdassarov, N. & Dingwell, D. B. 1992. A rheological investigation of vesicular rhyolite. J VOLCANOL GEOTHERM RES 50, 307–22.

Bagdassarov, N. & Dingwell, D. B. 1993a. Frequency-dependent rheology of vesicular rhyolite. J GEOPHYS RES 98, 6477–87.

Bagdassarov, N. & Dingwell, D. B. 1993b. Deformation of foamed rhyolites under internal and external stresses. BULL VOLCANOL 55, 147–54.

Bagdassarov, N., Dingwell, D. B. & Webb, S. L. 1994a. Viscoelasticity of crystal- and bubble-bearing rhyolite melts. PHYS EARTH PLANET INTER 83, 83–99.

Bagdassarov, N., Dorfman, A. & Dingwell, D. B. 1994b. Effect of alkalis on surface tension of haplogranite melts. EOS, TRANS AM GEOPHYS UNION 75, 724.

Burnham, C. W. 1963. Viscosity of a water-rich pegmatite. SPEC PAP GEOL SOC AM 76, 26.

Burnham, C. W. & Davis, N. F. 1971. The role of water in silicate melts: I. P–V–T relations in the system NaAlSi₃O₈–H₂O to 10 kilobars and 1100°C. AM J SCI 270, 54–79.

Burt, D. M., Bikun, J. V. & Christiansen, E. H. 1982. Topaz rhyolites: distribution, origin and significance for exploration. ECON GEOL 77, 1818–36.

Cerny, P., Meintzer, R. E. & Anderson, A. J. 1985. Extreme fractionation in rare-metal pegmatites: selected examples of data and mechanisms. CAN MINERAL 23, 381–421.

Chakraborty, S., Dingwell, D. B. & Chaussidon, M. 1993. Chemical diffusion of boron in melts of haplogranitic composition. GEOCHIM COSMOCHIM ACTA 57, 1741–52.

Dingwell, D. B. 1987. Melt viscosities in the system NaAlSi₃O₈–H₂O–F₂O₋₁. In Mysen, B. O. (ed.) Magmatic processes: physicochemical principles. GEOCHEM SOC SPEC PUBL 1, 423–33.

Dingwell, D. B. 1989. The effect of fluorine on the viscosity of diopside melt. AM MINERAL 174, 333–8.

Dingwell, D. B. 1990. Shear viscosities of galliosilicate liquids. AM MINERAL 75, 1231–7.

Dingwell, D. B. 1993. Experimental strategies for the investigation of low temperature properties in granitic and pegmatitic melts. CHEM GEOL 108, 19–30.

Dingwell, D. B. 1995a. Relaxation in silicate melts: some applications. MIN SOC AM REV MINERAL 32, 21–66.

Dingwell, D. B. 1995b. Viscosity and anelasticity of melts and glasses. In Ahrens, T. (ed.) Mineral physics and crystallography. A handbook of physical constants. AGU REF SHELF 2, 209–217.

Dingwell, D. B. The glass transition in hydrous granitic melts. PHYS EARTH PLANET INT, in press.

Dingwell, D. B. & Webb, S. L. 1989. Structural relaxation in silicate melts and non-Newtonian melt rheology in geologic processes. PHYS CHEM MINERAL 16, 508–16.

Dingwell, D. B. & Webb, S. L. 1990. Relaxation in silicate melts. EUR J MINERAL 2, 427–49.

Dingwell, D. B., Knoche, R., Webb, S. L. & Pichavant, M. 1992. The effect of B₂O₃ on the viscosity of haplogranitic melts. AM MINERAL 77, 457–61.

Dingwell, D. B., Knoche, R. & Webb, S. L. 1993. The effect of fluorine on the density of haplogranitic melts. AM MINERAL 78, 325–30.

Dingwell, D. B., Romano, C. & Hess, K.-U. The effect of water on the viscosity of a haplogranitic melt under P–T–X–conditions relevant to silicic volcanism. CONTRIB MINERAL PETROL, in press.

Dorfman, A., Hess, K.-U. & Dingwell, D. B. Centrifuge-assisted falling sphere viscometry. EUR J MINERAL, in press.

Hess, K.-U., Dingwell, D. B. & Webb, S. L. 1995. The influence of excess alkalis on the viscosity of a haplogranitic melt. AM MINERAL 80, 297–304.

Hess, K.-U., Dingwell, D. B. & Webb, S. L. 1996. The influence of alkaline earth oxides on the viscosity of granitic melts: systematics of non-Arrhenian behaviour. EUR J MINERAL 8, 371–81.

Hess, K.-U., Dingwell, D. B. & Rössler, E. Parameterization of viscosity temperature relationships of aluminosilicate melts. CHEM GEOL, in press.

Holtz, F., Behrens, H., Dingwell, D. B. & Taylor, R. 1992. Water solubility in aluminosilicate melts of haplogranitic composition at 2 kbar. CHEM GEOL 96, 289–302.

Holtz, F., Behrens, H. & Dingwell, D. B. 1995. Water solubility in haplogranitic melts. Compositional pressure and temperature dependence. AM MINERAL 80, 94–108.

Hurwitz, S. & Navon, O. 1994. Bubble nucleation in rhyolitic melts: experiments at high pressure temperature and water content. EARTH PLANET SCI LETT 122, 267–80.

Jahns, R. H. & Burnham, C. W. 1969. Experimental studies of pegmatite genesis I. A model for the derivation and crystallisation of granitic pegmatites. ECON GEOL 64, 843–64.

Knoche, R., Webb, S. L. & Dingwell, D. B. 1992. A partial molar volume for B₂O₃ in haplogranitic melts. CAN MINERAL 30, 561–9.

Knoche, R., Dingwell, D. B. & Webb, S. L. 1995. Leucogranitic and pegmatitic melt densities: partial molar volumes for SiO₂, Al₂O₃, Na₂O, K₂O, Rb₂O, Cs₂O, Li₂O, BaO, SrO, CaO, MgO, TiO₂, B₂O₃, P₂O₅, F₂O₋₁, Ta₂O₅, Nb₂O₅, and WO₃. GEOCHIM COSMOCHIM ACTA 59, 4645–52.

Kushiro, I. 1978a. Density and viscosity of hydrous calkalkaline andesite magma at high pressures. CARNEGIE INST WASHINGTON YEARB 77, 675–7.

Kushiro, I. 1978b. Viscosity and structural changes of albite (NaAlSi₃O₈) melt at high pressures. EARTH PLANET SCI LETT 41, 87–90.

London, D. 1992. The application of experimental petrology to the genesis and crystallisation of granitic pegmatites. CAN MINERAL 30, 499–540.

Manning, D. A. C. 1981. The effect of fluorine on liquidus phase relationships in the system qz–ab–or with excess water at 1 kb. CONTRIB MINERAL PETROL 76, 206–15.

Mungall, J., Bagdassarov, N., Romano, C. & Dingwell, D. B. Numerical modelling of stress generation and microfracturing of vesicle walls in glassy rocks. J VOLCANOL GEOTHERM RES, in press.

Mungall, J., Dingwell, D. B. & Chaussidon, M. 1996. Trace element diffusion in synthetic granite and granitic pegmatite melts. GAC/MAC WINNIPEG ABSTR PROGRAM.

Naney, M. T. & Swanson, S. E. 1980. The effect of Fe and Mg on crystallisation in granitic systems. AM MINERAL 65, 639–53.

Nowak, M. & Behrens, H. 1995. The speciation of water in haplogranitic glasses and melts determined by in situ near-infrared spectroscopy. GEOCHIM COSMOCHIM ACTA 59, 3445–50.

Persikov, E. S., Zharikov, V. A., Bukhtiyarov, P. G. & Polskoy, S. F. 1990. The effects of volatiles on the properties of magmatic melts. EUR J MINERAL 2, 621–42.

Pichavant, M. 1981. An experimental study of the effect of boron on a water-saturated haplogranite at 1 kbar pressure: geological applications. CONTRIB MINERAL PETROL 76, 430–9.

Pichavant, M., Valencia Herrera, J., Boulmier, S., Briqueu, L., Joron, J.-L., Juteau, M., Marin, L., Michard, A., Sheppard, S. M. F., Treuil, M. & Vernet, M. 1987. The Macusani glasses, SE Peru: evidence of chemical fractionation in peraluminous magmas. GEOCHEM SOC SPEC PUBL 1, 359–73.

Richet, P. 1984. Viscosity and configurational entropy of silicate melts. GEOCHIM COSMOCHIM ACTA 48, 471–84.

Richet, P., Lejeune, A-M., Holtz, F. & Roux, J. Water and the viscosity of andesite melts. CHEM GEOL, in press.

Romano, C., Bagdassarov, N., Dingwell, D. B. & Mungall, J. Strength and explosive behaviour of vesicular glassy lavas: experimental constraints. AM MINERAL, in press.

Rosenhauer, M., Scarfe, C. M. and Virgo, D. 1979. Pressure dependence of the glass transition in glasses of diopside, albite and sodium trisilicate composition. CARNEGIE INST WASHINGTON YEARB 78, 556–9.

Sakuyama, M. & Kushiro, I. 1979. Vesiculation of hydrous andesitic melt and transport of alkalis by separated vapour phase. CONTRIB MINERAL PETROL 71, 61–6.

Schairer, J. F. & Bowen, N. L. 1956. The system Na₂O–Al₂O₃–SiO₂. AM J SCI 254, 129–95.

Schulze, F., Behrens, H., Holtz, F., Roux, J. & Johannes, W. The influence of water on the viscosity of a haplogranitic melt. AM MINERAL, in press.

Shaw, H. R. 1963. Obsidian–H₂O viscosities at 1000 and 2000 bars in the temperature range 700 to 900°C. J GEOPHYS RES 68, 6337–43.

Shaw, H. R. 1972. Viscosities of magmatic silicate liquids: an empirical method of prediction. AM J SCI 272, 870–89.

Shen, A. & Keppler, H. Infrared spectroscopy of hydrous silicate melts to 1000°C and 10 kbars: direct observation of water speciation in a diamond anvil cell. AM MINERAL **80,** 1335–8.

Sparks, R. S. J. 1978. The dynamics of bubble formation and growth in magmas. J VOLCANOL GEOTHERM RES **3,** 1–37.

Sparks, R. S. J., Barclay, J., Jaupart, C., Mader, H. M. & Phillips, J. C. 1994. Physical aspects of magmatic degassing I. Experimental and theoretical constraints on vesiculation. REV MINERAL **30,** 413–45.

Stevenson, R. J., Dingwell, D. B., Webb, S. L. & Bagdassarov, N. S. 1995. The equivalence of enthalpy and shear stress relaxation in rhyolitic obsidians and quantification of the liquid–glass transition in volcanic processes. J VOLCANOL GEOTHERM RES **68,** 297–306.

Thomas, R. 1995. Assessment of water content in granitic melts using melt inclusion homogenisation data: method–results–problems. *In* Brown, M. & Piccoli, Ph. M. (eds) *The origin of granites and related rocks. Third Hutton Symposium—Abstracts.* US GEOL SURV CIRC **1129,** 145–6.

Tuttle, O. F. & Bowen, N. L. 1958. Origin of granite in the light of experimental studies in the system $NaAlSi_3O_8$–$KAlSi_3O_8$–SiO_2–H_2O. GEOL SOC AM MEM **74,** 1–154.

Webb, S. L., Knoche, R. & Dingwell, D. B. 1992. Determination of liquid expansivity using calorimetry and dilatometry. EUR J MINERAL **14,** 95–104.

Wong, J. & Angell, C. A. 1976. *Glass structure by spectroscopy.* New York: Dekker, 864 pp.

D. B. DINGWELL, K.-U. HESS and R. KNOCHE, Bayerisches Geoinstitut, Universität Bayreuth, 95440 Bayreuth, Germany.

Transactions of the Royal Society of Edinburgh: Earth Sciences, **87**, 73–83, 1996

Melt segregation in the lower crust: how have experiments helped us?

Tracy Rushmer

ABSTRACT: The rheological and chemical behaviour of the lower crust during anatexis has been a major focus of geological investigations for many years. Modern studies of crustal evolution require significant knowledge, not only of the potential source regions for granites, but also of the transport paths and emplacement mechanisms operating during granite genesis. We have gained significant insights into the segregation and transport of granitoid melts from the results of experimental studies on rock behaviour during partial melting. Experiments performed on crustal rock cores under both hydrostatic conditions and during deformation have led, in part, to two conclusions. (1) The interfacial energy controlling melt distribution is anisotropic and, as a result, the textures deviate significantly from those predicted for ideal systems—planar solid–melt interfaces are developed in addition to triple junction melt pockets. The ideal dihedral angle model for melt distribution cannot be used as a constraint to predict melt migration in the lower crust. (2) The 'critical melt fraction' model, which requires viscous, granitic melt to remain in the source until melt fractions reach >25 vol%, is not a reliable model for melt segregation. The most recent experimental results on crustal rock cores which have helped advance our understanding of melt segregation processes have shown that melt segregation is controlled by several variables, including the depth of melting, the type of reaction and the volume change associated with that reaction. Larger scale processes such as tectonic environment determine the rate at which the lower crust heats and deforms, thus the tectonic setting controls the melt fraction at which segregation takes place, in addition to the pressure and temperature of the potential melting reactions. Melt migration therefore can occur at a variety of different melt fractions depending on the tectonic environment; these results have significant implications for the predicted geochemistry of the magmas themselves.

KEY WORDS: melting reactions, melt migration, static experiments, experimental rock deformation, stress, strain, geochemistry.

The Third Hutton Symposium on the Origin of Granites and Related Rocks comes during a time of major re-evaluation of melt segregation and melt migration in the lower crust. Knowledge of the rheological and chemical behaviour of the partially molten crust forms the basis of current models of crustal growth. The present thinking on melt segregation in the lower crust has come from integrating the results from several fronts: (1) compaction and fracture modelling studies (Clemens & Mawer 1992; Brown *et al.* 1995a; Petford 1995); (2) detailed field work in well-known areas such as the Grenville Front, the Himalayas and the Ivrea Zone (Sawyer 1991; Sinigoi *et al.* 1994; Harris *et al.* 1995); and (3) several experimental investigations of crustal rock behaviour during partial melting.

This paper reviews the experimental results in some detail and evaluates the impact of these results on two melt segregation models: the dihedral angle model, which uses equilibrium partial melt distribution in a solid–melt system to assess melt segregation potential; and the critical melt fraction (CMF) melt segregation model, which suggests that viscous melt can only migrate from its source at melt fractions greater than *c.* 25 vol%. All the experiments described here use realistic crustal assemblages, either as cores of solid rock or mechanically made aggregates, to study the partial melt behaviour in the lower crust. The experimental results suggest that melt connectivity and segregation are possible at very low melt fractions, much lower than considered possible by the CMF model, and are controlled by a variety of factors.

The experiments investigate the importance of such variables as depth (pressure) and type of mineral melting reaction, interfacial energy anisotropy and strain rate on controlling melt behaviour in the lower crust. In addition, they attempt to address the role of viscosity and water content of melts, the importance of fractures, different permeabilities and the effect of hydrostatic versus dynamic tectonic settings, all of which have been major issues in understanding melt segregation processes in the past. This focus has provided a solid base of information which enables us to think in new ways about melt segregation and transport mechanisms active in the crust. For example, the geochemistry and emplacement of low-melt fraction granites (e.g. Himalaya leucogranites), originally considered too viscous to segregate efficiently, are much better understood.

I consider here experiments performed under both hydrostatic conditions, where there is no applied stress to the sample, and non-hydrostatic or dynamic conditions, where samples are deformed during partial melting. Most of the hydrostatic experiments specifically characterise the distribution of granitoid melt in different crustal assemblages. The distribution data provide information on the fraction of melt which is necessary before interconnectivity can be achieved. Ideally, in monomineralic systems, the equilibrium melt distribution is similar to the liquid sintering processes described in publications on ceramics. Low wetting angles (<60°, e.g. Jurewicz & Watson 1984, 1985) suggest that melts can achieve interconnectivity by wetting the grain edges and can therefore achieve high enough permeabilities to segregate, whereas high wetting angles will hinder melt interconnectivity and the melt will form pools and not wet grain edges. Permeability estimates to model melt segregation are also provided by hydrostatic experiments which quantify melt distribution at the grain to grain scale. However, the application of this idealised model to the crust has been minimal because of the polymineralic, highly non-ideal nature of the lower crust (Laporte & Watson 1995). Alternative grain-scale melt segregation processes in natural rocks are currently being investigated in a hydrostatic experimental study on muscovite–quartz rock cores. This ongoing study is investigating the effect of positive volume change on the production of microcracks as a means of generating permeability during melting (Rushmer et al. 1995). The preliminary results are also briefly discussed here.

Dynamic segregation and transport processes in an active tectonic environment are better simulated in the deformation experiments. Deformation creates high melt pore pressures in the partially molten system (Davidson et al. 1994) and this high pore pressure can produce an extensive fracture network unless the melt reaches too high a fraction and the rock deforms ductily. Dynamic experiments are more easily scaled to the outcrop and to the lower crust because the microstructures are not limited to the individual grain scale, but are observed throughout the sample. There is, however, continued debate on the probability of melt-induced fractures propagating from the melting site into overlying, unmelted rocks and to what extent the rock deformation experiments can be applied to large-scale processes in the lower crust where deformation is occurring at slow strain rates (Brown et al. 1995b). These problems can be further evaluated with the successful integration of melt-induced fracturing and the compaction/fracture modelling.

Finally, I briefly discuss how the geochemistry of the melts is affected by the type of segregation and transport mechanism active during melting. Future experimental work and studies of natural systems must tie in the geochemical data on magmas with physical melt segregation models. These kinds of studies will provide important information on the rates of melt formation and melt extraction in different tectonic environments (e.g. Harris et al. 1995).

1. Hydrostatic experiments

Hydrostatic experiments on both mafic and felsic crustal assemblages have been used over the last few years to describe melt distributions in polymineralic rocks. Most of these experiments focus on characterising the granitoid (viscous) melt distribution among the different residual minerals.

All hydrostatic studies use either a cold-seal bomb or piston-cylinder apparatus. Pressure–temperature conditions of the experiments range between 0·1 and 1·5 GPa, with most experiments performed at 1·0 GPa, which is equal in pressure to the lowermost continental crust. Natural rock cores of partially molten mafic amphibolite are described by Hacker (1990) and Wolf and Wyllie (1995), powdered amphibole–silicic melt assemblages by Lupulescu and Watson (1995) and powdered biotite–hornblende–silicic melt assemblages by Laporte and Watson (1995). Early experiments on quartz–feldspar crustal assemblages by Jurewicz and Watson (1984, 1985) complement the more recent studies of quartz aggregates with granitic melt at different a_{H_2O} by Laporte (1994). Muscovite melting in natural quartz + muscovite rock cores investigated by Brearley and Rubie (1990) and Rubie and Brearley (1991) consider kinetically controlled disequilibrium melting processes. These experiments, plus several new ones

on the same starting material, have been used to study melt-induced fracturing during melting (Rushmer et al. 1995).

Although not all of these experiments produce equilibrium textures readily, the experimental results provide an understanding of the textures and reaction mechanisms at the grain scale in granitoid source regions.

1.1. Studies of equilibrium melt distribution

Hacker (1990) inferred slow reaction kinetics and perhaps disequilibrium melting in a set of hydrostatic experiments on amphibolite (hornblende + plagioclase + quartz) cores with no added water. These experiments produce ≤1% melt volume and melt pools at amphibole–quartz–plagioclase triple junctions and recrystallisation and growth of new plagioclase and amphibole nucleates on relict, or unreacted, grains of plagioclase and amphibole. With increased temperature, melting produces liquid as thin films between reactant quartz and amphibole, with pyroxene found along grain edges. Additional experiments by Hacker (1990) with water added increase the melt fraction, but the observed textures were very similar. Wolf and Wyllie (1995) also observed in their fluid-absent experiments (no water added) on hornblende + calcic plagioclase cores that low melt volumes (between 1 and 5 vol%) wetted hornblende–plagioclase grain boundaries. The melt distribution is interpreted by these workers to be initially crystallographically controlled by the hornblende in the amphibolite. Melt interconnectivity was achieved at low melt fractions, although precise melt fraction estimates were difficult because the cores were out of textural equilibrium. The low viscosity of the melt may also play a major part in the melt distribution observed. The initial melting is hydrous with >4 wt% water and calculated to have a viscosity of 10^{3-4} Pa s. Wolf and Wyllie (1995) estimated that the interconnectivity observed in the amphibolite assemblage occurred at melt fractions <5 vol%. Lupulescu and Watson (1995) reported the results of diffusion-couple piston-cylinder experiments, where powdered amphibole, saturated with hydrated tonalitic or granitic glass, was placed next to a hydrated amphibole-saturated glass, which acted as an infinite reservoir. The glass was doped with [151]Sm or ^{14}C, which was used as a tracer of diffusion to determine melt interconnectivity. The measured dihedral angles in the amphibolite were 46–48° (mean value), although the dihedral angle distribution was broad. Lupulescu and Watson (1995) estimated that interconnectivity was achieved at <10 wt% melt (see Table 1 for summary).

In granitic felsic systems at 1·0 GPa and 1000°C, Jurewicz and Watson (1985) found that the alkali feldspar–silicic melt wetting angles were moderate (44°), as were alkali feldspar–quartz–melt angles (49°). Laporte and Watson (1995) reported wetting angles of 28° for anorthite–silicate melt. Laporte's (1994) experiments on the quartz–granitic melt system at 1·0 GPa and between 800 and 900°C using quartz aggregates produced between 30 and 40 vol% melt and measured low dihedral angles between 12 and 16°. These latter results are particularly interesting as the experimental sets contained different amounts of water, causing the melt phase to be either dry, undersaturated or saturated. Surprisingly, the dihedral angles did not vary significantly as a function of the melt–water content. The water content causes profound structural changes in granitic melt, so greater angle changes were expected. One possibility is that the local reorganisation of the melt structure occurs at the melt–quartz interface, so bonding is facilitated. This would result in a low quartz–melt interfacial energy (Laporte, pers. comm.). These results suggest that the melt–water content does not influence the resulting interfacial energies, at least in the quartz–granitic melt system (Table 1).

Table 1 Summary of interconnectivity parameters obtained in hydrostatic partial melting experiments

Study	Assemblage	Pressure (GPa)	Temperature (°C)	Water present	Dihedral angle measured	Interconnectivity
Hacker (1990)	Amphibole+plagioclase+quartz	0·5–1·5	700–1000	H₂O-added and fluid-absent conditions	Not measured	Not estimated, but possible at >875°C
Wolf & Wyllie (1995)	Amphibole+plagioclase	1·0	750–1000	Fluid-absent conditions	Not measured	Estimated to occur at >875°C, at 2 vol% melt
Lupulescu & Watson (1995)	Amphibole+tonalitic or granitic melt (used as melt reservoir)	1·0	800	Starting melt phase is a 5 wt%-added hydrated glass	46–48°	Determined by tracer to occur at <10 wt% melt
Vicenzi et al. (1988)	Amphibole	0·8	800	Fluid-absent conditions	33°	Interconnectivity possible at low melt fractions
Laporte & Watson (1995)	Biotite+granite to granodioritic melt	1·0–1·2	800–1050	Starting melt phase is 3·5 wt%-added hydrated glass	23–32° (at 975–1050°C) 34–39° (at 800°C)	Estimated at low melt volumes for crustal assemblages in general (<5 vol%), but suggests very biotite-rich layers may be impermeable
Laporte & Watson (1995)	Amphibole+granite to granodioritic melt	1·2	975	Starting melt phase is 3·5 wt%-added hydrated glass	25° median	Estimated at ≤3–4 vol%
Jurewicz & Watson (1985)	Quartz–feldspar aggregate+granitic melt	1·0	1000	Dry	44° for feldspar–feldspar–melt	Interconnectivity possible at low melt fractions
Laporte (1994)	Quartz aggregate+granitic melt of different water contents	1·0	800–900	Glasses are H₂O-saturated, H₂O-undersaturated, or dry	12–16° (wet to dry, respectively)	Anisotropy controls melt distribution

The most complex crustal assemblages were studied in piston-cylinder experiments by Laporte and Watson (1995), who estimated wetting angles in biotite-bearing and hornblende-bearing crustal assemblages. The experiments were carried out at 1·0–1·2 GPa at different temperatures (800–1050°C), so the resulting parageneses of the assemblages differ as a function of temperature and pressure. Garnet, clinopyroxene, magnetite, ilmenite and metastable orthopyroxene are all observed in the run products from this set of experiments. The wetting angles measured, however, are relatively consistent and range between 23 and 39°. Low hornblende–hornblende–melt angles of 25° measured at 1·2 GPa and 975°C confirm that interconnectivity can be achieved at low melt fractions in amphibolite (<5 vol%). The study by Laporte and Watson (1995) illustrates the importance of considering anisotropy and the chemical and textural heterogeneity of the lower crust. This high degree of variability essentially eliminates the possibility of characterising the distribution of melt in the crust by a single dihedral angle. Laporte and Watson (1995) conclude, for example, that there is no unique wetting angle for a given source which can describe the melt distribution. Laporte and Watson (1995) also suggest that melt viscosity may be a limiting factor in granitic melt segregation and that there could be a range of melt fractions where melt is interconnected, but does not migrate.

Table 1 summarises these hydrostatic experimental results for melt distribution. All the experiments on crustal assemblages show moderate to low dihedral angles, but, more importantly, show the effect of anisotropy. Melt distribution textures in these studies are controlled by the interfacial energy anisotropy and appear to have dihedral angles of less than 60° in all crustal assemblages investigated to date, but this does not directly suggest that there will be interconnectivity starting at 0 vol% melt. The connectivity threshold is a low value (<5 vol% melt) because the melt distribution is sensitive to the solid–solid to solid–liquid interfacial energy ratios, as discussed by Laporte and Watson (1995). This was also observed in basalt melt–olivine experiments, in which it is no longer viewed as optimal to characterise these systems by dihedral angles. Instead, basaltic melt is better characterised by thin, elongated melt pockets (Faul et al. 1994). Mantle workers now describe anisotropy in these systems on a grain to grain scale. In general, the presence of anisotropy in residual mineral assemblages in the lower crust and in the mantle appears to enhance melt interconnectivity, at least for silicic melt systems.

1.2. Melt-induced fracturing experiments

Brearley & Rubie (1990) and Rubie & Brearley (1990) discuss disequilibrium melting processes during muscovite breakdown. Experimental results performed at 0·1 GPa and 757°C for different lengths of time (11·7, 96·0 and 214·2 hours) establish the kinetic control on the muscovite + quartz melting reaction. Localised melting initiated at muscovite–quartz grain boundaries and the progressive consumption of muscovite by the melting reaction eventually led to the pseudomorphic replacement of this phase by melt + mullite + biotite. The volume change of the reaction has a ΔV of $+2·7\%$. Also observed in these experiments are melt-filled cracks (Fig. 1).

In contrast with studies which characterise equilibrium melt distribution textures in rocks, experiments which produce melt by a positive volume reaction provide transient permeability data at the grain scale. At the onset of reaction, reactions with $\Delta V > 0$ can create significant melt overpressure. The subsequent microstructural evolution, deformation mechanisms and rate of melt extraction may depend on the rates and mechanisms

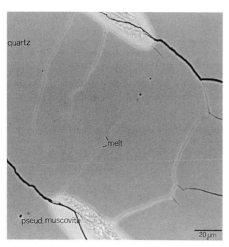

Figure 1 Example of melt-filled cracks produced during the break-down reaction of muscovite + quartz as shown by a scanning electron microscopy backscattered electron image. This experiment was reacted at 802°C and 0·3 GPa hydrostatic pressure for 285·9 hours. No water was added to the experiment. 'Pseud. muscovite' is muscovite reacted to biotite + mullite + melt. Melt is found also along cracks as indicated. Quartz is the other phase present. Note that the black fractures that cross-cut the photomicrograph are produced after the quench and are due to take down of the experiment.

of melting during this early stage, regardless of the equilibrium textural behaviour of the system. Determining the conditions under which melting reactions can create permeability (through fractures) in the surrounding rock is important in estimating the overall permeability in partially molten systems.

The study of Rushmer et al. (1995) uses the partial melting results of the muscovite-bearing quartzite experiments (Brearley & Rubie 1990) in addition to results from new experiments at 0·3 GPa at 802–853°C for 285·9 and 48·0 hours, respectively. The focus of the investigation is to measure changes in crack density as a function of time and reaction type (water-saturated or fluid-absent). The experiments allow an estimation of the permeability generated during melting reactions as a function of both melting rate and time, based on melting reactions having different $+\Delta V$ and melt viscosities. The preliminary results show that the length scale of melt migration, or crack length, as well as the width of the melt-filled fractures, increase with the extent of the melting reaction. Shorter term experiments (<12 hours) contain melt-filled fractures which are typically 1 μm wide and can extend up to 150 μm from the melting sites. In longer duration experiments (>12 hours and at least 50% reaction) the fractures have further propagated, have a length of several hundred micrometres, a width of 1–5 μm and form a network which connects many of the melting sites. The experiment, performed under fluid-absent conditions, shows that crack spacing and crack length may be dependent on not only reaction volume change, but on the viscosity of the melt. The melt phase under fluid-absent conditions (experiment performed at 802°C and 0·3 GPa) has a lower water content than the water-saturated experiments and therefore a higher melt viscosity. Crack lengths are shorter (cracks are <100 μm after 285·9 hours compared with reaching >200 μm in the water-saturated experiment performed for 214·2 hours) and are closer together.

The results from the early set of experiments (0·1 GPa, 757°C) were used to calculate the permeability of this system. Preliminary data suggest that crack growth slowed down and then levelled off as the wet quartz + muscovite reaction approached 100% completion. The change in crack growth rate is interpreted as the point at which interconnectivity was attained in the system, so no new crack was produced during the experiment. Early calculations based on the total crack density and the crack area result in permeabilities ranging between 10^{-14} and 10^{-13} m², depending on the point during the reaction at which interconnectivity was assumed to be achieved (Rushmer et al. 1995). To obtain the total crack density (number of cracks) and crack area, it was assumed that the cracks were randomly oriented so the crack lengths were approximately the same and a true average length in three dimensions could be estimated. The experimental data (crack length estimated in three dimensions and cracks/mm²) are then used to calculate the crack area per unit volume. The individual crack area (assumed to be the area for a penny-shaped crack) multiplied by the total number of cracks can then be set equal to the crack area per unit volume calculated from the data. This provides a value for the total number of cracks. A crack width is calculated by assuming that the total volume of cracks (equal to the volume of reaction of the given experiment) is equal to the individual crack volume (function of crack width) multiplied by the total number of cracks. The permeabilities calculated with this approach are similar to those of fractured crystalline rocks as given by Hanson (1995), 10^{-12} to 10^{-15} m², and are higher than those of McKenzie's (1985) molten, 10% porosity layer: $1·2 \times 10^{-12}$ m². Further experiments need to be performed to pinpoint the extent of reaction necessary to have interconnectivity in the system. In particular, further experiments are needed to calculate the potential permeabilities generated during fluid-absent reactions, as the crack spacing and density may be significantly different than under water-saturated conditions. Ultimately, a set of permeability data could be produced for typical reactions in the crust. These data could then be used in compaction/extraction models for melt segregation. These results are especially applicable to environments which undergo fast melting rates, such as during magmatic underplating. Melting reactions may produce transient periods of increased permeability on the grain scale. This small-scale process may be the most important first step in the development of a large-scale fracture network necessary for efficient melt segregation.

2. Non-hydrostatic (dynamic) experiments

Non-hydrostatic or rock deformation experiments on amphibolite by Hacker and Christie (1990) and Rushmer (1995) and on granite by Dell'Angelo and Tullis (1988) and Rutter and Neumann (1995) show that melting reactions can induce fracture and cataclastic deformation (see Tables 2 and 3 for summary). The effect of high fluid pore pressure on rock embrittlement has been previously investigated by Raleigh and Paterson (1965) and Paquet et al. (1981), among others, but the melting experiments at high pressures show that fracture-forming pathways for melt migration can occur at deep levels in the crust, even in previously ductile rock (Table 2). In addition, two experiments have produced melt fractions greater than 20 vol% (Rushmer 1995; Rutter & Neumann 1995). These experiments also produced microstructures consistent with viscous deformation and homogeneous flow. Significantly, Rutter and Neumann (1995) showed that a sharp decrease in effective viscosity is not observed at high melt volume, but rather the strength decreases more or less linearly as a function of increasing melt fraction.

Interpretation of the experiments suggests that melt may be able to migrate out of the system by either fracture or flow depending on the type of melting reaction (water-present versus water-absent), the depth of the melting reaction (brittle or ductile regime) and the strain rate (Dell'Angelo & Tullis 1988), among other factors. This is in contrast with a critical melt fraction segregation model in which viscous melt can

Table 2 Summary of deformation results obtained from dynamic, partial melting experiments on crustal rock types under fluid-absent conditions

Fluid-absent studies	Crustal assemblage	Pressure and temperature	Melt fraction	Strain rate (s^{-1})	Deformation regimes observed	Comments
Hacker & Christie (1990)	Amphibolite (amphibole + plagioclase)	0·5–1·5 GPa, 650–950°C	≪1 vol%	10^{-4}–10^{-7}	Deformation by fracturing and sliding along sample-scale fault zones	Brittle behaviour under subsolidus conditions. Cataclasis observed at the onset of melting (850–900°C)
Rushmer (1995)	Amphibolite (amphibole + plagioclase + quartz)	1·8 GPa, 650–1000°C	0–20 vol% 0–≥20 vol%	10^{-5}	Semi-brittle/ductile (no throughgoing fractures) 0 vol%; brittle with sample-scale cracks 2–5 vol%; cataclasis 5–20 vol%; ductile flow ≥20 vol%	Transition from ductile to brittle behaviour at the onset of melting is observed at this confining pressure
Rutter & Neumann (1995)	Westerly Granite (plagioclase + potassium feldspar + quartz + biotite + muscovite + minor chlorite)	250 MPa (0·25 GPa), 800–1100°C	3–50 vol%	10^{-5}	Brittle (throughgoing fractures) 0–10 vol% melt; cataclasis 10–40 vol%; ductile flow >40 vol%	Strength of rock decreases linearly as a function of increasing melt fraction

Table 3 Summary of deformation results obtained from dynamic, partial melting experiments on crustal rock types with water added

Water-added studies	Crustal assemblage	Pressure and temperature	Melt fraction	Strain rate (s^{-1})	Deformation regimes observed	Comments
van der Molen & Paterson (1979)	Granite	300 MPa (0·30 GPa), 800°C	0–24 vol% (water added in different amounts to produce desired melt fraction)	10^{-5}	Cataclasis and cataclastic flow (microcracking of grains with rotational frictional sliding of fragments)	Melt viscosity changes at different melt fractions due to added water; dramatic weakening observed at 24 vol% melt
Arzi (1978)	Granite	250 MPa (0·25 GPa), 860–1020°C	0–20 vol% (water added to produce melt)	10^{-5}	Formation of pervasive melt film fractures; cataclasis and cataclastic flow	Microstructures and weakening behaviour similar to that observed in van der Molen & Paterson (1979)
Paquet et al. (1981)	Aplite (quartz + K–feldspar + plagioclase)	250 MPa (0·25 GPa), 560–900°C	Not stated explicitly, but observed as several vol% (≤5 vol%)	10^{-5}	Free water induces frctures; cataclasis. Melt in fractures parallel to σ_1	Water content varies from dry (≈0·25 wt%) to 1·0 wt%
Dell'Angelo & Tullis (1988)	Aplite (quartz + potassium feldspar + oligoclase + minor biotite)	1500 MPa (1·5 GPa), 900°C	5–15 vol% (water added to produce melt)	10^{-5}–10^{-6}	Crystal plasticity at low melt volumes and slow strain rates; cataclasis at higher melt volumes	Low melt volumes did not produce high pore pressures at the slow strain rate, so cataclasis did not occur

only separate from its source at a pre-determined, specific melt fraction interval.

This section summarises these experiments and shows the need to re-evaluate our thinking with respect to melt segregation during deformation. For many years the experimental results on partially molten granite by van der Molen and Paterson (1979) were used to interpret different granitic terranes. Their results showed that at approximately 25 vol% melt there was a dramatic decrease in the effective viscosity, or strength, of the partially molten granite. Although there is a rheological transition from a mineral-supported framework to a melt-dominated system as a function of increasing melt fraction, this transition does not necessarily correlate with melt segregation. Many geologists have used these results to suggest that granitoid melt cannot segregate from its source unless it has reached high enough volumes (>25 vol%), so that the melt phase dominates the highly viscous magma structure, not the mineral framework, thereby dramatically weakening the system and allowing melt segregation. Wickham (1987) has pointed out that the melt fraction at which segregation can occur might change for a given system depending on such variables as strain rate (see also Miller *et al.* 1988), grain shape, melt viscosity and macroscopic heterogeneities; thus these more recent experimental results provide significant new data to consider when examining the concept of a 'critical' melt fraction with respect to melt segregation.

2.1. Deformation of amphibolite

Hacker and Christie (1990) performed a series of experiments on natural and synthetic hot-pressed cores of amphibolite in a solid-media Grigg's deformation apparatus. The amphibolite samples were deformed at 0·5–1·5 GPa confining pressure, between 650 and 950°C and at strain rates ranging from 10^{-4} to 10^{-7} s^{-1} with no added water. The natural amphibolite, composed of hornblende + plagioclase + minor quartz, deformed by fracturing and sliding along sample-scale fault zones under all conditions. Metamorphic reactions occurred in all experiments >650°C and the fault zones exhibited a crystal-plastic to cataclastic transition in deformation which coincided with the beginning of melting. Melt (similar to tonalite in composition) volumes are low, remaining at ≤1 vol% between 850 and 900°C. Some changes are observed in the slower strain rate and higher temperature experiments. Greater ductility is observed at 10^{-6} s^{-1} at 750°C and at 10^{-5} s^{-1} at 850°C and the fault zones are wider. The synthetic amphibolite exhibits more ductile behaviour than the natural cores, which is attributed to the hot pressing procedure, which may introduce a high density of defects into the mineral grains.

The solid-media deformation experiments by Rushmer (1995) used a metamorphosed alkali basalt from the Ivrea Zone as the starting amphibolite. It is similar to the Hacker and Christie (1990) assemblage, composed of Mg-hornblende + plagioclase + quartz, in addition to the accessory phases apatite, zoisite and titanite. The grain size of the amphibolite is smaller, averaging 200–250 μm. Macroscopically, the samples are unweathered, but in thin section some plagioclase grains are altered to sericite along the grain boundaries, providing some additional water to the experiments in addition to the breakdown of amphibole. No free water was added to the experiments. Previous piston-cylinder experiments, performed on a powder of the same amphibolite, provided a base for the deformation experiments. Temperatures between 650 and 1000°C were chosen so the amphibolite could deform at conditions spanning the fluid-absent solidus (between 800 and 850°C). The confining pressure in this set of experiments was 1·8 GPa so that the

garnet-producing melting reaction hornblende + plagioclase ± quartz = garnet + clinopyroxene + albitic plagioclase + trondjemitic melt would be observed at temperatures >850°C (Rushmer 1993). This higher confining pressure, plus perhaps a slight change in bulk chemistry, caused the deformation of the amphibolite to be macroscopically ductile at 650 and 750°C. No throughgoing faults like those of Hacker and Christie (1990) were observed under subsolidus conditions (Table 2).

The sample deformed at 650°C shows homogeneous flattening of plagioclase and quartz. Slip along cleavage planes in hornblende and plagioclase grains is also observed, but no localised fracturing is present. At 750°C, plagioclase again deforms by slip along the cleavage planes and most grains show undulatory extinction. In the experiments, hornblende is always more brittle than the plagioclase and grains often have small cracks oriented near-parallel to the compression direction. At higher temperatures, close to the solidus (800–850°C), the beginning of the hornblende breakdown reaction is observed. At these temperatures deformation is brittle, which is documented by micrometre-wide fractures offsetting grains of hornblende and plagioclase. Melt (1–2 vol%) is observed in cracks. The change in microstructure is most likely due to the onset of melting, shifting deformation from macroscopically ductile to fracture in the areas where melt is present (Rushmer 1995).

At 935°C, melt fractions reached ≈10–15 vol% and melt-enhanced embrittlement localised deformation in the sample. A conjugate set of ≈0·5 mm wide shear zones oriented at an approximate angle of 45° to the compression direction was observed. The increased melt fraction was probably due to the more extensive breakdown of hornblende + plagioclase at this temperature and perhaps due to the release of water from the dehydration of altered plagioclase. Abundant evidence for reaction is found in the shear zones and hornblende is strongly cataclastically deformed in these zones, which may in turn have promoted more reaction (Fig. 2). The melt fraction increases in the shear zones, reaching ≈17 vol%. Overall, the deformation observed within the shear zones is considered to be mainly brittle. Outside the sheared areas, garnet, melt,

Figure 2 Backscattered scanning electron photomicrograph showing detail of a fractured hornblende. This embrittlement texture is observed in the cataclastic shear zones produced at 935°C and 1·8 GPa confining pressure with 10–15 vol% melt (up to 17 vol% in the shear zones themselves). New phases are garnet (not shown), melt, clinopyroxene (cpx), which replaces hornblende [hbd] or grows separately, and albitic plagioclase (plag). These new phases are produced by the fluid-absent reaction of hornblende + plagioclase. The embrittlement texture is due to the increase in melt fraction combined with deformation (10^{-5} s^{-1}) of the sample. Black fractures which cross-cut the sample are unloading cracks after the quench.

accessory titanite, zoisite and hornblende (altered to clinopy-roxene in many instances) are observed throughout the sample. Melt is commonly found in dilatant cracks in the hornblende and at grain boundaries between plagioclase and hornblende.

At the highest melt fraction observed in this study ($\approx 20\%$), additional weakening of the sample was noted, but shearing was no longer observed. Instead, 'hornblende' grains, now pyroxene, separated passively from each other and were found in pools of melt (Fig. 3). The sample deformed ductilely and shortened homogeneously (by 25%) by viscous flow and no fracturing occurred. This shift in deformation did not coincide with a sharp decrease in strength. As the temperature and melt fraction increased, the strength of the amphibolite decreased to about one-third of its original strength. The largest relative decrease in strength, however, occurred at the onset of melting. Here, the shift from macroscopically ductile to fracture due to the melt phase can also be documented. Figure 4a schematically shows the changes in deformation behaviour as the melt fraction increases. The changes in microstructures at higher melt fractions are comparable with those of Rutter and Neumann (1995) discussed in the following section.

2.2. Deformation of granite

Rutter and Neumann (1995) performed a set of experiments using the Paterson gas apparatus on partially molten Westerly granite at temperatures between 800 and 1100°C (Table 2). A low confining pressure of 0·25 GPa was used and the samples were deformed at a constant strain rate of $10^{-5}\,\mathrm{s}^{-1}$. The granite was composed of plagioclase + alkali feldspar + quartz + biotite + minor muscovite and chlorite and no water was added to the sample to increase the melt fraction. Under these experimental conditions, melt is produced by the breakdown of the hydrous phases and ranged from 3 vol% at 800°C to 50 vol% at 1100°C. At low melt fractions (<10%) deformation is brittle, documented by shearing along fault zones. Melt is found in dilatant cracks and the surrounding broken grains in the fault zone. At higher melt fractions, between 10 and 45 vol%, these workers describe deformation as shear-enhanced compaction. Grains are pervasively cataclas-tic and melt fills pores which then collapse during deformation.

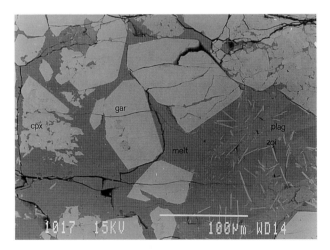

Figure 3 Backscattered scanning electron photomicrograph of experimental results at 1000°C and 1·8 GPa confining pressure. The sample was shortened by 25%, but shear zones are not produced. Instead, melt (≈ 20 vol%) surrounds new grains of garnet (gar), albitic plagioclase (plag) and clinopyroxene (cpx). Zoisite (zoi) occurs as prisms with the plagioclase and may be due to a water-present reaction of the anorthite component in the plagioclase. Deformation is by viscous flow and fracturing of grain boundaries is not observed. Again, black fractures which penetrate the sample are due to unloading of the sample after the quench.

Melt appears to have moved out from the specimens' centre and is often found at the sample's ends and between the jacket and sample. At melt fractions >45 vol%, grains (relatively unfractured) are found in melt pools and the samples are fully ductile, deforming by viscous flow. Figure 4b (modified from figure 9 in Rutter and Neumann 1995) summarises the observed microstructural textures and interprets deformation mechanisms as a function of increasing melt fraction.

Rutter and Neumann (1995) did not observe a 'critical' melt fraction in the sense that there was no dramatic weakening as the melt fraction increased above 20 vol%. The partially molten system did weaken significantly, but not at the transition between fracture and ductile flow. Although a shift from fracture and mineral-dominated microstructures to melt-dominated structures at high melt fractions is observed, melt segregation is not controlled by this rheological transition. This transition (which does occur as the melt fraction increases), however, influences the emplacement mechanism of granite and the deformability of magma as it cools. In addition, the transition between fracture and viscous flow will probably occur at different melt fractions for different magmas depending on the water content and composition (which influence the melt viscosity) and the deformation rate.

The effect of deformation rate on melt migration is shown by Dell'Angelo and Tullis (1988). In their experiments on an aplite deformed at 1·5 GPa and 900°C, the microstructures show that, without melt, the quartz and feldspar deform by dislocation creep. In one set of experiments they added water to the samples to form melt. These workers found that the partially molten aplite deformed at slow strain rates ($10^{-6}\,\mathrm{s}^{-1}$) still showed microstructures which indicated that the active deformation mechanism is crystal plasticity. The melt appears to have been 'squeezed' out laterally from the grain boundaries and did not influence the deformation behaviour. They observed cataclasis at faster strain rates because the melt could not move out from the deforming matrix fast enough to keep the pore pressure low. In these experiments the melt viscosity was also low, approximately $2\cdot 7 \times 10^3$, because of the water added to the charges. The melt viscosity was lower than in the fluid-absent experiments on amphibolite described by Rushmer (1995), which yielded melts with viscosities of $7\cdot 7 \times 10^4$ Pa s [both viscosities calculated by the empirical equation of Shaw (1972)]. Experiments at slower strain rates on the amphibolite (Rushmer, unpublished data) found that fracture still dominated the deformation microstructures at melt fractions between approximately 5 and 20 vol%.

Melt viscosity will probably have a significant effect on the ability of the melt to migrate along grain boundaries, or along micrometre-sized fractures produced during melting. Melt viscosity may also influence the rate at which a partially molten system weakens. This is pointed out by Rutter and Neumann (1995), who show that the dramatic weakening observed in the experiments of van der Molen and Paterson (1979) at 20 vol% melt is in fact due to a combination of a lower melt viscosity and high melt fraction, not due to the increase of melt fraction alone. In the earlier sets of experiments, water is added to produce melt of a desired fraction. This, however, changes the melt viscosity and in the case of van der Molen and Paterson (1979), the melt viscosity itself decreases by almost four orders of magnitude by 20 vol% melt, helping to cause the observed sharp decrease in strength observed in the partially molten granite (Rutter & Neumann 1995).

3. Discussion

These experimental results have significant consequences for models of melt segregation in the lower crust. The results

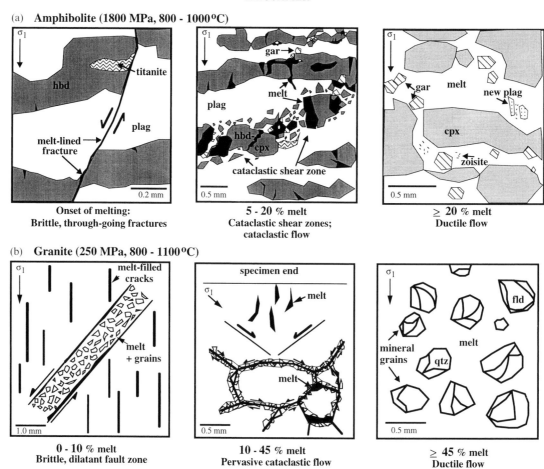

Figure 4 Schematic illustration of the microstructures observed at high confining pressure (amphibolite) and low confining pressure (granite) as a function of increasing melt fraction. The strain rate for both experimental sets is 10^{-5} s^{-1}. (a) Summary of amphibolite microstructures from the onset of melting (800°C) to the highest melt fraction observed at 1·8 GPa (1800 MPa) confining pressure. Deformation is macroscopically ductile with no throughgoing faults below the solidus (650 and 750°C). Left-hand panel: as melting begins (800–850°C), micrometre-wide fractures brittlely deform the sample and melt is found along some of these cracks (plag, plagioclase; hbd, hornblende). Middle panel: melt is found in dilatant cracks in the hornblende (partially reacted to clinopyroxene, cpx) and in cataclastic shear zones which transect the sample (see also Fig. 2 for detail). Product phases such as garnet (gar), new albitic plagioclase, titanite and zoisite are also present. Right-hand panel: at 20 vol% melt and greater (in the centre of the sample at 1000°C), deformation is again ductile and no throughgoing fracture is observed. Garnet grains have increased in size, new albitic plagioclase (new plag) and zoisite prisms are present in melt pools (see also Fig. 3). (b) Interpretation by Rutter and Neumann (1995; modified from their fig. 9) of the evolution of the microstructures observed in the experiments on partially molten Westerly Granite at 0·25 GPa (250 MPa) confining pressure. Left-hand panel: at the lower confining pressure, deformation is brittle below and above the solidus. As the melt fraction increases, dilatant melt-filled cracks, which lead to faulting, are observed. Middle panel: cataclastic deformation dominates the sample. The formation and collapse of melt-filled pores leads to the proposed shear-enhanced compaction model for moving melt in this regime. Right-hand panel: a transition between fracture and viscous flow is observed in these experiments at melt fractions ≥40 vol%, which is a higher melt fraction than that observed in the mafic experiments at higher pressures (≥20 vol%).

suggest that the segregation and migration of melt from the grain scale to crustal scale is controlled by several variables, mainly melt viscosity, deformation rate, mineral melting reactions and, when applicable, interfacial energies. The hydrostatic experiments show that melt interconnectivity on the grain scale can be readily achieved in crustal assemblages, as the dihedral angles are low due to the high ratio of grain boundary energy to the solid–melt interfacial energy. Melt movement on the grain scale is also controlled by other factors besides surface energy, such as melt hydrofracturing when the melting rates are high. The deformation experiments provide information on melt segregation processes at low melt fractions and show that the efficient extraction of viscous melt can occur, thus changing our view of a 'critical' melt fraction for the segregation of granite.

On the grain scale, the viscosity of the melt may initially inhibit melt movement under hydrostatic conditions, even if interconnectivity is achieved (Laporte & Watson 1995). Viscosity is a function of the type of reaction, fluid-absent reactions producing melts of higher viscosity. However, these reactions induce fractures more readily than wet melting because of the larger positive volume changes. This can increase the permeability of the rock. Even if higher viscosity melts cannot escape as quickly as wetter melts under hydrostatic conditions, deformation of the source rock will produce pressure gradients which can get the melt to move along the developed fracture network. Local changes in melt distribution may also create pressure gradients capable of inducing melt migration (see later). The depth of melting will also place constraints on melt segregation because pressure and temperature determine how minerals themselves deform, either by crystal plasticity at deeper levels in the crust or by more brittle deformation at shallower levels.

The link between grain and outcrop scale fracture networks and large-scale melt migration from the lower crust needs to be better understood. Rushmer *et al.* (1995) have begun to develop two-dimensional heat–melt flow models to determine the major factors that control melt fracturing at different scales in the crust, based on earlier work by Connolly and Ko (1995) concerning the evolution of metamorphic fluid pressure and fluid flow. They find that the transition from grain-scale to large-scale melt hydrofracturing in an otherwise hydrostatic environment will be dictated by three interdependent time-related processes. Firstly, once large-scale failure occurs, drainage of melt into the fracture will reduce the overpressure in the vicinity of the fracture. Secondly, the local reduction in melt overpressure can, in turn, raise the rate of melt production and increase the pressure gradients responsible for driving melt flow into the fracture. As the fracture grows, the lower portion of the crack becomes underpressured, causing compaction of the permeable microcrack network responsible for feeding melt into the fracture. Thirdly, the fracture becomes isolated from the melt source when the dilational strain rate within the fracture exceeds the volumetric rate of melt supply. The experimentally determined microfracture permeabilities from the muscovite–quartz experiments are currently being developed to determine the fracture dimensions and spacing for which this will occur (Rushmer *et al.* 1995).

Larger scale processes, such as deformation rates in the lower crust, also have a significant effect on melt segregation. Slower rates cause wet melts to be passively squeezed from grain boundaries, but may allow them to accumulate to high melt fractions locally. Slower strain rates in the crust, however, enhance the impact of the rate of melting in fracturing the source rock. Fast rates of melting cause a rapid build up of pore pressure and reduce the strength of the rock by inducing embrittlement, even if the strain rates are slow. In contrast, fast rates of deformation induce large-scale fracturing at a much lower melt fraction. Clearly, scaling of experiments to the lower crust is a significant problem. Currently, the style of fracturing on the sample scale is considered to be transferable to larger scales; the experimental results will provide constraints in developing more realistic numerical models.

Most importantly, the tectonic environment determines the melt fraction at which segregation takes place because the rate at which the crust is heated and deformed, in addition to the pressure and temperature of the potential melting reactions, are dictated by the imposed conditions. Melt migration can occur at a variety of different melt fractions depending on the tectonic environment, which, in turn, has significant implications for the resulting geochemistry of the magmas themselves.

3.1. Melt migration and melt geochemistry

Future work must tie in geochemical data with physical melt segregation models. Information is required, however, on the partitioning of trace and rare earth elements (REEs) during hydrostatic melting and during deformation. Fortunately, we are now able to obtain, with new ion probe technology and experiments, some of these necessary data.

In an experimental study by Cavallini *et al.* (1995) major and trace element data have been collected using secondary ion mass spectrometry from a partially molten metapelite under equilibrium conditions. At 850°C and 0·85 GPa, the melt is enriched in LREEs (La/Sm), depleted in HREEs (Gd/Yb = 8·3) and is very fractionated (La/Yb = 103·2), reflecting the presence of garnet in the residual assemblage. Also

present in the assemblage are sillimanite, alkali feldspar, quartz, ilmenite and titanite. The melt is also enriched in Rb and K, but depleted in Ba, Sr and Y and has an Eu anomaly, as expected in an equilibrium situation. Ion probe data have also been collected from the deformed partially molten amphibolite sample at 1·8 GPa and 1000°C which contains garnet + clinopyroxene + plagioclase + trondhjemitic glass (Rushmer *et al.* 1994). The accessory phases, titanite and zoisite were also analysed. The REE pattern of the trondhjemitic melt is steep (La/Sm = 4·8–7·5; Sm/Yb = 16–20) owing to the presence of garnet, but the high field strength elements (HFSEs) have unpredicted bulk distributions. The mass balance between the experimental charges and the starting composition shows that the abundances of some elements are less than unity for the major residual assemblages in the experiments. This means, theoretically, that the melt phase should be enriched in these elements relative to the starting composition, but the data show that the melt is not enriched in important elements, such as the HFSEs. This behaviour is because the minor phases, in particular titanite, have not completely dissolved and have retained many of the elements which should be in the melt. In general, the data show that major trends, such as a depletion in Y and HREEs when garnet is present, are as expected. However, even though local major element chemical equilibrium appears to have been achieved in the melted rock cores, HFSEs such as Zr are not fully equilibrated. This suggests that the slow dissolution rates of accessory phases such as apatite and titanite may easily result in extracted melt compositions which have lower abundances of these elements than predicted by equilibrium partition coefficients.

Studies of orogenic granites (Deniel *et al.* 1987; Barbero *et al.* 1995; Harris *et al.* 1995) have shown evidence of chemical disequilibrium. In the study of Harris *et al.* (1995), low-temperature fluid-absent melting of muscovite appears to have produced some of the leucogranitic magmas of the Himalayas. These results have found that melt did achieve some level of equilibration (e.g. Sr), but further show that detailed work on product phases (such as garnet) revealed some disequilibrium between the melt and restite with respect to Y and the HREEs. As a result, the LREE/HREE ratios are low. This suggests that in some crustally derived granites low LREE/HREE ratios may not be due to a garnet-free source, but to slow diffusion rates of REEs at the low temperature end of crustal melting. These kinds of results, combined with the dissolution rates of accessory phases such as apatite (Harrison & Watson 1984; Wolf & London 1995), monazite (Rapp & Watson 1986; Montel 1993) and zircon (Harrison & Watson, 1983) in felsic compositions, provide important constraints on the rates of melt formation and segregation in the lower crust. Further work on crustal granites must consider the dissolution rates and diffusion rates of trace elements given the lower temperatures of most crustal melting, plus the mode of melt segregation so geochemical modelling can be used most effectively.

4. Conclusions

As we continue to update our geochemical and physical models of melt segregation and migration, our overall understanding of the diverse possibilities of granite generation and potential emplacement mechanisms in the earth's crust will advance. Future melt segregation models must consider both permeability/porosity changes that occur during melting and the rate of melting, which is a major factor in increasing permeability during reaction. These variables need to be quantified by specific key experiments. When melting rates are relatively fast, such as during magmatic underplating, transient periods of increased permeability may be produced in previously impermeable rock. This process may be one of the most important first steps in the development of the fracture propagation mechanism for melt segregation. During deformation-enhanced melt segregation in the lower crust, REEs may have lower abundances than predicted by equilibrium partition coefficients because of the incomplete dissolution of accessory phases.

5. Acknowledgements

I thank the Department of Geology at University of Vermont for continued support since March 1995. Schweizerischer Nationalfonds project 2-77-590-92 funded the amphibolite experimental work in Dr Steve Kirby's rock deformation laboratory at the U.S.G.S. in Menlo Park. I thank Kim Hannula at Middlebury College, Vermont for use of the SEM facility and Djordje Grujic and Dorotea Dietrich at the Department of Geology, ETH-Zurich, Switzerland for their help with the SEM facility there. Helpful discussions with Jamie Connolly, Mike Brown, Dave Rubie, Afina Lupulescu and Barry Doolan are gratefully acknowledged. Very constructive reviews by Didier Laporte, Simon Inger and Barry Doolan greatly improved the manuscript.

6. References

Arzi, A. A. 1978. Critical phenomena in the rheology of partially melted rocks. TECTONOPHYSICS **44**, 173–84.

Barbero, L., Villaseca, C., Rogers, G. & Brown, P. E. 1995. Geochemical and isotopic disequilibrium in crustal melting: an insight from the anatectic granitoids from Toledo, Spain. J GEOPHYS RES **100**, 15,745–66.

Brearley, A. J. & Rubie, D. C. 1990. Effects of H$_2$O on the disequilibrium breakdown of muscovite+quartz. J PETROL **31**, 925–56.

Brown, M., Averkin, Y. A., McLellan, E. L. & Sawyer, E. W. 1995a. Melt segregation in migmatites. J GEOPHYS RES **100**, 15,655–80.

Brown, M., Rushmer, T. & Sawyer, E. W. 1995b. Segregation of melts from crustal protoliths: mechanisms and consequences. J GEOPHYS RES **100**, 15,551–64.

Cavallini, M., Vielzeuf, D., Bottazzi, P., Mazzucchelli, M. & Ottolini, L. 1995. Direct measurement of rare earth contents in partial melts from metapelites. TERRA ABSTR **7**, 344–5.

Clemens, J. D. & Mawer, C. K. 1992. Granitic magma transport by fracture propagation. TECTONOPHYSICS **204**, 339–60.

Connolly, J. A. D. & Ko, S.-C. 1995. Development of excess fluid pressure during dehydration of the lower crust. TERRA ABSTR **7**, 312.

Davidson, C., Schmid, S. M. & Hollister, L. S. 1994. Role of melt during deformation in the deep crust. TERRA NOVA **6**, 133–42.

Dell'Angelo, L. N. & Tullis, J. 1988. Experimental deformation of partially melted granitic aggregates. J METAMORPH GEOL **6**, 495–516.

Deniel, C., Vidal, P., Fernandez, A., Le Fort, P. & Peucat, J. J. 1987. Isotopic study of the Manaslu granite (Himalaya, Nepal): inferences of the age and source of Himalayan leucogranites. CONTRIB MINERAL PETROL **96**, 78–92.

Faul, U. H., Toomey, D. R. & Waff, H. S. 1994. Intergranular basaltic melt is distributed in thin, elongated inclusions. GEOPHYS RES LETT **21**, 29–32.

Hacker, B. 1990. Amphibolite-facies-to-granulite-facies reactions in experimentally deformed, unpowdered amphibolite. AM MINERAL **75**, 1349–61.

Hacker, B. & Christie, J. M. 1990. Brittle/ductile and plastic/cataclastic transitions in experimentally deformed and metamorphosed amphibolite. In: Durham, W., Duba, A., Handin, J. and Wang, H. (eds) *Brittle–ductile transitions. The Heard volume.* GEOPHYS MONOGR SER AM GEOPHYS UNION **20**, 127–47.

Hanson, R. B. 1995. The hydrodynamics of contact metamorphism. GEOL SOC AM BULL **107**, 595–611.

Harris, N., Ayres, M. & Massey, J. 1995. Geochemistry of granitic melts produced during the incongruent melting of muscovite: implications for the extraction of Himalayan leucogranite magmas. J GEOPHYS RES **100**, 15, 767–78.

Harrison, T. M. & Watson, E. B. 1983. Kinetics of zircon dissolution and zirconium diffusion in granitic melts of variable water content. CONTRIB MINERAL PETROL **84**, 66–72.

Harrison, T. M. & Watson, E. B. 1984. The behavior of apatite during crustal anatexis: equilibrium and kinetic considerations. GEOCHIM COSMOCHIM ACTA **48**, 1468–77.

Jurewicz, S. R. & Watson, E. B. 1984. Distribution of partial melt in a felsic system: the importance of surface energy. CONTRIB MINERAL PETROL **85**, 25–9.

Jurewicz, S. R. & Watson, E. B. 1985. The distribution of partial melt in a granitic system: the application of liquid phase sintering theory. GEOCHIM COSMOCHIM ACTA **49**, 1109–22.

Laporte, D. 1994. Wetting behavior of partial melts during crustal anatexis: the distribution of hydrous silicic melts in polycrystalline aggregates of quartz. CONTRIB MINERAL PETROL **116**, 489–99.

Laporte, D. & Watson, E. B. 1995. Experimental and theoretical constraints on melt distribution in crustal sources: the effect of crystalline anisotropy on melt interconnectivity. CHEM GEOL **124**, 161–84.

Lupulescu, A. & Watson, E. B. 1995. Tonalitic melt connectivity at low-melt fraction in a mafic crustal protolith. EOS, TRANS AM GEOPHYS UNION **76**, 299–300.

McKenzie, D. 1985. The extraction of magma from the crust and mantle. EARTH PLANET SCI LETT **74**, 81–91.

Miller, C. F., Watson, E. B. & Harrison, T. M. 1988. Perspectives on source, segregation and transport of granitic magms. TRANS R SOC EDINBURGH **79**, 135–56.

Montel, J. M. 1993. A model for monazite/melt equilibrium and application to the generation of granitic magmas. CHEM GEOL **110**, 127–46.

Paquet, J., Francois, P. & Nedelec, A. 1981. Effect of partial melting on rock deformation: experimental and natural evidence on rocks of granitic composition. TECTONOPHYSICS **78**, 545–65.

Petford, N. 1995. Segregation of tonalitic–trondhjemitic melts in the continental crust: the mantle connection. J GEOPHYS RES **100**, 15,735–44.

Raleigh, C. B. & Paterson, M. S. 1965. Experimental deformation of serpentine and its tectonic implications. J GEOPHYS RES **70**, 3965–85.

Rapp, R. P. & Watson, E. B. 1986. Monazite solubility and dissolution kinetics: implications for the thorium and light rare earth chemistry of felsic magmas. CONTRIB MINERAL PETROL **94**, 304–16.

Rubie, D. C. & Brearley, A. J. 1990. A model for rates of disequilibrium melting during metamorphism. In: Ashworth, J. R. & Brown, M. (eds) *High temperature metamorphism and crustal anatexis*, 57–86. London: Unwin Hyman.

Rushmer, T. 1993. Experimental high-pressure granulites: some applications to mafic xenoliths and Archean terranes. GEOLOGY **21**, 411–4.

Rushmer, T. 1995. An experimental deformation study of partially molten amphibolite: application to low-fraction melt segregation. J GEOPHYS RES **100**, 15,681–96.

Rushmer, T., Pearce, J. A., Ottolini, L. & Bottazzi, P. 1994. Trace element behavior during slab melting: experimental evidence. EOS, TRANS AM GEOPHYS UNION **75**, 746.

Rushmer, T., Rubie, D. C. & Connolly, J. A. D. 1995. Melt-induced fracturing as a function of rate and time: implications for melt migration at the onset of reaction. GEOL SOC AM ANNU MEET ABSTR PROGRAMS **27**, 431.

Rutter, E. & Neumann, D. 1995. Experimental deformation of partially molten Westerly granite under fluid-absent conditions with implications for the extraction of granitic magmas. J GEOPHYS RES **100**, 15,697–715.

Sawyer, E. W. 1991. Disequilibrium melting and the rate of melt–residuum separation during migmatization of mafic rocks from the Grenville Front, Quebec. J PETROL **32**, 701–38.

Shaw, H. R. 1972. Viscosities of magmatic silicate liquids: an empirical method of prediction. AM J SCI **272**, 870–93.

Sinigoi, S., Quick, J. E., Clemens-Knott, D., Mayer, A., Demarchi, G., Mazzucchelli, M., Negrini, L. & Rivalenti, G. 1994. Chemical evolution of a large mafic intrusion in the lower crust, Ivrea-Verbano Zone, northern Italy. J GEOPHYS RES **99**, 21, 575–90.

van der Molen, I. & Paterson, M. S. 1979. Experimental deformation of partially-melted granite. CONTRIB MINERAL PETROL **98**, 7–22.

Vicenzi, E. P., Rapp, R. P. and Watson, E. B. 1988. Crystal/melt wetting characteristics in partially molten amphibolite. EOS, TRANS AM GEOPHYS UNION **69**, 482.

Wickham, S. M. 1987. The segregation and emplacement of granitic magmas—some examples from the Pyrenees. J GEOL SOC LONDON **144,** 281–97.

Wolf, M. B. & London, D. 1995. Incongruent dissolution of REE- and Sr-rich apatite in peraluminous granitic liquids: differential apatite, monozite and xenotime solubilities during anatexis. AM MINERAL **80,** 765–75.

Wolf, M. B. & Wyllie, P. J. 1995. Liquid segregation parameters from amphibolite dehydration melting experiments. J GEOPHYS RES **100,** 15,611–22.

TRACY RUSHMER, Department of Geology, Perkins Hall, University of Vermont, Burlington, VT 054505-0122, U.S.A.

Transactions of the Royal Society of Edinburgh: Earth Sciences, **87,** 85–94, 1996

Melt segregation and magma flow in migmatites: implications for the generation of granite magmas

E. W. Sawyer

ABSTRACT: To form a granite pluton, the felsic melt produced by partial melting of the middle and lower continental crust must separate from its source and residuum. This can happen in three ways: (1) simple melt segregation, where only the melt fraction moves; (2) magma mobility, in which all the melt and residuum move together; and (3) magma mobility with melt segregation, in which the melt and residuum move together as a magma, but become separated during flow. The first mechanism applies to metatexite migmatites and the other two to diatexite migmatites, but the primary driving forces for each are deviatoric stresses related to regional-scale deformation. Neither of the first two mechanisms generates parental granite magmas. In the first mechanism segregation is so effective that the resulting magmas are too depleted in FeO_T, MgO, Rb, Zr, Th and the REEs, and in the second no segregation occurs. Only the third mechanism produces magmas with compositions comparable with parental granites, and occurs at a large enough scale in the highest grade parts of migmatite terranes, to be considered representative of the segregation processes occurring in the source regions of granites.

KEY WORDS: anatexis, diatexite, melt–residuum separation, metatexite, restite.

Since the early studies of water-saturated melting (Winkler & von Platen 1961), a large body of experimental data has been gathered to show that anatexis of a wide range of crustal lithologies, e.g. pelite, greywacke, metabasite, granite and tonalite, produces granitic melt in the middle and lower continental crust. In contrast with earlier work, it is now believed that to have generated the large volumes of granite found in the upper crust, most melting must have occurred at temperatures in excess of 850°C, and under low a_{H_2O} conditions (Le Breton & Thompson 1988; Vielzeuf & Holloway 1988; Stevens & Clemens 1993). Although much is now known about the nature of the sources and conditions of melting during granite formation, comparatively little is known about the processes by which granitic melts and magmas separate from their residuum and their source rocks.

The transition from high-grade metamorphic rocks through migmatite to granite with increasing temperature proposed by Read (1957) is intuitively attractive. Because of the position migmatites occupy in this scheme, they have received much attention as analogues of the processes occurring in the source regions of granites. Progress has been made in understanding melt segregation at small scales, such as in migmatite leucosomes (Wickham 1987a; Sawyer 1991, 1994; Brown *et al.* 1995; Williams *et al.* 1995). However, a clarification of if, and how, these results apply to the large-scale problem of the generation of granite magmas is needed, because the intermediate position of migmatites in the origin of granites has now been questioned (Le Breton & Thompson 1988; White & Chappell 1990).

Migmatites display a wide range of morphologies—some of which are under-reported in published work—so it seems premature to deny a link between migmatites and granites. A pertinent question is therefore which migmatites might represent the source regions of granites. To answer this, the approach will be first to examine what conditions are necessary for the separation of melt from its residuum and the formation of a granite magma. The field characteristics of migmatites are

outlined to determine: (1) in which types of rock suitable conditions existed at the required scale for the formation of granite magmas; and (2) which migmatites have suitable melt, or magma, compositions to be the source material of parental granite magmas. The final section examines which driving forces and mechanisms may apply to the formation of granite magmas.

1. Conditions for melt segregation and magma mobility

The essential step in passing from a partially molten protolith to a granite magma is the physical separation of a large volume of the melted fraction from its residuum and source. This magma can then rise through the continental crust. Here, melt segregation is defined as the separation of the melt fraction from its residuum and any early formed crystals. In contrast, magma mobility is defined as the movement of the melt fraction plus entrained crystallisation products and residuum. Melt segregation and magma mobility are considered in terms of the relative movement of the principal components—melt and residuum—because these are the most easily observed features in terranes where anatexis has occurred.

For either melt segregation or magma mobility to occur there must be a differential movement between the melt and residuum, or between melt plus entrained crystals (in magma) and some pre-melting reference frame in the source region. This differential movement can be considered in terms of the velocity of the melt-rich part, (v_m), and the residual solids (v_s). The condition $v_m = v_s = 0$ is an end-member case in which both the melt fraction and the solid residuum remain at the site of melting; consequently, there is neither melt segregation nor magma mobility. This condition prevails at the onset of melting when small volumes of melt, a few centimetres across, form patch migmatites (Fig. 1a) and, at larger volumes, nebulitic migmatites. Some mantle–core structures in granulites, such as melt rims on garnet or orthopyroxene cores, are

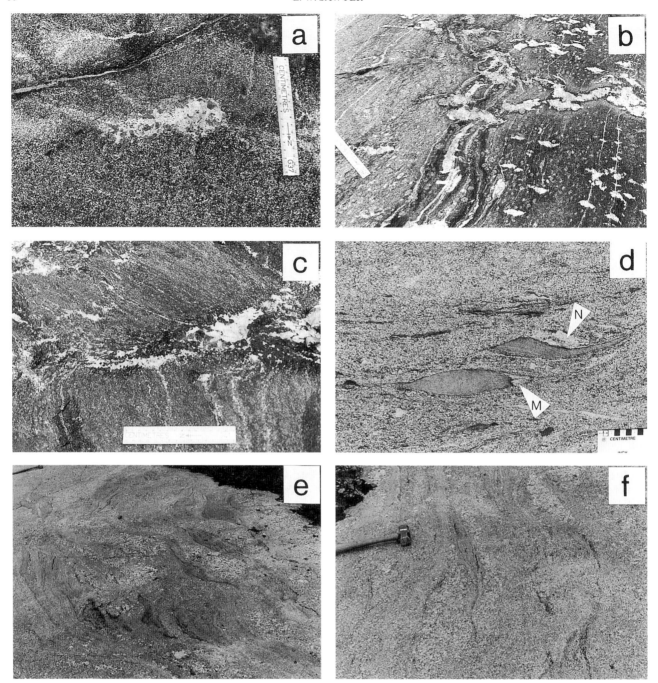

Figure 1 Appearance of migmatites with increasing metamorphic grade. (a) Initial melting yields a migmatite with small patches of unsegregated melt and residuum. (b) Typical metatexite migmatite with a small melt volume; note structurally controlled leucosomes and the well-preserved pre-migmatisation structure. (c) Typical metatexite leucosome–melanosome pair located in a small shear band. (d) Diatexite migmatite containing magma flow structures with melt segregation (M) at the tail of enclaves and at other sites of heterogeneities in the flow (N); note lack of preserved pre-migmatisation structures. (e) Diatexite with asymmetrical enclaves indicating the sense of magma flow; note also coarse-grained, cross-cutting and banding-parallel melt segregations with diffuse borders. (f) Wispy banding in diatexite due to segregation of residuum crystals (mostly biotite) during magma flow; note also subtle variations in mafic mineral (residuum) content within the coarser grained parts. Scale is 15 cm long and hammer handle is 50 cm long. Metatexites derived from mafic rocks in (a), (b) and (c) are from the Central Metasedimentary Belt of the Grenville Province near Gracefield, Quebec; the diatexite derived from metapelite in (d) is from the Quetico Subprovince near Atikokan, Ontario; and the diatexites derived from leucotonalites in (e) and (f) are from the Opatica Belt, Superior Province, Quebec.

also examples. It is unlikely that $v_m = v_s = 0$ can be maintained at large (greater than metre) scales unless the melt fraction present is below that required for permeability. At higher melt fractions deformation will create instabilities that initiate and drive melt segregation or magma mobility, thus $v_m > 0$. There are three cases which lead to melt segregation or magma mobility.

1. $v_m > v_s = 0$. Simple melt segregation in which the residuum

remains *in situ*, but the melt moves. This describes those migmatites that have leucosomes (former melt) with adjacent melanosomes (residuum). Published accounts of leucosomes formed under this condition indicate that the segregation process was particularly effective at separating melt from residuum; typically, the leucosomes contain very few crystals of the major residuum phases (Sawyer 1991; Whitney & Irving 1994), but the entrainment of small accessory phase crystals, such as zircon, is common (Gupta & Johannes 1985; Brouand et al. 1990). Thus granite magmas formed where this condition prevailed might be recognised by their very leucocratic nature.

2. $v_m = v_s > 0$. This constitutes magma mobility without melt segregation. Such a magma consists of the melt fraction plus all of the residuum and any crystallisation products. It is unlikely that this condition can remain stable; heterogenieties in the magma flow regime will soon develop and generate pressure and velocity gradients that will drive melt segregation. Nevertheless, some of the Quetico (Sawyer & Barnes 1988) and Ashuanipi (Percival 1991) migmatites that have high melt fractions are examples of this type of magma mobility.

3. $v_m > v_s > 0$. This is magma mobility with melt segregation. Both the melt and the residuum move (magma mobility), but because the melt fraction has the greater velocity, melt segregation also occurs. Regions where this condition applied should show structures indicating both bulk flow and segregation of the melt from its residuum. Furthermore, the parts showing flow structures should also show variable degrees of residuum separation; some parts will contain a larger residuum fraction than others, although in general much of the residuum remains in, or near to, the source region.

1.1. How much melt is needed?

Although the conditions just outlined provide a framework for describing melt segregation and magma mobility, they tell us nothing about when these processes begin, or what drives them. Melt segregation can begin once there is enough melt in the matrix to form an interconnected network; the melt fraction needed for this has been termed the permeability threshold (Maaløe 1982) and the first percolation threshold (FPT) by Vigneresse et al. (1991). Estimates of the melt fraction at the FPT range from <1% for basalt melts in the mantle (Daines & Richter 1988) to about 5% for granitic melt in a partially molten aplite (Dell'Angelo & Tullis 1988).

An estimate of the FPT for felsic melts in migmatites formed under condition 1 is given by the degree of melting (F), provided the leucosome–melanosome–palaeosome set is a closed system; the lowest value of F provides an upper limit on the FPT. Estimates of F range from 0·1 to 0·64 for such migmatites, hence melt segregation can begin when melt fractions are fairly small (<10%).

In contrast, magma mobility (conditions 2 and 3) can only occur when there is sufficient melt present that the matrix framework breaks down—a limit variously termed the rheologic critical melt percentage (Arzi 1978), the critical melt fraction (Wickham 1987a) and the second percolation threshold (SPT) by Vigneresse et al. (1991). The principal factors that determine the SPT for a particular case of melting include: (1) the range of grain sizes and shapes in the palaeososome: (2) the mineralogy (Brown et al. 1995); (3) the presence or absence of a shape-preferred orientation in the rock (Nicolas et al. 1993); and (4) the presence or absence of shearing stresses during melting (Nicolas et al. 1993; Sawyer 1994). Melting experiments on Westerly granite (Rutter & Neumann 1995) suggest that >40% melt must be present for magma flow to occur in rocks of low anisotropy. In contrast, Nicolas et al. (1993) report that in systems with a strong shape anisotropy undergoing shearing, magma mobility can occur with <20% melt and perhaps even with <10% melt present. Estimates of F from migmatites showing evidence of magma flow are few, but could place an upper limit on the melt fraction at the SPT in natural melting. Migmatites formed from strongly foliated leucotonalites in the Opatica Belt of

Canada (Sawyer & Benn 1993) show abundant evidence of magma mobility and in most cases record values of F between 0·34 and 0·70, i.e. broadly in accord with the results obtained by Rutter and Neumann (1995). However, some of the Opatica migmatite samples experienced magma mobility when they had undergone less than 20% melting (Sawyer unpublished data)—a value similar to that proposed by Nicolas et al. (1993). Clearly, there is no fixed value of the SPT and considerable variation can be expected from place to place within the source regions of granite magmas, depending on local factors. In particular, the effect that anisotropy due to strong shape-preferred orientation and layering/foliation, two common properties of high-grade rocks, has on magma mobility during deformation requires further study.

It could be argued that, during shearing of migmatites containing a melt fraction between the FPT and the SPT, the higher strain rates due to melt-aided diffusive mass transfer (Dell'Angelo & Tullis 1988) or grain boundary sliding (Sawyer 1994) generates flow structures similar to those produced during magma mobility. However, the grain-scale textures are not like those of a crystallised magma. Because of the presence of the solid matrix framework during shearing, the grains should show evidence of deformation by intracrystalline plasticity. Moreover, melt fractions between the FPT and SPT may be transient in a shearing matrix. Shearing will drive melt segregation (Sawyer 1994; Rutter & Neumann 1995) and drain the matrix, resulting in a rock containing a melt fraction equal to the FPT, but with scattered domains (the leucosomes or veins) where the melt has collected. The melt fraction can only build up in the matrix if the melting rate exceeds the extraction rate.

2. The metatexite–diatexite division in migmatites

In this section, field observations are used to argue that a major rheological change takes place within migmatite sequences. Extensive mapping in the English River (Breaks et al. 1978), St Malo (Brown 1979; Martin 1980), Trois Seigneurs (Wickham 1987b; Vielzeuf, pers. comm.), the Quetico (Sawyer & Barnes 1988), Higo (Obata et al. 1994) and Opatica (Sawyer, unpublished data) migmatites show a similar, two-stage, systematic change with increasing metamorphic grade.

Typically, there is a lower grade part where the leucosomes representing crystallised melt are small (less than several thousands of cubic centimetres), and, although locally numerous, constitute only a small fraction of the rock volume. In this zone the pre-migmatisation fabric (layering, bedding, foliation or even folding) in the palaeosome is preserved between the leucosomes and is largely unaffected by the migmatisation (Fig. 1b). The leucosomes can have a wide range of morphologies, but basically there are two types: (1) more or less in situ leucosomes that have mafic selvedges (melanosomes) associated with them (Fig. 1c)—these leucosomes are typically, but not exclusively, located parallel to pre-existing planes of anisotropy (layering, foliation or bedding), or in syn-melting deformation structures, such as boudin necks, shear bands or fold closures; and (2) leucosomes consisting of melt that was injected from external sources and which form discordant (Sawyer 1987) or layer-parallel (Collins & Sawyer in press) veins—typically, these leucosomes lack mafic selvedges.

In contrast, the highest grade part of the migmatite terrane is marked by a wholesale textural modification resulting in a more homogeneous appearance. This involves a coarsening of grain size and the development of flow structures that overprint and destroy the pre-existing structures (Fig. 1d, 1e and 1f). Pre-migmatisation structures, if preserved at all, occur in small xenolith-like enclaves within a more homogeneous host (Sawyer & Barnes 1988). Leucosomes are less abundant in the higher grade zone and are of a different appearance to those in the lower grade migmatites. There are three main types: (1) small leucosomes located in low pressure sites within the magma flow structure (Fig. 1d); (2) veins or sheets with diffuse borders (Fig. 1e)—these can be both concordant or discordant to the flow banding; and (3) discordant veins with sharp boundaries that post-date the textural homogenisation and flow structures (Sawyer & Barnes 1988). In addition, there is a considerable variation in the proportion of residuum to melt from place to place in the migmatite; concentrations of residual mafic minerals produce characteristic mafic schlieren oriented parallel to the flow (Fig. 1f).

Thus there are two distinct parts in many migmatite terranes: a lower grade part where the palaeosome structure is preserved and the texture is metamorphic and a higher grade part showing widespread textural modification and the development of flow structures; these correspond to metatexite and diatexite migmatites, respectively, as described and defined by Brown (1973). In metatexites the bulk of the rock, where the pre-migmatisation structures are preserved, has not melted. Only a few layers or domains have melted, although in some instances these may have experienced fairly high degrees of partial melting (F between 0·4 and 0·6). In contrast, melting in diatexites is pervasive and affects almost the whole rock volume and, although the degree of partial melting may be higher, in some instances it is not (e.g. the Opatica diatexites where $F = 0·4$). Therefore the crucial factor in the transition from metatexite to diatexite is the development of a pervasive melt fraction throughout the whole rock volume as suggested by Brown (1973: 375) and not simply a higher degree of partial melting. The presence of pervasive melt enables two important processes to occur: (1) the coarsening of grain size and development of igneous textures (e.g. well-developed crystal faces on feldspar crystals) through more rapid diffusion of components in the melt and crystallization from the melt; and (2) the bulk flow (magma mobility) of melt and residuum by destroying the contiguity of the solid matrix framework. Thus the easily mapped transition from metatexite to diatexite marks a fundamental change in crustal rheology—at lower grades the crust has the rheology of solid rock, but across the transition the higher grade diatexites have the rheology of a magma.

Recalling the three conditions for melt segregation and magma mobility, the first applies exclusively to metatexites, but the second and the third (especially) apply to diatexites. From the aspect of melt separation, regions of diatexite migmatite are more likely to represent the source regions of granitic magmas than metatexites because the physical conditions there are conducive to the large-scale formation and movement of magma and the segregation of melt. Moreover, the scale of some diatexite terranes is comparable with plutonic belts (excluding plutonic arcs)—for example, the $> 30\,000$ km^2 area of the Ashuanipi diatexites and the even larger Minto diatexites (Percival 1991). Unfortunately, most published work on migmatites examines the origin of the leucosomes and melt segregation in metatexites.

2.1. Conditions of metatexite and diatexite formation
The metatexite and diatexite migmatites at Trois Seigneurs (Wickham 1987b), St Malo (Brown 1979, 1994) and the Opatica Belt (Sawyer, unpublished data) formed under amphibolite facies conditions where a_{H_2O} was externally buffered to a high value. Wickham (1987b) argues for the infiltration of water into the pelites as the cause of widespread melting at Trois Seigneurs. The presence of biotite, and the high degree

of melting at comparatively low temperatures (750–800°C) also suggest the presence of a water-rich volatile phase during melting in the Opatica Belt. For the upper amphibolite facies metatexites and diatexites in the English River (Breaks et al. 1978) and Quetico (Sawyer 1987) subprovinces, the initial stages of melting probably occurred in the presence of a water-rich volatile phase, but the bulk of melting occurred at progressively lower a_{H_2O} conditions, first with the breakdown of minor muscovite and then biotite. The Higo migmatite terrane contains the amphibolite–granulite facies transition with the 'orthopyroxene-in' isograd located in the metatexites, thus the diatexites represent low a_{H_2O} melting due to biotite breakdown (Obata et al. 1994). In many migmatite terranes both the metatexites and the diatexites are within the granulite facies and formed at low a_{H_2O}, e.g. Wuluma complex (Collins et al. 1989), Ashuanipi subprovince (Percival 1991; Lapointe & Chown 1993) and the Minto subprovince (Percival et al. 1990).

Clearly, diatexite migmatites have formed under a wide range of a_{H_2O} conditions, from water-rich volatile phase present to volatile phase absent. Given a fertile source, then the volume of melt formed depends on the availability of water and the temperatures reached. If water infiltrates the source rocks, then pervasive melting can occur at relatively low temperatures (amphibolite facies diatexites), but if water is only available through the breakdown of hydrous phases, then high temperatures ($> 850°C$) are required for pervasive melting (granulite facies diatexites). Although it is widely accepted that melting in the granulite facies leads to voluminous granite production (Stevens & Clemens 1993), amphibolite facies melting at high water activity can also, e.g. the Mancellian granites derived from the St Malo migmatites (Brown 1994).

3. Comparison of the compositions of migmatites and granites

3.1. Metatexite melts and parental granite magmas
The compositions of melts from metatexite migmatites used here includes only those leucosomes which have associated melanosomes and are believed to have formed in situ; this is to exclude injected exotic melts. The 93 leucosomes in the database come from many types of palaeosome, including clastic metasediments, metapelites, metabasites, felsic metavolcanics, granodiorites, tonalites and intermediate orthogneisses. Unfortunately, few of the samples include a full set of trace element determinations.

A large part of the compositional variation in granitic rocks has been attributed to three effects: fractional crystallisation (McCarthy & Groves 1979; Wall et al. 1987), restite separation (White & Chappell 1977) and magma mixing (Reid et al. 1983; Poli et al. 1989). Much of the published compositional data on granites represents either fractionated liquids or hybrid magmas and not the initial, or parental, magma composition. The data set compiled for this work includes only the compositions for granite magmas reported to be primary, i.e. parental, and derived by melting of the continental crust. It excludes oceanic plagiogranites and the fractionated felsic melts derived from the crystallisation of mafic magmas. The data set is not exhaustive (116 samples), but covers the major tectonic settings where granites occur, as outlined by Pearce et al. (1984), for example. It is believed to cover the likely range of parental granite magma compositions and so allow a meaningful comparison with migmatite melts from which general conclusions can be drawn.

Reported SiO_2 ranges in metatexite-derived melts are similar to those in parental granite magmas, but the distribution of

SiO_2 contents differs, despite a similarly wide range of protoliths. All but six of the metatexite melts have silica contents of between 71 and 77 wt%. Five of the six samples with <70 wt% SiO_2 are reported to contain fragments of melanosome material (Sawyer 1991; Whitney & Irving 1994). In contrast, parental granite magmas show a uniform distribution between 66 and 76 wt% SiO_2 (Fig. 2a). Melts from metatexites have lower $FeO_T + MgO$ (Fig. 2a) contents (also

TiO_2) than the parental granite magmas, although the compositional field of metatexite melts does overlap that of the most siliceous parental granite magmas. A significant point is that the low-SiO_2, residuum-enriched metatexite melts overlap the low-SiO_2 portion of the granite field—a point that will be discussed later. The K_2O contents of parental granitic magmas rises systematically from 1.5 to 6 wt% as SiO_2 increases, but there is no systematic increase in K_2O in the

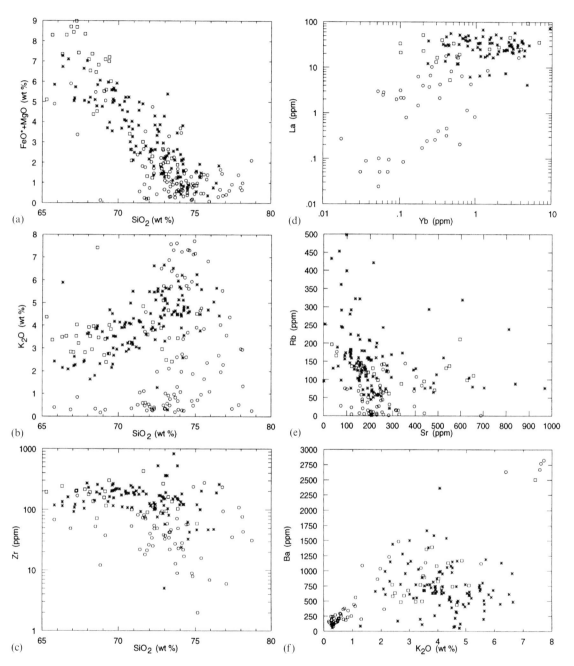

Figure 2 Selected element variation diagrams, (a) $FeO_T + MgO$ versus SiO_2, (b) K_2O versus SiO_2, (c) Zr versus SiO_2, (d) La versus Yb, (e) Rb versus Sr and (f) Ba versus K_2O, comparing parental granite magmas (asterisks) with the melts from metatexite migmatites (circles) and melt-rich parts of diatexites (open squares). In general, the metatexite-derived melts are depleted in most elements (except SiO_2) compared with both the parental granite magmas and the diatexite magmas. Note the overlap between the parental granite magmas and melt-rich diatexite magmas compositional fields, but poor overlap with the metatexite melts. Data sources: metatexite melts (leucosomes)—Barbey et al. (1989), Brouand et al. (1990), Collins et al. (1989), Fershtater (1977), Gupta and Johannes (1982), Henkes and Johannes (1981), Martin (1980), Mehnert and Büsch (1982), Nédélec et al. (1993), Sawyer (1991), Weber et al. (1985), White (1966), Whitney and Irving (1994); parental granite magmas— Albuquerque (1971, 1978), Arth and Hanson (1975), Bagby et al. (1981), Barker et al. (1986, 1992), Bickford et al. (1981), Castelli and Lombardo (1988), Chappell and White (1992), Crawford and Windley (1990), Crisci et al. (1979), Currie and Pajari (1981), Day and Weblen (1986), Jensen (1985), McKenzie and Clarke (1975), Neiva (1981), Norman et al. (1992), Pearce et al. (1984), Poli et al. (1989), Saavedra et al. (1985), Sawyer (1987), Weaver et al. (1992); melt-rich diatexites—Brown (1979), Martin (1980), Percival (1991), Torres-Roldan (1983), Sawyer (unpublished data), and Wickham (1987b). Database is available on request from the author.

trend of metatexite melts (Fig. 2b) and hence little overlap
between the two compositional fields.

The differences between metatexite melts and parental
granite magmas are more marked for the trace elements.
Metatexite melts have systematically low Zr (Fig. 2c), Hf, Th,
P_2O_5, Y and REE (Fig. 2d) contents; typically, there is only a
small overlap between the upper compositional range for
metatexite melts and the lower part of parental granite magma
field. The metatexite melts have systematically lower Rb
contents, but similar ranges for Ba and Sr compared with
parental granite magmas (Fig. 2e, 2f). However, most metatexite melts contain considerably less Ba (< 500 ppm) than the
parental granite magmas (> 500 ppm) and show a positive
correlation with K_2O, whereas the granites exhibit a weak
tendency for Ba to decrease with increasing K_2O content
(Fig. 2f). At equivalent SiO_2 contents the parental granite
magmas have higher P_2O_5 contents than the metatexite melts.

The results of this comparison indicate that there is little
overlap between the compositions of melts from metatexite
migmatites and parental granite magmas. Generally, the
metatexite melts are too siliceous and too depleted in TiO_2,
FeO_T, MgO, P_2O_5, Rb, Zr, Hf, Th, Y and the REEs to be
considered as the source melts for parental granitic magmas—
they are too leucocratic. Because there is some compositional
overlap the data allow that certain high-SiO_2, pelite-derived
granites, such as the Himalayan leucogranites, could be derived
from the amalgamation of metatexite melts. However, studies
of the zircons from the Tibetan slab migmatites and the
Himalayan leucogranites indicate that the leucosomes are not
the unsegregated equivalent of the Himalayan leucogranite
magma (Brouand et al. 1990), even though both have a
common source. Thus the Himalayan leucogranite magma
underwent a different melting and segregation process to that
which formed the observed metatexite leucosomes.

The overlap between parental granite magmas and the five
metatexites containing trapped residual material is en-
lightening. One implication is that the melt segregation process
operating in metatexites is effective and produces melts that
are too clean to be the source of typical granitic magmas. If
the segregation mechanism was less effective and allowed a
larger residuum component into the melt, then the resulting
magmas could be the source of parental granite magmas. So,
unless the sampling of metatexite is strongly biased towards
'clean' leucosomes, then to find the source melts for granites
we must look for parts of the crust where melt and residuum
are imperfectly separated. Based on the field observations,
diatexites are such an environment.

3.2. Comparison of diatexite and parental granite magma compositions

Diatexites cannot be compared with parental granite magmas
in the same way as the metatexite melts. This is because there
is no clear distinction between melt, melanosome and palaeosome in diatexites; there is simply a range between melt-rich
and residuum-rich portions and the end-members may not be
present (see, for example, the Trois Seigneurs data of Wickham
1987b, or the Higo data of Obata et al. 1994). Because it is
the melt-rich portions of diatexites that we wish to compare
with parental granite magmas, it is first necessary to distinguish
these from the residuum-rich diatexite samples.

Fortunately, predictions can be made about palaeosome
and felsic melt compositional trends which can be used to
identify the melt-rich diatexite samples. For example, the
Opatica diatexites (Sawyer, unpublished data) formed from a
leucocratic tonalite which shows an igneous trend of decreasing
(FeO_T + MgO), but increasing K_2O, as SiO_2 increases (Fig. 3).
Samples that are enriched in the leucogranite melt fraction

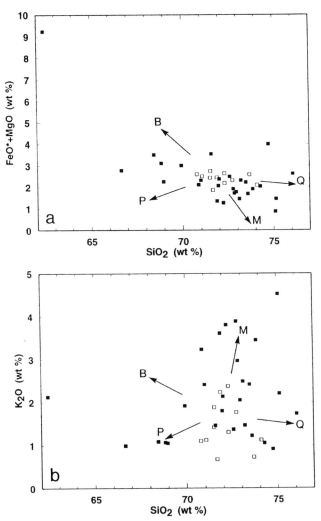

Figure 3 Harker diagrams (a) FeO_T + MgO and (b) K_2O for the
Opatica Belt migmatites: diatexites (closed squares) and palaeosome
(open squares). The arrows represent vectors for the main residuum
phases—biotite (B), plagioclase (P) and quartz (Q)—and the granitic
melt (M). During magma flow and melt segregation melt-rich diatexite
compositions move away from the palaeosome compositional field in
the direction of M.

can therefore be identified because they extend to lower
(FeO_T + MgO) and higher K_2O, relative to the palaeosome
field. The residuum-enriched samples extend either to lower
SiO_2, higher (FeO_T + MgO) and lower K_2O if biotite and
plagioclase are the principal residual phases, or to higher SiO_2
if quartz is a significant residual phase. For pelite-derived
diatexites, the palaeosome field shows increases in both K_2O
and (FeO_T + MgO) as SiO_2 decreases, because biotite is a
major reservoir for K_2O, FeO_T and MgO. In contrast, a
granitic melt fraction shows a decrease in (FeO_T + MgO), but
an increase in K_2O as SiO_2 increases, as K-feldspar contains
more K_2O than biotite. Thus on K_2O versus SiO_2 plots the
melt-rich and melt-depleted samples form a new trend that is
oblique to the pelite (palaeosome) trend; the melt-enriched
samples are identified because they contain higher K_2O and
SiO_2 (see fig. 8b of Obata et al. 1994) relative to the
palaeosome field.

All the diatexite data were plotted on K_2O versus SiO_2
plots and about 37 melt-rich diatexite samples were identified
(see Fig. 2 caption for data sources) for comparison with
parental granite compositions. Figure 2a shows that the melt-
rich diatexites have a similar range of SiO_2 and (FeO_T + MgO)
to the parental granite magmas. The gap in diatexite range
between 70 and 72% SiO_2 is attributed to a sampling bias;

five of six data sets are pelite-derived diatexites. As diatexite data from source rocks intermediate in composition between pelite and leucotonalite (e.g. granites or granodiorites) becomes available, this gap should disappear. The melt-rich diatexite and parental granite magma fields also overlap for the other major oxides (e.g. K_2O, Fig. 2b). The trace elements Zr (Fig. 2c), Hf, Th, Y and the REE (Fig. 2d) contents of diatexites and parental granite magmas also show a close correspondence. Consequently, the melt-enriched parts of diatexites are suitable sources for parental granite magmas, i.e. they could be parental granite magmas that remained more or less *in situ*.

4. Driving forces and mechanisms

Two principal driving forces have been proposed for melt-residuum separation: (1) gravity (McKenzie 1985; Wickham 1987a); and (2) differential or shearing stresses (Sawyer 1991, 1994; Nédélec *et al.* 1993; Brown *et al.* 1995). Because migmatite terranes are anisotropic and subjected to shearing during melting, Sawyer (1994) argued that deviatoric stresses, and not gravity, are the driving forces for the melt segregation process forming leucosomes in metatexites. In simple terms, the melt can be thought of as being squeezed out of the compacting matrix (melanosome) or being sucked into a low pressure sink (site of the leucosome). Consequently, this is called deformation-assisted melt segregation (Sawyer 1994) or shear-enhanced compaction (Rutter & Neumann 1995) to differentiate it from purely gravity-driven segregation.

Gravity is a viable driving force for melt segregation at scales where sufficiently large volumes of melt, or magma, have been generated (or accumulated) so that the buoyancy forces exceed the yield strength of the enclosing rocks. Gravity can then drive melt segregation by the buoyant rise of the melt-rich fraction either (1) along self-propagating cracks (Clemens & Mawer 1992) or pre-existing fractures (Petford *et al.* 1993) and shear zones (D'Lemos *et al.* 1992) or (2) as classical diapirs. Because the scale of melting is much larger than in metatexite migmatites and the bulk viscosity is lower (i.e. magma rheology), gravity may be an important driving force for melt–residuum separation in diatexites—field work is needed to test this.

4.1. Magma mobility and melt segregation in diatexites

Because magma mobility requires that pervasive melting advance to the stage where the matrix framework breaks down, the size and geometry of the melting layer is expected to exert some control on which driving force predominates. As with melt segregation in metatexites, there must be a sink to which the magma can move, otherwise magma mobility will not occur. If the sink is in the upper crust, then a connecting melt transfer path must be available (see Collins & Sawyer, in press).

The Opatica Belt contains examples of diatexite formation in a fertile layer between unmelted refractory layers. Diatexite magma flow is then confined between rigid walls (Fig. 4a) and so is analogous to magma flow in a dyke or sill; segregation of the melt fraction from the residuum occurs during this magma flow and will be considered later. The width of such a confined diatexite magma depends on local geology; it could, for example, be only a 1 m wide layer, or it could be a 500 m wide pelite layer sandwiched between massive quartzites. In the Opatica example, the sense of magma flow determined from the asymmetrical shape of enclaves in the diatexite corresponds to the shear sense determined for local and regional deformation. This suggests that magma mobility is driven by deviatoric stresses related to regional deformation, which in turn are related to arc collision following subduction

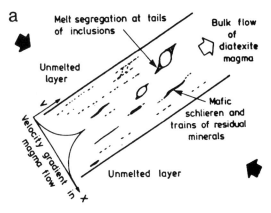

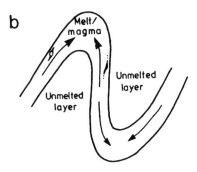

Figure 4 (a) Sketch based on a diatexite layer from the Opatica Belt confined between two unmelted, refractory wallrocks giving rise to magma mobility that is equivalent to flow in a dyke or sill. Closed arrows indicate far-field stress orientation and open arrow indicates the direction of bulk magma flow. (b) Diatexite layer from the Arunta Inlier, folded during melting, showing deformation-driven and possible buoyancy-driven magma mobility towards the antiformal hinge.

(Calvert *et al.* 1995). Thus magma mobility in the Opatica Belt is ultimately driven by plate movements and the mantle. Gravity may also drive magma mobility, especially if the diatexite layer is steeply inclined, and then flow is analogous to buoyant rise up a dyke.

Combined regional deformation and buoyancy effects are suggested for a folded diatexite layers between massive, unmelted, intermediate and mafic felses at Wuluma in the Arunta Inlier (fig. 4b). In this example the diatexite was folded during melting. Because there is more magma (containing less residuum) in the antiformal hinges than the synformal hinges, some magma flow occurred through buoyancy, but in general magma mobility is related to the stress that produced the folding.

The dominant driving force for magma mobility in examples of very large-scale melting in thick, laterally extensive fertile rocks (e.g. sediments in accretionary complexes such as the Ashuanipi and Quetico subprovinces; Percival 1991) may vary depending on the scale examined. Regional tectonics may drive magma mobility at the largest scale, but on an intermediate scale buoyancy effects may be important [e.g. fig. 8 of Percival (1990)]. Finally, at the outcrop scale magma mobility may be driven by shear stresses related to local structures.

Field observations suggest at least three mechanisms for the segregation of melt from its residuum during magma flow ($v_m > v_s > 0$) in diatexites. They arise because the magma is heterogeneous at all scales.

1. Small-scale segregation of melt typically occurs at perturbations in the flow regime. Two examples are: (a) at the tails of asymmetrical enclaves (Fig. 1d), in which case segregation can be inferred to be due to pressure gradients at the tails of the inclusions, by analogy to pressure shadows

(e.g. Stormgard 1973); and (b) elongate lenses at the borders of curviplanar flow banding. Diatexites commonly contain a banding (Fig. 1e) which, because it is coarse grained, has an igneous-like rather than a metamorphic texture (some feldspars with crystal faces) and contains asymmetrical flow structures, is interpreted to be magmatic. However, the bands exhibit layer to layer variations in geochemistry that mimic that in the protolith, which implies that the flow regime in the confined diatexite was approximately laminar and mixing between layers did not occur. Variations in the aspect ratio of enclaves from band to band indicate strain gradients in the magma and hence viscosity, velocity and pressure gradients. Thus the boundaries between flow bands provides the driving force (pressure gradients) and site (velocity discontinuities) for local melt segregation.

2. Many diatexites contain persistent bands or laminae enriched in scattered residuum minerals (i.e. biotite flakes) or trains of mafic schlieren. Magma flow between rigid walls can give rise to velocity gradients because of viscous drag near the bounding walls. Under these conditions melt–residuum separation due to dispersive stress between particles in a flow (Bagnold 1956) could occur in diatexites and trains of residuum material could form in the centre of the flow. In reality, the velocity structure in a diatexite is more complex, as indicated by the presence of banding in some instances. Thus there may be melt–residuum separation due to the Bagnold-type effects in several places across a confined flow.

3. The most common evidence melt–residuum separation in diatexites is a gradual change in the concentration of residuum phases versus melt fraction across an outcrop. Typically, this pattern of melt segregation is observed in the more homogeneous parts of diatexites. The homogeneity probably reflects a thick and uniform palaeosome. A possible explanation for this type of segregation is the difference in the flow characteristics due to a platy shape and greater density of the residuum phases, biotite, garnet, orthopyroxene and plagioclase, relative to the melt fraction.

Broadly, the three types of mechanisms might be collectively termed flow-assisted melt segregation. At present, only reconnaissance field observations of melt segregation in diatexites are available. Once detailed field studies have been completed to outline the scale, and some of the boundary conditions for these processes, they should become amenable to detailed examination using the techniques of fluid mechanics.

5. Conclusions

Metatexite migmatites have: (1) the bulk rheology of solid rock because melting was not pervasive: (2) a very effective segregation mechanism that results in very leucocratic melts depleted in TiO_2, FeO_T, MgO, P_2O_5, Hf, Zr, Th, Y and the REEs; and (3) a short melt transport scale. Consequently, they are not likely to represent the source of melts from which parental granite magmas formed. In contrast, diatexite migmatites have: (1) a magma-like rheology because of pervasive melting which enables large-scale magma transport; (2) segregation mechanisms which yield variable degrees of melt–residuum separation resulting in magma compositions that are comparable with those of parental granite magmas; and (3) areal extents comparable with plutons and plutonic belts and which, hence, appear to be suitable sources for granite magmas. If the source regions of granite magmas do resemble diatexite migmatites, then existing models of melt segregation developed using a metatexite migmatite analogy (i.e. based on the condition $v_m > v_s = 0$) are inappropriate for describing the melt–residuum separation process in the source regions of granite magmas.

Acknowledgements

I thank the organisers of the Third Hutton Symposium on Granites for the invitation to write this paper. I have benefited greatly from conversations with Scott Barboza, George Bergantz, Mike Brown, Anne Nedelec, Gary Solar and Jean-Louis Vigneresse, all of whom have shared their knowledge of melt segregation and migmatites with me—I thank them all. I also thank the reviewers J. Ashworth and M. Williams for their comments. This work was funded by a Natural Sciences and Engineering Research Council of Canada operating grant.

References

Albuquerque, C. A. R. 1971. Petrochemistry of a series of granitic rocks from Northern Portugal. GEOL SOC AM BULL **82**, 2783–98.

Alburquerque, C. A. R. 1978. Rare earth elements in 'younger' granites, northern Portugal. LITHOS **11**, 219–29.

Arth, A. G. & Hanson, G. N. 1975. Geochemistry and origin of the early Precambrian crust of northeastern Minnesota. GEOCHIM COSMOCHIM ACTA **39**, 197–241.

Arzi, A. A. 1978. Critical phenomena in the rheology of partly melted rocks. TECTONOPHYSICS **44**, 173–84.

Bagby, W. C., Cameron, K. L. & Cameron, M. 1981. Contrasting evolution of calc-alkali volcanic and plutonic rocks of western Chihuahua, Mexico. J GEOPHYS RES **86**, 10402–10.

Bagnold, R. A. 1956. The flow of cohesionless grains in a fluid. PHIL TRANS R SOC LONDON **A249**, 235–97.

Barbey, P., Bertrand, J. M., Angoua, S. & Dautel, D. 1989. Petrology and U/Pb geochronology of the Telohat migmatites, Aleksod, Central Hoggar, Algeria. CONTRIB MINERAL PETROL **101**, 207–19.

Barker, F., Arth, J. G. & Stern, T. W. 1986. Evolution of the Coast batholith along the Skagway Traverse, Alaska and British Columbia. AM MINERAL **71**, 632–643.

Barker, F., Farmer, G. L., Ayuso, R. A., Plafker, G. & Lull, J. S. 1992. The 50 Ma granodiorite of the eastern Gulf of Alaska: melting in an accretionary prism in the forearc. J GEOPHYS RES **97**, 6757–78.

Bickford, M. E., Sides, J. R. & Cullers, R. L. 1981. Chemical evolution of magmas in the Proterozoic terrane of the St. Francois mountains, southeastern Missouri 1. Field, petrographic and major element data. J GEOPHYS RES **86**, 10365–86.

Breaks, F. W., Bond, W. D. & Stone, D. 1978. Preliminary geological synthesis of the English River Subprovince, northwestern Ontario and its bearing upon mineral exploration. ONTARIO GEOL SURV MISC PAP **72**,

Brouand, M., Banzet, G. & Barbey, P. 1990. Zircon behaviour during crustal anatexis. Evidence from the Tibetan Slab migmatites (Nepal). J VOLCANOL GEOTHERM RES **44**, 143–61.

Brown, M. 1973. Definition of metatexis, diatexis and migmatite. PROC GEOL ASSOC **84**, 371–382.

Brown, M. 1979. The petrogenesis of the St. Malo Migmatite Belt, Armorican Massif, France, with particular reference to the diatexites. N JAHRB MINERAL ABH **135**, 48–74.

Brown, M. 1994. The generation, segregation, ascent and emplacement of granite magma: The migmatite-to-crustally-derived granite connection in thickened orogens. EARTH-SCI REV **36**, 83–130.

Brown, M., Averkin, Y. A., McLellan, E. L. & Sawyer, E. W. 1995. Melt segregation in migmatites. J GEOPHYS RES **100**, 15655–79.

Calvert, A. J., Sawyer, E. W., Davis, W. J. & Ludden, J. N. 1995. Archaean subduction inferred from seismic images of a mantle suture in the Superior Province, NATURE **375**, 670–4.

Castelli, D. & Lombardo, B. 1988. The Gophu La and western Lunana granites: Miocene muscovite leucogranites of the Bhutan Himalaya. LITHOS **21**, 211–25.

Chappell, B. W. & White, A. J. R. 1992. I- and S-type granites in the Lachlan Fold Belt. TRANS R SOC EDINBURGH EARTH SCI **83**, 1–26.

Clemens, J. D. & Mawer, C. K. 1992. Granitic magma transport by fracture propagation. TECTONOPHYSICS **204**, 339–60.

Collins, W. J. & Sawyer, E. W. Pervasive magma transfer through the lower–middle crust during non-coaxial compressional deformation: an alternative to dyking. J METAMORPH GEOL, in press.

Collins, W. J., Flood, R. H., Vernon, R. H. & Shaw, S. E. 1989. The Wuluma granite, Arunta Block, central Australia: an example of in situ, near-isochemical granite formation in a granulite facies terrane. LITHOS 23, 63–83.

Crawford, M. B. & Windley, B. F. 1990. Leucogranites of the Himalaya/Karakoram: implications for magmatic evolution within collisional belts and the study of collision related leucogranite genesis. J VOLCANOL GEOTHERM RES 44, 1–19.

Crisci, G. M., Maccarrone, E. & Rottura, A. 1979. Cittanova peraluminous granites (Calabri, southern Italy). MINER PETROGR ACTA 23, 279–302.

Currie, K. L. & Pajari, G. E. 1981. Anatectic peraluminous granites from the Carmanville area, northeastern Newfoundland. CAN MINERAL 19, 147–62.

Daines, M. J. & Richter, F. M. 1988. An experimental method for directly determining the interconnectivity of melt in a partially molten system. GEOPHYS RES LETT 15, 1459–62.

Day, W. C. & Weblen, P. W. 1986. Origin of Late Archaean granite: geochemical evidence from the Vermilion Granite Complex of northern Minnesota. CONTRIB MINERAL PETROL 93, 283–96.

Dell'Angelo, L. N. & Tullis, J. 1988. Experimental deformation of partially melted granitic aggregates. J METAMORPH GEOL 6, 495–516.

D'Lemos, R. S., Brown, M. & Strachan, R. A. 1992. The relationship between granite and shear zones: magma generation, ascent and emplacement within a transpressional orogen. J GEOL SOC LONDON 149, 487–90.

Fershtater, G. B. 1977. Isochemical migmatization and genesis of quartzofeldspathic rocks of the Taratash metamorphic complex (southern Urals). GEOCHEM INT 14, 63–72.

Gupta, L. N. & Johannes, W. 1982. Petrogenesis of a stromatic migmatite (Nelaug, southern Norway). J PETROL 23, 548–67.

Gupta, L. N. & Johannes, W. 1985. Effect of metamorphism and partial melting of host rocks on zircons. J METAMORPH GEOL 3, 311–23.

Henkes, L. & Johannes, W. 1981. The petrology of a migmatite (Arvika, Varmland, western Sweden). N JARHB MINERAL ABH 141, 113–33.

Jensen, I. S. 1985. Geochemistry of the central granitic stock in the Glitrevann cauldron within the Oslo Rift, Norway. NORSK GEOL TIDSSKR 65, 201–16.

Lapointe, B. & Chown, E. H. 1993. Gold-bearing iron-formation in a granulite terrane of the Canadian Shield: a possible deep-level expression of an Archean gold-mineralizing system. MINERAL DEPOSITA 28, 191–7.

Le Breton, N. & Thompson, A. B. 1988. Fluid-absent (dehydration) melting of biotite in metapelites in the early stages of crustal anatexis. CONTRIB MINERAL PETROL 99, 226–37.

Maaløe, S. 1982. Geochemical aspects of permeability controlled partial melting and fractional crystallization. GEOCHIM COSMOCHIM ACTA 46, 43–57.

Martin, H. 1980. Comportement de quelques elements en traces au cours de l'anatexie. L'exemple du massif de Saint Malo (Bretagne, France). CAN J EARTH SCI 17, 927–41.

McCarthy, T. S. & Groves, D. I. 1979. The Blue Tier Batholith, Northeastern Tasmania. CONTRIB MINERAL PETROL 71, 193–209.

McKenzie, D. 1985. The extraction of magma from the crust and mantle. EARTH PLANET SCI LETT 74, 81–91.

McKenzie, C. B. & Clarke D. B. 1975. Petrology of the South Mountain Batholith, Nova Scotia. CAN J EARTH SCI 12, 1209–18.

Mehnert, K. R. & Büsch, W. 1982. The initial stage of migmatite formation. N JARHB MINERAL ABH 145, 211–38.

Nédélec, A., Minyem, D. & Barbey, P. 1993. High–P–high–T anatexis of Archaean tonalitic grey gneisses: the Eseka migmatites, Cameroon. PRECAMBRIAN RES 62, 191–205.

Neiva, A. M. R. 1981. Geochemistry of hybrid granitoid rocks and their biotites from central northern Portugal and their petrogenesis. LITHOS 14, 149–63.

Nicolas, A., Freydier, Cl., Godard, M. & Vauchez, A. 1993. Magma chambers at oceanic ridges; how large? GEOLOGY 21, 53–6.

Norman, M. D., Leeman, W. P. & Mertzman, S. A. 1992. Granites and rhyolites from the northwestern U.S.A.: temporal variation in magmatic processes and relations to tectonic setting. TRANS R SOC EDINBURGH EARTH SCI 83, 71–81.

Obata, M., Yoshimura, Y., Nagakawa, K., Odawara, S. & Osanai, Y. 1994. Crustal anatexis and melt migrations in the Higo metamor-

phic terrane, west-central Kyushu, Kumamoto, Japan. LITHOS 32, 135–47.

Pearce, J. A., Harris, N. B. W. & Tindle, A. G. 1984. Trace element discrimination diagrams for the tectonic interpretation of granitic rocks. J PETROL 25, 956–83.

Percival, J. A. 1990. Archean tectonic setting of granulite terranes of the Superior Province, Canada: a view from the bottom. In Vielzeuf, D. & Vidal, Ph. (eds) Granulites and crustal evolution, 171–193. Amsterdam: Kluwer.

Percival, J. A. 1991. Granulite-facies metamorphism and crustal magmatism in the Ashuanipi Complex, Quebec–Labrador, Canada. J PETROL 32, 1261–97.

Percival, J. A., Card, K. D., Stern, R. A. & Bégin, N. J. 1990. A geological transect of northeastern Superior Province, Ungava Peninsula, Quebec: the Lake Minto area. CURR RES PART C GEOL SURV CAN 90–1c, 133–41.

Petford, N., Kerr, R. C. & Lister, J. R. 1993. Dike transport of granitoid magmas, GEOLOGY 21, 845–8.

Poli, G., Ghezzo, C. & Conticelli, S. 1989. Geochemistry of granitic rocks from the Hercynian Sardinia–Corsica batholith: implication for magma genesis. LITHOS 23, 247–66.

Read, H. H. 1957. The granite controversy. London: Murby.

Reid, J. B., Evans, O. C. & Fates, D. G. 1983. Magma mixing in granitic rocks of the central Sierra Nevada, California. EARTH PLANET SCI LETT 66, 243–61.

Rutter, E. H. & Neumann, D. H. K. 1995. Experimental deformation study of partly molten Westerly granite under fluid-absent conditions, with implications for the extraction of granitic magmas. J GEOPHYS RES 100, 15697–716.

Saavedra, J., Rossi de Toselli, J., Toselli, A. & Garcia-Sanchez, A. 1985. The origin of the two-mica granites of the Loma Pelada pluton. Tucuman, northwest Argentina. LITHOS 18, 179–85.

Sawyer, E. W. 1987. The role of partial melting and fractional crystallization in determining discordant migmatite leucosome compositions. J PETROL 28, 445–73.

Sawyer, E. W. 1991. Disequilibrium melting and the rate of melt-residuum separation during migmatisation of mafic rocks from the Grenville Front, Quebec. J PETROL 32, 701–38.

Sawyer, E. W. 1994. Melt segregation in the continental crust. GEOLOGY 22, 1019–22.

Sawyer, E. W. & Barnes, S.-J. 1988. Temporal and compositional differences between subsolidus and anatectic migmatite leucosomes from the Quetico metasedimentary belt, Canada. J METAMORPH GEOL 6, 437–50.

Sawyer, E. W. & Benn, K. 1993. Structure of the high-grade Opatica Belt and adjacent low-grade Abitibi Subprovince, Canada: an Archaean mountain front. J STRUCT GEOL 15, 1443–58.

Stevens, G. & Clemens, J. D. 1993. Fluid-absent melting and the roles of fluids in the lithosphere: a slanted summary? CHEM GEOL 108, 1–17.

Stormgard, K. E. 1973. Stress distribution during the formation of boudinage and pressure shadows. TECTONOPHYSICS 16, 215–48.

Torres-Roldan, R. L. 1983. Fractionated melting of metapelite and further crystal–melt equilibria—the example of the Blanca Unit migmatite complex, north of Estepona (southern Spain). TECTONOPHYSICS 96, 95–123.

Vielzeuf, D. & Holloway, J. R. 1988. Experimental determination of the fluid-absent melting relations in the pelitic system: consequences for crustal differentiation. CONTRIB MINERAL PETROL 98, 257–76.

Vigneresse, J. L., Cuney, M. & Barbey, P. 1991. Deformation assisted crustal melt segregation and transfer. GAC-MAC ABSTR 16, A128.

Wall, V. J., Clemens, J. D. & Clarke, D. B. 1987. Models for granitoid evolution and source compositions. J GEOL 95, 731–49.

Weaver, S. D., Adams, C. J., Pankhurst, R. J. & Gibson, I. L. 1992. Granites of Edward VII Peninsular, Marie Byrd Land: anorogenic magmatism related to Antarctic–New Zealand rifting. TRANS R SOC EDINBURGH EARTH SCI 83, 281–90.

Weber, C., Barbey, P., Cuney, M. & Martin, H. 1985. Trace element behaviour during migmatization: evidence for a complex melt-residuum–fluid interaction in the St. Malo migmatitic dome (France). CONTRIB MINERAL PETROL 90, 52–62.

White, A. J. R. 1966. Genesis of migmatites from the Palmer region of South Australia. CHEM GEOL 1, 165–200.

White, A. J. R. & Chappell, B. W. 1977. Ultrametamorphism and granitoid genesis. TECTONOPHYSICS 43, 7–22.

White, A. J. R. & Chappell, B. W. 1990. Per migma ad magma downunder. GEOL J 25, 221–5.

Whitney, D. L. & Irving, A. J. 1994. Origin of K-poor leucosomes in a metasedimentary migmatite complex by ultrametamorphism, syn-metamorphic magmatism and subsolidus processes. LITHOS **32**, 173–92.

Wickham, S. M. 1987a. The segregation and emplacement of granitic melts. J GEOL SOC LONDON **144**, 281–97.

Wickham, S. M. 1987b. Crustal anatexis and granite petrogenesis during low-pressure regional metamorphism: the Trois Seigneurs Massif, Pyrenees, France. J PETROL **28**, 127–69.

Winkler, H. G. F. & von Platten, H. 1961. Experimentelle Gesteinsmetamorphose—V. GEOCHIM COSMOCHIM ACTA **24**, 250–9.

Williams, M. L., Hanmer, S., Kopf, C. & Darrach, M. 1995. Syntectonic generation and segregation of tonalitic melts from amphibolite dykes in the lower crust, Striding–Athabasca mylonite zone, northern Saskatchewan. J GEOPHYS RES **100**, 15 717–34.

E. W. SAWYER, Sciences de la Terre, Sciences Appliquees, Université du Québec à Chicoutimi, Chicoutimi, Quebec, Canada, G7H 2B1. E-mail:ewsawyer@UQAC.UQuebec.CA

Transactions of the Royal Society of Edinburgh: Earth Sciences, **87**, 95–103, 1996

Ascent mechanism of felsic magmas: news and views

Roberto Ferrez Weinberg

ABSTRACT: Diapirism has been discredited as a transport mechanism for magmas partly because diapirs seem to be unable to bring magmas to shallow crustal levels (< 10 km) and partly because recent developments in the theory of dyke propagation have shown that sufficiently wide dykes are able to efficiently transport felsic magmas through the crust. However, it is still unclear how felsic dykes grow to widths that allow them to propagate faster than they close by magma freezing. Ultimately, it may be the ability of felsic dykes to grow within the source that controls which mechanism dominates ascent.

The ability of dykes to propagate from the top of rising diapirs depends among other factors on the changing temperature gradient of the wall rocks. The steep gradient around rapidly rising diapirs in the low viscosity lower crust will cause dykes to freeze. As diapirs rise to colder stiffer crust and decelerate, heat diffuses further from the diapir, resulting in shallower temperature gradients that favour dyke propagation. The mechanism may thus swap, during ascent, from diapirism to dyking. Calculations of the thermal evolution of diapirs and their surroundings show that basaltic diapirs may never form because they would be drained by dykes at a very early stage; felsic diapirs may be unable to give rise to successful dykes, whereas diapirs of intermediate magmas may propagate dykes during ascent.

KEY WORDS: diapirs, dykes, magma segregation, magma emplacement, magma transport.

There has been considerable discussion on the mechanisms of magma transport through the continental crust in recent publications (Cruden 1990; Emerman & Marrett 1990; Lister & Kerr 1991; Clemens & Mawer 1992; Cruden & Aaro 1992; Paterson & Fowler 1993, 1994; Weinberg 1994; Weinberg & Podladchikov 1994, 1995; Petford, this volume). Discussion concentrates mainly on two mechanisms: dykes and diapirs. The limiting parameter controlling the velocity of dyke propagation is magma viscosity. For diapirs magma viscosity plays only a minor part in determining the velocity, which is mainly controlled by the much higher viscosity of the diapir's surrounding. The models of Grout (1945) and Ramberg (1967) made diapirs a popular mechanism to explain the ascent of felsic magmas through the crust. This was because diapirs are buoyant and may give rise to voluminous plutons that show geometries similar to those resulting from the models, such as elliptical shape and concentric patterns of foliation (Sylvester 1964; Brun & Pons 1981; Bateman 1985; Courrioux 1987; Ramsay 1989; Cruden & Aaro 1992; Paterson & Fowler 1993). The idea that felsic magmas rise as diapirs was further strengthened by the observation that large granitic dyke swarms are rare compared with basic swarms, suggesting that most felsic magmas would be too viscous to rise rapidly enough to avoid freezing.

The controversy regarding the transport of felsic magmas started in the 1980s when several workers suggested that the crust was too viscous to allow diapirs to rise fast, thus they froze after travelling only a short distance (Marsh 1982; Morris 1982; Ribe 1983; Daly & Raefsky 1985; Mahon et al. 1988). Similar results were found when the effects of the thermal softening of wall rocks by hot diapirs (hot Stokes models) were included: the faster ascent due to lower viscosity was counteracted by faster magma cooling. These workers concluded that diapirism of magmas through the crust was a slow process limited by the thermal energy of diapirs, and that to be efficient diapirism requires either an anomalously hot crust or several diapirs heating up a path through the crust. Paterson & Vernon (1995) argued that the latter is just what is observed in nature, with several diapirs following the same path to nest into each other and form the commonly observed pluton zonation. Doubts about diapirs were strengthened by a test carried out by Schwerdtner (1990) in the Archaean crust of Ontario, which showed that diapirism was unlikely to give rise to the structures observed around the studied domes.

The situation at that time was such that it seemed that the viscosity of the crust was too high for diapirs, and that the viscosity of felsic magmas was too high to allow dykes to propagate without first freezing. In the beginning of the 1990s, Lister and Kerr (1991), Clemens and Mawer (1992) and Petford et al. (1993) showed that, _given a large enough initial dyke_, felsic magmas can rapidly and episodically rise across the crust through dykes and form large batholiths in the upper crust, despite their high viscosity. The critical width required for felsic dykes to rise without freezing was estimated to be from a few metres to a few tens of metres (Petford et al. 1993, 1994; Lister 1995), reasonable values for observed dyke widths (Wada 1994, 1995; Kerr & Lister 1995). The common flow structures observed in and around plutons and batholiths fed by dykes would be caused, not by ascent, but by the emplaced magma warming up the surrounding rocks which then flowed due to the expansion of the magma chamber.

The efficiency of magma transport through dykes, coupled with the common spatial association between plutons and large crustal-scale fault zones, have led several workers to suggest that faults and shear zones are pathways for magmas and are also responsible for creating space for magma emplacement through extension (e.g. Pitcher 1979;

Guineberteau *et al.* 1987; Hutton *et al.* 1990; Hacker *et al.* 1992; Hutton & Reavy 1992; Petford & Atherton 1992; Petford *et al.* 1993; Brown 1994; Ingram & Hutton 1994). Sleep (1975) showed that shallow basaltic magma chambers can be maintained at shallow depths along mid-ocean ridges due to the prevailing extension rates. Only very recently have we learned that crustal extensional rates are able to open space rapidly enough to give rise to a steady-state felsic magma chamber (Hanson & Glazner 1995). A steady-state chamber describes a chamber where magma slowly accumulates and interacts with older (still not totally solidified) magma, in contrast with a collection of small frozen magma batches or dykes that results when extension is too slow. Hanson & Glazner (1995) note, however, that accurate dating is needed to determine if magma chambers grow at appropriate extension rates or faster. The problem of space for the emplacement of large magma volumes is by no means solved (e.g. Paterson & Fowler 1993, 1994; Weinberg 1994; Schwerdtner 1995; Petford, this issue). That magmas find or open space for themselves does not seem to be a problem; the space problem lies in understanding how this is done. Once this is understood, we might be able to fully understand the ascent and emplacement of granites. Diapirs make their own space by imposing flow on the surrounding crust. The width of the strain aureole is controlled by the temperature distribution and the rheology of the wall rocks (Weinberg & Podladchikov 1995). Paterson and Fowler (1993), studying the aureoles of several plutons, suggested that the observed shortening was insufficient to account for the pluton volume and that therefore mechanisms other than the viscous flow of the wall rocks were necessary to create that space. However, Weinberg (1994) and Schwerdtner (1995) pointed out that volume estimations resulting from shortening measured at the margins of essentially two-dimensional plutons cannot be integrated into three dimensions with confidence, and that the method used by Paterson and Fowler (1993) is unable to yield reliable estimates of the volume displaced by diapirs. For plutons fed by dykes it is clear that space for magma emplacement needs to be opened and several mechanisms are available, such as faulting, stoping, doming of the roof, caldera subsidence and viscous flow of the wall rocks. As suggested by Petford *et al.* (1994), emplacement may depend on the temperature and rheology of the crust and the rate of magma flow into the chamber. Buddington (1959) and, more recently, Paterson and Fowler (1993) suggested that more than one mechanism may be required to explain the ascent and emplacement of shallow level batholiths and that different mechanisms may dominate ascent at different depths.

Counterbalancing the recent tendency in publications to favour dykes as the most effective ascent mechanism, Rubin (1993a, b, 1995) pointed out that a fundamental question remains to be answered by the proponents of the dykes-ballooning model: how do felsic dykes survive the early stages of propagation without freezing? As the velocity of dyke propagation increases linearly with dyke length (and decreases linearly with magma viscosity), small dykes starting from the magma source and progressing into rocks at subsolidus temperatures will propagate slowly and magma freezing may clog the dyke and stop propagation. Freezing will halt most rhyolitic dykes soon after intrusion into rocks at subsolidus temperature (Rubin 1995). However, hot rhyolites (100°C above its solidus) may successfully develop dykes given low magma viscosity (10^4 Pa s) or high magma pressure (> 10 MPa), or low temperature gradients in the surrounding rocks (< 5°C/km), or some combination of these (Rubin 1995). These conditions limit felsic dykes to narrow and rather anomalous situations. Regarding diapirism, recent work has shown that diapirs may be more efficient than previously thought (Weinberg & Podladchikov 1994, 1995). These workers showed how the strain rate softening of power law crust, rather than thermal softening (hot Stokes models), may allow diapirs to rise fast enough to reach upper crustal levels before freezing. Strain-rate softening relies on the diapir's velocity to decrease the crustal viscosity rather than the heat content of the rising magma, allowing diapirs to rise faster without considerably increasing their cooling rate (Weinberg & Podladchikov 1994). These workers showed, however, that individual diapirs would require anomalous crustal temperatures or extremely low viscosity upper crust to reach shallow crustal levels (< 10 km).

Although considerable progress has been made in understanding magma transport through the crust, the theories of dyking and diapirism are still unable to answer several fundamental questions. I start this paper by discussing the problem of dyke initiation, suggesting conditions that may favour or inhibit dyking. Results by Rubin (1993a, 1995) are applied to a rising diapir to determine if at any point during rise dykes would successfully leave the diapir and drain it of its magma. The results suggest that a swap in ascent mechanism may occur for magmas of viscosity of intermediate values ($\approx 10^3$–10^6 Pa s). Finally, I discuss a few other questions regarding dykes and diapirs. Firstly, the proximity of plutons and fault/shear zones is discussed and it is suggested that this may result from shear zones focusing melt migration in the source and thus controlling the site for magmatic ascent either as diapirs or dykes. Then, I discuss the difficulties in finding evidence of the pathways of diapirs and the question of why felsic plutons balloon (Rubin 1995).

1. Controls on dyke initiation from a partially molten zone

We know that large wide dykes (two or more metres in the direction perpendicular to the dyke walls) may be able to crack the surrounding rocks and transport rapidly large volumes of felsic magmas through the crust. We also know that wide felsic dykes are not uncommon (Wada 1994). Very small and narrow dykes will take infinitely long to grow because their velocity tends to zero as their length or width tends to zero. If the propagation velocity is slow the magma in the dyke will freeze if surrounded by rocks at subsolidus temperature (e.g. Bruce & Huppert 1989; Rubin 1995). Bruce and Huppert (1989) introduced the concept of 'critical initial width', which is the minimum width of a dyke which allows it to propagate without becoming clogged by freezing magma. If dykes are able to grow within the supersolidus magma source to widths beyond critical, they will survive when they propagate into and across the subsolidus crust. The question then is how melt segregation processes in the source are likely to influence or control dyke initiation. Is the process of porous flow of melt into a network of veins that drain into large dykes capable of draining the source efficiently, and are dykes within the source large enough to survive the freezing temperatures outside the magma source (Sleep 1988)? If segregation is unable to give rise to such dykes, then doming of the buoyant source with or without concomitant segregation may lead to diapirs (Fig. 1). The limited ability of stiff magmas to initiate dykes may ultimately control which transport mechanism dominates ascent. In the next section I discuss how magma segregation, tectonic stresses and viscosity ratios between magmas and wall rocks may favour or inhibit the initiation of dykes.

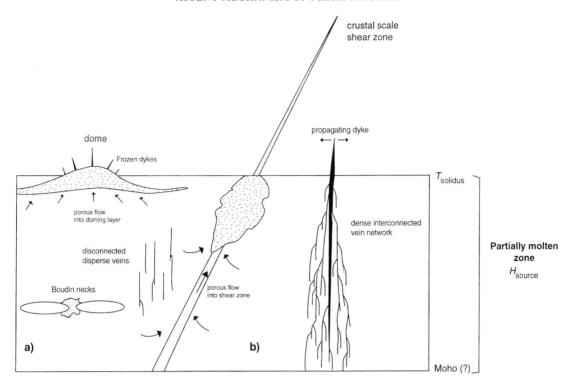

Figure 1 Schematic view of melt segregation, doming and dyking of a partially molten zone. Melt flows into low pressure sites such as boudin necks, low viscosity shear zones (active or inactive) and the hinge of a dome or regions previously enriched in melt to form veins (Stevenson 1989). Where dykes are unable to propagate, either the whole source may dome or, as depicted, melt accumulated close to the top of the source layer through porous flow (or in any low pressure site) may start to dome (a); where melt segregation gives rise to a dense network of interconnected veins, dykes may be able to propagate into colder surroundings without freezing (b). The shear zone cutting across the source may act as a magma sink and drain the surroundings. Independently of whether shear zones control the mechanism of ascent, by acting as a low pressure site, they may be an important control on the site of magma ascent.

1.1. Veins in the source

The starting point of this discussion is that creep is the natural response of the lower crust to applied stress and that granite veins only develop because of particular conditions prevailing in the molten zone. Dell'Angelo and Tullis (1988) found that in partially molten granite samples undergoing deformation in laboratory experiments, the presence of melt may result in the concurrent operation of rock cataclasis, diffusion creep and dislocation creep. These mechanisms are not normally operative for the same P–T conditions and their simultaneous occurrence may lead to incorrect interpretations regarding these conditions. Particularly relevant here is that their experiments showed that when the imposed strain rate is faster than the rate with which melt can flow laterally to low pressure sites (out of the system in their experiments), high pore pressures result in rock cataclasis with fractures parallel to the orientation of the maximum compressional axis (σ_1). Once cataclasis occurs, melt migrates to low pressure sites, the pore pressure drops and the rock resumes creep. The results of Dell'Angelo and Tullis (1988) suggest that cataclasis of the source would be enhanced by high strain rates or high magma viscosity. Owing to spatial variations in melt fractions and permeability, cataclasis may occur in some parts of the source while deformation is controlled by creep in others. As cataclasis at these high P–T conditions is triggered by the presence of melt, creep will still dominate the deformation of the surrounding solid rock. In other words, the fact that the viscous partially molten zone is able to crack does not imply that the stiffer subsolidus crust is able to do the same. Cataclasis is thus a brittle mechanism that may give rise to veins within the source zone when the strain rate is fast enough to increase the local pore pressure.

In a heterogeneous partially molten zone there are generally several natural low pressure sites that act as melt sinks, such as boudin necks (Brown 1994), fold hinge zones (Allibone & Norris 1992) or, less obviously, low pressure zones resulting from the layering of rocks of different viscosities (filter pressing; Robin 1979; Miller & McLellan 1986; Wickham 1987). Stevenson (1989) showed that a partial melt undergoing large-scale deformation is unstable, and melt migrates through pores in the direction of the minimum compressional axis σ_3 to low pressure regions slightly enriched in melt to give rise to veins parallel to the maximum compressional axis σ_1. This process may have a preferred length scale of the order of a metre (Stevenson 1989) and although it does not involve rock failure, it could give rise to a network of interconnected veins oriented parallel to σ_1. This mechanism is probably common to any source regions undergoing large-scale deformation and gives rise to veins without requiring the brittle failure of source rocks.

1.2. Melt segregation, tectonic stresses and magma stresses

Different segregation mechanisms and different stress orientations will influence the ability of dykes to propagate into subsolidus crust by controlling the magma pressure of initiating dykes. The pressure at the tip of a dyke is one of the main factors controlling the success of dyke propagation (Lister & Kerr 1991; Rubin 1993a, b). Magma pressure in the source depends on the dynamics of melt segregation and on dyke orientation, which is strongly controlled by tectonic stresses. If the segregation of magmas occurs by porous flow (McKenzie 1984) resulting in a magma layer close to the top of the partially molten zone (Fig. 1; Fountain *et al.* 1989), the maximum bouyancy stress (magma pressure, σ_m) on an ensuing crack will be

$$\sigma_m = \Delta\rho \, g \, H_{melt}$$

where $\Delta\rho$ is the density difference between the melt and its surroundings, g is the gravity acceleration and H_{melt} is the thickness of the molten layer ($H_{melt} < H_{source}$, where H_{source} is the source layer thickness). If, on the other hand, melt segregates to form veins, the maximum buoyancy stress results from an interconnected network of steeply dipping static (immobile melt) veins cutting across the entire source. In this ideal case

$$\sigma_m = \Delta\rho \, g \, H_{source}.$$

If magma is moving through this network (most likely), the pressure decreases significantly due to friction at the walls to

$$\sigma_m = \Delta\rho \, g \, L$$

where L is the compaction length (McKenzie 1987), characteristically much smaller than H_{source}.

Vein orientation is strongly controlled by crustal heterogeneities and the orientation of tectonic stresses. Although extensional regimes would tend to favour vertical or steeply dipping veins, shortening regimes would favour horizontal or shallow dipping veins. If the combination of crustal heterogeneities and stress results in horizontal or shallow dipping veins and dykes (e.g. compressional regimes), their orientation is perpendicular to the direction of the buoyancy force, which is distributed over a large area (low stress), and

$$\sigma_m = \Delta\rho \, g \, w_{dyke}$$

where w_{dyke} is the dyke width. (In this discussion, any overpressure caused by the exsolution of gases has been neglected, as the melt is assumed to be in equilibrium within the source).

In summary, for a given magma viscosity, the most favourable conditions for dyke propagation from the source arises in an extending crust in the case where segregation gives rise to an immobile interconnected network of steeply dipping veins where high pressure is concentrated at the dyke tips (Fig. 1b). Conversely, conditions least likely to favour dykes and most favourable to diapirs arise in shortening environments, when segregation is dominated by porous flow or when it gives rise to disconnected veins (Fig. 1a). The magma pressure in dyke tips required to enable propagation into cold subsolidus crust depends, among other parameters, on the temperature gradient in the subsolidus crust, magma viscosity and its excess temperature (i.e. the difference between magma temperature and solidus temperature, ΔT^*; see Rubin 1995 for detailed discussion). If segregation within the source gives rise to these high pressure dykes, we may have rapid melt extraction. On the other hand, if the source is unable to give rise to high pressure dykes or if this process is slow (e.g. high viscosity melt or low permeability), the buoyant source or those low pressure sites where magma converged during segregation may start to dome. Doming may in turn enhance melt segregation and focus melt flow towards the dome's hinges. Magma bodies thus formed may eventually detach from the source, forming a diapir and leaving restite-rich migmatites along its tail. The efficiency and velocity of this process depends on the efficiency of melt segregation, the magma buoyancy, and the viscosity of the wall rocks (a function of rock rheology, temperature and diapir buoyancy; see Weinberg & Podladchikov 1994).

1.3. Viscosity ratio between melt and solid rocks

Corriveau and Leblanc (1995) showed in the Grenville Province, Quebec how magmas that were ascending in dykes became trapped in a low viscosity marble-rich layer, through which they rose as diapirs. This change in ascent mechanism results from the faster viscous response of the marble to the applied magma stress compared with that of stiffer rocks. Two mechanisms were envisaged by Rubin (1993a) where the viscosity contrast between the magma and the surrounding rocks may prevent dykes propagating efficiently: (a) if the viscosity contrast is sufficiently small, the tip of an initiating dyke might become blunt in the time required for the dyke to inflate, so that the stress concentration at the tip (a function of tip sharpness) never becomes sufficient to further crack the rock; (b) in the case of magma invading a pre-existing fracture above a rising diapir, if the viscosity contrast is sufficiently small, the large-scale flow around the diapir widens the dyke faster than the dyke tip propagates. Whereas the diapiric flow of the wall rocks limits the dyke's ability to develop in the latter, it is the process of fracturing that limits dyking in the former. In the field example of Quebec, either of the two (or both) mechanisms may have prevented the dykes from propagating through the marble-rich layer.

Rubin's (1993a) study of the response of viscoelastic rocks to a propagating dyke showed that for expected magma pressures, elastic stiffness $G = (1-v)/\mu$ (where v is the Poisson's ratio and μ the shear modulus) and magma viscosities, the wall rock viscosities would have to be extremely low to respond as an essentially viscous fluid. For example, using Rubin's (1993a) figure 4, for magma viscosity of 10^8 Pa s, the surrounding crust would need viscosities less than 10^{15} Pa s to behave as a purely viscous fluid. Although the viscosity of a very soft marble submitted to magma stresses could be that low, most crustal rocks are likely to be stiffer. For this more general case the crust would respond to a propagating dyke either as a purely elastic medium or as a combination of viscous and elastic. The viscoelastic response requires a low, but geologically reasonable, viscosity ratio, and results in a wide dyke with the tip propagating through an essentially elastic medium, whereas the more central parts of the dyke encounter a viscous response from the surroundings and widens.

2. Magma ascent: from diapirs to dykes

Rubin (1993a, 1995) assumed steep temperature gradients around magma chambers ($dT/dz = 0.1–1°C/m$) that tend to cause early dyke freezing. In natural magma chambers the temperature gradient is likely to evolve in time from steep during and just after magma emplacement, to shallow as the magma heat diffuses into the surroundings. Dykes that would initially freeze in the cold surroundings might be able to propagate at a later stage through warmer wall rocks. Similarly, the fate of dykes initiating at the top of a rising diapir may change as the temperature gradient evolves. In low viscosity hot crust, fast diapirs impose a steep temperature gradient in the surroundings that causes early dyke freezing. As diapirs rise to stiffer and colder crust, the temperature gradient becomes shallower and dykes may successfully leave the diapir. In this section, we examine if, at some point, as a diapir progresses upwards from hot and soft to cold and stiff crust, magma-filled cracks initiating at the diapir's top are able to propagate without freezing and drain the diapir. This is carried out here by studying the evolution of rising diapirs of different sizes and different viscosity magmas using a modified version of the program Rise (Weinberg & Podladchikov 1994, 1995) that includes the results of Rubin (1993a).

2.1. Method

Rubin (1993a, b) showed that, for magmas at the solidus temperature, a single freezing parameter β predicts the ability of dykes leaving a chamber to propagate into subsolidus crust

$$\beta = \frac{2(3^{\frac{1}{4}}c|dT/dz|(\kappa\eta)^{\frac{1}{4}}}{\pi^{\frac{1}{4}}L(p/G)^2 p^{\frac{1}{4}}} < \approx 0.15 \qquad (1)$$

where c is the heat capacity, dT/dz is the temperature gradient away from the chamber walls, κ is the thermal diffusivity, η is the magma viscosity, L is the latent heat and G is the elastic stiffness. The magma pressure, p, at the dyke entrance is taken here to be the buoyancy stress of the diapir ($p \approx \Delta\rho g r$, where r is the diapir radius, Rubin 1993a). Rubin (1993b) showed that for $\beta < 0.15$ dykes would be able to propagate faster than the speed with which freezing would shut them. This critical value of β is weakly dependent on the suction at the cavity at the tip of the dyke and may be rewritten for magmas at solidus temperature and $c = 1 \text{ kJ kg}^{-1}\,^{\circ}\text{C}^{-1}$, $L = 400 \text{ kJ kg}^{-1}$, $\kappa = 10^{-6} \text{ m}^2 \text{ s}^{-1}$ and $G = 10$ GPa as

$$\beta^* = \frac{|dT_0/dz|^2\eta}{p^5} < 10^{-31} \text{ s}^{\circ}\text{C m}^{-2} \text{ Pa}^{-4} \qquad (2)$$

The computer code Rise calculates the velocity of spherical diapirs rising through crust of Westerly granite rheology (strain rate and temperature-dependent viscosity; Hansen & Carter 1982). As the diapir rises into colder and stiffer crust, the magma temperature and the temperature gradient above the diapir are calculated. The magma temperature is assumed to be homogeneous within the diapir and the magma viscosity and density difference to the surroundings is constant throughout the ascent. In this way β^* can be calculated at each step and compared with the critical value. When $\beta^* \leqslant \beta^*_{\text{crit}} = 10^{-31}$, magma-filled fractures starting at the top of the diapir will be able to propagate into the surrounding crust. The value of β^*_{crit} used assumes that the magma is at its solidus temperature. Although magma, in the calculations, is generally above the solidus temperature, this assumption should not greatly influence the results because, as shown by Rubin (1995), the addition of an excess temperature (ΔT^*) should not greatly enhance the ability of magmas to penetrate the subsolidus crust as dykes. For example, using figure 3 in Rubin (1995), for $\Delta T^* = 100\,^{\circ}$C, and dT/dz of 0.1 or $1\,^{\circ}$C/m, when β exceeds the critical value by three times, a dyke would propagate only a distance of $2.6l_0$ (where $l_0 = \Delta T^*/dT/dz = 1000$ or 100 m, respectively). If β exceeds the critical value by more than three this propagating distance becomes even smaller. (It is important to note that the maximum ΔT^* coincides with the maximum β at the first step of the calculation, so that ΔT^* will not greatly enhance dyke penetration.)

The solutions of Rubin (1993a, 1995) are for Newtonian fluids. Solutions for power law fluids are unavailable, but the concentration of stresses at the tip of dykes might decrease the viscosity of power law rocks considerably and allow a prompter viscous response to tip stresses so that cracking may become more difficult than for Newtonian fluids. The approach of this paper is to calculate the velocity and the temperature gradient in front of diapirs rising through power law fluids, but to assume Newtonian behaviour during dyke propagation and use Rubin's results. This approach favours dyke propagation. In Equation (2), for a fixed magma chamber, dT/dz does not depend on the magma pressure and $\beta \propto p^{-5}$. However, when considering diapirs, an increase in p causes an increase in diapir velocity proportional to p^n (where n is the power law exponent of crustal rocks; Weinberg & Podladchikov 1994). As $dT/dz \propto Nu$ (from Daly & Raefsky 1985) and $Nu \propto Pe^{\frac{1}{2}} \propto V^{\frac{1}{2}} \propto p^{n/2}$ (see Weinberg & Podladchikov 1994)

then $dT/dz \propto p^{-n/2}$ (where Nu is the Nusselt number). Thus for diapirs $\beta \propto p^{n-5}$.

As a diapir rises through the crust, it warms up its surroundings. The temperature gradient around the diapir is controlled by its size, velocity and magma temperature, as well as the temperature and thermal diffusivity of the wall rocks. The temperature gradient also varies around the diapir, being steepest at the top and shallowest at the tail, as rocks at the tail spent a longer time in the vicinity of the diapir. The size and velocity of the diapir control the width of the thermal boundary layer through the Peclet number ($Pe = Vr/\kappa$, where V is the diapir's velocity). As the diapir's velocity continuously decreases as the diapir rises into colder and stiffer crust, Pe decreases and the magma heat propagates further into the surroundings, decreasing the temperature gradient (dT/dz) and increasing the survival chances of initiating dykes (decrease in β^*). In this way, dykes that would have rapidly frozen when leaving a fast rising lower crustal diapir (high β^*) may become successful in stiffer shallower crust (low β^*).

The temperature gradient in front of the diapir is calculated at every step by first determining, based on Pe, the distance from the diapir's surface in which the temperature decays to $1/e$ of the diapir's temperature (δ_T as defined by Daly & Raefsky 1985), following the methodology described in Weinberg & Podladchikov (1994). In the calculations the initial excess temperature (ΔT^*) equals the initial temperature difference between magma and surrounding solid rock ($\Delta T^* = \Delta T_0$). In this instance, the width of the thermal aureole is generally larger than, but of the same magnitude, as Rubin's (1995) l_0, the distance that a dyke would propagate before it reached subsolidus crust. Therefore, it is assumed here that the temperature gradient within this thermal aureole effectively controls dyke propagation. Because the influence of variation in dT/dz in dyke propagation is unknown, dT/dz is assumed to be arbitrarily linear within the thermal aureole (Fig. 2) and is

$$\frac{dT}{dz} = \frac{T_m - T_c}{\delta_T}\left(1 + \frac{1}{e}\right) \qquad (3)$$

where T_m is magma temperature, T_c is undisturbed crustal temperature at the crustal depth δ_T above the top of the diapir. As the diapir rises from step to step, the diapir and the undisturbed surroundings cool and a new temperature gradient is calculated (see Weinberg & Podladchikov 1994 for the cooling of the diapir).

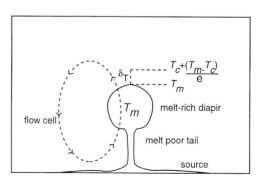

Figure 2 A melt-rich diapir leaves a melt-poor tail behind and causes convection of the surrounding crust. Convection feeds the source with new, fresh material from shallower crustal levels, that may warm up, melt and give rise to a new diapir. The temperature gradient above the diapir is found by determining δ_T (the distance from the top in which the magma temperature decays to $1/e$) and assuming a linear gradient within that zone (see text for details).

2.2. Results

The calculation starts with a spherical body of magma at 880°C, 100°C warmer than the surroundings, which are at the magma solidus temperature ($\Delta T^* = \Delta T_0 = 100°C$, $T = 780°C$) and a starting depth of 50 km. The temperature of the crust decreases upwards linearly by 15·6°C/km. The magma pressure is assumed to be constant, which disregards any pressure changes related to magma solidification and volatile exsolution. The spherical diapir rises and β is calculated at each step. If at any point during ascent $\beta \leqslant \beta_{crit}$ the calculation is stopped, because the diapir will cease to exist as its magma is drained rapidly through propagating dykes. The same procedure was carried out for a series of diapirs of different sizes and viscosities (Fig. 3). As expected from Equation (2), increasing magma pressures (increasing radii in Fig. 3) allow more viscous magmas to crack the crust and rise in dykes. Highly viscous magmas rise as diapirs without ever being drained by dykes. Conversely, low viscosity magmas will immediately go from the initial geometry into dykes. There is, however, a zone of intermediate viscosity in which ascent starts as diapirs and swaps to dykes at some depth. Surprisingly, this intermediate zone occurs at a range of viscosities that corresponds broadly to that of magmas of intermediate composition. (Note that wet, silica-rich magmas may have viscosities as low as 10^5 Pa s, but the tendency of wet magmas to solidify when decompressed is likely to cause them to remain close to their source whether they rise as diapirs or dykes.)

The influence of the initial temperature difference between magma and solid crust, ΔT_0, was studied. When $\Delta T_0 = 0$ (the case of a hot crust surrounding the magma), the temperature gradient above the magma body is the geothermal gradient, generally a small value that would greatly favour dyke propagation. The dynamics of magma segregation within the source is likely to give rise, through the flow of magma and advection of heat, to magma pools within the source region that are warmer than their immediate melt–depleted surrounding rocks ($\Delta T > 0$; e.g. Fountain *et al.* 1989). It was found here that, above some critical value, changes in ΔT_0 do not greatly influence either the depth at which the swap occurs or the viscosity that limits the fields in Figure 3. This is because for ΔT_0 above critical, warm magma diapirs will cool faster than cool diapirs and their temperatures tend to converge rapidly as they rise. For the input parameters of Figure 3 this critical value was found to be between 5 and 10°C. For temperatures below the critical value, the temperature gradient becomes so small that dykes will propagate from the very beginning of the calculations even for high viscosity magmas. For values around the critical value, small temperature differences may cause significant changes in the distance diapirs are able to rise before they go on to dykes, especially for small diapirs.

The rheology of Westerly granite used in these calculations is particularly suited for diapirs, as it yields a low viscosity when submitted to common magma buoyancy stresses. If stiffer rocks such as mafic granulites had been used, the domain in Figure 3 in which dykes dominate would have been larger, and perhaps only very viscous magmas would be unable to give rise to dykes. Similarly, the spherical initial geometry of the diapir gives rise to relatively fast initial velocities and steep temperature gradients that inhibit dyke propagation. If the initial geometry of the buoyant body was a layer or an oblate ellipsoid (of long horizontal dimensions), its initial velocity would be slow and the temperature gradient at the initial stages would be shallow and favour dykes. However, this may be partly counteracted by a decrease in magma pressure for these other geometries (remember $\beta^* \approx p^{n-5}$). Another unexplored possibility is the propagation of dykes from the sides of diapirs. These areas are warmer than the top (wider thermal boundary layer) and dykes could be favoured.

These results should be seen as a limited application of Rubin's results. Several important aspects have not been considered such as the increase in magma viscosity as the magma cools; the faster cooling of the magma close to the margins; the changes in the relation between excess magma temperature (ΔT^*) and temperature gradient (dT/dz) that controls the length of dykes as the diapir rises; changes in magma pressure due to solidification and gas exsolution; and the improved propagation of dykes by the transverse flow of magma within the dyke induced by rising bubbles or boundary roughness (Carrigan *et al.* 1992). Despite the simplifications the results corroborate what geological observations had already suggested, and what Rubin (1995) concluded: that the transport of felsic and mafic magmas may differ simply because of differences in magma viscosity; it also explains the often observed radial pattern of dykes around felsic, elliptical plutons (Paterson & Vernon 1995). These results also add that the ascent of magmas as diapirs may be suppressed at some point by the successful propagation of dykes.

3. Discussion

3.1. Shear zone control on emplacement and ascent location

The suggestion that faults control magma ascent (e.g. Hutton 1988; Hutton *et al.* 1990) is appealing because of the common association between regional-scale faults and plutons and because faults have the potential of creating, through extension, the necessary space for the emplacement of large magma volumes. However, there are still several aspects of this relation that remain unclear.

Before discussing unsolved problems, it is worth discussing

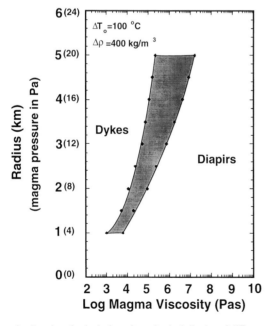

Figure 3 Results of calculations for spherical diapirs of different radii and magma viscosity rising through crust of Westerly granite rheology ($n = 1·9$, $A = 2 \times 10^{-4}$ MPa^{-n}/s and $E = 1·41 \times 10^5$ kJ/mol). 'Dykes' define conditions in which the magma leaves the spherical chamber at the initial set-up conditions; 'diapirs' defines conditions in which diapirs rise to their freezing depth without giving rise to successful dykes; and the intermediate field (shaded area) defines the conditions under which diapirs swap to dykes at some point during ascent. This field corresponds to viscosities of magmas of intermediate composition (10^3–10^6 Pa s, depending on magma pressure).

recent developments regarding space for magma emplacement. Although numerous plutons have been shown to be a collection of sheets (e.g. Hutton 1992; McCaffrey 1992), the largest volume of felsic magmas resides in plutons that better fit models of steady-state magma chambers. If extension or the heat supply to emplacement levels is not fast enough, each magma batch will freeze before the next one arrives, resulting in a collection of frozen sheets (Sleep 1975). It was unknown until recently whether extension along faults in the continental crust is sufficient to open space rapidly enough to give rise to a steady-state felsic magma chamber in cold crust. Hanson and Glazner (1995) showed that this is possible and that it depends on several parameters such as the heat content of the magma (the difference between the initial temperature and solidus temperature, latent heat and heat capacity), the undisturbed temperature of the surroundings at emplacement level, heat diffusivity, the extension rate and the volume of the initial magma chamber.

Fracture resistance of elastic (and viscoelastic) rocks can be neglected for wide, self-propagating dykes (Lister & Kerr 1991; Rubin 1993a). Short initiating dykes, however, may greatly depend on pre-existing cracks to grow to the size where they become independent and self-propagating (Rubin 1993a). However, it seems unlikely that cracks would remain open in the lower crust, particularly in the neighbourhood of hot, partially molten zones, because the stress around the fracture would cause rocks to flow and close it. Faults in the lower crust are therefore unlikely to provide cracks that could be used by growing dykelets.

As rocks tend to soften when strained (Drury *et al.* 1991), active or inactive shear zones cutting across a partially molten zone provide potential low pressure sites that may act as melt sinks, focusing magma migration (Fig. 1). The extent to which such low pressure sites might influence melt migration is unknown, but a shear zone a few kilometres wide may be able to drain magma from a large volume of surrounding rocks. Whatever the preferred mechanism of magma ascent, the accumulation of magma in the shear zone within the source can control the site where magma ascends, explaining the often observed spatial closeness between plutons and faults. If the magma is unable rise as dykes (as discussed earlier), the low viscosity of the shear zone associated with a high melt concentration may lead to early doming and diapirism close to the shear zone.

Finally, and perhaps most importantly, the interaction between the melt and an active shear zone is poorly known. Would an active shear zone attract or repel melt? Does the high strain rate during shearing expel melt out of the shear zone, as in the experiments of Dell'Angelo and Tullis (1988)? Or would the permeability within the zone be so much higher than the surroundings that melt flows towards the zone and is flushed vertically? If movement along the fault is not constant, what happens to the melt during faulting and during repose intervals?

In summary, the opening of magma chambers by faults is plausible, but is likely to be a long drawn out process and to generate sheeted intrusions when extension or heat input are slow. Although pre-existing fractures may help short dykes to get started, large dykes are independent of any crustal weaknesses. The main point here regarding the influence of shear zones is that rather than controlling the ascent mechanism, low viscosity shear zones cutting across the magma source may control the site of magma ascent by focusing melt accumulation in the source and providing a favourable site for the initiation of diapirs or dykes. The close spatial relationship between plutons and shear zones does not necessarily imply that the latter controlled either magma ascent or emplacement.

3.2. Diapir pathways

Diapirs are expected to leave along their path characteristic structures that should be easily recognisable in the field. An argument that has been raised against diapirism is the lack of field examples that indicate their paths through the crust (see Clemens & Mawer 1992; Petford, this issue). I argue here that the lack of such field examples may be partly due to the difficulty for structural geologists to assign observed structures to diapirs that are no longer there, and partly due to the simplified picture of diapirism emerging from models, compared with the structural complexities arising from diapirs travelling through heterogeneous crust that may undergo syn- or post-emplacement deformation.

As pointed out earlier, it seems that diapirs require very special (unusual?) crustal conditions to rise to shallow crustal levels (< 10 km). Typically, diapirs would rise to the 10–20 km depth range, so that their pathways may only be exposed in deep crustal sections. Not only are these sections rarely exposed, but they very often present complex deformation patterns. The results of simplified numerical and laboratory models suggest that the structures at the tail of a diapir should be a vertical narrow cylinder of intensely sheared rocks with subvertical lineations, showing a diapir-up sense of shear and an increase in metamorphic temperature towards the centre where granitic rocks could be found. The cylinder diameter should be of the order of the diapir diameter, i.e. a few kilometres or less depending on the power law exponent of the wall rocks, and sometimes a rim syncline could have developed (see Weinberg & Podladchikov 1995). The *P–T–t* history of a rock sample in the aureole should be one of fast heating followed by slow cooling, accompanied by accelerating decompression (as the diapir approaches the sample), followed by decelerating decompression (as the diapir leaves it behind). If the crust is undergoing deformation during diapirism, the hot and sheared tail of diapirs might localise and rapidly become overprinted by tectonic strains. Even after the tail cools, strain softening due to earlier shearing in the tail might localise later strains. A possible result of combining the deformation caused by the passage of diapirs and contemporaneous or later deformation is the development of large kilometre-scale steeply plunging sheath folds. A possible example of a diapir tail is the Vrådal granite in Norway (Fig. 4; Sylvester 1964), where several of these features have been described.

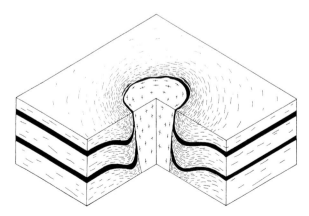

Figure 4 Vrådal pluton in Norway (redrawn from Sylvester 1964) showing structures typical of those expected inside and around diapirs or their tails, such as the circular shape, the concentric pattern of foliation increasing towards the contact and the rim synform in the wall rocks.

3.3. Why do felsic plutons balloon?

Apart from the main question as to how felsic dykes begin, two very puzzling questions remain to be answered by the proponents of felsic transport mainly through dyke. Firstly, as pointed out by Rubin (1995), why do granitic magmas balloon when many large basaltic magmas very often seem not to (e.g. Muscox, Canada and the Skaergaard, Greenland)? This is particularly puzzling because the heat content of basaltic magmas is much higher than that of granitic magmas, and they should therefore be able to soften and viscously deform their surroundings more easily than felsic magmas. A possible answer for that question suggested by R. Kerr (pers. comm.) is that the single magma batch that formed Skaergaard filled the chamber on too short a time-scale too allow viscous deformation of the surroundings. A second point, brought up by Paterson (pers. comm.) is why should granitic magmas be able to deform their surroundings at the initially cold emplacement level viscously, but be unable to deform the hot and soft lower crust viscously?

4. Conclusions

The main conclusion of this discussion is that the transport of felsic magmas through the crust is most likely to rely on a combination of mechanisms, which may simultaneously or sequentially control ascent and emplacement. A single mechanism seems to be unable to overcome all physical barriers as well as explaining the common features of plutons and batholiths (Paterson & Fowler 1993; Petford, this issue). A single diapir seems to be unable to reach shallow crustal levels and their pathways through the crust are yet to be clearly described. Felsic dykes, on the other hand, may be unable to get started because freezing may close the dykes faster than they propagate. The ideal conditions for dyke initiation occur in extensional environments, when melt segregation gives rise to a dense interconnected network of veins that drains within the source into a few large high pressure dykes. Dykes may be inhibited by shortening environments associated with segregation mechanisms that are unable to give rise to large dykes (e.g. porous flow or sparse disconnected veins). If dykes are unable to drain the source, or if the drainage is too slow compared with viscous deformation of the source and surroundings, then diapirs may dominate. The influence of shear zones in the initiation of dykes is unclear, but the low viscosity expected in those zones is likely to focus melt migration in the source and provide a favourable site for the initiation of both dykes and diapirs.

The results of calculations of the temperature evolution of diapirs and surroundings suggest that diapirs of intermediate composition may start their ascent as diapirs and swap to dykes as they slow down when reaching stiffer rocks. Large felsic diapirs may also undergo this swap, but more generally felsic magmas will tend to rise as diapirs whereas mafic magmas rise through dykes. Future work should concentrate on finding examples of diapir pathways by taking into consideration the possible interaction between diapirs and syn- or post-emplacement regional deformation. There is also a need to understand better the part that melt segregation may play in initialising dykes, and to understand the interaction of active and inactive shear zones with segregation and dyke initiation. Most importantly, we need to know more about the interplay between the several mechanisms that might enhance or inhibit the transport of felsic magma through the crust.

Acknowledgements

I thank Ross Kerr for numerous discussions on all aspects of granitic processes discussed in this paper. I also thank Scott Paterson and Nick Petford for careful revision of the paper.

References

Allibone, A. H. & Norris, R. J. 1992. Segregation of leucogranite microplutons during syn-anatectic deformation: an example from the Taylor Valley, Antarctica. J METAMORP GEOL 10, 589–600.

Bateman, R. 1985. Aureole deformation by flattening around a diapir during in situ ballooning: the Cannibal Creek Granite. J GEOL 93, 293–310.

Brown, M. 1994. The generation, segregation, ascent and emplacement of granite magma: the migmatite-to-crustally derived granite connection in thickened orogens. EARTH SCI REV 36, 83–130.

Bruce, P. M. & Huppert, H. E. 1989. Thermal control of basaltic fissure eruptions. NATURE 342, 665–7.

Brun, J. P. & Pons, J. 1981. Strain patterns of pluton emplacement in a crust undergoing non-coaxial deformation, Sierra Morena, Southern Spain. J STRUCT GEOL 3, 219–29.

Buddington, A. F. 1959. Granite emplacement with special reference to North America. GEOL SOC AM BULL 70, 671–747.

Carrigan, C. R., Schubert, G. & Eichelberger, J. C. 1992. Thermal and dynamical regimes of single- and two-phase magmatic flow in dikes. J GEOPHYS RES 97, 17 377–92.

Clemens, J. D. & Mawer, C. K. 1992. Granitic magma transport by fracture propagation. TECTONOPHYSICS 204, 339–60.

Corriveau, L. & Leblanc, D. 1995. Sequential nesting of magma in marble, southwestern Grenville Province, Quebec: from fracture propagation to diapirism. TECTONOPHYSICS 246, 183–200.

Courrioux, G. 1987. Oblique diapirism: the Criffel granodiorite/granite zoned pluton (southwest Scotland). J STRUCT GEOL 9, 313–30.

Cruden, A. R. 1990. Flow and fabric development during the diapiric rise of magma. J GEOL 98, 681–98.

Cruden, A. R. & Aaro, S. 1992. The Ljugaren granite massif, Dalarna, central Sweden. GFF 114, 209–25.

Daly, S. F. & Raefsky, A. 1985. On the penetration of a hot diapir through a strongly temperature-dependent viscosity medium. GEOPHYS J R ASTRON SOC 83, 657–81.

Dell'Angelo, L. N. & Tullis, J. 1988. Experimental deformation of partially melted granitic aggregates. J METAMORP GEOL 6, 495–515.

Drury, M. R., Vissers, R. L. M., Van der Wal, D. & Strating, E. H. H. 1991. Shear localisation in upper mantle peridotites. PURE APPL GEOPHYS 137, 439–60.

Emerman, S. & Marrett, R. 1990. Why dikes? GEOLOGY 18, 231–3.

Fountain, J. C., Hodge, D. S. & Shaw, R. P. 1989. Melt segregation in anatectic granites: a thermo-mechanical model. J VOLCANOL GEOTHERM RES 39, 279–96.

Grout, F. F. 1945. Scale models of structures related to batholiths. AM J SCI 243A, 260–84.

Guineberteau, B., Bouchez, J.-L. & Vigneresse, J. L. 1987. The Mortagne granite pluton (France) emplaced by pull-apart along a shear zone: structural and gravimetric arguments and regional implications. BULL GEOL SOC AM 99, 763–70.

Hacker, B. R., Yin, A., Christie, J. M. & Davis, G. A. 1992. Stress magnitude, strain rate, and rheology of extended middle continental crust inferred from quartz grain sizes in the Whipple mountains, California. TECTONICS 11, 36–46.

Hansen, F. D. & Carter, N. L. 1982. Creep of selected crustal rocks at a 1000 MPa. EOS, TRANS AM GEOPHYS UNION 63, 437.

Hanson, R. B. & Glazner, A. F. 1995. Thermal requirements for extensional emplacement of granitoids. GEOLOGY 23, 213–6.

Hutton, D. H. 1988. Granite emplacement mechanisms and tectonic controls: inferences from deformation studies. TRANS ROYAL SOC EDINBURGH EARTH SCI 79, 245–55.

Hutton, D. H. W. 1992. Granite sheeted complexes: evidence for the dyking ascent mechanism. TRANS ROYAL SOC EDINBURGH EARTH SCI 83, 377–82.

Hutton, D. H. W. & Reavy, R. J. 1992. Strike-slip tectonics and granite petrogenesis. TECTONICS 11, 960–7.

Hutton, D. H. W., Dempster, T. J., Brown, P. E. & Becker, S. D. 1990. A new mechanism of granite emplacement: intrusion in active extensional shear zones. NATURE 343, 452–55.

Ingram, G. M. & Hutton, D. H. W. 1994. The Great Tonalite Sill: emplacement into a contractional shear zone and implications for Late Cretaceous to early Eocene tectonics in southeastern

Alaska and British Columbia. BULL GEOL SOC AM **106,** 715–28.

Kerr, R. C. & Lister, J. R. Comment on 'On the relationship between dike width and magma viscosity' by Y. Wada. J GEOPHYS RES **100,** 15 541.

Lister, J. R. 1995. Fluid-mechanical models of the interaction between solidification and flow in dykes. *In* Baer, G. and Heimann, A. (eds) *Physics and chemistry of dykes,* 115–24. Rotterdam: Bakema.

Lister, J. R. & Kerr, R. C. 1991. Fluid-mechanical models of crack propagation and their application to magma transport in dykes. J GEOPHYS RES **96,** 10 049–77.

Mahon, K. I., Harrison, T. M. & Drew, D. A. 1988. Ascent of a granitoid diapir in a temperature varying medium. J GEOPHYS RES **93,** 1174–88.

Marsh, B. D. 1982. On the mechanics of igneous diapirism, stoping, and zone melting. AM J SCI **282,** 808–55.

McCaffrey, K. J. W. 1992. Igneous emplacement in a transpressive shear zone: Ox Mountains igneous complex. J GEOL SOC LONDON **149,** 221–35.

McKenzie, D. 1984. The generation and compaction of partially molten rock. J PETROL **25,** 713–65.

Miller, E. G. & McLellan, E. L. 1986. Textural controls on the viscosity and critical melt percentage (RCMP) of partially molten mushes: a model for filter-pressing in layered migmatites. GEOL SOC AM ABSTR PROGRAMS **18,** 696.

Morris, S. 1982. The effects of a strongly temperature-dependent viscosity on slow flow past a hot sphere. J FLUID MEC **124,** 1–26.

Paterson, S. R. & Fowler, T. K. Jr 1993. Re-examining pluton emplacement processes. J STRUCT GEOL **15,** 191–206.

Paterson, S. R. & Fowler, T. K. Jr 1994. Re-examining pluton emplacement processes: reply. J STRUCT GEOL **16,** 747–8.

Paterson, S. R. & Vernon, R. H. 1995. Bursting the bubble of ballooning plutons: a return to nested diapirs emplaced by multiple processes. GEOL SOC AM BULL **107,** 1356–80.

Petford, N. 1997. Granitoid transport mechanism: dykes or diapirs? TRANS R SOC EDINBURGH EARTH SCI **88,** 000–000.

Petford, N. & Atherton, M. P. 1992. Granitoid emplacement and deformation along a major crustal lineament: the Cordillera Blanca, Peru. TECTONOPHYSICS **205,** 171–85.

Petford, N., Kerr, R. C. & Lister, J. R. 1993. Dike transport of granitoid magmas. GEOLOGY **21,** 845–8.

Petford, N., Lister, J. R. & Kerr, R. C. 1994. The ascent of felsic magmas in dykes. LITHOS **32,** 161–8.

Pitcher, W. S. 1979. The nature, ascent and emplacement of granitic magmas. J GEOL SOC LONDON **136,** 672–62.

Ramberg, H. 1967. *Gravity, deformation and the Earth's crust.* London: Academic Press.

Ramsay, J. G. 1989. Emplacement kinematics of a granite diapir: the Chindamora batholith, Zimbabwe. J STRUCT GEOL **11,** 191–209.

Ribe, N. 1983. Diapirism in the earth's mantle: experiments on the motion of a hot sphere in a fluid with temperature dependent viscosity. J VOLCANOL GEOTHERM RES **16,** 221–45.

Robin, P.-Y. 1979. Theory of metamorphic segregation and related processes. GEOCHIM COSMOCHIM ACTA **43,** 1587–600.

Rubin, A. M. 1993a. Dikes vs. diapirs in viscoelastic rock. EARTH PLANET SCI LETT **119,** 641–59.

Rubin, A. M. 1993b. On the thermal viability of dikes leaving magma chambers. GEOPHYS RES LETT **20,** 257–60.

Rubin, A. M. 1995. Getting granite dikes out of the source region. J GEOPHYS RES **100,** 5911–29.

Schwerdtner, W. M. 1990. Structural tests of diapir hypothesis in Archean crust of Ontario. CAN J EARTH SCI **27,** 387–402.

Sleep, N. H. 1975. Formation of oceanic crust: Some thermal constraints. J GEOPHYS RES **80,** 4037–42.

Sleep, N. H. 1988. Tapping of melt by veins and dikes. J GEOPHYS RES **93,** 10 255–72.

Stevenson, D. J. 1989. Spontaneous small-scale melt segregation in partial melts undergoing deformation. GEOPHYS RES LETT **16,** 1067–70.

Sylvester, A. G. 1964. The Precambrian rocks of the Telemark area in south central Norway. III. Geology of the Vrådal granite. NORSK GEOL TIDSSKR **44,** 445–82.

Wada, Y. 1994. On the relationship between dike width and magma viscosity. J GEOPHYS RES **99,** 17 743–55.

Wada, Y. 1995. Reply 'On the relationship between dike width and magma viscosity' by Y. Wada. J GEOPHYS RES **99,** 17 743–55.

Weinberg, R. F. 1994. Re-examining pluton emplacement processes: discussion. J STRUCT GEOL **16,** 743–6.

Weinberg, R. F. & Podladchikov, Y. 1994. Diapiric ascent of magmas through power-law crust and mantle. J GEOPHYS RES **99,** 9543–59.

Weinberg, R. F. & Podladchikov, Y. Y. 1995. The rise of solid-state diapirs. J STRUCT GEOL **17,** 1183–95.

Wickham, S. M. 1987. The segregation and emplacement of granitic magma. J GEOL SOC LONDON **144,** 281–97.

ROBERTO FERREZ WEINBERG, Research School of Earth Sciences, The Australian National University, ACT 0200, Canberra, Australia.

Transactions of the Royal Society of Edinburgh: Earth Sciences, **87**, 105–114, 1996

Dykes or diapirs?

Nick Petford

ABSTRACT: Until the last few years, diapirism reigned supreme among granitoid ascent mechanisms. Granitoid masses in a variety of material states, from pure melt through semi-molten crystal mushes to solid rock, were believed to have risen forcefully through the continental crust to their final emplacement levels in a way analogous to salt domes. The structural analogy between granite plutons and salt diapirs, which gained acceptance in the 1930s, has clearly been attractive despite the pessimistic outcomes of thermal models and, at best, ambiguous field evidence.

In contrast with traditional diapiric ascent, dyke transport of granitoid magmas has a number of important implications for the emplacement and geochemistry of granites that have yet to be fully explored. Rapid ascent rates of $\approx 10^{-2}$ m/s predicted for granite melts in dykes (cf. m/a for diapirs) mean that felsic magmas can be transported through the continental crust in months rather than thousands (or even millions) of years, and that large plutons can in principle be filled in $< 10^4$ a. Granitic melts are likely to rise adiabatically from their source regions, leading to the resorption of any entrained restitic material. Ascending melts in dykes close to their critical minimum widths may have little opportunity to assimilate significant amounts of country rock, and if source extraction is sufficiently rapid, most crustal contamination will be restricted to the site of emplacement. Rates of pluton and batholith inflation will be determined by the amount and rate of melt extraction at source.

The construction of large plutons and batholiths piecemeal from a number of magma pulses separated by periods of relative quiescence provides a means of reconciling rapid ascent rates with times for batholith construction based on average rates. Field and seismic evidence that shows batholiths as large, sheet-like structures with flat roofs and floors is consistent with a general model for plutons and batholiths as laccolith-type structures, fed from depth by dykes. The overall geometry of this type of structure helps ameliorate the space problem, which developed as a consequence of the unrealistic volumes of upwelling granite associated with the classical diapir model.

KEY WORDS: granite ascent, magma pulsing, emplacement, laccoliths.

Although James Hutton first suggested in 1788 that granite magma developed at great depths in the earth and was made to invade and break (intrude) into the surrounding strata, before about 1930 geologists working on the origin of granitic rocks had not concerned themselves to any great extent with the mechanisms of magma ascent. Daly (1903) emphasised the importance of brittle processes such as stoping, but it was not until the detailed map work of Cloos (1925) and the structural studies of Balk (1937) that geologists became interested in the patterns of magmatic flow seen in and around plutons. Impressed by structural and geometrical similarities between granite plutons, anticlinal salt domes and mantled gneiss domes at outcrop, these features were taken as *prima face* evidence for a common mode of emplacement. Unfortunately, the waters at the time were being muddied by the ongoing debate between transformists and magmatists as to the formation and material status of granite in general. Thus, although Rast (1970), in reviewing the main factors involved in the ascent of granitic magmas, was able to cite diapirism, gas fluxation, tectonic squeezing and overpressure in the source region as viable mechanisms, the role of dykes was not addressed. It is interesting to note in the published work of the time a strong tendency to treat granite as somehow different from other igneous rocks, particularly basalt. This view is exemplified in the writings of H. H. Read, who was clearly unimpressed with the idea that basalt and granite have anything much in common:

'that (the one magma view proposed by Bowen and others) degrades granitic magma to the status of the dregs of the primary basaltic' (Read 1947: 6).

Reaction of this kind against the 'one magma' view may explain why the idea of granitic magmas ascending along dykes was not addressed until fairly recently.

This review is divided broadly into two sections. In the first, which is concerned with the mode of ascent, I begin by tracing the origin and subsequent development of ideas concerning granites as diapiric structures before assessing the role of dykes and fractures in transporting granitic melt in the crust. In the second, more open section, some of the problems of reconciling rapid dyke ascent with emplacement and pluton filling times are discussed, along with ancillary observations related to magma pulsing, the heat source and the shapes of plutons and batholiths with depth. Finally, a speculative model is proposed where batholiths are assembled from tabular sills or laccoliths fed from depth by granitic feeder dykes.

1. A brief history of granite diapirs

The word diapir, from the Greek verb $\delta\iota\alpha\pi\varepsilon\iota\rho\omega$ to pierce, was first introduced by Mrazec (1927) to describe anticlinal salt domes in Romania. Other contemporary studies that drew attention to the similarities in the structure of granites in orogenic belts and salt and other evaporite diapirs include those of Nicolesco (1929) and Wegmann (1930). Nicolesco (1929) describes what he calls volcanic anticlinal diapirs from Madagascar, and suggested an origin for viscous granitic rocks in the 'pyrosphere', a layer of volcanic source material located deep in the crust that supposedly ringed the earth.

Wegmann, like many continental European geologists of the time, was a transformist, who believed that granites were formed *in situ* in the solid state from other crustal rocks via the transforming effects of heat and 'rock juices' liberated during large-scale earth movements (the so-called *ichor* of Sederholm 1926). Wegmann considered that plutons were emplaced in the solid state after having being first made into a granite layer by metasomatism. He noted the occurrence of granite bodies in most orogenic belts and regarded the main driving force for ascent as due to tectonic squeezing. Wegmann also acknowledged that many granite plutons are multiple, and suggested that during orogenesis, granitic layers of different composition become mobilised successively and intrude each other. This line of reasoning was championed by Eskola (1949), who considered that two orogenic stages were necessary for the formation of granites and associated mantled gneiss domes. However, he was also of the opinion that granite may have intruded through the crust during orogenesis as convection currents made up solely of granitic material.

The first experiments on the dynamics of diapiric ascent are described by Grout (1945), who used mixtures of oil and corn syrup to simulate rising magmas. It is Grout who was responsible for introducing the popular concept of a diapir as an inverted teardrop piercing the crust (Fig. 1). Although the experiments of Grout do not stand up in the face of rigorous scaling, the later experiments of Ramberg (1967, 1970) and Dixon (1975) and the finite element modelling of Berner *et al.* (1972) are much more sophisticated. Using centrifuge techniques, Ramberg modelled the various geometric features of

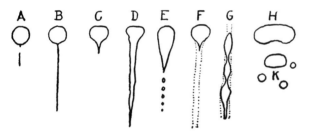

Figure 1 Shapes of fluid 'diapirs' rising in a viscous matrix (reproduced from Grout 1945, *American Journal of Science*, with permission). (A and B) Cylinder oil diapirs in corn syrup matrix at room temperature; (C and D) cold oil diapirs in cool and warm syrup matrix; (E) warm oil in cold syrup; (F) water diapir in syrup; (G) water diapirs following in the trail of (F); and (H and K) air bubble diapir in water. Note the distinctive trail left in the wake of diapir (F). Where is the evidence for similar structures in the crust?

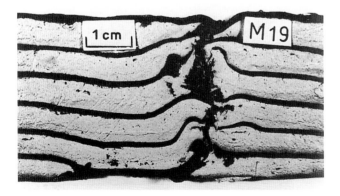

Figure 2 Modelling magma ascent: experiment M19 (reproduced from Ramberg 1970, with permission J. Wiley & Sons Ltd), showing the ascent path and geometry taken by a $KMnO_4$ solution (magma) rising under gravity through a layered overburden. These and similar structures were generated with a viscosity contrast of $c.\ 10^{10}$ and are thus more realistically scaled to reflect magma ascent than low viscosity contrast diapirs and domes.

diapiric development, from initial doming in the source region through upwards migrating bodies attached to the source by a thin stalk which eventually mushroomed out at shallow crustal levels.

Although the interaction between the evolving pluton and envelope is well demonstrated, it is applicable only where a low viscosity contrast ($< 10^3$) exists between the diapir and country rock, and as such is not valid for ascending melts (see also Castro 1987), although the strong similarity between experimentally produced structures and those seen in and around many salt domes, particularly their rounded shape at outcrop, continued to reinforce the idea that granite magmas ascend dipirically (e.g. Sorgenfrei 1971). When Ramberg (1981: 333–9) did try to model the ascent of magma with a viscosity contrast appropriate for magma ascent through the crust ($c.\ 10^{10}$) using a $KMnO_4$ solution, the 'melt' rose upwards rapidly through channels and fissures (Fig. 2).

One obvious criticism of the experiments of Grout and Ramberg is the lack of any thermal constraints. Marsh (1982) was able to assess the effects of heat transfer between the rising blob and the surrounding country rock numerically using a hot Stokes flow model, which describes the ascent a spherical body of uniform temperature in an isoviscous fluid (it should be noted that Marsh was not concerned specifically with granite in his study). In essence, Marsh (1982) found that the lithosphere is too cool to allow a hot Stokes magma diapir to rise to crustal levels, and diapirism was confined to the lower lithosphere and asthenosphere where the rocks were hotter and more ductile (see also Ribe 1983). A similar conclusion was reached in a numerical study by Mahon *et al.* (1988), who concluded that even allowing for a stress-free boundary condition during ascent through a temperature varying (non-isoviscous) wallrock, diapiric ascent velocities at normal crustal temperatures are so low that ascent is not possible to upper crustal levels in time-scales less than the freezing times of large plutons ($c.\ 10^5$ a).

1.2. Power law diapirs

Although classical models for diapiric ascent are based largely on historical comparisons between granite and salt, and despite the pessimistic outcome of thermal models, it may yet be premature to rule out diapirism totally as a granite transport mechanism. Work by Weinberg and Podladchikov (1994, 1995) has shown that heat transfer alone need not be the rate-limiting step controlling diapiric ascent and they argue that the process is more effective if the rheology of the crust is considered as non-Newtonian. Here, the mechanism required to soften the crust is not thermal, as in the hot Stokes model, but dependent on strain rate, with higher strains resulting in lower country rock viscosities in the vicinity of the rising diapir. Using as an example the Tara granodiorite described by Miller *et al.* (1988), their elegant modelling suggests that for a power law crust the Tara pluton would rise diapirically from the Moho and solidify at a depth of about 15 km from the surface. Although power law diapirs still have trouble in reaching shallow crustal depths of less than 10 km, under special conditions they may rise to within 6 km of the surface (Weinberg & Podladchikov 1994, 1995). Power law diapirism need not be confined to purely upwards motion, and some (several kilometres) downward movement of a magma body may occur after solidification if it is negatively buoyant (Weinberg & Podladchikov 1995).

The main parameters controlling the depth of diapir solidification in a power law crust are the pre-exponential factor A, and the activation energy, E. Values for E (kJ/mol) and A (MPa/s) used by Weinberg and Podladchikov (1994) to represent the mid–upper continental crust are taken from experimental data on the Westerly granite. However, as pointed out by these workers, estimates based on power law ascent are very sensitive to the experimental uncertainties inherent in A and E. For example, small changes in activation energy resulted in a variation in emplacement depth that spanned the entire thickness of the crust.

The claim by Weinberg and Podladchikov (1994) that the transport of granitic magmas through the crust as diapirs is as effective as dyke-fracture transport is discussed in Section 2.4. However, even allowing for a power law rheology, the model does not account for the many high-level granites and related acidic volcanic rocks emplaced less than 5 km from the surface. Furthermore, the assumption that diapiric ascent relies largely on the development of a Rayleigh–Taylor instability in the source is challenged by work on melt segregation that relates the initiation of magma ascent to pressure changes accompanying deformation and partial melting reactions (Sawyer 1991; Brown 1994; Petford 1995; Rutter & Neumann 1995; Section 2.5.1). Experiment work by Webb and Dingwell (1990) on the non-Newtonian rheology of silicate melts may also be relevant to power law ascent models. These workers found that for strains close to the onset of non-Newtonian (viscoelastic) behaviour, the melt underwent tensile failure and breakage. It would be interesting to know under what conditions (power law or otherwise) strain rates in ascending igneous melts result in brittle failure.

1.3. Field evidence

The first person to catalogue in detail the structures that should form in the vicinity of a rising granite diapir was Wegmann (1930). These include the now classic rim synform (later exemplified in the experiments of Ramberg), rounded outcrop shape and stretching lineations at the pluton margin. DeWaard (1949) provides a structural comparison between granite and salt domes and describes some typical diapiric phenomena including folding and doming of salt (and granite) in the diapir core, synthetic folding in the country rock and radial tension joints in the overlying strata.

Without doubt the major stumbling block for the diapiric ascent of granitoid magmas is the absence of vertical shear zones and vertical or upturned bedding in the mid- to lower crust that can be related conclusively to the passage of an ascending diapir (Clemens & Mawer 1992). This problem is particularly acute given that these structures are reproduced in both experimental and numerical simulation, and thus accepted universally as essential components of all diapiric ascent models. Pitcher (1979) has termed these elusive structures 'blind diapirs'. An example of the types of experimental structures formed in the vicinity of a diapiric upwelling are given by Dixon (1975, figs 12, 13). Perhaps the most remarkable result of Dixon's modelling was the large amount of flattening (>90%) and extension (>100%) in and around the diapir. Dixon (1975) states that the results of his method for estimating finite strain around a diapiric structure (in his example a mantled gneiss dome) could be used by structural geologists to assess the deformation paths followed by rocks in and around a rising diapir. Unfortunately, the vertical structures such as shear zones or mylonites extending through large sections of the crust predicted from experiment have not been found.

A similar lack of field evidence applies to Groutian blobs, where the inward dips expected in the vicinity of the 'tails' of hundreds of diapir-like structures reported from around the world are absent (Pitcher 1979). With the exception of the infamous rim syncline, many 'diapiric' structures such as highly strained margins, attributed to power law ascent (Weinberg & Podladchikov 1995), can be explained by other

means, most notably *in situ* ballooning as described by Bateman (1985) in his study of the Cannibal Creek pluton. Sadly, the use of emplacement-related structures such as deformed aureoles as indicators of magma ascent is still a source of confusion (Paterson & Fowler 1994; Weinberg 1994).

In the case of excessively hot crust, classical diapiric ascent may be possible, although explanations such as diapir swarms heating the crust or diapirs following the trails of previous diapirs are nothing more than special pleading. However, in a comparative study of salt stocks and Archaean gneiss diapirs, at a time when the crust was undoubtedly hotter than today, Schwerdtner (1982, 1990) still could not show conclusively using field evidence that the gneiss domes of western Ontario had formed in a way similar to that proposed by Ramberg. Similar studies of granites in Archean greenstone belts have failed to find the narrow umbilical necks expected from theory and models. Other structures in greenstone belts such as recumbent sheath folds considered as possible indicators of diapir tracks (Weinberg, this issue) are more likely due to intense deformation in large, low-angle shear zones (Coward 1976).

2. Granite dykes

The simple alternative to granite ascent by diapiric rise is that granite magmas ascend in dykes the same as basalts. Following earlier ideas about the funnelling and focusing of granitic melt along faults or dyke-like conduits proposed initially by Leake (1978) and Pitcher (1979), there are currently two general models for granitic ascent in dykes, one involving ascent by magma fracture (Clemens & Mawer 1992), and the other related to ascent in fault-related conduits (Petford et al. 1993, 1994). Although both models are different geologically, with the latter emphasising the important role of crustal extension in initiating dyke conduits, their physics are essentially the same.

The historical lack of enthusiasm for dyke transport of granitic melts is especially odd given that, at crustal levels and in the lithosphere in general, magma typically ascends seismically in dykes and fractures (e.g. Spera 1980; Turcotte 1982, Spence & Turcotte 1985; Lister & Kerr 1991). Shaw (1980) gives an extended review of fracture mechanisms, emphasising the importance of dyke propagation in magma transport. In particular, magma at lithostatic pressure will give rise to tensional (effective) stresses that enable magma-filled fractures to form and propagate throughout much of the lithosphere. Basaltic melts can flow in dykes and fractures at velocities in excess of 5 m/s, with dyke emplacement controlled fundamentally by the viscosity of the fluid in the crack as opposed to the rate of crack propagation (e.g. Spence & Turcotte 1985). Although the number of granite workers speculating that plutons may have been fed by granite ascending in fractures and dykes (Pitcher & Berger 1972; Leake 1978; Pitcher 1979; Bateman 1985; Hutton 1982, 1992; Brun et al. 1990; Emerman & Marrett 1990; Petford & Atherton 1992) has been increasing since the late 1970s, it was not until 1992 that the first in-depth study by Clemens and Mawer (1992) dedicated to the topic of granite dyking as an ascent mechanism was published.

2.1. Dyke propagation

There are two forces involved in dyke propagation: buoyancy forces (as exploited in the experiments of Ramberg) and excess pressure at source (the pressure in excess of the lithostatic load). Excess pressure can drive a dyke in any direction, while buoyancy forces result in vertical motion only (Wilson & Head 1981). During vertical transport, buoyancy forces must be larger than the excess pressure, with the latter used to initiate the fracture. Excess pressure will remain constant as long as the dyke volume remains small compared with the volume of magma in the source. Once fracturing has commenced, the orientation and shape of the dyke are influenced to a large degree by the regional stress regime (Rubin & Pollard 1987), although the final emplacement geometry may differ from that during active ascent, (e.g. Lister & Kerr 1991).

The most efficient initial geometry for a dyke is a planar fissure. Simple viscous flow of magma in a vertical dyke can be expressed generally as

$$V_{ave} = \frac{\Delta \rho g w^2}{12 \mu} \qquad (1)$$

where V_{ave} is the average flow velocity, $\Delta \rho$ is the density contrast between the magma and the wallrock, w is the dyke width, μ is the magmatic viscosity and g is the acceleration due to gravity. No excess pressure is assumed and ascent is driven by buoyancy alone. For this condition to apply, the excess (elastic) pressure $(w/H)\sigma$ must be significantly less than $\Delta \rho g H$, where w is the full dyke thickness, H is the height and σ is the elastic stiffness of the wallrock (Lister & Kerr 1991; Rubin 1995). Viscous flow is characterised by a parabolic flow profile that increases from zero at the dyke wall to a maximum velocity

$$V_{max} = \frac{3V_{ave}}{2} \qquad (2)$$

at the centre of the dyke (Batchelor 1967). Equations (1) and (2) hold only for laminar, low Reynolds number (Re) flow

$$Re = \frac{V_{ave} w}{v} \qquad (3)$$

where r is the dyke radius and v the kinematic viscosity of the melt (μ/ρ). This condition is satisfied even for the lowest viscosity granitoid melts, where maximum Re numbers estimated during dyke flow are several orders of magnitude less than those generally required for turbulent flow (e.g. Tritton 1988; Fig. 3).

2.2. Control of magma properties on dyke widths

A major difference between the granite diapir and granite dyke ascent models is the fundamental control of magma

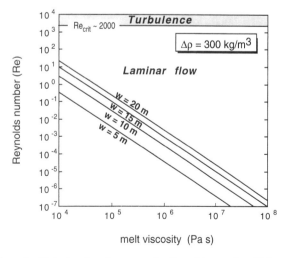

Figure 3 Plot showing the range in Reynolds number (Re) as a function of melt viscosity and dyke width. Re numbers increase with decreasing melt viscosity and increasing dyke width (w). For the wide range of conditions shown, Re numbers are subcritical, indicating that magma ascent in dykes will be governed by laminar flow.

viscosity on dyke widths and flow-rates, although excess (magmatic) pressure within the dyke will also be important if the wallrocks deform elastically. In a study of basic and felsic dykes from Peru and Japan, Wada (1994) has shown that dyke widths are strongly dependent on magma viscosity, with higher viscosities producing wider dykes. The viscosity of granitic melts has been the subject of both experimental and theoretical investigation by a variety of workers, most notably Shaw (1965). Silicate melt viscosity is related in a complex way to melt composition and structure, volatile content, temperature and to a lesser extent pressure (for a review, see Dingwell et al. 1993). Direct viscosity measurements of crystal-free granitic melts and estimated based on empirical models (Shaw 1972; Baker, in press) fall in the general range 10^4–10^8 Pa s, depending on the temperature and water content. Tonalites and granodiorites, the major pluton-forming magmas in Cordilleran batholiths, have estimated crystal-free viscosities in the range 10^4–10^5 Pa s (Petford 1995). It should not go unnoted that the viscosity of a given granitic melt is controlled primarily by the P–T–X conditions in the source region. Crystal loads are also known to strongly influence magma viscosity, although from the analysis of Wada (1994), magmas with phenocryst contents <35% have effective viscosities similar to crystal-free melts, whereas magmas with crystal contents generally in excess of 50% are too viscous to be intruded as dykes, regardless of melt composition (Kerr & Lister 1991).

2.3. Thermal considerations

Although up to six orders of magnitude more viscous that average basalt melts (c. 10^1–10^2 Pa s), granite magmas are still capable of flowing in narrow dykes and conduits without freezing. From the thermal analysis of Bruce and Huppert (1989, 1990), the flow of hot magma through cool lithosphere requires a minimum dyke thickness, below which the conductive heat loss of magma to the surrounding wallrock is greater than heat advected by magma flow and the dyke will freeze. This minimum thickness can be estimated by the following argument (Petford et al. 1994). During the early stages of dyke formation, conductive heat loss gives rise to a thermal boundary layer of thickness proportional to $(\kappa t)^{\frac{1}{2}}$ with rate of solidification $(\partial w / \partial t)$ approximated by

$$\frac{\partial w}{\partial t} \propto -\frac{\kappa}{S_\infty} \frac{1}{(\kappa t)^{\frac{1}{2}}} \qquad (4)$$

where κ is the thermal diffusivity and S_∞ is a Stefan number that relates the ratio of latent heat to the specific heat capacity of the dyke magma to the far-field temperature of the surrounding country rock. As flow continues and convection becomes dominant, a thermal boundary layer forms at height z with thickness proportional to $(wz\kappa/V_{ave})^{\frac{1}{3}}$. Thus after initial conductive cooling along the dyke margins, a thermal balance will be reached where the rate of melt-back along the margins is greater than the rate of freezing along the dyke walls, approximated by

$$\frac{\partial w}{\partial t} \propto \frac{\kappa}{S_m} \left(\frac{V_{ave}}{wz\kappa} \right)^{\frac{1}{3}} \qquad (5)$$

where S_m is a Stefan number for the dyke magma that allows for a migrating solidification front at the dyke walls. Substitution of (1), (4) and (5) into the condition that meltback is achieved at $z = H$ within the timescale $(WS_\infty)^2/\kappa$ of solidification leads (after some rearrangement) to the expression

$$w_c = C \left(\frac{S_m}{S_\infty^2} \right)^{\frac{3}{4}} \left(\frac{H\kappa\mu}{\Delta \rho g} \right)^{\frac{1}{4}} \qquad (6)$$

where w_c is the minimum critical dyke width for flow to occur without freezing, C is a constant of value 1·5 and H is the vertical height of the dyke (Petford *et al.* 1994: 166). More detailed calculations are given in Lister and Dellar (1996). Actual dyke widths may of course be much larger.

For basalt dykes, the critical minimum width, w_c, is of the order of 1 m for dykes kilometres in length. As w_c is proportional to viscosity raised to the $\frac{1}{4}$ power, granitic melts (generally 10^3–10^6 times more viscous than basalt) require dykes that are a factor of $10^{\frac{3}{4}}$–$10^{6/4}$ wider ($\approx$ 6–30 m) for dykes of similar length to prevent freezing. In contrast, maximum dyke widths are controlled, not by thermal effects, but by the rate at which melt can be supplied from depth (Lister & Kerr 1991). For granite dykes, this will depend on the amount of partial melting in the source region and the rate at which it is extracted.

The predicted relationship between magma viscosity and minimum dyke width is shown in Figure 4 as a function of density contrast between dyke melt and country rock. In general, the more mafic, higher temperature tonalites and granodiorites characteristic of Cordilleran magmatic arcs, including the Great Tonalite Sill and plutons from the Western Peninsula Ranges batholith, have the narrowest critical dyke widths and plot on the left-hand side of the diagram. Conversely, lower temperature, more viscous, silica-rich melts formed during crustal thickening events (e.g. Himalayan leucogranites) require correspondingly wider dykes to ascend without freezing and lie towards the right-hand side of the figure. Syn-tectonic plutons such as the Main Donegal lie in the middle of the diagram. Thus the variation in estimated minimum dyke width can be related in a general way to the tectonic setting in which these plutons are found. This, in turn, may reflect the type of source material being melted, with relatively mafic sources beneath the Cordilleras giving low viscosity tonalite melts, while predominantly metasedimentary(?) sources beneath the Himalaya yield higher viscosity leucogranites.

2.4. Rates of granite dyke ascent and pluton filling

Perhaps the most significant implication of dyke ascent of granitoid magmas is the extreme rapidity of magma ascent compared with diapiric rise. The analysis of Clemens and

Mawer (1992) showed that granite dyke velocities during active magma fracturing were typically of the order of 10^{-2} m/s. Similar ascent rates were obtained by Petford *et al.* (1993). These rates are orders of magnitude faster than even the most speedy diapirs, which have estimated ascent velocities, depending on their size, of between 10^{-10} and 10^{-7} m/s (Fig. 5).

To illustrate how fast dyke ascent of granitoid melt can be, I take as an example the Tara granodiorite, SE Australia, which has been the subject of previous studies on diapiric granitic ascent (Miller *et al.* 1988; Weinberg & Podladchikov 1994). Using the physical properties of crust and magma given by Miller *et al.* (1988) their table 3, and the compositional data provided by White *et al.* (1977), the Tara granodiorite, with an estimated magmatic viscosity of $10^{5·8}$ Pa s, would have a critical minimum dyke width of 5·2 m and could in principle ascend in a dyke 20 km through the crust in just 0·03 a. This is considerably faster than a diapiric ascent rate of 10^4 a given by Weinberg and Podladchikov (1994) and casts some doubt on their assertion that diapirism can be as effective in transporting magma as dyke ascent. Although the ascent time is not equal to the emplacement time, rapid ascent rates of granitic magma in dykes does allow relatively large plutons ($> 10^4$ km^3) to be filled extremely quickly (Clemens & Mawer 1992; Petford *et al.* 1993). The time (Δt) taken for a pluton of volume Q to be filled provided flow is continuous can be estimated from

$$\Delta t = \frac{Q}{V_{ave}wl} \qquad (7)$$

where V_{ave} is the flow-rate [from Equation (1)], w is the dyke width and l is the dyke length (Petford *et al.* 1993). Estimated pluton filling times are shown in Figure 6 for melt viscosities of 10^4 and 10^8 Pa s a function of density contrast ($\Delta\rho$) between the melt and wallrock responsible for driving vertical dyke flow. Typical values of $\Delta\rho$ between the granite melt and wallrocks (c. 200–400 kg/m^3) are shaded. For a 5000 km^3 pluton, a 5 m wide dyke ($l = 10$ km) and a mean $\Delta\rho$ of 300 kg/m^3, $\Delta t \approx 6$ a for a melt viscosity of 10^4 Pa s and $\approx 6 \times 10^4$ a for a melt viscosity of 10^8 Pa s.

2.5. Problems with dyke ascent

Rubin (1995) has criticised dyke transport of granitic melts as *a priori*, arguing that (unlike less viscous basaltic melts) it may not be possible for granite dykes to grow to the required critical minimum width in the source region, and although recognising the shortcomings inherent in classical models of diapiric ascent, nevertheless cautions against the notion that

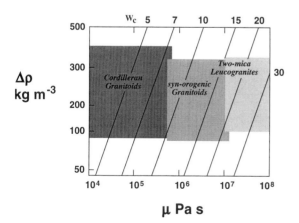

Figure 4 Plot of density contrast between magma and country rock ($\Delta\rho$) as a function of magma viscosity. Numbered contour lines show the critical minimum dyke width needed for magma ascent without freezing. Fields summarise the positions of granitoid plutons and batholiths in relation to known tectonic setting (after Petford *et al.* 1994). Small critical minimum widths define granitoids found in magmatic arc settings, whereas largest minimum widths are for high-silica leucogranites in contractional settings. The increase in critical dyke width from right to left across the diagram is due to an increase in melt viscosity of granitoid rocks found in each setting.

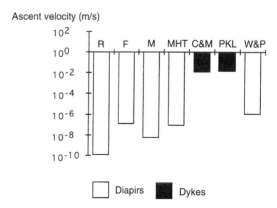

Figure 5 Histogram of ascent rates of granitic magmas from various sources in metres per second. Ascent rates are greatest for magmas transported in dykes. R, Ramberg (1967); F, Fyfe (1970); M, Marsh (1982); MHT, Mahon *et al.* (1988); C&M, Clemens and Mawer (1992); PKL, Petford *et al.* (1993); and W&P, Weinberg & Podladchikov (1994).

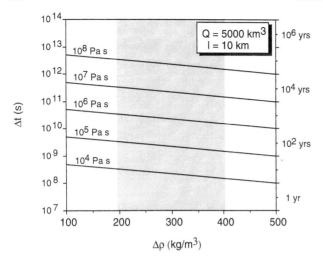

Figure 6 Estimated pluton filling times (Δt) for a range in melt viscosity (Pa s) and density contrast ($\Delta\rho = \rho_{\text{country rock}} - \rho_{\text{magma}}$). Typical crustal values of $\Delta\rho$ are shaded. Q is pluton volume and l is dyke length (both fixed). Filling times increase with increasing melt viscosity and decreasing density contrast from a minimum of ≈ 5 a (melt = 10^4 Pa s, $\Delta\rho = 500$ kg/m³) to a maximum of $\approx 10^5$ a (melt = 10^8 Pa s, $\Delta\rho = 100$ kg/m³).

all granite magmas are transported through the crust in dykes. Rubin (1995) argues that where ascent is driven by excess magma pressure at source instead of buoyancy, most (but not *all*) granite dykes would freeze shortly after the magma encountered subsolidus country rock (see also Weinberg, this issue). In his original formulation, Rubin (1993) assumed the dyke magma (at temperature T_m) entered the propagating dyke at its freezing temperature (T_s). The distance from the source at which the country rock temperature falls below T_m can be used to define a length scale $l_o = \Delta T^*|dT_o/dx|$, where $\Delta T^* = T_m - T_s$ and $|dT_o/dx|$ is the temperature gradient in the country rock. However, even for this end-member case, magma temperatures close to that of the country rock may not kill the dyke by freezing if the dyke height $H < l_o$. For the more realistic case where melt enters the dyke at some temperature above its solidus (i.e. $\Delta T^* > 0$), granite dyking from a deep source is possible at modest (c. 10 MPa) source pressures (Rubin 1995).

2.5.1. Source control on magma ascent. The conceptual model underpinning Rubin (1993, 1995) seems to be based on field observations of aplite and pegmatite veins in the vicinity of plutons, which form during the final stages of emplacement and crystallisation. Thus although this analysis may be valid for the formation of late-stage dykes and veins in the upper crust, it may not provide an accurate model of a granitic source region in the lower crust where partial melting above a mafic mantle-derived heat source has led to the formation of a mushy zone (e.g. Bergantz 1989; Petford 1995). Although the physics of melt segregation in the continental crust are still poorly understood (Brown *et al.* 1995), rates of melt production will depend on the type (i.e. fluid-present or fluid-absent) and kinetics of the partial melting reactions, themselves a product of the protolith composition. For example, Clemens & Mawer (1992) estimate volume changes during fluid-absent partial melting of quartzo-feldspathic rocks of between 2 and 20%. This raises the possibility that different source lithologies may give rise during partial melting to a range of excess source pressures, some of which are more effective than others in driving magma flow. Other factors considered by Rubin (1995) likely to aid dyke formation in a partially molten source region include viscous deformation of the wallrocks and the influx of melt through the dyke walls. Where melt is

exploiting a pre-existing fracture or fault zone at source, the thorny problem of crack initiation is removed, although for dykes longer than a few metres the difference in ascent velocity between a dyke following an existing fracture and one that is self-propagating is negligible as both are working against a confining pressure.

In summary, if the degree of partial melting and excess pressure in the source are governed by the type of melting reaction, then the physical state of the source is likely to have important knock-on effects, and may even control the final ascent mechanism (Weinberg, this issue).

2.6. Petrological consequences of dyke ascent

Rapid vertical flow in dykes may allow the ascending granitoid melts to rise adiabatically. Previous studies investigating the effects of isentropic adiabatic ascent of granitoid magmas (e.g. Sykes & Holloway 1987; Holtz & Johannes 1994) have shown that considerable resorption of entrained, restitic material will occur. For example, Holtz and Johannes (1994) have shown that rising magma with an initial 50:50 melt to solid ratio will resorb some 35% solid by volume, with the final ratio of melt to solid on emplacement as 85:15. Modelling by Clemens *et al.* (1996) has shown that for typical ascent velocities of 10^{-2} m/s, dykes may contain significant (up to 50°C) constitutional superheat. As a result, very little crystallisation will occur during ascent, and for dyke widths $\gg w_c$ negligible conductive heat loss should allow dyke magmas to be emplaced virtually free of entrained material, consistent with the observation that most granitoid magmas are non-restitic (Clemens 1989). Another consequence of the resorption of entrained material during ascent is that the reduction in effective viscosity (the ratio of suspended solids to melt) will lead to an increase in magma flow-rates, with a reduction in suspended load from 30 to 10% more than doubling the ascent velocity (Fig. 7).

Laminar (non-turbulent) flow during granite ascent (Fig. 3), combined with a lack of crystallisation and associated release of latent heat, may act to prevent significant assimilation of wallrock during ascent. This should be most pronounced for dykes close to their critical minimum width, where the presence of growing chilled margins reduces the amount of contact between the flowing magma and country rock. If melt extraction from the source is sufficiently rapid, the only significant magma reservoir will exist at the level of emplacement. It is here that most assimilation of country rock should

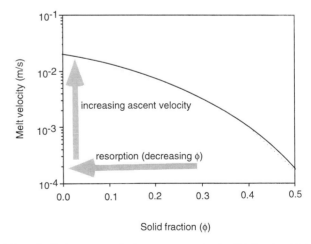

Figure 7 Dyke ascent velocity versus entrained solid fraction (ϕ) in an adiabatically ascending granitic magma. Resorption of entrained material leads to a reduction in effective viscosity and an increase in melt ascent velocity.

occur, along with the development of magmatic fabrics, as crystallisation proceeds.

2.7. Granite dykes in the field

An increasing number of field studies from different tectonic environments now show granite dykes to be commonplace. Granite dykes intruding either undeformed crust or exploiting faults and shear zones have been reported from the Himalaya (LeFort 1981; Inger & Harris 1992; Reddy et al. 1993), Alaska (Hutton & Ingram 1992), Ireland (Hutton 1982; McCaffrey 1992), the Andes (Pitcher 1979; Petford & Atherton 1992), Peru and Japan (Wada 1994), along with the extensive occurrence of composite dykes with granite cores in Iceland and the British Tertiary Igneous Province, and similar dyke swarms in the Channel Islands. Extensive granite dyke swarms are exposed in Namibia. In most instances, the widths of the granite dykes ranges from 10^{-2} to 10^2 m, in good agreement with theory (Wada 1994).

2.7.1. Pluton floors and feeder dykes. That granitic dykes exist is beyond doubt. However, their existence does not mean that they were involved necessarily in batholith construction, and a common criticism against the dyke ascent of granitoid melts is the apparent lack of field examples showing granitic dykes unambiguously feeding large plutons (e.g. Rubin 1995). Although the field evidence for dykes beneath plutons provides the best way of testing the dyke transport models, it is frustrating that the floors of plutons are rarely, if ever, exposed in any detail. Where pluton floors are seen in the field, they do appear to show evidence for inflation via dykes. The Himalayan Manaslu and gangotri leucogranites are examples (LeFort 1981; Scaillet et al. 1995), as is the Bergell pluton, Central Alps (Rosenberg et al. 1995). Although fed by a dyke, neither pluton shows any evidence for having been emplaced as a series of vertical sheets (cf. Paterson 1994). Similar observations can be made of many large mafic intrusions.

Most geophysical investigations of granite plutons involve gravity studies which, although failing to yield unique results (see Vigneresse 1990 for a review), often reinforce the classical picture of a batholith with outward dipping contacts extending to depth, creating as it does so the infamous (but largely imaginary?) space problem. In contrast, relatively few studies using seismic reflection imaging, a technique more likely to reveal the true nature of contacts between plutons and country rock at depth, have been made. One exception is a seismic reflection study of the internal structure of the English Lake District batholith (Evans et al. 1994). Their results show the presence at depth of a series of reflector-poor zones, interpreted as granitic sills 500–1000 m in thickness, separated by highly reflective lenses of country rock. These structures suggest that the batholith is made up of a series of horizontal sheets with flat tops and floors. Furthermore, the steep western margin lies parallel to the Lake District boundary faults, which may have aided the ascent and emplacement of the batholith magmas. On the basis of their seismic profiling, the overall structure of the batholith is regarded by Evans et al. (1994) as laccolithic. Similar sill-like geometries occur in sections of the Sierra Nevada batholith (Coleman et al. 1995) and the High Himalaya (Scaillet et al. 1995).

3. Relationship between source, ascent and emplacement of granitic magmas

3.1. Magma pulsing

A common feature of many large plutons and batholiths is that they are made up of discrete magma pulses or batches of differing composition. An early proponent of the multiple

pulse hypothesis of batholith formation was Noble (1950), who suggested that the Sierra Nevada batholith was built up by multiple, forceful intrusions of a number of small intrusive units. Pitcher (1979), based on his experiences of the Peruvian Coastal Batholith, made the intriguing observation that granitoid plutons of a given composition tend to be limited to a maximum size. This is borne out in numerous field studies of large (>1000 km^2) batholiths of the circum-Pacific region, which have been found to comprise multiple bodies. Within these intrusions there is a marked trend towards smaller sized plutons with increasing acidity, with tonalitic pulses (300 km^2) > granodiorite (150 km^2) > granite (80 km^2). Several explanations have been put forward to explain this apparent size limitation, including the thickness of the source region and buoyancy limitation (Pitcher 1979). Whatever the reason, compositional pulsing of this kind must reflect in some way a decrease in supply (or generation) of magma at source and/or changing source compositions with time (Fig. 8).

3.1.1. Heat source and pulse composition. An allied problem relates to the heat source required for melting to occur and although mantle-derived basaltic magma (e.g. Huppert & Sparks 1988; Bergantz 1989) is the only realistic possibility, the amounts of material added, and at what rates, are clearly fundamental. Given the potential for extreme heterogeneity in the lower crust, it may be argued (e.g. Miller et al. 1988) that a range of melt compositions produced during anatexis is inevitable, thus explaining the change in composition of magma pulses with time. However, the sequence from basic to acid is often repeated cyclically in many large batholiths (Pitcher et al. 1985) and it would be unlikely that the same sequence of source rocks could be melted with sufficient regularity to produce the observed zonations. Furthermore, there is growing recognition that the source rocks for many Cordilleran (tonalitic–granodioritic) granitoids is relatively mafic (Atherton 1990; Tepper et al. 1992). During partial melting of an intermediate to mafic protolith, the composition of the partial melt will become more mafic with increasing melt fraction, with felsic melts representing the smallest melt fraction. One possible explanation for the zonation and magma pulsing seen in the Coastal Batholith is that initial (mantle-derived) heat input caused relatively large amounts of partial melting, resulting in more mafic compositions, followed by smaller, more acidic melts as the heat source wanes.

3.1.2. Batholith assembly time. If plutons and batholiths are constructed piecemeal from many smaller pulses as outlined earlier, then it may be more convenient to express total batholith assembly time (T_i) as the sum of the time *between*

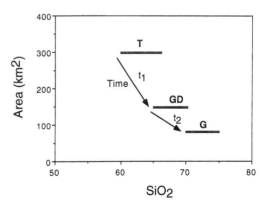

Figure 8 Plot of area (km^2) versus silica content for individual tonalite (T), granodiorite (GD) and granitic (G) facies in the Coastal Batholith, Peru. Each pulse (P) becomes smaller in volume and more evolved with time.

constituent pulses

$$\sum_{i=1}^{n} T_i = t_1 + t_2 + t_3 \ldots + t_n \qquad (8)$$

where t_1, t_2, etc. is the time gap separating one pulse from the next. Thus although individual magma batches may ascend rapidly to their emplacement levels in dykes, the total time required to build a batholithic mass may well be considerably longer. If we assume that, once formed, melt can readily escape its source, as seems to be the case generally (Brown *et al.* 1995), the rate-limiting step in pluton formation will not be controlled either by ascent or emplacement rates where space is made by faulting (Hanson & Glazner 1995), but by how fast (and how much) melt can be created in the source region at any one time. This raises the possibility that although individual events in batholith formation may be rapid geologically, in those instances where emplacement is fault-controlled batholith assembly overall may be dominated by long periods of inactivity. In this way, estimates of emplacement and cooling rates based on tectonic averages (e.g. Paterson & Tobisch 1992; Hanson & Glazner 1995) may be reconciled with the rapid bursts of activity predicted by models of dyke ascent.

4. Discussion

4.1. Towards a possible solution

Does a model currently exist that can unite the various requirements of source processes, pulsing, ascent rates, size, composition and observed emplacement geometries in one all-encompassing mechanism? In keeping with the results of Evans *et al.* (1994), laccoliths seem to fit most of the requirements for a united ascent and emplacement model. First described by Gilbert (1877), laccoliths are intruded horizontally as sill-like bodies. They have flat bases, with domed roofs, and are generally circular or ovoid in plan as exemplified by the Henry Mountains Group in Utah.

In contrast with plutons and batholiths, the emplacement of laccoliths in the crust has been the subject of a number of quantitative studies (e.g. Pollard & Johnson 1973; Jackson & Pollard 1988; Corry 1988). Three basic stages are involved in laccolith formation: (1) vertical ascent in narrow dykes; (2) emplacement and horizontal propagation as sills; and (3) *in situ* inflation and doming. Note that although fed by dykes, there is no requirement for the laccolith to be internally sheeted. At some point, vertical flow is translated to horizontal flow (due to factors such as neutral buoyancy or crustal anisotropy) and the laccolith begins to inflate, with the doming of roof rocks in excess of 2·5 km and bending strains in the roof of 10% reported by Jackson and Pollard (1988). Both ascent and emplacement are likely to be fast, with emplacement and cooling time measured in 10^0–10^2 a. As an example of the potential rapidity of the emplacement process, Kerr and Lister (1995) estimate that the silicic Tennant Creek porphyry sill, Australia, intruded into unconsolidated sediments, was able to flow 100 km laterally in about 1 a. The relatively small size of many laccoliths is consistent with a model for a pulsed supply of magma at source which ascends rapidly through the crust. Against this is the fact that most laccoliths are emplaced at high crustal levels and quickly cooled, resulting in a general lack of contact metamorphism. Estimates of the magmatic overpressure required to dome the roof rocks (an essential feature of laccolith geometry) apparently restrict magma emplacement to 2–3 km of the surface (Corry 1988), although the undomed geometry (sill) could in principle form at deeper crustal levels.

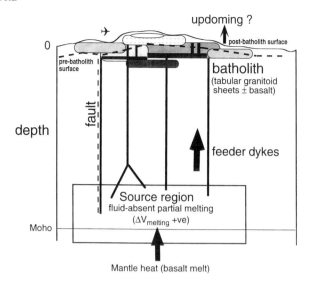

Figure 9 Summary cartoon showing the proposed relationship between magma source, ascent and emplacement (not to scale). Basaltic melts intrude the lower crust and initiate partial melting. Pulses of granitic and minor(?) basaltic melt ascend rapidly in dykes through the crust and are emplaced as tabular sills and sheets (e.g. Coleman *et al.* 1995; Pitcher *et al.* 1985). Space is created by a combination of faulting, ballooning and uplift at the surface.

Thus, although regarded in the classic sense as shallow level intrusions, the processes involved in laccolith formation fulfil in a general way the requirements of the dykists in that magma ascent is (a) dyke controlled and (b) shows that a clear difference exists between the mechanisms of ascent (vertical motion of magma) and emplacement (predominantly horizontal movement of magma). Their structural development also caters to the diapirists/ballooners in that laccolith outcrop patterns and associated structures, both internally and externally, appear to match closely those predicted from experiment. Depending on the depth of emplacement, initial brittle country rock structures would soon become more ductile as the region around the magma bodies heats up. As emplacement is likely to involve both brittle and ductile deformation of the crust in relative amounts that will vary with time, brittle deformation will be favoured by the emplacement of small and infrequent magma batches into cold crust (e.g. sheets), whereas more prolonged and continuous input will induce a more ductile response in the country rock, resulting in the familiar oblate, high-aspect ratio outcrop geometries characteristic of many plutons. The whole gambit of magmatic foliations and country rock structures associated with ballooning and diapiric 'emplacement' as cooling and crystallisation proceeds and ductility contrast decreases are thus rationalised with a granitic body that ascended the crust in a dyke or fissure.

It is easy to take this one step further and consider what would happen if dozens or hundreds of laccoliths coalesced at a similar level of emplacement. The result would be a composite batholithic body, fed by dykes with a flat bottom and a domed roof (vertically created space), comprised of tabular shaped intrusions whose size may relate ultimately to their composition (Fig. 9). Such a structure, built up in this way, where high-level magmas can also vent to the surface, would resemble the Coastal Batholith of Peru, and perhaps other Cordilleran batholith complexes.

5. Acknowledgements

I am indebted to Mike Atherton and Wally Pitcher for introducing me to granites in the first place, and to John Clemens and Ken McCaffrey for irreverent lunch time discussions on mechanisms of granite ascent. Ross Kerr, John Lister, Allan Rubin, Roberto Weinberg and Phil Candela are thanked for providing comment and critical reviews. I thank the Royal Society for the award of a University Research Fellowship that allowed me to pursue the ideas presented here.

6. References

Atherton, M. P. 1990. The Coastal Batholith of Peru: the product of rapid recycling of new crust formed within rifted continental margin. GEOL J **25**, 337–49.

Baker, D. R. 1996. Granitic melt viscosities: empirical and configuration entropy models for their calculation. AM MIN **81**, 126–34.

Balk, R. 1937. Structural behavior of igneous rocks. GEOL SOC AM BULL **5**, 177 pp.

Batchelor, G. K. 1967. *An introduction to fluid dynamics.* Cambridge: Cambridge University Press.

Bateman, R. 1985. Aureole deformation by flattening around a diapir during in situ ballooning: the Cannibal Creek granite. J GEOL **93**, 293–310.

Berganz, G. W. 1989. Underplating and partial melting: implications for melt generation and extraction. SCIENCE **254**, 1093–5.

Berner, H., Ramberg, H. & Stephasson, O. 1972. Diapirism in theory and experiment. TECTONOPHYSICS **15**, 197–218.

Brown, M. 1994. The generation, segregation ascent and emplacement of granite magma: the migmatite-to-crustally derived granite connection in thickened orogens. EARTH SCI REV **36**, 83–130.

Brown, M., Rushmer, T. & Sawyer, E. W. 1995. Mechanisms and consequences of melt segregation from crustal protoliths. J GEOPHYS RES **100**, 15 551–63.

Bruce, P. M. & Huppert, H. E. 1989. Thermal control of basaltic fissure eruptions. NATURE **342**, 665–7.

Bruce, P. M. & Huppert, H. E. 1990. Solidification and melting along dykes by the laminar flow of basaltic magma. In Ryan, M. P. (ed.) *Magma transport and storage,* 87–101. New York: Wiley.

Brun, J. P., Gapais, D., Cogne, J. P., Ledru, P. & Vingeresse, J. L. 1990. The Flamanville granite (northwestern France): an unequivocal example of a syntectonically expanding pluton. GEOL J **25**, 271–86.

Castro, A. 1987. On granitoid emplacement and related structures: a review. GEOL RUNDSCH **76**, 101–24.

Clemens, J. D. 1989. The importance of residual source material (restite) in granite petrogenesis: a comment. J PETROL **30**, 1313–6.

Clemens, J. D. & Mawer, C. K. 1992. Granitic magma transport by fracture propagation. TECTONOPHYSICS **204**, 339–60.

Clemens, J. D., Petford, N. & Mawer, C. K. 1996. Ascent mechanisms of granitic ascent: causes and consequences. MIN SOC SPEC PUB, in press.

Cloos, H. 1925. *Einfuhrung in die tektonische Behandlung magmatischer Erscheinungen (Granittektonik). I. Spezieller Teil. Das Riesengebirge in Schlesien. Bau, Bilding, und Oberflachengestaltung.* Berlin: Borntraeger.

Coleman, D. S., Glazner, A. F., Miller, J. S., Bradford, K. J., Frost, T. P., Joye, J. L. & Bachl, C. A. 1995. Exposure of a late Cretaceous layered mafic–felsic magma system in the central Sierra Nevada batholith, California. CONTRIB MINERAL PETROL **120**, 129–36.

Corry, C. E. 1988. Laccoliths: mechanisms of emplacement and growth. GEOL SOC AM SPEC PUB 220.

Coward, M. P. 1976. Archean deformation patterns in southern Africa. PHIL TRANS R SOC LONDON **283**, 313–31.

Daly, R. A. 1903. The mechanics of igneous intrusion. AM J SCI **16**, 107–26.

DeWaard, D. 1949. Tectonics of the Mt. Aigoual pluton in the southeastern Cevennes, France. PROC KON NED AKAD **52**, 389–402.

Dingwell, D. D., Bagdassarov, G. Y., Bussod, G. Y. & Webb, S. L. 1993. Magma rheology. In Scarfe, C. M. (ed.) Experiments at high pressure and applications to the earth's mantle. MINER ASSOC CAN SHORT COURSE HANDB **21**, 131–96.

Dixon, J. M. 1975. Finite strain and progressive deformation in models of diapiric structures. TECTONOPHYSICS **28**, 89–124.

Emerman, S. H. & Marrett, R. 1990. Why dikes? GEOLOGY **18**, 231–3.

Eskola, P. 1949. The problems of mantled gneiss domes. Q J GEOL SOC LONDON **54**, 461–76.

Evans, D. J., Rowley, W. J., Chadwick, R. A., Kimbell, G. S. & Millward, D. 1994. Seismic reflection data and the internal structure of the Lake District batholith, Cumbria, northern England. PROC YORKSHIRE GEOL SOC **50**, 11–24.

Fyfe, W. S. 1970. Some thoughts on granitic magmas. In Newall, G. N. & Rast, N. (eds) *Mechanism of igneous intrusion.* GEOL J SPEC ISSUE **2**, 201–16.

Gilbert, G. K. 1877. Geology of the Henry Mountains, Utah. *US geographical and geological survey of the Rocky Mountains Region.*

Grout, F. F. 1945. Scale models of structures related to batholiths. AM J SCI **243A**, 260–84.

Hanson, B. R. & Glazner, A. F. 1995. Thermal requirements for extensional emplacement of granitoids. GEOLOGY **23**, 213–6.

Holtz, F. & Johannes, W. 1994. Maximum and minimum water contents of granitic melts: implications for chemical and physical properties of ascending magmas. LITHOS **32**, 149–59.

Huppert, H. E. & Sparks, R. S. J. 1988. The generation of granitic magmas by the intrusion of basalt into continental crust. J PETROL **29**, 599–624.

Hutton, D. H. W. 1982. A tectonic model for the emplacement of the Main Donegal granite, NW Ireland. J GEOL SOC LONDON **139**, 615–31.

Hutton, D. H. W. 1992. Granite sheeted complexes: evidence for the dyking ascent mechanism. TRANS R SOC EDINBURGH EARTH SCI **83**, 377–82.

Hutton, D. H. W. & Ingram, G. M. 1992. The Great Tonalite Sill of southeastern Alaska and British Columbia: emplacement into an active contractional high angle reverse shear zone (extended abstract). TRANS R SOC EDINBURGH EARTH SCI **83**, 383–6.

Hutton, J. 1788. Theory of the Earth. TRANS R SOC EDINBURGH **1**, 209.

Inger, S. & Harris, N. B. W. 1992. Geochemical constraints on leucogranitic magmatism in the Langtang Valley, Himalaya. J PETROL **34**, 345–68.

Jackson, M. D. & Pollard, D. D. 1988. The laccolith stock controversy: new results from the southern Henry mountains, Utah. GEOL SOC AM BULL **100**, 117–39.

Kerr, R. C. & Lister, J. R. 1991. The effects of shape on crystal settling and on the rheology of magmas. J GEOL **99**, 457–67.

Kerr, R. C. & Lister, J. R. 1995a. The lateral intrusion of silicic magmas into unconsolodated sediments: the Tennant Creek porphyry revisited. AUST J EARTH SCI **42**, 223–4.

Kerr, R. C. & Lister, J. R. 1995b. Comment on 'On the relationship between dike width and magma viscosity' by Y. Wada. J GEOPHYS RES **100**, 15 541.

Leake, B. E. 1978. Granite emplacement; the granites of Ireland and their origin. In Bowes, D. R. & Leake, B. E. (eds) *Crustal evolution of northwestern Britain and adjacent regions.* GEOL J SPEC PUBL **10**, 221–48.

LeFort, P. 1981. Manaslu leucogranite: a collisional signature of the Himalaya, a model for its genesis and emplacement. J GEOPHYS RES **86**, 10 545–68.

Lister, J. R. & Kerr, R. C. 1991. Fluid-mechanical models of crack propagation and their application to magma transport in dykes. J GEOPHYS RES **96**, 10 049–77.

Lister, J. R. & Dellar, P. J. 1996. Solidification of pressure driven flow in a finite rigid channel. J FLUID MECH (submitted).

Mahon, K. I., Harrison, T. M. & Drew, D. A. 1988. Ascent of a granitoid diapir in a temperature varying medium. J GEOPHYS RES **93**, 1174–88.

Marsh, B. D. 1982. On the mechanics of igneous diapirism, stoping and zone melting. AM J SCI **282**, 808–55.

McCaffrey, K. J. W. 1992. Igneous emplacement in a transpressive shear zone: the Ox Mountains igneous complex. J GEOL SOC LONDON **149**, 221–35.

Miller, C. F., Watson, E. B. & Harrison, M. T. 1988. Perspectives on the source, segregation and transport of granitoid magmas. TRANS R SOC EDINBURGH EARTH SCI **79**, 135–56.

Mrazec, L. 1927. Les plis diapirs et le diapirisme en general. C R SEANCES INST GEOL ROUMANIE VI (1914–1915), 226–70.

Nicolesco, C. P. 1929. Anticlimax diapirs sedimentaries, volcaniques et plutoniques. BULL SOC GEOL FR **29**, 21–4.

Noble, J. A. 1950. Evaluation of criteria for the forcible intrusion of magma. J GEOL **60**, 34–57.

Paterson, S. R. & Fowler, T. K. 1994. Re-examining pluton emplacement processes: reply. J STRUCT GEOL **16**, 747–8.

Paterson, S. R. & Tobisch, O. T. 1992. Rates of processes in magmatic arcs: implications for the timing and nature of pluton emplacement and wall rock deformation. J STRUCT GEOL **14**, 291–300.

Petford, N. 1995. Segregation of tonalitic and trondhjemitic melts in the continental crust: the mantle connection. J GEOPHYS RES **100**, 15 735–43.

Petford, N. & Atherton, M. P. 1992. Granitoid emplacement and deformation along a major crustal lineament: the Cordillera Blanca, Peru. TECTONOPHYSICS 205, 171–85.

Petford, N., Kerr, R. C. & Lister, J. R. 1993. Dike transport of granitoid magmas. GEOLOGY 21, 845–8.

Petford, N., Lister, J. R. & Kerr, R. C. 1994. The ascent of felsic magmas in dykes. LITHOS 32, 161–8.

Pitcher, W. S. 1979. The nature, ascent and emplacement of granitic magmas. J GEOL SOC LONDON 136, 627–22.

Pitcher, W. S. & Berger, A. R. 1972. *The geology of Donegal: a study of granite emplacement and unroofing.* London: Wiley-Interscience.

Pitcher, W. S., Atherton, M. P., Cobbing, E. J. & Beckinsale, R. D. 1985. *Magmatism at a plate edge: the Peruvian Andes.* Glasgow: Blackie Halsted Press.

Pollard, D. D. & Johnson, A. M. 1973. Mechanisms of growth of some laccolith intrusions in the Henry Mountains, Utah II. Bending and failure of overburden layers and sill formation. TECTONOPHYSICS 18, 311–45.

Ramberg, H. 1967. *Gravity deformation and the earth's crust as studied by centrifuge models.* Academic Press, London.

Ramberg, H. 1970. The initiation, ascent and emplacement of magmas. *In* Newall, G. N. & Rast, N. (eds) *Mechanism of igneous intrusion.* GEOL J SPEC ISSUE 2, 261–86.

Ramberg, H. 1981. Gravity, Deformation and the earth's Crust in theory, experiments and geological application, 2nd edn. London: Academic Press.

Rast, N. 1970. The initiation, ascent and emplacement of magmas. *In* Newall, G. N. & Rast, N. (eds) *Mechanism of igneous intrusion.* GEOL J SPEC ISSUE 2, 332–69.

Read, H. H. 1947. Granites and granites. GEOL SOC AM MEM 28, 1–19.

Reddy, S. M., Searle, M. P. & Massey, J. A. 1993. Structural evolution of the High Himalayan gneiss sequence, Langtamng Valley, Nepal. *In* Treloar, P. J. & Searle, M. P. (eds) *Himalayan tectonics.* GEOL SOC LONDON SPEC PUBL 74, 375–89.

Ribe, N. M. 1983. Diapirism in the Earth's mantle: experiments on the motion of a hot sphere in a fluid with temperature dependent viscosity. J VOLCANOL GEOTHERM RES 16, 221–45.

Rosenberg, C., Berger, A. & Schmid, S. M. 1995. Observations from the floor of a granitoid pluton: a constraint on the driving force for final emplacement. GEOLOGY 23, 443–6.

Rubin, A. M. 1993. On the thermal viability of dykes leaving magma chambers. GEOPHYS RES LETT 20, 257–60.

Rubin, A. M. 1995. Getting granite dikes out of the source region. J GEOPHYS RES 100, 5911–29.

Rutter, E. & Neumann, D. 1995. Experimental deformation of partially molten Westerly Granite under fluid-absent conditions, with implications for the extraction of granite magma. J GEOPHYS RES 100, 15 697–715.

Sawyer, W. E. 1991. Disequilibrium melting and the rate of melt-residuum separation during migmatisation of mafic rocks from the Grenville Front, Quebec. J PETROL 32, 701–38.

Schwerdtner, W. M. 1982. Salt stocks as natural analogues of Archean gneiss diapirs. GEOL RUNDSCH 71, 370–9.

Schwerdtner, W. M. 1990. Structural tests of diapir hypotheses in Archean crust of Ontario. CAN J EARTH SCI 27, 387–402.

Sederholm, J. J. 1926. On migmatites and associated Precambrian rocks of southwest Finland. II. BULL COMM GEOL FINLANDE 77, 141 pp.

Shaw, H. R. 1965. Comments on viscosity, crystal settling and convection in granitic magmas. AM J SCI 263, 120–53.

Shaw, H. R. 1972. Viscosities of magmatic silicate liquids: an empirical method of prediction. AM J SCI 272, 870–93.

Shaw, H. R. 1980. Fracture mechanisms of magma transport from the mantle to the surface. *In* Hargraves, R. B. (ed.) *Physics of magmatic processes*, 201–264. Princetown: Princetown University Press.

Sorgenfrei, T. 1971. On the granite problem and the similarity of salt and granite structures. GEOL FORH 93, 371–435.

Spence, D. A. & Turcotte, D. L. 1985. Magma-driven propagation of cracks. J GEOPHYS RES 90, 575–80.

Spera, F. 1980. Aspects of magma transport. *In* Hargraves, R. B. (ed.) *Physics of magmatic processes*, 263–323. Princetown: Princetown University Press.

Sykes, M. L. & Holloway, J. R. 1987. Evolution of granitic magmas during ascent: a phase equilibrium model. *In* Mysen, B. O. (ed.) *Magmatic processes: physicochemical principles.* GEOCHEM SOC SPEC PUBL 1, 447–61.

Tepper, J. H., Nelson, G. W., Bergantz, G. W. & Irving, A. J. 1992. Petrology of the Chilliwack Batholith, North Cascades, Washington: generation of calc-alkaline granitoids by melting of mafic lower crust with variable water fugasity. CONTRIB MINERAL PETROL 70, 299–318.

Tritton, D. L. 1988. *Physical fluid dynamics.* Oxford: Oxford University Press.

Turcotte, D. L. 1982. Magma migration. ANNU REV EARTH PLANET SCI 10, 397–408.

Turcotte, D. L. 1987. Physics of magma segregation processes. *In* Mysen, B. O. (ed.) *Magmatic processes: physicochemical principles.* GEOCHEM SOC SPEC PUBL 1, 69–74.

Vigneresse, J. L. 1990. Use and misuse of geophysical data to determine the shape at depth of granitic intrusions. GEOL J 25, 249–60.

Wada, Y. 1994. On the relation between dike width and magma viscosity. J GEOPHYS RES 99, 17 743–55.

Webb, S. L. & Dingwell, D. B. 1990. The onset of non-Newtonian rheology of silicate melts, a fiber elongation study. PHYS CHEM MINERAL 17, 125–32.

Wegmann, C. E. 1930. Uber Diapirismus. BULL COMM GEOL FINLANDE 92, 58–76.

Weinberg, R. F. 1994. Re-examining pluton emplacement processes: discussion. J STRUCT GEOL 16, 743–6.

Weinberg, R. F. 1996. The ascent mechanism of felsic magmas: news and views. TRANS R SOC EDINBURGH EARTH SCI 87, 000–000.

Weinberg, R. F. & Podladchikov, Y. Y. 1994. Diapiric ascent of magmas through power-law crust and mantle. J GEOPHYS RES 99, 9543–59.

Weinberg, R. F. & Podladchikov, Y. Y. 1995. The rise of solid-state diapirs. J STRUCT GEOL 17, 1183–95.

White, A. J. R., Williams, I. S. & Chapell, B. W. 1977. *Geology of the Berridale 1:100,000 sheet.* GEOL SURV NEW SOUTH WALES 8625.

Wilson, L. & Head, J. W. 1981. Ascent and eruption of basaltic magma on the earth and the moon. J GEOPHYS RES 86, 2971–3001.

NICK PETFORD, School of Geological Sciences, Kingston University, Kingston-Upon-Thames, Surrey KT1 2EE, UK. E-mail: N.PET@kingston.ac.uk

Transactions of the Royal Society of Edinburgh: Earth Sciences, **87**, 115–123, 1996

Pluton emplacement in arcs: a crustal-scale exchange process

Scott R. Paterson, T. Kenneth Fowler Jr and Robert B. Miller

ABSTRACT: Buddington (1959) pointed out that the construction of large crustal magma chambers involves complex internal processes as well as multiple country rock material transfer processes (MTPs), which reflect large horizontal, vertical and temporal gradients in physical conditions. Thus, we have attempted to determine the relative importance of different magmatic and country rock MTPs at various crustal depths, and whether country rock MTPs largely transport material vertically or horizontally, rather than seeking a single model of magma ascent and emplacement.

Partially preserved roofs of nine plutons and in some cases roof–wall transitions with roof emplacement depths of 1·5–11 km were mapped. During emplacement, these roofs were not deformed in a ductile manner, detached or extended by faults, or significantly uplifted. Instead, sharp, irregular, discordant contacts are the rule with stoped blocks often preserved immediately below the roof, even at depths of 10 km. The upper portions of these magma chambers are varied, sometimes preserving the crests of more evolved magmas or local zones of volatile-rich phases and complex zones of dyking and magma mingling. Magmatic structures near roofs display a wide variety of patterns and generally formed after emplacement. Transitions from gently dipping roofs to steep walls are abrupt. At shallow crustal levels, steep wall contacts have sharp, discordant, stepped patterns with locally preserved stoped blocks indicating that the chamber grew sideways in part by stoping. Around deeper plutons, an abrupt transition (sometimes within hundreds of metres) occurs in the country rock from discordant, brittle roofs to moderately concordant, walls deformed in a ductile manner defining narrow structural aureoles. Brittle or ductile faults are not present at roof–wall joins.

Near steep wall contacts at shallow to mid-crustal depths (5–15 km), vertical and horizontal deflections of pre-emplacement markers (e.g. bedding, faults, dykes), and ductile strains in narrow aureoles (0·1–0·3 body radii) give a complete range of bulk strain values that account for 0–100% of the needed space, but average around 30%, or less, particularly for larger batholiths. A lack of far-field deflection of these same markers rules out significant horizontal displacement outside the aureoles and requires that any near-field lateral shortening is accommodated by vertical flow. Lateral variations from ductile (inner aureole) to brittle (outer aureole) MTPs are typically observed. Compositional zoning is widespread within these magma bodies and is thought to represent separately evolved pulses that travelled up the same magma plumbing system. Magmatic foliations and lineations commonly cross-cut contacts between pulses and reflect the strain caused either by the late flow of melt or regional deformation.

Country rocks near the few examined mid- to deep crustal walls (10–30 km) are extensively deformed, with both discordant and concordant contacts present; however, the distinction between regional and emplacement-related deformation is less clear than for shallower plutons. Internal sheeting is more common, although elliptical masses are present. Lateral compositional variations are as large as vertical variations at shallower depths and occur over shorter distances. Magmatic foliations and lineations often reflect regional deformation rather than emplacement processes.

The lack of evidence for horizontal displacement outside the narrow, shallow to mid-crustal aureoles and the lack of lateral or upwards displacement of pluton roofs indicate that during emplacement most country rock is transported downwards in the region now occupied by the magma body and its aureole. The internal sheeting and zoning indicate that during the downwards flow of country rock, multiple pulses of magma travelled up the same magma system. If these relationships are widespread in arcs, magma emplacement is the driving mechanism for a huge crustal-scale exchange process.

KEY WORDS: crustal-scale exchange processes, pluton emplacement, arcs

In a comprehensive review of plutons, Buddington (1959) proposed a three-tiered classification for emplacement styles based on crustal level (Fig. 1). According to Buddington, plutons of the shallow epizone (0–10 km) are characterised by almost wholly discordant contact relations and were emplaced by brittle processes such as stoping, block foundering during cauldron subsidence and sometimes roof uplift. The mesozone (6–16 km) represents a complex transition from dominantly brittle to dominantly ductile processes, with plutons having both concordant and discordant features. Emplacement mechanisms include stoping, block foundering, ductile flow during radial expansion and assimilation. Plutons of the catazone

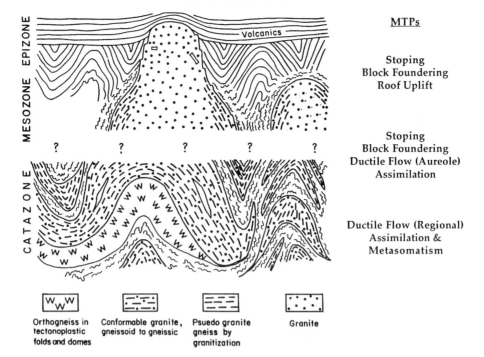

Figure 1 Diagram redrafted from Buddington (1959) summarising changing mechanisms of emplacement with depth. MTPs, Material transfer processes that Buddington (1959) thought were dominant at different crustal levels. Note the discordant stepped contact of epizonal plutons, that brittle processes such as stoping operated well into the mesozone, and that pluton emplacement was intimately associated with folding but not faulting. Also note that these magmatic systems were viewed as being vertically extensive, consisting of multiple pulses of magma and sometimes without clear-cut floors.

(>10 km) are dominantly concordant and were inferred by Buddington to be emplaced largely by ductile country rock flow during diapirism and regional deformation with assimilation/metasomatism playing a secondary part. Emplacement during horizontal extension by faulting was not called upon at any crustal level. Buddington understood the importance of such variable quantities as temperature, bulk composition and volatile content on rheology, and regarded the depth estimates as only a first approximation. He also viewed these systems as commonly vertically extensive (Fig. 1), composed of multiple batches of magma, potentially without clear-cut floors, and he repeatedly noted examples where magmatic foliations/lineations cross-cut internal contacts.

However, following this review, single solution models of pluton ascent and emplacement continued to gain in popularity. Such models include hot Stokes or power law diapirism (Marsh 1982; Mahon *et al.* 1988; Weinburg & Podladchikov 1994), in which the country rock is displaced only by viscous flow, extensional models (Hutton 1988), in which country rock is displaced laterally along faults, and dyking (Shaw 1980; Delaney & Pollard 1981; Lister & Kerr 1991; Clemens & Mawer 1992; Rubin 1993), in which country rock is displaced laterally by brittle failure and elastic strain. These models are clearly applicable under certain circumstances. However, the volumetrically huge magmatic systems described by Buddington (1959) involved the interaction of multiple magmatic and country rock transfer processes that resulted in many features not described by existing mechanical models. In this paper, we briefly summarise recent mapping around the roofs and walls of magmatic systems that typically occur in subduction-related arcs to re-emphasize Buddington's conclusions that complex horizontal and vertical changes in both magmatic and country rock transfer processes occur during the construction of these systems.

1. Pluton roofs

We have now mapped the partially preserved roofs of nine plutons in the Cordillera with roof emplacement depths of 1·5–11 km and country rock hosts of unmetamorphosed plutonic rock, weakly metamorphosed clastic, calc-silicate, carbonate and volcanic rock, and amphibolite grade schist and gneiss (Table 1). All of these roof contacts are roughly planar with average initial dips of <20°. However, in detail the contacts are highly irregular (e.g. Fig. 2), with numerous steps or jogs at the metre to hundreds of metres scale (e.g. Crossland 1994; Fowler *et al.* 1995). These contacts sharply truncate all pre-existing structures (e.g. Fig. 2) even at the scale of individual grains (e.g. Fowler 1994a), are typically not associated with ductile strains or large-scale brittle faulting, and are thus mode I extensional fractures. The general characteristics of these roofs are displayed in Figure 3.

In every instance stoped blocks are preserved at or immediately below roof contacts even in the deeper plutons (e.g. Paterson 1992; Crossland 1994, Fowler & Paterson, in press). The blocks range in size from <1 m to hundreds of metres across, generally have irregular rectangular shapes and sometimes display structures that have been overprinted by hornfelsic textures and have been rotated with respect to equivalent structures in the roof. Near roof contacts these blocks are commonly separated from the country rock by zones of igneous layering (e.g. schlieren, petrographic banding), implying that during or after the blocks dropped, multiple pulses of magma flowed in behind them. Further from roof contacts (50–400 m), magmatic foliations are typically not deflected around stoped blocks as would be expected during the settling and rotation of large blocks, and thus must have formed after the blocks ceased to move—that is, after the magma chamber was constructed (Fowler & Paterson, in press).

Table 1 Summary of the characteristics of pluton roofs and corners mapped in this work. References cited summarise the likely emplacement depths.

Pluton name	Country rock	Emplaced depth (km)	Corners and walls	Dykes	Roof uplift	Roof extension	Stoped blocks	Ductile shortening	Reference
Chita	Sandstone	1·5–3	Brittle	Rare	Minor	No	Yes	0	Yoshinobu et al. (1995)
Agua Negra	Sandstone	1·5–3?	Brittle	Rare	Minor?	No	Yes	?	Yoshinobu et al. (1995)
Mt Powell	Sedimentary and plutonic rocks	>3·0	?	Common	None	No	Yes	0	Hyndman et al. (1987)
Yerington	Volcanic and sedimentary rocks	1·0–2·0	Brittle to ductile	?	Minor	No	Yes	0	Dilles (1982)
Gaudalupe	Slate and sandstone	4–7	Brittle to ductile	Rare	Minor	No	Yes	?	Paterson et al. (1991b)
Castle Creek	Plutonic	5–8	Brittle	Common	?	No	Yes	0	Ague and Brimhall (1988)
Mitchell	Plutonic	5–8	Brittle	Common	None	No	Yes	0	Ague and Brimhall (1988)
Mt Stuart	Schist	8–11	Ductile	Rare	?	?	Yes	20% ? maximum	Paterson et al. (1994)
Hall Canyon	Schist, dolomite	≈10	None	Rare	None	No	Yes	0	Crossland (1994)

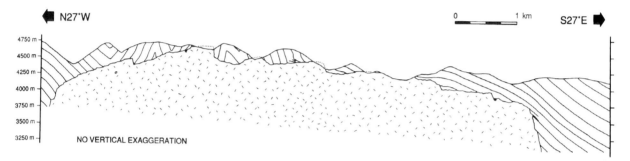

Figure 2 Cross-section of the roof and walls of the Chita pluton, Argentina. Lines in the country rock define patterns of actual mapped bedding. Note the irregular stepped contact along both the roof and walls, the sharp truncation of previously folded bedding and the continuity of roof and wall structures. As the roof and wall markers are continuous (i.e. not displaced relative to one another) and not deflected during emplacement, all country rock displaced during emplacement must have been transported downwards in the region now occupied by the magma chamber.

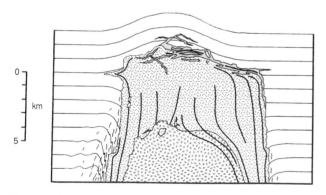

Figure 3 Cartoon summary of most common characteristics of roofs and roof–wall transitions in plutons mapped in this work (see Table 1). The relatively flat-lying roof contact rolls over abruptly into steep, but locally stepped, wall contacts. Internal patterns show nesting of two diapirs, a small amount of heterogeneous intrusive rock along the roof, and magmatic foliations (dark lines) that are locally truncated but sometimes continuous across internal contacts. Solid lines in the country rock show the behaviour of pre-emplacement passive markers, which are only rarely bowed up above the pluton (laccoliths), commonly abruptly truncated along shallow sides, and typically deflected in narrow aureoles at slightly deeper levels. Short curved lines represent foliation. Note that none of these features needs be developed symmetrically around a pluton.

During emplacement, pre-emplacement markers in roof rocks (e.g. bedding, dykes, faults) indicate that these roofs were not, in general, detached or extended by faults, or significantly displaced upwards or laterally with respect to other parts of the roof (e.g. Fig. 2). In some instances, well-preserved markers (e.g. conglomerates, bedding thickness) allow us to document that no detectable ductile strain occurred in roofs during emplacement (Crossland 1994; Fowler et al. 1995).

The crestal portions of these magma chambers are varied, sometimes preserving regions of less dense and/or more evolved magmas (e.g. Guadalupe Igneous Complex and Hall Canyon pluton), sometimes local zones of volatile-rich phases (e.g. Chita pluton) and sometimes complex zones of dyking, schlieren and magma mingling (e.g. Mitchell and Mt Stuart plutons). Comb layering, a few centimetres thick, formed directly at the roof of the Chita pluton; otherwise there is no evidence of early quenched phases or 'roofwall' crystallisation resulting in strong vertical zoning at roof contacts. Dykes rooted in the magma chamber are present, but are not particularly abundant in roof rocks of the studied plutons. Where present, dykes sometimes cut across roof rock structures and other times take numerous jogs reminiscent of the pluton–roof rock contacts, with the direction of jogs only occasionally controlled by pre-emplacement roof structures.

2. Roof–wall transitions

In a few instances we have been able to map roof–wall transitions where roof rocks above plutons can be followed continuously into wall rocks adjacent to steep-sided pluton contacts (Fig. 3). These 'corners' are preserved around the Chita pluton, Argentina (Fig. 2), the Guadalupe Igneous Complex, California (Fig. 7a), the Yerington batholith, Nevada

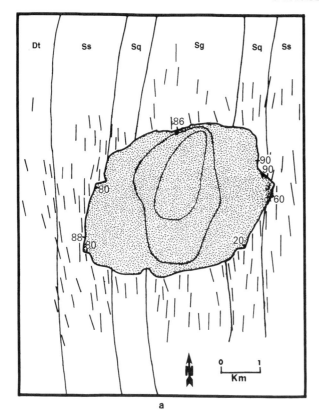

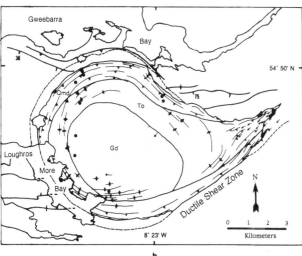

(Fig. 7b), the Mitchell Intrusive Suite, California and the Mt Stuart batholith, Washington. In every instance these transitions are abrupt with flat-lying roof contacts (<20° dip) rapidly changing into steep-sided (>70° dip) wall contacts over vertical distances of tens to hundreds of metres, as previously noted for plutons in the Andes (Cobbing & Pitcher 1972) and in North America (Buddington 1959). At shallow crustal levels the steep wall contacts have sharp, discordant, stepped contacts. In some plutons, stoped blocks are locally preserved along these wall contacts (e.g. Chita; Fowler *et al.* 1995). Around the more deeply emplaced plutons, a relatively abrupt transition occurs, again within hundreds of metres, from discordant, brittle roof contacts to concordant, ductilely deformed wall contacts (e.g. Mt Stuart, Guadalupe Igneous Complex). Field relationships and the scale over which the transition occurs argue against control by rock type or depth. Instead, we argue that these transitions represent an increasing magnitude of downwards flow, with respect to undeformed roof rocks, along the sides of the plutons.

These 'corners' also provide us with important constraints on emplacement models. In every instance pre-emplacement markers in the roofs can be followed continuously into the wall rocks. Except in the narrow ductile aureoles, these markers are not deflected from their regional orientations and show no evidence of being ductilely stretched or shortened. We also have not found brittle or ductile faults at roof–wall joins, although pervasive fractures, which accommodated minor roof uplift, do occur near the Yerington 'corner'. This relationship indicates that roof rocks and wall rocks outside narrow aureoles have not been displaced with respect to one another during emplacement, nor have they been displaced with respect to rocks immediately outside the aureoles. For the plutons examined it also directly rules out some emplacement models (lateral or vertical extension by faulting, significant roof uplift), places severe constraints on others (e.g. hot Stokes diapirism, ballooning, dyking) and requires that by whatever means country rock was displaced, the overall direction was downwards and in the region now occupied by the magma chamber and its aureole (Fowler *et al.* 1995; Paterson & Vernon 1995).

Igneous layering and complex patterns of overprinting magmatic foliations and lineations are common near these roof–wall contacts. The layering, defined by compositional and textural differences, occurs on scales ranging from centimetres to tens of metre widths, with the layering being commonly subparallel to the margins. The magmatic foliation/lineation patterns are so varied from one pluton to the next that no simple summary can be provided.

3. Shallow to mid-crustal systems

Extensive studies have been completed of shallow to mid-crustal (5–15 km) plutons with steep wall contacts by other workers (see summaries by Buddington 1959; Pitcher 1979; Barton *et al.* 1991), and much of our recent work on such plutons has been reported elsewhere (Miller & Bowring 1990;

Figure 4 Examples of mid-crustal plutons. (a) Bruinbin pluton, Australia. Weak internal zonation from a granite margin to a granodiorite core was mapped by Bateman (1982) and inferred by Bateman to reflect two magma pulses. The contact aureole mineral assemblages indicate a depth of emplacement between 2 and 4 kbar. Note the irregular wall margin of the pluton. The numbers along the contact represent the dips of the contact (dip direction shown with short lines perpendicular to the contact) where the exposure and topographic relief allowed three-dimensional calculations. The external lithological contacts (solid lines) and bedding-parallel foliation (short lines) mapped by Paterson (unpublished data) are largely undeflected and sharply truncated by the pluton contact even at the scale of a few centimetres. Dt, Tuff, dacite, slate; Ss, slate, siltstone, greywacke; Sq, quartzite, slate; and Sg, quartz-rich greywacke. Note that the pluton occurs within, but sharply cuts across, the hinge of a regionally extensive fold. (b) Ardara pluton, Ireland, after Paterson and Vernon (1995). Within the pluton solid lines represent compositional contacts; Gd, granodiorite; To, tonalite or quartz monozodiorite; and Qmd, quartz monzodiorite. Note cross-cutting and asymmetrical nature of different magmatic pulses; thinner short lines, magmatic foliations with solid triangles showing dip directions; short arrows, shallowly plunging mineral lineations; black dots, steep-plunging lineations; short broken line external to the pluton marks the position of an abrupt decrease in emplacement-related strain/structures in the

outer aureole; note how the foliation trend lines cut the internal contacts and that foliations steepen inwards (a pattern suggestive of nested diapirs); also note the change from vertical magmatic lineations in the NW part of the pluton to subhorizontal lineations in the SE tail; pre-emplacement country rock markers shown externally; line with teeth, thrust fault (teeth on upper plate); lines with dip symbols, lithological contacts (numbers = average dips of contact). Note that both steeply and shallowly dipping external markers do not show significant deflection as they approach the pluton except for moderate deflections near the NW margin. Sources of data listed in Vernon and Paterson (1995).

Paterson *et al.* 1991a, 1991b; Paterson & Fowler 1993; Paterson & Vernon 1995). We will thus only briefly summarise a few pertinent results. Most of these plutons are elliptical in map view and steep-sided (Fig. 4a, 4b), although some are sheet-like (e.g. Paterson *et al.* 1991a).

Xenoliths occur, but always make up less than a few per cent of the magma chamber area. Asymmetrical internal compositional zoning is common in some magma chambers (Fig. 4b) and elsewhere we and others have argued that this zoning usually reflects the nesting of magma pulses rather than *in situ* fractional crystallisation (e.g. Bouchez & Diot 1990; Stephens 1992; Paterson & Vernon 1995). In other chambers (e.g. the Guadalupe Igneous Complex), vertical compositional layering is much more pronounced than lateral zoning. Magmatic foliation and lineation almost always cross-cut internal contacts and other igneous features (Fig. 4b) and are believed to form late, usually after the construction of the magma chamber (Buddington 1959; Paterson & Vernon 1995; Fowler & Paterson, in press). We (Yuan & Paterson 1993; Paterson & Vernon 1995) and others before us (e.g. Mackin 1947; Bateman *et al.* 1963; Berger & Pitcher 1970; Pitcher, 1993) have argued that these structures reflect the late strain of crystal–rich magma and not necessarily magma flow planes and flow directions.

Vertical and lateral deflections of pre-emplacement markers (e.g. bedding, faults, dykes) and ductile strains in narrow aureoles (0·1–0·3 of magma chamber radii) give a complete range of bulk flow values that can account for 0–100% of the needed magma chamber space, but average around 30%, or less, particularly for larger plutons and batholiths (Paterson & Fowler 1993; Fowler 1994b). A lack of deflection of these same markers outside the narrow aureoles rules out significant lateral far-field displacements as required by simple extensional and ballooning models. This lack of far-field deflection also requires that any flow reflected by near-field lateral shortening is accommodated by vertical flow within the aureoles. Finally, the combined lack of near-field bulk strain and far-field deflections implies that on average 70% or more of the country rock originally located where the magma chamber now resides has been transported vertically out of the exposed map surface.

Lateral and/or temporal variations from ductile to brittle country rock MTPs are sometimes observed (e.g. Compton 1955; Bykerk-Kauffman 1990; Stein 1994), but commonly remain hard to document because of the difficulty in clearly tying brittle structures in the outer parts of aureoles to emplacement. Our work in progress also suggests that it may be common to have both brittle (e.g. rigid rotation and slip on pre-emplacement anisotropies) and ductile (crystal-plastic flow) processes operating simultaneously during emplacement at these crustal levels (e.g. Stein 1994).

4. Mid- to deep-crustal systems

We are presently mapping wall rocks near mid- to deep crustal (15–30 km) plutons in the Cascades Mountains, Washington (e.g. Fig. 5). These plutons are highly elongate with length to width ratios of 3:1 to 15:1, consist internally of numerous metre to hundreds of metres wide sheets and sometimes larger elliptical bodies, and are extensively deformed. Lateral compositional and textural variations are widespread, are as large as the vertical variations observed in other chambers at shallower depths, and occur over shorter distances. Magmatic foliations define complex patterns and overprint contacts between the above-described compositional variations. They often define structural patterns that are similar to and/or continuous with those in nearby country rock (Fig. 5a, 5b),

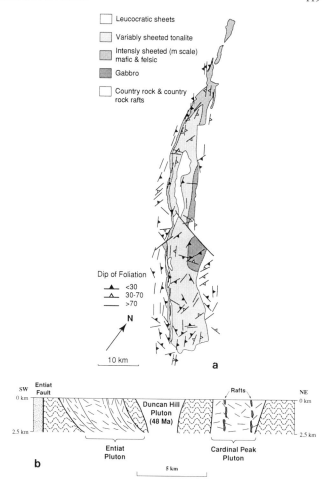

Figure 5 (a) Simplified map and (b) cross-section of the Entiat and nearby plutons, Washington. Al-in-hornblende indicates a depth of emplacement of 6–7 kbar for the Entiat pluton (Dawes 1993). Map of Entiat pluton shows location of mappable sheets and more irregular masses. Most of these units are internally complex and, particularly in the NW third of the pluton, consist of hundreds of metre–scale sheets. Foliation patterns (largely magmatic internally with an increasing component of subsolidus deformation near the margins) in (a) and (b) are even more complex than displayed, define large- and small-scale folds, sometimes cross-cut internal sheeting and are typically continuous with foliations in the country rock. Country rock units (shown with an upright fold pattern in cross-section) around the Entiat pluton largely consist of meta-amphibolite and meta-chert (Napeequa Formation), Swakane biotite gneiss and various orthogneisses. The structures in these units are complex, with older recumbent folds and strong axial planar foliations folded into the upright folds displayed in the cross-section.

including patterns of large- and small-scale folding. Solid-state fabrics are also common, particularly near the pluton margins or the margins of individual sheets. Interestingly, even at these depths the pluton margins are sometimes discordant to pre-emplacement markers even though magmatic foliation and lineation orientations may be parallel to equivalent country rock structures across the contacts (Fig. 5b).

Xenoliths and/or raft trains are more common in these deeper plutons than in higher level plutons, but still do not make up more than a few per cent of the total exposed areas of the plutons. It is often difficult to determine if these pieces of country rock are stoped blocks that have settled from above or relatively *in situ* country rock preserved between magmatic sheets (e.g. the raft trains of Pitcher & Berger 1972). As noted in summaries by Buddington (1959) and Barton *et al.* (1991), metamorphic aureoles are difficult to recognise and the differentiation between regional and emplacement-related deformation is less clear than at shallower levels.

Timing relationships in the Cascades indicate that regional

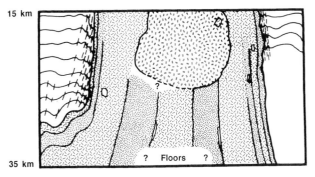

Figure 6 Summary of most common characteristics of the few steep-sided mid- to deep crustal plutons examined in this work (e.g. Fig. 5). Various random patterns represent plutonic rocks showing sheeting and one elliptical mass in the magma chamber. Visible sheeting decreases at shallower levels and may or may not feed into elliptical masses. Country rock and stoped blocks/raft trains shown with no pattern. Solid lines in the country rock show a range of behaviour of pre-emplacement markers. These plutons are typically pervasively foliated, with foliation sometimes subparallel to sheeting, but can also cut across internal contacts and mimic regional country rock patterns; only two such foliation trends are shown for the sake of clarity. Note that relatively narrow, sometimes discordant aureoles still occur at these crustal levels although regional, gentle to tight, possibly emplacement-related folds are more widespread than in Figure 3.

deformation occurred before, during and after emplacement, making it extremely difficult to evaluate emplacement mechanisms. However, the features, summarised in Fig. 6, indicate that emplacement scenarios at these crustal levels must involve the vertical transport of magma in numerous sheet-like and elliptical magma batches, regional deformation and some means of making steep magma chamber contacts that in places cut sharply across country rock structure. Regional deformation typically includes (1) arc-perpendicular contraction and shortening of the wall rocks and magma chamber perpendicular to the long dimension of the chamber, (2) crustal thickening and (3) arc-parallel extension. We believe the steep, sometimes discordant contacts again suggest that significant

amounts of country rock have been transported vertically out of the presently exposed map surface.

5. Tilted sections

Another valuable means of constraining depth changes in magma chamber processes is to examine tilted plutons. We have examined two, the Guadalupe Igneous Complex, California (Fig. 7a; Best 1963; Paterson *et al.* 1991b; Haeussler & Paterson 1993) and the Yerington batholith, Nevada (Fig. 7b; Proffett & Dilles 1984). Other examples include those described by Barnes *et al.* (1986) in the Klamath Mountains, California, by Hopson and co-workers (Hopson & Dellinger 1987) in the Cascade Mountains, Washington, by Flood and Shaw (1979) in the New England batholith, Australia, in the Newberry Mountains, Nevada (Hopson *et al.* 1994), the Chemehuevis Mountains, California (John 1988) and the Bergell massif, Italy (Rosenberg *et al.* 1995). Some of these sections represent oblique sections through the top and sides of the body (e.g. they are tilted <60°), but still provide important constraints on vertical changes in magma chambers.

The Guadalupe Igneous Complex, California (Best 1963; Paterson *et al.* 1991b) has been tilted approximately 30° with the SW side up (Haeussler & Paterson 1993). Restoration to the pre-tilt configuration brings the internal contacts and bedding in overlying volcanic rocks approximately back to the horizontal and corrects for the rotation of palaeomagnetic data (Haeussler & Paterson 1993). In this configuration (Fig. 7a), the Guadalupe Igneous Complex consists, from bottom to top, of roughly horizontal units of layered gabbro, diorite, mingled diorite and granite, and granophyre, and is overlain by felsic volcanic rocks, slate and greywacke (Best 1963; Paterson *et al.* 1991b). Cumulate-like layering in the gabbro is subhorizontal at some localities, but dips of 30–40° are common. Felsic and mafic dykes have variable orientations from subhorizontal to subvertical. Enclaves have irregular shapes and are generally not aligned, even near the margins. Magmatic foliation and lineation are rarely visible.

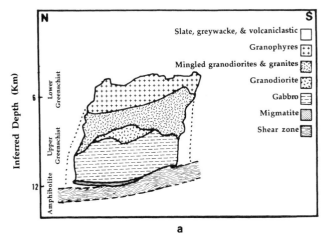

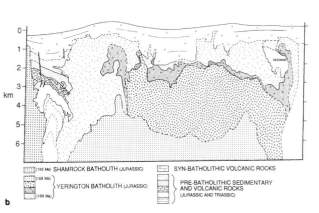

Figure 7 Crustal sections of two restored tilted plutons examined in this work (vertical = horizontal scale). (a) Oblique section through the Guadalupe Igneous Complex, Sierra Nevada, California after untilting of 30° and the removal of post-emplacement deformation in the country rock (see Haeussler & Paterson 1993; Paterson *et al.* 1991b for full discussion). The country rock is a Late Jurassic slate–greywacke sequence with local volcaniclastic-rich zones. Dotted lines represent the inferred extent of emplacement-related deformation. These regions display complex folding with the bedding often showing steep dips and at deeper levels some margin-parallel foliation. Bedding in the roof typically has dips of less than 20° and only a bedding-parallel compaction foliation. Note that the ductile shear zone at the base of this pluton has possibly cut off the lower parts of the chamber and that this shear zone still dips 30–40° to the east in this reconstruction. (b) Yerington batholith, Nevada after Dilles (1987). Depths and structural reconstructions based on excellent stratigraphic and structural control. Note that volcanic roof rocks show only small deflections. In the narrow aureole along the left-hand side of the Yerington batholith, the stratigraphic units show increasing deflection and amounts of ductile shortening with depth. Internally, this batholith consists of several nested elliptically-shaped bodies (diapirs?).

From the bottom to top, the country rock (after the removal of tilt and the effects of post-emplacement regional deformation) displays the following: (1) a large, syn-emplacement, ductile shear zone (Bear Mountains fault zone) and local migmatite in amphibolite grade wall rocks at the deepest exposed levels; (2) a small amount of ductile strain in a narrow, discontinuous aureole in upper greenschist grade rocks along the exposed sides; (3) an irregular, often stepped, discordant contact in lower greenschist facies rocks with no evidence of regional faulting or extension near the roof–wall joins; and (4) a relatively flat-lying roof of weakly deformed volcanic and volcaniclastic rocks (Fig. 7a). This horizontally layered chamber contrasts sharply with that of the Yerington Batholith described in the following.

Cenozoic extensional faulting has dismembered and tilted the Jurassic Yerington batholith 60–110° westwards, exposing a natural cross-section of the batholith roof and walls with >5 km of structural relief (Proffett 1977; Geissman et al. 1982; Proffett & Dilles 1984). When restored to it's original orientation (Fig. 7b), the roughly horizontal roof of this piston-shaped body occurs in shallowly dipping, genetically-related volcanic rocks. Along its steeply dipping walls, the batholith intrudes folded and metamorphosed Triassic and Jurassic volcanic and sedimentary rocks (Proffett & Dilles 1984). The roof is not extended or detached from the walls. The batholith is composite, consisting of four compositionally distinct pulses that were intruded between 169 and 168 Ma (Dilles & Wright 1988). In contrast with the Guadalupe Igneous Complex, these compositional units are nested, piston-shaped bodies. The youngest unit (granite) intruded the oldest (quartz monzodiorite) after the older unit was substantially crystallised; the

granite truncates fabric trends in the quartz monzodiorite (Fowler, unpublished data).

The emplacement style of the batholith changes from purely brittle to both ductile and brittle mechanisms (Fowler, unpublished data). At the structurally highest levels ($\approx$1–2 km paloeodepth), batholith emplacement is associated with modest, localised doming of quartzite and volcanic roof rocks, accommodated by penetrative brittle fracture along numerous small faults and breccia zones. Immediately below the roof ($\approx$2–3.5 km), the sharp, irregularly stepped wall contact truncates bedding and bedding-parallel foliation in the folded host rocks (Fig. 7b). At the deepest structural levels ($\approx$3.5–6 km), pluton emplacement is associated with the tightening of regional(?) folds and the development of penetrative cleavage with moderate to strong shortening strains (about −30 to −70%) in a narrow ductile aureole (0–300 m). Interestingly, emplacement-related host rock cleavage that deforms early dykes and is cut by late dykes is, in turn, locally cut by the intrusive contact, indicating a temporal transition from ductile to brittle behaviour.

6. Summary and conclusions

Our studies support Buddington's (1959) general conclusions regarding the complexity and changing internal and external characteristics of magmatic systems with depth. In general, the systems we have examined have flat roofs with steep sides. Internally, the crestal portions of magma chambers generally consist of less dense and/or more evolved magmas with little evidence of lateral zoning or early quenched margins, but on a smaller scale can be petrologically fairly varied. These grade

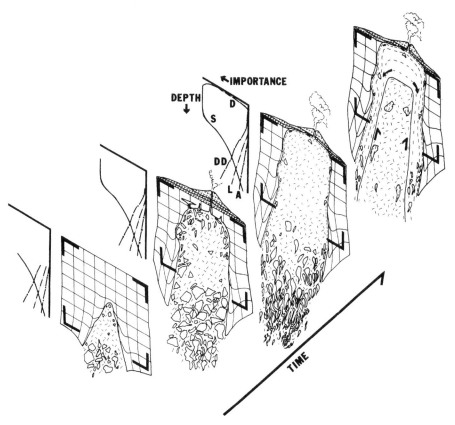

Figure 8 Time slice cartoons showing one possible emplacement scenario. Redrafted from Paterson and Vernon (1995). Slices show behaviour of country rock and are areally balanced. The graphs at the back show the magnitude of different MTPs with depth (these will vary from pluton to pluton). D, Doming of roof; S, stoping; DD, ductile deformation; L, lateral translation; and A, assimilation. The final time slice shows the nesting of diapirs and a summary of country rock constraints on the magnitude of and displacement paths for the transfer of country rock during emplacement. Note that although local upward transport may occur, most materials are transported downwards in the narrow aureole or within the region now occupied by the pluton.

downwards into steep-sided systems in which internal characteristics define two distinct mid-crustal patterns (Fig. 7): (1) lateral normally zoned or less commonly reversely zoned (e.g. Ayuso 1984; Nabelek *et al.* 1986; Allen 1992) bodies that we believe to be nested diapirs; and (2) horizontally layered systems in which lateral zoning is at best only weakly developed (e.g. Guadalupe Igneous Complex). At deeper levels, the plutons we are familiar with are petrologically complex systems typically dominated by many sheets and small elliptical masses. At all crustal levels magmatic foliations and lineations form late during freezing rather than early during the construction of these systems.

Our composite view of the plutons examined clearly supports the operation of vertically, laterally and temporally changing country rock material transfer processes during the upwards and sideways growth of magma chambers. Potentially more importantly, however, the evidence against significant upwards or lateral displacement of pluton roofs and evidence against horizontal displacement outside the narrow aureoles indicate that during chamber construction most, if not all, of the country rock in the mid- to upper crust is transported downwards in the region now occupied by the magma chamber and its aureole (Fig. 8). We emphasise that this statement seems to hold even for depths of 20–25 km, although the magnitude of downwards transported material is poorly constrained because of the less certain role of regional deformation during emplacement.

The internal sheeting at deeper levels and along margins at shallow levels and the nested elliptical magma batches at midcrustal levels indicate that during the downwards flow of country rock, multiple pulses of magma usually travelled up the same magma system (see also Bergantz & Dawes 1994). Thus if these relationships are widespread in arcs, as is suggested by our initial studies (e.g. Yoshinobu & Paterson 1995), and those of others (Buddington 1959; Cobbing & Pitcher 1972; Pitcher 1979; Saleeby 1990; Bateman 1992), magma emplacement in arcs is the driving mechanism for a huge crustal-scale exchange and sometimes a recycling process. As magma plumbing systems are constructed by the rise of multiple batches of magma, country rock is transported downwards by a variety of material transport processes in the laterally migrating margins of these systems. Presumably an increasing fraction of this downwards transported country rock could become assimilated by or chemically interact with magmas at deeper crustal levels (Bergantz & Dawes 1994; Reiners *et al.* 1995) and thus potentially rise again in crustally contaminated magmas.

7. Acknowledgements

This research was supported by NSF Grants EAR-9218741 and EAR-9304058 awarded to Paterson and EAR-8917343 and EAR-9219536 awarded to Miller. We thank Sandy Cruden and Barbara John for constructive reviews and Dallas Peck for editorial assistance.

8. References

Ague, J. J. & Brimhall, G. H. 1988. Magmatic arc asymmetry and distribution of anomolous plutonic belts in the batholiths of California: effects of assimilation, crustal thickness, and depth of crystallization. BULL GEOL SOC AM **100**, 912–27.

Allen, C. M. 1992. A nested diapir model for the reversely zoned Turtle pluton, southeastern California. TRANS SOC EDINBURGH EARTH SCI **83**, 179–90.

Ayuso, R. A. 1984. Field relations, crystallization, and petrography of reversely zoned granitic pluton in the Bottle Lake Complex, Maine. US GEOL SURV PROF PAP **1320**.

Barnes, C. G., Rice, J. M. & Gribble, R. F. 1986. Tilted plutons in the

Klamath Mountains of California and Oregon. J GEOPHYS RES **91**, 6059–71.

Barton, M. D., Staude, J. M., Snow, E. A. & Johnson, D. A. 1991. Aureole systematics, in contact metamorphism. AM MINERAL SOC REV MINERAL **26**, 723–847.

Bateman, P. C. 1992. Plutonism in the central part of the Sierra Nevada Batholith, California. US GEOL SURV PROF PAP **1483**.

Bateman, P. C., Clark, L. D., Huber, N. K., Moore, J. G. & Rinehart, C. D. 1963. The Sierra Nevada batholith—a synthesis of recent work across the central part. US GEOL SURV PROF PAP **414D**, D1–46.

Bateman, R. 1982. The zoned Bruinbin granitoid pluton and its aureole. J GEOL SOC AUST **29**, 253–65.

Bergantz, G. W. & Dawes, R. 1994. Aspects of magma generation and ascent in the continental lithosphere. In Ryan, M. P. (ed.) *Magmatic systems*, 291–317. San Diego: Academic Press.

Berger, A. R. & Pitcher, W. S. 1970. Structures in granite rocks: a commentary and critique on granite tectonics. PROC GEOL ASSOC **81**, 441–61.

Best, M. G. 1963. Petrology and structural analysis of metamorphic rocks in the southwestern Sierra Nevada Foothills, California. UNIV CALIFORNIA PUBL GEOL SCI **42**, 111–58.

Bouchez, J. L. & Diot, H. 1990. Nested granites in question: contrasted emplacement kinematics of independent magmas in the Zaer pluton, Morocco. GEOLOGY **18**, 966–9.

Buddington, A. F. 1959. Granite emplacement with special reference to North America. GEOL SOC AM BULL **70**, 671–747.

Bykerk-Kauffman, A. 1990. *Structural evolution of the northeastern Santa Catalina Mountains, Arizona: a glimpse of the pre-extension history of the Catalina Complex*. Unpublished Ph.D. Dissertation, University of Arizona.

Clemens, J. D. & Mawer, C. K. 1992. Granitic magma transport by fracture propagation. TECTONOPHYSICS **204**, 339–60.

Cobbing, E. J. & Pitcher, W. S. 1972. The coastal batholith of Peru. J GEOL SOC LONDON **128**, 421–60.

Compton, R. 1955. Trondhjemite batholith near Bidwell Bar, California. GEOL SOC AM BULL **66**, 9–44.

Crossland, A. 1994. Implications of roof structures for the emplacement of the Hall Canyon pluton, Panamint Mtns, California. GEOL SOC AM ABST PROGRAMS **26**, A-134.

Dawes, R. L. 1993. *Mid-crustal, Late Cretaceous plutons of the North Cascades: petrogenesis and implications for the growth of continental crust*. Unpublished Ph.D. Dissertation, University of Washington, Seattle.

Delaney, P. T. & Pollard, D. D. 1981. Deformation of host rocks and flow of magma during growth of minette dikes and breccia bearing intrusions near Ship Rock, New Mexico. US GEOL SURV PROF PAP **1202**.

Dilles, J. H. 1987. The petrology of the Yerington batholith, Nevada: evidence for the evolution of porphyry copper ore fluids. ECON GEOL **72**, 769–95.

Dilles, J. H. & Wright, J. E. 1988. The chronology of early Mesozoic arc magmatism in the Yerington District of western Nevada and its regional implications. GEOL SOC AM BULL **100**, 644–52.

Flood, R. H. & Shaw, S. E. 1979. K-rich cumulate diorite at the base of a tilted granodiorite pluton from the New England Batholith, Australia. J GEOL **87**, 417–25.

Fowler, T. K. Jr. 1994a. Granitoid emplacement into older plutonic host-rocks. GEOL SOC AM ABSTR PROGRAMS **26**, A-134.

Fowler, T. K. Jr. 1994b. Using geologic maps to constrain pluton emplacement mechanisms. GEOL SOC AM ABSTR PROGRAMS **26**, 52.

Fowler, T. K. Jr. & Paterson, S. R. Timing and nature of magmatic fabrics from structural relations around stoped blocks. J STRUCT GEOL, in press.

Fowler, T. K. Jr, & Paterson, S. R., Crossland, A. & Yoshinobu, A. 1995. Pluton emplacement mechanisms: a view from the roof. In Brown, M. & Piccoli, P. M. (eds) *The origin of granites and related rocks* US GEOL SURV CIRC **1129**, 57.

Fowler, T. K. Jr, Yoshinobu, A., Paterson, S. R., Tickyj, H., Llambias, E. J. & Sato, A. M. 1995. Chita pluton, San Juan Province, Argentina: 3D constraints on pluton emplacement by magmatic stoping. GEOL SOC AM ABSTR PROGRAMS **27**, 7–125.

Geissman, J. W., Van Der Voo, R. & Howard, K. L. Jr 1982. A paleomagnetic study of the structural deformation in the Yerington district, Nevada. AM J SCI **282**, 1042–109.

Haeussler, P. J. & Paterson, S. R. 1993. Post-emplacement tilting and burial of the Guadalupe Igneous Complex, Sierra Nevada, California. GEOL SOC AM BULL **105**, 1310–20.

Hopson, C. A. & Dellinger, D. A. 1987. Evolution of four-dimensional compositional zoning, illustrated by the diapiric Duncan Hill pluton, north cascades, Washington. GEOL SOC AM ABSTR PROGRAMS **19**, 707.

Hopson, C. A., Gans, P. B., Baer, E., Blythe, A., Calvert, A. & Pinnow, J. 1994. Spirit Mountain Pluton, Southern Nevada: a progress report. GEOL SOC AM ABSTR PROGRAMS **26**, 60.

Hutton, D. H. W. 1988. Granite emplacement mechanisms and tectonic controls: inferences from deformation studies. TRANS R SOC EDINBURGH **79**, 245–55.

Hyndman, D. W., Silverman, R. E., Benoit, W. R. & Wold, R. 1982. The Phillipsburg batholith, western Montana. BUR MINES GEOL MEM **49**.

John, B. E. 1988. Structural reconstruction and zonation of a tilted midcrustal magma chamber; the felsic Chemehuevi Mountains Plutonic Suite. GEOLOGY **16**, 613–7.

Lister, J. R. & Kerr, R. C. 1991. Fluid-mechanical models of crack propogation and their application to magma transport in dikes. J GEOPHYS RES **96**, 10 049–77.

Mackin, J. H. 1947. Some structural features of the intrusions in the Iron Springs district. UTAH GEOL SOC GUIDE GEOL UTAH **2**, 62.

Mahon, K. I., Harrison, T. M. & Drew, D. A. 1988. Ascent of a granitoid diapir in a temperature varying medium. J GEOPHYS RES **93**, 1174–88.

Marsh, B. D. 1982. On the mechanics of igneous diapirism, stoping, and zone melting. AM J SCI **282**, 808–55.

Miller, R. B. & Bowring, S. A. 1990. Structure and chronology of the Oval Peak batholith and adjacent rocks: implications for the Ross Lake fault zone, North Cascades, Washington. GEOL SOC AM BULL **102**, 1361–77.

Nabelek, P. I., Papike, J. J. & Laul, J. C. 1986. The Notch Peak Granite stock, Utah: origin of reverse zoning and petrogenesis. J PETROL **27**, 1035–9.

Paterson, S. R. 1992. Pluton emplacement processes: implications of the structural characteristics of pluton roofs. GEOL SOC AM ABSTR PROGRAMS **24**, 73.

Paterson, S. R. & Fowler, T. K. Jr 1993. Re-examining pluton emplacement processes. J STRUCT GEOL **15**, 191–206.

Paterson, S. R. & Fowler, T. K. Jr 1995. Construction of magma chambers in arcs: a perspective from the country rock. GEOL SOC AM ABSTR PROGRAMS **27**, 80.

Paterson, S. R. & Vernon, R. H. 1995. Bursting the bubble of ballooning plutons: a return to nested diapirs emplaced by multiple processes. GEOL SOC AM BULL **107**, 1356–80.

Paterson, S. R., Vernon, R. H. & Fowler, T. K. Jr 1991a. Aureole tectonics. *In* Kerrick, D. M. (ed.) *Contact metamorphism.* MINERAL SOC AM REV MINERAL **26**, 673–722.

Paterson, S. R., Tobisch, O. T. & Vernon, R. H. 1991b. Emplacement and deformation of granitoids during volcanic arc construction in the Foothills terrane, central Sierra Nevada, California. TECTONOPHYSICS **191**, 89–110.

Paterson, S. R., Miller, R. B., Anderson, L. A., Lund, S., Bendixen, J., Taylor, N. & Fink, T. 1994. Emplacement and evolution of the Mt. Stuart batholith. *In* Swanson, D. A. & Haugerud, R. H. (eds) *Geologic field trips in the Pacific Northwest: 1994 Geological Society of America Annual Meeting,* 2F-1–47.

Pitcher, W. S. 1979. The nature, ascent and emplacement of granite magmas. J GEOL SOC LONDON **136**, 627–62.

Pitcher, W. S. 1993. *The nature and origin of granite.* Glasgow: Blackie Academic and Professional.

Pitcher, W. S. & Berger, A. R. 1972. *The geology of Donegal: a study of granite emplacement and unroofing.* New York: Wiley.

Proffett, J. M. Jr 1977. Cenozoic geology of the Yerington district, Nevada, and implications for the nature and origin of basin and range faulting. GEOL SOC AM BULL **88**, 247–66.

Proffett, J. M. & Dilles, J. H. 1984. Geologic map of the Yerington district, Nevada. *Nevada Bureau of Mines and Geology Map 77, 1:24,000 scale.*

Reiners, P. W., Nelson, B. K. & Ghiorso, M. K. 1995. Assimilation of the felsic crust by basaltic magma: thermal limits and extents of crustal contamination of mantle-derived magmas. GEOLOGY **23**, 563–66.

Rosenberg, C. L., Berger, A. & Schmid, S. M. 1995. Observations from the floor of a granitoid pluton: inferences on the driving force of final emplacement. GEOLOGY **23**, 443–6.

Rubin, A. M. 1993. Getting granite dikes out of the source region. J GEOPHYS RES **100**, 5911–29.

Saleeby, J. B. 1990. Progress in tectonic and petrogenetic studies in an exposed cross-section of young ($\approx$ 100 Ma) continental crust, southern Sierra Nevada, California. *In* Salisbury, M. H. & Fountain, D. M. (eds) *Exposed cross-sections of the continental crust,* 137–58. Dordrecht: NATO Advanced Studies Institute/ Kluwer Academic.

Sato, A. M. 1987. Chita granitic stock: a closed system crystallization. *In 10 Congreso Geologico, Tucuman,* Vol. 4, 69–99.

Shaw, H. R. 1980. The fracture mechanism of magma transport from the mantle to the surface. *In* Hargraves, R. B. (ed.) *Physics of magmatic processes,* 201–64. Princeton: Princeton University Press.

Stein, E. 1994. Structures in the aureole of the Joshua Flat Pluton: implications for country rock flow patterns during emplacement. GEOL SOC AM ABSTR PROGRAMS **26**, 95.

Stephens, W. E. 1992. Spatial, compositional, and rheological constraints on the origin of zoning in the Criffell pluton, Scotland. TRANS R SOC EDINBURGH EARTH SCI **83**, 191–9.

Weinberg, R. F. & Podladchikov, Y. 1994. Diapiric ascent of magmas through power-law crust and mantle. J GEOPHYS RES **99**, 9543–60.

Yuan, E. S. & Paterson, S. R. 1993. Evaluating flow from structures in plutons. GEOL SOC AM ABSTR PROGRAMS **25**, 305.

Yoshinobu, A. S. & Paterson, S. R. 1995. Multiple space-making mechanisms and vertical material transfer in the Sierra Nevada batholith. GEOL SOC AM ABSTR PROGRAMS **27**, 85.

Yoshinobu, A. S., Okaya, D. A., Paterson, S. R. & Fowler, T. K. 1995. Testing fault-controlled magma emplacement mechanisms. US GEOL SURV CIRC **1129**.

SCOTT R. PATERSON and T. KENNETH FOWLER Jr University of Southern California, Los Angeles, CA 90089-0740, U.S.A. E-mail: Paterson@usc.edu
ROBERT B. MILLER Department of Geology, San Jose State University, San Jose, CA 95192, U.S.A.

Transactions of the Royal Society of Edinburgh: Earth Sciences, **87**, 125–138, 1996

Status of thermobarometry in granitic batholiths

J. Lawford Anderson

ABSTRACT: Most granitic batholiths contain plutons which are composed of low-variance mineral assemblages amenable to quantification of the P–T–f_{O_2}–f_{H_2O} conditions that characterise emplacement. Some mineral thermometers, such as those based on two feldspars or two Fe–Ti oxides, commonly undergo subsolidus re-equilibration. Others are more robust, including horn-blende–plagioclase, hornblende–clinopyroxene, pyroxene–ilmenite, pyroxene–biotite, garnet–horn-blende, muscovite–biotite and garnet–biotite. The quality of their calibration is variable and a major challenge resides in the large range of liquidus to solidus crystallisation temperatures that are incompletely preserved in mineral profiles. Further, the addition of components that affect K_d relations between non-ideal solutions remains inadequately understood. Estimation of solidus and near-solidus conditions derived from exchange thermometry often yield results $> 700°C$ and above that expected for crystallisation in the presence of an H_2O-rich volatile phase. These results suggest that the assumption of crystallisation on an H_2O-saturated solidus may not be an accurate characterisation of some granitic rocks.

Vapour undersaturation and volatile phase composition dramatically affect solidus temperatures. Equilibria including hypersthene–biotite–sanidine–quartz, fayalite–sanidine–biotite, and annite–sanidine–magnetite (ASM) allow estimation of f_{H_2O}. Estimates by the latter assemblage, however, are highly dependent on f_{O_2}. Oxygen fugacity varies widely (from two or more log units below the QFM buffer to a few log units below the HM buffer) and can have a strong affect on mafic phase composition. Ilmenite–magnetite, quartz–ulvospinel–ilmenite–fayalite (QUILF), annite–sanidine–magnetite, biotite–almandine–muscovite–magnetite (BAMM), and titanite–magnetite–quartz (TMQ) are equilibria providing a basis for the calculation of f_{O_2}.

Granite barometry plays a critical part in constraining tectonic history. Metaluminous granites offer a range of barometers including ferrosilite–fayalite–quartz, garnet–plagioclase–hornblende–quartz and Al-in-hornblende. The latter barometer remains at the developmental stage, but has potential when the effects of temperature are considered. Likewise, peraluminous granites often contain mineral assemblages that enable pressure determinations, including garnet–biotite–muscov-ite–plagioclase and muscovite–biotite–alkali feldspar–quartz. Limiting pressures can be obtained from the presence of magmatic epidote and, for low-Ca pegmatites or aplites, the presence of subsolvus versus hypersolvus alkali feldspars.

As with all barometers, the influence of temperature, f_{O_2}, and choice of activity model are critical factors. Foremost is the fact that batholiths are not static features. Mineral compositions imperfectly record conditions acquired during ascent and over a range of temperature and pressure and great care must be taken in properly quantifying intensive parameters.

KEY WORDS: thermometry, barometry, granites.

Much remains to be learned about the intensive parameters characterising the formation and emplacement of granitic rocks. In many instances, this is due to high–variance phase assemblages, resetting during slow cooling, or the lack of well characterised, experimentally calibrated reactions. However, portions of most granitic batholiths contain low-variance phase assemblages that are amenable to detailed quantification of P–T–f_{O_2}–f_{H_2O} conditions of final equilibration.

Only a portion of mineral compositions in igneous rocks represent solidus conditions, which makes the retrieval of meaningful estimates of pressure, volatile fugacities and temperature a challenge. Some mineral phases continue to react during subsolidus cooling. Other minerals resist chemical change and partially record conditions recorded during ascent. Only through careful petrography and analysis of zoning profiles is it possible to obtain reliable estimates of the intensive parameters that characterise solidus and hypersolidus conditions.

Most orogenic terranes contain a protracted history of plutonism. Estimates for P–T conditions of well-dated plutons

provide direct evidence for the ascent or descent of exposed crustal sections through time, thus providing fundamental information about tectonic processes leading to the orogenic construction of new crust. This is a common goal in thermochronological studies of metamorphic terranes (Spear *et al.* 1984) and in multiply intruded terranes, plutons serve as crustal 'nails' providing important constraints complementary to those achieved through other means (Anderson *et al.* 1988).

1. Factors affecting thermometry and barometry

A number of problems can lead to erroneous estimates of intensive parameters in igneous rocks. The preferred thermometer or barometer is one that is grounded in reversed experimental data. Yet the quality of an experimental calibration can vary depending of the ease of chemical change or experimental design. Other calibrations are empirical and thus carry significant added uncertainty. Although most thermometers are not very sensitive to pressure, barometers are fundamentally temperature dependent. Thus an inadequate assessment of temperature can lead to erroneous pressure estimates.

The major source of error in thermobarometric calculations rests with the fact that the original experimental calibration may have been conducted for an end-member reaction with mineral compositions far removed from that observed in granitic systems. Whereas some calibrations are 'adjusted' on empirical grounds, the preferred method uses extrapolation based on activity models. Activity models likewise vary in quality due to uncertainties related to solid solution non-ideality and the effects of extra components. Analytical uncertainty magnifies errors caused by extrapolation necessitating the evaluation of the full extent of uncertainty through the propagation of error.

Finally, and given that igneous minerals crystallise over a range of P–T conditions and remain in contact well into the subsolidus realm, there can be no assurance that mineral phases have 'locked in' solidus compositions. It would be rare to find a granite, for example, that did not have compositionally zoned minerals and much of the zoning can be related to mineral growth above the solidus. Other minerals easily change their compositions during subsolidus cooling.

For further discussion of uncertainties in thermobarometry, the reader is referred to Powell and Holland (1988), McKenna and Hodges (1988) and Kohn and Spear (1991a, b). There are several computer programs available to aid in calculations involving P–T–f_{H_2O}–f_{O_2}, including EQUILI (Valley & Essene 1980), THERMO (Perkins *et al.* 1987), THERMOCALC (Powell & Holland 1988), TWQ (Berman 1988), WEBINVEQ (Gordon 1992) and QUILF (Andersen *et al.* 1993). The programs have the benefit of offering internally consistent data sets, which is important when different thermobarometers are applied to a given assemblage. However, they can also have the drawback of preventing a choice of activity model or the incorporation of more recent parameters that may better describe a particular solid solution. The following sections summarise the current status of thermometers and barometers applicable to granites and, for purposes of comparison, the original calibrations are used.

2. Thermometers

There are many excellent reviews of thermometers used in metamorphic systems (e.g. Essene 1982, 1989; Bohlen & Lindsley 1987; Spear 1993) and many of these thermometers are also applicable to granitic rocks. If minerals in granitic magmas all equilibrated on a water-saturated solidus, there would be little reason to determine the temperature as solidus temperatures for most igneous systems are well known. For pressures >3 kbar, for example, the water-saturated granite solidus is nearly isothermal at ≈650–675°C (Wyllie 1984). Yet thermobarometric data for many granitic plutons yield temperatures above 700°C (see review in Anderson & Smith 1995) either due to the preservation of compositions acquired above the solidus or to the effects of a mixed H_2O–CO_2 vapour phase on the solidus or vapour undersaturation. Because the determination of pressure, f_{O_2} and volatile composition are temperature dependent, an accurate assessment of crystallisation temperature is mandated. How accurate any one thermometer measures solidus conditions remains an open question given the concerns expressed earlier. Thus it should be clear that petrologists should use many different thermometers to assess the crystallisation temperatures of magmas. It should also be recognised that a given thermometer may be providing information about only a portion of the liquidus to subsolidus temperature range experienced by a pluton. Naturally, it should not be assumed that the mineral compositions obtained reflect equilibrium conditions. On a graphical basis, all involve the rotation of temperature-dependent tie lines. Thus K_d relations among minerals can be used as a guide in the assessment of the nature and extent of equilibrium.

2.1. Thermometers applicable to a range of granitic rocks

Almost all granites contain two feldspars and many contain two Fe–Ti oxides. Typically one or both minerals of these two pairs may have changed in composition during subsolidus cooling and therefore the derived temperatures are either meaningless or reflect the cooling path. In some instances, however, magmatic information can be retrieved by an integrated form of analysis.

Two-feldspars. Two-feldspar thermometry has had a long history of development and all formulations but that of Whitney and Stormer (1977) are based on disordered structural states. Early calibrations were based on the binary exchange of the albite component, including Stormer (1975), Whitney and Stormer (1977) and Haselton *et al.* (1983). Brown and Parsons (1981) noted that models for feldspar thermometry should use ternary components and more recent calibrations, including Green and Usdansky (1986), Fuhrman and Lindsley (1988) and Elkins and Grove (1990) offer three calibrations for each feldspar pair based on the exchange of albite, anorthite and orthoclase components, respectively. The results using anorthite and orthoclase components are subject to analytical error given that the amount of anorthite component in K-feldspar and orthoclase component in plagioclase are usually very low.

Figure 1 compares each of the six calibrations for albite tested against experimental feldspar data of Elkins and Grove (1990) at temperatures from 650 to 900°C at 1 to 3 kbar. The

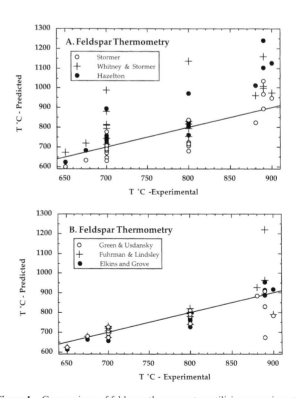

Figure 1 Comparison of feldspar thermometry utilising experimental data of Elkins and Grove (1990). (a) Calibrations of Stormer (1975), Whitney and Stormer (1977) and Haselton *et al.* (1983). (b) Calibrations of Green and Usdansky (1986), Fuhrman and Lindsley (1988) and Elkins and Grove (1990). The solid lines delineate a 1 : 1 correspondence between calculated and experimental temperature.

binary calibrations exhibit the some scatter of estimated versus known temperature. The ternary thermometer calibrations utilise improved solution parameters extended to include minor components (Or in plagioclase and An in K-feldspar). In the 650–800°C range, the three calibrations agree within a range of ±50°C. The only serious discrepancy is for pairs having high An plagioclase at 900°C. Thus any of the three recent ternary calibrations can be expected to yield reliable estimates of solidus temperature for granitic plutons. The effect of the structural state remains unknown, however, as there has been no calibration with partially ordered orthoclase.

Fe–Ti oxides. The ilmenite–magnetite thermometer is classic to petrology. Dating back to Buddington and Lindsley (1964), the thermometer offers independent estimates of T and f_{O_2}. The recent formulation of Ghiorso and Sack (1991) combines the spinel solution model of Sack and Ghiorso (1991) with the ilmenite–hematite solution model of Ghiorso (1990). Lindsley *et al.* (1990) present a different formulation which combines the projection algorithm of Spencer and Lindsley (1981) with the thermodynamic models for oxides of Andersen and Lindsley (1988). These build on earlier methods of projection, including that of Stormer (1983).

Fe–Ti oxides in many granitic plutons have re-equilibrated at subsolidus conditions; see Frost (1991) for a discussion of how ilmenite and magnetite may or may not change in composition during cooling. However, a few granites retain compositions of primary Fe–Ti oxides, particularly those that were emplaced under shallower and/or hotter conditions.

Figure 2 depicts T–f_{O_2} estimates for three plutons that contain primary Fe–Ti oxides, emphasising the differences in the formulations of Lindsley and Spencer (1982), Stormer (1983), Lindsley *et al.* (1990) and Ghiorso and Sack (1991). Rutherford and Hill (1993) found that the Lindsley *et al.* (1990) model reproduced temperatures within ±10°C for oxide pairs synthesised at 850°C.

2.2. Thermometers applicable to hornblende-bearing granitic rocks

Most batholiths contain hornblende-bearing intrusions and, as a result, the hornblende–plagioclase thermometer is expected to be widely applied. A few deep–seated intrusions contain garnet + hornblende, although the two phases are often not in textural equilibrium.

Hornblende–plagioclase. Spear (1980, 1981) presented early calibrations of the hornblende–plagioclase thermometer based on pairs in metamorphic rocks for which temperature was known by other means. For rocks having a restricted range of plagioclase composition, Nabelek and Lindsley (1985) pro-

posed an experimentally based thermometer using variations in Al^{IV} of hornblende. Using a combined set of experimental and empirical data, Blundy and Holland (1990) formulated a calibration for the equilibrium

$$\text{Edenite} + 4 \text{ quartz} = \text{tremolite} + \text{albite} \qquad \text{(A)}$$

They used an ideal mixing-on-sites model for hornblende and a quadratic formalism to describe non-ideality in plagioclase. The calibration was criticised by Poli and Schmidt (1992) on thermodynamic grounds and for yielding erroneously high temperatures for amphibolites containing aluminous hornblendes, particularly those that coexist with garnet.

Subsequently, Holland and Blundy (1994) revised the calibration for the edenite–tremolite reaction and introduced a new calibration based on the equilibrium

$$\text{Edenite} + \text{albite} = \text{richterite} + \text{anorthite} \qquad \text{(B)}$$

For both equilibria they have discarded the ideal solution model for hornblende and have used a symmetrical formalism to account for non-ideality.

These two calibrations are sensitive to the Fe^{3+} content of hornblende which, in turn, affects the occupancy of several sites. Holland and Blundy (1994) follow an amphibole normalisation scheme similar to that of Spear and Kimball (1984) and, as it was used as a basis for calibrations, their procedure must be followed in the application of the thermometers. Cosca *et al.* (1991) have presented evidence that a 13 cation normalisation is a better method for estimating Fe^{2+}/Fe^{3+} ratios in amphibole which will usually lead to higher estimates of Fe^{3+} than the method used by Holland and Blundy.

Figure 3 depicts the sensitivity of both of the Holland and Blundy (1994) thermometers (labelled reaction A and reaction B) to estimated Fe^{3+} concentrations for hornblende–plagioclase pairs from a tonalite. In contrast, the older Blundy and Holland thermometer is less affected by the hornblende normalisation scheme.

Figure 4 compares results for all three calibrations for two groups of granitic plutons, one having low-alumina hornblende ($Al_2O_3 < 8$ wt%) and the other having high-alumina hornblende ($Al_2O_3 > 10$ wt%) as a function of emplacement depth. The results for shallow intrusions show a high correspondence for temperatures derived from all three calibrations. In contrast, temperature estimates for higher pressure, mid-crustal intrusions differ significantly with the Blundy and Holland (1990) calibration yielding unusually high tempera-

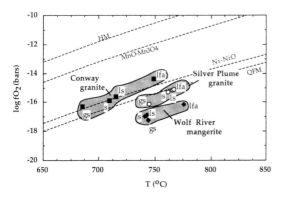

Figure 2 Comparison of calibrations used for ilmenite–magnetite thermometry. gs = Ghiorso and Sack (1991); lfa = Lindsley *et al.* (1990); ls = Lindsley and Spencer (1982); s = Stormer (1983). Fe–Ti oxide data from Stormer (1983), Anderson and Thomas (1985) and Anderson (1980).

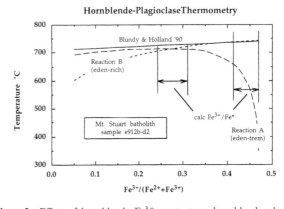

Figure 3 Effect of hornblende Fe^{3+} content on hornblende–plagioclase thermometry. Calibrations are those of Blundy and Holland (1990) and Holland and Blundy (1994). Two estimates of Fe^{3+}/Fe are depicted, an upper bracket based on 13 cations (Cosca *et al.* 1991) and a lower bracket based on the method of Holland and Blundy (1994). Amphibole sample is from the Mt Stuart batholith (Paterson *et al.* 1994; Anderson and Smith 1995).

Hornblende-Plagioclase Thermometry

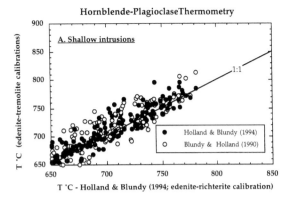

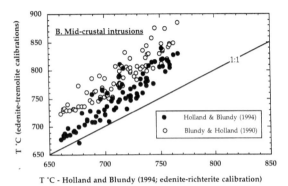

Figure 4 Comparison of hornblende–plagioclase thermometers including calibration of Blundy and Holland (1990) and calibrations of reactions A and B of Holland and Blundy (1994) using coexisting hornblende–plagioclase rims from (A) the Mt Stuart batholith (Anderson and Smith 1995), Eagle Mountains pluton (Mayo 1994) and the Teutonia batholith (Beckerman *et al.* 1982) and (B) Whipple (Anderson 1988), San Gabriel (Barth 1989), Cargo Muchacho (Hayes 1992) and Granite Mountains (Young 1990).

tures. The results derived from the calibration of reaction B (edenite–richterite) of Holland and Blundy (1994) yield the lowest and are considered to be the most reliable because they more precisely reproduce the temperatures derived from other thermometers.

Garnet–hornblende. The garnet–hornblende thermometer of Graham and Powell (1984) is based on the exchange of Fe^{2+}/Mg and was empirically calibrated against the garnet–clinopyroxene thermometer of Ellis and Green (1979). The thermometer has not been experimentally investigated, nor has it been re-calibrated against newer versions of the garnet–clinopyroxene thermometer (Green and Adam 1991). Thus the thermometer can only be expected to offer qualified estimates of temperature.

Dawes and Evans (1991) used the thermometer for magmatic epidote-bearing dykes from which coexisting phenocrysts of garnet and hornblende yielded temperatures (740–780°C) in the range observed from garnet–biotite and plagioclase–hornblende thermometry.

2.3. Thermometers applicable to pyroxene-bearing granitic rocks

Many calc-alkaline batholiths contain pyroxene-bearing dioritic to mafic tonalitic intrusions. The pyroxene is often mantled by hornblende and, based on textural observations, may not have been stable at solidus conditions. In contrast, pyroxene-bearing granitic rocks are common in the Proterozoic (Anderson 1983; Anderson & Morrison 1992) and occur in association with anorthosite–rapakivi granite intrusions. As a result of low f_{H_2O} and high temperature, pyroxenes in these

rocks are inferred to have been stable during solidus crystallisation (Emslie & Stirling 1993).

Clinopyroxene–orthopyroxene. Two-pyroxene thermometry has long been applied to granulites, other high-temperature metamorphic rocks and gabbros. The thermometer is also applicable to many intermediate to felsic igneous rocks. Early forms of the thermometer (Wood & Banno 1973; Wells 1977) were based on a binary model of Ca–Mg exchange. Kretz (1982) offered two calibrations, one based on a ternary model and the other for Fe–Mg exchange. The most complete formulations are those of Lindsley (1983) and Davidson and Lindsley (1985), who offer corrections for non-quadrilateral components.

For dioritic and quartz dioritic rocks of the Mt Stuart batholith, Paterson *et al.* (1994) observed that temperatures obtained with the Davidson and Lindsley (1985) calibration averaged between 830 and 860±25°C compared with 905±35°C derived from the older calibrations. Cotkin and Medaris (1993) determined similar temperatures (845±80°C) for intermediate plutons of the Russian Peak complex.

Clinopyroxene–hornblende and clinopyroxene–biotite. Clinopyroxene–hornblende and clinopyroxene–biotite thermometers are presented together as both are applicable in many intermediate composition rocks. Based on the partitioning of Mg/Fe, the thermometers are presented in graphical form by Perchuck *et al.* (1985) with no information about the nature or quality of calibration. Figure 5 shows the graphical form of the clinopyroxene–hornblende thermometer applied to dioritic and quartz dioritic rocks of the Mt Stuart batholith of Washington. The author has fitted both thermometers to an algebraic form that is available on request. Although the two thermometers can yield similar temperatures, these results can only be viewed with caution as the data in support of the calibration have not been published in a referred journal and nor have the effects of variable Fe^{3+}/Fe^{2+} been reported. Kretz and Jen (1978) have also investigated the hornblende-clinopyroxene thermometer and have reported the substantial effect of Al^{IV} in hornblende on the Mg/Fe K_d between the phases.

Orthopyroxene–biotite. Sengupta *et al.* (1990) have calibrated an Fe–Mg exchange thermometer for orthopyroxene-biotite using the experimental data of Fonarev and Konilov (1986). Included are terms correcting for Fe–Mg non-ideality in orthopyroxene and the effects of mixing of Al and Ti in the octahedral sites of biotite. The calibration is also minimally pressure dependent ($\approx 4°C\ kbar^{-1}$).

Application of the thermometer to five pyroxene–biotite pairs from a diorite from the Mt Stuart batholith yields

CPX-HB Thermometry

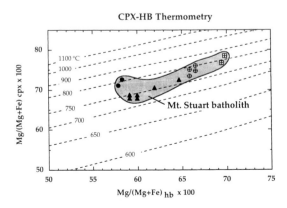

Figure 5 Results of clinopyroxene–hornblende thermometry (Perchuck *et al.* 1985) applied to the Mt Stuart batholith (Paterson *et al.* 1984). Open symbols represent diorites and closed symbols tonalites.

average temperature of $1042 \pm 25°C$ (Paterson *et al.* 1994). A controversial aspect of the calibration is the Fe–Mg nonideality term for orthopyroxene as previous workers have concluded that the Fe–Mg mixing is ideal (Harley 1984; Lee & Ganguly 1987). Removing this term lowers temperatures for the Mt Stuart diorites down to $970 \pm 21°C$.

Pyroxene–ilmenite. $Mg–Fe^{2+}$ partitioning between clinopyroxene–ilmenite and orthopyroxene–ilmenite has been calibrated experimentally by Bishop (1980). These thermometers are applicable to many mafic and intermediate rocks, but only those with measurable Mg in ilmenite. Although reversed, the calibrations remain of uncertain quality due to a lack of homogeneous run products. Moreover, the reported value for the derived mixing parameter W_G^{Cpx} (2.728 ± 0.166 kcal mol^{-1}) is at variance with that obtained by Perkins and Vielzeuf (1992; 0.950 ± 0.101 kcal mol^{-1}).

2.4. Thermometers applicable to peraluminous granitic rocks

Garnet–biotite. The garnet–biotite thermometer initially calibrated by Ferry and Spear (1978) was developed for garnets with low Ca and Mn ($X_{Ca} + X_{Mn} < 0.2$) and biotites with low Ti and Al^{VI}. Peraluminous granites commonly have garnet and biotite, but the compositions of both phases often exceed the boundary conditions recommended by Ferry and Spear (1978), particularly in terms of X_{Mn}. Perchuck and Lavrent'eva (1983) present an independent calibration for the thermometer which generally yields lower temperatures (Chipera & Perkins 1988). However, the problem for garnet–biotite thermometry in granites is commonly one of temperatures being too low rather than too high due largely to the effect of garnet Mn on the Mg/Fe K_d between garnet and biotite (Dallmeyer 1974). This can be seen in Figure 6, which documents how the temperatures of Ferry and Spear (1978) decrease with increasing $X_{Mn, garnet}$.

There have been several attempts to re-formulate the Ferry and Spear (1978) thermometer to correct for substitutions that affect Fe–Mg mixing in both phases, particularly Al^{VI} and Ti in biotite and Mn and Ca in garnet (Ganguly & Saxena 1984; Hodges & Crowley 1985; Indares & Martignole 1985; Williams & Grambling 1990; Dasgupta *et al.* 1991; Kullerud 1995). Figure 6 compares garnet–biotite temperatures derived from granites with a wide range of garnet and biotite compositions. Garnets in granitic rocks are commonly Mn-rich and the derived temperatures can be expected to be in the sequence $T_{Ferry-Spear} < T_{Hodges-Crowley} < T_{Ganguly-Saxena} < T_{Williams-Grambling} < T_{Indares-Martignole}$. Notably, the Ganguly and Saxena (1984) formulation consistently estimates temperatures in the range 650–750°C and it may be the most robust version to account for the effects of high Mn.

Not shown in Figure 6 are temperatures derived for granites with calcic garnets ($X_{Ca} > 0.10$). The Whipple Wash granites (Anderson 1988) contain garnets with rim compositions of $X_{Mn} = 0.16–0.21$ and $X_{Ca} = 0.23–0.25$. For these rocks, the Ferry and Spear (1978) temperatures are $712 \pm 62°C$ and all other formulations, including that of Ganguly and Saxena (1984), yield temperatures that are much higher, ranging from 777 ± 33 to $862 \pm 41°C$. Although these formulations contain parameters to compensate for Ca in garnet, it appears that the results are overcorrected.

In summary, no one formulation of the garnet–biotite thermometer is sufficiently robust to be applied to all compositions of garnet and biotite in granites. The original Ferry and Spear calibration is suitable for rocks containing low Mn and low Ca garnet and may be the preferred formulation for those with high Ca with modest Mn abundances. The most common occurrence involves high-Mn garnet, for which the Ganguly and Saxena (1984) formulation appears to yield the most appropriate results. This conclusion is in contrast with studies by Berman (1990) and Koziol (1990), who have inferred that Mn mixing in Ca–Fe–Mg garnets is ideal.

2.5. Apatite–zircon saturation thermometry

Most granites contain accessory zircon and apatite. Watson and Harrison (1983) and Harrison and Watson (1984) present models to estimate the saturation temperatures of zircon and apatite, respectively. A basic assumption is that neither phase is cumulate, xenocrystic or restitic in origin. These models can be used to estimate liquidus conditions, where one or both phases are early in the crystallisation sequence, or solidus conditions, where one or both phases are late. As a derived temperature may be indicative of conditions throughout the sequence of crystallisation, careful petrography is needed to determine the timing of zircon and apatite crystallisation relative to that of other phases. As both minerals also carry a strong trace element signature, inspection of whole-rock elemental data can further aid in determining if either is of cumulate or foreign origin.

Figure 7 shows zircon saturation temperatures for several high-Zr Proterozoic granite batholiths calculated from data in Cullers *et al.* (1992, 1993). Portions of these granites yield

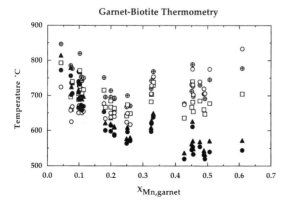

Figure 6 Effect of the Mn content of garnet on garnet–biotite thermometry of 23 peraluminous granites. Source of data include Anderson *et al.* (1988), Cullers *et al.* (1993) and Anderson (unpublished data). Symbols: ⊕, Williams and Grambling (1990); ○, Indares and Martignole (1985); □, Ganguly and Saxena (1984); ▲, Hodges and Crowley (1985); and ●, Ferry and Spear (1978).

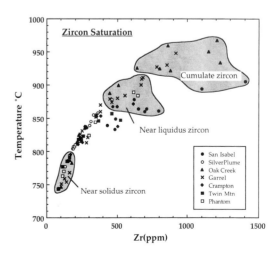

Figure 7 Zircon saturation temperatures estimated for Proterozoic granites of Colorado. Data from Anderson and Thomas (1985), Cullers *et al.* (1992, 1993), Cullers (1994, pers. comm.).

much higher zircon saturation temperatures, in part due to crystal accumulation (evidenced by anomalously high whole-rock Zr, Sr and Ba). Excluding these samples, the other rocks with high Zr offer an estimate of liquidus conditions at >850–900°C. For samples with lower Zr, near-solidus conditions are estimated at ≈740–790°C, consistent with reported temperatures based on hornblende–plagioclase and two-feldspar thermometry.

2.6. Comparison of thermometers

The Mt Stuart batholith (Anderson 1992; Anderson & Paterson 1991; Paterson et al. 1994) has a sufficient range in rock composition and mineral assemblage to allow the comparison of a large number of thermometers. Figure 8 shows the results for the batholith applying up to nine thermometers in rock types ranging from diorite to tonalite and granodiorite.

Dioritic rocks allow intrasample comparison of the largest number of thermometers. All indicate final equilibration at $T \geqslant 750°C$. The pyroxene-based exchange thermometers may indicate hypersolidus conditions as much of the pyroxene is mantled by late crystallising hornblende. The very high orthopyroxene–biotite temperature appears anomalous. Estimated garnet–biotite and hornblende–plagioclase temperatures for intermediate to felsic rocks are appropriately lower, including in $700 \pm 20°C$ for quartz diorite, $670 \pm 30°C$ for tonalite and $630 \pm 20°C$ for granodiorite. It is common for minerals in these rocks to be zoned and, as a result, the core compositions yield higher temperatures than those derived for adjacent or touching rims.

3. Oxygen and water fugacity

The fugacities of water and oxygen exert far more control on the composition of minerals in granitic rocks than any other set of intensive parameters, including pressure and temperature. Many arc-related batholiths crystallise near the experimental Ni–NiO buffer (Czamanske et al. 1981; Speer 1987). Others, particularly ilmenite series and fayalite-bearing granites (Anderson 1983; Emslie & Stirling 1993), crystallise near the quartz–fayalite–magnetite buffer and a few high f_{O_2}, magnetite series granites (Barth et al. 1995) crystallise a few log units below the haematite–magnetite buffer. Thus the observed range of f_{O_2} in granites spans seven orders of magnitude and

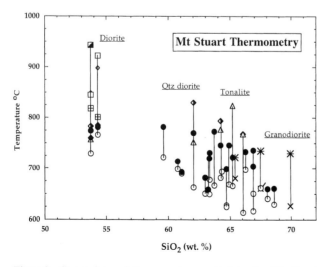

Figure 8 Comparison of thermometers applied to the Mt Stuart batholith (Anderson & Paterson 1991; Paterson et al. 1994). Symbols: ◩, OPX-Bio; ◈, CPX-BiO; △, CPX-Hb; ◇, CPX-Ilm; ◆, OPX-Ilm; ⊞, 2PX solvus; □, 2PX exchange; ●, Pl-Hb cores; ○, Pl-Hb rims; ∗, Gar-Bio core; X, Gar-Bio rim.

this duplicates the range of f_{O_2} measurements for rapidly erupted volcanic rocks (Ewart et al. 1975; Luhr et al. 1984; Carmichael 1991).

Experimental investigations on various rock–water systems (Wyllie 1984) show that for a range of granitic compositions, isobaric crystallisation will reach water saturation where $P_{H_2O} = P_{total}$. If this were true in nature, then the solidus for granites emplaced at pressures >3 kbar would be nearly isothermal at temperatures of $650 \pm 30°C$. Yet, as reviewed by Anderson and Smith (1995), the estimated temperatures for many granitic plutons are much higher than 700°C, implying that the thermometers have not recorded solidus conditions, the presence of a mixed fluid phase or volatile phase undersaturation. A granitic magma can crystallise without fluid saturation if a magmatic process such as accumulation or filter pressing segregates the more evolved portion of a magma away from a crystallised residue. At present, there is no method to calculate reduced fugacities for the fluid-undersaturated portions of magmas. For the purposes of calculation, we can only assume mixed fluid saturation to estimate f_{H_2O} or X_{H2O}.

The calculation of volatile fugacities is highly dependent on a knowledge of temperature and, for some reactions, pressure. As all involve displaced equilibria; they are also dependent on the activity model. For a summary of activity models, see Berman (1988), Essene (1989) and Spear (1993).

3.1. H₂O barometry in pyroxene and fayalite granites

Whereas many granites have only hydrous mafic phases, the occurrence of pyroxene and/or fayalitic olivine is an indication of elevated temperature and/or low f_{H_2O}. The temperatures estimated for such rocks are, in fact, often high (Fuhrman et al. 1988) relative to granitic rocks containing hydrous phases. There are two equilibria that can be used to constrain f_{H_2O}. The equilibrium

$$\text{Phlogopite} + 3 \text{ quartz} = 3 \text{ enstatite} + \text{sanidine} + H_2O$$

can be modelled for an assemblage that includes biotite–orthopyroxene–alkali feldspar–quartz based on experiments by Bohlen et al. (1983), Vielzeuf and Clemens (1992) and Clemens (1995). Using EQUILI and reduced mineral activities, Cotkin and Medaris (1993) determined an X_{H2O} of 0·55 at 780°C and 3·0 kbar for the Russian Peak quartz diorite.

Crystallisation conditions for fayalite granites can be modelled based on the equilibrium

$$\text{Annite} + \tfrac{3}{2} \text{ quartz} = \text{sanidine} + \text{fayalite} + H_2O$$

Emslie and Stirling (1993) present a calibration of this equilibrium derived from the experimental data of Wones (1981) and Myers and Eugster (1983), where

$$\log f_{H_2O}(\text{bar}) = -7402/T(K) + 10·84 + 0·036(P_{bar} - 1)/T(K)$$
$$+ \log a_{annite} + \tfrac{3}{2} \log a_{SiO_2} - \log a_{san} - \tfrac{3}{2} \log a_{fa}$$

For rocks of the Umiakovik batholith, Emslie and Stirling (1993) have estimated f_{H_2O} to be 700–900 bar at emplacement conditions of ≈740°C and 3·5 kbar.

3.2. Annite–sanidine–magnetite (ASM)

Wones and Eugster (1965) advanced the idea of using the equilibrium among annite, sanidine and magnetite (ASM) to estimate f_{H_2O} in granitic magmas. In practice, however, the equilibrium

$$\text{Annite} + \tfrac{1}{2}O_2 = \text{sanidine} + \text{magnetite} + H_2O$$

is dependent on f_{O_2} and is more appropriately used as an oxybarometer where P, f_{H_2O} and T have been independently derived. The calibration of Wones (1981) is

$$\log f_{O_2}(\text{bar}) = -2[4819/T(\text{K}) + 6\cdot69 - \log f_{H_2O}$$

$$- 0\cdot011(P_{\text{bar}} - 1)/T(\text{K}) + \log a_{\text{annite}} - \log a_{\text{san}} - \log a_{\text{Mt}}]$$

which can be used at f_{O_2} levels $\leqslant$ NNO. The equilibrium has also been investigated by Cygan *et al.* (1991) and Dachs (1994).

Figure 9 shows T–f_{O_2} estimates for 12 Proterozoic granite batholiths of the SW U.S.A. based on the ASM oxybarometer and assuming $P_{H_2O} \approx P_{\text{total}}$ (note error bars for the effects of P_{H_2O} uncertainty). Biotite in these plutons (Anderson & Bender 1989) exhibits marked differences in composition leading to f_{O_2} estimates ranging from $10^{-16\cdot3}$ bar (at 700°C) in the most reduced pluton to $10^{-13\cdot4}$ bar for the one that crystallised under the most oxidised conditions. As magnetite and K-feldspar in many granites vary little in composition, much of the control on calculated f_{O_2} comes from the activity of annite in biotite, for which an ideal ionic model can be approximated as

$$a_{\text{annite}} = (\text{Fe}^{2+}/3)^3 (\text{OH}/2)^2 \text{ K}/(\text{K} + \text{Na} + \text{Ca}).$$

3.3. Quartz–ulvospinel–ilmenite–fayalite (QUILF)
Like the ASM equilibrium, the equilibrium

$$\text{Quartz} + 2 \text{ ulvospinel} = 2 \text{ ilmenite} + \text{fayalite}$$

emphasises the relationship between f_{O_2}, Fe–Ti oxides and mafic silicates. With magnetite and haematite being added to the oxides in this equilibrium, the equilibrium is shifted to higher f_{O_2}, the mafic silicates become Mg-bearing and iron–rich orthopyroxene substitutes for fayalite + quartz. The utility of the QUILF equilibrium was first recognised by Frost *et al.* (1988) and was subsequently expanded by Lindsley and Frost (1992) and Frost and Lindsley (1992). More recently, Andersen *et al.* (1993) have provided a computer program (termed QUILF) using solution models and internally consistent standard state data to search for equilibrium by free energy minimisation.

The above equilibrium is one of several which can also be expressed in the four component system Fe–O–TiO$_2$–SiO$_2$. Only three equilibria can be independent in this system, which thus becomes overdetermined when additional equilibria are considered. The above equilibrium can be expressed by the four additional equilibria

$$6 \text{ Haematite} = 4 \text{ magnetite} + O_2$$

$$2 \text{ Magnetite} + 3 \text{ quartz} = 3 \text{ fayalite} + O_2$$

$$2 \text{ Haematite} + 2 \text{ quartz} = 2 \text{ fayalite} + O_2$$

$$\text{Ulvospinel} + \text{haematite} = \text{magnetite} + \text{ilmenite}.$$

Because the QUILF assemblage is overdetermined, it can be used in part or whole and constitutes a powerful tool to see through the later re-equilibration of the oxides and/or to calculate the equilibrium composition of missing or altered phases (Andersen *et al.* 1993). Frost and Lindsley (1992) show calculations for a range of pyroxene-bearing granitoids with f_{O_2} estimates ranging from 2 log units above to 1 log unit below QFM.

Fayalite-bearing granites containing magnetite can also be used to constrain f_{O_2} based on the activity-reduced displacement of the QFM buffer, a subset of the QUILF equilibria. Frost (1991) provides the following expression for QFM, modified for impure olivine and magnetite as follows

$$\log f_{O_2}(\text{bar}) = -25\,096\cdot3/T(\text{K}) + 8\cdot735 + 0\cdot110(P_{\text{bar}} - 1)/T$$
$$- \log a^3_{\text{Fa}} + \log a^2_{\text{Mt}}$$

Activity models for fayalite are provided by Davidson and Mukhopadhyay (1984), which at $T > 700°C$ approach the ideal model of $a_{\text{Fa}} = X^2_{\text{Fe}}$ where $X_{\text{Fe}} = \text{Fe}^{2+}/(\text{Fe}^{2+} + \text{Mg} + \text{Mn} + \text{Ca})$. Activity models for magnetite are described by Andersen and Lindsley (1988), Andersen *et al.* (1991) and Ghiorso and Sack (1991).

3.4. Titanite–magnetite–quartz (TMQ)
Ishihara (1977) noted that granites can be subdivided into magnetite series (high f_{O_2}) and ilmenite series (low f_{O_2}) with the boundary located approximately between the NNO and QFM buffers. More recently, Wones (1989) presented a calibration for the equilibrium

$$3 \text{ Titanite} + 2 \text{ magnetite} + 3 \text{ quartz}$$

$$= 3 \text{ hedenbergite} + 3 \text{ ilmenite} + O_2$$

Based on thermodynamic data, Wones (1989) offers the expression

$$\log f_{O_2}(\text{bar}) = -30\,930/T(\text{K}) + 14\cdot98 + 0\cdot142(P_{\text{bar}} - 1)/T$$

which, for pure phases, trends at a steeper angle in T–f_{O_2} space than the experimental buffers. It intersects the QFM buffer at 650°C and crosses above the Ni–NiO buffer at 720°C. The expression can be corrected for reduced activities by adding

$$- 3 \log a_{\text{hd}} - 3 \log a_{\text{ilm}} + 3 \log a_{\text{titanite}} + 2 \log a_{\text{mt}}$$

In natural rocks, the equilibrium constitutes a fundamental boundary for ilmenite versus magnetite series granites. In practice, most granites will lack the full assemblage, but the above expression, modified for reduced activities, may serve as a minimum or maximum constraint in terms of T and f_{O_2}.

3.5. Biotite–almandine garnet–muscovite–magnetite (BAMM)
Often representing the only means to determine oxygen fugacity in peraluminous granites, the equilibrium

$$\text{Annite} + \text{almandine} + O_2$$

$$= \text{muscovite} + 2 \text{ magnetite} + 3 \text{ quartz}$$

has been calibrated by Zen (1985) from thermodynamic data

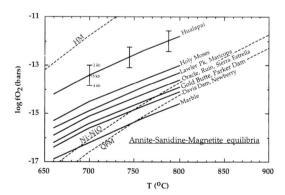

Figure 9 Estimation of T–f_{O2} contours based on the Wones (1981) calibration of the ASM oxybarometer for 12 Proterozoic plutons. Estimates recalculated at 3 kbar from Anderson and Bender (1989) with representative effect of ± 1 kbar depicted. Experimental buffers calculated from data in Frost (1991).

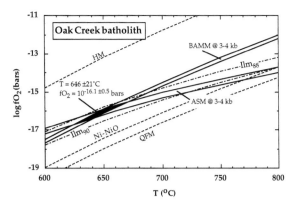

Figure 10 Application of the BAMM (Zen 1985) and ASM (Wones 1981) oxybarometers to the Oak Creek batholith using data in Cullers *et al.* (1993). Ilmenite isopleths from Andersen *et al.* (1993) and experimental buffers calculated from data in Frost (1991).

leading the following equation

$$\log f_{O_2}(\text{bar}) = 10{\cdot}29 - 26\,284/T(\text{K}) + 0{\cdot}148(P_{\text{bar}} - 1)/T$$
$$- 4 \log X_{\text{si,bio}} - 3 \log X^{2+}_{\text{Fe,bio}} - 3 \log X^{2+}_{\text{Fe,gar}}$$
$$+ 2 \log X^{\text{VI}}_{\text{Al,mu}} + 4 \log X_{\text{si,mu}}$$

It is intended for rocks with low-Ca garnet and assumes that the magnetite is of an ideal end-member composition. If the magnetite is impure, then the above expression can be modified by adding the term $+2 \log a_{\text{mt}}$ using the magnetite activity models referenced in the previous section.

As many peraluminous granitoids also contain alkali feldspar, BAMM and ASM can be solved simultaneously for T and f_{O_2}. Figure 10 shows the results for the Oak Creek batholith (Cullers *et al.* 1993). At 3–4 kbar (pressure determined from the garnet–plagioclase–biotite–muscovite barometer described in the following), BAMM and ASM intersect at $T = 646 \pm 21°C$ and $f_{O_2} = 10^{-16 \cdot 1 \pm 0 \cdot 5}$ bar. This f_{O_2} estimate is consistent with the observed composition of ilmenite (Ilm$_{88-90}$), indicating that the Fe–Ti oxide has not undergone subsolidus re-equilibration.

3.6. Temperature dependence of F–Cl–OH partitioning
Munoz and Ludington (1974), Ludington (1974) and Munoz and Swenson (1981) proposed models relating the halogen content of biotite and apatite to temperature and the relative fugacities of HF/HCl and HF/H_2O. Ague and Brimhall (1988) and Czamanske *et al.* (1991) have applied these calibrations to determine variations in f_{HF}/f_{H_2O} and $f_{\text{HF}}/f_{\text{HCl}}$ within batholiths. These models have been superseded by those of Zhu and Sverjensky (1991), who have calculated internally consistent standard-state thermodynamic data for a range of F and Cl end-members, including annite, phlogopite, muscovite, tremolite and apatite, that are consistent with the thermodynamic data base of Berman (1988). Their data represent a powerful tool to investigate the temperature and pressure dependence of F–Cl–OH partitioning among minerals and fluids. For example, Zhu and Sverjensky (1991) demonstrate that increasing temperature favours partitioning of F into the fluid and Cl into annite and that, unlike F, Cl partitioning is markedly pressure dependent.

4. Barometers for granitic plutons

4.1. Fe-rich orthopyroxene–fayalite–quartz
The occurrence of ferrosilite-rich orthopyroxene provides a constraint on the minimum pressure. Under most crustal conditions, this phase is absent and is replaced by the lower pressure assemblage fayalite + quartz. Experiments on the stability of iron-rich pyroxene, fayalite and quartz have been provided by Bohlen *et al.* (1980), Bohlen and Boettcher (1981) and Davidson and Lindsley (1989). Mg and Mn substitution into these phases lowers the pressure of the equilibria and, as a result, the assemblage occurs in many granites, particularly those from anorogenic settings. For example, compositions of olivine (Fa$_{94-96}$) and orthopyroxene (Fs$_{79}$En$_{18}$Wo$_{02}$Rh$_{01}$) restrict pressure in the granites of the Umiakovik batholith (Emslie and Stirling 1993) to 3·5–4·5 kbar at 800°C.

4.2. Al-in-hornblende
The Al-in-hornblende barometer has received considerable attention since its initial empirical calibration by Hammarstrom and Zen (1986) and Hollister *et al.* (1987). The central thesis of both of these studies is that the alumina content in hornblende increases with pressure and that for low-variance granites containing seven solid phases (quartz, K-feldspar, An$_{25-35}$ plagioclase, biotite, hornblende, titanite, Fe–Ti oxide) + melt + vapour, the Al content could be used as a barometer. A potential barometric equilibrium, suggested by Hollister *et al.* (1987), is

2 Quartz + 2 anorthite + biotite = orthoclase + tschermak

where tschermak refers to the component $Ca_2(Mg,Fe)_3Al^{VI}_2Si_6$ $Al^{IV}_2O_{22}(OH)_2$. With increased pressure, the equilibrium shifts to the right, causing the amphibole to change in composition by the exchange vector $Si + R^{2+} = Al^{IV} + Al^{VI}$.

Experimental confirmation of the pressure dependence has been reported by Johnson and Rutherford (1989), Thomas and Ernst (1990) and Schmidt (1992). Unfortunately, no experimental study has fully investigated the barometric reaction nor reported the composition of other involved phases. All barometers have a temperature dependence and are often affected by variations in other intensive parameters, including f_{O_2}. Throughout the last decade, however many petrological studies have reported pressures derived from this barometer without any correction for temperature. As a result, all of these determinations remain suspect.

Anderson and Smith (1995) have incorporated the experimental data of Johnson and Rutherford (1989) and Schmidt (1992) to derive an expression for the barometer that incorporates temperature. Their calibration is

$$P(\text{kbar}) = 4{\cdot}76\text{Al} - 3{\cdot}01 - [(T(°C) - 675)/85]$$
$$\times [0{\cdot}530\text{Al} + 0{\cdot}005294(T(°C) - 675)]$$

where Al is the sum of $Al^{IV} + Al^{VI}$ per 13 cations. The temperature dependence increases with pressure, from 1·3 kbar per 100°C at 2 kbar to >2 kbar per 100°C at 8 kbar. The lack of temperature constraints can lead to erroneous estimates of pressure, leading to the inaccurate determination of emplacement depth, pluton thickness and tilt.

Anderson and Smith (1995) also document that the barometer fails by yielding elevated pressures for low-f_{O_2} plutons with iron-rich hornblendes coexisting with the full barometric assemblage. They recommend that the barometer is used for hornblende with Fe/(Fe + Mg) < 0·65.

Figure 11 offers an example of hornblende barometry using mineral rim compositions for 14 Mesozoic plutons of the Mojave Desert. The range of observed crystallisation pressures, all corrected for temperature (hornblende–plagioclase) using the calibration of Anderson and Smith (1995), reveal that at least three crustal levels are exposed in this region. Notably, P–T estimates for the deeper plutons do not conform to a water-saturated solidus.

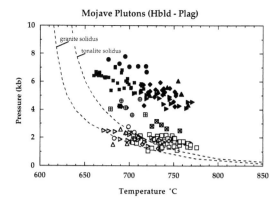

Figure 11 Hornblende–plagioclase thermobarometry applied to 14 Mesozoic and Tertiary plutons of the Mojave Desert. Calibrations include hornblende–plagioclase reaction B (edenite–richterite) of Holland and Blundy (1994) and hornblende barometry calibration of Anderson and Smith (1985). *P–T* solutions calculated from mineral rim data in Beckerman *et al.* (1982), Anderson *et al.* (1988), Anderson (1988), Barth (1989), Young (1990), Hayes (1992) and Anderson *et al.* (1992). Mid-crustal plutons: ●, Axtel qd; ■, Cargo Muchacho qm; ▲, San Gabriel grd; ◆, Granite J spgr; ▶, Granite K gr. Intermediate level plutons: ⊞, 29 Palms qm; ⬦, Queen Mtn grd; ⊕, Sacramento grd; ⊞, Rock spgs qmd. Shallow intrusions: □, Eagle Mtns grd; ▷, War Eagle qd; ○, War Eagle grd; △, Mid Hills gr; and Providence qm.

4.3. Garnet–hornblende–plagioclase–quartz

Kohn and Spear (1989, 1990) have empirically calibrated a barometer for the assemblage garnet + hornblende + plagioclase + quartz based on an Mg and an Fe end-member water-conservative reaction. The Mg end-member equilibrium is

6 Anorthite + 3 tremolite = 2 grossularite + 1 pyrope

+ 3 tschermakite + 6 quartz

The high-pressure assemblage (written on the product side) predicts that garnet will become more calcic and hornblende more aluminous with increasing pressure. The calibrations, one for the Mg end-member equilibrium and the other for the Fe end-member reaction are

$$P_{Mg} = [79\,507 + T(29 \cdot 14 + 8 \cdot 3144 \ln K_{Mg})]/10 \cdot 988 (\text{bar})$$

$$P_{Fe} = [35\,327 + T(56 \cdot 09 + 8 \cdot 3144 \ln K_{Fe})]/11 \cdot 906 (\text{bar})$$

where

$$K_{Mg} = (a_{Gr}^2 a_{py} a_{tsch}^3)/(a_{an}^6 a_{tr}^3)$$

and

$$K_{Fe} = (a_{gr}^2 a_{alm} a_{Fetsch}^3)/(a_{an}^6 a_{FeAct}^3).$$

Temperature is in Kelvin and it is assumed that $a_{SiO_2} = 1$. Activity models are given by Kohn and Spear (1990) for plagioclase (An$_{15-70}$), garnet ($X_{Sp} < 0 \cdot 15$) and hornblende (Al = $1 \cdot 90 – 3 \cdot 75$ atoms/pfu and Fe^{2+}/(Fe^{2+} + Mg) between $0 \cdot 4$ and $0 \cdot 6$). Accuracy is estimated to be within $\pm 0 \cdot 4$ kbar.

By all respects, this appears to be a well-calibrated barometer. Unfortunately, there seem to be few opportunities to evaluate its application to granite plutons. The assemblage occurs in several plutons, including the Bushy Point tonalite (Zen & Hammarstrom 1984b), Mt Lowe pluton (Barth 1990) and the Mt Stuart batholith (Paterson *et al.* 1994), but in each instance garnet and hornblende are not in textural equilibrium. For the Bushy Point tonalite, Zen and Hammarstrom (1984b) suggest that the garnet formed at deeper levels and was not in equilibrium with hornblende at the solidus.

4.4. Garnet–biotite–plagioclase–muscovite

Many peraluminous granites contain the assemblage garnet + biotite + muscovite + plagioclase, which allows utilisation of the pressure-sensitive equilibria

Pyrope + grossular + muscovite = 3 anorthite + phlogopite

Almandine + grossular + muscovite = 3 anorthite + annite

Ghent and Stout (1981) empirically calibrated these reactions utilising the thermodynamic data of Helgeson *et al.* (1978) and found that the derived pressures compared well with independently derived pressures from garnet–plagioclase–Al$_2$SiO$_5$–quartz (GASP) equilibria. This apparent success should not be considered an endorsement as the GASP barometer can carry significant errors, in part due to long extrapolation to the low grossular contents of most crustal rocks and uncertainty in its calibration (Bohlen & Lindsley 1987; McKenna & Hodges 1988; Koziol & Newton, 1988). Hodges and Crowley (1985) revised the calibration for the second reaction using an expanded database and garnet end-member activity models of Newton and Haselton (1981) and Hodges and Spear (1982). However, these models do not account for the effect of Mn on grossular mixing and thus the Hodges and Crowley formulation should not be used for garnets with $X_{Mn} > 0 \cdot 15$. Hoisch (1991) has also reformulated the barometer, but for a restricted range of mineral compositions that do not correspond well to that observed in most two-mica granites.

Figure 12 depicts results using the original Ghent and Stout (1981) calibration for several granitic plutons. These calculations are indistinguishable, within error, from results using the Powell and Holland (1988) database. As a result of mineral zoning, *P–T* estimates using the interior and core compositions are typically higher than those derived from mineral rim data and this full range of calculated pressure and temperature is depicted for each pluton. *P–T* estimates based on rim compositions are considered to be the most reliable and these roughly correspond to the location of wet solidii for granite and tonalite. However, these pressures are likely to be imprecise. There is considerable new thermodynamic data for garnet solutions (Anovitz and Essene 1987; Berman, 1990; Koziol 1990) and the barometer needs to be recalculated.

4.5. Phengite barometry

The phengite barometer was first proposed by Velde (1965), who concluded that the celadonite content of muscovite–celadonite micas (phengite) increases with pressure.

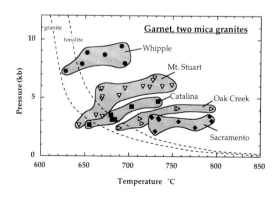

Figure 12 Garnet–plagioclase–biotite–muscovite barometry (Ghent & Stout 1981) and garnet–biotite thermometry applied to five granitic plutons. *P–T* solutions were calculated from data in Anderson (1988), Cullers *et al.* (1993) and Paterson *et al.* (1984). The field of *P–T* solutions for each is a consequence of mineral zoning and, in each instance, results using paired rims are at the lower temperature end of each range.

Unfortunately, the experiments were conducted at $P < 4.5$ kbar and largely at low temperature. Powell and Evans (1983) subsequently used Velde's data to calibrate the barometer based on the equilibrium

$$3 \text{ Celadonite } [\text{KMgAlSi}_4\text{O}_{10}(\text{OH})_2]$$

$$= \text{phlogopite} + 2 \text{ sanidine} + 3 \text{ quartz} + 2\text{H}_2\text{O}$$

The white mica in peraluminous granites is characteristically celadonitic (Miller *et al.* 1981; Anderson & Rowley 1981; Zen 1988) and commonly it coexists with biotite, K-feldspar and quartz. Massonne and Schreyer (1987) presented new experimental data for this assemblage at conditions up to 22 kbar and 700°C. However, their experiments were not reversed, the celadonite content was estimated only from X-ray measurements and the compositions of biotite and feldspar were not determined. Massonne and Schreyer (1987) present their calibration in graphical form using only the Si content (per 11 oxygens) as a function of P and T. Their graphical solutions can be expressed analytically where

$$P(\text{kbar}) = -2.6786\text{Si}^2 + 43.975\text{Si} + 0.01253\ T(°\text{C}) - 113.9995$$

This expression fits their data with high precision ($r^2 = 0.999$), but the utility of the barometer is limited. The calibration is temperature sensitive (a $\pm 50°$C error corresponds to an uncertainty of ± 0.6 kbar) and is strongly affected by analytical error (± 0.05 atoms Si pfu yields an additional error of ± 1.3 kbar), leading to an overall uncertainty of ± 2 kbar. In addition, the white mica in peraluminous granites is typically titaniferous (> 0.4 wt% TiO_2) and the effects of Ti and other components (Fe^{3+}, F, Cl) remain unknown. Thus, like the Al-in-hornblende barometer, the phengite barometer remains imperfect, because it is based on a single element's abundance in one phase for a restricted mineral assemblage.

Figure 13 compares the averaged pressures obtained from this barometer with those obtained from garnet–plagioclase–biotite–muscovite for 46 samples from six peraluminous plutons. This comparison shows a general correspondence between the two barometers, although both need to be re-calibrated. The phengite barometer probably has greater potential in application to metamorphosed two-mica granites where Ti and other substitutions are more limited.

4.6. Pressure-limiting assemblages
Some information about the relative pressures of crystallisation can be determined from restrictions based on phase assemblages. Magmatic epidote (Zen & Hammarstrom 1984; Schmidt 1983) has a lower limit of stability of 4–6 kbar,

depending on f_{O_2}. Likewise, magmatic muscovite in granite has a lower limit near 4 kbar (Chatterjee & Johannes 1974; Anderson & Rowley 1981). The presence of two (subsolvus) alkali feldspars in low-Ca granitic rocks, such as aplite or pegmatite, indicates pressures greater than about 2.2 kbar (Smith 1974). These conclusions assume crystallisation on a water-saturated solidus and ignore the effects of extra components in the first two examples and the complexities of Al–Si ordering on the latter.

5. Conclusions

The mineralogy of granitic plutons enables the use of a large number of thermometers and barometers that vary significantly in the quality of calibration. The fact that the solid phases in granitic magmas crystallise over a range of temperatures during ascent further complicates the problems of constraining conditions leading to emplacement. Judicious application can, however, yield accurate information when a full evaluation of error and uncertainty are used. Figure 14 shows thermobarometric results for three metamorphic core complexes from the SW U.S.A. using data recalculated from Anderson (1988) and Anderson *et al.* (1988). Barometric constraints include the Al-in-hornblende and garnet–plagioclase–biotite–muscovite barometers for various plutons and the phengite barometer for mylonitised granitic lithologies utilising ages based on U–Pb dating. Despite the relative imprecision of these barometers, the estimated depths of emplacement consistently shallow with decreasing age, providing direct evidence for crustal ascent during stages of core complex evolution.

The quality of thermometry and barometry of granitic rocks is at a state where considerable information can be gained about relative levels of emplacement and T–f_{O_2} conditions of crystallisation. The volatile content of magmas remains the most difficult parameter to evaluate. Substantial improvement is needed in the quality of many calibrations and far more attention needs to be devoted to a consideration of the error associated with activity models and the effects of additional components on mixing and phase stability. Problems associated with metastability and subsolidus equilibration on a cooling path are intrinsic features to plutons that simply require more thorough forms of analysis preceded by careful and insightful methods of petrography.

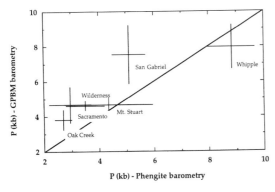

Figure 13 Comparison of phengite barometry (from calibration of Massonne & Schreyer 1987) to garnet–plagioclase–biotite–muscovite barometry (calibration of Ghent & Stout 1981) applied to Proterozoic, Mesozoic and Tertiary peraluminous granites. *P–T* solutions calculated from data in Anderson (1988), Cullers *et al.* (1993) and Paterson *et al.* (1994).

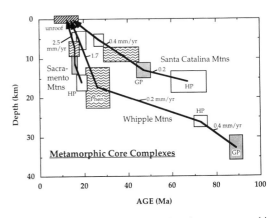

Figure 14 Depth–age tracks estimated for three metamorphic core complexes (recalculated from Anderson 1988 and Anderson *et al.* 1988, with additional age data from Pease *et al.* 1995). Boxes represent 2σ error in estimated age and depth. Thermobarometers include hornblende–plagioclase (light shade, labelled 'HP'; calibrations of reaction B of Holland & Blundy (1994) and hornblende barometry calibration of Anderson and Smith (1995); garnet–plagioclase–biotite–muscovite barometry (dark shade, labelled 'GP'; calibration of Ghent & Stout 1981); and, for mylonitic rocks, phengite barometry (wavy pattern, labelled 'Phen'; calibration of Massonne & Schreyer 1987).

6. Acknowledgements

The author acknowledges critical reviews by Philip Candela, Jane Hammarstrom, Jean Morrison and Malcolm Rutherford. This work was supported by NSF grant 92-19347.

References

Ague, J. J. & Brimhall, G. H. 1988. Magmatic arc asymmetry and distribution of anomalous plutonic belts in the batholiths of California: effects of assimilation, crustal thickness, and depth of crystallization. BULL GEOL SOC AM **100**, 912–27.

Andersen, D. J., Bishop, F. C. & Lindsley, D. H. 1991. Internally consistent solution models for Fe–Mg–Mn–Ti oxides: Fe–Mg–Ti oxides and olivine. AM MINERAL **76**, 427–44.

Andersen, D. J., Lindsley, D. H. & Davidson, P. M. 1993. QUILF: a Pascal program to assess equilibria among Fe–Mg–Mn–Ti oxides, pyroxenes, olivine, and quartz. COMPUT GEOSCI **19**, 1333–50.

Andersen, D. J. & Lindsley, D. H. 1988. Internally consistent solution models for Fe–Mg–Mn–Ti oxides: Fe–Ti oxides. AM MINERAL **73**, 714–26.

Anderson, J. L. 1980. Mineral equilibria and crystallization conditions in the late Precambrian Wolf River rapakivi massif, Wisconsin. AM J SCIENCE **280**, 289–332.

Anderson, J. L. 1983. Proterozoic anorogenic granite plutonism of North America. MEM GEOL SOC AM **161**, 133–52.

Anderson, J. L. 1988. Core complexes of the Mojave–Sonoran Desert: conditions of plutonism, mylonitization, and decompression. *In* Ernst, W. G. (ed.) *Metamorphism and crustal evolution of the western United States, Rubey Volume VII*, 503–25. Englewood Cliffs: Prentice Hall.

Anderson, J. L. 1992. Compositional variation within the high-Mg, tonalitic Mount Stuart batholith, north Cascades, Washington. GEOL SOC AM ABSTR PROGRAM **24**, 3.

Anderson, J. L. & Bender, E. E. 1989. Nature and origin of Proterozoic A-type granitic magmatism in the southwestern United States. LITHOS **23**, 19–52.

Anderson, J. L. & Morrison, J. 1992. The role of anorogenic granites in the Proterozoic crustal development of North America. *In* Condie, K. C. (ed.) *Proterozoic crustal evolution*, 263–99. New York: Elsevier.

Anderson, J. L. & Paterson, S. R. 1991. Emplacement of the Cretaceous Mt. Stuart Batholith, central Cascades, Washington. GEOL SOC AM ABSTR PROGRAM **23**, A387.

Anderson, J. L. & Rowley, M. C. 1981. Synkinematic intrusion of two-mica and associated metaluminous granitoids, Whipple Mountains, California. CAN MINERAL **19**, 83–101.

Anderson, J. L. & Smith, D. R. 1995. The effect of temperature and oxygen fugacity on Al-in-hornblende barometry. AM MINERAL **80**, 549–59.

Anderson, J. L. & Thomas, W. M. 1985. Proterozoic anorogenic two-mica granites: Silver Plume and St. Vrain batholiths of Colorado. GEOLOGY **13**, 177–80.

Anderson, J. L., Barth, A. P. & Young, E. D. 1988. Mid-crustal roots of Cordilleran metamorphic core complexes. GEOLOGY **16**, 366–9.

Anderson, J. L., Barth, A. P., Young, E. D., Davis, M. J., Farber, D., Hayes, E. M. & Johnson, K. A. 1992. Plutonism across the Tujunga–North American terrane boundary: a middle to upper crustal view of two juxtaposed arcs. *In* Bartholomew, M. J., Hyndman, D. W., Mogk, D. W. & Mason, R., (eds) *Characterization and comparison of ancient and Mesozoic continental margins*, 205–30. Dordrecht: Kluwer Academic.

Anovitz, L. M. & Essene, E. J. 1987. Compatibility of geobarometers in the system CaO–FeO–Al₂O₃–SiO₂–TiO₂ (CFAST); implications for garnet mixing models. J GEOL **95**, 635–45.

Barth, A. P. 1989. *Mesozoic rock units in the upper plate of the Vincent thrust fault, San Gabriel Mountains, southern California*. Ph.D. Thesis, University of Southern California.

Barth, A. P. 1990. Mid-crustal emplacement of Mesozoic plutons, San Gabriel Mountains, California, and implications for the geologic history of the San Gabriel Terrane. *In* Anderson, J. L. (ed.) *The nature and origin of cordilleran magmatism*. GEOL SOC AM MEM. **174**, 33–45.

Barth, A. P., Wooden, J. L., Tosdal, R. M. & Morrison, J. 1995. Crustal contamination in the petrogenesis of a calc-alkalic rock series: Josephine Mountain intrusion, California. BULL GEOL SOC AM **107**, 201–12.

Beckerman, G. M., Robinson, J. P. & Anderson, J. L. 1982. The Teutonia batholith: a large intrusive complex of Jurassic and Cretaceous age in the eastern Mojave Desert, California. *In* Frost, E. G. & Martin, D. L. (eds) *Mesozoic–Cenozoic tectonic evolution of the Colorado River region, California, Arizona, and Nevada*, 205–221. San Diego: Cordilleran Publishers.

Berman, R. G. 1988. Internally consistent thermodynamic data for minerals in the system Na₂O–K₂O–CaO–MgO–FeO–Fe₂O₃–Al₂O₃–SiO₂–TiO₂–H₂O–CO₂. J PETROL **29**, 445–522.

Berman, R. G. 1990. Mixing properties of Ca–Mg–Fe–Mn garnets. AM MINERAL **75**, 328–44.

Bishop, F. C. 1980. The distribution of Fe²⁺ and Mg between coexisting ilmenite and pyroxene with applications to geothermometry. AM J SCI **280**, 46–77.

Blundy, J. D. & Holland, T. J. B. 1990. Calcic amphibole equilibria and a new amphibole–plagioclase geothermometer. CONTRIB MINERAL PETROL **104**, 208–24.

Bohlen, S. R. & Boettcher, A. L. 1981. Experimental investigation and geological application of orthopyroxene barometry. AM MINERAL **66**, 951–64.

Bohlen, S. R. & Lindsley, D. H. 1987. Thermometry and barometry of igneous and metamorphic rocks. ANNU REV EARTH PLANET SCI **15**, 397–420.

Bohlen, S. R., Essene, E. J. & Boettcher, A. L. 1980. Reinvestigation and application of olivine–quartz–orthopyroxene barometry. EARTH PLANET SCI LETT **47**, 1–10.

Bohlen, S. R., Boettcher, A. L., Wall, V. J. & Clemens, J. D. 1983. Stability of phlogopite–quartz and sanidine–quartz: a model for melting in the lower crust. CONTRIB MINERAL PETROL **83**, 270–7.

Brown, W. L. & Parsons, I. 1981. Towards a more practical two-feldspar thermometer. AM MINERAL **70**, 356–61.

Buddington, A. L. & Lindsley, D. H. 1964. Iron–titanium oxide minerals and synthetic equivalents. J PETROL **5**, 310–57.

Carmichael, I. S. E. 1991. The redox states of basic and silicic magmas: a reflection of their source regions? CONTRIB MINERAL PETROL **106**, 129–41.

Chatterjee, N. D. & Johannes, W. 1974. Thermal stability and standard thermodynamic properties of synthetic 2M1 muscovite, KAl₂AlSi₃O₁₀(OH)₂. CONTRIB MINERAL PETROL **49**, 89–114.

Chipera, S. J. & Perkins, D. 1988. Evaluation of biotite–garnet geothermometers: application to the English River subprovince, Ontario. CONTRIB MINERAL PETROL **98**, 40–8.

Chou, I.-M. 1978. Calibration of oxygen buffers at elevated P and T using the hydrogen fugacity sensor. AM MINERAL **63**, 690–703.

Clemens, J. D. 1995. Phlogopite stability in the silica-saturated portion of the system KAlO₂–MgO–SiO₂–H₂O: new data and a reappraisal of phase relations to 1·5 GPa. AM MINERAL **80**, 982–97.

Cosca, M. A., Essene, E. J. & Bowman, J. R. 1991. Complete chemical analyses of metamorphic hornblendes: implications for normalizations, calculated H₂O activities, and thermobarometry. CONTRIB MINERAL PETROL **108**, 472–84.

Cotkin, S. J. & Medaris, L. G. 1993. Evaluation of the crystallization conditions for the calc-alkaline Russian Peak intrusive complex, Klamath Mountains, Northern California. J PETROL **34**, 543–71.

Cullers, R. L., Griffin, T., Bickford, M. E. & Anderson, J. L. 1992. Origin and chemical evolution of the 1360 Ma-old San Isabel batholith, Wet Mountains, Colorado, U.S.A.: A mid-crustal granite of anorogenic affinities. BULL GEOL SOC AMER **104**, 316–28.

Cullers, R. L., Stone, J., Anderson, J. L., Sassarini, N. & Bickford, M. E. 1993. Petrogenesis of Mesoproterozoic Oak Creek and West McCoy Gulch plutons, Colorado: an example of cumulate unmixing of mid-crustal, two mica granite of anorogenic affinity. PRECAMBRIAN RES **62**, 139–69.

Cygan, G. L., Chou, I.-Ming & Sherman, D. M. 1991. Reassessment of the annite breakdown reaction using new hydrothermal experimental techniques. EOS, TRANS AM GEOPHYS UNION **72**, 313.

Czamanske, G. K., Ishihara, S. & Atkin, S. A. 1981. Chemistry of rock-forming minerals of the Cretaceous–Paleocene batholith in southwestern Japan and implications for magma genesis. J GEOPHYS RES **86**, 10431–69.

Dachs, E. 1994. Annite stability revised. 1. Hydrogen-sensor data for the reaction annite = sanidine + magnetite + H₂. CONTRIB MINERAL PETROL **117**, 229–40.

Dallmeyer, R. D. 1974. The role of crystal structure in controlling the partitioning of Mg and Fe²⁺ between coexisting garnet and biotite. AM MINERAL **59**, 201–3.

Dasgupta, S., Sengupta, P., Guha, D. & Fukuoka, M. 1991. A refined garnet–biotite Fe–Mg exchange thermometer and its application in amphibolites and granulites. CONTRIB MINERAL PETROL **109**, 130–7.

Davidson, P. M. & Lindsley, D. H. 1985. Thermodynamic analysis of quadrilateral pyroxene. Part II. Model calibration from experiments and applications to geothermometry. CONTRIB MINERAL PETROL **91**, 390–404.

Davidson, P. M. & Lindsley, D. H. 1989. Thermodynamic analysis of pyroxene–olivine–quartz equilibria in the system CaO–MgO–FeO–SiO$_2$. AM MINERAL **74**, 18–30.

Davidson, P. M. & Mukhopadhyay, D. K. 1984. Ca–Fe–Mg olivines: phase relations and a solution model. CONTRIB MINERAL PETROL **86**, 256–63.

Dawes, R. L. & Evans, B. W. 1991. Mineralogy and geothermobarometry of magmatic epidote-bearing dikes, Front Range, Colorado. BULL GEOL SOC AM **103**, 1017–31.

Elkins, L. T. & Grove, T. L. 1990. Ternary feldspar experiments and thermodynamic models. AM MINERAL **75**, 544–59.

Ellis, D. J. & Green, D. H. 1979. An experimental study of the effect of Ca upon garnet–clinopyroxene Fe–Mg exchange equilibria. CONTRIB MINERAL PETROL **71**, 13–22.

Emslie, R. F. & Stirling, J. A. R. 1993. Rapakivi and related granitoids of the Nain plutonic suite: geochemistry, mineral assemblages, and fluid equilibria. CAN MINERAL **31**, 821–47.

Essene, E. J. 1982. Geologic thermometry and barometry. *In* Ferry, J. M. (ed.) *Characterization of metamorphism through mineral equilibria.* MINERAL SOC AM REV MINERAL **10**, 153–206.

Essene, E. J. 1989. The current status of thermobarometry in metamorphic rocks. *In* Daly, J. S., Cliff, R. A. & Yardley, B. W. D. (eds) *Evolution of metamorphic belts.* SPEC PUBL GEOL SOC LONDON **43**, 1–44.

Ewart, A., Hildreth, W. & Carmichael, I. S. E. 1975. Quaternary acid magmas in New Zealand. CONTRIB MINERAL PETROL **51**, 1–27.

Ferry, J. M. & Spear, F. S. 1978. Experimental calibration of the partitioning of Fe and Mg between biotite and garnet. CONTRIB MINERAL PETROL **66**, 113–7.

Fonarev, V. I. & Konilov, A. N. 1986. Experimental study of Fe–Mg distribution between biotite and orthopyroxene at P = 490 MPa. CONTRIB MINERAL PETROL **93**, 227–35.

Frost, B. R. 1991. Introduction to oxygen fugacity and its petrologic importance. *In* Lindsley, D. H. (ed.) *Oxide minerals.* MINERAL SOC AM REV MINERAL **25**, 1–9.

Frost, B. R. & Lindsley, D. H. 1992. Equilibria among Fe–Ti oxides, pyroxenes, olivine, and quartz: part II. Application. AM MINERAL **77**, 1004–20.

Frost, B. R., Lindsley, D. H. & Andersen, D. J. 1988. Fe–Ti oxide–silicate equilibria: assemblages with fayalitic olivine. AM MINERAL **73**, 727–40.

Fuhrman, M. L. & Lindsley, D. H. 1988. Ternary feldspar modeling and thermometry. AM MINERAL **73**, 201–15.

Fuhrman, M. L., Frost, B. R. & Lindsley, D. H. 1988. Crystallization conditions of the Sybille monzosyenite, Laramie anorthosite complex, Wyoming. J PETROL **29**, 699–729.

Ganguly, J. & Saxena, S. 1984. Mixing properties of aluminosilicate garnets: constraints from natural and experimental data, and applications to geothermo-barometry. AM MINERAL **69**, 88–97.

Ghent, E. D. & Stout, M. Z. 1981. Geobarometry and geothermometry of plagioclase–biotite–garnet–muscovite assemblages. CONTRIB MINERAL PETROL **76**, 92–7.

Ghiorso, M. S. 1990. Thermodynamic properties of hematite–ilmenite–geikielite solid solutions. CONTRIB MINERAL PETROL **104**, 645–67.

Ghiorso, M. S. & Sack, R. O. 1991. Fe–Ti oxide thermometry: thermodynamic formulation and estimation of intensive variables in silicic magmas. CONTRIB MINERAL PETROL **108**, 485–510.

Gordon, T. M. 1992. Generalized thermobarometry; solution of the inverse chemical equilibrium problem using data for individual species. GEOCHIM COSMOCHIM ACTA **56**, 1793–800.

Graham, C. M. & Powell, R. 1984. A garnet–hornblende geothermometer: calibration, testing, and application to the Pelona schist, southern California: J METAMORPH GEOL **2**, 13–31.

Green, N. L. & Usdansky, S. I. 1986. Ternary-feldspar mixing relations and thermobarometry. AM MINERAL **71**, 1100–8.

Green, T. H. & Adam, J. 1991. Assessment of the garnet–clinopyroxene Fe–Mg exchange thermometer using new experimental data. J METAMORPH GEOL **9**, 341–7.

Hammarstrom, J. M. & Zen, E-an. 1986. Aluminum in hornblende, an empirical igneous geobarometer. AM MINERAL **71**, 1297–313.

Harley, S. L. 1984. Comparison of the garnet–orthopyroxene geobaro-meter with recent experimental studies and applications to natural assemblages. J PETROL **25**, 697–712.

Harrison, T. M. & Watson, E. B. 1984. The behavior of apatite during crustal anatexis: equilibrium and kinetic considerations. GEOCHIM COSMOCHIM ACTA **48**, 1467–77.

Haselton, H. T., Hovis, G. L., Hemingway, B. S. & Robie, R. A. 1983. Calorimetric investigation of the excess entropy of mixing in analbite–sanidine solid solutions: lack of evidence for Na, K short-range order and implications for two-feldspar thermometry. AM MINERAL **68**, 398–413.

Hayes, E. M. 1992. *Petrology of Jurassic plutons and older crystalline units, the Cargo Muchacho Mountains, southeastern California.* M.S. Thesis, University of Southern California.

Helgeson, H. C., Delaney, J. M., Nesbitt, H. W. & Bird, D. K. 1978. Summary and critique of the thermodynamic properties of rock-forming minerals. AM J SCI **278A**, 1–229.

Hodges, K. V. & Crowley, P. D. 1985. Error estimation and empirical geothermobarometry for pelitic systems. AM MINERAL **70**, 702–9.

Hodges, K. V. & McKenna, L. W. 1987. Realistic propagation of uncertainties in geologic thermobarometry. AM MINERAL **72**, 671–80.

Hodges, K. V. & Spear, F. S. 1982. Geothermometry, geobarometry, and the Al$_2$SiO$_5$ triple point at Mt. Moosilauke, New Hampshire. AM MINERAL **67**, 1118–34.

Hoisch, T. D. 1991. Equilibria with the mineral assemblage quartz = muscovite + biotite + garnet + plagioclase. CONTRIB MINERAL PETROL **108**, 43–54.

Holland, T. & Blundy, J. 1994. Non-ideal interactions in calcic amphiboles and their bearing on amphibole–plagioclase thermometry. CONTRIB MINERAL PETROL **116**, 433–47.

Hollister, L. S., Grissom, G. C., Peters, E. K., Stowell, H. H. & Sisson, V. B. 1987. Confirmation of the empirical correlation of Al in hornblende with pressure of solidification of calc-alkaline plutons. AM MINERAL **72**, 231–9.

Indares, A. & Martignole, J. 1985. Biotite–garnet geothermometry in the granulite facies: the influence of Ti and Al in biotite. AM MINERAL **70**, 272–8.

Ishihara, S. 1977. The magnetite-series and ilmenite-series granitic rocks. MIN GEOL **27**, 293–305.

Johnson, M. C. & Rutherford, M. J. 1989. Experimental calibration of an aluminum-in-hornblende geobarometer with application to Long Valley caldera (California) volcanic rocks. GEOLOGY **17**, 837–41.

Kohn, M. J. & Spear, F. S. 1989. Empirical calibration of geobarometers for the assemblage garnet + hornblende + plagioclase + quartz. AM MINEAL **74**, 77–84.

Kohn, M. J. & Spear, F. S. 1990. Two new geobarometers for garnet amphibolites, with applications to southeastern Vermont. AM MINERAL **75**, 89–96.

Kohn, M. J. & Spear, F. S. 1991a. Error propagation for barometers: 1. Accuracy and precision of experimentally located end-member reactions. AM MINERAL **76**, 128–37.

Kohn, M. J. & Spear, F. S. 1991b. Error propagation for barometers: 2. Application to rocks. AM MINERAL **76**, 138–47.

Koziol, A. M. 1990. Activity-composition relationships of binary Ca–Fe and Ca–Mn garnets determined by reversed, displaced equilibrium experiments. AM MINERAL **75**, 319–27.

Koziol, A. M. & Newton, R. C. 1988. Redetermination of the anorthite breakdown reaction and improvement of the plagioclase–garnet–Al$_2$SiO$_5$–quartz geobarometer. AM MINERAL **73**, 216–23.

Kretz, R. 1982. Transfer and exchange equilibria in a portion of the pyroxene quadrilateral as deduced from natural and experimental data. GEOCHIM COSMOCHIM. ACTA **46**, 411–22.

Kretz, R. & Jen, L. S. 1978. Effect of temperature on the distribution of Mg and Fe^{2+} between calcic pyroxene and hornblende. CAN MINERAL **16**, 533–7.

Kullerud, K. 1995. Chlorine, titanium, and barium rich biotites: factors controlling biotite composition and implications for garnet–biotite geothermometry. CONTRIB MINERAL PETROL **120**, 42–59.

Lee, H. & Ganguly, J. 1987. Equilibrium compositions of coexisting garnet and orthopyroxene: experimental determinations in the system FeO–MgO–Al$_2$O$_3$–SiO$_2$ and applications. J PETROL **29**, 93–114.

Lindsley, D. H. 1983. Pyroxene thermometry. AM MINERAL **68**, 477–93.

Lindsley, D. H. & Frost, B. R. 1992. Equilibria among Fe–Ti oxides, pyroxenes, olivine, and quartz: part I. Theory. AM MINERAL **77**, 987–1003.

Lindsley, D. H. & Spencer, K. J. 1982. Fe–Ti oxide geothermometry: reducing analyses of coexisting Ti–magnetite (Mt) and ilmenite (Ilm). EOS, TRANS AM GEOPHYS UNION 63, 471.

Lindsley, D. H., Frost, B. R., Andersen, D. J. & Davidson, P. J. 1990. Fe–Ti oxide equilibria: assemblages with orthopyroxene. In Spencer, R. J. & Chou, I.-M (eds) Fluid-mineral interactions. SPEC PUBL GEOCHEM SOC 2, 103–19.

Luhr, J. F., Carmichael, I. S. E. & Varekamp, J. C. 1984. The 1982 eruptions of El Chichon volcano, Chiapas, Mexico: mineralogy and petrology of the anhydrite-bearing pumices. J VOLCANOL GEOTHERM RES 23, 69–108.

Ludington, S. 1978. The biotite–apatite geothermometer revisited. AM MINERAL 63, 551–3.

Massonne, H. J. & Schreyer, W. 1987. Phengite geobarometry based on the limiting assemblage with K–feldspar, phlogopite, and quartz. CONTRIB MINERAL PETROL 96, 212–24.

Mayo, D. P. 1994. Estimating crystallization conditions in metaluminous granodiorite using microtextures and equilibria involving mafic silicates and oxides. GEOL SOC AM ABSTR PROGRAM 26, 71.

McKenna, L. W. & Hodges, K. V. 1988. Accuracy versus precision in locating reaction boundaries: implications for the garnet–plagioclase–aluminum silicate–quartz geobarometer. AM MINERAL 73, 1205–8.

Miller, C. F., Stoddard, E. F., Bradfish, L. J. & Dollase, W. A. 1981. Composition of plutonic muscovite: genetic implications. CAN MINERAL 19, 25–34.

Munoz, J. L. & Ludington, S. D. 1974. Fluoride–hydroxyl exchange in biotite. AM MINERAL 274, 396–413.

Munoz, J. L. & Swenson, A. 1981. Chloride–hydroxyl exchange in biotite and estimation of relative HCL/HF activities in hydrothermal fluids. ECON GEOL 76, 2212–21.

Myers, J. & Eugster, H. P. 1983. The system Fe–Si–O: oxygen buffer calibrations to 1500 °K. CONTRIB MINERAL PETROL 82, 75–90.

Nabelek, C. R. & Lindsley, D. H. 1985. Tetrahedral Al in amphibole: a potential thermometer for some mafic rocks. GEOL SOC AM ABSTR PROGRAM 17, 673.

Newton, R. C. & Haselton, H. T. 1981. Thermodynamics of the garnet–plagioclase–Al$_2$SiO$_5$–quartz geobarometer. In Newton, R. C. (ed.) Thermodynamics of minerals and melts, 131–47. New York: Springer Verlag.

O'Neil, J. R. 1986. Theoretical and experimental aspects of isotopic fractionation. In Valley, J. W., Taylor, H. P. & O'Neil, J. R. (eds) Stable isotopes in high-temperature geologic processes. MINERAL SOC AM REV MINERAL 167, 1–40.

Paterson, S. R., Miller, R. B., Anderson, J. L., Lund, S., Bendixen, J., Taylor, N. & Fink, T. 1994. Emplacement and evolution of the Mt. Stuart batholith. In Swanson, D. A. & Haugerud, R. A. (ed.) Geologic field trips of the Pacific Northwest, Vol. 2. GEOL SOC AM GUIDE 2F1–47.

Pease, V., Foster, D., Wooden, J., O'Sullivan, P. & Argent, J. 1995. Tertiary plutonism and extension in the Sacramento Mountains, SE California, U.S.A. EOS, TRANS AM GEOPHYS UNION 76, 639.

Perkins, D. & Vielzeuf, D. 1992. Experimental investigation of Fe–Mg distribution between olivine and clinopyroxene: implications for mixing properties of Fe–Mg in clinopyroxene and garnet–clinopyroxene thermometry. AM MINERAL 77, 774–83.

Perkins, D., Essene, E. J. & Wall, V. J. 1987. THERMO: a computer program for calculation of mixed volatile equilibria. AM MINERAL 72, 446–7.

Perchuck, L. L. & Lavrent'eva, I. V. 1983. Experimental investigation of exchange equilibria in the system cordierite–garnet–biotite. In Saxena, S. K. (ed.) Kinetics and equilibrium in mineral systems, 199–239. New York: Springer Verlag.

Perchuck, L. L., Aranovich, L. Y., Podlesskii, K. K., Lavrant'eva, I. V., Gerasimov, V. Y., Fed'kin, V. V., Kitsul, V. I., Karsakov, L. P. & Berdnikov, N. V. 1985. Precambrian granulites of the Aldan shield, eastern Siberia, USSR. J METAMORPH GEOL 3, 265–310.

Poli, S. & Schmidt, M. W. 1992. A comment on 'calcic amphibole equilibria and a new amphibole–plagioclase geothermometer'. CONTRIB MINERAL PETROL 111, 273–82.

Powell, R. & Evans, J. A. 1983. A new geobarometer for the assemblage biotite–muscovite–chlorite–quartz. J METAMORPH GEOL 1, 331–6.

Powell, R. E. & Holland, T. J. B. 1988. An internally consistent dataset with uncertainties and correlations: 3. Applications to geobarometry, worked examples and a computer program. J METAMORPH GEOL 6, 173–204.

Rutherford, M. J. & Hill, P. M. 1993. Magma ascent rates from amphibole breakdown: an experimental study applied to the 1980–1986 Mount St. Helens eruption. J GEOPHYS RES 98, 19 667–85.

Sack, R. O. & Ghiorso, M. S. 1991. An internally consistent model for the thermodynamic properties of Fe–Mg titanomagnetite-aluminate spinels. CONTRIB MINERAL PETROL 106, 474–505.

Schmidt, M. W. 1992. Amphibole composition in tonalite as a function of pressure: an experimental calibration of the Al-in-hornblende barometer. CONTRIB MINERAL PETROL 110, 304–10.

Schmidt, M. W. 1993. Phase relations and compositions in tonalite as a function of pressure: an experimental study at 650°C. AM J SCI 293, 1011–60.

Sengupta, P., Dasgupta, S., Bhattacharya, P. K. & Mukherjee, M. 1990. An orthopyroxene–biotite geothermometer and its application in crustal granulites and mantle-derived rocks. J METAMORPH GEOL 8, 191–7.

Smith, J. V. 1974. Feldspar minerals. Heidelberg: Springer.

Spear, F. S. 1980. NaSi=CaAl exchange equilibria between plagioclase and amphibole, an empirical model. CONTRIB MINERAL PETROL 72, 33–41.

Spear, F. S. 1981. Amphibole–plagioclase equilibria: an empirical model for the reaction albite + tremolite = edenite + 4 quartz. CONTRIB MINERAL PETROL 77, 355–64.

Spear, F. S. 1993. Metamorphic phase equilibria and pressure-temperature time paths. MINERAL SOC MONOGR.

Spear, F. S. & Kimball, K. L. 1984. Recamp—a fortran IV program for estimating Fe^{3+} contents in amphiboles. COMPUT GEOSCI 10, 317–25.

Spear, F. S., Selverstone, J., Hickmott, D., Crowley, P. & Hodges, K. 1984. P–T paths from garnet zoning: A new technique for deciphering tectonic processes in crystalline terranes. GEOLOGY 12, 87–90.

Speer, J. A. 1987. Evolution of magmatic AFM mineral assemblages in granitoid rocks: the hornblende + melt = biotite reaction in the Liberty Hill pluton, South Carolina. AM MINERAL 72, 863–78.

Spencer, J. J. & Lindsley, D. H. 1981. A solution model for coexisting iron-titanium oxides. AM MINERAL 66, 1189–202.

Stormer, J. C. 1975. A practical two-feldspar thermometer. AM MINERAL 60, 667–74.

Stormer, J. C. 1983. The effects of recalculation on estimates of temperature and oxygen fugacity from analyses of multicomponent iron-titanium oxides. AM MINERAL 68, 586–94.

Thomas, W. M. & Ernst, W. G. 1990. The aluminum content of hornblende in calc-alkaline granitic rocks: a mineralogic barometer calibrated experimentally to 12 kbars. In Spencer, R. J. & Chou, I.-M. (ed) Fluid–mineral interactions: a tribute to H. P. Eugster. GEOCHEM SOC SPEC PUBL 2, 59–63.

Valley, J. W. & Essene, E. J. 1980. Calc-silicate reactions in Adirondack marbles: the role of fluids and solid solution. BULL GEOL SOC AM 91, 114–7.

Velde, B. 1965. Phengitic micas: synthesis, stability, and natural occurrence. AM J SCI 263, 886–913.

Vielzeuf, D. & Clemens, J. D. 1992. The fluid absent melting of phlogopite + quartz: experiments and models. AM MINERAL 77, 1206–22.

Watson, E. B. & Harrison, T. M. 1983. Zircon saturation revisited: temperature and compositional effects in a variety of crustal magma types. EARTH PLANET SCI LETT 64, 295–304.

Wells, P. R. A. 1977. Pyroxene thermometry in simple and complex systems: CONTRIB MINERAL PETROL 62, 129–139.

Whitney, J. A. & Stormer, J. C. Jr 1977. The distribution of NaAlSi$_3$O$_8$ between coexisting microcline and plagioclase and its effect on geothermometric calculations. AM MINERAL 62, 687–91.

Williams, M. L. & Grambling, J. A. 1990. Manganese, ferric iron, and the equilibrium between garnet and biotite. AM MINERAL 75, 886–908.

Wones, D. R. 1981. Mafic silicates as indicators of intensive parameters in granitic magmas. MIN GEOL 31, 191–212.

Wones, D. R. 1989. Significance of the assemblage titanite + magnetite + quartz in granitic rocks. AM MINERAL 74, 744–9.

Wones, D. R. & Eugster, H. P. 1965. Stability of biotite: experiment, theory, and application. AM MINERAL 50, 1228–72.

Wood, B. J. & Banno, S. 1973. Garnet–orthopyroxene and orthopyroxene–clinopyroxene relationships in simple and complex systems. CONTRIB MINERAL PETROL 42, 109–27.

Wyllie, P. J. 1984. Constraints imposed by experimental petrology on possible and impossible maga sources and products. PHIL TRANS R SOC LONDON A310, 439–56.

Young, E. D. 1990. *Geothermobarometric and geochemical studies of two crystalline terrains of the eastern Mojave Desert, USA*. Ph.D. Thesis, University of Southern California.

Zen, E-an. 1985. An oxygen buffer for some peraluminous granites and metamorphic rocks. AM MINERAL **70**, 65–73.

Zen, E-An. 1988. Phase relations of peraluminous granitic rocks and their petrogenetic implications. ANNU REV EARTH PLANET SCI **16**, 21–51.

Zen, E-an & Hammarstrom, J. M. 1984a. Magmatic epidote and its petrologic significance. GEOLOGY **12**, 515–8.

Zen, E-an & Hammarstrom, J. M. 1984b. Mineralogy and a petrogenetic model for the tonalite pluton at Bushy Point, Revillagigedo Island, Ketchikan $1° \times 2°$ quadrangle, southeastern Alaska, U.S. GEOL SURVEY CIRC **939**, 118–23.

Zhu, C. & Sverjensky, D. A. 1991. Partitioning of F–Cl–OH between minerals and hydrothermal fluids. GEOCHIM COSMOCHIM ACTA **55**, 1837–58.

J. LAWFORD ANDERSON, Department of Earth Sciences, University of Southern California, Los Angeles, CA 90089-0740, U.S.A.

Transactions of the Royal Society of Edinburgh: Earth Sciences, **87**, 139–146, 1996

Experimental constraints on the compositional evolution of crustal magmas

Hanna Nekvasil and William Carroll

ABSTRACT: Recent water-undersaturated phase equilibrium data on the subsystems of the granite–H_2O system have provided important new constraints on the topology of the cotectic surfaces and hence on the compositional evolution of felsic magmas. The effect of water on phase relations can be deduced from a comparison of anhydrous and H_2O-saturated data or from data obtained in the presence of a CO_2-bearing fluid. However, although new experimental evidence indicates that the silica enrichment of evolving H_2O-undersaturated, H_2O-unbuffered melts during the co-precipitation of quartz and feldspar is as previously thought for orthoclase-rich compositions, it suggests that such a trend is considerably less for Ab-rich compositions. For water-poor trachytic melts, the newly recognised strong destabilisation of the sanidine melt component relative to the anorthite melt component with increasing water content indicates that the co-precipitation of two feldspars will result in saturation of the melt with ternary alkali feldspar at an earlier stage (i.e. higher melt anorthite content) than previously thought. This, in turn, implies that the melt differentiation path will have a greater component of anorthite depletion during the equilibrium co-precipitation of ternary feldspars and that the melt will remain in the peritectic region of the two feldspar plus liquid surface over a greater interval of crystallisation, thereby enhancing the possibility that the resoption of plagioclase during the early stages of equilibrium with alkali feldspar may go to completion. Comparison of CO_2-free and CO_2-bearing haplogranitic phase equilibrium data suggests that CO_2 may be playing an independent part in the modification of phase equilibria and may induce a significant destabilisation of the orthoclase melt component.

KEY WORDS: compositional evolution, crustal magmas, experimental constraints, water saturation.

The ubiquitous presence of water in crustal melts, its high solubility and its major effect on phase relations have made water a main focus of numerous studies of natural and synthetic crustal magmas. The more recent recognition that H_2O-undersaturated conditions may be common during most of the crystallisation history of silicic magmas has led to variety of studies of H_2O-undersaturated phase equilibria, from experimental studies on compositionally simplified synthetic systems (e.g. Holtz *et al.* 1992; Pichavant *et al.* 1992) and natural compositions (e.g. London 1986; Webster *et al.* 1987) to theoretical studies on simplified systems (e.g. Nekvasil & Burnham 1987; Nekvasil 1992) that are pinned by experimental data. Obtaining H_2O-undersaturated phase equilibria experimentally is not, however, a trivial task. Investigations in which an amount of water less than that required for water saturation is added to the starting material (e.g. Luth 1969; Whitney 1972; Steiner *et al.* 1975; Huang & Wyllie 1976) are hampered by the slow structural relaxation times of the highly viscous melts and, hence, slow diffusion rates. The resulting problems include nucleation delay, the non-homogeneous distribution of water and disequilibrum solid solution compositions. For these reasons, most recent investigations (e.g. Keppler 1989; Ebadi & Johannes 1991; Holtz *et al.* 1992; Pichavant *et al.* 1992; Bohlen *et al.* 1995) have instead attained H_2O-undersaturated conditions through the diffusion-enhancing presence of a vapour phase in which the activity of water is less than unity.

Because the solubility of CO_2 in felsic magmas appears to be low (e.g. Wyllie & Tuttle 1959; Fogel & Rutherford 1990),

many workers have used CO_2 as an inert fluid dilutant for lowering the activity of water in the fluid. In such fluid-present experiments the composition of the H_2O–CO_2 fluid is considered to remain fixed during crystallisation. The fixed water content of the fluid, in turn, fixes the activity of water in the melt (and its water content) where the value of the activity of water in the melt is dependent on the activity–composition relations of H_2O and CO_2 in the fluid (and, of course, the choice of standard state for water in the melt and fluid). [This condition, in which the activity of water in the melt remains dictated by that in the fluid, will hereafter be referred to as 'H_2O-buffered'. Importantly, however, the activity of water in the melt will vary slightly as a function of temperature and composition.]

There are several caveats in the use of this method. (a) The $H_2O/(H_2O + CO_2)$ ratio of the fluid will not be that loaded at elevated pressures because of the significant solubility of water. Instead, it may be a value considerably less, depending on the pressure. It is possible to overcome this problem to a certain extent by pre-saturating the melt with water at a pressure below that of the investigation pressure and using the quenched glass as the starting material. Alternatively, this can be overcome by investigating phase relations only near the solidus (i.e. in the presence of only small amounts of melt). (b) The crystallisation of anhydrous phases will result in the exsolution of water from the melt into the fluid in an attempt to preserve the activity of water imposed by the fluid. This, in turn, will change the $H_2O/H_2O + CO_2$ ratio of the fluid, thereby imposing a new activity on the melt. The evolution of water activity in both fluid and melt phases can in this way be minimised if the molar ratio of the fluid to the melt is large. Large ratios, however, present their own problems because of the increasing likelihood of changes in the bulk melt composition by the dissolution of melt components into the vapour phase. (c) Because the activity of water in the melt is composition-dependent (as can be readily seen by differing H_2O solubilities in the individual component melts, e.g. Holtz et al. 1995), the assemblage at each projected anhydrous composition within a phase diagram (e.g. within the haplogranite system), while having nominally the same water activity, will have been determined for a different water content. (d) The use of the activity of water in the fluid to obtain the value of the activity of water in the melt is dependent on our knowledge of the mixing behaviour of the fluid components. Joyce and Holloway (1993) determined that H_2O and CO_2 mix ideally at 2 kbar relative to their chosen standard states, although this is not the case at and above 5 kbar. The use of this result by equating the activity of water in the melt to the mole fraction of water in the fluid forces the standard state of water in the melt to be pure water vapour. As this is not the standard state used by the Burnham (1975) model for water in the melt, for example, conversion of H_2O activities to mole fractions in the melt cannot be easily carried out with available models. (e) The composition of the fluid lies within the C–O–H system and the actual fluid species may not remain H_2O and CO_2 unless the runs remain at oxygen fugacities greater than NNO–0·5 (Holloway & Blank 1994). Any H_2 gain through the capsule, as may be common in CO_2-rich experiments, will result in the formation of H_2O. In addition to these problems, the question remains about whether or not CO_2 can significantly affect phase relations; this will be discussed in more detail later.

Qualitatively, phase relations obtained at a constant water content or water activity will be similar as even the constant activity experiments should see only small changes in the water content with temperature and melt composition (as long as the abundances of crystalline phases is minor). The anhydrous projections of constant activity data are slightly more difficult to interpret than the constant water data in that the water content of each composition varies within each section because of the compositional dependence of the water activity. Isoactivity data can directly yield differentiation paths under H_2O-buffered conditions. Such paths, however, would only be relevant for the rare cases of a magma differentiating either in the presence of large volumes of fluid or of a hydrous crystalline phase. Neither the H_2O isoactivity or isoplethal data of these types of experiments can directly indicate differentiation paths for felsic magmas in which the water content increases during crystallisation. For such information we must rely on experiments such as those of Whitney (1969, 1972) or thermodynamic models based on available data that allow a systematisation of data that can be used for interpolation, coupled with a methodology for differentiation path calculations (e.g. Nekvasil 1988).

Since the original work of Burnham and Nekvasil (1986) in which thermodynamic models were developed based on evaluating the effect of water through a comparison of the available H_2O-saturated and anhydrous data, more recent experimental data have put additional constraints on H_2O-unbuffered crystallisation paths. The effect of water on the eutectic compositions in the bounding quartz-bearing binaries of the haplogranite–H_2O system, for example, provides important constraints on the topology of the alkali feldspar + quartz cotectic surface. In addition, such data, when combined with data on the effect of water on the eutectic in the system An–Or(–H_2O), provide constraints on the topology of the two feldspar + liquid (L) surface in the feldspar–H_2O system; all, in combination, aid in constraining the two feldspar + quartz + L surface in the granite–H_2O system. The topological nature of these surfaces dictate the compositional evolution of melts across the surfaces during the co-precipitation of two or three phases.

1. Experimental constraints

The haplogranite system remains of great importance to our understanding of melt compositional evolution. Ebadi and Johannes (1991) and Holtz et al. (1992) experimentally determined the isobaric effect of water in the haplogranite system using H_2O–CO_2 fluids as a means of reducing the activity of water in the melt in the presence of a vapour phase. Figure 1 shows phase relations in a portion of the haplogranite system at 5 kbar. The solid line shows the location of the H_2O-saturated solidus as determined by Holtz et al. (1992) and Nekvasil and Holloway (1989). The data for $a_{H_2O}^{Fl} = 0.5$ from Ebadi and Johannes (1991) are plotted for comparison.

The cotectic isoactivity curves in Figure 1 indicate the topology of the alkali feldspar + quartz liquidus surface. This surface extends from the anhydrous haplogranite base towards the Qz–H_2O sideline of the haplogranite–H_2O tetrahedron. Importantly, however, unlike the implications of the earlier anhydrous and H_2O-saturated data, the fluid-present data indicate that the effect of water is very different in each of the Qz-bearing bounding binaries and this difference carries over into the ternary. However, as the positions of the fluid-present eutectics have not been determined at low water activities, the question remains about what happens at low water contents. If the effect of water on the eutectic is very non-linear on the Ab–Qz side, there is still the possibility that the surface extends still further away from Qz with decreasing water content. This would imply that the slope of this surface steepens with increasing water content to a greater extent near the Ab–Qz sideline than the Or–Qz sideline. To investigate this possibility, an experimental investigation of the eutectic shift from dry to H_2O-saturated conditions was conducted at high pressure where the H_2O solubility is high and hence any shift in the eutectic with water would be magnified relative to that at lower pressures.

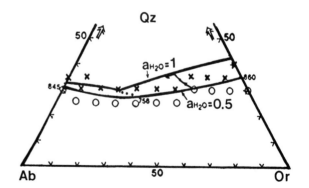

Figure 1 Quartz + alkali feldspar cotectic surface in the haplogranite system at 5 kbar based on the H_2O-saturated data of Holtz et al. (1992) and Nekvasil and Holloway (1989) ($a_w^{fl} = 1.0$) and the H_2O-undersaturated data and resulting cotectic H_2O-isoactivity curve of Ebadi and Johannes (1991) ($a_w^{fl} = 0.5$). Symbols: crosses, quartz + L; circles, alkali feldspar + L. The arrowed thin curve schematically indicates the compositional evolution of the melt during H_2O-unbuffered crystallisation. The dotted curve schematically indicates the locus of isoactivity thermal minima.

1.1. The bounding binary Ab–Qz(–H₂O)

Experimental technique. Efforts were made to enhance the chances of obtaining reversed equilibrium data in the anhydrous experiments. The starting material for all experiments was cryptocrystalline material obtained from the hydrothermal crystallisation of gel mixtures (using 10 wt% added H_2O and synthesis conditions of 5 kbar and 600°C for 4 days). Optical examination of the products showed very fine-grained crystalline material and no evidence of glass, and X-ray diffraction (XRD) revealed well-crystallised albite and quartz. The resulting powder was reground for several hours in alcohol, dried at 500°C for 1 hour and then stored at 120°C.

For each anhydrous experiment, the starting material was loaded into small Pt capsules that were welded at the bottom and then lightly crimped at the top. The capsules were heated to 500°C for 30 minutes to dry the sample thoroughly and were welded shut immediately thereafter. All ceramic spacers in the furnace assembly, as well as the furnace and BN furnace sleeve were dried at 1050°C for 30 minutes in an N_2 atmosphere. The thermocouple was separated from the Pt capsule by a thin ceramic disc; a NaCl sleeve was used in the experiments. The experiments were conducted in a solid-medium piston-cylinder apparatus with the temperatures controlled using $W_{96}Re_4$–$W_{74}Re_{26}$ thermocouples. The friction correction for NaCl was calibrated for the high-temperature runs using the reaction anhydrous Mg-cordierite = sapphirine + quartz (Newton et al. 1990, pers. comm.) and indicated no need for friction correction in the experiments. Triple-headed thermocouples were used to calibrate the temperature gradient within the capsules, which was found to be ≈13°C. The recorded temperature of each run is that at the thermocouple. All runs were held at the final temperature for 4–5 days.

Melting experiments were conducted by taking the crystalline starting material directly to temperature. Crystallisation experiments involved holding the charge above the liquidus temperature until completely molten (as determined by both XRD and optical examination of the quenched melt) and then rapidly dropping the temperature to the final value. For these crystallisation experiments the samples were superheated at the minimum temperature needed to render the sample completely molten within 24 hours. This was done in the hope of preserving nuclei within the melt to facilitate crystallisation at the subliquidus temperatures. All experiments critical to the location of the eutectic and solidus temperatures were reversed by conducting both melting and crystallisation experiments. Run products were identified from their optical properties and electron microprobe analysis.

After 4 days all the melting experiments above the eutectic temperature showed significant reaction and production of liquid. The crystalline material remained fairly fine-grained, however, and could only be identified by XRD. In contrast, the crystallisation experiments yielded large (up to 1 mm long), homogeneous, euhedral crystals. These experiments gave no indication of the significant nucleation delay or sluggish crystal growth rates that have been considered characteristic of anhydrous silicic systems. The subsolidus crystallisation experiments, however, did exhibit the metastable persistance of a few vol% glass, which was not seen in the subsolidus melting experiments. Both the use of cryptocrystalline starting material and the elevated temperatures of the runs may have contributed significantly to the success of the crystallisation experiments. Additionally, the care taken to superheat only as much as necessary to obtain complete melting within 24 hours may have been important.

The H_2O-saturated experiments were conducted in a similar manner, but with the omission of the drying of the sample before welding. Also, an all-NaCl cell was used instead of the BN sleeve. After each run the capsule was punctured and heated to check for H_2O loss.

Results. Tables 1 and 2 give the experimental data for both anhydrous and H_2O-saturated conditions at 11·3 kbar. Figure 2 shows the experimental results plotted within the

Table 1 Anhydrous experimental data in the system Ab–Qz at 11.3 kbar (the Ab content is the composition of the starting material in wt%).

Temperature (°C)	Run products
Ab₁₀₀	
1400; 1240	Gl
1400; 1230	Gl
1400; 1225	Ab
1400; 1200	Ab
Ab₇₈	
1290	Gl
1300; 1270	Qz + Gl
1300; 1250	Qz + Gl
1300; 1230	Qz + Gl
1300; 1225	Qz + Gl
1300; 1215	Qz + Gl
1300; 1190	Qz + Gl
1300; 1175	Qz + L
1300; 1169	Ab + Qz
1300; 1162	Ab + Qz
Ab₈₂	
1225	Gl
1225; 1210	Gl
1225; 1190	Gl
1225; 1175	Ab + Gl
Ab₈₆	
1325	Gl
1325; 1225	Gl
1325; 1215	Gl
1325; 1210	Gl
1325; 1200	Ab + Gl

Note: Single temperature entries denote melting experiments. Double entries denote crystallisation experiments (i.e. the sample was held at the first temperature for 24 hours and then at the second temperature for 4 days) and melting experiment reversals.

Table 2 H_2O-saturated experimental data for the system Ab–Qz–H₂O at 11.3 kbar (the Ab content is the composition of the starting material in wt%).

Temperature (°C)	Run products
Ab₇₅	
710	Gl + V
700; 658	Qz + Gl + V
700; 653	Qz + Gl + V
700; 642	Qz + Ab + V
Ab₇₈	
700	Gl + V
700; 655	Gl + V
700; 650	Gl + V (tr. Ab + Qz)
700; 647	Ab + Qz + V
700; 645	Ab + Qz + V
Ab₈₂	
800	Gl + V
800; 670	Gl + V
800; 665	Ab + L + V
800; 660	Ab + L + V

Note: Single temperature entries denote melting experiments. Double entries denote crystallisation experiments (i.e. the sample was held at the first temperature for 24 hours and then at the second temperature for 4 days) and melting experiment reversals.

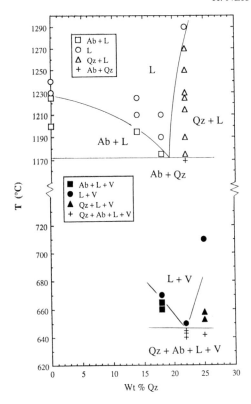

Figure 2 Reversed experimental data on the anhydrous and H_2O-saturated eutectics in the system Ab–Qz(–H_2O).

systems Ab–Qz(–H_2O). The H_2O-saturated eutectic lies at $\approx 650°C$ and Ab_{78} (wt%). The eutectic temperature is in agreement with that of Boettcher and Wyllie (1969), but the composition is several wt% higher in Ab than that determined by Luth et al. (1964). This latter difference is in keeping with the re-determinations of Holtz et al. (1992) and Nekvasil and Holloway (1989). The anhydrous eutectic was determined to lie at 1170°C, over 40°C higher than the determination of Boettcher et al. (1984). The eutectic composition of Ab_{81} lies several wt% Ab higher than that of Luth (1969). The solidus of pure albite is determined to lie at 1225°C, about 25°C higher than that determined by Boettcher et al. (1982). Importantly, the differences between the existing data and these new data indicate the importance obtaining reversed data for these systems in which kinetics makes it easy to obtain depressed solidus temperatures.

Comparison of the anhydrous and H_2O-saturated eutectic compositions allows an assessment of the effect of water in this system. Because of the small shift, the direction was further ascertained by using the composition of the H_2O-saturated cotectic (Fig. 2) for anhydrous experiments in which the liquidus phase was determined. The presence of Qz+L in the reversed anhydrous experiments and the large temperature interval between the liquidus and solidus all indicate the expansion of the Qz+L field with decreasing water content. As over the full spectrum of water contents, from dry to H_2O-saturated, the shift of the eutectic remains small, the topology of the whole surface must be similar to the water-rich part shown in Figure 1. It is interesting to note that the magnitude of the contraction of the Qz+L field at 11·3 kbar from dry to H_2O-saturated conditions is close to that obtained when a_w^{fl} changes from 0·5 to 1 in the CO_2-bearing studies at lower pressures. This is surprising as we might anticipate a greater shift in the higher pressure study because of the much greater range of water activities studied and, because of the higher

pressure, of the water contents of the melts. Certainly all of the data indicate a contraction of the liquidus field of quartz with increasing water content.

1.2. The haplogranite system

It is reasonable to anticipate that the shift in the position of the thermal minimum in the haplogranite–H_2O system with increasing water content of the melt would generally follow the shift in the isoactivity cotectic curves. However, Figure 1 shows that in the CO_2-bearing experiments there appears to be an additional component to the shift direction, i.e. towards the Ab–Qz sideline with increasing water content. As any cotectic H_2O–isoactivity curve is a product of the intersection of two liquidus surfaces and the minimum is the lowest temperature of this intersection, the easiest way in which the observed shift of the minimum with increasing water content can be induced is if the contraction of the Qz+L field is accompanied by an increase in the Ab : Or ratio of the thermal valley on the alkali feldspar liquidus surface. The compositions of coexisting melt and alkali feldspar at 2 kbar for $a_w^{fl}=1$ and 0·5 from Holtz et al. (1992) at 2 kbar (Fig. 3) and 5 kbar indeed indicate such a shift in the alkali feldspar + L thermal valley. However, these workers conclude that the shift component towards Qz is much smaller than is indicated by the data of Ebadi and Johannes (1991); they suggest instead that the minimum shifts almost along a Qz isopleth with increasing water content. In opposition to this, calculations by Nekvasil (1988), which were based on the behaviour in the binaries, mainly predicted a shift of the thermal minimum towards Qz with only a very slight component of the shift towards Ab–Qz. Additionally, the calculations did not show the strong shift in the alkali feldspar thermal valley towards the Ab–Qz sideline with increasing water content found in the CO_2-bearing experiments. Reconnaissance work of Johannes (1994, pers. comm.) on the location of the anhydrous minimum at 5 kbar has indicated an intriguing result. His data are consistent with the calculations, i.e. he found no strong shift component in the Ab–Qz direction when comparing his anhydrous data with H_2O-saturated data, implying that the anhydrous minimum is not continuous with the trend of H_2O–isoactivity minima observed from the H_2O–CO_2 experiments (Fig. 4). This could be readily explained if the effects of CO_2 and water are different. If this is true, then CO_2 cannot be considered as an inert vapour phase dilutant and the effect of water and CO_2 must be individually considered even for Fe- and Mg-poor magmas.

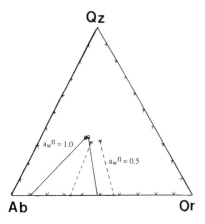

Figure 3 Compositions of coexisting feldspar and melt on the alkali feldspar liquidus surface at 2 kbar for $a_w^{fl}=1·0$ (solid tielines from Tuttle and Bowen 1958) and $a_w^{fl}=0·5$ (broken tie-lines from Holtz et al. 1992) showing the shift in position of the alkali feldspar thermal valley near the ternary minimum with increasing CO_2 content.

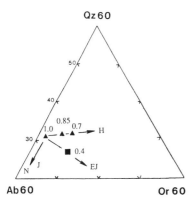

Figure 4 Shift in position of the thermal minimum in the haplogranite system at 5 kbar for various H_2O–CO_2 ratios of the fluid as indicated by the data of Holtz *et al.* (1992) [H], Ebadi and Johannes (1991) [EJ] and Johannes (1994, pers. comm.) [J]. Also shown for comparison is the calculated shift in thermal minimum of Nekvasil (1988) [N] with decreasing water content. Numerical values indicate mole fraction of water in the fluid.

The possible role of CO_2. Solubility studies indicate less than 1 wt% (≈ 5 mol%) CO_2 solubility in albite melts (Stolper *et al.* 1987) and in rhyolite melts below 20 kbar (Fogel & Rutherford 1990), with a general decrease in CO_2 solubility with decreasing mol% SiO_2 (Blank & Brooker 1994). This implies that even for CO_2-saturated conditions, the effect on liquidus temperatures should be minor. Furthermore, because the dissolution occurs mainly as molecular CO_2 for these compositions (Fogel & Rutherford 1990), there should be little differential activity lowering effect on any of the melt components beyond the differences imposed by differing ΔS°_{fus}. However, Bohlen *et al.* (1982) and Boettcher *et al.* (1987) found a significant solidus lowering effect in albite and, particularly, in sanidine melts. It is possible that the $\approx 30^{\circ}C$ solidus depression that Boettcher *et al.* (1982) found at 10 kbar in albite–CO_2 melts was a result of H_2 diffusion into the capsule and the production of water and graphite. Fine and Stolper (1985) indicate that the water contents from CO_2-bearing albite glasses can be expected to be about 0·2 to 0·5 wt%, despite drying, the use of a double capsule with a buffered assembly and the limitation of run duration to only a few hours. This, however, cannot readily explain the >100°C lowering of the sanidine solidus observed by Boettcher *et al.* (1987). These data, however, are more difficult to evaluate because of the incongruent behaviour of sanidine melts below 16 kbar.

There has been some suggestion that the presence of H_2O increases the solubility of CO_2. The data of Mysen (1976) indicate a 50% increase in CO_2 solubility at 20% H_2O relative to pure CO_2 in albite melts at 20 kbar (from 7·7 to 11·1 mol%); this was not seen, however, at low pressure by Stolper *et al.* (1987). In spite of such possible increases in CO_2 solubility with water content at low water contents, the solubility of CO_2 in silicic melts remains low and it is reasonable to expect that, based on solubility alone, it should therefore only nominally affect the phase relations.

Low solubility, however, may be necessary, but insufficient, to validate the conclusion that CO_2 does not affect phase relations. The solubility of N_2 in albite melts, for example, is only ≈ 400 ppm at 960°C and 2 kbar (Kessen & Holloway 1974) and decreases with temperature and f_{O_2}. [More recent results suggest that the solubility of N_2 may be even significantly lower than this (Shilobreyeva *et al.* 1994, quoted in Carroll & Webster 1994).] In contrast, CO_2 solubility is significantly higher (≈ 1150 ppm at 2 kbar and 900°C in

rhyolite, which is similar to that in albite melts—see figure 1, Blank & Brooker 1994). It would therefore be anticipated that the effect of CO_2 on solidus temperatures should be greater than that of N_2 for a given H_2O activity. Keppler (1989) experimentally determined the solidus temperature of haplogranite compositions with Na : K ratios of 1 : 1 in the presence of both H_2O–CO_2 and H_2O–N_2 melts for the same initial fluid H_2O content. His results indicate that N_2 has a much stronger effect on solidus depression than CO_2. As this difference must be caused by a factor independent of melt solubility, it must be the vapour phase that is controlling the behaviour. In the case of N_2 versus CO_2, stronger positive deviations from the ideal mixing of fluid species in N–O–H mixtures relative to C–O–H fluid mixtures would raise the activity of water imposed on the melt to a greater extent in the N-bearing system. Even small increases in melt water activity could result in major changes in phase equilibria.

As in the case just described, it is likely that we must look to the vapour phase to provide an answer to the question regarding the preferential destabilisation of Or in the haplogranite system. If melt components dissolve to different extents in C–O–H fluids, then the composition of the coexisting melt could be significantly affected in spite of the low melt solubilities of CO_2. For example, if there is an increasing extent of congruent dissolution of the Or melt component relative to Ab (and Qz) with increasing CO_2 content of the fluid, then the melt composition would be more albitic than the bulk composition even in the superliquidus region. This would result in a shift of the thermal minima in the Or–Ab system increasingly towards Or with increasing CO_2 content of the fluid, and this shift would be reflected by a similar shift of the thermal minimum in the haplogranite system, similar to that shown in Figure 1.

Incongruent dissolution of Or into the fluid may instead result in a shift of thermal minima similar to that described by Holtz *et al.* (1992). If potassium (and silica) were preferentially removed from the melt, then in addition to lowering the activity of the Or melt component, the melt would become peraluminous. (If the solubility of alkalis is strongly temperature-dependent, then such peraluminocity would not be quenchable in a closed system.) Voigt and Joyce (1991) showed that even small amounts of normative sillimanite result in significant destabilisation of quartz in both the systems albite–quartz and sanidine–quartz and at the haplogranite minimum. At sillimanite saturation (as little as 3·6 wt% normative sillimanite), quartz is preferentially destabilised to the extent of 30°C at 2 kbar (H_2O-saturated) in the Or–Qz system relative to 15°C in the Ab–Qz system (and to an intermediate extent at the haplogranite minimum). This destabilisation of quartz would counteract the quartz stabilisation that would occur by lowering the Or melt activity and could induce an almost isoplethal shift of the thermal minimum.

1.3. The granite system

The kinetic difficulties encountered in investigations of anhydrous phase relations in plagioclase-free systems are compounded when plagioclase becomes part of the phase assemblage because of the strong temperature dependence of its composition and the need for coupled diffusion for compositional adjustment. However, from data in the bounding systems some constraints can be placed on the evolution of melts during H_2O-undersaturated, H_2O-unbuffered crystallisation. In the Ab-free system An–Qz, anhydrous data are still needed to evaluate the magnitude of the shift. However, because of the much higher $\Delta H°$ of fusion of An (e.g. Ghiorso et al. 1983), there is little by way of interaction that can be envisioned to counter the large shift of the eutectic towards Qz that is anticipated on the addition of water in this system.

Experimental investigation of the effect of water in An-bearing systems has in part been conducted by Ai and Green (1989) and Nekvasil and Carroll (1993) on the systems An–Or. All experimental data on the anhydrous phase relations and available H_2O-saturated data (Yoder et al. 1957) indicate that there is a significant preferential destabilisation of Or by water. The reversed data of Nekvasil & Carroll (1993) indicate a shift of the eutectic of 10 mol% towards Or from anhydrous to H_2O-saturated conditions (which is about half of that suggested by the data of Ai and Green 1989) at 10 kbar. This is still considerably larger than the effect of water on the Ab–Qz eutectic and larger than the shift of the Or–Qz eutectic. These data in combination allow the general topology of all cotectic surfaces to be constructed in the subsystems of the granite–H_2O system and in turn provide information on the effect of water on the four-phase surface within the granite–H_2O hyperspace.

2. Implications for differentiation paths

Recent experimental investigations have shed new light on the topologies of the cotectic surfaces in the granite system. Although the possibility of a significant CO_2 effect remains unresolved, the surface topologies remain robust and can still be used to infer compositional trends during differentiation of H_2O-undersaturated, H_2O-unbuffered magmas. Such differentiation trends should in turn be fairly robust even for natural rocks, as feldspars and quartz are the most abundant phases in most felsic rocks and the presence of ferromagnesian minerals results in only a second-order modification of such trends. The differentiation trends of Nekvasil (1988) for granitic/rhyolitic magmas and Nekvasil (1992) for syenitic/trachytic magmas remain good descriptors of magma differentiation. The new data, however, indicate a few additional second-order changes in the paths of the melt evolution that will be discussed in the following.

Nekvasil (1992) showed how H_2O-unbuffered equilibrium crystallisation of Qz-poor trachytic magmas with low bulk water contents will have markedly different evolution paths than magmas with high bulk water contents. The former paths will include the Or enrichment of alkali feldspar during cooling and the partial resorption of plagioclase during the early stages of alkali feldspar crystallisation. This can be followed by the co-precipitation of plagioclase and alkali feldspar, and plagioclase compositional evolution along An-isopleths. Figure 5, modified from Nekvasil (1992), for a hypothetical Qz-free trachyte shows a calculated equilibrium path for 1·2 wt% H_2O. The experimental data of Nekvasil and Carroll (1993) indicate that the plagioclase + alkali feldspar + L surface in the feldspar–H_2O system extends away from Or with decreasing water content to a much greater extent than previously thought. This implies than when plagioclase is on the liquidus, alkali feldspar will precipitate much earlier (i.e. at a higher An content of the melt and higher melt fraction for a given bulk water content and bulk composition) and will be more ternary than indicated by the calculated path in Figure 5a. As shown schematically in Figure 5, the path of melt evolution will have a greater component of An depletion and less of Or depletion during the initial stages of coexistence of two feldspars than previously realised. The melt will thus remain in the peritectic region of the two feldspar + L surface over a greater crystallisation interval. This could lead to the

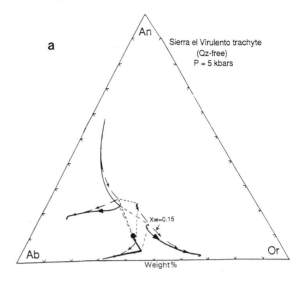

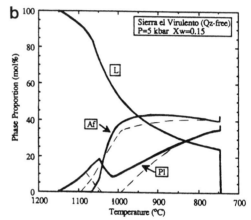

Figure 5 (a) Calculated equilibrium crystallisation paths (solid curves) for a trachyte composition (closed circle) for 1·2 wt% bulk water and 2 kbar (modified from Nekvasil 1992) compared with a schematic projection of the crystallisation path in light of the recent data in the An–Or system (broken curves). The broken triangles indicate the three-phase assemblage at the onset of coexistence of two feldspars. (b) Variations in phase abundances during the crystallisation paths in (a). Broken curves indicate schematically the possibility of compete resorption of plagioclase followed by re-precipitation at a later stage.

fascinating situation of the complete resorption of plagioclase followed by re-precipitation as the melt evolves to a higher water content and a lower An content, as shown schematically in Figure 5b! Although quantitative investigation of this specific possibility must await model recalibration in the light of the new experimental data, the experimental data certainly suggest that peritectic behaviour of feldspars should be expected more commonly than previously thought.

As concluded by Nekvasil (1988), for Qz-rich melts (e.g. rhyolites), if quartz is the liquidus phase, the precipitation of quartz will continue to a greater extent (and hence the melt will see a greater extent of silica depletion) before either plagioclase or alkali feldspar is stabilised if the bulk water content is low. Importantly, however, the extent of this silica depletion will be significantly less for Ab-rich than Or-rich melts. Once quartz and a feldspar co-precipitate, the melt will undergo silica enrichment as the water content continues to build up and the H_2O-saturated cotectic surface is approached. The resulting path could readily explain the variations between the early and late phases of the Bishop Tuff found by Hildreth (1979). For most An-bearing compositions, however, the H_2O-saturated cotectic will not be reached because the melt will be simultaneously evolving away from the Ab–An sideline if plagioclase is precipitating (or away from the Ab–Or sideline if alkali feldspar is precipitating) and towards the four-phase surface where saturation with a second feldspar will occur. This intersection will occur at a higher An and Qz content if the magma is drier at this stage. Importantly, once co-precipitation begins it will lead to a faster depletion of An in the melt than previously thought as the melt follows a curved path across this surface towards the H_2O-saturated, two feldspar + L curve.

Orthoclase-poor rocks such as diorites, with their high abundances of ferromagnesian minerals, are less easily compositionally modelled by the granite system. Importantly, however, the new data indicate that because of the minor shift of the quartz + plagioclase cotectic surface towards Qz with increasing water content for Or-poor compositions, there will be little difference between the melt compositions at the onset of co-precipitation of these two phases (for a given bulk composition), regardless of the water content. Additionally, only a minor amount of silica enrichment will occur during co-precipitation and evolution of the melt across the surface.

In summary, experimental investigations have defined more clearly the effect of water on the cotectic surfaces in the granite system and have important implications for the compositional evolution of felsic magmas. Further experimental investigations and model development should allow more quantitative analysis of the paths within the system. Additional investigation is also needed to define the role that CO_2 may be playing in both synthetic and natural systems.

References

Ai, Y. & Green, D. H. 1989. Phase relations in the system anorthite–potassium feldspar at 10 kbar with emphasis on their solid solutions. MINERAL MAG **53**, 337–45.

Blank, J. G. & Brooker, R. A. 1984. Experimental studies of carbon dioxide in silicate melts: solubility, speciation, and stable isotope behavior. *In* Carroll and Holloway, (eds) *Volatiles in magmas.* MINERAL **30**, 157–86.

Boettcher, A., Burnham, C. W., Windom, K. E. & Bohlen, S. R. 1982. Liquids, glasses, and the melting of silicates to high pressures. J GEOL **90**, 127–38.

Boettcher, A., Guo, Q., Bohlen, S. R. & Hanson, B. 1984. Melting in feldspar-bearing systems at high pressures and the structure of aluminosilicate liquids. GEOLOGY **12**, 202–4.

Boettcher, A., Luth, R. W. & White, B. S. 1987. Carbon in silicate liquids: the systems $NaAlSi_3O_8$–CO_2, $CaAlSi_2O_8$–CO_2 and $KAlSi_3O_8$–CO_2. CONTRIB MINERAL PETROL **97**, 297–304.

Boettcher, A. L. & Wyllie, P. 1969. Phase relationships in the system $NaAlSiO_4$–SiO_2–H_2O to 35 kilobars pressure. AM J SCI **267**, 875–909.

Bohlen, S. R., Boettcher, A. L. & Wall, V. J. 1982. The system albite–H_2O–CO_2: a model for melting and activities of water at high pressures. AM MINERAL **67**, 451–62.

Bohlen, S. R., Boettcher, A. L., Wall, V. J. & Clemens, 1983. Stability of phlogopite–quartz and sanidine–quartz: a model for melting in the lower crust. CONTRIB MINERAL PETROL **83**, 270–7.

Bohlen, S. R., Eckert, J. O. & Hankins, W. B. 1995. Experimentally determined solidi in the Ca-bearing granite system $NaAlSi_3O_8$–$CaAl_2Si_2O_8$–$KAlSi_3O_8$–SiO_2–H_2O–CO_2. AM MINERAL **80**, 752–6.

Burnham, C. W. 1975. Water in magmas: a mixing model. GEOCHIM COSMOCHIM ACTA **39**, 1077–84.

Burnham, C. W. & Nekvasil, H. 1986. Equilibrium properties of granite pegmatite magmas. AM MINERAL **71**, 239–63.

Carroll, M. D. & Webster, J. D. (1994) Solubilities of sulfur, noble gases, nitrogen, chlorine and fluorine in magmas. *In* Carroll, M. R. & Holloway, J. R. (eds) *Volatiles in magmas.* REV MINERAL **30**, 231–279.

Ebadi, A. & Johannes, W. 1991. Experimental investigation of composition and beginning of melting in the system $NaAlSi_3O_8$–$KAlSi_3O_8$–SiO_2–H_2O–CO_2. CONTRIB MINERAL PETROL **106**, 286–95.

Fine, G. & Stolper, E. 1985. The speciation of carbon dioxide in sodium aluminosilicate glasses. CONTRIB MINERAL PETROL **91**, 105–21.

Fogel, R. A. & Rutherford, M. J. 1990. The solubility of carbon dioxide on rholitic melts: a quantitative FTIR study. AM MINERAL **75**, 1311–26.

Ghiorso, M. S., Carmichael, I. S. E., Rivers, M. L. & Sack, R. L. 1983. The Gibbs free energy of mixing of natural silicate materials, an expanded regular solution approximation for calculation of magmatic intensive variables. CONTRIB MINERAL PETROL **84**, 107–45.

Hildreth, W. 1979. The Bishop Tuff: evidence for the origin of compositional zonation in silicic magma chambers. GEOL SOC AM SPEC PAP **180**, 43–75.

Holloway, J. R. & Blank, J. 1994. Application of experimental results to C–O–H species in natural melts. *In* Carroll, M. R. & Holloway, J. R. (eds) *Volatiles in magmas* REV MINERAL **30**, 187–230.

Holtz, F., Pichavant, M., Barbey, P. & Johannes, W. 1992. Effects of H_2O on liquidus phase relations in the haplogranite system at 2 and 5 kbar. AM MINERAL **77**, 1223–41.

Holtz, F., Behrens, H., Dingwell, D. B. & Johannes, W. 1995. H_2O solubility in haplogranitic melts: compositional, pressure, and temperature dependence. AM MINERAL **80**, 94–108.

Huang, L. & Wyllie, P. J. 1976. Phase relationships of gabbro–tonalite–granite–water at 15 kbar with applications to differentiation and anatexis. AM MINERAL **71**, 301–16.

Joyce, D. B. & Holloway, J. R. 1993. An experimental determination of the thermodynamic properties of H_2O–CO_2–NaCl fluids at high pressures and temperatures. GEOCHIM COSMOCHIM ACTA **57**, 733–46.

Keppler, H. 1989. The influence of fluid phase composition on the solidus temperatures in the haplogranite system $NaAlSi_3O_8$–$KAlSi_3O_8$–SiO_2–H_2O–CO_2. CONTRIB MINERAL PETROL **102**, 321–7.

Kesson, S. E. & Holloway, J. R. 1974. The generation of N_2–CO_2–H_2O fluids for use in hydrothermal experimentation II. Melting of albite in a multispecies fluid. AM MINERAL **59**, 598–603.

London, D. 1986. The magmatic–hydrothermal transition in the Tanco rare-element pegmatite: evidence from fluid inclusions and phase equilibrium experiments. AM MINERAL **71**, 376–95.

Luth, W. C. 1969. The system Ab–Qz and Sa–Qz to 20 kbar and the relationship between H_2O content, P_{H_2O}) and P_T in granitic magmas. AM J SCI **267A**, 325–41.

Luth, W. C., Jahns, R. & Tuttle, F. 1964. The granite system at pressures of 4 to 10 kbar. J GEOPHYS RES **69**, 759–73.

Mysen, B. O. (1976) The role of volatiles in silicate melts: solubility of carbon dioxide and water in feldspar, pyroxene, and feldspathoid melts to 30 kbar and 1625°C. AM J SCI 969–96.

Nekvasil, H. 1988. The effect of anorthite on the crystallization paths of H_2O-undersaturated haplogranitic melts. AM MINERAL **73**, 966–81.

Nekvasil, H. 1992. Ternary feldspar crystallization in high-temperature felsic magmas. AM MINERAL **77**, 592–604.

Nekvasil, H. & Burnham, C. W. 1987. The calculated individual effects of pressure and H_2O content on phase equilibria in the granite system. *In* *Magmatic processes: physicochemical principles.* GEOCHEM SOC SPEC PUBL **1**, 433–46.

Nekvasil, H. & Carroll, W. 1993. Experimental constraints on the high temperature termination of the 2 feldspar + L curve in the system Ab–Or–An–H_2O. AM MINERAL **78**, 601–6.

Nekvasil, H. & Holloway, J. R. 1989. H_2O-undersaturated phase relations in the system Ab–Or–Qz–H_2O: new considerations. EOS, TRANS AM GEOPHYS UNION **70**, 506.

Pichavant, M., Holtz, F. & McMillan, P. F. 1992. Phase relations and compositional dependence of H_2O solubility in quartz feldspar melts. CHEM GEOL **96**, 303–19.

Steiner, J. C., Jahns, R. H. & Luth, W. C. 1975. Crystallization of alkali feldspar and quartz in the haplogranite system at 5 kbar. GEOL SOC AM BULL **86**, 83–98.

Stolper, E. M., Fine, G. J., Johnson, T. & Newman, S. 1987. The solubility of carbon dioxide in albitic melt. AM MINERAL **72**, 1071–85.

Voigt, D. E. & Joyce, D. B. 1991. Depression of the granite minimum by the addition of sillimanite. AM GEOPHYS UNION SPRING 1991 MEET PROGRAM ABSTR 304.

Webster, J. D., Holloway, J. R. & Hervig, R. L. 1987. Phase equilibria of a Be, U, and F-enriched vitrophyre from Spor Mountain, Utah. GEOCHIM COSMOCHIM ACTA, 389–402.

Whitney, J. A. 1969. *Partial melting relationships of three granitic rocks*. M.S. Dissertation. Massachussetts Institute of Technology.

Whitney, J. A. 1972. History of granodioritic and related magma systems: an experimental study. Ph.D. Dissertation, Stanford University.

Wyllie, P. J. & Tuttle, O. F. 1959. Effect of carbon dioxide on the melting of granite and feldspars. AM J SCI **257**, 548–655.

Yoder, H. S., Stewart, D. B. & Smith, J. R. 1957. Ternary feldspars. CARNEGIE INST WASHINGTON YEARB **55**, 206–14.

HANNA NEKVASIL and WILLIAM CARROLL, Department of Earth and Space Sciences, State University of New York, Stony Brook, NY 11794-2100, U.S.A. E-mail: HNEKVASIL@ccvm.sunysb.edu

Transactions of the Royal Society of Edinburgh: Earth Sciences, **87**, 147–157, 1996

Insights from igneous reaction space: a holistic approach to granite crystallisation

John P. Hogan

ABSTRACT: Petrological investigations of granite commonly reveal multiple periods of growth punctuated by resorption for many of the constituent minerals. Complementary to such textures are mineral compositional heterogeneity manifested by zoning or grain to grain variability. These features ultimately reflect changes in the intensive parameters or activities of components during melt solidification. Such complexities of granite crystallisation can be simultaneously modelled in a reaction space constructed from the set of linearly independent reactions describing the equilibria among all phases and components in the system of interest.

The topology of the linearly independent reactions that define the reaction space for garnet–muscovite–biotite granites yields the following insights: (1) there is no one unique reaction that produces or consumes aluminous minerals (e.g. garnet); (2) minerals can alternate as reactants or products in different reactions accounting for textures indicating multiple periods of crystallisation separated by resorption; (3) mineral compositions are regulated by the reaction(s) producing them and vary as the stoichiometry of the reaction(s) producing them varies; (4) resorption of early crystallising garnet is likely to reflect decreasing pressure, presumably during magma ascent; (5) late crystallisation of garnet, at the expense of biotite, reflects an increase in melt aluminosity and does not necessarily require high Mn activities for the melt and (6) increasing melt H_2O, at H_2O-undersaturated conditions, favours the formation of biotite–muscovite granite.

Application of the reaction space method to other granite types holds considerable promise for elucidating reactions that regulate mineral assemblages and compositions during crystallisation.

KEY WORDS: reaction space, granite, compositional variability, garnet, muscovite, biotite.

Petrological studies reveal complex crystallisation histories for granite. Many of the constituent minerals can exhibit evidence of multiple periods of growth punctuated by episodes of resorption. Rapakivi feldspars are one common example. Complementary to such textures are compositional heterogeneities exhibited by minerals (e.g. growth zoning in garnet). These features serve as witness that most granites are the integrated product of crystallisation over a range in values for P, T, f_{H_2O}, f_{O_2}, as well as other components of interest.

The complex phase relationships of biotite–muscovite–garnet granites are discussed in this paper as an example. These granites typically exhibit the following crystallisation sequence: (1) early crystallising garnet, when observed, is replaced by biotite and muscovite; (2) near the end of crystallization, garnet rejoins the crystallising assemblage of biotite and muscovite; and (3) eventually biotite is eliminated from the final solidus assemblage, which crystallises as muscovite–garnet granite. Reaction relationships among these phases during crystallisation have been discussed in detail by numerous workers (Miller & Stoddard 1981a; Abbott & Clarke 1979; Abbott 1981a, 1985; Clemens & Wall 1988; Speer 1981; Speer & Becker 1992; Zen 1988). Peritectic reactions that involve garnet, biotite and muscovite have previously been inferred from textural relationships, modal variation, compositional zoning and graphical analyses of AFM projections. Although the exact nature of the reactions involved remains contentious (e.g. Abbott 1981b; Clemens & Wall 1982; Miller & Stoddard 1981b, 1982), most workers concur with Hall (1965) that an increase in the Mn content of the liquid as a result of fractional crystallisation stabilises the crystallisation of late garnet from felsic magmas (Miller & Stoddard 1981a; Abbott & Clarke 1979; Abbott 1981a, 1985; Zen 1988).

For large variance systems, the equivocal nature of textural interpretations makes it difficult, if not impossible, to determine specific reaction(s) responsible for changes in the crystallising assemblage and mineral composition by observations alone (Bowen 1912; Flood & Vernon 1988; Zen 1988). However, Thompson *et al.* (1982) and Thompson (1982a, b) developed a method that overcomes this obstacle by algebraic determination of the set of linearly independent reactions which fully describe the system of interest. A subset of these reactions is selected as basis vectors for the *reaction space*. Although this selection can be entirely arbitrary, *net-transfer reactions* are commonly chosen as these reactions affect the modal abundance of minerals, and thus mineral textures, and are also commonly sensitive to intensive parameters (e.g. pressure, temperature) or the chemical potential of some component of interest. The path through reaction space is reconstructed from mineral textures, compositional variation and modal variations. More importantly, the reactions controlling these variations can be completely described by linear combination of the basis vectors and, as noted by Thompson (1982b), 'Reactions such as these often give surprisingly simple interpretations for baffling textural relationships.'

The reaction space technique has been applied extensively to understanding the reaction histories of metamorphic rocks and, as demonstrated by Thompson (1982b), can be applied equally well to igneous systems. Although traditional petrological tools (e.g. AFM projections) are effective for analysing relationships among reduced subsets of the crystallising assemblage, such methods must commonly appeal to additional components not accounted for in the analysis (e.g. MnO) to achieve compatibility between the predicted and observed reaction histories. The reaction space method overcomes this limitation through its ability to treat complex mineral solid solutions and through its consideration of all phases and components within the system of interest. The holistic approach of the reaction space method is applied here to the crystallisation of garnet, biotite and muscovite in granitic magmas, using the Northport monzogranite as a typical example, to elucidate potential driving forces for changes in the crystallising assemblage.

1. Geological setting

The Northport monzogranite, previously described by Hogan (1984, 1993), is associated with the mid-Palaeozoic plutonic and volcanic rocks of the Coastal Maine Magmatic Province (Hogan & Sinha 1989). The pluton is an elliptical stock ($\approx 14\,km^2$) that intruded a regionally metamorphosed (andalusite–sillimanite grade) graphitic to sulphidic schist comprised of thinly interbedded pelite and quartz-rich layers. The schist was re-metamorphosed along the contact with the pluton to a cordierite–andalusite hornfels. Conventional and ion microprobe studies of zircon from the granite yielded an age of 430 Ma without much evidence for an inherited component (Eriksson et al. 1989).

Detailed petrographic descriptions, modal analysis and the composition of whole rock samples from the Northport pluton are given in Hogan (1984, 1993) and are briefly summarised here. The Northport pluton is comprised predominantly of relatively homogeneous, medium- to coarse-grained, hypidomorphic to allotriomorphic, inequigranular seriate, garnet–muscovite–biotite monzogranite. Plagioclase commonly occurs as inclusions in subhedral to anhedral alkali feldspar crystals or in aggregates comprised of several subhedral to anhedral plagioclase crystals. The colour index varies from 8 to 13, with biotite as the dominant mafic phase (≈ 6–11% modal) and lesser primary and secondary muscovite (≈ 2–3·0%) and trace garnet (≈ 0.2%). Biotite is more abundant in scarce mafic enclaves (15–22 vol%), whereas it is rare to absent in scarce thin (centimetre scale) muscovite–garnet pegmatitic granite dykes that cross-cut the main phase of the granite. These pegmatite dykes appear to be confined to the interior of the granite and are interpreted here to be consanguineous with the Northport granite and represent the crystallisation of residual magma. Bulk rock chemical analyses indicate that the granite is peraluminous, with ≈ 2.1 wt% normative corundum and an $Al_2O_3/CaO + K_2O + Na_2O$ ratio of ≈ 1.6 (mol%) after correction for apatite. The Northport pluton is part of a larger suite of strongly peraluminous granites distributed throughout the Coastal Maine Magmatic Province (Hogan & Wones 1984; Hogan & Sinha, unpublished data).

Pressure and temperature estimates for the emplacement of the granite remain inadequately constrained. Geothermometry on a late-garnet–biotite pair recorded a temperature of 730°C (Hogan 1984). Comparison of the normative composition of whole rock samples with experimentally determined phase relationships in the systems SiO_2–$NaAlSi_3O_8$–$KAlSi_3O_8$–H_2O and SiO_2–$CaAl_2Si_2O_8$–$KAlSi_3O_8$–$NaAlSi_3O_8$–$KAlSi_3O_8$–H_2O suggest crystallisation at conditions of 1·0 kbar H_2O and temperatures of c. 730°C (Hogan 1993). The absence of miarolitic cavities and the scarcity of granite, pegmatite or aplite dykes within, or immediately adjacent to, the pluton suggests that the bulk of crystallisation proceeded at H_2O-undersaturated conditions. Thus pressure estimates based on comparisons with experimentally determined minima under H_2O-saturated conditions represent minimum values for emplacement. Nevertheless, such values are consistent with the low-pressure mineral assemblage of the contact aureole (cordierite–andalusite) and indicate an epizonal level of emplacement for this pluton.

2. Petrography

2.1. Quartz

Euhedral quartz crystals, with rectangular to hexagonal outlines suggestive of crystallisation as β-quartz, occur as inclusions in plagioclase and alkali feldspar. More typically, quartz is found in the matrix as interstitial, interlocking anhedral grains. Fine-grained anhedral quartz occurs with myrmekitic feldspar, $\pm$muscovite in pockets, or in cross-cutting veins. Hibbard (1979) interpreted such textures to represent a final quench or late-stage vapour saturation of residual magma. Alternatively, these features may be deuteric in origin. Locally, the presence of recrystallised quartz with well-developed subgrains indicates that portions of the pluton were subjected to varying degrees of post-crystallisation deformation.

2.2. Feldspars

Plagioclase crystals commonly display complex zonation. This includes: (1) dendritic or boxwork patterns in the core (e.g. Hibbard 1981); (2) reverse to normally zoned overgrowths on cores; and (3) unzoned crystals. For example, one plagioclase exhibited the following trend: a normally zoned core of An_{34-32} is mantled by a thin zone of An_{39} and overgrown by a normally zoned rim An_{39-24}. However, most plagioclase analyses are within a narrow range of An_{24-17}, with the total ranged observed being An_{40-10} (Hogan 1984).

Alkali feldspar occurs as prominent coarse subhedral to anhedral poikilitic grains that variably merge with the interstitial matrix. They are exsolved and form macro- to microperthite. String, vein and braided varieties of perthite are observed in thin section. Most grains are microcline and display well-developed 'tartan' patterns characteristic of combined albite and pericline twins. Carlsbad twins are also commonly observed. Compositions range from $Or_{91.61}Ab_{8.19}An_{0.21}$ to $Or_{96.59}Ab_{3.26}An_{0.15}$ and reflect post-crystallisation re-equilibration (Hogan 1984).

2.3. Biotite

Biotite occurs as subhedral to anhedral grains exhibiting light to dark brown pleochroism. It is found as isolated grains or inclusions in other minerals (e.g. feldspar). More typically biotite forms aggregates comprised of several biotite grains as well as other minerals (e.g. muscovite, garnet). These aggregates commonly occur along the margins of larger grains of feldspar or quartz. Zircon, monazite and apatite are common inclusions in biotite. In the main phase of the granite biotite grains typically appear unaltered and chloritised biotite is scarce or localised along obvious zones of fluid alteration defined by fractures and sericite. In granite pegmatite dykes, biotite grains are rare and ragged and may represent the remnants of incompletely reacted biotite incorporated from the main phase of the granite.

Biotite from the Northport monzogranite is Fe-rich (Table 1). Fe^{2+}/Fe^{3+} determinations on biotite separates indicate that 98% of the total Fe is present as Fe^{2+} (Hogan 1984). They are characterised by excess Al^{IV}, Al^{VI} and Ti^{4+} and contain vacancies in the octahedral and interlayer sites and exhibit considerable solid solution off the phlogopite–annite join towards siderophyllite while maintaining a constant Fe/Fe + Mg ratio of ≈ 0.72 (Fig. 1). Additional Al may be incorporated by a Tschermak exchange $Al^{IV}Al^{VI}Si_{-1}Fm_{-1}$ where $Fm = \sum Fe^{2+}$, Mg^{2+}, Mn^{2+}. However, an excess of Al^{VI}, beyond which can be accounted for by a Tschermak

Table 1 Representative electron microprobe analyses of biotite.

Sample:	36 Np25	37 Np25	83 Np102	85 Np102	108 Np17-1	109 Np17-1	130 Np32	196 Np108
SiO$_2$	32·51	32·02	35·19	35·39	34·33	34·15	34·57	34·59
TiO$_2$	1·93	1·75	3·33	3·18	1·44	1·60	2·32	2·72
Al$_2$O$_3$	18·89	18·64	16·99	16·60	19·22	18·96	19·23	18·39
FeO*	28·31	28·85	25·42	25·09	27·48	27·51	25·11	24·26
MnO	0·71	0·96	0·52	0·39	0·46	0·53	0·97	0·49
MgO	5·81	6·58	5·24	5·83	3·77	4·11	5·11	5·99
CaO	0·07	0·07	0·12	0·06	0·03	0·03	0·03	0·08
Na$_2$O	0·05	0·09	0·07	0·07	0·05	0·04	0·04	0·10
K$_2$O	7·74	6·67	8·57	8·82	8·77	8·78	8·89	9·35
Total	96·02	95·63	95·45	95·43	95·55	95·71	96·27	95·97

Notes: 36 and 37 biotite associated with L-type garnet; 83 and 85 early crystallising biotite; 108 and 109 late crystallising biotite in granite pegmatite; 130 and 196 biotite associated with other biotite grains. See Hogan (1984) for details of the analytical scheme.

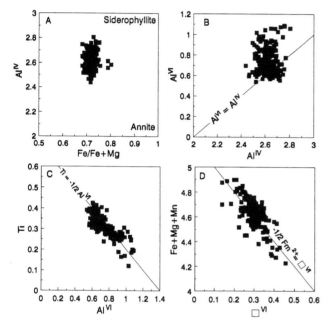

Figure 1 Composition of biotite from the Northport monzogranite, Maine. Lines represent expected compositional variation from the operation of exchange mechanisms discussed in the text. Concentrations are pfu (per formula unit).

exchange, requires the operation of additional exchanges involving Al^{VI} (Fig. 1B). Exchange mechanisms for the incorporation of Ti^{4+} into biotite also involve Al^{VI} or octahedral vacancies. The negative correlation between Ti^{4+} and Al^{VI} suggests Fm$_{+1}$ Ti$_{+1}$ Al$^{VI}_{-2}$ as a possible exchange (Fig. 1C). The inverse correlation between $\sum$ Fe, Mg, Mn and octahedral vacancies with a slope of $\approx 1/2$ suggests Fm$_{-2}$ Ti$_{+1}$ $\square^{VI}_{+1}$ as another possible exchange (Fig. 1D).

2.4. Muscovite

Muscovite is interpreted to be of both primary (i.e. magmatic) and secondary (i.e. subsolidus) origin. Petrographic criteria used to discern primary and secondary muscovites are those discussed by Miller et al. (1981). Primary muscovite is commonly associated with biotite, but also occurs as isolated grains or inclusions in other minerals. In contrast, inclusions in muscovite are rare and include garnet and zircon. Muscovite and garnet can exhibit clean planar terminations along crystal faces or they can be more irregularly intergrown. In late-stage granite pegmatite dykes, large (≈ 3.0 mm) muscovite grains form monomineralic aggregates or are associated with late-crystallising garnet. Secondary white mica varies from 0·1 to 2·7% modal. It is present as small grains along margins of

biotite, partially recrystallised primary muscovite, within plagioclase (i.e. sericite) and in late-stage cross-cutting veins.

Primary muscovite from the Northport monzogranite is similar in composition to plutonic muscovite reported by Miller et al. (1981) (Table 2). They deviate considerably from the composition of ideal muscovite due to the presence of excess Ti, Fe, Mg, Mn and Si, and a deficiency in Al (Fig. 2). Vacancies in A-site occupancy are greater than those reported by Miller et al. (1981) and, although the analytical scheme did not include Ba, work by Miller et al. (1981) suggests that Ba is present in negligible quantities [<0.01 per formula unit (pfu)] in primary muscovites. Calculated octahedral site occupancies for $\sum$ Fe, Mg, Mn and Ti both exhibit positive correlations with the magnitude in deviation of Al^{VI} from ideal muscovite (Fig. 2B, 2C). However, the Al^{VI} deficiency exceeds that which can be accounted for by a Tschermack exchange alone (Fig. 2A). Thus a linear combination of Si$_{+1}$Fm$_{+1}$Al$^{IV}_{-1}$ Al$^{VI}_{-1}$ and the Fm$_{+1}$Ti$_{+1}$ Al$^{VI}_{-2}$ exchange component may account for the presence of Ti and the additional Al^{VI} deficiency in these muscovites. Plotted in Figure 2D are muscovite compositions adjusted for the effects of Fm$_{+1}$Ti$_{+1}$ Al$^{VI}_{-2}$ by reduction of the $\sum$ Fe, Mg, Mn by a quantity equal to the abundance of Ti^{4+} (pfu) and reduction of the Al^{VI} deficiency by twice this quantity. The scatter exhibited by muscovites with the largest Al^{VI} deficiencies (cf. Fig. 2C) is significantly reduced. The adjusted compositions define an array consistent with operation of the Tschermack exchange. Displacement of this array above the line representing the Tschermack exchange, as well as a portion of the A-site vacancies, may reflect operation a dioctahedral–trioctahedral exchange Fm$_{+3}$ $\square_{-1}$Al$^{VI}_{-2}$.

2.5. Garnet

Two garnet populations are recognised from petrographic observations. Small (0·2–0·75 mm) early crystallising ('E-type') garnet, with very rare inclusions of opaque oxide (ilmenite?) or zircon/monazite, occurs almost exclusively as inclusions within plagioclase (Fig. 3A, 3B). E-type garnet is commonly euhedral, but may show rounding or scalloping of some faces. Late-crystallising ('L-type') garnet exhibits a larger range in size (0·15–6·0 mm) and occurs both in the main phase of the granite and in scarce cross-cutting pegmatite dykes. It occurs as euhedral crystals and anhedral grains and typically exhibits abundant irregular cracks (Fig. 3C, 3D). Alkali feldspar, muscovite, biotite and quartz are all found as inclusions. In the main phase of the granite, L-type garnet is commonly associated with biotite–muscovite aggregates. L-type garnet in pegmatite dykes typically forms larger crystals (0·75–6·0 mm) and is associated with muscovite and quartz.

The presence of two garnet populations is supported by subtle compositional differences (Table 3). Both garnet types

Table 2 Representative electron microprobe analyses of muscovite.

Sample:	43 Np25	84 Np17-1	29 Np17-1	53 Np17	100 Np69	69 Np102	131 Np105	154 Np108
SiO_2	46·92	46·71	47·09	47·78	46·92	45·11	43·90	47·00
TiO_2	0·41	0·18	0·11	1·06	0·64	1·89	2·71	1·80
Al_2O_3	32·31	36·33	35·78	32·78	34·68	31·63	28·64	31·40
FeO*	3·68	1·59	1·63	2·34	1·78	3·27	5·32	2·97
MnO	0·11	0·03	0·01	0·06	0·04	0·07	0·26	0·02
MgO	1·06	0·47	0·61	1·08	0·72	1·02	1·58	1·08
CaO	0·01	0·01	0·01	0·02	0·00	0·01	0·03	0·01
Na_2O	0·33	0·75	0·45	0·25	0·31	0·21	0·18	0·22
K_2O	10·42	9·38	8·97	10·79	9·91	10·51	10·05	10·73
Total	95·25	95·45	94·66	96·16	95·00	93·72	92·67	95·23

Notes: 43 muscovite associated with L-type garnet; 84 and 29 muscovite associated with L-type garnet in granite pegmatite; 53, 100, 69, 131 and 154 muscovite associated with biotite. See Hogan (1984) for details of the analytical scheme.

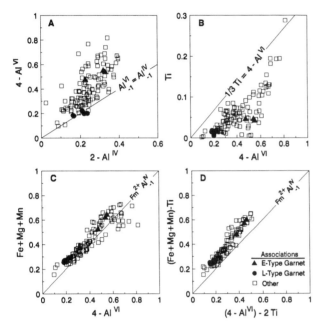

Figure 2 Composition of muscovite from the Northport monzogranite, Maine. Lines represent expected compositional variation from operation of exchange mechanisms discussed in the text. Concentrations are pfu. Compositions of muscovite associated with an E- and L-type garnet are indicated by a closed triangle and closed circle, respectively.

are predominantly almandine–spessartine solid solutions (Fig. 4). However, E-type garnet on average has higher almandine, pyrope and grossular contents than L-type garnet (Table 3). Compositional profiles of E-type garnets are relatively uniform. A narrow zone of increasing Mn and decreasing Fe (i.e. reverse zonation) is present along the rims (Fig. 5). L-type garnet displays larger compositional variations and more complex zoning profiles than E-type garnet. Both reverse or normally zoned (i.e. decreasing Mn and increasing Fe from core to rim) crystals occur (Fig. 6). In addition, a change from reverse to normal zonation has been documented within individual L-type garnet crystals (Fig. 6). Similar compositional zoning has been reported for igneous garnet elsewhere (e.g. Allan & Clarke 1981; Manning 1983).

2.6. Other minerals

Accessory minerals include apatite, zircon, monazite, scarce tourmaline and rare fluorite. Opaque minerals are present in extremely low modal abundance and include extremely rare ilmentite, of which two grains could be analysed yielding composition of $(Fe^{2+}_{0.90}Mn_{0.10})$ $(Fe^{3+}_{0.0}Ti_{1.03})O_3$ and $(Fe^{2+}_{0.71}Mn_{0.29})$ $(Fe^{3+}_{0.30}Ti_{0.97})O$, as well as pyrrhotite, pyrite, molybdenite and possible graphite. Chlorite, sericite, epidote, clinozoisite, titanite, rutile, opaque oxide and minor carbonate formed during subsolidus alteration.

3. Crystallisation sequence

The sequence in which minerals crystallise is sensitive to variations in P, T, f_{H_2O}, f_{O_2}, and bulk composition. Comparison

Table 3 Representative electron microprobe analyses of garnet.

	E-type garnet				L-type garnet			
Sample:	Core 123 Np108	Rim 125 Np108	Core 67 Np27	Rim 63 Np27	Core 84 Np69	Rim 85 Np69	Core 37 Np17-1	Rim 40 Np17-1
SiO_2	37·46	36·98	37·14	36·92	36·50	36·86	35·29	36·43
TiO_2	0·15	0·12	0·17	0·32	0·14	0·07	0·22	0·12
Al_2O_3	20·14	20·36	20·70	20·48	20·43	20·65	20·68	20·04
FeO*	30·39	30·18	30·33	30·64	25·52	28·14	19·73	28·32
MnO	9·32	9·56	9·47	9·80	15·99	12·75	22·79	12·79
MgO	1·59	1·64	1·55	1·36	0·91	0·92	0·51	1·17
CaO	1·43	1·45	0·83	1·09	0·51	0·53	0·32	0·48
Na_2O	0·04	0·00	0·01	0·00	0·02	0·04	0·02	0·00
K_2O	0·03	0·04	0·03	0·04	0·00	0·05	0·03	0·03
Total	100·55	100·33	100·23	100·65	100·02	100·01	99·59	99·38
Al	68·30	67·60	69·33	69·03	58·01	64·88	44·71	64·40
Py	6·37	6·55	6·31	5·46	3·69	3·78	2·06	4·74
Sp	21·21	21·69	21·92	22·36	36·81	29·77	52·30	29·46
Gr	4·12	4·16	2·43	3·15	1·49	1·57	0·93	1·40

Notes: Al, almandine; Py, pyrope; Sp, spessartine; Gr, grossular; in mol%. See Hogan (1984) for details of the analytical scheme.

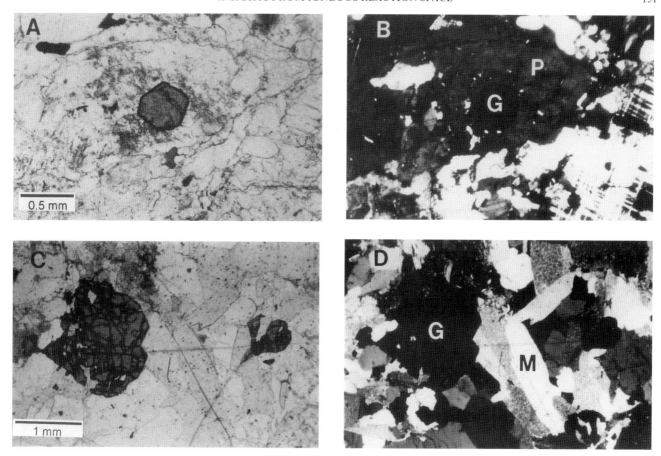

Figure 3 Photomicrographs of typical garnet habits in the Northport monzogranite, Maine. (A) Plane polarised light and (B) cross-polarised light of an early crystallising 'E-type' garnet (G) armoured in plagioclase (P). (C) Plane polarised light and (D) cross-polarised light of a late crystallising 'L-type' garnet (G) associated with primary muscovite (M).

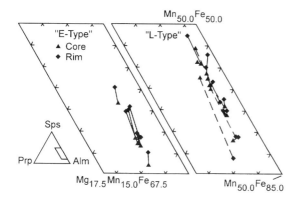

Figure 4 Core and rim compositions of E- and L-type garnets plotted in the spessartine–pyrope–almandine ternary system. Solid lines connect core and rim compositions of reversely zoned garnets; broken lines connect core and rim compositions of normally zoned garnets.

of observed crystallisation sequences with those predicted by experimental studies on similar bulk compositions can yield insights into the variation of these parameters during crystallisation (e.g. Naney 1983). However, petrographic criteria commonly used to interpret the order in which minerals *initiated* crystallisation (e.g. relative sizes, inclusion relationships) are equivocal, and may simply define the order in which minerals *ceased* crystallisation (Bowen 1912). This limitation is overcome through the use of compositional zoning and reaction relationships to recognise crystallisation events, which can be correlated among thin sections, to determine crystallisation sequences (Wiebe, 1968; Flood & Vernon 1988).

Peritectic reactions, in conjunction with mineral inclusion relationships, have been used to define the crystallisation

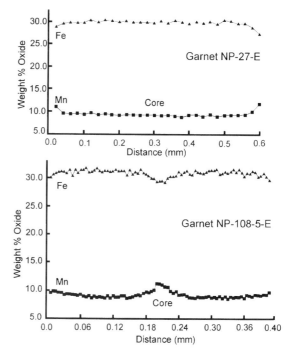

Figure 5 Compositional profile of two E-type garnets from the Northport monzogranite, Maine.

sequence for the Northport monzogranite (Fig. 7). Plagioclase glomerocrysts, a distinctive texture of this granite, are interpreted to have formed as a result of plagioclase resorption along the peritectic portion of the two-feldspar boundary line and were discussed in detail by Hogan (1993). Evidence for

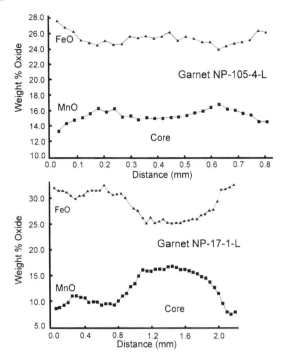

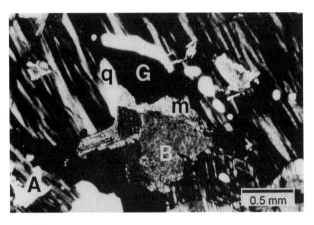

Figure 6 Compositional profile of two L-type garnets from the Northport monzogranite, Maine.

Figure 7 Cross-polarised photomicrographs of (A) anhedral partially resorbed E-type garnet (G) surrounded by quartz (q), muscovite (m) and biotite (B) included in alkali feldspar and (B) partially resorbed primary muscovite (M) embayed by quartz (q) included in alkali feldspar.

the co-crystallisation of plagioclase and alkali feldspar have not been observed. Thus this reaction readily subdivides the crystallisation history into two distinct periods: an early period of crystallisation when plagioclase was precipitating from the

melt and a late period of crystallisation when alkali feldspar was precipitating from the melt.

E-type garnet and plagioclase are both early crystallising phases, with E-type garnets occurring almost exclusively as armoured inclusions in the cores of partially resorbed plagioclase (Fig. 3). Anhedral E-type garnet fragments, either re-exposed during plagioclase resorption or initially too large to have been entirely consumed, can exhibit reaction relationships with biotite, muscovite and quartz and occur as inclusions in alkali feldspar (Fig. 7A). The reaction terminating E-type garnet crystallisation occurred before plagioclase resorption. Petrographic evidence for the co-crystallisation of E-type garnet with biotite and muscovite has not been observed, whereas L-type garnet, biotite and muscovite commonly occur in close spatial association in the main phase of the granite. The common occurrence of isolated E-type garnet inclusions within plagioclase, without inclusions of either biotite or muscovite, argues against the prolific crystallisation of garnet–biotite–muscovite early in the crystallisation history of the Northport pluton. This may reflect an initially low a_{H_2O} for this magma. In contrast, the topologies of both reaction spaces constructed for the Northport pluton, as shown later in the paper, allow reaction paths for which garnet, biotite and muscovite can co-precipitate throughout the entire crystallisation history. However, the crystallisation sequence observed for the Northport pluton, one with distinct early and late periods of garnet growth, is typical of many, if not most, garnet–muscovite–biotite granites.

Most of the crystallisation history appears to have been dominated by the co-crystallisation of biotite and muscovite. Both minerals are found along the margins of plagioclase glomerocrysts. Muscovite inclusions in alkali feldspar exhibit anhedral embayed margins that are typically rimmed by quartz (Fig. 7B). Arguably, the replacement of muscovite by alkali feldspar and quartz may have taken place in the subsolidus. However, recrystallisation of primary muscovite as well as feldspar in the Northport monzogranite, and other granitoids, typically forms finer grained secondary muscovite rather than alkali feldspar and quartz. Hogan (1984) suggested that this texture may represent the reaction of muscovite with the melt to form alkali feldspar + quartz as a result of decrease in μHK_{-1} in the melt just before the crystallisation of alkali feldspar (e.g. see Burt 1976, figs 3,4). If muscovite decomposition did occur in the presence of melt as suggested by Hogan (1984), then this reaction was of limited significance as primary muscovite later co-crystallised, in the presence of alkali feldspar and quartz, with L-type garnet and biotite in the main phase of the granite and with L-type garnet in granite pegmatite dykes.

L-type garnet crystallisation post-dates the resorption of plagioclase and the initiation of alkali feldspar crystallisation. In the main phase of the granite, L-type garnets exhibit sharp contacts and mutual inclusion relationships with biotite, muscovite and alkali feldspar (e.g. Fig. 3). However, biotite is extremely rare to noticeably absent from L-type garnet–muscovite granite pegmatite dykes. Thus the peritectic reaction eliminating biotite occurred after the initiation of L-type garnet crystallisation and before complete solidification of the magma.

4. Development of a general reaction space for garnet–muscovite–biotite granite

The crystallisation history of the garnet–muscovite–biotite Northport monzogranite is treated in two parts: (1) an early period in which alkali feldspar is absent from the crystallising assemblage; and (2) a later period in which alkali feldspar is part of the crystallising assemblage. A reaction space for each period of crystallisation was constructed following the method and examples described in Thompson (1982a,b), to which the reader is referred to as only a brief description of the method and the assumptions involved follows. Phase components, exchange components and system components used to describe this system are listed in Tables 4 and 5. The abbreviations used are those of Kretz (1983). Minerals and components present in extremely low to negligible abundance and concentrations (e.g. ilmenite, ZrO₂) are excluded. Complex mineral compositions are represented by an additive component and the dominant exchange components (Table 5). MgO and MgFe₋₁ are excluded because the pyrope contents of E- and L-type garnets are almost homogeneous and similar, and because Mg^{2+}, Mn^{2+} or Fe^{2+} are not discriminated by the dominant exchange components operating in the sheet silicates (cf. Figs 1, 2). Similarly, NaK₋₁ is excluded as this exchange is significant only among feldspars and melt and is a trivial component in the sheet silicates. Both of these assumptions will not influence the main conclusions of this paper as projections from exchange components, which alter the composition of phases without affecting the modal abundances, preserve the general form of the reaction space (Thompson 1982a,b).

MnO and MnFe₋₁ are included in the analysis because of substantial variations in MnFe₋₁ in L-type garnet and because of the presumed importance of MnO to garnet stability in granitic magmas (e.g. Hall 1965; Miller & Stoddard 1981; Zen 1988). The MnFe₋₁ exchange operates between garnet and melt and is considered insignificant in other phases (e.g. sheet silicates). Operation of the FeCa₋₁ exchange in garnet is included in the analysis as this component is sensitive to pressure. Measured oxide abundances of whole rock samples of Northport monzogranite (Hogan 1993), with a nominal amount of H₂O (≈ 0.5 wt%) added, are normalised to 24 oxygens and used to represent the initial melt composition (Table 5). Although the overall homogeneity of the granite as represented by the analysed samples (Hogan 1984) suggests crystallising minerals have not been removed from the system, mineral compositional heterogeneities (Figs 1, 2, 4–6) indicate that equilibrium with the residual melt has not been maintained during crystallisation. Thus the initial melt composition has been adjusted for the effects of fractional crystallisation by the removal of 34% plagioclase (An₂₅), 0.25% E-type garnet, 26.5% quartz, 7.0% biotite and 1.5% muscovite and the resulting melt composition, normalised to 24 oxygens, is used in the second reaction space (Table 6). The Northport pluton is interpreted to have crystallised entirely under H₂O-undersaturated conditions, as even the presence of pegmatite textures is not unequivocal evidence for vapour saturation (London 1992). Therefore, H₂O is included in the calculations as a system component, but not as a phase component. Thus the stoichiometry of reactions involving melt are sensitive to the H₂O content of the melt as biotite and muscovite are the only two hydrous minerals present. The importance of this will be discussed later during the examination of the second reaction space. All exchange components can operate on the melt.

The early period of crystallisation consists of 23 phase components (C_p) and nine system components (C_s). The number of linearly independent reactions (N_r) required to fully described this system is 14, as given by $N_r = C_p - C_s$ (Thompson 1982b). Twelve of these reactions are exchange reactions (N_{ex}). The other two reactions are net transfer reactions (N_t). Following Thompson (1982b), the specific net transfer reactions are solved by the algebraic manipulation of a matrix constructed from the composition of the phase components written in terms of the system components. The two linearly independent net transfer reactions for the early period of

Table 4 Matrix for the early reaction space.

Phase components	System components								
	SiO₂	AlO₁.₅	KO₀.₅	NaO₀.₅	CaO	FeO	MnO	TiO₂	H₂O
Quartz	1	0	0	0	0	0	0	0	0
Melt	9.44	2.28	0.67	0.84	0.25	0.32	0.01	0.02	0.67
Annite	3	1	1	0	0	3	0	0	1
NaSiCa₋₁Al₋₁	1	−1	0	1	−1	0	0	0	0
FeCa₋₁	0	0	0	0	−1	1	0	0	0
Almandine	3	2	0	0	0	3	0	0	0
MnFe₋₁	0	0	0	0	0	−1	1	0	0
Al₂Fe₋₁Ti₋₁	0	2	0	0	0	−1	0	−1	0
Muscovite	3	3	1	0	0	0	0	0	1
Anorthite	2	2	0	0	1	0	0	0	0
Al₂Fe₋₁Si₋₁	−1	2	0	0	0	−1	0	0	0

Table 5 Matrix for the late reaction space.

Phase components	System components								
	SiO₂	AlO₁.₅	KO₀.₅	NaO₀.₅	CaO	FeO	MnO	TiO₂	H₂O
Quartz	1	0	0	0	0	0	0	0	0
Almandine	3	2	0	0	0	3	0	0	0
Orthoclase	3	1	1	0	0	0	0	0	0
NaSiCa₋₁Al₋₁	1	−1	0	1	−1	0	0	0	0
FeCa₋₁	0	0	0	0	−1	1	0	0	0
Annite	3	1	1	0	0	3	0	0	1
MnFe₋₁	0	0	0	0	0	−1	1	0	0
Al₂Fe₋₁Ti₋₁	0	2	0	0	0	−1	0	−1	0
Muscovite	3	3	1	0	0	0	0	0	1
Anorthite	2	2	0	0	1	0	0	0	0
Al₂Fe₋₁Si₋₁	−1	2	0	0	0	−1	0	0	0
Melt	9.76	2.27	0.71	0.55	0.01	0.13	0.02	0	0.54

crystallisation in the Northport monzogranite are:

$$6.27 \text{ Melt} + 0.34 \text{ Alm} + 0.013 \text{ Al}_2\text{Fe}_{-1}\text{Ti}_{-1}$$

$$= \text{An} + 0.527 \text{ NaSiCa}_{-1}\text{Al}_{-1} + 0.299 \text{ Ann} + 0.121 \text{ Ms}$$

$$+ 3.152 \text{ Qtz} + 0.317 \text{ FeCa}_{-1} + 0.06 \text{ MnFe}_{-1} \qquad (R1)$$

and

$$1.254 \text{ Melt} + 1.367 \text{ FeCa}_{-1} + 0.025 \text{ Al}_2\text{Fe}_{-1}\text{Ti}_{-1}$$

$$= 1.053 \text{ NaSiCa}_{-1}\text{Al}_{-1} + 0.599 \text{ Ann} + 0.241 \text{ Ms} + \text{Al}_2\text{Fe}_{-1}\text{Si}_{-1}$$

$$+ 8.304 \text{ Qtz} + 0.320 \text{ Alm} + 0.013 \text{ MnFe}_{-1} \qquad (R2)$$

The addition of a new phase component, alkali feldspar, requires three linearly independent net transfer reactions to completely describe the later period of crystallisation. These reactions are as follows:

$$\text{Melt} + 0.560 \text{ FeCa}_{-1} + 0.139 \text{ Ann}$$

$$= 0.55 \text{ NaSiCa}_{-1}\text{Al}_{-1} + 0.02 \text{ MnFe}_{-1}$$

$$+ 0.679 \text{ Ms} + 0.376 \text{ Alm} + 0.170 \text{ Or} + 5.952 \text{ Qtz} \qquad (R3)$$

$$\text{An} + \text{FeCa}_{-1} + 0.333 \text{ Ann} = 0.333 \text{ Ms} + 0.667 \text{ Alm} \qquad (R4)$$

and

$$2 \text{ Qtz} + 0.667 \text{ Ann} + \text{Al}_2\text{Fe}_{-1}\text{Si}_{-1} = 0.667 \text{ Ms} + 0.333 \text{ Alm}$$
$$\qquad (R5)$$

Reactions (R1) and (R2) serve as the basis vectors for the early reaction space and (R3), (R4) and (R5) are the basis vectors for the late reaction space as these reactions regulate the modal abundance and composition of mineral phases.

Following transformation of the reaction coefficients to *oxy-equivalents*, the physical appearance of the early and late reaction spaces is calculated following Thompson *et al.* (1982). Reaction progress ceases when a reactant, product or the exchange capacity of a phase component is exhausted. Thus the portion of reaction space that is physically accessible is determined by the modal abundance and exchange capacity of the phases present. Assumed modal abundances, representing some point in the crystallisation history, are used to calculate the bounding [phase] planes to the reaction spaces. Each plane represents the limit to which a reaction path can progress before exhausting either one of the reactants or one of the products involved. Reaction progress away from or towards a [phase] plane results in the precipitation or resorption of that phase, respectively. For example, as crystallisation proceeds melt is typically consumed and reaction progress is towards the [Melt] plane (i.e. the solidus). Varying the initial modal abundance of phases shifts the position of [Phase] planes with respect to the origin while preserving their orientation relative to the axes of the reaction space. The exchange capacity of minerals has not been considered as a limiting factor due to the large exchange capacity of the melt. However, as a result of fractional crystallisation, Ti abundances in the late crystallising melt closely approach zero, resulting in projection of the late reaction space from $\text{Al}_2\text{Fe}_{-1}\text{Ti}_{-1}$.

4.1. Description of reaction space for early crystallisation

The reaction space for the early period of crystallisation is bounded by four [Phase] planes that define a polygon (Fig. 8). The origin passes through the [Ms] and [Ann] planes as the mode was selected to focus on the resorption of E-type garnets. A construction line subdivides the reaction space into two domains: one in which reaction progresses from the origin towards the [Melt] plane results in the precipitation of garnet along with biotite and muscovite, and the other results in the resorption of garnet and precipitation of biotite and muscovite. The modal abundance of a phase (e.g. garnet) can

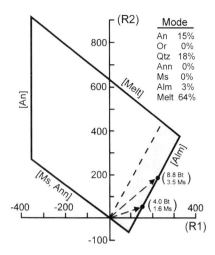

Figure 8 Reaction space polygon for modelling the early period of crystallisation of the Northport monzogranite, Maine. Short-dashed line represents the garnet construction line. The shaded area indicates the region for which garnet will begin to be resorbed by the melt for reaction paths emmanating from the origin. Two possible reaction paths are shown by long-dashed arrows, with the modal amount of biotite and muscovite produced shown in parentheses.

remain unchanged during crystallisation by reaction progress parallel to its [Phase] plane (i.e. along the construction line).

Based on the observed crystallisation sequence, reaction paths must conform to the following: (1) reaction progress is towards the [Melt] and [Alm] planes and away from the [An], [Ann] and [Ms] planes; (2) the reaction path progresses a limited distance along the [Alm] plane but does not reach the [Melt] plane. Several possible reaction paths, along with the modal proportion of muscovite and biotite produced, are shown in Figure 8. One possible reaction path from the origin to the [Alm] plane is $1.5*\text{R1} + 0.577*\text{R2}$ or

$$17.47 \text{ Melt} + 3 \text{ Alm} + 0.02 \text{ FeCa}_{-1} + 0.0016 \text{ Al}_2\text{Fe}_{-1}\text{Ti}_{-1}$$

$$= 6.3 \text{ An} + 0.607 \text{ NaSiCa}_{-1}\text{Al}_{-1} + 8.1 \text{ Qtz} + 4.2 \text{ Ann}$$

$$+ 1.7 \text{ Ms} + 1.9 \text{ Al}_2\text{Fe}_{-1}\text{Si}_{-1} + 0.006 \text{ MnFe}_{-1} \qquad (R6)$$

Similarly, the reaction path describing the [Alm] plane is $\text{R1} + 1.9231*\text{R2}$ or

$$27 \text{ Melt} + 0.7 \text{ FeCa}_{-1} + 0.27 \text{ Al}_2\text{Fe}_{-1}\text{Ti}_{-1}$$

$$= 4.2 \text{ An} + 1 \text{ NaSiCa}_{-1}\text{Al}_{-1} + 13.9 \text{ Qtz} + 6.5 \text{ Bi}$$

$$+ 2.7 \text{ Mu} + 0.63 \text{ Al}_2\text{Fe}_{-1}\text{Si}_{-1} + 0.01 \text{ MnFe}_{-1} \qquad (R7)$$

Forward reaction progress along (R6) represents a peritectic reaction in which garnet is resorbed as the liquid precipitates plagioclase, quartz, biotite and muscovite. Reactions of this form, where garnet is replaced by anorthite and annite, are sensitive to pressure, with garnet occurring on the high pressure side (e.g. Hoisch 1990). Experimental P–T–$x\text{H}_2\text{O}$ relationships of peraluminous granite show almandine-rich garnet to be a near-liquidus phase at H_2O-undersaturated conditions and $P \geqslant 4$–5 kbar (Clemens & Wall 1981). However, at 2·0 kbar the stability field for almandine-rich garnet shrinks significantly to near-solidus conditions. Resorption of E-type garnet in felsic igneous rocks is consistent with a decrease in pressure (Clemens & Wall 1988) and may mark the rise of this magma through the crust to its emplacement level.

Reactions (R6) and (R7) describe possible variations in mineral compositions, as well as modes, during crystallisation. For example, application of the $\text{NaSiCa}_{-1}\text{Al}_{-1}$ exchange entirely to anorthite produced by Reaction (R6) yields a plagioclase of composition An_{23}, which is in good agreement with measured plagioclase compositions from the Northport pluton. The calculation of other mineral compositions is complicated by the fact that exchange components can simultaneously operate on several phases, including melt. However, the general form of Reactions (R6) and (R7) predicts that sheet silicates produced during this reaction should be more titaniferous and more aluminous as a result of operation of the $Al_{+2}Ti_{-1}Fm_{-1}$ and $Al^{IV}Al^{VI}Si_{-1}Fm_{-1}$ substitutions. Biotite from the Northport pluton is titaniferous and aluminous (cf. Fig. 1). In particular, early crystallising muscovite exhibits higher titanium concentrations than late crystallising muscovite. However, early crystallising muscovite is less aluminous than pure muscovite and becomes more aluminous as crystallisation proceeds (cf. Fig. 2). Similar compositional trends have been reported for individually zoned muscovite and biotite crystals from peraluminous granites of the South Mountain Batholith (Ding *et al.*, unpublished data).

4.2. Description of reaction space for late crystallisation

Reactions (R3), (R4) and (R5) serve as the basis vectors for the reaction space used to model the later stages of crystallisation. This reaction space has a form similar to a trigonal prism (Fig. 9). The top of the prism, the [An] plane, parallels Reactions (R3) and (R5). The three sides to the prism are: (1) the [Melt] plane which parallels Reactions (R4) and (R5); (2) the [Ms] plane in the back; and (3) the [Ann] plane in the front; the last two intersect all three axes. The [Alm] plane forms the sloping bottom to the prism.

The portion of the reaction volume relevant to the crystallisation of garnet, biotite and muscovite from peraluminous granitic melts has been enlarged (Fig. 9). The [Melt]

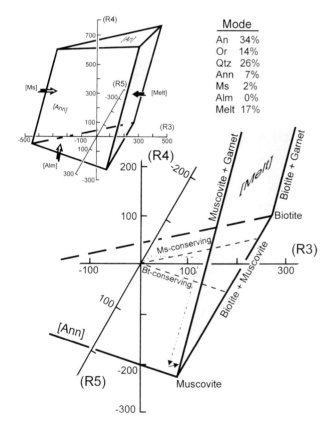

Figure 9 Reaction space polyhedron (upper left) and an enlargement of a portion of this polyhedron (lower right) for modelling the late period of crystallisation of the Northport monzogranite, Maine. The biotite and muscovite construction lines are shown as long dashed lines. A possible reaction path for the final stages of crystallisation is shown as a dotted line where the reaction path is on the [Alm] plane, a short dashed and dotted line where the path is within the reaction space volume and a solid line where the reaction path is on the [Ann] plane, finishing at the muscovite–garnet solidus.

plane is the solidus for biotite–muscovite–garnet granites. Intersection of the various [Phase] planes with the [Melt] plane defines lines and points that form the solidi for the following granite types: (1) biotite–muscovite granite [Alm]–[Melt]; (2) muscovite–garnet granite [Ann]–[Melt]; (3) biotite–garnet granites [Ms]–[Melt] planes; (4) muscovite granite [Ann]–[Alm]–[Melt]; and (5) biotite granite [Ms]–[Alm]–[Melt]. Reaction paths leading to the formation of a particular granite assemblage become more generalised as the area of the solidus for that assemblage increases. For this reaction space there is a greater probability for reaction paths to terminate on the [Melt] plane than on an edge or corner. This is consistent with the majority of the Northport pluton having formed as biotite–muscovite–garnet granite, with minor muscovite–garnet granite restricted to rare late dykes.

The driving forces behind the introduction of late crystallising garnet and the elimination of biotite from the crystallising assemblage are elucidated through consideration of the likely reaction path from the origin to the appropriate solidus (Fig. 9). At this point in the crystallisation history the reaction path is advancing along the [Alm] plane towards the [Melt] plane. E-type garnet has already been consumed, and biotite and muscovite are part of the stable crystallising assemblage. The reintroduction of garnet to the crystallising assemblage requires movement of the reaction path of the [Alm] plane into the interior of the reaction space volume. Reaction progress in this direction is regulated by Reaction (R4). Linear combinations involving slightly negative ($\leqslant -0.567$) to positive reaction coefficients for Reaction (R4) and positive reaction coefficients for Reaction (R3) will result in the precipitation of L-type garnet during crystallisation. Negative reaction progress along Reaction (R4) allows for continued crystallisation of plagioclase, whereas positive reaction progress predicts the resorption of plagioclase, a feature that has been previously documented for this pluton (Hogan 1993).

Reaction (R4) is more commonly written as

$$3\ CaAl_2Si_2O_8 + KFe_3AlSi_3O_{10}(OH)_2$$
$$= Fe_3Al_2Si_3O_{12} + Ca_3Al_2Si_3O_{12} + KAl_2AlSi_3O_{10}(OH)_2$$

and has been utilised as a geobarometer for quartz, muscovite, biotite, plagioclase, and garnet bearing assemblages (e.g. Hoisch 1990). Back-reaction of plagioclase and biotite, the low-pressure high-temperature assemblage, to form L-type garnet and muscovite, is interpreted to reflect decreasing temperature, rather than a sudden increase in pressure, as conditions at the emplacement level are likely to have been isobaric. More importantly, $MnFe_{-1}$ is not regulated by Reaction (R4), indicating that higher Mn concentrations are not necessary for the crystallisation of late garnet from granitic melts. However, reaction progress towards the solidus (i.e. [Melt] plane) requires a positive reaction along Reaction (R3), which also results in the precipitation of garnet and muscovite at the expense of biotite. Reaction (R3) regulates the $MnFe_{-1}$ exchange component; reaction progress towards [Melt] and along the [Alm] plane will increase the $MnFe_{-1}$ capacity of the residual melt, whereas reaction progress towards [Melt] and off the [Alm] plane will result in this exchange component being partitioned into late crystallising garnet. The higher average spessartine contents of L-type garnet compared with E-type garnet more than likely reflects this increase in the $MnFe_{-1}$ exchange capacity of later melt fractions due to fractional crystallisation. At H_2O-undersaturated conditions, the stoichiometry of Reaction (R3) is sensitive to the H_2O content of the melt as muscovite and biotite are the only two hydrous minerals. For the melt bulk composition used in these calculations, H_2O contents greater than ≈ 1.5 wt% result in almandine garnet and annite changing positions as either product or reactant in Reaction (R3). Thus at H_2O-undersaturated conditions, increasing the H_2O content of the melt eliminates almandine garnet from the crystallising assemblage and favours the formation of biotite–muscovite granite. The composition of garnet participating in such a reaction would be predicted to become increasingly more spessartine-rich as the almandine component is consumed in making biotite.

Reaction paths that eliminate biotite and give rise to muscovite–garnet granites must finish along the intersection of the [Ann] and [Melt] planes. The efficient elimination of biotite, without excessive increases in the modal abundance of garnet, is best accomplished by reaction paths involving large positive coefficients for Reaction (R5). Again, $MnFe_{-1}$ is not regulated by Reaction (R5), suggesting that high Mn contents are of secondary significance to biotite and garnet stability. In contrast, Reaction (R5) indicates that biotite and quartz are replaced by muscovite and garnet in response to increasing $Al_{+2}Fe_{-1}Si_{-1}$ activity in the melt. Thus the formation of muscovite–garnet granite late in the crystallisation history can be attributed to increasing melt aluminosity as a result of fractional crystallisation (e.g. Zen 1986) rather than to higher Mn activities. Correspondingly, the composition of muscovite forming progressively later in the crystallisation history becomes increasingly more aluminous and approaches the composition of ideal muscovite (cf. Fig. 2).

Compositional variation exhibited by 'L-type' garnet corresponds to growth along different segments of the reaction path. L-type garnet growth initiates with reaction progress off the [Alm] plane and towards the [Melt] plane. This segment of the reaction path is described by a linear combination of Reactions (R3) and (R4), which predicts an increase in both the grossular and spessartine components of L-type garnet towards the rims, as $MnFe_{-1}$ and $CaFe_{-1}$ are regulated by Reactions (R3) and (R4), respectively. The grossular contents of L-type garnet rims are consistently higher than the cores (Table 3). The spessartine content of L-type garnet may initially increase towards the rim, as predicted, but then decreases (e.g. Fig. 6). More typically, the spessartine contents of L-type garnet decrease towards their rim. This change in compositional zonation pattern marks a change in the reaction path. The resorption of biotite and co-precipitation of garnet and muscovite requires a sharp redirection of the reaction path towards the [Ann] plane. Although this segment of the reaction path can be described by a linear combination of all three net transfer reactions, the coefficient for Reaction (R5), which produces almandine garnet and muscovite at the expense of biotite, will be significantly greater than the coefficient for Reaction (R3), which regulates the spessartine content of garnet. Thus the commonly observed increase in almandine content of L-type garnet towards their rim corresponds with biotite breakdown and increasing melt aluminosity as described by reaction progress along Reaction (R5). In addition, the higher Al contents observed for biotite and muscovite associated with L-type garnet (e.g. Figs 1, 2) are predicted by reaction progress towards the [Ann] plane as well.

5. Conclusions

A reaction space approach to the crystallisation of garnet–muscovite–biotite granites provides new insight into the paragenesis of garnet during the solidification of these melts. Minerals can alternate as reactants or products in different reactions, accounting for textures indicating multiple periods of crystallisation separated by resorption, with early and late crystallising garnet being just one example. There is no one unique reaction that produces or consumes aluminous minerals during crystallisation. An infinite number of reaction paths, each described by a different linear combination of the net transfer reactions that serve as the basis vectors to the reaction space, results in the precipitation of garnet late in the crystallisation history of such granites. However, the topology of this reaction space indicates that the late crystallisation of garnet at the expense of biotite, and thus the formation of garnet–muscovite granites, is in response to increasing melt aluminosity during fractional crystallisation rather than being entirely attributed to the increasing Mn content of the residual melt as previously proposed. This conclusion is in agreement with the experimental results of Clemens and Wall (1981), which define a stability field for almandine-rich garnet ($Alm_{79}Sp_{10}Py_9Gr_2$) in peraluminous granite melt to pressures as low as 1 kbar. Furthermore, late crystallising garnet in the Northport pluton, as well as late crystallising garnet elsewhere (cf. Miller & Stoddard 1981a; Manning 1983), can have rims with *lower* Mn and *higher* Fe concentrations than their cores. Although such compositional profiles conflict with models requiring increasing Mn contents to promote garnet growth, the topology of this reaction space predicts that late crystallising garnet should become increasingly more almandine-rich, as observed, along the segment of the reaction path where biotite is rapidly being consumed. Complex compositional profiles, where a component such as $MnFe_{-1}$ initially increases and then decreases, have been shown to reflect growth along different segments of the reaction path. Thus, to a large extent mineral compositions during crystallisation are regulated by the reactions producing them and will vary as the stoichiometry of the reactions producing them varies.

Traditional petrological tools, such as Schreinmaker's analysis or AFM diagrams, determine reaction relationships applicable only to significantly reduced subsets of the crystallising assemblage. In contrast, the reaction space method of Thompson (1982a,b) determines the phase equilibria that regulate mineral assemblages and compositions for systems with a complexity of components and phases that more closely approaches natural crystallising assemblages. The topology of these equilibria reveal a previously unrecognised connectivity among phases and components that leads to new insights concerning interpretations of (1) textural observations, (2) compositional profiles and (3) forces dictating changes in the crystallising assemblage during the solidification of granitic melts.

6. Acknowledgements

I thank J. Laird for introducing me to the reaction space approach. I also acknowledge the helpful comments of M. P. Dickerson, D. R. Wones, D. A. Hewitt and A. K. Sinha in the initial stages of this study. Careful and thoughtful reviews by J. A. Speer and an anonymous reviewer are greatly appreciated. I thank M. Brown, P. A. Candela, and E. Zen for providing me with the opportunity to present this paper at the Third Hutton Symposium. Many thanks go to M. C. Gilbert for his insight, support and encouragement during completion of this work.

References

Abbott, R. N. Jr. 1981a. AFM liquidus projections for granitic magmas with special reference to hornblende, biotite, and garnet. CAN MINERAL **19**, 103–10.
Abbott, R. N. Jr 1981b. The role of manganese in the paragenesis of

magmatic garnet: an example from the Old Woman–Puite Range, California: a discussion. J GEOL **89**, 767–9.

Abbott, R. N. Jr 1985. Muscovite-bearing granites in the AFM liquidus projection. CAN MINERAL **23**, 553–61.

Abbott, R. N. & Clarke, D. B. 1979. Hypothetical liquidus relationships in the subsystem Al_2O_3–FeO–MgO projected from quartz, alkali feldspar, and plagioclase for $a(H_2O) \leqslant 1$. CAN MINERAL **17**, 549–60.

Allan, B. D. & Clarke, D. B. 1981. Occurrence and origin of garnets in the South Mountain Batholith, Nova Scotia. CAN MINERAL **19**, 19–24.

Bowen, N. L. 1912. The order of crystallization in igneous rocks. J GEOL **20**, 457–68.

Burt, D. M. 1976. Hydrolysis equilibria in the system K_2O–Al_2O_3–SiO_2–Cl_2O_{-1}: comments on topology. ECON GEOL **71**, 665–71.

Clemens, J. D. & Wall, V. J. 1981. Origin and crystallization of some peraluminous (S-type) granitic magmas. CAN MINERAL **19**, 111–31.

Clemens, J. D. & Wall, V. J. 1982. The role of manganese in the paragenesis of magmatic garnet: an example from the Old Woman–Puite Range, California: a discussion. J GEOL **89**, 339–341.

Clemens, J. D. & Wall, V. J. 1988. Controls on the mineralogy of S-type volcanic and plutonic rocks. LITHOS **21**, 53–66.

Eriksson, S. C., Hogan, J. P. & Williams, I. S. 1989. Ion microprobe resolution of age details in heterogeneous sources of plutonic rocks from a comagmatic province. GEOL SOC AM ABSTR PROG **21**, 361.

Flood, R. H. & Vernon, R. H. 1988. Microstructural evidence of orders of crystallization in granitoid rocks. LITHOS **21**, 237–45.

Hall, A. 1965. The origin of accessory garnet in the Donegal Granite. MINERAL MAG **35**, 628–33.

Hibbard, M. J. 1979. Myrmekite as a marker between preaqueous and postaqueous phase saturation in granitic systems. GEOL SOC AM BULL **90**, 1047–62.

Hibbard, M. J. 1981. The magma mixing origin of mantled feldspars. CONTRIB MINERAL PETROL **76**, 158–70.

Hogan, J. P. 1984. *Petrology of the Northport pluton Maine: a garnet-bearing muscovite–biotite granite.* Unpublished Masters Thesis, Virginia Polytechnic Institute and State University.

Hogan, J. P. 1993. Monomineralic glomerocrysts: textural evidence for mineral resorption during crystallization of igneous rocks. J GEOL **101**, 531–40.

Hogan, J. P. & Sinha, A. K. 1989. Compositional variation of plutonism in the coastal Maine magmatic province: mode of origin and tectonic setting. *In* Tucker, R. D. & Marvinney, R. G. (eds) *Igneous and metamorphic geology: studies in Maine geology* **4**, 1–33. Augusta, Maine: Maine Geological Survey, Department of Conservation.

Hogan, J. P. & Wones, D. R. 1984. Peraluminous plutonic rocks of the Penobscot Block, eastern Maine. GEOL SOC AM ABST PROG **16**, 24.

Hoisch, T. D. 1990. Empirical calibration of six geobarometers for the mineral assemblage quartz + muscovite + biotite + plagioclase + garnet. CONTRIB MINERAL PETROL **104**, 225–34.

Kretz, R. 1983. Symbols for rock forming minerals. AM MINERAL **68**, 277–9.

London, D. 1992. The application of experimental petrology to the genesis and crystallization of granitic pegmatites. CAN MINERAL **30**, 499–540.

Manning, D. A. C. 1983. Chemical variation in garnets from aplites and pegmatites, peninsular Thailand. MINERAL MAG **47**, 353–8.

Miller, C. F. & Stoddard, E. F. 1981a. The role of manganese in the paragenesis of magmatic garnet: an example from the Old Woman–Puite Range, California. J GEOL **89**, 233–46.

Miller, C. F. & Stoddard, E. F. 1981b. The role of manganese in the paragenesis of magmatic garnet: an example from the Old Woman–Puite Range, California: a reply. J GEOL **89**, 770–2.

Miller, C. F. & Stoddard, E. F. 1982. The role of manganese in the paragenesis of magmatic garnet: an example from the Old Woman–Puite Range, California: a reply. J GEOL **90**, 341–3.

Miller, C. F., Stoddard, E. F., Bradfish, L. J., & Dollase, W. A. 1981. Composition of plutonic muscovite: genetic implications. CAN MINERAL **19**, 25–34.

Naney, M. T. 1983. Phase equilibria of rock-forming ferromagnesian silicates in granitic systems. AM J SCI **283**, 993–1033.

Novak, G. A. & Gibbs, G. U. 1971. The crystal chemistry of silicate garnets. AM MINERAL **56**, 791–825.

Speer, J. A. 1981. Petrology of cordierite- and almandine + cordierite-bearing biotite granitoid plutons of the southern Appalachian Piedmont, USA. CAN MINERAL **19**, 35–46.

Speer, J. A. & Becker, S. W. 1992. Evolution of magmatic and subsolidus AFM mineral assemblages in granitoid rocks: biotite, muscovite, and garnet in the Cuffytown Creek pluton, south Carolina. AM MINERAL **77**, 821–33.

Thompson, J. B. Jr 1982a. Composition space: an algebraic and geometric approach. *In* Ferry, J. M. (ed.) *Characterization of metamorphism through mineral equilibria*, Vol. 10, 1–31. Washington: Mineralogical Society of America.

Thompson, J. B. Jr. 1982b. Reaction space: an algebraic and geometric approach. *In* Ferry, J. M. (ed.) *Characterization of metamorphism through mineral equilibria*, Vol. 10, 33–53. Washington: Mineralogical Society of America.

Thompson, J. B. Jr, Laird, J. & Thompson, A. B. 1982. Reactions in amphibolite, greenschist and blueschist. J PETROL **23**, 1–27.

Wiebe, R. A. 1968. Plagioclase stratigraphy: a record of magmatic conditions and events in a granitic stock. AM J SCI **266**, 690–703.

Zen, E-an. 1986. Aluminium enrichment in silicate melts by fractional crystallization: some mineralogical and petrographic constraints. J PETROL **27**, 1095–117.

Zen, E-an. 1988. Phase relations of peraluminous granitic rocks and their petrogenetic implications. ANNU REV EARTH PLANET SCI **16**, 21–51.

JOHN P. HOGAN, University of Oklahoma, School of Geology and Geophysics, Rm 818 Sarkey's Energy Center Building, 100 East Boyd Street, Norman, OK 73019-0628, U.S.A. E-mail: jhogan@uoknor.edu

Transactions of the Royal Society of Edinburgh: Earth Sciences, **88**, 159–170, 1997

Compositional variation within granite suites of the Lachlan Fold Belt: its causes and implications for the physical state of granite magma

B. W. Chappell

ABSTRACT: Granites within suites share compositional properties that reflect features of their source rocks. Variation within suites results dominantly from crystal fractionation, either of restite crystals entrained from the source, or by the fractional crystallisation of precipitated crystals. At least in the Lachlan Fold Belt, the processes of magma mixing, assimilation or hydrothermal alteration were insignificant in producing the major compositional variations within suites. Fractional crystallisation produced the complete variation in only one significant group of rocks of that area, the relatively high temperature Boggy Plain Supersuite. Modelling of Sr, Ba and Rb variations in the I-type Glenbog and Moruya suites and the S-type Bullenbalong Suite shows that variation within those suites cannot be the result of fractional crystallisation, but can be readily accounted for by restite fractionation. Direct evidence for the dominance of restite fractionation includes the close chemical equivalence of some plutonic and volcanic rocks, the presence of plagioclase cores that were not derived from a mingled mafic component, and the occurrence of older cores in many zircon crystals. In the Lachlan Fold Belt, granite suites typically evolved through a protracted phase of restite fractionation, with a brief episode of fractional crystallisation sometimes evident in the most felsic rocks. Evolution of the S-type Koetong Suite passed at about 69% SiO_2 from a stage dominated by restite separation to one of fractional crystallisation. Other suites exist where felsic rocks evolved in the same way, but the more mafic rocks are absent. In terranes in which tonalitic rocks formed at high temperatures are more common, fractional crystallisation would be a more important process than was the case for the Lachlan Fold Belt.

KEY WORDS: assimilation, crystal fractionation, fractional crystallisation, granite suites, hybrids, magmas, magma mingling, magma mixing, partition coefficients, restites, restite fractionation.

In studying the granites in eastern Australia it has been helpful to subdivide them into suites (Hine *et al.* 1978; White 1995), where particular suites comprise one or more plutons that can be grouped together because they share common compositional (chemical, modal, mineralogical and isotopic) features. Because of the implied similarities in origin, suites often share similar textural features, which may be useful distinguishing criteria in the field. Individual suites are thought to correspond to specific source rock compositions so that their compositions reflect or image features of those source rocks (Chappell 1979). Within the eastern Australian fold belts, the primary subdivision of granites and related volcanic rocks has been into the I- and S-types (Chappell & White 1992) and the much less abundant A-types. Within those groups, different levels of compositional fine structure can be recognised, through supersuites that group together I- or S-type granites which share many compositional features, to suites representing the finest degree of subdivision that can be made.

This paper evaluates the compositional variation within granite suites. Such variation can potentially be produced in several ways, which in some instances may be combined together either at the same stage of evolution, or sequentially. The different potential causes of within-suite variation are: (1) magma mixing and/or mingling; (2) assimilation or contamination; (3) fractional crystallisation; (4) restite separation; (5) hydrothermal alteration; and (6) variation inherited from heterogeneous source rocks. The parts played by these processes, particularly 1, 3 and 4, will be discussed here,

drawing principally on data from the Lachlan Fold Belt (LFB) of southeastern Australia.

Although the question of why sequences of related granites vary in composition is a basic petrogenetic problem that must be addressed, there are broader implications. A knowledge of the precise processes that lead to the compositional variation in a particular suite would enhance our understanding of the physical nature and temperature of the magma. If the process causing the variation is understood, then it may be possible to look through the chemical compositions of the granites to those of the source rocks, and it will enhance our knowledge of the mechanisms by which the crust is internally fractionated. It is critical in deducing whether individual samples of granite represent the compositions of magmas, or mixtures of magmas, or accumulations of minerals, or of liquids resulting from the removal of cumulates. The H_2O content of a bulk magma and the H_2O budget during granite evolution both depend on whether the magma is completely or only partly molten. Certain processes are more likely to lead to the concentration of incompatible trace metals at the magmatic stage, enhancing the probability of the later formation of a significant mineral deposit. These and other implications of a better understanding of the fractionation processes that produced granite suites are discussed in Section 8.

1. Magma mixing and/or magma mingling

1.1. Magma, mixing and mingling
The general term magma has long been used to refer to a substance that is a fluid made up of solid and liquid matter.

The term has passed from pharmaceutical to chemical to petrological usage, where magma was introduced by Scrope in 1862 as being 'composed of crystalline or granular particles to which a certain mobility is given by an interstitial fluid' (Shand 1950: 4). The great debate between the magmatists and metasomatists of 50 years ago to some extent revolved around a misunderstanding of the term magma, shown by Read's (1948) statement 'The *igneous* rocks are those produced by consolidation of *magma*, which is a completely fluid rock substance', presumably meaning completely molten. Grout (1948), at the same symposium, gave what is perhaps the pre-eminent definition of magma, stating among other criteria that it is a natural fluid. This paper will show that many granites form from magmas that were initially completely liquid, but not all do and probably most do not, which has many important implications.

The term mixing refers to the complete combination of two components in such a way that they are completely homogenised. Mingling refers to the process in which such components retain some of their identity, as, for example, in the frequently cited example of basalt mingling with a granite magma to produce mafic enclaves. Despite the distinction, the term mixing has often been used in an inclusive way, and that will be generally be done here. Sparks and Marshall (1986) have proposed that the term hybrid is restricted to rocks formed by true magma mixing resulting in a completely blended homogeneous mixed magma, and they used the term co-mingled magma for magmas that are mingled physically while compositional heterogeneities are preserved.

1.2. General evidence for magma mixing/mingling
Magma mixing or mingling is a popular hypothesis, generally proposed in terms of blending between a crustal melt and the mafic material from the mantle that caused the melting. For example, Barbarin (1991) proposed that the mafic enclaves of the Sierra Nevada batholith represent mafic magma that was injected into and subsequently mingled with already hybrid magmas just before or during ascent. He considered that those hybrid magmas had earlier been produced by thorough mixing between mantle-derived mafic magma and crustal felsic magma whose production was induced by the mafic component. The development of local hybrids has frequently been used to infer the operation of such a process on a larger scale, although Pitcher (1993: 136) has pointed out that although such outcrop evidence provides important clues, 'mingling and mixing in the higher levels of the crust represent but second-order processes in the diversification of the granitic rocks'. Wall *et al.* (1987) stated that 'mixing is the classic cause of linear variation in major and trace element Harker Diagrams'. Likewise, regular variations in isotopic compositions have been interpreted as resulting from the mixing of relatively homogeneous end members (e.g. DePaolo 1981b).

The above lines of evidence for magma mixing have been considered more fully by Chappell (1996) and are summarised here in Sections 1.3, 1.4 and 2.

1.3. Evidence from the Bega Batholith
The Bega Batholith (Beams 1980; see also Chappell 1996) is the largest I-type granite complex (8940 km²) in the LFB and it shows many features that have been interpreted both there and elsewhere in support of a magma mixing/mingling model. These include the development of local hybrids, common mafic enclaves in the more mafic rocks, linear patterns of elements on variation diagrams and systematic changes in isotopic composition across the batholith. Referring to mafic inclusions, linear variations on Harker diagrams and the presence of pseudo-isochrons or isotope mixing lines, Wall

et al. (1987) stated that 'if a granitic pluton has these field, geochemical and isotopic features, it is reasonable to conclude that magma mixing has played a major role in its chemical evolution'. All of these features are exhibited by the components of the Bega Batholith and it therefore provides an opportunity to examine the patterns of chemical variation in detail, to test their compatibility with an origin by large-scale magma mixing or mingling.

The Bega Batholith extends meridionally through a distance of more than 300 km, with a maximum width of 75 km. There are some strong compositional asymmetries transverse to the axis of the batholith, with the most significant changes being decreases in Na and Sr, and increases in Ca and Sc towards the west (Beams 1980). These correlate with distinct changes in the isotopic composition of Sr, Nd and O (McCulloch *et al.* 1982; Chappell *et al.* 1990).

Harker diagrams for Na, Sr, Ca, Rb and P for various suites of the Bega Batholith are plotted in Figure 1. Strong correlations exist between those elements and SiO_2, individually of the kind that would be generated by magma mixing. There are, however, features of the chemical variation not compatible with mixing on a scale that would have produced the major variations in those suites. Specifically, in comparing pairs of suites it is found that any difference in compositions at either end of the range in composition is also seen at the other end, so that both the most mafic and felsic rocks show similar relative abundances of particular elements. Such observations are not consistent with magma mixing or mingling, as such a process would imply that felsic granite melts containing different relative amounts of a range of elements would have to know in some way that they were required to mix with mafic components showing the same relative abundances, and conversely. Hence, although there is some field evidence that can be interpreted as such a mixing process taking place on a small scale, it cannot have been responsible for the major compositional variations within the suites of the Bega Batholith. This observation is more widely applicable to the granites of southeastern Australia, both I- and S-type.

Although the information presented here can be used rigorously to constrain the role of magma mixing/mingling only in the case of the Bega Batholith, it is noted again that the batholith shows those features that are generally thought to be the result of such mixing. Also, for all other suites of the LFB in which linear variations are seen, including some S-type granites, similar relationships are observed. For magma mixing to be established as a credible model for the generation of compositional changes on a large scale, the type of data reported here and by Chappell (1996) is needed, but showing patterns consistent with such mixing.

1.4. S-type granites and a magma mixing hypothesis
The S-type granites of the LFB will always present difficulties for magma mixing models because, as pointed out by White and Chappell (1988), the mafic cordierite-bearing S-type suites such as Bullenbalong become more peraluminous as they become more mafic. These rocks cannot be the products of simple magma mixing as the mafic end-member would have to be strongly peraluminous and more complex mixing scenarios would be required. Gray (1984) proposed such a scenario, whereby the mixing or mingling of basalt and crust-derived granite melt produced a mafic magma which fractionated to yield the more felsic and less peraluminous rocks.

The Cooma Granodiorite chosen by Gray (1984) as the crustal melt end-member contains 39·9% normative quartz, only 12·3% normative albite and 6·7% normative corundum [average of sample C1 of White and Chappell (1988) and five

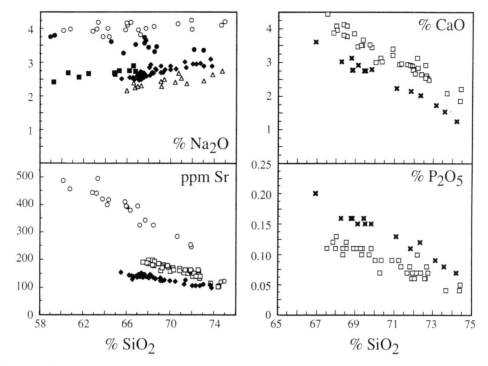

Figure 1 Harker diagrams for various elements in suites of the Bega Batholith. The suites represented are Moruya (open circles), Cobargo (closed circles), Kameruka (crosses), Candelo (closed squares), Bemboka (open squares), Glenbog (closed diamonds) and Tonghi (open triangles). These seven suites are located in a general sequence from east to west across the batholith.

analyses of Munksgaard (1988)]. Such a quartz-rich composition would require an extremely high temperature to completely melt, even at high H_2O contents, and it would be difficult to dissolve that amount of excess Al_2O_3 in such a melt. Because of problems such as these, a degree of consensus seems to have developed that the S-type granites of the LFB were produced by the partial melting of sedimentary source rocks with the major compositional variations within a suite resulting from varying degrees of retention of unmelted source material, according to the restite model of Chappell *et al.* (1987). Pitcher (1993: 112) stated that there are various features of the S-type granites of the LFB which 'when taken together, offer strong support for the restite model as an explanation for the origin of the Lachlan S-type granites.'

1.5. Mixed source rocks

The arguments used here against the hypothesis of magma mixing apply only to such a process producing the variations *within* granite suites. Mixing of components may still be a viable model for the generation of the source rocks or source magmas that later fractionated in some way to produce the variation observed within suites. Indeed, sometimes such mixing seems to be required. The common occurrence of older cores in zircons from all suites of the Bega Batholith apart from the Moruya Suite (Williams 1995) means that it is necessary to postulate mixing of some sedimentary material into the source materials of those granites (Section 4.2.1). The suggestion by Collins (1995) that there was an increasing metasediment component towards the west of the Bega Batholith is not new (see Chappell & McCulloch, 1990); also such an added component alone would not account for the observed changes. For example, Na and Sr are both relatively low in amount in sedimentary rocks that contain a pelitic component, and both elements decrease in amount westwards across the Bega Batholith. However, Na has its lowest abundance in the Tonghi Suite and Sr in the Glenbog Suite (Fig. 1), so their decreases are not exactly correlated. Moreover, Ca, which also has a low abundance in such sedimentary

rocks, and notably in the Ordovician rocks of the LFB (Wyborn & Chappell 1983), increases significantly in abundance, as can be seen by comparing the more westerly Bemboka Suite with the Kameruka Suite in Figure 1. These patterns can only be accounted for by the dominant igneous component of those western suites itself being different from that of the Moruya Suite. Hence the question of the number of source components and their proportions in the source material for each suite is not yet resolved. What is certain is that the mixing that did occur took place before the commencement of the fractionation processes that produced the variation within individual granite suites.

2. Assimilation or contamination

2.1. Assimilation by I-type magmas

The arguments developed in Section 1.3 against magma mixing or mingling based on the correlations between the relative abundances of elements in mafic and felsic rocks of a suite apply equally to any simple process of assimilation being responsible for the compositional variations within I-type granite suites of the LFB.

The most felsic rocks of the normally zoned Boggy Plain pluton show isotopic evidence of the slight assimilation of country rock material (Wyborn 1983). In that case, initial $^{87}Sr/^{86}Sr$ ratios change from an average of 0·70441 in the marginal diorites and the granodiorites (29% of area), to an average of 0·70479 in the dominant monzogranites (70%), to 0·70554 in one sample from the central aplitic rocks (0·9%). That I-type granite body is part of the Boggy Plain Supersuite (Section 3.1), which apparently formed at the highest temperatures of any granites in the LFB; it also underwent extended fractional crystallisation and hence potentially provided heat for assimilation, but the amount of assimilation relative to the total pluton was still very small. The effects of country rock assimilation by other I-type magmas in the LFB have not been detected. Although such processes would be accompanied by fractional crystallisation (AFC process) and therefore be more complex than the variations expected with simple mixing or mingling (Bowen 1928; DePaolo 1981a), the within-suite correlations between isotopic and chemical compositions that would be expected have not been observed. For the Bega Batholith there are distinct variations in isotopic compositions that take place systematically across the batholith and which correlate with the changes in chemical composition between suites (McCulloch *et al.* 1982). There are, however, no such correlations within suites and it must be concluded that the incorporation of sedimentary material took place before the fractionation that produced the compositional dispersion within the suites (Chappell *et al.* 1990).

2.2. Assimilation by S-type magmas

Those granites of the LFB that have been called S-type could in the first instance be regarded as prime candidates for the assimilation of sedimentary material from the country rocks because they have peraluminous compositions and contain widespread cordierite. Those features have long been recognised, most notably by Baker (1940), who described an occurrence of cordierite granite and favoured an origin involving the assimilation of shales. The more mafic S-type granites also contain a distinctive group of metasedimentary enclaves which would be interpreted as part of the assimilated material. The arguments against assimilation in the development of these rocks have developed in two ways. Firstly, as first pointed out by Stevens (1952) for the Cowra Granodiorite, and now recognised throughout the LFB, the enclaves are of higher grade than the country rocks. For example, enclaves in granites that intruded to shallow levels and may be emplaced into co-magmatic volcanic rocks sometimes contain almandine that formed at depths ≈ 15 km. Secondly, although the cordierite-bearing S-type granites do become more peraluminous as they become more mafic, except for the volumetrically minor Cooma Supersuite the observed compositions are inconsistent with country rock contamination, which have particularly low abundances of Ca and Na. These arguments do not rule out assimilation at depths where both the mineralogical and chemical compositions of the sedimentary rocks differ from those exposed at the surface. However, such a process is unlikely for several reasons. Firstly, different suites of S-type granite would have had to assimilate compositionally distinct packages of sedimentary materials. Secondly, although examples are less frequent than for the I-type granites, there is a strong indication that there are compositional correlations between felsic and mafic S-type granites analogous to those discussed in Sections 1.3 and 2.1 for the I-type granites, which would again rule out assimilation. Thirdly, there is a clear separation between the more mafic S- and I-type granites of the LFB (fig. 2 in Chappell & White 1992) and little or no

evidence of assimilation of sedimentary material by the latter (Section 2.1). Fourthly, although the relevant data are limited to one pair of samples, the relatively felsic KB31 (3·01% FeO$_t$; ^{87}Sr/^{86}Sr = 0·71528) and mafic KB32 (4·64% FeO$_t$; ^{87}Sr/^{86}Sr = 0·71504) from the Bullenbalong Suite, they do suggest that the suite has no significant difference in isotopic composition correlated with changes in chemical composition, which would generally be expected from a significant degree of assimilation. In general, it would seem that the incorporation of unmelted source material rather than assimilation is to be preferred (Section 4.1).

Collins (1995) stated that 'S-type granites of the LFB are heavily contaminated I-types'. This is not consistent with the observation that the two types are compositionally distinct at mafic compositions (fig. 2 in Chappell & White 1992); towards felsic compositions both types converge towards minimum temperature melt compositions. Collins (1995) proposed that the S-type Bullenbalong Suite resulted from the fractional crystallisation of a parent magma produced by a 70:30 mix of a Cooma granite composition and a mafic I-type Jindabyne Tonalite composition. There are problems with such a model. Firstly, granites with the composition of the Cooma Granodiorite (Cooma Supersuite) are uncommon. Secondly, those granites are not compositionally representative of the exposed Ordovician sedimentary rocks which might be a more appropriate assimilant because of their widespread occurrence, with the Na and Ca contents of those granites at the upper limit of values for the sediments. Thirdly, the compositional variation within the Bullenbalong Suite cannot be the result of fractional crystallisation (Sections 3.3.4 and 3.3.5). Fourthly, if that variation *had* resulted from such a process, as Collins (1995) suggested, then the mafic Bullenbalong compositions used in his modelling are those of cumulate rocks, not of parent magma, and certainly not of melts (Section 8.7); the same comment can be made about use of the mafic Jindabyne Tonalite composition. Finally, the Jindabyne Tonalite is itself isotopically evolved (Section 2.1), which means that it was either derived from source materials that were much older (Compston & Chappell 1979), or it contains a major sedimentary component (Gray 1984).

3. Fractional crystallisation

3.1. The Boggy Plain Supersuite

One group of granites in southeastern Australia, the Boggy Plain Supersuite (Wyborn *et al.* 1987), shows textural and compositional features that clearly indicate that the granites formed progressively as cumulates from an originally liquid or largely liquid magma. This is shown by the variation of Ba in the Boggy Plain pluton, which shows an inflexion at 66%

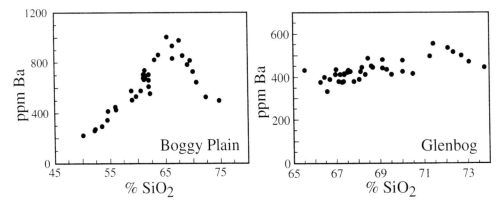

Figure 2 Harker diagrams for Ba in the Boggy Plain pluton and the Glenbog Suite.

SiO$_2$ corresponding to the appearance of biotite as a liquidus phase (Fig. 2). However, most of the granites of the LFB do not exhibit the recognisable fractional crystallisation trends of the Boggy Plain rocks (Chappell *et al.* 1987), so that such an origin is not self-evident.

3.2. Significance of plutonic–volcanic rock suites

The Boggy Plain Supersuite comprises both plutonic and volcanic rocks and this is true of other suites of the LFB, both I- and S-type. These plutonic–volcanic associations are critical in assessing the role of fractional crystallisation. Firstly, with that process, many plutonic rocks would be cumulates and the volcanic rocks would represent some of the complementary liquid compositions. Secondly, it is instructive to compare the mineral compositions of plutonic and volcanic rocks with similar chemical compositions, as the rapidly cooled volcanic rocks, rather than the granites, may have preserved the assemblage of minerals, either restitic, precipitated, or both, that was responsible for fractionation in the granites.

3.2.1. Chemical equivalence of plutonic and volcanic rocks. Wyborn *et al.* (1981) pointed out that the general compositions of some plutonic and volcanic suites of the LFB can be matched fairly closely, with differences between plutonic suites also being present in the volcanic suites. Wyborn and Chappell (1986) showed that co-magmatic plutonic and volcanic rocks of the LFB can be divided into two groups, one comprising those plutonic and volcanic rocks that can be matched closely in composition, and a second in which the volcanic rocks are more felsic than their related plutonic rocks. In the first instance, the most mafic I- and S-type granites correspond in composition to the most phenocryst-rich volcanic rocks and they must therefore represent true magma compositions and cannot be cumulate rocks produced during fractional crystallisation (Wyborn & Chappell 1986). Those mafic rocks, both volcanic and plutonic, must therefore be interpreted as restite-rich magmas. In the second instance, the plutonic rocks represent crystal cumulates and the volcanic rocks the complementary fractionated liquid. In the LFB this second situation is illustrated by the Boggy Plain Supersuite which includes rhyolitic lavas, the Mountain Creek Volcanics. This supports the earlier observation (Section 3.1) that the plutonic rocks of that supersuite show textural and compositional evidence for their formation as cumulates from a largely liquid magma.

3.3. Significance of Sr, Ba and Rb variations

Except within the most felsic compositions, there are some specific patterns of chemical variation for the I-type suites of the Bega Batholith and for the S-type granites of the Bullenbalong Suite, which preclude an origin by fractional crystallisation. In the following discussion it is first assumed that the granites did form by fractional crystallisation and that they are cumulate rocks. This will then be shown to be inconsistent with the observed compositional features of these granites.

The weight percentages and partition coefficients of the minerals that would remove Sr, Ba and Rb from a melt by crystallisation are shown in Table 1. The former are calculated from the modes of the cumulate minerals (vol.%) given in Sections 3.3.1 to 3.3.4. A fraction of 35 vol.% trapped melt is assumed in each instance, approximating to the critical melt fraction (Section 8.1). The partition coefficients for K-feldspar and biotite, and for Rb in plagioclase, are averages of published volcanic phenocryst–groundmass pairs (Chappell *et al.* unpublished data); for Sr and Ba in plagioclase the coefficients have been calculated from the equations of Blundy and Wood (1991) using plagioclase compositions and temperatures given in Sections 3.3.1 to 3.3.3. The trapped melt is treated as a separate fractionating phase with a partition coefficient of unity. In the context of this paper, it is not appropriate to set out fractionation models in full, and in any case the general conclusions do not require such detail. Detailed information will be provided by Chappell *et al.* (unpublished data).

3.3.1. Sr variation in the Glenbog Suite. Except for a few samples at the felsic end of the Moruya and Bemboka suites, the variation for Sr in the Bega Batholith suites is very regular and is strongly correlated with SiO$_2$ (Fig. 1). The observed range in normative plagioclase compositions in the Glenbog Suite is 19 mol.% An (An$_{34}$ to An$_{15}$). A range from An$_{50}$, the phenocryst composition of the comagmatic Kadoona Dacite, to An$_{30}$, and a modest drop in temperature from 800 to 750°C, corresponds to an increase in the partition coefficient for Sr in plagioclase relative to the melt, $D_{Sr}^{plag/L}$, from 4·5 to 9·1 using the equation of Blundy and Wood (1991). A comparable range in $D_{Sr}^{plag/L}$ values exists for other intervals of 20 mol.% An and 50°C within possible magmatic ranges.

Even if the bulk partition coefficients of the crystallising minerals plus trapped melt for Sr relative to the melt ($D_{Sr}^{Glenbog/L}$) remained relatively constant throughout a process of fractional crystallisation that produced the Glenbog Suite,

Table 1 Calculation of some bulk partition coefficients. %biot is the fraction of total biotite in the final cumulate assemblage that originally precipitated as biotite, rather than as pyroxene; % is the weight percentage of the mineral in each cumulate assemblage; and D is the partition coefficient used for each mineral or calculated for the bulk rock.

Sample	%biot	K-feldspar %	K-feldspar D	Plagioclase %	Plagioclase D	Biotite %	Biotite D	Melt %	Melt D	Bulk rock D
Sr, Mafic Glenbog Suite	50	4	3·7	28	4·5	6	0·2	32	1	1·74
Sr, Felsic Glenbog Suite	100	20	3·7	20	9·1	3	0·2	33	1	2·90
Ba, Mafic Glenbog Suite	100	4	6·7	20	0·37	12	7·5	32	1	1·55
Ba, Mafic Glenbog Suite	50	4	6·7	28	0·37	6	7·5	32	1	1·14
Ba, Mafic Glenbog Suite	25	4	6·7	28	0·37	3	7·5	32	1	0·92
Ba, Mafic Glenbog Suite	0	4	6·7	28	0·37	0	7·5	32	1	0·69
Ba, Felsic Glenbog Suite	50	20	6·7	20	0·86	3	7·5	33	1	2·00
Ba, Felsic Glenbog Suite	0	20	6·7	20	0·86	0	7·5	33	1	1·84
Ba, Mafic B'balong Suite	50	0	6·7	20	0·39	12	7·5	32	1	1·30
Ba, Felsic B'balong Suite	100	14	6·7	20	0·79	10	7·5	33	1	2·18
Ba, Felsic B'balong Suite	50	14	6·7	20	0·79	5	7·5	33	1	1·80
Rb, Mafic B'balong Suite	50	0	0·4	20	0·04	12	3·7	32	1	0·78
Rb, Felsic B'balong Suite	100	14	0·4	20	0·04	10	3·7	33	1	0·70
Rb, Felsic B'balong Suite	50	14	0·4	20	0·04	5	3·7	33	1	0·52

it is not certain that the observed linear variation in Sr would have resulted. If those coefficients changed significantly away from a value of unity with falling temperature and changing compositions, then the production of such a variation would have been impossible. Plagioclase would have been the major crystallising phase throughout the solidification of the Glenbog Suite, but increasing $D_{Sr}^{plag/L}$ values would have been partly balanced by a decreasing modal abundance, from ≈ 40 to $\approx 30\%$. However, K-feldspar modal abundances increase within the suite from ≈ 10 to $\approx 30\%$ and as $D_{Sr}^{Kspar/L}$ is high (≈ 3.7), that would have contributed to an increase in $D_{Sr}^{Glenbog/L}$. Because precise temperatures, plagioclase compositions and the proportion of trapped intercumulus melt and its compositions are not known, accurate modelling is not possible; however, a model calculation can be made. The six most mafic granites of the Glenbog Suite have an average mode of $Q_{29}Or_{11}Pl_{40}Bi_{14}Ho_6$ and the average of the six most felsic granites is $Q_{34}Or_{27}Pl_{31}Bi_7Ho_1$. The phenocrysts in the Kadoona Dacite indicate that pyroxenes rather than biotite and hornblende were initially the dominant mafic FeMg minerals in the mafic phases of the Glenbog Suite, but for calculations of Sr partitioning, only the feldspars are significant. If it is assumed that temperatures fell from 800 to 750°C, that plagioclase compositions ranged from An_{50} to An_{30}, and that there was 35 vol.% of trapped melt of composition $Q_{32}Or_{20}Pl_{35}Bi_{10}Ho_3$, then values of $D_{Sr}^{Glenbog/L}$ of 1.7 and 2.9 are obtained for the mafic and felsic extremes. This clearly indicates that a process of fractional crystallisation in the Glenbog Suite would have been associated with a progressive increase in $D_{Sr}^{Glenbog/L}$. This is not consistent with the linear variation of Sr seen in Figure 1.

3.3.2. Sr variation in the Moruya Suite. The Moruya Suite (SiO_2 from 60 to 75%) encompasses almost twice the compositional range of the Glenbog Suite (SiO_2 from 66 to 74%). The Sr variation is essentially linear, except for the two most Si-rich (MG20 and MG44) samples, which are displaced to relatively low Sr values (Fig. 1). Also, apart from MG20 and MG44, K-feldspar is much less abundant than in the Glenbog Suite, so that in most samples, under the fractional crystallisation scenario, that mineral would have crystallised only from trapped liquid. Also, relative to Glenbog, the plagioclase compositions are more sodic, with normative compositions between An_{26} and An_{12}, to An_7 for MG20 and MG44. Plagioclase is dominant, ranging between 60 and 50% of the mode, falling sharply to 35% in the two most felsic samples. If the Sr abundances in the Moruya Suite resulted from fractional crystallisation, then its removal would have been dominated by the separation of plagioclase in approximately constant amount, with a small contribution from K-feldspar in the most felsic rocks. If the formation of this suite involved temperatures falling from 850 to 750°C with plagioclase compositions changing from An_{30} to An_{15}, then the $D_{Sr}^{plag/L}$ values would have increased from 7.5 to 14.6. This would have lead to an approximate doubling of the value of $D_{Sr}^{Moruya/L}$. Again, this is not consistent with the observed variation. The relatively high concentrations of Na on the eastern side of the Bega Batholith (Section 1.3) means that plagioclase is both more abundant and more sodic than in the Glenbog Suite and both factors contribute to high values of $D_{Sr}^{Moruya/L}$. This makes the operation of fractional crystallisation, given the observed pattern of Sr variation, even less likely.

The data for Sr show that the more felsic compositions in the Glenbog and Moruya suites, apart from MG20 and MG44, cannot have been produced by fractional crystallisation. However, the effects of such crystallisation are seen in those two most felsic Moruya samples (Fig. 1), in which the Sr abundances were displaced from the previously established trend by the precipitation of sodic plagioclase at low temperatures, both factors contributing to high $D_{Sr}^{plag/L}$ values.

3.3.3. Ba variation in the Glenbog Suite. The removal of Ba from a melt by fractional crystallisation provides another view of that process because the partition coefficients for both K-feldspar and biotite are high (values of 6.7 and 7.5 are used here), whereas plagioclase is significant only if both of those other minerals have low abundances. The contrasting behaviour of Ba in the Boggy Plain pluton and the Glenbog Suite is shown in Figure 2. In the former case there is an inflexion in Ba abundances at about 66% SiO_2, corresponding to the appearance of biotite as a liquidus mineral in that zoned pluton (Wyborn 1983). The variation in Ba shown for the Glenbog Suite is typical of suites of the Bega Batholith in which that element generally increases slightly, but sometimes decreases slowly, in both instances without any hint of an inflexion.

Estimating the values of $D_{Ba}^{Glenbog/L}$ are subject to the uncertainties in the composition and proportion of trapped intercumulus melt, and temperature, encountered in Section 3.3.1 for calculating $D_{Sr}^{Glenbog/L}$. However, in this instance the composition of plagioclase is not critical as $D_{Ba}^{Plag/L}$ has a relatively small value. The fraction of the biotite now observed in the rocks that was originally precipitated as orthopyroxene and converted to biotite by reaction with the appropriate components in the trapped melt is not known and is critical for the calculation of a precise value of $D_{Ba}^{Glenbog/L}$. The Kadoona Dacite is very similar in composition to the most mafic rocks of the Glenbog Suite and pyroxenes comprise 75% of the ferromagnesian minerals. With the modes previously used in Section 3.3.1 for the Glenbog Suite, and fractions of 1, 0.5, 0.25 and 0 of the observed biotite originally precipitated as that mineral rather than pyroxene, $D_{Ba}^{Glenbog/L}$ values of 1.55, 1.12, 0.92 and 0.69 are obtained. Although these values are not precise estimates, the value of 0.91 which corresponds to the observed volcanic phenocryst abundances is close to a value that would, by fractional crystallisation, account for the pattern of Ba variation shown in Figure 2. Precipitation of biotite alone ($D_{Ba}^{Glenbog/L} = 1.55$) is clearly ruled out, in agreement with the volcanic phenocryst data, as the Ba abundances would then decrease with fractionation, contrary to what is observed.

The amount of K-feldspar in granites of the Glenbog Suite increases from ≈ 11 to $\approx 27\%$ as the rocks become more felsic, so that the mineral would have changed from a minor to a significant cumulate phase. Assuming that it comprised 20% of the cumulate minerals in the more felsic rocks, values for $D_{Ba}^{Glenbog/L}$ of 1.84 (no biotite) and 2.07 (no pyroxene) are calculated. These are substantially higher than values for the most mafic rocks and show that the magnitude of $D_{Ba}^{Glenbog/L}$ would have changed significantly during the evolution of the Glenbog Suite by fractional crystallisation. That that did not happen is shown clearly by the variation in Ba for that suite in Figure 2 and hence derivation of the suite by that process can again be ruled out.

3.3.4. Ba variation in the Bullenbalong suite. The variation of Rb and Ba in the Bullenbalong Suite is represented by the closed symbols in Figure 3. Although there is more scatter in this S-type suite, the trend in the variation for Ba is approximately constant, except for the most felsic sample. For that sample (KB45), Rb is a little higher than the trend for the suite (Fig. 3) and Sr is lower, so that the fractional crystallisation of feldspars and quartz from a felsic melt was involved in that instance. That process did not have a significant role in producing the other rocks of the Bullenbalong Suite, as will now be shown.

Modal data for the more mafic rocks of the Bullenbalong

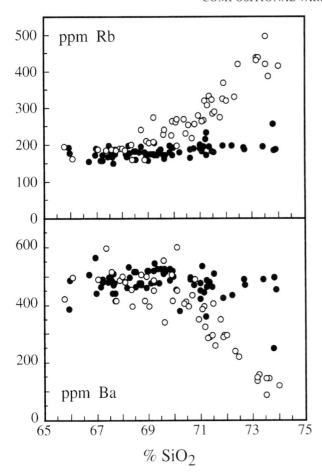

Figure 3 Harker diagrams for Rb and Ba in the Bullenbalong (closed circles) and Koetong (open circles) suites.

nature of the high values for $D_{Ba}^{Bullenbalong/L}$ that have just been considered.

3.3.5. Rb variation in the Bullenbalong suite. Figure 3 shows that the variation in Rb in the Bullenbalong Suite is rather flat; a regression line through all of the data except KB45, passes through 170 ppm Rb at 66% SiO_2 and 199 ppm at 74% SiO_2. Such variation could be consistent with fractional crystallisation with a value of the bulk partition coefficient $D_{Rb}^{Bullenbalong/L}$ a little less than unity, that did not change significantly with fractionation. Using the modal data in Section 3.3.4 and the mineral partition coefficients listed in Table 1, a value for $D_{Rb}^{Bullenbalong/L}$ of 0·78 is calculated for the case in which half of the final cumulate biotite was originally precipitated as that mineral. The error in that value could place it within the limits which would produce the observed variation. However, the value of $D_{Rb}^{Bullenbalong/L}$ must fall with fractionation as K-feldspar becomes the dominant potassic cumulate phase, and a value of 0·52 has been calculated for the most felsic granites apart from KB45. Such a value is too low to maintain the flat Rb trend that is observed, again ruling out the operation of fractional crystallisation. If the amount of biotite precipitating were increased to move the value of $D_{Rb}^{Bullenbalong/L}$ closer to unity, that would have a deleterious effect on the bulk partition coefficient for Ba, as far as fractional crystallisation is concerned.

3.3.6. Contrasting behaviour of the Bullenbalong and Koetong suites. The Koetong Suite is an extensive development of S-type granites with an area of 6740 km^2 in the Wagga Batholith, west of the Bullenbalong Suite. Granites of the Koetong Suite cover a similar range in SiO_2 contents to those of the Bullenbalong Suite, but show some striking differences in trace element compositions. Although the most mafic rocks of both suites are fairly similar in composition, above about 69% SiO_2 the abundances of many elements diverge, as illustrated by Rb and Ba in Figure 3. For the Koetong Suite, they and other trace elements show patterns of variation that would be expected to result from fractional crystallisation (Sections 3.3.4 and 3.3.5). If there were any question of fractional crystallisation not having a significant role in producing the compositional variation in the Bullenbalong Suite, that would be resolved by comparing those rocks with the Koetong Suite. This is done in Figure 3, where it can be seen that the variation for Rb and Ba in the Koetong Suite above 69% SiO_2 is precisely as would be expected from fractional crystallisation.

4. Restite separation

The restite model (Chappell *et al.* 1987) proposes that many granite suites contain crystals residual from melting, or restite, and that it is differences in the proportion of this restite that account for the variation in composition within those suites. The difficulties of generating most of the granite suites of the LFB by magma mixing or assimilation, and in many instances by fractional crystallisation, provide indirect support for the model.

4.1. Resolution of the problems of magma mixing and fractional crystallisation

Problems that have been addressed above which can be readily accounted for by using the restite model, include the following.

Firstly, the mixing or mingling of two end-members would produce linear correlations between elements. So too will varying degrees of separation or 'unmixing' of mafic residual material and felsic melt produced together during partial melting of the crust. That second mechanism offers the

Suite are incomplete, but the approximate mode of those rocks is $Q_{38}Or_5Pl_{32}Bi+Cord_{25}$. For the less mafic sample SV51 from the Hawkins Volcanics of the Bullenbalong Supersuite (FeO_t of 4·34% compared with an average of 4·90% for the six most mafic granites of the Bullenbalong Suite), the approximate mode of the phenocrysts is $Q_{20}Pl_{20}Biot_{10}Hy+Cord_{10}$. These data provide an indication of the abundances of the cumulate minerals in mafic granites of the Bullenbalong Suite, assuming fractional crystallisation. Taking a value of $Q_{20}Pl_{20}Biot_{12}Hy+Cord_{13}$ for the cumulate minerals in the most mafic granites of the Bullenbalong Suite, a value for $D_{Ba}^{Bullenbalong/L}$ of 1·30 is calculated (Table 1). That value is sensitive to the values chosen for the fraction of biotite in the cumulate minerals and the value of $D_{Ba}^{Biotite/L}$. It is therefore perhaps not in conflict with the data plotted in Figure 3, in which Ba abundances are constant through most of the range, implying a value of $D_{Ba}^{Bullenbalong/L}$ close to unity. There is less uncertainty about $D_{Ba}^{Bullenbalong/L}$ calculated for the felsic members of the suite as the value of $D_{Ba}^{Kspar/L}$ is better known and the fraction of K-feldspar in the cumulate minerals more precisely defined. Accordingly, the magnitude of $D_{Ba}^{Bullenbalong/L}$ is much greater in those felsic rocks, values of 2·18 and 1·80 being calculated for biotite alone or biotite comprising 50% of the mafic minerals, respectively. In any case, there is a substantial increase in the value of $D_{Ba}^{Bullenbalong/L}$ in passing from the mafic to the felsic rocks, so that once again the operation of fractional crystallisation in producing most of the range in composition of this suite is precluded. It has been noted earlier that one sample lies well below that main trend at the most felsic composition and it clearly does that because of fractional crystallisation and the advantage of accounting for the observation that distinct suites have both different mafic and felsic 'end-members', as those extremes of composition are each related to specific source compositions. Furthermore, it would account for the observed correlations in the relative abundances of different elements between the more mafic restite-rich and more felsic melt-rich rocks in a suite. It does that because the ratio of the abundances of any element between coexisting restite and melt is determined, within limits imposed by disequilibrium, by the bulk distribution coefficient of that element. For any element, the values of those coefficients will generally be similar between various suites.

Secondly, the observation that many S-type granites become more peraluminous as they become more mafic is not consistent with either the presence of a mafic basaltic component, or assimilation of the observed sediments, but is consonant with the presence of larger quantities of a sedimentary component, as restite, in the more mafic rocks.

Thirdly, the observed variations in Sr abundances in the Glenbog and Moruya suites, which cannot be accounted for by fractional crystallisation, can be readily explained as due to variations in the proportions of melt and restite in the different samples of granite. This also accounts for the variation within plutons being the same as the suite as a whole, with the necessary condition being that all units were derived from source materials of effectively the same composition.

Fourthly, the restite model can likewise account for the Ba variation in the Glenbog Suite, with that element continuing to increase in amount as the rocks become more felsic. Those compositions are a blend of felsic melt that formed in equilibrium with restite in which biotite and/or K-feldspar were absent or present in small amount and which therefore had a bulk distribution coefficient of less than unity.

4.1. Importance of restite minerals

The examples in Section 4.1 are those in which intrinsic difficulties of the magma mixing and fractional crystallisation models can be readily resolved by using the restite model. Other independent arguments to support that as the preferred model were presented by Chappell et al. (1987). Of those arguments, later information makes it appropriate to mention here those dealing with evidence for restite minerals. Two examples of what are considered to be primary restite minerals were discussed by Chappell et al. (1987): zircon crystals showing age inheritance and plagioclase cores. Both minerals show two stages of magmatic development, restite cores and melt precipitated rims.

4.2.1. Zircon crystals showing age inheritance. Fractions of zircon separated from granites and showing pre-emplacement ages when analysed in bulk by thermal ionisation mass spectrometry were interpreted by Chappell et al. (1987) as being restite material. Specific data on the ages of inherited zircons in LFB granites and elsewhere are now available from the SHRIMP ion microprobe (e.g. Williams 1992, 1995; Pidgeon & Compston 1992). SHRIMP analyses have confirmed the earlier observation by identifying inherited cores in many zircon crystals. Such old cores are extremely common in the S-type granites. The abundance of inherited zircon in the I-type granites is very low by comparison with the S-types, but has been found in all but one of the rocks that have been studied.

4.2.2. Plagioclase cores. Plagioclase crystals with cores that are uniform or almost uniform in composition, generally rounded and surrounded by strongly zoned melt precipitated plagioclase overgrowth are an important feature of many granites. The outer parts of such crystals are zoned because,

as Johannes (1978) pointed out, plagioclase is a particularly difficult mineral to re-equilibrate with a melt at temperatures below $1000°C$. For the same reason, Chappell et al. (1987) interpreted these cores as forming during the slow prograde ultrametamorphism or melting of the granite source rocks, so that they represent crystals of restite. These workers did not consider the alternative possibility, that the cores represent crystals from a mingled mafic magma that originally precipitated and homogenised at high temperatures. New considerations of the magma mixing/mingling hypothesis (Section 1 and Chappell in press), which show that it cannot be the cause of large-scale compositional variation within these granite suites, also imply that these cores cannot be accounted for as crystals of mingled mafic magma.

5. Hydrothermal alteration

Subsolidus hydrothermal alteration has the potential to greatly alter the composition of a granite. Such a process would be dominated by the replacement of feldspars by sheet silicates and the loss of those elements that cannot be accommodated in the latter, principally Na, Ca and Sr, whereas K and Rb could be added from circulating solutions. This process of feldspar destruction would have compositional effects analogous to those of feldspar separation at higher temperatures (Section 3.3), as in both instances feldspars are being removed from the system. The primary evidence for alteration is generally petrographic; however, chemical data generally show that its effects are overrated. Many rocks in which there are clear petrographic signs of alteration plot in a very tight field or array on chemical diagrams, which would not be expected if their compositions were in part the result of low temperature alteration. Also, for most felsic granites of this type, the compositions are generally very close to the minimum temperature compositions for hydrous melts in equilibrium with quartz and feldspars (Tuttle & Bowen 1958). For the Koetong Suite, in which the development of the more felsic rocks is ascribed to fractional crystallisation (Section 3.3.6), five analysed granites have a sum of normative $Q + Ab + Or$ greater than 90%. CIPW norms, modified to assign normative Hy and some of the normative or to the crystalline biotite presumed to have been present, have been calculated for those samples. Their average composition of $Q_{42}Ab_{29}Or_{29}$ is very close to the 50 MPa minimum of Tuttle and Bowen (1958) of approximately $Q_{40}Ab_{29}Or_{31}$, confirming the origin by fractional crystallisation inferred from other compositional features (Section 3.3.6).

It must be concluded that although petrographic evidence clearly points to the mobility of elements during the latest stage of formation of some granites, particularly the more felsic granites, the process occurs over a scale much smaller than that at which the rocks are sampled for chemical analysis. This process therefore does not contribute significantly to the chemical variations observed within suites of granites, although locally it can be very significant.

The data of Blundy and Wood (1991), showing that the partition coefficients for Sr and Ba for plagioclase increase substantially at albite-rich compositions, may make it possible to account for the very low Sr and Ba contents of many strongly fractionated granites for which it had been thought that hydrothermal alteration was an essential agent.

6. Variation inherited from heterogeneous source rocks

Suites of I-type granites in the LFB often show a remarkable compositional coherence, e.g. Sr in the Glenbog Suite (Fig. 1) shows regular variation throughout 12 plutons that occur over a distance of more than 250 km. Although homogeneity in a single pluton could be ascribed to thorough mixing at an early stage, for the Glenbog Suite this implies a relatively homogeneous distribution of Sr in the source rocks that must have contributed separately to the different plutons. The capacity to recognise suites depends on the internal variations within two source materials being small relative to the overall differences between them. A corollary of the very precise suite definitions that can be made in many instances in southeastern Australia is that either the source materials were very homogeneous or they were mixed thoroughly at an early stage in the production of the suite, the latter situation being unlikely when a suite comprises several dispersed plutons. Because sedimentary source rocks are more heterogeneous than those of I-type granites, S-type suites individually show more scatter in element concentrations about the dominant trend for a suite. Also, when the more mafic members of S-type suites have compositions close to those of their source rocks, they can apparently show variations inherited from those sources and examples of such variations are currently being investigated.

7. Relation between restite fractionation and fractional crystallisation

Compositional variation in the granites of the LFB resulted dominantly from processes of crystal fractionation, both of restite crystals and by fractional crystallisation. Those two processes both involve the removal of crystals, in one instance carried from the source and for the other precipitated from a melt, and are complementary. The compositional variation that is observed in individual suites may be completely dominated by one of those processes, or they may act together to produce the total range in composition. Chappell *et al.* (1987) referred to those two stages of evolution as the *restite fractionation regime* and the *fractional crystallisation regime*.

Rocks of the Boggy Plain Supersuite comprise ≈5% of the exposed area of granites in the LFB and they appear to be the only rocks which range widely in composition whose composition was determined exclusively by fractional crystallisation. The other situation, in which all of the observed variation must be ascribed to restite fractionation, is much more common and includes large numbers of both I-type and S-type plutons and suites. In other instances, dominant restite fractionation lead to a relatively minor episode of fractional crystallisation, apparently after all of the restite had been cleared from the magma, so that the original partial melt itself underwent fractional crystallisation; examples of such behaviour include two samples from each of the Moruya and Bemboka suites (Section 3.3.2 and Fig. 1), and the one sample (KB45) collected from the Snowy Gap granite of the Bullenbalong Suite (Section 3.3.4).

The Koetong Suite provides the only certain example in the LFB of a suite in which there is a distinct interval dominated by restite separation, in that case from 66% to about 69% SiO_2, followed by extended fractional crystallisation. There are other felsic granites in the LFB with trace element abundances that resulted from extended fractional crystallisation of a magma that had previously approached or reached minimum temperature compositions through the loss of restite, but more mafic rocks representing those earlier stages of evolution are not seen. In the LFB this is particularly shown by some S-type granites, and by two I-type suites, in most instances from Tasmania. There are distinct differences between the I- and S-type granites at that stage of extreme fractionation, as discussed by Chappell and White (1992).

The question remains: how typical is the Lachlan belt with its complete range of granite magma type, from completely molten magmas to those bearing maximum restite, of granite terranes in general? Although every belt will have some unique characters, there is no reason for the LFB to differ too significantly in this regard from some other S-type granite terranes and also from those containing what Pitcher (1982) termed the I- (Caledonian) type granites. High temperature tonalitic rocks, the I- (Cordilleran) type of Pitcher (1982) are absent from the LFB, and it is terranes that contain such rocks in abundance which would contain granites more typically derived from completely molten magmas than is the case for the LFB.

8. Broader implications

It can be concluded that restite fractionation and fractional crystallisation are dominant but complementary processes in producing variation within granite suites, and that in many instances fractional crystallisation only operates after a substantial degree of compositional variation has been induced by restite separation. Some of the areas in which these conclusions have broader implications will now be considered.

8.1. Implications for the physical state of granite magma
Rock magma is a fluid substance that is not necessarily completely liquid (Section 1.1). The observations listed in this paper have shown that, for granite magmas, all possibilities exist between pure melts and magmas charged with the maximum amount of solid material consistent with fluid behaviour and corresponding to the critical melt fraction (CMF). However, in studying the extraction of granite magma from its source and its movement through the crust, it seems to be widely assumed that those magmas are, at least initially, completely liquid (e.g. Clemens & Mawer 1992; Rushmer 1995). The CMF as defined by van der Molen and Paterson (1979) separates framework-controlled flow behaviour from suspension-like behaviour, at approximately 30–35% melt, so that for fractions of melt beyond that point the whole mass may move as a magma. Rushmer (1995) has used the CMF to refer to 'the minimum amount of melt needed before it can effectively segregate', which is a different usage that does not correspond with the loss of a crystal framework. That difference is significant in considering the movement from the source of magmas charged with restite and with a melt content which may just exceed the CMF. Granite magmas may contain any fraction of melt between the CMF and 100% and, although studies such as that of Rushmer (1995) are important for melt-rich conditions, evidence presented here shows that the other case, in which the whole mass moves bodily once the restite framework is broken and the whole mass becomes a crystal-rich magma, must also be important.

8.2. Implications for the ascent of granite magma
The view that granite magmas ascend as dykes to high levels in the crust to produce batholiths within periods ≈100–10 000 a after the initiation of a thermal anomaly in the source has been presented by Clemens and Mawer (1992). Such a mechanism involves the initial segregation of a partial melt followed by its movement upwards through propagating dykes. Modelling of melt-rich magmas in that way is a valuable contribution to understanding one end-member of a spectrum of magmas that extends away to those containing restite at the CMF. The empirical evidence that such crystal-rich (restite + early crystallising minerals) magmas are common, at least in the LFB and most likely elsewhere, is unambiguous, and a comprehensive account of the transport of granite

magmas must account for all instances. The statement of Clemens and Mawer (1992) that the 'Gravitational rise of an homogenized restite–melt system may not be a factor of any major significance in the process of granitoid magma movement', does not accord with the observations on granites discussed in Sections 3 and 4.

8.3. Implications for temperatures of granite magmas

Completely molten magmas are hotter than those of similar bulk composition containing entrained source material. In assessing the temperature of a particular magma it is critical to know the fraction of solid material that was present when it left the source. This is particularly important in considering one of the outstanding problems in granite genesis, that of the source of heat for melting.

The amount of heat required to produce a partly molten magma is lessened in two ways. Firstly, the amount of melt needed to produce a magma is reduced by up to two-thirds in the limiting case of approximately one-third melt at the CMF. Secondly, as a fraction of the mafic components, and perhaps a large fraction, are present as solids, and the melt itself is felsic relative to the bulk magma, lower temperatures of melting are required.

8.4. Implications for relative contributions of crust and mantle to granite magmas

An implication of the magma mixing hypothesis is that significant amounts of mafic material are introduced into the crust at the time of magmatism, so that granites contain a significant proportion of juvenile crust. If the observations from the LFB that magma mixing is not a significant process in generating the large-scale variations within granite suites (Section 1) is more widely applicable, this limits the amount of new crust formed at the time of granite magmatism. An implication of the partial melting–restite hypothesis is that granitic magmatism tends to recyle the crust (Section 8.6) rather than modify its bulk composition. This is likely to be generally the case in terranes, such the LFB, that contain the I- (Caledonian) type granites of Pitcher (1982).

8.5. Implications for source rock compositions

Granites form a compositional image of their source rocks. For isotopic compositions that image is relatively clear, but for chemical compositions its quality depends on the nature of the magma that transported material from the source region to the zone of final emplacement, the fraction of restite in the magma and the degree of fractionation after the magma leaves its source. Chappell and Stephens (1988) examined this in terms of two contrasting end-members, the Tuross Head Tonalite and the Boggy Plain Supersuite. The Tuross Head Tonalite is the most mafic unit of the Moruya Suite. The most mafic parts of that body represent compositions in which there was no separation of melt from restite, and hence the compositions are close to those of their source rocks. More felsic rocks of the Moruya Suite, from which there has been fractionation of some restite from the magma, provide a less distinct image of the source rocks, although a degree of modelling of source compositions only from more felsic compositions is possible. In the Boggy Plain Supersuite, derived from a completely molten magma, fractional crystallisation was the dominant process leading to compositional variation; the most mafic rocks are cumulates and it is difficult to make any precise statement about the major element composition of the original magma. For trace elements, the situation is better as many of these can be considered as abundances relative to specific major elements and in that way they can be compared with other suites. Hence the

relatively high abundances of Cu and incompatible elements such as K, Sr and Ba in the Boggy Plain Supersuite must reflect correspondingly high abundances in the source rocks. Between the extremes represented by the Moruya Suite and the Boggy Plain Supersuite, the mafic components of the source will be carried in the magma both as entrained restite crystals and in solution in the melt, and the image of the source would be of intermediate quality.

An important consequence of the occurrence of restite-rich magmas is that there must be some instances in which source rock compositions have been replicated in granites or volcanic rocks without fractionation.

8.6. Implications for chemical fractionation of the crust and the nature of the deep crust

The exposed continental crust is granodioritic or dacitic in composition because its composition has largely been determined by the movement of magmas up through the crust. An understanding of the mechanism by which the crust is vertically fractionated is therefore synonymous with understanding how granitic and related volcanic magmas form, move upwards and fractionate. The fact that many such magmas carry significant amounts of unmelted deeper crust material has had a significant effect on the composition of the upper crust, so that it is granodioritic rather than more haplogranitic.

8.7. Implications for whether granite compositions are those of magmas

It is often assumed that granite compositions represent the compositions of magmas, with the further presumption often made that those magmas were at some time completely molten. This can only be the case for granites that crystallised from melts that crystallised *in situ* after all restitic and/or precipitated crystals had been removed. This applies only to relatively felsic granite magmas that have undergone prior fractionation of such crystals in another place. More mafic granites that solidify from completely or largely molten magmas must generally be cumulate rocks.

Magmatic compositions may be retained by granite magmas that contain restite and completely solidify. This is evidenced by the coincident compositions of plutonic and volcanic rocks (Section 3.2.1) in such instances.

8.8. Implications for H₂O content of granite magmas and its flux from cooling magmas

Granite magmas carry incompatible components into the upper crust, of which the most prominent is H_2O. If such a magma is never completely molten, then the amount of H_2O required to form the magma, and hence its upward flux, is much less. A granite magma comprising one-third melt that contains 3% H_2O will contain only 1% H_2O, which is much less than the figures generally assumed to be present on the assumption that magmas are largely molten. Hence the amount of H_2O required to form granite magmas, and the flux of that component associated with such magmatism, is overall much less than is generally supposed.

8.9. Implications for abundance of compatible elements

Chappell *et al.* (1987) pointed out that an implication of the restite model is that some elements may be present at a significant concentration in granite magmas despite their very low solubility in granite melts. For example, much of the light rare earth elements and Th in mafic S-type granites of the LFB are present in monazite, which is probably restite. Also most of the Sn in oxidised I-type granites may be present in restite minerals such as biotite and sphene, rather than the melt.

8.10. Implications for mineralisation

Granite suites that evolve by fractional crystallisation will concentrate incompatible trace metals at that magmatic stage, enhancing the probability of the later significant mineralisation (Blevin & Chappell 1992). An excellent example of this is provided by the contrasting Bullenbalong and Koetong suites of the LFB (Section 3.3.6). In the first instance, Sn variation is rather constant throughout the range in granite compositions that resulted from the fractionation of melt from restite and rises to 10 ppm only in the one sample (KB45) that underwent fractional crystallisation. For the Koetong Supersuite, Sn levels are progressively higher above 68% SiO_2, to a maximum of 47 ppm; this occurred as extensive feldspar fractionation took place (Section 3.3.6). These differences correlate with the reported amounts of Sn obtained from the areas in which those two groups of rocks are found, < 10 t for Bullenbalong and ≈ 5 kt for Koetong. The higher temperatures of magmas undergoing fractional crystallisation (Section 8.3) would also enhance the formation of mineral deposits for other reasons (Blevin & Chappell 1992).

Acknowledgements

I thank the organisers of the Third Hutton Symposium for their invitation to present this paper at that conference. Discussions with many people over many years have helped to develop my thoughts on this subject, particularly Wally Pitcher, Lee Silver, Ed Stephens, Allan White, Ian Williams and Doone Wyborn. The stimulus provided by the numerous critics of the restite model is also acknowledged. Much of this paper was prepared while visiting the Department of Geology of the University of St Andrews and Peter Brown and Ed Stephens are thanked for their hospitality. Financial support from the Australian Research Council grant A39232908 is acknowledged. This is publication number 41 in the Key Centre for the Geochemical Evolution and Metallogeny of Continents.

References

Baker, G. 1940. Cordierite granite from Terip Terip, Victoria. AM MINERAL **25**, 543–8.

Barbarin, B. 1991. Enclaves of the Mesozoic calc-alkaline granitoids of the Sierra Nevada Batholith, California. In Didier, J. & Barbarin, B. (eds) *Enclaves and Granite Petrology*, 135–53. Amsterdam: Elsevier.

Beams, S. D. 1980. *Magmatic evolution of the Southeast Lachlan Fold Belt, Australia*. Ph.D. Thesis, La Trobe University, Melbourne.

Blevin, P. L. & Chappell, B. W. 1992. The role of magma sources, oxidation states and fractionation in determining the granite metallogeny of eastern Australia. TRANS R SOC EDINBURGH EARTH SCI **83**, 305–16.

Blundy, J. D. & Wood, B. J. 1991. Crystal-chemical controls on the partitioning of Sr and Ba between plagioclase feldspar, silicate melts, and hydrothermal solutions. GEOCHIM COSMOCHIM ACTA **55**, 193–209.

Bowen, N. L. 1928. *The Evolution of the Igneous Rocks*. Princeton: Princeton University Press.

Chappell, B. W. 1979. Granites as images of their source rocks. GEOL SOC AM ABSTR PROGRAM **11**, 400.

Chappell, B. W. 1996. Magma mixing and the production of compositional variation within granite suites: evidence from the granites of southeastern Australia. J PETROL **37**.

Chappell, B. W. & McCulloch, M. T. 1990. Possible mixed source rocks in the Bega Batholith: constraints provided by combined chemical and isotopic studies. GEOL SOC AUST ABSTR **27**, 17.

Chappell, B. W. & Stephens, W. E. 1988. Origin of infracrustal (I-type) granite magmas. TRANS R SOC EDINBURGH EARTH SCI **79**, 71–86.

Chappell, B. W. & White, A. J. R. 1992. I- and S-type granites in the Lachlan Fold Belt. TRANS R SOC EDINBURGH EARTH SCI **83**, 1–26.

Chappell, B. W., White, A. J. R. & Wyborn, D. 1987. The importance of residual source material (restite) in granite petrogenesis. J PETROL **28**, 1111–38.

Chappell, B. W., Williams, I. S., White, A. J. R. & McCulloch, M. T. 1990. Granites of the Lachlan Fold Belt. ICOG 7 Field Guide Excursion A-2. REC BMR GEOL GEOPHYS **1990/48**.

Clemens, J. D. & Mawer, C. K. 1992. Granitic magma transport by fracture propagation. TECTONOPHYSICS **204**, 339–60.

Collins, W. J. 1995. S- and I-type granitoids of the eastern Lachlan Fold Belt: three-component mixing, not restite unmixing. In Brown, M. & Piccoli, P. M. (eds) *The Origin of Granites and Related Rocks. Third Hutton Symposium Abstracts*. US GEOL SURV CIRC **1129**, 37–8.

Compston, W. & Chappell, B. W. 1979. Sr-isotope evolution of granitoid source rocks. In McElhinny, M. W. (ed.) *The Earth: its Origin, Structure and Evolution*, 377–426. London: Academic Press.

DePaolo, D. J. 1981a. Trace element and isotopic effects of combined wallrock assimilation and fractional crystallization. EARTH PLANET SCI LETT **53**, 189–202.

DePaolo, D. J. 1981b. A neodymium and strontium isotopic study of the Mesozoic calc-alkaline granitic batholiths of the Sierra Nevada and Peninsular Ranges, California. J GEOPHYS RES **86**, 10470–88.

Gray, C. M. 1984. An isotopic mixing model for the origin of granitic rocks in southeastern Australia. EARTH PLANET SCI LETT **70**, 47–60.

Grout, F. F. 1948. Origin of granite. In Gilluly, J. (ed.) *Origin of Granite*. GEOL SOC AM MEM **28**, 45–54.

Hine, R., Williams, I. S., Chappell, B. W. & White, A. J. R. 1978. Contrasts between I- and S-type granitoids of the Kosciusko Batholith. J GEOL SOC AUST **25**, 219–34.

Johannes, W. 1978. Melting of plagioclase in the system Ab–An–H_2O and Qz–Ab–An–H_2O at $P_{H2O} = 5$ kbars, an equilibrium problem. CONTRIB MINERAL PETROL **66**, 295–303.

McCulloch, M. T. & Chappell, B. W. 1982. Nd isotopic characteristics of S- and I-type granites. EARTH PLANET SCI LETT **58**, 51–64.

McCulloch, M. T., Chappell, B. W. & Hensel, H. D. 1982. Nd and Sr isotope relations in granitic rocks of the Tasman Fold Belt, eastern Australia. ABST ICOG **5**, 246–7.

Munksgaard, N. C. 1988. Source of the Cooma Granodiorite, New South Wales—a possible role of fluid–rock interactions. AUST J EARTH SCI **35**, 363–77.

Pidgeon, R. T. & Compston, W. 1992. A SHRIMP ion microprobe study of inherited and magmatic zircons from four Scottish Caledonian granites. TRANS R SOC EDINBURGH EARTH SCI **83**, 473–83.

Pitcher, W. S. 1982. Granite type and tectonic environment. In Hsu, K. J. (ed.) *Mountain Building Processes*, 19–40. London: Academic Press.

Pitcher, W. S. 1993. *The Nature and Origin of Granite*. Glasgow: Blackie Academic & Professional.

Read, H. H. 1948. Granites and granites. In Gilluly, J. (ed.) *Origin of Granite*. GEOL SOC AM MEM **28**, 1–19.

Rushmer, T. 1995. An experimental deformation study of partially molten amphibolite: application to low melt fraction segregation. J GEOPHYS RES **100**, 15681–95.

Shand, S. J. 1950. *Eruptive Rocks*, 4th edn. London: Murby.

Sparks, R. S. J. & Marshall, L. A. 1986. Thermal and mechanical constraints on mixing between mafic and silicic magmas. J VOLCANOL GEOTHERM RES **29**, 99–124.

Stevens, N. C. 1952. The petrology of the Cowra intrusion and associated xenoliths. PROC LINN SOC NSW **77**, 132–41.

Tuttle, O. F. & Bowen, N. L. 1958. Origin of granite in the light of experimental studies in the system $NaAlSi_3O_8$–$KAlSi_3O_8$–SiO_2–H_2O. GEOL SOC AM MEM **74**.

van der Molen, I. & Paterson, M. S. 1979. Experimental deformation of partially-melted granite. CONTRIB MINERAL PETROL **70**, 299–318.

Wall, V. J., Clemens, J. D. & Clarke, D. B. 1987. Models for granitoid evolution and source compositions. J GEOL **95**, 731–49.

White, A. J. R. 1995. Suite concept in igneous geology. In Leon T. Silver 70th Birthday Symposium and Celebration. 113–6. The California Institute of Technology.

White, A. J. R. & Chappell, B. W. 1988. Some supracrustal (S-type) granites of the Lachlan Fold Belt. TRANS R SOC EDINBURGH EARTH SCI **79**, 169–81.

Williams, I. S. 1992. Some observations on the use of zircon U–Pb geochronology in the study of granitic rocks. TRANS R SOC EDINBURGH EARTH SCI **83**, 447–58.

Williams, I. S. 1995. Zircon analysis by ion microprobe: the case of the eastern Australian granites. *In Leon T. Silver 70th Birthday Symposium and Celebration* 27–31. The California Institute of Technology.

Wyborn, D. 1983. *Fractionation processes in the Boggy Plain zoned pluton.* Ph.D. Thesis, The Australian National University, Canberra.

Wyborn, D. & Chappell, B. W. 1986. The petrogenetic significance of chemically related plutonic and volcanic rock units. GEOL MAG **123,** 619–28.

Wyborn, D., Chappell, B. W. & Johnston, R. M. 1981. Three S-type volcanic suites from the Lachlan Fold Belt, southeast Australia. J GEOPHYS RES **86,** 10335–48.

Wyborn, D., Turner, B. S. & Chappell, B. W. 1987. The Boggy Plain Supersuite: a distinctive belt of I-type igneous rocks of potential economic significance in the Lachlan Fold Belt. AUST J EARTH SCI **34,** 21–43.

Wyborn, L. A. I. & Chappell, B. W. 1983. Chemistry of the Ordovician and Silurian greywackes of the Snowy Mountains, southeastern Australia: an example of chemical evolution of sediments with time. CHEM GEOL **39,** 81–92.

B. W. CHAPPELL, Key Centre in Geochemistry and Metallogeny of the Continents (GEMOC), Department of Geology, The Australian National University, Canberra, ACT 0200, Australia.

Transactions of the Royal Society of Edinburgh: Earth Sciences, **87**, 171–181, 1996

Lachlan Fold Belt granitoids: products of three-component mixing

W. J. Collins

ABSTRACT: The paradox of Lachlan Fold Belt (LFB) granitoids is that although contrasted chemical types (S- and I-types) imply melting of distinct crustal sources, the simple Nd–Sr–Pb–O isotopic arrays indicate a continuum, suggesting mixing of magmatic components. The paradox is resolved by the recognition that the previously inferred, isotopically primitive end-member is itself a crust–mantle mix, so that three general source components, mantle, lower crust and middle crust, comprise the granitoids. Based on Nd isotopic evidence, mantle-derived basaltic magmas melted and mixed with Neoproterozoic–Cambrian, arc–backarc-type material to produce primitive I-type, parental granitoid magmas in the lower–middle crust. Ordovician metasediment, locally underthrust to mid-crustal levels, was remobilised under the elevated geotherms and is most clearly recognised as diatexite in the Cooma complex, but it also exists as gneissic enclaves in S-type granites. The diatexite mixed with the hybrid I-type magmas to produce the parental S-type magmas. Unique parent magma compositions of individual granite suites reflect variations within any or all of the three major source components, or between the mixing proportions. For example, chemical tie-lines between Cooma diatexite and mafic I-type Jindabyne suite magma encompass almost all mafic S-type granites of the vast Bullenbalong supersuite, consistently in the proportion Jindabyne : Cooma, 30 : 70. The modelling shows that LFB S-type magmas are heavily contaminated I-type magmas, produced by large-scale mixing of hot I-type material with lower temperature diatexite in the middle crust. The model implies a genetic link between migmatite and pluton-scale, crustally derived (S-type) granites.

Given the chemical and isotopic contrasts of the crustally derived source components, and their typically unequal proportions in the magmas, it is not surprising that the LFB granitoids are so distinctive and have been categorised as S- and I-type. The sublinear chemical trends of the granitoid suites are considered to be secondary effects associated with crystal fractionation of unique parental magmas that were formed by three-component mixing. The model obviates the necessity for multiple underplating events and Proterozoic continental basement, in accordance with the observed tectono-stratigraphy of the Lachlan Fold Belt.

KEY WORDS: granitoid genesis, crystal fractionation, gabbros, S-type, I-type, enclaves, crustal evolution.

The vast, 700 km wide, Palaeozoic Lachlan Fold Belt (LFB) of eastern Australia is characterised by a ubiquitous pile of Ordovician flysch into which granitoids of contrasted geochemical character have intruded. The granitoids, called I- and S-types (Chappell & White 1974), were considered to be derived from lower crustal igneous and sedimentary precursors, respectively, and the contrasts were highlighted in the Kosciusko Batholith by distinct linear chemical trends (Hine et al. 1978). Such chemical trends were considered to indicate a consanguinity of rock types, defined as a granitoid suite, and the chemical variation within suites was ascribed to restite 'unmixing' (White & Chappell 1977; Chappell et al. 1987).

A paradox exists for the eastern LFB granitoids because the marked geochemical differences are not reflected isotopically: the hyperbolic Sr–Nd–Pb–O isotopic array for S- and I-type granitoids (McCulloch & Chappell 1982; McCulloch & Woodhead 1993) suggests a crust–mantle mixing system (Gray 1984, 1990). Gray (1984) argued for a two-component mixing model involving the mantle and Ordovician sediments, but it was rejected using chemical arguments (McCulloch & Chappell 1982; Chappell & White 1992; Chappell 1994). This paper will show that a more complex mixing system is the likely solution to the paradox.

This paper seeks to determine the source components for

the granitoids of the eastern LFB by focusing on the I-type Jindabyne and S-type Bullenbalong suites of the Kosciusko Batholith and the various supersuites of the Bega Batholith (Fig. 1). Over 300 LFB granitoid analyses were obtained from White et al. (1977), White and Chappell (1989) and Beams (1980), and compared with analyses of high- and low-grade Ordovician (450–500 Ma) sediments from the LFB (Wyborn & Chappell 1983; Munksgaard 1988) and Neoproterozoic–Cambrian (530–850 Ma) sediments from the adjacent Adelaide Fold Belt (Turner et al. 1993). The data indicate that S-type granitoids do not image any of the analysed metasediments, but that they bear many similarities to Ordovician sediments of the LFB.

1. General features of Lachlan Fold Belt I- and S-type granitoids

S- and I-type granites of the LFB form an overlapping chemical and isotopic array, and have similar average compositions (Chappell & White 1992; McCulloch & Woodhead 1993). The distinction is partly mineralogical, in that mafic S-type granitoids are peraluminous and contain cordierite, and mafic I-type granites are metaluminous and contain hornblende. At higher silica compositions (>70%),

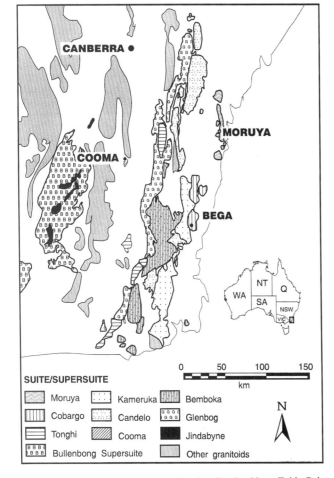

Figure 1 Distribution of granitoids in the Lachlan Fold Belt. Supersuites in the Bega Batholith, and the Jindabyne and Bullenbalong supersuites of the Kosciusko Batholith, are highlighted.

the differences are far more subtle. Another important, understated similarity is that I-type granites contain similar populations of inherited zircons to the S-type granites (Williams *et al.* 1992).

Nonetheless, at low silica contents the two granite types are distinctive and a chemical bimodality is apparent (Chappell & White 1992), which does not accord well with a simple crust–mantle mixing system. Also, many I-type suites include granites with $\approx 60\%$ SiO_2, whereas S-type granites rarely have less than $\approx 64\%$ SiO_2. Furthermore, S-types commonly contain gneissic enclaves, whereas I-types do not, but both contain microgranitoid enclaves (Vernon 1983; Elburg & Nicholls 1995), a feature commonly understated in petrogenetic models. Any successful model must explain the similarities as well as the differences between S- and I-type granites.

2. Linear chemical variations

Chemical variations in LFB granitoid suites has been considered to result from restite unmixing, two–component magma mixing or crystal fractionation. These possibilities are assessed in the following sections.

2.1. Restite unmixing

Linear variation in granitoid suites was used by White and Chappell (1977) and Chappell *et al.* (1987) as a diagnostic criterion of restite unmixing. According to the restite model, the source partitions into felsic melt and mafic residue (restite) during partial melting, and rises *en mass* as a magma. During ascent and emplacement, the restite is progressively removed,

causing the geochemical variation in granite suites. In this model, the source composition is inferred to lie on the chemical trends and can be constrained by back extrapolation. Thus the granites are inferred to be 'images' of their source rocks (Chappell 1979, 1994).

Direct evidence for restite was initially cited as the widespread mafic 'xenoliths' (enclaves) and xenocrystic 'clots' in granitoids (White & Chappell 1977). However, Phillips *et al.* (1981), Wall *et al.* (1987) and Vernon (1983, 1990) have shown that the enclave textures of so-called 'restite' are the result of crystallisation from a melt and the unequivocal evidence for restite is reduced to inherited zircon and possibly calcic plagioclase cores.

2.2. Two-component magma mixing

Simple two-component isotopic mixing models were constructed to account for the $\varepsilon_{Nd–Sr}$ hyperbolic array for LFB granitoids by Gray (1984, 1990), who argued that the end-member components were defined by the low-Si and high-Si extremes of the linear chemical variation trends of granitoid suites. This model contrasts with that of McCulloch and Chappell (1982), and has been rejected by Chappell *et al.* (1987) and Chappell (1994), among others, largely on geochemical grounds, because the specific crust–mantle mix required by the isotopes could not be matched with the bulk–rock chemical composition of the granitoids. Gray (1990, 1995) has repudiated some of these claims, but the general criticism is still valid.

These arguments do not preclude all mixing models. They simply imply that if mixing has occurred, the mixing line must be defined by a vector other than the variation within granitoid suites. If the linear variation is not restite-controlled, nor the result of simple crust–mantle mixing, what is it?

2.3. Crystal fractionation

Many workers have demonstrated that chemical variation in some LFB granitoid suites is a result of crystal fractionation, including the S-type Strathbogie Granite (Phillips *et al.* 1980) and the Violet Town Volcanics (Clemens & Wall, 1984), and the I-type Jindabyne Suite (Wall *et al.*, 1987).

Chemical variation in the I-type Moruya Suite can also be explained by crystal fractionation. The suite is a closed isotopic system (Compston & Chappell 1979). An arbitrary parent (MG 17) and daughter (MG 20) were taken as representative end-members of the suite (Griffin *et al.* 1978), along with mineral analyses from the tonalite (Table 1). Removal of plagioclase (29%), hornblende (20%) and minor biotite (3·7%) accounts for the fractionation, leaving a cumulate with composition comparable with some gabbroic diorites in the Bingie Complex. Using these mass balance constraints and mineral liquid partition coefficients for dacitic magmas (Rollinson 1993), calculated Ba, Rb and Sr variation matches the observed trends (Fig. 2), indicating that crystal fractionation is a viable mechanism for trace element variation in the Moruya Suite. The chemical similarity of the calculated cumulate (Table 1) with some Bingie Bingie gabbroic diorites is independent evidence that the mafic Moruya tonalite is a parent magma, not a cumulate.

Excluding the zoned intrusion of the Boggy Plains Suite (Wyborn *et al.* 1987), a chemical hiatus exists between the gabbroids and related granitoid suites of the LFB at 55–60% SiO_2. Given this common lower SiO_2 limit (60%), it is likely that the mafic tonalites are parent magmas rather than cumulates. Furthermore, cumulus textures are not common in the tonalites, unlike many gabbroic and dioritic (appinitic) rocks. Thus many LFB granitoid suites could be the products

Table 1 Major element modelling of Moruya Suite granites. Residual sum of squares, 0·261.

	(Parent) Tuross Tonalite	(Daughter) Bodalla Adamellite	Plagioclase (An₄₇)	Biotite (diorite)	Hornblende (diorite)	Apatite	Magnetite	Cumulate
SiO₂	60·78	74·91	56·62	36·79	45·55	0	0	50·19
TiO₂	0·84	0·15	0	4·1	2·49	0	0	1·23
Al₂O₃	16·26	13·37	27·6	14·01	8·33	0	0	18·97
Σ FeO	5·18	1·58	0	17·97	12·72	0	100	8·36
MnO	0·11	0·07	0	0·15	0·23	0	0	0·1
MgO	3·51	0·24	0	13·36	13·98	0	0	6·2
CaO	5·55	1·1	9·56	0·1	12·19	56·84	0	9·79
Na₂O	3·98	4·21	6·21	0·83	2·27	0	0	4·26
K₂O	1·75	3·5	0	9·44	0·53	0	0	0·86
P₂O₅	0·3	0·06	0	0	0	43·16	0	0·04
Proportion used (%)	100	(f) 45·91	29·1	3·68	20·01	0·05	1·25	54·09

Note: f = Fraction of liquid remaining.

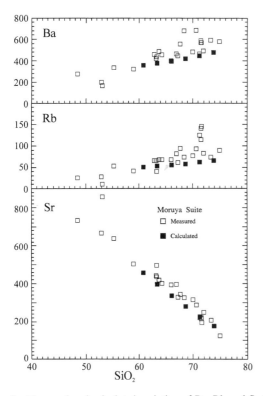

Figure 2 Measured and calculated variation of Ba, Rb and Sr in the Moruya Suite. MG 17 (Griffin *et al.* 1978) is the arbitrary parent and bulk distribution coefficients were calculated from mass balance modelling. Mineral–liquid partition coefficients are from Rollinson (1993).

of fractional crystallisation from parent magmas containing ≥60% SiO₂.

2.4. Partial melting

Granitoid suites which contain numerous spatially separate plutons that may extend over several hundred kilometres, such as some Bega Batholith supersuites (Fig. 1), are unlikely to have formed by crystallisation from a single magma. Nonetheless, many of these suites show coherent chemical trends, suggesting derivation by different degrees of melting from a uniform, extensive source. Each partial melt would then produce a pluton that has chemical similarity with the adjacent pluton, provided that the source rock is not affected by prior production of that adjacent pluton. If affected, the source will become dehydrated and A-type granites are likely to result (Landenberger & Collins 1996). A further restriction is that the individual plutons cannot fractionate extensively, as that is likely to produce chemical arrays that are divergent to the partial melting trend, which is not observed.

In summary, chemical variation in both S- and I-type granite suites in the LFB can be explained equally well by crystal fractionation, restite unmixing or partial melting. However, experimental, textural and numerical analysis, and the remarkably linear chemical trends, suggest that fractionation is the most likely alternative. If so, the linear chemical trends cannot be used to constrain source compositions or components, but extrapolate back to parental magma compositions.

3. Low-K, I-type granitoids and coeval mafic rocks

3.1. The Moruya Suite

The Moruya Suite has particular significance as it is the most isotopically primitive suite in the LFB, and was used initially to infer restite-controlled chemical variation (White & Chappell 1977). Also, it is associated with the well-exposed Bingie gabbroid complex, contains abundant microgranitoid enclaves and closely resembles 'Cordilleran-type' low-K granites (Chappell & Stephens 1988).

The Bingie Complex is coeval with, and has close chemical similarities to, the Moruya Suite (Keay *et al.*, unpublished data). Both have low TiO₂, Na₂O, Sr and Zr, relative to other suites of the Bega Batholith (Beams 1980). The only other studied gabbroid complex in the LFB is the Blind gabbro. Like the adjacent Jindabyne Suite, the Blind Gabbro has high Al₂O₃, CaO, K₂O, P₂O₅ and large ion lithophile elements (LILEs), relative to other suites (Griffin *et al.* 1978). These rare gabbroid complexes are particularly important because they have geochemical features that are reflected in the adjacent, low-K granitoid suites, which indicates a fundamental link between mantle-derived magmas and I-type granitoid petrogenesis.

Field and petrographic evidence of enclaves from the Bingie gabbroic complex indicate that they were mingled globules of magma, which was more mafic than the solidifying host tonalite of the Moruya Suite (Vernon *et al.* 1988). Isotopic evidence shows that the larger enclaves are isotopically more primitive than the granitoids and are similar to the mantle-derived microdiorite dykes (Keay *et al.*, unpublished data), indicating that the enclaves were hybrid magmas of gabbroid and tonalite. Taken with the presence of mantled plagioclase and quartz ocelli, the enclaves are interpreted as hybridised crust–mantle magma that has mingled with solidifying granite magma in a high-level pluton.

Granites and gabbroids at Bingie Bingie Point define an ε_{Nd–Sr} isotopic array, indicating crust–mantle mixing (Keay *et al.*, unpublished data). The array encompasses and extends the primitive or 'mantle' end of the LFB granitoid

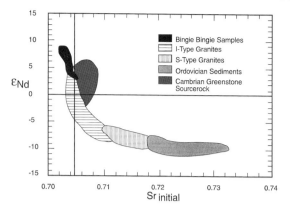

Figure 3 ε_{Nd-Sr} array for LFB S- and I-type granitoids and associated rocks. Note that the Bingie Bingie array defines a crust–mantle mix. As this mix encompasses the primitive isotopic array of the LFB granitoid spectrum, the evolved end-member is the third component of the mixing system (from Keay *et al.*, unpublished data).

Nd–Sr spectrum (Fig. 3) and indicates that the previously considered, single, primitive end-member of LFB granitoids is a two-component system, comprising mantle-derived, hydrous basaltic magmas that mixed with an older meta-igneous crustal component in the lower crust (Keay *et al.*, unpublished data). The typical, more isotopically evolved I-type granitoid suites must contain a third source component.

3.2. Implications for I-type granite petrogenesis
The Moruya–Bingie data indicate that this most isotopically primitive, low-K, I-type suite in the LFB is a crust–mantle mix that produced a parent tonalite magma with $\approx 60\%$ SiO_2. The unique chemical character of the suite (e.g. high TiO_2, high Na_2O and high Sr) reflects the mantle contribution in the granitoid parent magma, as these features are also seen in the basaltic dykes. The crustal end-member, inferred to be derived by the partial melting of older mafic rocks in the lower crust, produced a much more silicic partial melt (70–71% SiO_2; Keay *et al.*, unpublished data), which is incapable of strongly influencing the chemistry of the parent magma. The linear chemical variation that characterises the suite could result from either crystal fractionation or from various mixes of the basaltic magmas with the silicic partial melts derived from the lower crust. Because chemical variation in the low-K Jindabyne Suite also reflects the unique character of the Blind Gabbro, it is suggested that the mafic Jindabyne tonalites are also crust–mantle mixes. Given the similar silica content for many mafic members of I-type granitoid suites ($\approx 60\%$ SiO_2), it is considered that crust–mantle mixing is a general petrogenetic process in the LFB. Thus I-type granites appear to 'image' the mantle component of their parentage, not a particular crustal source, contrary to the restite model.

4. S-type granites

4.1. Images of their sources?
If S-type granites are images of their supracrustal sources (cf. Chappell 1979, 1994), the protoliths should resemble some of the older supracrustal rocks exposed on the Australian continent. Likely candidates for S-type source material include the Neoproterozoic sediments of the Adelaide Fold Belt (AFB), which separates the LFB from the older cratonic interior, and Ordovician sediments of the LFB. Available analyses include 'shales' of varying ages (530–900 Ma) from the AFB (Turner *et al.* 1993) and low- and high-grade Ordovician sediments of

the Cooma Complex (Munksgaard 1988). The 'typical S-type granite' is taken as the Bullenbalong Supersuite, which represents one-third of all exposed granites in the eastern LFB and 94% of the exposed S-types described by Chappell (1984).

Figure 4 indicates that the LFB sediments, including the high-grade gneisses at Cooma, cannot be exclusive geochemical parents to the S-type granites, as they lack sufficient CaO and Na_2O. Analyses of metasediments from the Palaeoproterozoic Arunta Inlier, central Australia, are very similar to the LFB sediments (Collins, unpublished data), and also cannot be sole parents to the S-type granites. The only possible candidates are some high-Ca sediments from the AFB, which are a disparate group of variably aged samples from South Australia, Antarctica and western Victoria (Turner *et al.* 1993). Figure 4 also demonstrates that the MgO and Al_2O_3 contents of some high–Ca sediments plot at the mafic extrapolation of the LFB S-type granite trend, but that the remaining major elements hold no such correlation. Indeed, not one individual sample 'images' the LFB granites for all major elements. Therefore, no sediment from eastern or central Australia, ranging in age from Ordovician to Palaeoproterozoic, is the direct source for LFB S-type granites.

Nonetheless, the S-type granites must contain a major metasedimentary component. Given that the Cooma Granite and LFB sediments plot as one ε_{Sr-Nd} isotopic end-member of the LFB S-type granite system (McCulloch & Chappell 1982; McCulloch & Woodhead 1993), these rocks are now re-evaluated as possible source components.

4.2. Ordovician LFB sediment as a source component
Ordovician LFB sediments form a linear array for most elements on Harker diagrams, which is distinctly different from the S-type granite trends (Fig. 4). Elements such as Fe, Ti, P, V, Cr and Ni, which are compatible in granitic magmas, all decrease more rapidly at higher SiO_2 in the granitoids, whereas many of the LILEs, such as K, Rb, Pb and Th, increase more rapidly. Al_2O_3 and Ga form subparallel, but slightly offset, trends, whereas only Na and Ca form divergent trends at low-Si compositions. The different linear trends of the sediments reflect clay–sand mixing (Wyborn & Chappell 1983), whereas the array for the S-types is considered to reflect crystal fractionation (e.g. Clemens & Wall 1981), as discussed in Section 2.3.

Mafic granites of the Bullenbalong Suite most closely approximate the S-type parent magma, if the variation results from fractionation. Most of the granites plot in the LFB sediment array at 65–70% SiO_2, indicating that the S-type parental magmas are chemically related to the sediments. However, the CaO and Na_2O contents are markedly different, so the relation is not simple.

4.3. Ordovician metasedimentary xenoliths in S-type granites
Figure 4 shows the chemical relation between metasedimentary xenoliths in S-type granites and the LFB sediments. Mica-rich, schistose inclusions and foliated quartzo-feldspathic inclusions from the Bullenbalong and Koetong suites (Chen *et al.* 1989) plot within the well-defined linear chemical array of LFB sediments, with most at the low-Si end of the spectrum. Only the three analysed silicic xenoliths contain more CaO than typical LFB sediments, which may be the result of local diffusion. Many xenoliths also encompass the poorly defined, wide ranging AFB sediment field, but their MgO and CaO contents are lower, and their Al_2O_3 and K_2O contents higher than the AFB sediments. On the other hand, the tight array defined by the xenoliths indicates their much closer affinity with the Ordovician turbidites of the LFB.

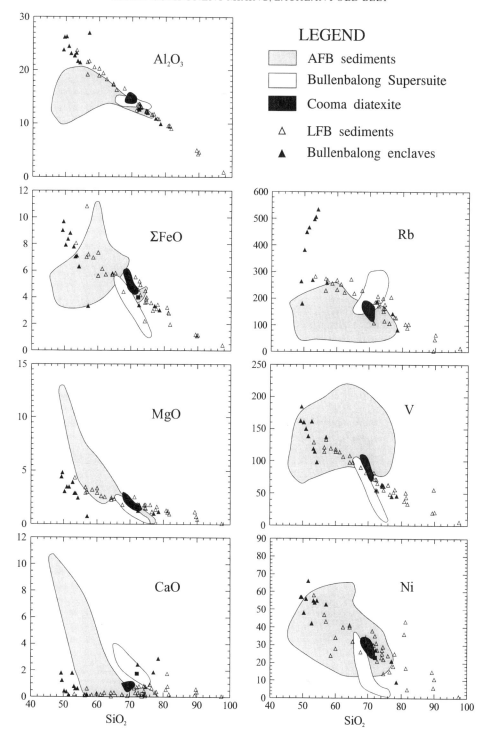

Figure 4 Chemical fields for LFB and AFB sediments, and for S-type granites and gneissic enclaves from the Bullenbalong supersuite. Note that the mafic S-types and Cooma diatexite plot in or close to the LFB sediment field, except for CaO, and that most of the enclaves plot within the LFB sediment field.

Recognition of Ordovician metasedimentary xenoliths in the Bullenbalong and Koetong suites has strong implications for the origin of LFB S-type granites. The xenoliths have been interpreted as refractory lithologies from the source region (Chen *et al.* 1989). The absence of LFB xenoliths between 58 and 72% SiO_2 (Fig. 4) suggests preferential melting of metasediments within this compositional range, which is the 'fertile window' described by White and Chappell (1988). The similarity of zircon inheritance patterns between S-type granites and Ordovician metasediment (Williams *et al.* 1992) and their strong chemical similarities (Section 4.2) is also consistent with a major component of LFB sediment in the S-type granites.

4.4. Remobilisation of Ordovician sediments

Ordovician sediments have partially melted to form granitoid magma, evidenced from field, chemical and isotopic relations in the migmatitic Cooma Complex (Fig. 1). Lithological contacts grade over several metres from metatexite, where compositional layering or stromatic structure is still apparent, into diffuse zones of schlieric diatexite where the original structure is outlined by biotite schlieren. Locally, the structure is lost and small diffuse pods of homogeneous diatexite formed, which contain millimetre- to metre-scale restitic and refractory components of metatexite.

Chemical analyses of the Cooma 'diatexite' yield a granitoid composition which plots on the chemical trends defined by the surrounding metasediments (Munksgaard 1988). Considerable isotopic overlap between 'diatexite' and metasediments also exists (Munksgaard 1988). Accordingly, the granitoid is an 'image' of, and presumably was derived from equivalents of, the surrounding LFB sediments and can be regarded as a restite-rich granitoid magma or a homogeneous diatexite. The term diatexite is preferred as it highlights the link with the surrounding sediments and serves to distinguish the rock from the typical higher level *contact aureole* S-type granites of the LFB (cf. White *et al.* 1974).

The average Cooma diatexite corresponds to the 'fertile window' of S-type granites (cf. White & Chappell 1988) and is similar to that produced experimentally by melting New Zealand greywackes (Conrad *et al.* 1988). The Cooma diatexite indicates that Ordovician sediment is capable of being 'remobilised' by partial melting and incorporation of restite and refractory material, to produce a homogeneous granitoid magma that corresponds to the evolved isotopic end-member of the ε_{Nd-Sr} LFB granitoid spectrum.

5. Three-component magma mixing in the Lachlan Fold Belt

The isotopic data for LFB granitoids define a strongly covariant ε_{Nd-Sr} array, which has been previously interpreted as a two-component mixing system. However, detailed investigation of the isotopically 'primitive' end-member reveals that it is a two-component system of mantle magmas and partially melted mafic lower crust (Keay *et al.*, unpublished data). Therefore, three components are evidently involved in LFB granitoid genesis.

5.1. The 'primitive' end-members

The ε_{Nd-Sr} isotopic array defined at Bingie Bingie Point indicates that the most isotopically primitive granitoid suite in the LFB, the Moruya suite, was derived by the mixing of depleted mantle melts with silicic partial melts derived from a dominant mafic lower crustal source component. The mantle component is under-represented in the LFB because of deep crustal mixing (see Section 5.4), but the Palaeozoic age coincidence of inferred widespread mantle metasomatism beneath the LFB (Griffin *et al.* 1988) is consistent with large-scale mantle activity during LFB granite genesis. Furthermore, lithospheric mantle xenoliths from SE Australia almost entirely fall within the field defined by isotopically primitive LFB granitoids (Keay *et al.*, unpublished), suggesting that mantle magmas with isotopic compositions similar to the xenoliths could have contributed to the end-member granites.

Nd model ages from the primitive Moruya suite suggests that the crustal source component was <700 Ma old (Keay *et al.*, unpublished data), which overlaps with the 450–630 Ma age of abundant inherited zircons in granites and mafic enclaves from the Bega Batholith (Chen & Williams 1990). Thus two independent lines of isotopic evidence suggest that the lower crustal I-type source component of many eastern LFB granites is ≤630 Ma old. The occurrence of fault–bound slices of Cambrian boninitic, tholeiitic and andesitic greenstones in western Tasmania (Crawford & Berry 1992), central and western Victoria (Crawford *et al.* 1984) and as ophiolite complexes at the eastern margin of the LFB (Aitchison *et al.* 1992) strongly suggests the presence of widespread arc, back-arc crust beneath the ubiquitous Ordovician sedimentary pile. This arc-like material could be the lower crustal component of LFB granitoids. Experimental evidence indicates that dehydration melting of similar mafic, hydrated crust should produce magmas with ≈70% SiO_2 (Beard & Lofgren 1993).

Mixing of 60% silicic, lower crustal derived melts with 40%

of basaltic magmas, represented by the most primitive basaltic dykes at Bingie Bingie Point, satisfies the isotopic and chemical constraints of the parental Moruya tonalite (Keay *et al.*, unpublished data). As the isotopic array at Bingie Bingie Point defines the entire isotopically primitive spectrum, it is considered that Neoproterozoic–Cambrian lower crustal and depleted mantle components were generally involved in I-type magma genesis in the LFB.

5.2. The 'evolved' end-member

The narrow ε_{Nd-Sr} isotopic array for evolved LFB granitoids projects to a well-defined crustal end-member, with isotopic and geochemical characteristics similar to the Ordovician metasediments (McCulloch & Woodhead 1993, table 1; McCulloch and Chappell 1982). Furthermore, field, geochemical and isotopic evidence suggests that the sediments were 'remobilised', yielding melts such as the Cooma diatexite. Therefore this sedimentary material is the probable third component within LFB granitoids.

5.3. Three-component mixing model

The Cooma diatexite is chemically distinct from typical S-type granites of the LFB (Fig. 4). It typically contains higher TiO_2, FeO^T, MgO, Pb, Zr, Nb, Ni, Cr, Zn, Ga and much lower CaO and Na_2O than other LFB S-types of similar silica range (68–71% SiO_2). Al_2O_3, K_2O, P_2O_5, Ba, Rb, Sr, U, Th, Sr, Y and Cu are similar. However, the diatexite defines one of the isotopic end–members of the LFB spectrum. How can the chemical and isotopic data be reconciled?

Mafic S-type granites of the Bullenbalong Suite plot in a tightly defined field between the Cooma diatexite and mafic tonalites of the Jindabyne Suite (Fig. 5). The field is a mixing line and the mafic S-type granites consistently plot as a 70:30 mix between average Cooma diatexite and average mafic Jindabyne tonalite, with mixing values ranging between 0·63 for Sr and 0·79 for Rb. Most elements can be modelled as a mix of 68–75% 'Cooma diatexite' and mafic I-type tonalite (Figs 5, 6). Accordingly, S-type granites of the vast Bullenbalong suite are considered to represent hybrids of ≈70% 'Cooma-diatexite' and 30% I-type magma, represented by the mafic Jindabyne Suite.

The chemical variation of S- and I-type granites of the Kosciusko Batholith can thus be resolved into mixing and fractionation trends, which are summarised in Figure 6. The mixing line (m) lies between mafic I-type granites of the Jindabyne suite (J) and the Cooma diatexite (C). Following the conclusion of Section 2.4, the mafic tonalites of the I-type Jindabyne and S-type Bullenbalong suites are considered to be parent magmas which fractionate to higher SiO_2 along the vector (f). The mixing line has a positive slope against silica when the diatexite contains a greater abundance of a particular element relative to the Jindabyne parent magma (e.g. Fig. 6b, 6d, 6e) and negative slope for the reverse (Fig. 6a, 6c, 6f). Crystal fractionation trends of the S- and I-type granite suites are subparallel and can be subdivided into the incompatible elements that increase with silica (Fig. 6d, 6e) and the compatible elements that decrease (Fig. 6a, 6b, 6c, 6f). Thus six patterns of chemical variation involving mixing and fractionation are identified.

5.4. Melting and mixing dynamics

A lack of S-type granitoids with compositions more mafic than the mafic Jindabyne tonalite–Cooma diatexite tie-line suggests that the S-types were *not* produced by direct interaction with basalt. This reinforces the argument against simple crust–mantle mixing models and suggests that mixing of geochemically contrasted granitoid magmas (Jindabyne and

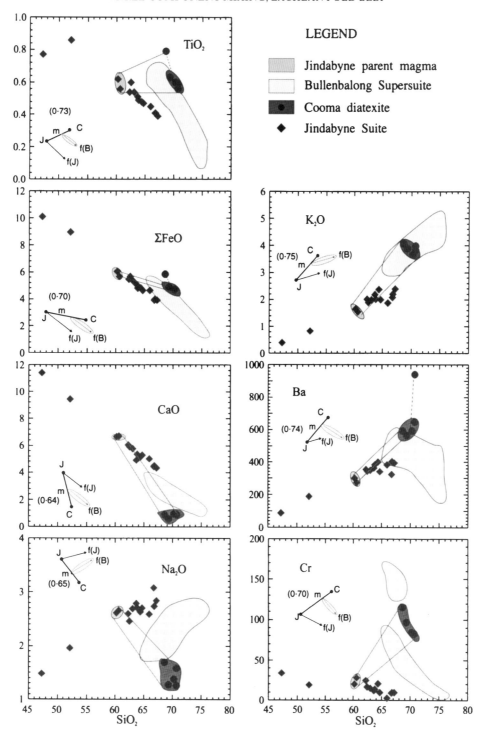

Figure 5 Chemical fields for I-type Jindabyne and S-type Bullenbalong suites of the Kosciusko batholith and the average Cooma diatexite. Mafic S-types consistently plot on a mixing line (m) between Jindabyne mafic magma (J) and Cooma diatexite (C). The mixing proportion for each element is given in parentheses. f(B) and f(J) represent the fractionation trends with increasing SiO_2 for the Bullenbalong and Jindabyne supersuite, respectively.

Cooma) produces the typical S-type magmas of the LFB. In effect, typical S-types represent heavily contaminated I-type magmas, similar to the regional-scale crustal contamination of some Californian I-type granites (Ague & Brimhall 1987).

LFB S-type granites formed at ≈4–6 kbar and ≥800°C (Wyborn et al. 1981; Clemens & Wall 1981, 1984) and were emplaced as *contact aureole* granites, having moved from their source region (White et al. 1974). Thus S-type granites formed at hotter temperatures and slightly greater depths (≈15–20 km) than the Cooma-type diatexite, which equilibrated at 3·5–4·0 kbar and 670–730°C (Ellis & Obata 1992). Therefore,

migmatite complexes such as Cooma are considered to represent the top of a major zone of partial melting in the middle crust of the LFB, below the presently exposed S-type granite batholiths. This is consistent with the structural interpretation of Glen (1992) that the Cooma Complex is an upthrust zone of LFB middle crust.

A likely melting scenario for the LFB is that 400 Ma ago, upwelling mantle-derived magmas intra- and underplated the Neoproterozoic–Cambrian lower crust, which began to melt. The resultant crustally derived partial melts mixed with basaltic material to produce the low-K, isotopically primitive

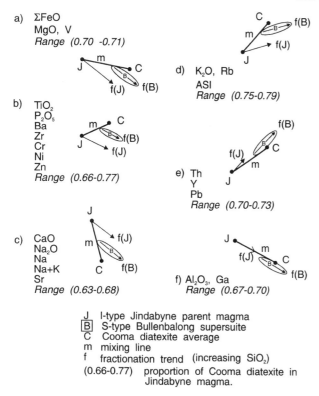

J   I-type Jindabyne parent magma
B   S-type Bullenbalong supersuite
C   Cooma diatexite average
m   mixing line
f   fractionation trend (increasing SiO_2)
(0.66-0.77)   proportion of Cooma diatexite in
              Jindabyne magma.

Figure 6 Summary of Harker diagrams showing that the same relation holds for almost all analysed elements, with the mafic Bullenbalong mafic S-types representing a consistent mix of 63–79% Cooma: 21–37% Jindabyne parent. Note that the Jindabyne parent is a crust–mantle mix, so that the Cooma diatexite is the third component.

tonalites of the eastern LFB. Experimental evidence suggests that crustally derived I-type granites, produced under fluid-absent conditions, will form at 900–950°C (e.g. Roberts & Clemens 1993), so that the crust–mantle mixes are at least this hot. The continued intrusion of mantle-derived melts and advective heating by rising I-type magmas extended the zone of partial melting to mid-crustal levels, where underthrust Ordovician sediment existed locally.

Extensive melting of Ordovician metasediment occurred to levels as shallow as 10–12 km, indicated by the widespread development of diatexite in the Cooma and Omeo complexes. The abundance of diatexite implies fluid-present melting, as suggested for Cooma by Ellis and Obata (1992). This is consistent with the experimental results of Conrad et al. (1988), who indicated that ≈60% melt is produced at temperatures <800°C in the middle–lower crust, for a_{H_2O} ≈0·5; so at higher water activity even greater abundances of melt should be produced at relatively low (<800°C) temperatures.

Hot (>900°C) rising crust–mantle mixes of tonalitic I-type magma would readily incorporate the cooler (≤800°C) Cooma-type diatexite in the middle crust (4–6 kbar) by elevating its temperature and lowering the viscosity and water activity, producing typical S-type granites which are capable of rising as separate contact aureole plutons. Accordingly, the S-types should initially contain restitic and refractory material, but only the larger, more refractory fragments are likely to survive as enclaves and these preserve their original Ordovician chemical signature (Section 4.3). On a smaller scale, only the most refractory minerals will survive, such as zircon, which explains why the S-type granites and more isotopically evolved I-types have similar populations of inherited zircons, and why these populations match those of the Ordovician turbidites (Williams et al. 1992). This model also provides a connection

between migmatites and large-scale bodies of crustally derived (S-type) granite (cf. Brown 1994).

The mantle-derived basaltic component of the granites is likely to be under-represented in the LFB, as indicated by rare coeval gabbroic complexes. According to the three-component mixing model, this material must first interact with lower crustal greenstones, and possibly with fertile metasediments, before reaching the upper crust. Thus it should be efficiently mixed within the middle and upper crust, but is likely to be a major component at deeper levels if underplating continued after the removal of I-type partial melts. The underplated material will eventually conductively cool to form a thick mafic lower crust with much of the pre-existing deep crust removed during I-type magma generation (Collins 1994; Collins & Vernon 1994). This helps to account for the large component of mafic lower crust in the LFB, identified from seismic profiling (Finlayson et al. 1980).

Microgranitoid enclaves are rare to non-existent in the middle crustal exposures (migmatite complexes) of the LFB. According to the model just outlined, enclaves would not be expected at these levels because the lower crustal levels are dynamic mixing zones, where mantle-derived melts are readily disaggregated and assimilated into the newly formed, I-type granitoid magmas, possibly similar to the MASH zones postulated by Hildreth and Moorbath (1988). The mixing zone may extend to higher crustal levels (10–20 km depth), where extensive migmatites exist and S-type granites are produced. Only in the later stages of granite generation, when large-scale mixing is completed and the lower and middle crust is relatively depleted in fertile material, and therefore felsic magma, will the mafic mantle–crust hybrids rise into solidifying high-level plutons to quench as microgranitoid enclaves.

5.5. Bimodality of S- and I-type granites

Chappell and White (1992) highlighted the bimodal character of *mafic* S- and I-type granites of the LFB to argue against a mixing model. The bimodality is also regional, with I-types abundant and S-types absent east of the I–S line, and S-types abundant and I-types rare to the west (White et al. 1976). According to the three-component mixing model, the bimodality is related to the ability to form S-type magmas and is probably affected by two independent factors: (1) the area of deeply underthrust Ordovician sediment is limited and bounded by faults; and (2) reaction kinetics, whereby the formation of diatexite and mixing with I-type magmas is a rapid process. Once the water-saturated solidus is exceeded, fluid-present melting to produce Cooma diatexite (Ellis & Obata 1992) proceeds rapidly and abundant melt is produced, within 50°C of the solidus if the results of Conrad et al. (1988) can be extrapolated to mid-crustal depths. The hot rising I-type magmas enter the zone of migmatites and rapidly mix with them to produce S-type granites. Therefore, in areas of deeply underthrust Ordovician sediment (5–6 kbar), where mid-crustal geotherms exceed the water-saturated granite solidus, typical S-type granites are rapidly generated. In these areas, such as the Kosciusko Batholith, I-types are rare and typically younger, because they can only intrude once the mid-crustal melt zone is depleted, and mixing with sediment is minimal. Nonetheless, some early I-types might be expected if they were emplaced before the sediments became migmatitic.

In areas where the crust was insufficiently thickened, geotherms in the Ordovician turbidite pile may not have exceeded the granitoid solidus, so that diatexites did not form and contamination by metasediments was minimal, such as the Bega Batholith (see Section 5.6). At the edges of thickened crustal welts, along the I–S line, I- and S-type granites are likely to occur in equal abundance, as within the Berridale Batholith. Nonetheless, subequal mixes of I- and Cooma-type diatexite will be rare; they are represented along the I–S line by the two small plutons of the Murrumbucka Suite, and were recognised as mixes by Chappell and White (1992).

5.6. Significance of the Bega Batholith

Granitoids of the Bega Batholith show a remarkable chemical and isotopic asymmetry from east to west. Beams (1980) subdivided the Bega Batholith into seven meridional-trending supersuites (Fig. 1), which show a systematic westward increase in CaO and Sc, concomitant with a decrease in Sr, Al_2O_3 and Na_2O contents (Fig. 7). Westwards, strong isotopic asymmetry in the Bega Batholith is indicated by the decrease of ε_{Nd} from $+3.5$ to -6.6, the increase of initial $^{87}Sr/^{86}Sr$ from 0.7040 to 0.7094, and the increase of δO^{18} from 7.9 to 9.5 (Chappell et al. 1991). SHRIMP analysis of inherited zircon in the Bega Batholith indicates an increase in inheritance westwards, which Williams et al. (1992) interpreted to result from a westerly increase in the amount of a sedimentary component in the granites. What is the sedimentary component?

Williams et al. (1992) also stated that the inherited zircon populations in S- and I-type granites are similar and that they mimic the pattern of the Ordovician metasediments, although the proportions are different. McCulloch and Woodhead (1993) demonstrated a strongly covariant ε_{Nd-Sr} array for S- and I-type granites of the Berridale and Kosciusko batholiths, and of the I-type Bega Batholith further east. A colinear array of ε_{Nd} with $^{207}Pb/^{206}Pb$ also exists for the Bega and Berridale batholiths (McCulloch & Woodhead 1993), and of δO^{18} with ε_{Nd} and ε_{Sr} in the Berridale and Kosciusko batholiths (McCulloch & Chappell 1982). Thus the three batholiths appear to be part of a similar isotopic mixing system and the westwards transition in isotopic systems reflects a progressive increase in Ordovician metasedimentary material within the deep crust of the LFB.

6. Tectonic implications

The three-component mixing model obviates the need for successive underplating events beneath the LFB required to accommodate the 'remagmatism' model for I-type granite genesis (Chappell & Stephens 1988). It also overcomes the severe problem of requiring Cordilleran tonalites (and presum-

ably a Cordilleran arc) beneath the Ordovician turbidites, as these are the presumed source for the LFB I-type granitoids (Chappell & Stevens 1988, Chappell 1994).

The postulated Proterozoic granite protoliths in the deep crust of the LFB, based on Nd model ages of the S-type granites (McCulloch & Woodhead 1993), is also probably incorrect. Rather, the Ordovician sediments are recycled Proterozoic detritus with Nd model ages of ≈ 1700 Ma (cf. Turner et al. 1993), consistent with palaeocurrent and paleogeographical data which indicate derivation from the Australian craton (e.g. Cas 1983). Variable incorporation of the metasediments into the I-type magmas causes the range of Mesoproterozoic Nd model ages for the LFB granites. It follows that the concentration of inherited cores at ≈ 1000–800 and 650–450 Ma in LFB granites reflect peaks of detrital populations from the remobilised Ordovician metasediment, combined with a dominant $\leqslant 630$ Ma lower crustal component derived from partial melts of the underlying, deeper crustal greenstone succession and perhaps the lithospheric mantle. Accordingly, a Proterozoic basement of continental character under the LFB is not required, consistent with the lack of field evidence for its existence.

The model also removes the necessity for crustal-scale Pb isotopic homogenisation of the crust immediately before LFB granite genesis (cf. McCulloch & Woodhead 1993), as the positive correlation of Pb–Pb and Nd model ages can be ascribed to three-component mixing. In this case, the Ordovician turbidites represent a relatively homogeneous, high-Pb crustal component that dominated the Pb isotopic mixing system in the granitoids.

Melting of Ordovician sediments to produce S-type granites requires that the thin turbidite pile was underthrust to mid-crustal levels (≈ 15 km) in the Omeo–Kosciusko–Cooma region before melting began. The P–T–deformation path for the Cooma complex indicates that peak metamorphism (and granite emplacement) was late in the structural history (Johnson & Vernon 1995), after crustal thickening. Given that the Cooma Granodiorite is one of the oldest in the LFB (Williams 1995, pers comm.), the general presence of S-type granites implies prior crustal thickening of the Ordovician sediments in the LFB.

A review of structural evidence for the 'Benambran Orogeny' indicates that deformation diminishes in extent east of the Cooma Complex, in the vicinity of the western margin of the Bega Batholith (Collins & Vernon 1992; Fergusson & Coney 1992). A consequence is that the amount of Ordovician metasediment in the mid-crust would decrease eastwards. This accords well with the chemical and isotopic asymmetry in the Bega Batholith, which suggests an easterly decrease of Ordovician sediment in the granitoids. The restriction of S-types to the 'Kosciusko and Wagga basement terranes' (Chappell et al. 1988), largely coincides with the extent of the Benambran Orogeny, and the I–S line is considered to be the eastern edge of deeply underthrust Ordovician sediment. Thus structural evidence provides independent evidence for localised crustal thickening of Ordovician sediments before S-type granite generation, as required by the three-component mixing model.

The lower crust beneath the Ordovician turbidite pile is inferred to be arc–back-arc material with relatively primitive isotopic signature (e.g. Nelson et al. 1984). It is comparable, isotopically, with the mantle end-member, but contrasts strongly with the chemical and isotopic signature of the sediments, which are largely Proterozoic detritus. Accordingly, unequal mixing of the various components will produce a strongly covariant isotopic array and obvious chemical contrasts in the LFB. This contrast differs from granitoids

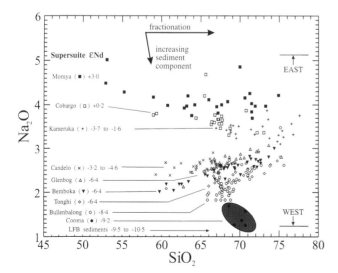

Figure 7 Harker diagram showing the systematic westward decrease of Na_2O and ε_{Nd} in supersuites across the I-type Bega Batholith. Note that the S-type Bullenbalong suite Cooma diatexite and LFB sediments, are part of this chemical system.

produced in many active continental margins because arc detritus is commonly rapidly recycled back into the source region by subduction–accretion processes, so the chemical distinction between meta-igneous and metasedimentary is minimal and the contrasts of subsequently produced I- and S-type granites is subdued. Thus the contrasting S- and I-type granites of the LFB might not be the product of typical continental arc magmatism.

7. Conclusions

A major thermal anomaly in the LFB, at ≈ 400 Ma, generated by underplating and intraplating of crust by mafic magmas, caused extensive partial melting and remobilisation of LFB crust. The mantle-derived magmas melted and mixed with Neoproterozoic–Cambrian, arc–back-arc crust to produce tonalitic I-type magmas. These hot ($>900°C$) rising magmas readily mixed with cooler ($\leqslant 800°C$) Cooma-type diatexite, derived from Ordovician sediment in the middle crust (4–6 kbar), producing typical S-type granites. Thus LFB granitoids are the product of mixing three general components: mantle-derived magmas, partially melted mafic lower crust and supracrustal middle crust. The crustal components were the two major pre-existing components of the LFB, and direct evidence of the 400 Ma mantle contribution are the coeval gabbroic complexes and basaltic dykes and the widespread microgranitoid enclaves within the granitoids. Therefore, the LFB granites reflect crustal-scale reworking processes, triggered by mantle advection, leading to the production of stable continental crust from rocks that initially formed a thin oceanic crust.

Acknowledgements

This work was supported by ARC grant (A39230976). Bill Landenberger and Hope Nesmith are sincerely thanked for help in manipulation of data and drafting of figures. Simon Beams allowed access to his PhD thesis on the Bega batholith. I also thank Ron Vernon, Bob Pankhurst and an anonymous referee for clarification of some ideas and an enhanced presentation of the manuscript.

9. References

Ague, J. J. & Brimhall, G. H. 1987. Granites of the batholiths of California: products of local assimilation and regional-scale crustal contamination. GEOLOGY 15, 63–6.

Aitchison, J. C., Ireland, T. R., Blake, M. C. Jr & Flood, P. G. 1992. 530 Ma zircon for ophiolite from the New England Orogen: oldest rocks known from eastern Australia. GEOLOGY 20, 125–8.

Beams, S. D. 1980. *Magmatic evolution of the southeast Lachlan Fold Belt, Australia.* Ph.D. Thesis. La Trobe University.

Beard, J. S. and Lofgren, G. E. 1991. Dehydration melting and water-saturated melting of basaltic and andesitic greenstones and amphibolites at 1, 3 and 6·9 kb. J PETROL 32, 365–401.

Brown, M. 1994. The generation, segregation, ascent and emplacement of granite magma: the migmatite-to-crustally-derived granite connection in thickened orogens. EARTH-SCI REV 36, 83–130.

Cas, R. A. F. 1983. A review of the palaeogeographic and tectonic development of the Palaeozoic Lachlan Fold Belt of southeastern Australia. SPEC PUBL GEOL SOC AUST 10.

Chappell, B. W. 1979. Granites as images of their source rocks. GEOL SOC AM ABSTR PROGRAMS 11, 400.

Chappell, B. W. & Stephens, W. E. 1988. Origin of infracrustal (I-type)

granite magmas. TRANS R SOC EDINBURGH EARTH SCI 79, 71–86.

Chappell, B. W. & White, A. J. R. 1974. Two contrasting granite types. PACIFIC GEOL 8, 173–4.

Chappell, B. W. & White, A. J. R. 1992. I- and S-type granites in the Lachlan Fold Belt. TRANS R SOC EDINBURGH EARTH SCI 83, 1–26.

Chappell, B. W., White, A. J. R. & Wyborn, D. 1987. The importance of residual source material (restite) in granite petrogenesis. J PETROL 28, 1111–38.

Chappell, B. W., White, A. J. R. & Hine, R. 1988. Granite provinces and basement terranes in the Lachlan Fold Belt, southeastern Australia. AUST J EARTH SCI 35, 505–21.

Chappell, B. W., White, A. J. R. & Williams, I. S. 1991. A transverse section through granites of the Lachlan Fold Belt. *In Second Hutton Symposium Excursion Guide, Canberra.* BMR GEOL GEOPHYS REC 1991 p. 122.

Chen, Y. & Williams, I. S. 1990. Zircon inheritance in mafic inclusions from Bega Batholith granites, southeastern Australia: an ion microprobe study. J GEOPHYS RES 95, 17787–96.

Chen, Y., Price, R. C. & White, A. J. R. 1989. Inclusions in three S-type granites from southeastern Australia. J PETROL 30, 1181–218.

Clemens, J. D. & Wall, V. J. 1981. Origin and crystallization of some peraluminous (S-type) granitic magmas. CAN MINERAL 19, 111–31.

Clemens, J. D. & Wall, V. J. 1984. Origin and evolution of a peraluminous silicic ignimbrite suite: the Violet Town volcanics. CONTRIB MINERAL PETROL 88, 354–71.

Collins, W. J. 1994. Upper and middle crustal response to delamination: an example from the Lachlan Fold Belt, eastern Australia. GEOLOGY 22, 143–6.

Collins, W. J. & Vernon, R. H. 1992. Palaeozoic arc growth, deformation and migration across the Lachlan Fold Belt, southeastern Australia. TECTONOPHYSICS 214, 381–400.

Collins, W. J. & Vernon, R. H. 1994. A rift–drift–delamination model of crustal growth: Phanerozoic tectonic development of eastern Australia. TECTONOPHYSICS 235, 249–75.

Compston, W. & Chappell, B. W. 1979. Sr-isotope evolution of granitoid source rocks. *In* McElhinny, M. W. (ed.) *The earth, its origin, structure and evolution,* 377–426. London: Academic Press.

Conrad, W. K., Nicholls, I. A. & Wall, V. J. 1988. Water-saturated and undersaturated melting of metaluminous and peraluminous crustal compositions at 10 kb: evidence for the origin of silicic magmas in the Taupo Volcanic Zone, New Zealand, and other occurrences. J PETROL 29, 765–803.

Crawford, A. J. and Berry, R. F. 1992. Tectonic implications of Late Proterozoic–Early Palaeozoic igneous rock associations in western Tasmania. TECTONOPHYSICS 214, 37–56.

Crawford, A. J., Cameron, W. E. & Keays, R. R. 1984. The association boninite, low T andesite, tholeiite in the Heathcote Greenstone Gelt, Victoria. AUST J EARTH SCI 31, 161–75.

Elburg, M. A. and Nicholls, I. A. 1995. Origin of microgranitoid enclaves in the S-type Wilson's Promontory Batholith, Victoria: evidence for magma mingling. AUST J EARTH SCI 42, 423–35.

Ellis, D. J. & Obata, M. 1992. Migmatite and melt segregation at Cooma, New South Wales. TRANS R SOC EDINBURGH EARTH SCI 83, 95–106.

Fergusson, C. L. & Coney, P. J. 1992. Convergence and intraplate deformation in the Lachlan fold Belt of southeastern Australia. TECTONOPHYSICS 214, 417–39.

Finlayson, D. M., Collins, C. D. N. & Denham, D. 1980. Crustal structure under the Lachlan Fold Belt, eastern Australia. PHYS EARTH PLANET INTER 21, 321–42.

Glen, R. A. 1992. Thrust, extensional and strike-slip tectonics in an evolving Palaeozoic orogen—a structural synthesis of the Lachlan Orogen of southeastern Australia. TECTONOPHYSICS 214, 341–80.

Gray, C. M. 1984. An isotopic mixing model for the origin of granitic rocks in southeastern Australia. EARTH PLANET SCI LETT 70, 47–60.

Gray, C. M. 1990. A strontium isotopic traverse across the granitic rocks of southeastern Australia: petrogenetic and tectonic implications. AUST J EARTH SCI 37, 331–49.

Gray, C. M. 1995. Discussion of 'Lachlan and New England: fold belts of contrasting magmatic and tectonic development'. J PROC R SOC NSW 128, 29–32.

Griffin, T. J., White, A. J. R. & Chappell, B. W. 1978. The Moruya Batholith and geochemical contrasts between the Moruya and Jindabyne suites. J GEOL SOC AUST 25, 235–47.

Hildreth, W. & Moorbath, S. 1988. Crustal contributions to arc

Chappell, B. W. & White, A. J. R. 1984. Source rocks of S- and I-type granites in the Lachlan Fold Belt, southeastern Australia. PHIL TRANS R SOC LONDON A310, 693–707.

Chappell, B. W. 1994. Lachlan and New England: fold belts of contrasting magmatic and tectonic development. J PROC R SOC NSW 127, 47–59.

magmatism in the Andes of central Chile. CONTRIB MINERAL PETROL **98**, 455–89.

Hine, R. H., Williams, I. S., Chappell, B. W. & White, A. J. R. 1978. Geochemical contrasts between I- and S-type granitoids of the Kosciusko Batholith. J GEOL SOC AUST **25**, 215–34.

Johnson, S. & Vernon, R. H. 1995. Stepping stones and pitfalls in the determination of an anticlockwise P–T–t–deformation path: the low-P, high-T Cooma Complex, Australia. J METAMORPH GEOL **13**, 165–83.

Landenberger, B. & Collins, W. J. Derivation of A-type granites from a dehydrated charnockitic lower crust: evidence from the Chaelundi complex, eastern Australia. J PETROL **37**, 145–70.

McCulloch, M. T. & Chappell, B. W. 1982. Nd isotopic characteristics of S- and I-type granites. EARTH PLANET SCI LETT **58**, 51–64.

McCulloch, M. T. & Woodhead, J. D. 1993. Lead isotopic evidence for deep crustal–scale fluid transport during granite petrogenesis. GEOCHIM COSMOCHIM ACTA **57**, 659–74.

Munksgaard, N. C. 1988. Source of the Cooma Granodiorite, New South Wales—a possible role of fluid–rock interactions: AUST J EARTH SCI **35**, 263–378.

Nelson, D. R., Crawford, A. J. & McCulloch, M. T. 1984. Nd–Sr isotopic and geochemical systematics in Cambrian boninites and tholeiites from Victoria, Australia. CONTRIB MINERAL PETROL **88**, 164–72.

Phillips, G. N., Wall, V. J. & Clemens, J. D. 1980. Petrology of the Strathbogie Batholith: a cordierite-bearing granite. CAN MINERAL **19**, 47–63.

Roberts, M. P. & Clemens, J. D. 1993. Origin of high-potassium, calc-alkaline, I-type granitoids. GEOLOGY **21**, 825–8.

Rollinson, H. R. 1993. *Using geochemical data: evaluation, presentation, interpretation.* Harlow: Longman Scientific and Technical.

Turner, S. P., Foden, J. D., Sandiford, M. & Bruce, D. 1993. Sm–Nd isotopic evidence for the provenance of sediments from the Adelaide Fold Belt and southeastern Australia with implications for episodic crustal additions. GEOCHIM COSMOCHIM ACTA **57**, 1837–56.

Vernon, R. H. 1983. Restite, xenoliths and microgranitoid enclaves in granites. J PROC R SOC NSW **116**, 77–103.

Vernon, R. H. 1990. Crystallization and hybridism in microgranitoid enclave magmas: microstructural evidence. J GEOPHYS RES **95**, 17 849–59.

Vernon, R. H., Etheridge, M. E. & Wall, V. J. 1988. Shape and microstructure of microgranitoid enclaves: indicators of magma mingling and flow. LITHOS **22**, 1–11.

Wall, V. J., Clemens, J. D. & Clarke, D. B. 1987. Models for granitoid evolution and source compositions. J GEOL **95**, 731–49.

White, A. J. R. & Chappell, B. W. 1977. Ultrametamorphism and granitoid genesis. TECTONOPHYSICS **43**, 7–22.

White, A. J. R. & Chappell, B. W. 1988. Some supracrustal (S-type) granites of the Lachlan Fold Belt. TRANS R SOC EDINBURGH EARTH SCI **79**, 169–82.

White, A. J. R. & Chappell, B. W. 1989. *Geology of the Numbla 1 : 000 000 sheet (8624).* Sydney: New South Wales Geological Survey.

White, A. J. R., Chappell, B. W. & Cleary, J. R. 1974. Geologic setting and emplacement of some Australian Paleozoic batholiths and implications for intrusion mechanisms. PACIFIC GEOL **8**, 159–71.

White, A. J. R., Williams, I. S. & Chappell, B. W. 1976. The Jindabyne Thrust and its tectonic, physiographic and petrogenetic significance. J GEOL SOC AUST **23**, 105–12.

White, A. J. R., Williams, I. S. & Chappell, B. W. 1977. *Geology of the Berridale 1 : 000 000 sheet (8625).* Sydney: New South Wales Geological Survey.

Williams, I. S., Chappell, B. W., Chen, Y. & Crook, K. A. W. 1992. Inherited and detrital zircons—vital clues to the granite protoliths and early igneous history of southeastern Australia. TRANS R SOC EDINBURGH EARTH SCI **83**, 503.

Wyborn, D., Chappell, B. W. & Johnston, R. W. 1981. Three S-type volcanic suites from the Lachlan Fold Belt, southeast Australia. J GEOPHYS RES **86**, 10 335–48.

Wyborn, D., Turner, B. S. & Chappell, B. W. 1987. The Boggy Plains Supersuite: a distinctive belt of I-type igneous rocks of potential economic significance in the Lachlan Fold Belt. AUST J EARTH SCI **34**, 21–43.

Wyborn, L. A. I. & Chappell, B. W. 1983. Geochemistry of the Ordovician and Silurian greywackes of the Snowy Mountains, southeastern Australia: an example of chemical evolution of sediments with time. CHEM GEOL **39**, 81–92.

W. J. COLLINS, Department of Geology, University of Newcastle, Newcastle, NSW, 2308, Australia.

Transactions of the Royal Society of Edinburgh: Earth Sciences, **87**, 183–191, 1996

Cretaceous granitoids in SW Japan and their bearing on the crust-forming process in the eastern Eurasian margin

Takashi Nakajima

ABSTRACT: The Cretaceous granitic rocks and associated regional metamorphic rocks in SW Japan were formed by a Cordilleran-type orogeny. Southwest Japan is regarded as a hypothetical cross-section of the upper to middle crust of the Eurasian continental margin in the Cretaceous, comprising (1) high-level granitoids (called San-yo type) and weakly to unmetamorphosed accretionary complexes that are exposed on the back-arc side and (2) low-level (Ryoke type) granitoids with high-grade metamorphites up to migmatitic gneisses on the forearc side. All these granitoids are of the ilmenite series, and predominantly I-type, with a subordinate amount of garnet- or muscovite-bearing varieties in the Ryoke zone, but none of these contains cordierite. These mineralogical variations are likely to depend more on their slightly peraluminous chemistry rather than the pressure differences during crystallisation.

In the eastern part of SW Japan, the granitoids of both levels give K–Ar biotite ages of approximately 65 Ma, whereas the magmatic age of high-level granitoids is approximately 70 Ma, 15 Ma younger than the nearly 85 Ma old lower level granitoids. This implies that the formation of the middle crust started approximately 15 Ma before that of the upper crust. The middle crust material was kept over 500°C for 15–20 Ma after solidification, then it cooled together with the upper crust to 300°C, 6–7 Ma after the formation of the upper crust. The coincidence of cooling history below 500°C of the upper and middle crust may reflect the regional uplift of the crust.

The low-level granitoids have higher $^{87}Sr/^{86}Sr$ initial ratios than those of high-level granitoids in the middle-western part (Chugoku district), but the relationship appears to be opposite in the eastern part. This may imply that the two plutonic series formed by separate magmatic pulses at an interval of *c.* 15 Ma, even though they are not independent, but rather part of a larger episode of crustal growth.

KEY WORDS: SW Japan, crustal formation process, Cretaceous, high-level granites, low-level granites, initial $^{87}Sr/^{86}Sr$ ratios, middle crust, upper crust.

Much of the continental crust world-wide consists of granitic rocks and this crust grows as a result of granitic magmatism. Active convergent plate margins are one of the most important granite-forming sites. They comprise Cordilleran-type orogens which are composed of various kinds of accretionary complex, granitic rocks and their related rocks. More than 60% of the Cordilleran-type orogens in the world are now exposed in the Circum-Pacific regions, called Circum-Pacific plutonic terranes (Roddick 1983), where Middle to Late Mesozoic continental margin-type granitic rocks are widespread.

The Japanese Islands, located on the NW Pacific rim (Fig. 1a), have been an active convergent plate margin from the Late Palaeozoic to the present, so their basement geology is composed of the products of the Cordilleran-type orogeny, such as subhorizontal nappe piles of accretionary complexes of Late Palaeozoic to Cenozoic age and their metamorphosed equivalents (Faure 1985; Isozaki *et al.* 1990). They are intruded by Cretaceous granitic rocks and overlain by related volcanic rocks.

Before the opening of the Japan Sea during the Miocene (Otofuji & Matsuda 1984; Tosha & Hamano 1988; Tamaki *et al.* 1992), the Japanese Islands were a part of the Eurasian continent. Thus these orogenic rocks in the Japanese Islands were formed at the continental margin of Eurasia or its constituent land mass before its amalgamation. This Cordilleran-type orogen is best exposed in SW Japan, where it comprises a hypothetical crustal cross-section from the near-surface to presumably mid-crustal level at the Cretaceous Eurasian margin (Nakajima 1994).

This paper delineates the history of crustal growth of the eastern Eurasian margin based on the temporal, lithological, petrographical and chemical variations of the granitic rocks in the major part of SW Japan.

1. Granitoids in SW Japan

Approximately 30% of the ground surface of SW Japan is occupied by granitoids and related volcanic rocks (Geological Survey of Japan 1982). The isotopic ages of the granitoids in SW Japan show a remarkable concentration between 50 and 100 Ma and a small, but sharp, sub-peak between 10 and 20 Ma. The 10–20 Ma granitoids occur as relatively small bodies, intruding the Cretaceous accretionary complex exposed on the Pacific side of SW Japan. The large number of 50–100 Ma granitoids reflects the huge surface exposure area of Cretaceous granitoids in SW Japan. The exposure area covers nearly all of SW Japan from Kyushu Island to central Japan, extending for nearly 800 km (Fig. 1b) and is similar in size to that of the Sierra Nevada and Peninsular Ranges batholiths in western North America. It is noteworthy that the granitoids were generated not constantly, but at particular times, although SW Japan has been an active margin for more than 200 Ma. This implies that continental crust was formed episodically in SW Japan. The Cretaceous granitoids in SW Japan are calc-alkaline in chemistry (Takahashi *et al.* 1980;

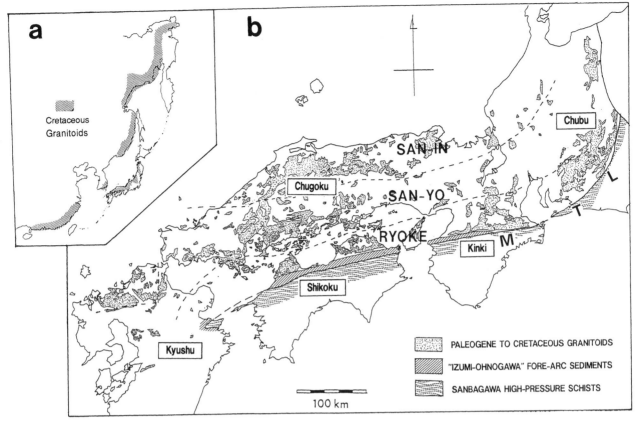

Figure 1 (a) Cretaceous granitic provinces in the NW Pacific rim including the Japanese Islands. Modified after Takahashi (1983). (b) Three granitic provinces in SW Japan. MTL denotes the Median Tectonic Line, a large strike-slip fault which runs through the centre of SW Japan along the arc trend.

Takahashi 1983). They are designated to be arc-type magmatism, like the granitoids from the Mesozoic Andes, which occupy a similar area on the (Y + Nb)–Rb diagram (Fig. 2).

The granitic province of SW Japan is subdivided into three zones, which are arranged parallel to the trend of the arc, the San-in zone, the San-yo zone and the Ryoke zone, from back-arc side to forearc side, respectively (Fig. 1b). The granitoids

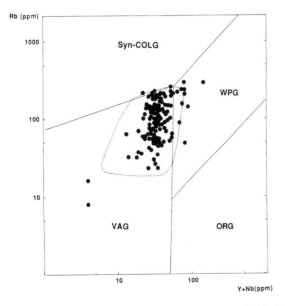

Figure 2 Chemical composition of the Cretaceous granitoids from SW Japan on the (Y + Nb)–Rb discrimination diagram of Pearce *et al.* (1984). The compositional field of the granitoids in the Andes (Pichowiak *et al.* 1990; Cobbing 1990; Cingolani *et al.* 1991) is shown with a dotted line. Abbreviations: VAG, volcanic arc granitoids; Syn-COLG, syn-collisional granitoids; WPG, within-plate granitoids; and ORG, oceanic ridge granitoids.

in each zone have typical characteristics (Table 1), so we refer to these granitoids herein as of San-in type, San-yo type and Ryoke type. The areal distribution boundary between magnetite series and ilmenite series plutons (Ishihara 1977) defines the San-in and the San-yo zones. Granitoids of all three zones are predominantly of the I-type. Some of the Ryoke-type granitoids carry garnet or muscovite, but never cordierite, and have an accessory mineralogy similar to the Chemehuevi Mountains Suite (John & Wooden 1990) and Old Woman–Piute Range batholith (Miller *et al.* 1990), both in SE California, whereas the San-in and San-yo granitoids do not contain these peraluminous minerals. Accordingly, in his review paper, Pitcher (1983) designated the Ryoke-type granitoids as 'intermediate I-type'. However, the surface exposure area of the garnet-bearing and two-mica granitoids is less than 10% of the total Ryoke granitoids. They are common in the Ryoke zone, but are not a representative lithology. In the Kyushu district, the San-yo/Ryoke definition of each pluton has not yet been established well enough to compare with other areas in SW Japan.

These three zones are not independent tectonic units such as juxtaposed suspect terranes because the intrusive contact relationships of the two types of granitoids at the zone boundaries have been documented. All the Ryoke and San-yo granitoids were emplaced during the Cretaceous, whereas the ages of the San-in granitoids range from Cretaceous to Neogene, forming several age clusters (Murakami 1979; Iizumi *et al.* 1985), with the main activity during the Palaeogene.

2. Crustal section of the eastern Eurasian continental margin

In the San-in and San-yo zones, the granitoids are massive to porphyritic and they are sometimes granophyric, lithologically

Table 1 Characteristics of the three types of granitic rocks in Southwest Japan. Pressure estimates with amphibole geobarometer by Hollister *et al.* (1987) are referred from Takahashi (1993).

	San-in zone	San-yo zone	Ryoke zone
Lithology	Porphyritic and granophyric	Massive to porphyritic	Gneissose to massive
I/S type	I	I	I
Magnetite/ilmenite series	Magnetite	Ilmenite	Ilmenite
Ore deposit	Mo	W	None
Accompanied by	Acidic volcanic rocks	Acidic volcanic rocks	Gneiss
Facies	High level	High level	Low level
Estimated pressure (kbar)	<2–3·4	2·9–4·2	4·0–8·5
Age	Palaeogene to Cretaceous	Cretaceous	Cretaceous

indicating high-level plutons. They cross-cut the structure of the unmetamorphosed accretionary complex and the overlying Cretaceous felsic volcanic rocks. On the other hand, Ryoke granitoids are often foliated and accompany low-pressure regional 'Ryoke metamorphic rocks' of various grades up to migmatites of upper amphibolite facies. The foliated granitoids intruded concordantly with the gneissosity of the host metamorphic rocks. The pressure of the highest grade metamorphic rocks is estimated to be 3·5–4 kbar, corresponding to a 12–14 km mid-crustal depth (Nakajima *et al.* 1992; Nakajima 1994). The maximum depth of the Ryoke granitoids is presumably around 12–14 km as they are closely associated with these metamorphic rocks. The pressure estimated with the amphibole geobarometer of Hollister *et al.* (1987) for these granitoids is consistent with this 12–14 km depth range (Table 1).

In the Yanai district in western Japan (Fig. 1b), the Ryoke metamorphic rocks grade down continuously from migmatite of upper amphibolite facies to the virtually unmetamorphosed fossiliferous Jurassic accretionary complex (Ikeda 1991; Nakajima 1994). Metamorphic mineral zones are defined as the muscovite zone, biotite zone, cordierite zone and sillimanite zones I and II in ascending order of metamorphic grade (Nakajima 1994). The metamorphic rocks are intruded by various granitoids. They have narrow contact aureoles of less than 200 m and the mineral zone boundaries are independent of the granitoid bodies, indicating that they are regional metamorphic isograds. The granitoids are of San-yo type in the low-grade and unmetamorphosed zones, whereas those of Ryoke type are in intermediate- to high-grade zones. The high-level granitoids occur with high-level host rocks whereas the low-level granitoids are associated with low-level wallrocks. The independently estimated maximum depths of the granitoids and host rocks are similar. The relationship allows us to interpret the rocks to represent a hypothetical cross-section of upper to middle crust. Therefore, we can view the across-arc transect on the present surface from the San-yo granitoids and weakly to unmetamorphosed accretionary complex to the Ryoke granitoids and high-grade migmatites to be a crustal cross-section of the Cretaceous Eurasian continental margin (Fig. 3). A contemporaneous accretionary prism was being formed at the trench. Cretaceous granitic magmas were intruding Jurassic and pre-Jurassic accretionary complexes—that is, one or two generation older accretionary complexes—to form new granitic continental crust.

As shown in Fig. 4, SW Japan is characterised by a relatively low pressure greenschist–amphibolite facies transition compared with that of many of the documented cross-sections of continental crust in the world, such as Skagit Methow in the western USA (Kriens & Wernicke 1990) or Hidaka, N Japan (Komatsu *et al.* 1986). The SW Japanese metamorphism is similar in pressure regime to the Seward Peninsula, Alaska

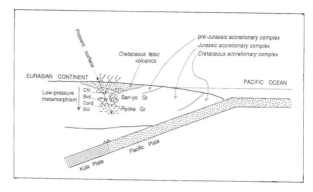

Figure 3 Schematic cross-section of the Cretaceous Eurasian continental margin including the granitic and metamorphic province. Not to scale.

(Hannula *et al.* 1995). This implies a remarkably high effective geotherm of 45–60°C/km at the Cretaceous Eurasian margin.

3. San-yo versus Ryoke granitoids as the upper versus middle crust

Nakajima (1994) presented a model upper to mid-crustal section of SW Japan, in which the upper and middle crust were represented by the San-yo and Ryoke granitoids, respectively. In this chapter, we will have a closer look at each component of the crust in terms of various depths. In the following section, we will examine the geochronological and geochemical differences and similarities between the two granitoid groups and discuss their relationships.

3.1. Geochronology

It has long been known that the K–Ar ages of the granitoids from SW Japan are relatively old in the W and young towards the E (e.g. Teraoka 1977). Nakajima *et al.* (1990) examined the Rb–Sr whole rock isochron ages and biotite ages of the San-yo type granitoids in SW Japan. They concluded that their emplacement ages and cooling ages at about 300°C (Dodson & McClelland-Brown 1984) showed an along-arc lateral variation of eastward younging of the emplacement and cooling ages of the San-yo granitoids from *c.* 100 Ma on Kyushu Island to 70 Ma in the Chubu district, with a coherent time lag of 5–10 Ma from emplacement to cooling. Nakajima *et al.* (1990) concluded that the magmatism of San-yo granitoids shifted along the Eurasian continental margin from W to E, and presented a ridge subdution model in which the Kula–Pacific ridge encountered the Eurasia plate and the RTT triple junction shifted along the continental margin with time. Similar along-arc age shifts were documented in the Early Tertiary granitic rocks from southern Alaska (Bradley *et al.*

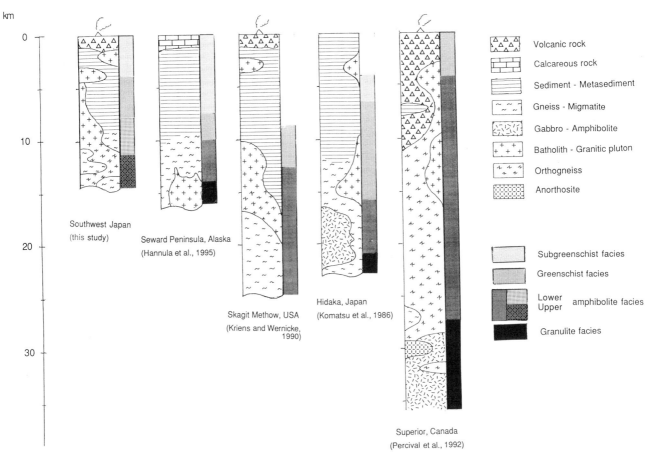

Figure 4 Cross-sections of the documented continental crust of the world.

1993) and the Late Tertiary volcanic rocks in coastal California (Cole & Basu 1995), and were used as an evidence for ridge subduction.

The K–Ar and Rb–Sr biotite ages of the Ryoke granitoids and metamorphic rocks show a similar along-arc variation to the San-yo granitoids. They are nearly parallel with the along-arc variation trend of the Rb–Sr whole rock isochron ages of the San-yo granitoids (Fig. 5). However, there are few reliable Rb–Sr whole rock isochron ages of the Ryoke granitoids, not enough to compare with the San-yo granitoids. Nakajima

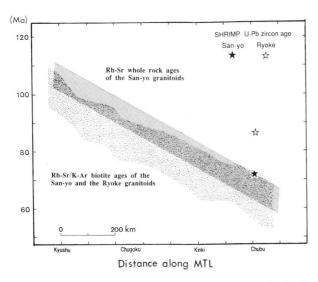

Figure 5 Along-arc age variation of the San-yo and the Ryoke granitoids with preliminary SHRIMP zircon ages. Along-arc position of the plots are defined as the normal projection from the sample locality points onto the Median Tectonic Line (MTL).

(1994) compiled paired K–Ar hornblende and biotite age data from the same rock samples of Ryoke granitoids, which are regarded as the cooling ages of $c.\,500°C$ (Harrison 1981; Dodson & McClelland-Brown 1984; Nakajima et al. 1990) and 300°C, respectively, and found that the cooling rate of the Ryoke granitoids is similar to that of the San-yo granitoids. He extended his ridge subduction model to involve the Ryoke granitoid and metamorphic rocks as well as the San-yo granitoid rocks. Strictly speaking, however, we can conclude only that the cooling rate of both granitoid types below 500°C were similar. The problem of the real magmatic ages of the Ryoke granitoids is still unsolved.

U–Pb zircon ages of a Ryoke granitoid and a San-yo granitoid from the Chubu district (Fig. 5) were preliminarily determined using the SHRIMP in the Australian National University to be $86·1 \pm 1·4$ and $71·3 \pm 1·6$ Ma (Fig. 6a, 6b). The full documentation of these data will appear elsewhere in the near future, but some necessary description will be presented here. Sampling sites for these rocks were chosen such that age discrepancies due to along-arc variation could be avoided (Fig. 5). Although based on only two data points, the Ryoke and the San-yo granitoids have a $c.\,15$ Ma age difference in central Japan. This result may indicate that the middle crust of SW Japan was formed some time before the upper crust. The K–Ar biotite ages were determined to be 64–66 Ma for the Ryoke granitoid (Shibata et al. 1979) and 65–72 Ma for the San-yo granitoid (Kawano & Ueda 1966), both from the same body from which SHRIMP ages were obtained. The Ryoke granitoid had a longer crustal residence time than the San-yo granitoid. Thus it can be presumed that the upper and middle crust of SW Japan were formed from separate magmatic pulses with an interval of approximately 15 Ma between them.

a

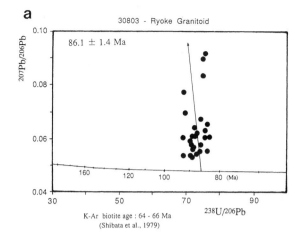

b

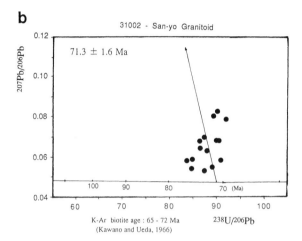

Figure 6 Tera & Wasserburg (1972) concordia diagram for the SHRIMP U–Pb zircon ages of (a) Ryoke granitoid and (b) San-yo granitoid.

3.2. Ryoke granitoids with garnet or muscovite: rock chemistry or crystallisation pressure?

There is no typical S-type granitoid in the Cretaceous granitoids of SW Japan. Both the Ryoke and San-yo granitoids are predominantly I-type (Takahashi *et al.* 1980). Some Ryoke granitoids have garnet and/or muscovite, but not cordierite. The most representative two-mica granites in the Ryoke type granitoids are exposed as relatively large plutons (>10 km in size) of fine-grained felsic granite of $SiO_2 > 70\%$wt. Garnet-bearing granitoids occur as relatively small bodies. Except for concentrations in aplites or pegmatites, the modal amount of garnet is less than 3%. The occurrence of these peraluminous minerals is independent of the areal distribution of the sedimentary host rock, implying that they were not affected by the local assimilation of sediments as in eastern Sierra Nevada (Ague & Brimhall 1987).

In contrast, none of the San-yo granitoids contains these minerals. However, the average whole rock chemistry of the Ryoke and San-yo granitoids does not show any clear difference in ACF type diagrams (Takahashi *et al.* 1980). In general, garnet and muscovite occur in peraluminous granitoids such as S-type rocks (White & Chappell 1977; Chappell & White 1992), as shown graphically in Fig. 7, but it is also widely known that high-pressure crystallisation favours these minerals as well. Many experimental studies have shown that the stability field of garnet expands as the pressure increases (Vielzeuf & Holloway 1988; Green 1992). The Ryoke granitoids are regarded as being from a relatively deeper level than the San-yo granitoids. Thus we cannot simply ascribe the S-type-

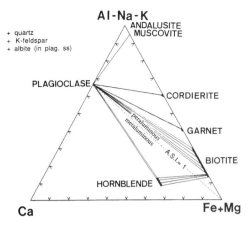

Figure 7 ACF diagram showing the mineral assemblages of peraluminous and metaluminous granitoids (modified after fig. 15 of White *et al.* 1977). Plagioclase–(Mg + Fe) line defines A.S.I. (alumina saturation index) = 1, which divides peraluminous and metaluminous granitoids.

like mineralogy of some Ryoke granitoids to their having a more peraluminous composition than the San-yo granitoids. The effect of whole rock chemistry and pressure conditions on the occurrence of these S-type-like minerals should be examined before we discuss the origin of the Ryoke and San-yo granitic magmas and the relationships between them.

The chemical composition of 46 petrographically well-examined Cretaceous granitoids and a few metamorphic rocks from SW Japan were analysed using a Philips PW-1404 XRF in the Geological Survey of Japan to examine the compositional difference between the Ryoke and San-yo granitoids. Representative analysis data are shown in Table 2. Analytical errors are less than 1% relative for SiO_2, less than 2% relative for the other major elements and below 5% for trace elements (Togashi 1989; Ujiie & Togashi 1992). They are plotted on the SiO_2–A.S.I. [$= Al_2O_3/(CaO + Na_2O + K_2O)_{mol}$] diagram with some published reference data on which petrographic information is available (Fig. 8). In spite of a small overlap, it appears that the Ryoke granitoids are slightly more Al-rich than the San-yo granitoids in most of the SiO_2 range, and that garnet-bearing and two-mica granitoids plot within the relatively Al-rich part of the Ryoke cluster. We could thus conclude that the whole rock chemistry of the Ryoke granitoids spreads towards the garnet-bearing field beyond the anorthite–

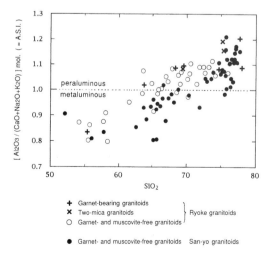

Figure 8 Chemical composition of the San-yo and the Ryoke granitoids in the SiO_2–A.S.I. diagram.

Table 2 Representative chemical of San-yo granitoids (Nos. 1–3), Ryoke granitoids (Nos. 4–7) and a Ryoke pelitic metamorphic rock (No. 8).

	San-yo Granitoid			Ryoke Granitoid				Ryoke pelite
	1	2	3	4	5	6	7	8
SiO_2	68·75	76·76	75·00	75·25	65·96	61·57	66·33	70·92
TiO_2	0·45	0·11	0·15	0·15	0·61	0·86	0·54	0·75
Al_2O_3	15·57	12·70	13·52	14·03	16·62	16·28	16·62	15·66
Fe_2O_3*	4·10	1·44	1·78	1·63	4·50	6·71	4·31	4·10
MnO	0·07	0·03	0·05	0·03	0·07	0·11	0·07	0·04
MgO	0·91	0·05	0·25	0·29	1·85	3·57	0·87	1·81
CaO	3·64	1·22	1·32	1·06	5·01	5·46	3·54	0·53
Na_2O	3·73	3·27	3·27	3·14	3·71	3·01	3·68	1·03
K_2O	2·68	4·42	4·62	4·37	1·52	2·25	3·88	5·03
P_2O_5	0·10	<0·02	0·04	0·05	0·13	0·17	0·15	0·16
Total	100·00	100·00	100·00	100·00	100·00	100·00	100·00	100·00
Zr	180	113	105	70	110	171	200	254
Y	25	45	43	20	10	24	18	27
Sr	236	75	94	172	472	323	230	87
Rb	79	215	194	130	32	87	117	175
Ba	591	329	347	801	464	491	943	793
Zn	58	19	31	55	62	81	54	72
Cu	4	<3	<3	<3	4	17	5	3
Ni	3	3	3	<2	12	27	<2	22
Cr	13	<5	<5	<5	51	119	8	80
V	27	<5	<5	<5	42	137	29	78
Nb	9	11	11	15	8	12	16	17
A.S.I.	1·002	1·029	1·061	1·191	0·995	0·949	1·003	1·937

Notes: Major elements in wt-% and trace elements in ppm by weight. Total Fe given as Fe_2O_3*.

biotite join in the ACF-type diagram (Fig. 7), but not too far, and that the chemistry of the San-yo granitoids is restricted to within the garnet-free field. It seems probable that the whole rock chemistry of the granitic magma controlled the varieties of minerals in the Ryoke granitoids more strongly than the pressure. In other words, the Ryoke and San-yo granitoids are chemically different enough to have a different mineralogy.

3.3. Variation of initial $^{87}Sr/^{86}Sr$ ratios

The $^{87}Sr/^{86}Sr$ and $^{143}Nd/^{144}Nd$ initial ratios of the Cretaceous granitoids in SW Japan have been determined to be in the range 0·705–0·711 and −8·0 to +1·2 (in εNd), respectively (Terakado & Nakamura 1984; Kagami *et al.* 1992). According to their data, $^{87}Sr/^{86}Sr$ and $^{143}Nd/^{144}Nd$ ratios show a systematic negative correlation and the $^{87}Sr/^{86}Sr$ ratio seems to be more sensitive, so the $^{87}Sr/^{86}Sr$ ratio will be mainly dealt with in this paper. The initial $^{87}Sr/^{86}Sr$ ratio of 0·705–0·711 in SW Japan is similar to other I-type granitic provinces such as the Lachlan Fold Belt in eastern Australia (McCulloch & Chappell 1982) and the Sierra Nevada (Bateman 1980). This value seems consistent with their chemical I-type affinity with weakly peraluminous composition.

Figure 9 shows an along-arc variation of initial $^{87}Sr/^{86}Sr$ ratio of the Ryoke and San-yo granitoids, covering the same area as that of the isotopic age diagram (Fig. 5). As a whole, the initial $^{87}Sr/^{86}Sr$ ratios of the Cretaceous granitoids in SW Japan have an along-arc lateral variation: relatively low (0·705–0·708) in the western part and high (0·707–0·711) in the eastern part of SW Japan. Viewed more closely, the initial Sr isotopic ratio of the Ryoke granitoids is higher than the San-yo granitoids in the middle-western part, but lower in the eastern part. In the westernmost part, which corresponds to Kyushu Island, the San-yo/Ryoke definition of each pluton has not yet been well established compared with other areas in SW Japan. Kagami *et al.* (1992) stated that the isotopic features in Kyushu Island are different from other areas and regarded it as a separate segment. If we exclude the four data

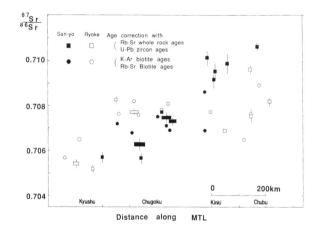

Figure 9 Along-arc distribution of the $^{87}Sr/^{86}Sr$ initial ratios of the San-yo and the Ryoke granitoids.

from the Kyushu Island, the initial $^{87}Sr/^{86}Sr$ ratio of the San-yo granitoid systematically increases eastwards along arc, whereas that of the Ryoke granitoids is nearly flat over the entire area. The two trends cross in central SW Japan. Therefore, the areal distribution patterns of the Sr isotopic compositions of the San-yo and the Ryoke granitoids are different.

4. Discussion

4.1. Origin of the San-yo and the Ryoke magma

On the basis of the differences discussed here in major element chemistry and Sr isotopes between the Ryoke and San-yo granitoids, we could assume that they were derived from separate source magmas or different magmatic events, or both. The magmatic source of these granitoids is discussed and the hypothesis of different magmatic events is presented.

Migmatitic gneisses are widespread in the high-grade zone

of the 'Ryoke regional metamorphic belt', occasionally accompanying small-scale leucocratic pods which could be fossil melt pockets (e.g. Obata *et al.* 1994). Phase relations in the highest grade zone of the Ryoke metamorphic belt indicate near-solidus conditions and dehydration melting could have occurred (Brown & Nakajima 1994). The age-corrected $^{87}Sr/^{86}Sr$ ratio of the Jurassic mudstones in the accretionary complex, which are unmetamorphosed equivalents of those migmatites, were 0·715–0·720 at 80 Ma in eastern SW Japan (Yuhara & Kagami 1995) and 0·715–0·716 at *c.* 90 Ma for the Ryoke metamorphic rocks in the Yanai area (Fig. 1), western SW Japan (Shigeno & Yamaguchi 1976). They are all distinctly higher than those of both the Ryoke and San-yo granitoids. It is concluded that these migmatic gneisses were not the main source of these granitic magmas. The strontium isotopic nature and the predominantly metaluminous whole rock chemistry of these granitoids suggest that the source of these granitic magmas was mainly igneous rocks, not sedimentary. An anatectic melt fraction would have existed and would have mixed with the magmatic melts, but it was not enough to control the chemical composition of the final granitic magmas. The sporadic occurrences of the slightly peraluminous varieties of some Ryoke granitoids may depend on the local variation of the composition of melt influx into the magma.

The lateral variation patterns of initial $^{87}Sr/^{86}Sr$ ratios of the Ryoke and San-yo granitoids provide an important constraint on the generation model of these granitoids. The magmatic ages of the San-yo granitoids young towards the east and the initial $^{87}Sr/^{86}Sr$ ratio is expected to increase slightly because of the age effect of isotopic evolution. However, the actual eastwards increase of the initial $^{87}Sr/^{86}Sr$ is far beyond the correction range for the age effect for the actual Rb/Sr ratio of the granitoids. Moreover, in spite of the eastwards increase of the initial $^{87}Sr/^{86}Sr$ of the San-yo granitoids, their A.S.I. does not increase eastwards nor does it surpass the A.S.I. of the Ryoke granitoids in eastern SW Japan. This indicates that the variation of the initial $^{87}Sr/^{86}Sr$ ratio of the San-yo granitoids cannot be attributed to interaction with a sedimentary component, but is more likely derived from their source magma. Therefore the source region of the San-yo granitoid magmas had a lateral variation with $^{87}Sr/^{86}Sr$ increasing eastwards.

Differences in the lateral variation pattern of the initial $^{87}Sr/^{86}Sr$ ratio between the San-yo and the Ryoke granitoids indicate different magma sources for these granitoids. There are two possible interpretations for their different sources: (1) different sites of magma generation or (2) different times of magma generation. (1) In the San-yo–Ryoke crustal cross-section model (Fig. 3), the different source regions may be interpreted as different depths. The present surface plane is assumed to be subvertical in Fig. 3 for simplicity, but it can also be assumed to be less steep. The across-arc chemical and isotopic variation should also be taken into account. (2) The Ryoke granitoids were formed *c.* 15 Ma before the San-yo granitoids in the Chubu district based on the recent SHRIMP data. Fifteen million years may be long enough to reset the strontium isotopic nature throughout the source region by fluid doping from the subducting slab or input of magmatic melt from beneath. However, it should be noted that the reset event must have worked to reduce the $^{87}Sr/^{86}Sr$ ratio in the western part while increasing it in the eastern part of SW Japan, which does not seem likely. Furthermore, the age data on the time lag between the San-yo and Ryoke magmatism is based on few data and not yet thoroughly resolved, nor

confirmed laterally. This must await additional geochronological information from these granitoids.

4.2. History of the crust-forming process of the SW Japan

The granitoids of both high and low levels of SW Japan have similar cooling ages (given by K–Ar and Rb–Sr biotite ages) and show a systematic along-arc variation of eastwards younging, but the magmatic ages (given by SHRIMP U–Pb zircon ages) of different levels are different. In the eastern part of SW Japan, the granitoids of both levels give K–Ar biotite ages of approximately 65 Ma, whereas the magmatic age of the high-level San-yo granitoids is approximately 70 Ma, 15 Ma younger than the 85 Ma age of the lower level Ryoke granitoids. This implies that the formation of the middle crust had started approximately 15 Ma before the upper crust. Generally, the Cretaceous granitoids in SW Japan have a similar cooling rate of 20–40°C/Ma during cooling from 500°C to 300°C, judging from the K–Ar hornblende and biotite ages (Nakajima 1994). This cooling rate indicates that it took from 5 to 10 Ma to cool from complete crystallisation to 300°C, as opposed to 15–20 Ma implied by the difference between U–Pb SHRIMP zircon age and K–Ar biotite age on the low-level Ryoke granitoids. Therefore, it is concluded that the middle crust material was at over 500°C for 15–20 Ma after solidification, presumably foliated and recrystallised at depth, and then cooled together with the upper crust to 300°C, 6–7 Ma after the formation of the upper crust. The coincidence of cooling history below 500°C of the upper and middle crust may reflect regional uplift. The timing of uplift as determined by thermochronological work on the Ryoke granitoids in the Chubu district with the ^{40}Ar–^{39}Ar method (Dallmeyer & Takasu 1991) and fission track (Tagami *et al.* 1988) accords well with this model. It may imply that the upper and middle crust were formed by different magmatic pulses separated by *c.* 15 Ma in eastern SW Japan, even though the pulses were not independent and fall within a large single geological event.

5. Acknowledgements

The author expresses his sincere thanks to Professor M. Brown for inviting the submission of this paper to this proceedings volume as well as giving critical comments. Dr S. Togashi of GSJ helped during use of the XRF facility and provided helpful discussions and permission to access her unpublished data, for which thanks are due. The author is also grateful to Dr H. Kamioka of the GSJ for his instruction and help on mass spectrometry. Careful review of the manuscript by Professor S. Ishihara of the Hokkaido University, Dr Paul Tomascak of the University of Maryland, Dr C. Mandeville of the GSJ and Dr D. L. Peck of the USGS are gratefully acknowledged. This work was partly supported by the AIST (Agency of Industrial Science and Technology) Research Project 'Behavior and concentration mechanism of the trace elements in the island are crust'.

6. References

Ague, J. J. & Brimhall, G. H. 1987. Granites or the batholiths of California: products of local assimilation and regional-scale crustal contamination. GEOLOGY **15**, 63–6.

Bateman, P. C. 1980. Geologic and geophysical constraints on models for the origin of the Sierra Nevada Batholith, California. *In* Ernst, W. G. (ed.) *The geotectonic development of California*, 71–86. London: Prentice Hall.

Bradley, D. C., Haeussler, P. J. & Kusky, T. M. 1993. Timing of Early Tertiary ridge subduction in Southern Alaska. US GEOL SURV BULL **2068**, 163–77.

Brown, M. & Nakajima, T. 1994. High-T–low-P metamorphism in

the Ryoke belt of Japan: consequence of ridge subduction. ABSTR ANNU MEET GEOL SOC AM **26**, A-214.

Chappell, B. W. & White, A. J. R. 1992. I- and S-type granites in the Lachlan Fold Belt. TRANS R SOC EDINBURGH EARTH SCI **83**, 1–26.

Cingolani, C., Salda, L. D., Herve, F., Munizaga, F., Punkhurst, R. J., Parada, M. A. & Rapela, C. W. 1991. The magmatic evolution of northern Patagonia; new impressions of pre-Andean and Andean tectonics. GEOL SOC AM SPEC PAP **265**, 29–44.

Cobbing, E. J. 1990. A comparison of granites and their tectonic settings from the South American Andes and the Southeast Asian tin belt. GEOL SOC AM SPEC PAP **241**, 193–204.

Cole, R. B. & Basu, A. R. 1995. Nd–Sr isotopic geochemistry and tectonics of ridge subduction and middle Cenozoic volcanism in western California. GEOL SURV AM BULL **107**, 167–79.

Dallmeyer, R. D. & Takasu, A. 1991. Middle Paleocene terrane juxtaposition along the Median Tectonic Line, Southwest Japan: evidence from ^{40}Ar–^{39}Ar mineral ages. TECTONOPHYSICS **200**, 281–97.

Dodson, M. H. & McClelland-Brown, E. 1984. Isotopic and paleomagnetic evidence for rates of cooling, uplift and erosion. MEM GEOL SOC LONDON **10**, 47–64.

Faure, M. 1985. Microtectonic evidence for eastward ductile shear in the Jurassic orogen of SW Japan. J STRUCT GEOL **7**, 175–86.

Geological Survey of Japan 1982. *Geological atlas of Japan*. Tsukuba: Geological Survey of Japan.

Green, T. H. 1992. Experimental phase equilibrium studies of garnet-bearing I-type volcanics and high-level intrusives from Northland, New Zealand. TRANS R SOC EDINBURGH EARTH SCI **83**, 429–38.

Hannula, K. A., Miller, E. L., Dumitru, T. A., Lee, J. & Rubin, C. M. 1995. Structure and metamorphic relations in the southwest Seward Peninsula, Alaska: crustal extension and the unroofing of blueschists. GEOL SOC AM BULL **107**, 536–53.

Harrison, T. M. 1981. Diffusion of ^{40}Ar in hornblende. CONTRIB MINERAL PETROL **78**, 324–31.

Hollister, L. S., Grissom, G. C., Peters, E. K., Stowell, H. H. & Sisson, V. B. 1987. Confirmation of the empirical correlation of Al in hornblende with pressure of solidification of calc-alkaline plutons. AM MINERAL **72**, 231–9.

Iizumi, S., Sawada, Y., Sakiyama, T. & Imaoka, T. 1985. Cretaceous to Paleogene magmatism in the Chugoku and Shikoku district, Japan. CHIKYU-KAGAKU (EARTH SCIENCE) **39**, 372–84 [in Japanese with English abstract].

Ikeda, T. 1991. Heterogeneous biotite from the Ryoke metamorphic rocks in the Yanai district, southwest Japan. J GEOL SOC JPN **97**, 537–47.

Ishihara, S. 1977. The magnetite-series and ilmenite-series granitic rocks. MIN GEOL **27**, 293–305.

Isozaki, Y., Maruyama, S. & Furuoka, Y. 1990. Accreted oceanic materials in Japan. TECTONOPHYSICS **181**, 179–205.

John, B. E. & Wooden, J. 1990. Petrology and geochemistry of the metaluminous to peraluminous Chemehuevi Mountains Plutonic Suite, southeastern California. MEM GEOL SOC AM **174**, 71–98.

Kagami, H., Iizumi, S., Tainosho, Y. & Owada, M. 1992. Spatial variations of Sr and Nd isotopic ratios of Cretaceous–Paleogene granitoid rocks, Southwest Japan arc. CONTRIB MINERAL PETROL **112**, 165–77.

Kawano, Y. & Ueda, Y. 1966. K–Ar dating of the Japanese igneous rocks; (V) the granitic rocks of southwest Japan. J JPN ASSOC PETROL MINERAL ECON GEOL **56**, 191–211 [in Japanese with English abstract].

Komatsu, M., Miyashita, S. & Arita, K. 1986. Composition and structure of the Hidaka metamorphic belt, Hokkaido—historical review and present status. MONOGR ASSOC GEOL COLLAB JPN **31**, 487–93 [in Japanese with English abstract].

Kriens, B. & Wernicke, B. 1990. Characteristics of a continental margin magmatic arc as a function of depth. In Salisbury, M. H. and Fountain, D. M. (eds) *Exposed cross-sections of the continental crust*. NATO ASI SER **C317**, 159–73.

Miller, C. J., Wooden, J. L., Bennett, V. C., Wright, J. E., Solomon, G. C. & Hurst, R. W. 1990. Petrogenesis of the composite peraluminous–metaluminous Old Woman-Piute Range batholith, southeastern California; isotopic constraints. MEM GEOL SOC AM **174**, 99–109.

McCulloch, M. T. & Chappell, B. W. 1982. Nd isotopic characteristics of S- and I-type granites. EARTH PLANET SCI LETT **58**, 51–64.

Murakami, N. 1979. Outline of the longitudinal variation of late Mesozoic to Paleogene acid igneous rocks in eastern Chugoku, Southwest Japan. MEM GEOL SOC JPN **17**, 3–18 [in Japanese with English abstract].

Nakajima, T. 1994. The Ryoke plutonometamorphic belt: Cretaceous crustal section of the Eurasian continental margin. LITHOS **33**, 51–66.

Nakajima, T., Shirahase, T. & Shibata, K. 1990. Along-arc variation of Rb–Sr and K–Ar ages of Cretaceous granitic rocks in Southwest Japan. CONTRIB MINERAL PETROL **104**, 381–9.

Nakajima, T., Ishiwatari, A., Sano, S., Kunugiza, K., Okamura, M., Sohma, T., Kano, T. & Hayasaka, Y. 1992. Geotraverse across the Southwest Japan arc: an overview of tectonic setting of Southwest Japan. In *Metamorphic belts and related plutonism in the Japanese Islands IGC '92 Kyoto, Field Trip Guidebook*, Vol. 5, 171–253. Tsukuba: Geological Survey of Japan.

Obata, M., Yoshimura, Y., Nagakawa, K., Odawara, S. & Osanai, Y. 1994. Crustal anatexis and melt migrations in the Higo metamorphic terrane, west-central Kyushu, Kumamoto, Japan. LITHOS **32**, 135–47.

Otofuji, Y. & Matsuda, T. 1984. Timing of rotational motion of Southwest Japan inferred from paleomagnetism. EARTH PLANET SCI LETT **70**, 373–82.

Pearce, J. A., Harris, N. B. W. & Tindle, A. G. 1984. Trace element discrimination diagrams for the tectonic interpretation of granitic rocks. J PETROL **25**, 956–83.

Percival, J. A., Fountain, D. M. & Salisbury, M. H. 1992. Exposed crustal cross sections as windows on the lower crust. In Fountain, R. J., Arculus, R. & Kay, R. W. (eds) *Continental lower crust*. 317–62. Amsterdam: Elsevier.

Pichowiak, S., Buchelt, M. & Damm, K.-W. 1990. Magmatic activity and tectonic setting of the early stages of the Andean cycle in northern Chile. GEOL SOC AM SPEC PAP **241**, 127–44.

Pitcher, W. S. 1983. Granite type and tectonic environment. In Hsü, K. J. (ed.) *Mountain building processes*, 19–40. London: Academic Press.

Roddick, J. A. 1983. Circum-Pacific plutonic terranes: an overview. MEM GEOL SOC AM **159**, 1–3.

Shibata, K., Uchiumi, S. & Nakagawa, T. 1979. K–Ar dating data— 1. BULL GEOL SURV JPN **30**, 675–86 [in Japanese with English abstract].

Shigeno, H. & Yamaguchi, M. 1976. Study on the metamorphism and plutonism of the Ryoke belt in the Yanai area with Sr isotopic ratio and Rb and Sr contents. J GEOL SOC JPN **82**, 687–98 [in Japanese with English abstract].

Tagami, T., Lal, L., Sorkhabi, R. B. & Nishimura, S. 1988. Fission track thermochronologic analysis of the Ryoke belt and the Median Tectonic Line, Southwest Japan. J GEOPHYS RES **93B**, 13705–15.

Takahashi, M. 1983. Space–time distribution of Late Mesozoic to Early Cenozoic magmatism in east Asia and its tectonic implications. In Hashimoto, M. & Uyeda, S. (eds) *Accretion tectonics in the Circum-Pacific regions*, 69–88. Tokyo: TERRAPUB.

Takahashi, M., Ishihara, S. & Aramaki, S. 1980. Magnetite series/ Ilmenite series vs. I-type/S-type granitoids. In Ishihara, S. & Takenouchi, S. (eds) *Granitic magmatism and related mineralization.* MIN GEOL SPEC ISSUE NO 8, 13–28.

Takahashi, Y. 1993. Al in hornblende as a potential geobarometer for granitoids: a review. BULL GEOL SURV JPN **44**, 597–608 [in Japanese with English abstract].

Tamaki, K., Suyehiro, K., Allan, J., Ingle, J. C. & Pisciotto, K. A., 1992. Tectonic synthesis and implications of Japan Sea ODP drilling. PROC ODP SCI RES **127/128**, 1333–48.

Tera, F. & Wasserburg, G. J. 1972. U/Pb systematics in lunar basalts. EARTH PLANET SCI LETT **17**, 65–78.

Terakado, Y. & Nakamura, N. 1984. Nd and Sr isotopic variations in acidic rocks from Japan: significance of upper mantle heterogeneity. CONTRIB MINERAL PETROL **87**, 407–17.

Teraoka, Y. 1977. Cretaceous sedimentary basin on the Ryoke and Sambagawa belts. In Hide, K. (ed.) *The Sambagawa belt*, 419–31. Hiroshima: Hiroshima University Press [in Japanese with English abstract].

Togashi, S. 1989. Determination of major elements in igneous rocks using Sc/Mo dual anode tube, XRF analytical report 1/89. OPEN-FILE REP GEOL SURV JPN NO 132.

Tosha, T. & Hamano, Y. 1988. Paleomagnetism of Tertiary rocks from the Oga Peninsula and the rotation of the Northeast Japan. TECTONOPHYSICS **7**, 653–62.

Ujiie, M. & Togashi, S. 1992. Determination of Rb, Sr, Y, Zr and Ba in igneous rocks using Sc/Mo tube, XRF analytical report 2/92. OPEN-FILE REP GEOL SURV JPN No 183.

Vielzeuf, D. & Holloway, J. R. 1988. Experimental determination of the fluid-absent melting relations in the pelitic systems. CONTRIB MINERAL PETROL **98**, 257–76.

White, A. J. R. & Chappell, B. W. 1977. Ultrametamorphism and granitoid genesis. TECTONOPHYSICS **43,** 7–22.

White, A. J. R., Williams, I. S. & Chappell, B. W. 1977. *Geology of the Berridale 1:100,000 sheet (8625).* Sydney: New South Wales Geological Survey.

Yuhara, M. & Kagami, H. 1995. Cooling history of the Katsuma quartz diorite in the Ina district of the Ryoke belt, Southwest Japan Arc. J GEOL SOC JPN **101,** 434–42 [in Japanese with English abstract].

TAKASHI NAKAJIMA, Geochemistry Department, Geological Survey of Japan, Higashi, Tsukuba, Ibaraki 305, JAPAN. E-mail: tngeoch@gsj.go.jp

Transactions of the Royal Society of Edinburgh: Earth Sciences, **87,** 193–203, 1996

Monzonite suites: the innermost Cordilleran plutonism of Patagonia

C. W. Rapela and R. J. Pankhurst

ABSTRACT: In Patagonia a Triassic–Early Jurassic Cordilleran interior magmatic belt preceded the widespread eruption of Middle Jurassic syn-extensional rhyolites. Two plutons (La Calandria and La Leona) represent the easternmost plutonic rocks of this belt, >750 km east of the present oceanic trench. They define a high-K calc-alkaline monzonite series in contrast with the main Andinotype arc magmatism of the Pacific margin: they are enriched in large ion lithophile elements (K, Rb, Ba, Sr and Th), LREE and P_2O_5 and depleted in HREE and Y, with low FeO*/MgO ratio. The range of observed compositions (56–76% SiO_2) resulted from high-level fractionation of plagioclase, hornblende, biotite, K-feldspar and accessories (sphene, apatite and zircon).

Initial $^{87}Sr/^{86}Sr$ ratios, average εNd_t and mean depleted-mantle Nd model ages of the two plutons are 0·70487, −0·5 and 1050 Ma for La Calandria and 0·70509, −1·4 and 1125 Ma for La Leona, respectively. They are thus isotopically more primitive than the Middle Jurassic rhyolites, previously attributed to partial melting of Mesoproterozoic mafic lower crust. The preferred model for the origin of the monzonites is remelting of an amphibole- + garnet-bearing, plagioclase-poor, high-K mafic source (?underplating). This occurred in a distal sector of a dying oblique subduction regime, immediately preceding the extensional silicic volcanism.

KEY WORDS: granites, Jurassic, geochemistry, isotopes, petrogenesis, modelling, subduction, South America, Andes, Deseado Massif.

The Mesozoic–Recent Cordilleran plutonism of the western Americas provides the classical example of granitoid batholiths associated with active continental margins. More specifically, that of South America is considered an archetype, defined as *Andinotype* by Pitcher (1979). Despite the many similarities between the *Cordilleran* batholiths of North and South America, there is a major difference in the width of the crustal section affected by coeval or quasi-coeval plutonism. In South America, the Upper Jurassic–Recent Andean batholiths are typically only ≈ 100 km wide, outcropping more or less continuously along the Andean continental margin, and inland plutonic activity is mostly restricted to subparallel belts or satellite plutons emplaced less than 200 km from the axis of the main Cordillera. On the other hand, the North American inland equivalents occupy a broad zone, the *Cordilleran interior*, extending more than 600 km eastwards from the coastal batholiths (Miller & Barton 1990). This mid-Mesozoic to early Cenozoic plutonic activity, which sometimes shows isotopic evidence of ancient crustal signature, was roughly synchronous with compressional tectonics (Miller & Barton 1990). The Late Cretaceous–Tertiary igneous rocks of the Colorado Mineral Belt (Stein & Crock 1990), which include monzonite suites, represent the easternmost zone of the Cordilleran interior, 1000 km inland from the continent edge.

Although the direct role of subduction and mafic underplating in the development of the Cordilleran interior is a matter of debate (see Miller & Barton 1990 and Pitcher 1993 for differing interpretations), it is clear that in the western USA a wide zone along the edge of the adjacent craton underwent crustal deformation, thrusting and contemporaneous magmatism, either directly or indirectly associated with mid-Mesozoic–early Tertiary subduction. It is curious that such plutonism is not common in an equivalent position behind the Andean arc. However, an exception to the typical Andean

batholithic setting occurred in the southern Andes during late Triassic to early Jurassic times. Northern Patagonia is characterised by an oblique calc-alkaline belt cropping out from 41° to 44° 30′ S. This reaches into central parts of the continent in the Gastre area and the south-easternmost outcrops occur in the Deseado Massif, near the Atlantic coast (c. 48° 00′ S; 67° 20′ W). Recent Rb–Sr geochronological studies indicate a predominantly Early Jurassic age for these granitoids, previously assigned to the Upper Palaeozoic (Rapela *et al.* 1991; Varela *et al.* 1991; Gordon & Ort 1993; Pankhurst *et al.* 1993). Rb–Sr data have also provided more precise ages for the overlying Jurassic rhyolites of this area, which show a southward decrease in age from *c.* 190 Ma at 40° S to 168 Ma at 47° S (Pankhurst & Rapela 1995). Both absolute ages and the southwards younging are confirmed by unpublished Ar–Ar dating (Alric *et al.* 1995). The widespread continental extensional regime associated with the eruption of these rhyolites has been interpreted as a precursory stage of Gondwana break-up (e.g. Storey *et al.* 1992).

This paper describes petrological and geochemical aspects of the Early Jurassic suites in the Deseado Massif, >750 km east of the present oceanic trench. The Jurassic magmatic evolution of Patagonia is used to evaluate the role of plate convergence and continental extension during the early stages of supercontinent break-up.

1. Late Triassic–Early Jurassic Cordilleran interior of Patagonia

During Jurassic times, a first-order magmatic segmentation of the Patagonian Andes occurred at about latitude 39° 30′–40° S (Fig. 1). South of this line, the main magmatic units are: (1) widespread mid-Jurassic silicic volcanic rocks in the extra-Andean region; (2) the Patagonian batholith in the coastal Cordillera and sub-Cordilleran areas; and (3) an oblique

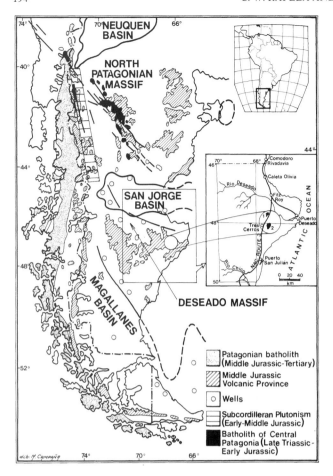

Figure 1 Simplified geological and location map of the Mesozoic magmatic units of Patagonia. Inset shows location of the monzonite suites in the Deseado Massif: (1) La Calandria; and (2) La Leona.

(NNW–ESE) belt of Late Triassic–Early Jurassic batholiths and plutons. The latter comprises three distinctive groups: (a) the Batholith of Central Patagonia in the southwestern sector of the North Patagonian Massif; (b) Sub-Cordilleran plutonism between 40 and 44° S and (c) the plutonic suites of the Deseado Massif in SE Patagonia.

1.1. The Batholith of Central Patagonia

Early Jurassic to Late Triassic batholiths and plutons extend southeastwards from Lake Panguipulli at 40° S in Chile to the Gastre area in central Patagonia (Fig. 1). These rocks are closely associated with the Gastre Fault System, a dextral transcurrent strike-slip fault that displaced southern Patagonia during the early stages of the Gondwana break-up (Rapela & Pankhurst 1992). Very shallow emplacement at less than 1 kbar is inferred for many of these units, based on field and textural relations and on hornblende geobarometry (Rapela *et al.* 1991). Roughly coeval rhyolites and ash-flow tuffs suggest that these bodies had an associated volcanic counterpart. Compared with Cretaceous–Tertiary granitoids of the Patagonian batholith, the central Patagonia plutons are predominantly leuco-monzogranites and granodiorites rather than metaluminous tonalites, diorites and granodiorites (Rapela & Alonso 1991). However, the geochemistry of the Batholith of Central Patagonia does show typical calc-alkaline trends, with no intraplate signature (Rapela *et al.* 1991). Its emplacement has been ascribed to Late Triassic–Early Jurassic subduction along the southwestern margin of Gondwana (Rapela & Pankhurst 1992).

1.2. Sub-Cordilleran plutonic belt

The Late Jurassic–Tertiary Patagonian batholith and its associated volcanic rocks define the axis of the Andean chain between 39° 30′–56° S (Fig. 1). Between 40° S and 42° S, Jurassic plutonic units form the eastern border of the Patagonian batholith (González Díaz 1982; Rapela & Kay 1988). Further south, between 42° 30′ and 44° 30′ S, the Jurassic plutons occur as a NNW–SSE line of isolated plutons along the Pre cordillera of SW Chubut province, informally grouped as a Sub-Cordilleran batholith (Gordon & Ort 1993). This is regarded as separate from the Patagonian batholith, from which no ages older than 160 Ma have been recorded south of 45° S.

The Sub-Cordilleran belt and the Batholith of Central Patagonia may have been at least partially coeval. Gordon & Ort (1993) report Rb–Sr whole rock ages of 200 ± 24 and 183 ± 13 Ma in the sector 41° 45′ S–71° 15′ W, although the Leleque granite at 42° 30′ S–71° 00′ W, considered as part of this suite, gives an age of 164 ± 4 Ma (Pankhurst & Gordon, unpublished data). K–Ar ages range 180–207 Ma in granitoids of the Chubut Pre cordillera and 170–218 Ma in granitoids from deep wells in the central-western sector of the San Jorge basin (see compilation by Linares & González 1990). Isolated plutons of gabbro and diorite along the Sub-Cordilleran sector between 40° S and 44° 30′ S might also be related to this episode, with Jurassic K–Ar ages extending up to 206 ± 10 and 211 ± 10 Ma (Franchi & Page 1980).

The lithology of the Sub-Cordilleran belt is very similar to that of the Cretaceous–Tertiary Patagonian batholith, with a predominance of zoned plutons and metaluminous associations ranging from gabbros, diorite, quartz diorite and tonalite to granodiorite and granite. The inferred tectonic environment is an easterly dipping pre-'Andean' subduction zone along the margin of Gondwana, which seems to be consistent with the lithological associations, and typical calc-alkaline Cordilleran-type geochemistry (e.g. Franchi & Page 1980; Gordon & Ort 1993).

2. Plutonic suites of the Deseado Massif

2.1. Field relations and depth of emplacement

The southeastermost exposures of early Jurassic intrusive rocks occur in the NE of the Deseado Massif at 48° S (Fig. 1). Formerly described as the La Leona formation (Arrondo 1972), these granitoid suites crop out in erosional windows through the widespread silicic units of the Mid-Jurassic Volcanic Province and younger cover. The largest outcrops occur in the La Calandria–La Juanita and Bajo de la Leona districts, where the intrusive rocks were emplaced into Permian–Triassic continental sediments and volcaniclastic rocks (Godeas 1985). An irregular 500–1000 m wide aureole exhibits contact metamorphic assemblages including andalusite–cordierite–alkali feldspar–biotite–quartz (Godeas 1985). The absence of muscovite in the andalusite stability field (Guowei Xu 1994) suggests emplacement at pressures below 2·5 kbar.

A shallow depth of emplacement is also indicated by abundant miarolitic cavities in the more evolved facies and by coeval dacitic dykes with pilotaxitic groundmass. Late aplogranitic bodies and aplites show normative compositions clustering around Q : Ab : Or = 38 : 30 : 32 for La Calandria and Q : Ab : Or = 44 : 33 : 23 for La Leona, suggesting minimum crystallisation pressures between 1 and 0·5 kbar (based on minimum melt compositions at different P_{H_2O} after Luth *et al.* 1964).

2.2. Age of the intrusives

Close control on the ages of these monzonitic suites is provided by Rb–Sr whole-rock isochrons (Pankhurst et al. 1993). Representative granitoids and enclaves from La Calandria (14 samples) give an age of 203 ± 2 Ma (MSWD = 1·9). Eighteen samples from Bajo de la Leona give 202 ± 2 Ma (MSWD = 0·7). Full data and plots are available from the second author. The isochrons include the full compositional range of exposed rocks and demonstrate rapid emplacement of the two complexes in earliest Jurassic times.

2.3. Lithology and petrography

The granitoid complexes of the Deseado Massif are lithologically distinct from those of most Cordilleran batholiths, including that of north Patagonia (Fig. 2). They display a continuous low-quartz trend from diorite to syenogranite, with hornblende–biotite quartz monzonite and quartz monzodiorite as the dominant facies. Such suites have been described as *calc-alkaline monzonitic* or *monzonitic series* (Lameyre 1987). The absence of tonalitic compositions is characteristic.

The *main facies*, quartz monzonite and quartz monzodiorite, contains plagioclase, quartz, K-feldspar, amphibole/clinopyroxene and biotite, with sphene, zircon, apatite, allanite and magnetite as accessory minerals, and secondary chlorite. Idiomorphic to sub-idiomorphic plagioclase (41–54%) is strongly zoned: An_{34} to acid oligoclase at the rim. Large idiomorphic hornblende crystals, occasionally showing resorbed cores of clinopyroxene, are as abundant as biotite (3–10% each). Anhedral K-feldspar (23–33%), with the development of thin myrmekite along the contacts with plagioclase and with inclusions of small individual plagioclase crystals, suggests late crystallisation. Sphene, both as idiomorphic and skeletal crystals, is the most conspicuous accessory (0·2–0·8%). Fine-grained, ubiquitous, disc-shaped enclaves of dark quartz monzodiorite (generally 5–20 cm, but occasionally 50 cm to 2 m) are common. They show a porphyritic texture with plagioclase phenocrysts (An_{44-34}), hornblende and poikilitic biotite, set in a groundmass mosaic of small prismatic plagioclase and allotriomorphic K-feldspar with inclusions of plagioclase, hornblende, biotite and anhedral quartz.

A dark dioritic to quartz monzodioritic *border facies* is developed against the SE contact with the Triassic sediments in La Calandria district (see Pankhurst et al. 1993; Fig. 2). The mineralogy and texture are essentially the same as in the main facies, but with more basic plagioclase (An_{38-40} in the cores), a higher colour index (20–30%) and the absence of K-feldspar in the diorites. Large hornblende phenocrysts have a poikilitic texture enclosing earlier formed plagioclase and pyroxene, suggesting orthocumulate crystallisation.

The intermediate *main facies* is transitional to a *biotite monzogranite facies* in both districts. The colour index decreases to 2–9% and hornblende is usually absent or present only in minor amounts. Modal quartz is low (21–27%); zoned plagioclase (An_{30}–An_{23}; 24–36%) and twinned perthitic K-feldspar (37–46%) make up most of the rock. The most extreme acid compositions are represented by thin (15 cm thick) *leuco-syenogranitic aplites* at La Calandria and 20 cm to 10 m bodies of *aplo-leucomonzogranites* and *syenogranites* at La Leona. Miarolitic cavities and fractures, filled in some instances with copper oxides, are usually associated with the aplogranitic bodies, which often display a mesoscopic graphic texture.

Small bodies of quartz monzodioritic porphyries and *dacitic dykes* with a phenocryst assemblage composed of corroded zoned plagioclase (An_{42}–An_{25}), chloritised hornblende, resorbed quartz and biotite are probably the latest igneous products associated with the monzonitic suites. The phenocryst phases are 'swimming' in a groundmass with a trachytic texture, composed of plagioclase, hornblende, quartz and anhedral K-feldspar. High–temperature hydrothermal alteration particularly affected these later porphyries and dacitic dykes, which together with the aplogranitic bodies carry a widespread Cu- and Mo–mineralisation (Godeas 1985; Márquez 1994).

3. Isotope results and geochemistry

Forty fresh granitoid samples of the monzonitic suites were collected for analysis, avoiding zones of hydrothermal alteration. Rb–Sr data for the same samples showed highly coherent behaviour (Pankhurst et al. 1993), indicating little isotopic or chemical disturbance. Selected chemical analyses of representative sample are shown in Table 1 and new whole-rock Sm–Nd data are given in Table 2. Technical details are supplied in the tables.

The initial $^{87}Sr/^{86}Sr$ ratios of the two monzonite suites (recalculated with slight adjustment from Pankhurst et al. 1993) are distinct, just outside analytical error at the 2σ level (La Calandria $0·70509 \pm 0·00005$; La Leona $0·70487 \pm 0·00006$). La Calandria has εNd_t $-0·8$ to $-1·5$, with one diorite, probably a cumulate, at $-2·5$ (overall mean of $-1·4$), whereas La Leona has εNd_t $-0·3$ to $-0·8$ (mean $-0·5$). These small but significant isotopic differences suggest that each body represents independent evolution from similar, but not identical, parent magmas.

Harker plots show a continuous range of SiO_2 from 56 to 76%, encompassing granitoids, basic enclaves and dykes (Fig. 3). The monzonite suites are calc-alkaline (alkali–lime index $\approx 60·5$) and plot in the subalkaline field of the total alkalis–silica (TAS) diagram (Fig. 3). The *main* and *border* facies are metaluminous, evolving to moderately peraluminous compositions above 70% SiO_2. With increasing SiO_2 content, Al_2O_3, TiO_2, FeO(total), MnO, MgO, CaO and Na_2O decrease, whereas K_2O and the aluminium saturation index (molecular $Al_2O_3/CaO + Na_2O + K_2O$) increase. K_2O is a notable exception to the almost continuous trends, with a gap between 2 and 3% at 65% SiO_2, separating the low-K_2O

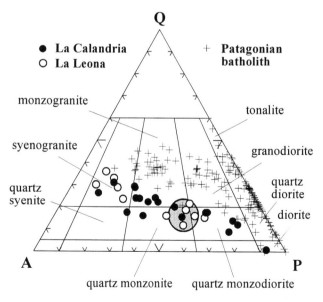

Figure 2 Modal composition of the La Calandria and La Leona monzonite suites in terms of quartz (Qz), alkali feldspar (A) and plagioclase (P), based on a minimum count of 1200 points in thin sections. The solid line outlines the field of the most abundant rocks in each district (main facies). The modal composition of the Patagonian batholith at latitudes 47° and 54° S, shown for comparison, is after Weaver (1988) and Bruce (1988).

Table 1   Selected geochemical analyses of the Early Jurassic monzonite suites, Deseado Massif.

| | La Calandria district | | | | | | | La Leona district | | | | | |
| | Border facies | | Main facies | | Granite | Aplite | Enclave | Enclave | Main facies | | Granites | | Dacite dyke |
Sample	LC-2	LC-22	LC-5	LC-24	LC-29	LC-8	LC-21	LL-44	LL-12	LL-37	LL-40	LL-33	LL-46
Major oxides (wt%)													
SiO_2	55·90	62·90	66·50	68·40	71·50	75·70	61·38	60·81	66·51	69·02	72·50	76·51	65·08
TiO_2	0·84	0·55	0·40	0·40	0·31	0·07	0·65	0·75	0·35	0·26	0·17	0·10	0·45
Al_2O_3	16·60	15·50	14·85	15·30	14·57	13·15	14·20	16·95	14·95	15·35	14·30	13·21	15·91
Fe_2O_3	3·41	2·65	1·27	1·66	0·83	0·22	2·74	2·22	1·91	1·54	0·65	0·56	2·39
FeO	4·14	2·68	1·48	1·04	0·92	0·25	3·49	3·23	1·54	1·17	0·51	0·22	2·26
MnO	0·10	0·10	0·04	0·04	0·01	0·01	0·10	0·11	0·06	0·02	0·01	0·01	0·05
MgO	4·51	2·83	1·88	1·84	0·87	0·16	3·86	3·48	2·28	1·59	0·25	0·04	2·67
CaO	7·21	5·58	3·53	3·56	1·91	0·69	6·82	5·98	3·68	2·86	2·98	0·68	3·48
Na_2O	4·14	3·60	3·65	3·60	3·49	3·36	3·86	4·24	3·54	3·75	3·84	2·66	3·71
K_2O	0·91	1·76	3·41	3·18	3·86	5·01	1·71	1·68	3·24	3·77	4·21	5·24	2·91
P_2O_5	0·42	0·30	0·21	0·21	0·12	0·01	0·33	0·33	0·21	0·12	0·06	0·01	0·28
H_2O^+	0·70	0·85	1·50	1·50	0·48	0·28	0·70	0·43	0·58	0·48	0·28	0·35	0·62
H_2O^-	0·28	0·18	0·52	0·52	0·25	0·18	0·20	0·21	0·38	0·13	0·13	0·18	0·28
Total	99·19	99·48	99·24	99·82	99·12	99·09	100·04	100·42	99·23	100·06	99·89	99·77	100·09
Trace elements (ppm)													
Cs	1·5	1·8	2·6	1·9	3·0	2·5	1·3	4·1	6·5	5·9	4·3	3·1	15·9
Rb	24·1	59	240	100	169	189	34·5	76·6	126	146	134	173	153
Sr	960	861	564	590	492	63·1	963	665	519	466	307	17·6	565
Ba	458	610	823	910	0·0	134	470	421	638	476	421	52	746
La	29·6	32·6	40·9	30·5	33·8	39·5	21·5	23·9	28·7	17·9	22·9	37·7	19·2
Ce	63·0	67·4	79·3	59·0	60·1	56·0	43·0	49·9	54·0	37·4	39·6	67·4	42·7
Nd	35·2	28·8	32·1	23·0	19·7	12·0	20·0	25·3	22·0	14·9	12·4	17·6	23·2
Sm	6·71	5·02	4·85	3·27	3·17	1·07	3·98	4·58	3·35	2·31	1·81	2·34	4·29
Eu	1·95	1·38	1·05	0·99	0·58	0·21	1·24	1·16	0·92	0·77	0·57	0·39	1·01
Gd	5·57	4·04	3·88	3·36	2·40	n.a.	0·0	3·86	2·91	2·01	1·50	1·89	3·97
Tb	0·9	n.a.	0·7	0·5	0·4	0·3	0·4	0·5	0·5	0·3	0·3	0·4	0·6
Er	n.a.	1·28	1·98	1·45	1·21	1·20	n.a.	1·64	1·44	1·22	n.a.	2·10	1·95
Yb	2·04	1·30	1·41	0·88	1·07	0·96	1·07	1·35	1·12	0·91	0·67	1·80	1·54
Lu	0·27	1·20	0·19	0·13	0·15	0·15	0·14	0·19	0·15	0·13	0·08	0·26	0·23
U	0·9	1·1	4·0	2·9	5·1	6·3	0·6	1·7	2·0	2·5	1·4	15·7	4·5
Th	2·7	5·5	22·6	12·9	25·0	37·3	2·8	9·6	16·0	19·4	25·8	50·5	15·5
Y	23	13	13	10	n.a.	11	18	22	12	9	6	17	16·0
Nb	2	8	12	4	n.a.	2	5	3	4	3	3	6	6·0
Zr	296	240	246	205	237	111	220	222	195	168	110	108	136
Hf	6·1	3·9	5·9	4·1	4·9	4·3	3·4	4·1	4·4	3·6	2·9	4·5	4·0
Ta	1·1	1·4	2·3	1·6	2·2	2·3	0·6	1·3	0·8	1·6	1·9	2·7	1·3
Sc	18·8	8·3	6·2	5·7	3·3	1·0	11·0	11·7	7·7	4·8	1·7	1·4	9·2
Mo	5	1·0	7	2	7	4	1	2	3	2	4	11	4
Co	37·7	32·5	41·8	27·3	30·5	31·0	27·5	30·3	31·4	26·5	28·2	29·0	28·7
Cr	60·2	15·5	27·8	42·2	27·5	12·5	22·1	20·5	25·8	8·2	19·6	21·5	23·8

Notes: Analyses mostly carried out at Centro de Investigaciones Geológicas, La Plata. Al, total Fe, Ca, Mg, Mn, Na and K determined by atomic absorption spectrometry; Si, P and Ti by colorimetry; and Fe^{2+} by volumetry. Y, Zr and Nb were determined by X-ray fluorescence (Miniussi *et al.* 1980) and the REEs by inductively coupled plasma techniques. Reproducibility for the REE contents of the standard granite AC-E varies 3–5% (1σ). Ba, U, Th Hf, Ta, Sc, Mo, Co and Cr were determined by instrumental neutron activation analysis in a selected group of samples by ACTLABS (Canada).

basic border facies and basic enclaves from a high-K_2O group that includes the main facies, the monzogranites and the aplite and dacite dykes. Rb mimics the K_2O trend and gap.

Major element trends are typical of the Sierran intrusive suites of the western Americas and reflect decreasing abundances of hornblende, biotite and calcic plagioclase and increasing proportions of K-feldspar and quartz from the metaluminous to the peraluminous rocks. However, compared with the Patagonian batholith, the monzonite suites show a low and relatively constant FeO/MgO ratio, and higher K_2O and P_2O_5 (Fig. 3). The low FeO/MgO ratio seems to be a general characteristic of monzonite suites with actinolitic hornblendes and Mg-rich biotites (Sabatier 1980; Lameyre 1987), and of Andean shoshonites, which also show high P_2O_5 (Deruelle 1991).

Trace element trends are generally as smooth as those of major elements: Rb, Ta, Th and U increase, whereas Sr, Sc, Y, Hf, Zr, Zn, Sm, Nd, Eu, Gd, Tb, Yb and Lu decrease towards acid compositions. The Cs level is different in the two suites (La Calandria 1–3 ppm, La Leona 3–7 ppm). Compared with the rocks of the Patagonian batholith at a given SiO_2 content, the monzonite suites are enriched in large ion lithophile (LIL) elements, such as La, Ce, Rb, Sr and Ba, but depleted in HREEs and Y.

The chemistry of the main facies is similar in both districts: SiO_2, 65–69%; K_2O, 3·0–3·8%; Rb, 100–140 ppm; Sr, 470–600 ppm; $[La/Yb]_N$, 16–23; Eu/Eu^*, 0·75–1·1). They show enrichment of LIL elements and depletion of high field strength (HFS) elements such as Zr, Hf and HREEs, relative to ocean ridge granites (Fig. 4). Ta is slightly enriched, a feature that the monzonite suites share with the Chilean arc granites (Pearce *et al.* 1984). The dark enclaves in both districts show a fairly uniform composition (SiO_2, 60–61·5%; K_2O, 1·7%; Rb, 35–77 ppm; Sr, 660–960 ppm; $[La/Yb]_N$, 12–13·5; Eu/Eu^*, 0·85–1·0). The most primitive rocks are the diorite to dark quartz monzodiorite border facies of La Calandria (SiO_2, 56–64%; K_2O, 0·9–1·80%; Rb, 25–70 ppm; Sr, 700–1000 ppm; $[La/Yb]_N$, 10–17; Eu/Eu^*, 0·95–1·1).

Table 2 Rb–Sr and Sm–Nd isotopic data for the monzonite suites of the Deseado Massif.

Sample		Sm (ppm)	Nd (ppm)	$\frac{^{147}Sm}{^{144}Nd}$	$\frac{^{143}Nd}{^{144}Nd}$	εNd_t	T_{DMM}	Rb (ppm)	Sr (ppm)	$\frac{^{87}Rb}{^{86}Sr}$	$\frac{^{87}Sr}{^{86}Sr}$	εSr_t
La Calandria (203 Ma)												
LC-1	QMD	5·48	27·5	0·1204	0·512412	−2·5	1209	42·9	797	0·1556	0·705668	14
LC-2	D	7·04	34·4	0·1237	0·512465	−1·5	1133	24·1	960	0·0726	0·705377	13
LC-8	A	0·92	8·9	0·0622	0·512404	−1·1	1100	63·1	8·7	8·694	0·730235	13
LC-27	QM	4·60	27·9	0·0995	0·512471	−0·8	1071	149	540	0·7974	0·707368	12
LC-29	MG	3·10	19·8	0·0944	0·512440	−1·3	1111	189	324	1·6897	0·709965	12
La Leona (202 Ma)												
LL-12	QM	3·69	21·6	0·1033	0·512499	−0·3	1035	126	519	0·7001	0·706854	8
LL-33	SG	2·48	17·9	0·0839	0·512452	−0·8	1070	173	17·6	28·691	0·787296	9
LL-39	QMD	3·50	19·8	0·1071	0·512494	−0·5	1051	118	613	0·5557	0·706547	10
LL-44	ME	4·91	25·1	0·1182	0·512494	−0·8	1075	76·6	665	0·3335	0·705784	8
LL-46	DD	4·79	23·0	0·1259	0·512532	−0·3	1030	153	565	0·7840	0·707082	8
Standard												
BHVO-1		6·10	24·5	0·1504	0·512969	6·4	251					
BHVO-1		6·16	25·0	0·1489	0·512987	6·8	212					

Notes: Isotope data obtained at the NERC Isotope Geosciences Laboratory, Keyworth, as in Pankhurst & Rapela (1995). 1-Sigma precision is 0·5% on $^{87}Rb/^{86}Sr$, 0·01% on $^{87}Sr/^{86}Sr$, 0·2% on $^{147}Sm/^{144}Nd$ and 0·005% on $^{143}Nd/^{144}Nd$.
Key to rock type: D, diorite; QMD, quartz monzodiorite; QM, quartz monzonite; MG, monzogranite; SG, syenogranite; A, syenogranitic aplite; DD, dacitic dyke; and ME, mafic enclave. CHUR values: $^{143}Nd/^{144}Nd = 0.51264$; $^{147}Sm/^{144}Nd = 0.1967$. bulk earth values: $^{87}Sr/^{86}Sr = 0.70450$; $^{87}Rb/^{86}Sr = 0.0863$. T_{DMM}, Multistage depleted mantle model age, after DePaolo et al. (1991).

A fresh dacitic dyke from La Leona is compositionally similar to the more basic rocks of the main facies (SiO_2, 65·0%; K_2O, 2·9%; Rb, 153 ppm; Sr, 565 ppm; $[La/Yb]_N$, 8; $Eu/Eu*$, 0·75). In both districts the biotite monzogranites are evolved (SiO_2, 70–74%; K_2O, 3·8–4·08%; Rb, 180–190 ppm; Sr, 280–330 ppm; $[La/Yb]_N$, 21–23; $Eu/Eu*$, 0·6–0·7). Strong negative Eu anomalies are only found in aplite and small aplogranitic bodies (SiO_2, 74·5–76·5%; K_2O, 4·7–5·2%; Rb, 135–210 ppm; Sr, 20–150 ppm; $[La/Yb]_N$, 26–28; $Eu/Eu*$, 0·4–0·7).

4. Crystallisation sequences

Petrographic observations indicate that intermediate plagioclase, augite and hornblende (plus zircon, apatite, sphene and magnetite) were the first minerals to crystallise, suggesting that the parent liquids were water-undersaturated. As the amount of water increased, augite became unstable, remaining only as relics in hornblende phenocrysts. The interstitial geometry of quartz and alkali feldspar in the main facies implies that they did not start to crystallise until the framework of plagioclase, hornblende and biotite had developed. Water saturation was probably reached only in the late-stage peraluminous aplogranitic bodies and aplite dykes.

Normative compositions of representative samples of La Calandria (Table 1) are compared with experimental results in the felsic tetrahedron Qz–Ab–Or–An at $P_{H_2O} = 0.5$ kbar (Fig. 5). Again, an evolutionary trend is defined that encompasses all facies from diorite to aplogranite, supporting the assumption that the rocks are consanguineous and consistent with the perfect fit of Rb–Sr isochron systematics for each body. Samples of the main facies (LC-5), the border facies (LC-1, LC-2) and the basic enclaves (LC-21) all lie within the plagioclase volume, which is concordant with the observed early crystallisation of plagioclase. Compared with the border facies and enclaves, the composition of the main facies (LC-5) is closest to the divariant cotectic surface Qz–plag–melt. Crystallisation of plagioclase should have shifted the composition of the main facies melt towards this surface, along which monzogranite crystallises. Only later, water-saturated liquids represented by syenogranitic aplites reached the univariant cotectic line describing the equilibrium Qz–plag–alkali feldspar–melt (LC-8). The inferred order

of crystallisation of the felsic minerals (plagioclase; plagioclase+quartz; plagioclase+quartz+alkali feldspar) is otherwise characteristic of many metaluminous Cordilleran magmas.

5. Trace element fractional–crystallisation model

Fractional crystallisation models using trace elements that are mainly contained in the major minerals, such as Rb, Sr and Ba, are shown in Figure 6a, 6b. The same models for elements that might be partially contained in accessory minerals (e.g. Th and REEs) are depicted in Figure 6c and 7. A working assumption in these models is that the composition of the main facies (65–67% SiO_2) is close to the parental magma composition. This seems reasonable in that (a) the main facies is by far the most abundant rock composition in both districts, (b) the most primitive compositions (56–60% SiO_2) appear only as a subordinate border facies interpreted as cumulate and (c) dykes with a composition similar to the main facies suggest the periodic input of intermediate magmas at higher levels.

Rayleigh fractionation expressions were used to model two stages of crystallisation. The proportions of solids crystallising in the plagioclase volume of the quaternary system (Fig. 5) were inferred from the modal composition of the dark diorite facies of La Calandria (66% plagioclase An_{40}, 27% hornblende, 6% biotite, 1% quartz). A second stage of crystallisation using the modal proportion of quartz monzonite (43% plagioclase An_{25}, 32% alkali feldspar, 18% quartz, 3% biotite, 4% hornblende) was used to describe the evolution once quartz and alkali feldspar appeared on the liquidus and hornblende had drastically decreased.

The Rb and Sr fractional crystallisation path of a parent melt similar in composition to the main facies is shown in Figure 6a. Up to 30% crystallisation of the parent melt, the trends are controlled by the joint precipitation of intermediate plagioclase, hornblende and biotite. At 70% crystallisation, alkali feldspar and acid plagioclase dominate the partition coefficient, resulting in a trend with a gentle increase in Rb and a drastic decrease in Sr. Figure 6a shows that the rocks of La Calandria define a trend with an inflexion point at ≈ 500 ppm Sr. The basic border facies and the basic enclaves define a low-Rb sub-trend leading to cumulus-rich differen-

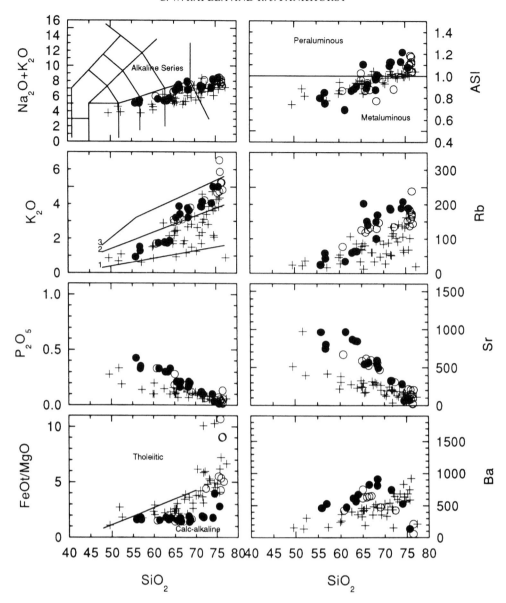

Figure 3 Harker variation diagrams for the monzonite suites and the Patagonian batholith. Symbols as in Fig. 2. Major element concentrations in weight per cent, trace elements in parts per million. Lines 1, 2 and 3 in K₂O plot delimit medium-K, high-K and shoshonite fields (Le Maitre *et al.* 1989; Peccerillo & Taylor, 1976).

tiates, with the gap (60–110 ppm) previously noted. The scatter observed in the Sr versus SiO₂ plot (Fig. 3) at low SiO₂ is probably due to varying proportions of modal plagioclase in the cumulates. The basic enclaves are consistently associated with the early differentiates (Fig. 6b, 6c), suggesting a cognate origin. They are most probably early crystal aggregates formed in isotopic equilibrium with the enclosing monzonitic magma.

The main facies, monzogranites, aplogranites and aplites from both complexes describe a trend similar to the modelled liquids L2–L3 (Fig. 6a). The inflexion in the Rb–Sr line of liquid descent in granitic magmas, coupled with a compositional hiatus in the trend of solids when alkali feldspar and acid plagioclase become liquidus phases, was predicted by McCarthy and Robb (1978). Concomitant variations in K, Ca and Na, and a K-gap, are also expected in this crystallisation sequence (Rapela & Shaw 1979).

As the Rb/Sr ratio increases through more than two orders of magnitude from the most primitive to the most evolved rocks, this ratio has been used as an index of differentiation against Ba and Th. The modelled fractional crystallisation path for Ba predicts an inflexion point in the liquid line of descent when K-feldspar and acid plagioclase come to

dominate the bulk partition coefficient, changing the $D_{Ba}^{solid\ liquid}$ from 0·5 to 4·0 (Fig. 6b). This is well seen in the more evolved rocks of the main facies (approximately 69–70% SiO₂). The suites of La Calandria and La Leona define similar, but not identical magma lineages. The distribution of Th versus Rb/Sr follows the model remarkably well (Fig. 6c), suggesting early stage control by hornblende and intermediate plagioclase, with the final stage of crystallisation being dominated by K-feldspar and acid plagioclase.

Rare earth element patterns for La Calandria show an uniform increase in [La/Yb]ₙ from 10 in the border facies to 19 (main facies), 22 (granite) and 27 (aplite), coupled with a depletion in MREEs and HREEs and an increase in the negative Eu anomaly with increasing SiO₂ (Fig. 7a). Figure 7b depicts a model based on fractionation of the essential minerals as described earlier, with accessory minerals estimated from thin section analysis (1200 points; sphene, 0·6%; apatite, 0·25%; zircon, 0·15%; magnetite, 1%). The modelled patterns are strongly affected by the accessory assemblage, especially sphene, which, together with hornblende, produces the observed depletion in MREEs and HREEs (Fig. 7b). The model is a semiquantitative approximation, as the amount of

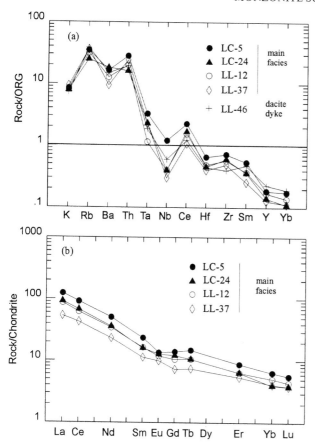

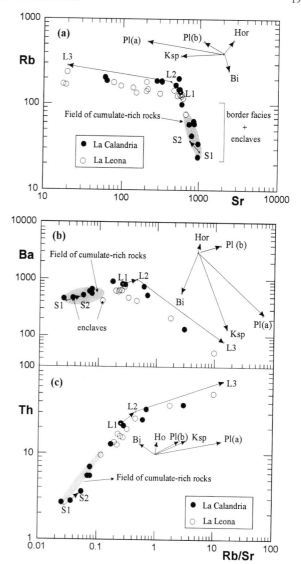

Figure 4 (a) Ocean ridge granite (ORG) normalized variation diagrams (Pearce *et al.* 1984) for rocks of the main facies of the monzonite suites and a dacitic dyke (analyses in Table 1). (b) Chondrite normalized REE patterns of the main facies. Note that the Eu anomaly varies from slightly negative to slightly positive.

Figure 6 Modelled distribution of Rb, Sr, Ba and Th during fractional crystallisation. Solid lines show trends of liquids (L1–L2) and solids (S1–S2) removed continually from an initial liquid of composition equal to L1 (sample LC-5, Table 1), both representing fractionation of an intermediate plagioclase, hornblende and biotite up to 30% crystallisation of melt L1. Solid line L2–L3 illustrates the trend of liquid calculated up to 70% crystallisation, as a quartz monzonitic solid assemblage begins to separate. Vectors for fractional crystallisation are after the Rayleigh law for $f = 0.7$. Partition coefficients chosen from published work as in Pankhurst and Rapela (1995). (a) Distribution of Rb and Sr in the monzonite suites of the Deseado Massif. Solid lines describe the path of liquid (L1–L2–L3) and solids (S1–S2) when a melt with 140 ppm Rb and 560 ppm (L1) Sr undergoes fractional crystallisation. (b) Plot of Rb/Sr versus Ba for monzonite suites of the Deseado Massif and fractional crystallisation model identical to that in (a). Initial compositions: Rb/Sr = 0.25; Ba = 820 ppm. (c) Plot of Rb/Sr versus Th for monzonite suites of the Deseado Massif and fractional crystallisation model identical to that in (a). Initial compositions: Rb/Sr = 0.25; Th = 23 ppm.

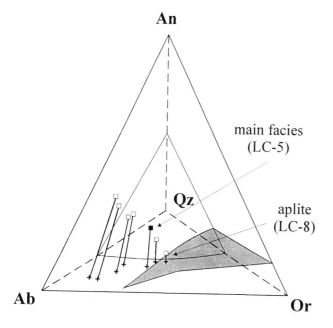

accessories can only be roughly estimated, but suggests that these minerals played a key part in the fractionation history. This seems to be an unavoidable conclusion when modelling REEs in granitic rocks (e.g. Mittlefehldt & Miller 1983).

6. Genesis of the monzonite suites

6.1. Possible source materials

Given that the monzonite suites resulted from high-level crystal fractionation of primary magmas corresponding to the least differentiated main facies rocks, evidence for the composi-

Figure 5 Qz–Ab–Or–An (–H₂O) diagram at $P_{H_2O} = 0.5$ kbar for rocks of La Calandria district. The position of the cotectic line at $P_{H_2O} = 0.5$ kbar in the ternary system Qz–Ab–Or is from Luth *et al.* (1964). The approximate position of the divariant cotectic surface plagioclase–alkali feldspar–melt at $P_{H_2O} = 0.5$ kbar (shaded surface) is based on interpolations between the configuration at low pressure (Carmichael 1963) and that at 2 kbar (Winkler & Lindemann 1972). Representative chemical analyses of La Calandria rocks are in Table 1.

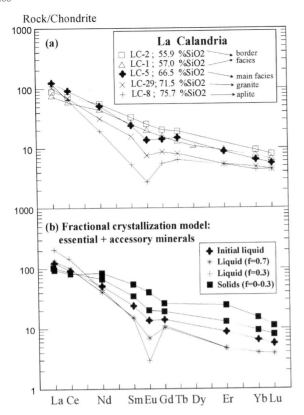

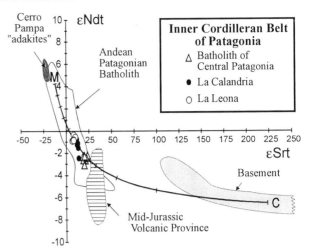

Figure 8 Variation of εNd_t versus εSr_t for the monzonite suites of the Deseado Massif and other Patagonian igneous rocks. Data sources: Batholith of Central Patagonia, Gastre area, Rapela *et al.* (1991); Mid-Jurassic Volcanic Province, Pankhurst and Rapela (1995); Patagonian batholith, Pankhurst, Weaver and Hervé (unpublished data); Cerro Pampa dacite, Kay *et al.* (1993); Deseado Massif basement, Pankhurst & Rapela (unpublished data). The assumed mantle end-member for the mixing model (M) has 1000 ppm Sr ($^{87}Sr/^{86}Sr = 0.7034$ at 200 Ma) and 30 ppm Nd ($\varepsilon Nd_t = +4.2$); the crustal end-member (C) has 100 ppm Sr (0.720) and 15 ppm Nd (-6.5). Ticks on the binary mixing line represent 10% increments.

Figure 7 (a) Chondrite normalised REE patterns of rocks from La Calandria district. Note the increase of La/Yb ratios and the negative Eu anomaly, coupled with the MREE depletion with increasing SiO_2. (b) Modelled patterns resulting from crystallisation of the essential and accesory minerals: 0–30% crystallisation with hornblende (26%); intermediate plagioclase (65%); biotite (6%); quartz (1%); sphene 0.6%; apatite (0.25%); zircon (0.15%); magnetite (1%): 30–70% crystallization with acid plagioclase (42%); alkali feldspar (31%); quartz (18%); hornblende (4%); biotite (3%); sphene 0.6%; apatite (0.25%); zircon (0.15%); and magnetite (1%). Partition coefficients as in Pankhurst and Rapela (1995).

tion of the source must be based on the characteristics of the latter. In this context, the following points are considered relevant:

1. The low ASI and initial $^{87}Sr/^{86}Sr$ of the monzonite suites preclude derivation from a primarily sedimentary crustal source with a previous history of weathering which would have increased its alumina saturation and Rb/Sr ratio.

2. The Early Jurassic monzonite suites of the Deseado Massif and granites of the Central Patagonia Batholith are isotopically more primitive than the widespread Middle Jurassic silicic magmas (initial $^{87}Sr/^{86}Sr \approx 0.7068$ and $\varepsilon Nd_{175} \approx -4.0$; see Fig. 8). The latter have been attributed to partial melting of a basaltic lower crust first formed in Grenvillian times, on account of their Nd model ages of 1150–1600 Ma (Pankhurst & Rapela 1995). Mean multi-stage depleted mantle Nd model ages for La Calandria and La Leona are significantly younger at 1040–1200 Ma (Table 2), indicating a different source and ruling out a genetic relationship with the silicic volcanism of extra-Andean Patagonia. The monzonite source is either younger than that of the Mid-Jurassic rhyolites, or it has a higher time-averaged Sm/Nd ratio (and lower Rb/Sr), i.e. a more LIL element depleted character.

3. Sr and Nd isotope systematics for the Early Jurassic granites and monzonites fall within the field of subduction-related Andean granites (Cretaceous to Tertiary) and lavas from the Southern and Austral Volcanic Zones of the modern Andes (Fig. 8). Some other aspects of the trace

element data of the main facies are also consistent with a volcanic arc signature, e.g. high LIL element/HFS element ratios (Fig. 4a), Ba/La ratios between 20 and 30 and low Rb/Zr ratios (0.5–0.9). A relatively LIL element enriched source is indicated by the high Th/La (0.4–0.7), Th/Ta (8–20) and Rb/Sr (0.2–0.4) ratios of the main facies rocks. With respect to their high LIL element and P_2O_5 contents, and low FeO/MgO ratios, the monzonites of Patagonia are transitional from high-K calc-alkaline to shoshonitic series, although the latter is usually restricted to rocks with less than 56% SiO_2. The geochemistry and isotopic composi-tions of the Patagonian monzonites are closer to those found in orogenic continental high-K and shoshonitic rocks than to those located above the shallow dipping subduction zone in the central Andes, which have a depleted source signature with low La/Yb ratio and high FeO/MgO ratios (Kay & Gordillo 1994).

4. Miocene dacite erupted in the Pre cordilleran region, 250 km to the west of La Calandria and La Leona, has been interpreted as a direct melt of the hot oceanic plate, before the collision of the Chile Rise (Kay *et al.* 1993). The monzonite suites are isotopically very different from these andesite–dacite (63–68% SiO_2) 'adakite' melts, which show very low $^{87}Sr/^{86}Sr$ (0.70285–0.70309) and positive $\varepsilon Nd = +6.9$ to $+5.5$ (Kay *et al.* 1993). The high LIL element concentrations of the monzonites also rules out a depleted N-MORB source, even if the melting percentage was low. Nevertheless, the main facies of the monzonites do show some geochemical affinities with the 'adakites', such as depletion in HREEs Yb and Y (Yb = 0.9–1.45 ppm, Y = 9–15 ppm, La/Yb = 22–35), low HFS elements (e.g. negative Nb and Ta anomalies), as well as relatively high Sr (>500 ppm) and a lack of a significant Eu anomaly (Defant *et al.* 1991). These features are consistent with a plagioclase-poor, garnet+amphibole-bearing residual assemblage (Defant *et al.* 1991), that might be well inferred for the source of the monzonite magmas. However, such character-istics might also result from the partial melting of basalt at pressures high enough for garnet to become stable, e.g.

where partial melting of a new mafic underplating takes place under a thick crust, well inboard of coastal Cordilleran batholiths (Atherton & Petford 1993; Muir et al. 1995).

5. The high P_2O_5 and K_2O are primary characteristics that differentiate the monzonite suites from both the 'adakite' suites and the Na-rich magmas emplaced inboard of the coastal batholiths (Atherton & Petford 1993). Negative correlation of P_2O_5 with SiO_2 in metaluminous granitoid suites is taken to reflect the temperature-dependent solubility of apatite (Bea et al. 1992). The slope of the correlation increases with increasing saturation temperature of the melts, and implies a higher temperature of crystallisation for the monzonite melts (850–900°C) than for their Cordilleran counterparts (<850°C). The low FeO/MgO ratios might indicate interaction with mantle peridotite (Kay et al. 1993), consistent with a high temperature for the monzonite magmas.

Altogether, these constraints suggest the subcontinental mantle as the most probable ultimate source for the monzonite suites. The enrichment of the main facies in K, Rb, Ba, Th and light REEs (Fig. 4) suggests that the source had most of the mantle-derived geochemical and isotopic features associated with arc rocks. However, peridotitic source materials, whether chondritic or enriched, lack an appropriate bulk composition to generate large volumes of high-Sr intermediate magmas in one-stage melting. Hence, our preferred model for the origin of the monzonite suites is a remelting of amphibole + garnet-bearing, high-K basaltic materials, formed in turn by partial melting of an enriched mantle wedge in a distal sector of the subduction zone.

6.2. Crustal contamination

The main facies rocks show an increase in Cs from La Calandria (1–3 ppm) to La Leona (2–7 ppm), with the late dacite dyke at La Leona showing an even higher value (16 ppm). High levels of Cs are sometimes considered to be an indication of crustal contamination (e.g. Kay et al. 1993). In principle, the Sr–Nd isotopic variations within the Triassic–Jurassic extra-Andean province could be described by mixing curves between a primitive mafic end-member in the mantle field and a crustal end-member equivalent to basement rocks of the Deseado Massif (Fig. 8). A high proportion of crust is required for binary mixing (about 50%), although this would be much less if an AFC process were invoked.

However, the perfect isochron relationships in the monzonite suites preclude any contamination at the level of fractionation. The highest Cs sample from La Leona (LL-46) lies on the same isochron as the rest of the rocks from this intrusion and has an equivalent εNd_t value (Table 2). Individually, this rock cannot be explained by AFC contamination. Different degrees of contamination during magma ascent (either binary mixing or AFC) are also unlikely to result in such perfectly homogeneous, but distinct, magma batches. In terms of mixing, this really only leaves the possibility of heterogeneity within the source region of the melts, e.g. partial melting in a zoned crust. For remelting of a mafic source (?underplate), either the heterogeneities were inherited from an earlier stage or they arose through the long-term isotopic evolution of the source. If the Nd model ages are taken to indicate a maximum possible age of source differentiation from normal mantle of, say 1000 Ma, then it is likely that its Rb/Sr ratio was about 0·05 and the differences between the two magma batches could be explained by a 10% variation in Rb/Sr in this source. The dyke with 16 ppm Cs should represent a lower degree of partial melting (by a factor of about three) than the main facies monzonites, but as its Rb content is not similarly

increased, this may point to a Rb-retaining phase (perhaps phlogopite) in the residue.

7. Tectonic setting

Previous studies in the Batholith of Central Patagonia, the Sub-Cordilleran plutonic belt and the Deseado Massif have related their Early Jurassic plutonism to easterly dipping subduction at the margin of Gondwana (e.g. Franchi & Page 1980; Rapela et al. 1991; Godeas, 1993; Márquez 1994). The geochemical and isotope data are also consistent with a mantle origin and arc-dominated signature for the Deseado monzonites. Furthermore, monzonite suites are frequently recognised as the deepest-seated melts in subduction zones (Pitcher 1993), commonly emplaced in distal zones of the orogene (Lameyre 1987).

The unusual NW–SE to NNW–SSE trend of the Sub-Cordilleran plutonic belt and the Batholith of Central Patagonia (Fig. 1) suggests oblique northeastwards convergence of the palaeo–Pacific plate relative to the Patagonian continental margin. In this interpretation, the Sub–Cordilleran belt is considered to be a remnant of an outer, Cordilleran segment of the arc, whereas the Batholith of Central Patagonia and the monzonite suites in the Deseado Massif are progressively more distal products of the oblique subduction. Northeastwards convergence is also consistent with the dextral displacement of the Gastre Fault System (Fig. 1), which controlled the emplacement of the Batholith of Central Patagonia (Rapela & Pankhurst 1992). The oblique subduction model also agrees well with the location of a narrow and elongated NW trending Liassic marine sedimentary basin between the Sub-Cordilleran plutonic belt and the Batholith of Central Patagonia, which has also been related to a convergent regime (González Bonorino 1990).

There is little evidence of Early Jurassic subduction along the Andean palaeo-margin to the north of Patagonia. Late Triassic–Early Jurassic magmatic activity in central Chile and the northern Andes has been related to a southwards propagating extensional or transtensional tectonic environment, during a subduction-free interval (e.g. Parada et al. 1991; Gana 1991). In the island of South Georgia, recognised as a displaced part of the southern Andes, Alabaster and Storey (1990) favour a transition from subduction to strike-slip tectonics during the Jurassic.

Another important point is that crustal stretching and normal faulting accompanied the eruption of the Middle Jurassic volcanism over a vast area immediately after pluton emplacement in the Cordilleran interior. As pointed out in the introduction, there is little evidence of coastal Cordilleran granitoids or coeval subduction-related volcanism for the period 190–170 Ma in Patagonia. Thus a tectonic scenario consistent with the new evidence is a southward cessation or slowing of subduction, along a highly oblique strike-slip Patagonian margin. The Mid-Jurassic Volcanic Province would have developed in a similar setting to the syn-extensional magmatism of the Basin and Range Province of western USA (e.g. Gans et al. 1989). This could also explain the remarkable similarities between the Patagonian monzonites and the shoshonite series in island and continental arcs, where these rocks appear during the termination of a subduction phase and coincident with the onset of arc extension, or during the transition between two subduction regimes of different orientation (Morrison 1980; Kepezhiskas 1995).

8. Conclusions

The monzonite suites of southern Patagonia are considered as the extreme interior of a Triassic–Jurassic belt of granitoids related to oblique (NE-directed) subduction along the Gondwana margin. They were emplaced early during the sequence of tectono-magmatic events that culminated in the breakup of Gondwana during the early Cretaceous. The overall tectonic scenario probably developed through the following stages.

1. 220–195 Ma. Development of a wide Cordilleran interior in the Patagonian segment of the proto-Andes. Emplacement of calc-alkaline granitoids in central Patagonia and the monzonite suites in the most distal position.

2. 190–168 Ma. Slowing or cessation of subduction accompanied by a large-scale continental extension and lower crust-derived acid volcanism.

3. 165 Ma. Initiation of a normal 'Chilean-type' subduction along the Patagonian margin. Start of the emplacement of a coastal Cordilleran batholith.

4. 133 Ma. Massively increased subduction-related magmatism in the Cordillera and opening of the South Atlantic behind the crustal region that had been thinned during Early Jurassic extension and inundated by thick acid volcanic products—this region continued to sink during formation of the Cretaceous sedimentary basins.

The monzonite suites of the Deseado Massif, like other granitoids of the Cordilleran interior, show a dominant mantle-type signature, although we interpret the immediate source of the main facies magmas as a mafic intermediate rather than ultramafic. The lack of significant negative Eu anomalies in the REE patterns of the monzonites rules out a large proportion of residual plagioclase in the source, whereas the high LREEs and low HREEs suggest a role for garnet. This would imply that any mafic source would have been in the eclogite facies. The monzonite magmas were high-temperature melts produced at greater depth than the Cordilleran granitoids, from a garnet-bearing source enriched in LIL elements. Small isotopic differences between the two observed magma batches are most likely to be due to small degrees of long-term heterogeneity in Rb/Sr and Sm/Nd in the source (either the mafic underplate itself or the mantle region from which it was derived).

9. Acknowledgements

This study was carried out with financial support from the National Geographic Society and CONICET, Argentina (Grant PID 3661/92). In particular we are indebted to Anne Grunow for her help in collecting the granite sampling and Marcelo Márquez for sharing his knowledge of the area. V. Posadas, C. Cavarozzi and J. Wlasiuk assisted with the sample preparation and chemical analyses. We thank Suzanne Kay and John Foden for their useful reviews. This paper is classified as NERC Isotope Geosciences Laboratory Publication No. 148 and is a contribution to IGCP Project No. 345 ('Andean Lithospheric Evolution').

10. References

Alabaster, T. & Storey, B. C. 1990. Modified Gulf of California model for South Georgia, north Scotia Ridge, and implications for the Rocas Verdes back-arc basin, southern Andes. GEOLOGY 18, 497–500.

Alric, V., Feraud, G., Bertrand, H., Haller, M., Labudia, C. & Zubia, M. 1995. ^{40}Ar/^{39}Ar dating of Patagonian Jurassic volcanism: new constraints on Gondwana break-up. TERRA NOVA ABSTR SUPPL 7(1), 353.

Arrondo, O. G. 1972. Estudio geológico y paleontológico en la estancia La Juanita y alrededores, provincia de Santa Cruz, Argentina. REV MUS LA PLATA VIII PALEONTOL 43.

Atherton, M. P. & Petford, N. 1993. Generation of sodium-rich magmas from newly underplated basaltic crust. NATURE 362, 144–6.

Bea, F., Fershtater, G. & Corretgé, L. G. 1992. The geochemistry of phosphorus in granite rocks and the effect of aluminium. LITHOS 29, 43–56.

Bruce, R. M. 1988. Petrochemical evolution and physical construction of an Andean arc: evidence from the southern Patagonian batholith at 53° S. Ph.D. Thesis, Colorado School of Mines.

Carmichael, I. S. E. 1963. The crystallization of feldspar in volcanic acid rocks. Q J GEOL SOC LONDON 119, 95–131.

Defant, M. J., Richerson, P. K., De Boer, J. Z., Stewart, R. H., Maury, R. C., Bellon, H., Drummond, M. S., Feigenson, M. D. & Jackson, T. E. 1991. Dacite genesis via both slab melting and differentiation: petrogenesis of La Yeguada Volcanic Complex, Panama. J PETROL 32, 1101–42.

DePaolo, D. J., Linn, A. M. & Schubert, G. 1991. The continental crustal age distribution: methods of determining mantle separation ages from Sm–Nd isotopic data and application to the Southwestern United States. J GEOPHYS RES 96B, 2071–88.

Deruelle, B. 1991. Petrology of the Quaternary shoshonitic lavas of northwestern Argentina. In Harmon, R. S. & Rapela, C. W. (eds) Andean magmatism and its tectonic setting. GEOL SOC AM SPEC PAP 265, 201–16.

Franchi, M. R. & Page, R. F. N. 1980. Los basaltos cretácicos y la evolución magmática del Chubut occidental. REV ASOC GEOL ARGENTINA 35(2), 208–29.

Gana, P. 1991. Magmatismo bimodal del Triásico superior-Jurásico inferior, en la Cordillera de la Costa, provincias de Elqui y Limari, Chile. REV GEOL CHILE 18(1), 55–67.

Gans, P. B., Mahood, G. A. N. & Schermer, E. 1989. Synextensional magmatism in the Basin and Range Province; a case study from the eastern Great Basin. GEOL SOC AM SPEC PAP 233, 53pp.

Godeas, M. 1985. Geología en el Bajo de La Leona y su mineralización asociada, provincia de Santa Cruz. REV ASOC GEOL ARGENTINA 40(3–4), 262–78.

Godeas, M. 1993. Geoquímica y marco tectónico de los granitoides en el Bajo de La Leona (Formación la Leona), provincia de Santa Cruz. REV ASOC GEOL ARGENTINA 47(3), 347–47.

González Bonorino, G. 1990. Cambios relativos en el nivel del mar y su posible relación con magmatismo en el Jurásico temprano, Formación Lepa, Chubut noroccidental, Argentina. REV ASOC GEOL ARGENTINA 45(1–2), 129–35.

González Díaz, E. F. 1982. Chronological zonation of granitic plutonism in the northern Patagonian Andes of Argentina: the migration of intrusive cycles. EARTH SCI REV 18, 365–93.

Gordon, A. & Ort, M. H. 1993. Edad y correlación del plutonismo subCordillerano en las provincias de Río Negro y Chubut (41°–42° 30′ L.S.). XII CONGR GEOL ARGENTINO, MENDOZA ACTAS IV, 120–7.

Guowei Xu, T., Wil, T. M. & Powel, R. 1994. A calculated petrogenetic grid for the system K$_2$O–FeO–MgO–Al$_2$O$_3$–SiO$_2$–H$_2$O, with particular reference to contact metamorphosed pelites. J METAMORPHIC GEOL 12, 99–119.

Kay, S. M. & Gordillo, C. E. 1994. Pocho volcanic rocks and the melting of depleted continental lithosphere above a shallowly dipping subduction zone. CONTRIB MINERAL PETROL 117, 25–44.

Kay, S. M., Ramos, V. A. & Márquez, M. 1993. Evidence in Cerro Pampa volcanic rocks for slab-melting prior to ridge-trench collision in southern South America. J GEOL 101, 703–14.

Kepezhinskas, P. 1995. Diverse shoshonite magma series in the Kamchatka arc: relationships between intra-arc extension and composition of alkaline magmas. In Smellie, J. L. (ed.) Volcanism associated with extension at consuming plate margins. SPEC PUBL GEOL SOC LONDON 81, 249–64.

Lameyre, J. 1987. Granites and evolution of the crust. REV BRAS GEOCIÊNC 17, 349–59.

Le Maitre, R. W., Bateman, P., Dubek, A., Keller, J., Lameyre, J., Le Bas, M. J., Sabine, M. A., Schmid, R., Sorensen, H., Streckeisen, A., Woolley, A. R. & Zanettin, B. (eds) 1989. A classification of igneous rocks and glossary of terms: recommendations of the International Union of Geological Sciences Subcommission on the Systematic of Igneous Rocks. Oxford: Blackwell, 193 pp.

Linares, E. & González, R. R. 1990. Catálogo de edades radimétricas de la República Argentina 1957–1987. ASOC GEOL ARGENTINA PUBL ESP SERIE "B" 19, 628 pp.

Luth, W. C., Jahns, R. H. & Tuttle, O. F. 1964. The granite system at pressures of 4 to 10 kilobars. J GEOPHYS RES 69, 759–73.

McCarthy, T. S. & Robb, L. J. 1978. On the relation between cumulus mineralogy and trace and alkali element chemistry in the Archean granite from the Barberton region, South Africa. GEOCHEM COSMOCHIM ACTA **42**, 21–6.

Márquez, M. J. 1994. El plutonismo Mesozoico en el Macizo del Deseado y su vinculación con mineralización tipo cobre diseminado, Prov. de Santa Cruz, Argentina. ZENTRALBL GEOL PALÄONTOL **Teil I, H.1/2,** 115–32.

Miller, C. F. & Barton, M. D. 1990. Phanerozoic plutonism in the Cordilleran interior, U.S.A. *In* Kay, S. M. & Rapela, C. W. (eds) *Plutonism from Antarctica to Alaska.* GEOL SOC AM SPEC PAP **241**, 213–32.

Miniussi, C., Merodio, J. C., Posadas, V. G. & Meda, J. 1980. Determinación de elementos minoritarios y traza por fluorescencia de rayos-X. REV ASOC ARGENTINA MINERAL PETROL SEDIMENTOL **11**, 15–21.

Mittlefehldt, D. W. & Miller, C. F. 1983. Geochemistry of the Sweetwater Wash Pluton, California: implications for 'anomalous' trace element behaviour during differentiation of felsic magmas. GEOCHEM COSMOCHIM ACTA **47**, 109–24.

Morrison, G. W. 1980. Characteristics and tectonic setting of the shosonitic rock association. LITHOS **13**, 97–108.

Muir, R. J., Weaver, S. D., Bradshaw, J. D., Eby, G. N. & Evans, J. E. 1995. The Cretaceous Separation Point batholith, New Zealand: granitoid magmas formed by melting of mafic lithosphere. J GEOL SOC LONDON **152**, 689–701.

Pankhurst, R. J. & Rapela, C. W. 1995. Production of Jurassic rhyolite by anatexis of the lower crust of Patagonia. EARTH PLANET SCI LETT **134**, 23–36.

Pankhurst, R. J., Rapela, C. W. & Márquez, M. 1993. Geocronología y petrogénesis de los granitoides jurásicos del noreste del Macizo del Deseado. XII CONGR GEOL ARGENTINO MENDOZA ACTAS IV, 134–41.

Parada, M. A., Levi, B. & Nystrom, J. O. 1991. Geochemistry of the Triassic to Jurassic plutonism of central Chile (30 to 33° S); petrogenetic implications and a tectonic discussion. *In* Harmon, R. S. & Rapela, C. W. (eds) *Andean magmatism and its tectonic setting.* GEOL SOC AM SPEC PUBL **265**, 99–112.

Pearce, J. A., Harris, N. B. W. & Tindle, A. G. 1984. Trace element discrimination diagrams for the tectonic interpretation of granitic rocks. J PETROL **25**, 956–83.

Peccerillo, A. & Taylor, S. R. 1976. Geochemistry of Eocene calc-alkaline volcanic rocks from the Kastamonu area, northern Turkey. CONTRIB MINERAL PETROL **58**, 63–81.

Pitcher, W. S. 1979. The nature, ascent and emplacement of granitic magmas. J GEOL SOC LONDON **136**, 627–62.

Pitcher, W. S. 1993. *The nature and origin of granite.* London: Blackie Academic & Professional, 321 pp.

Rapela, C. W. & Alonso, G. 1991. Litología y geoquímica del Batolito de la Patagonia Central. VI CONGR GEOL CHILENO VIÑA DEL MAR ACTAS **1**, 236–40.

Rapela, C. W. & Kay, S. M. 1988. The Late Paleozoic to Recent magmatic evolution of northern Patagonia. EPISODES **11**, 175–82.

Rapela, C. W. & Pankhurst, R. J. 1992. The granites of northern Patagonia and the Gastre Fault System in relation to the break-up of Gondwana. *In* Storey, B. C., Alabaster, T. & Pankhurst, R. J. (eds) *Magmatism and the causes of continental break-up.* SPEC PUBL GEOL SOC LONDON **68**, 209–20.

Rapela, C. W., Pankhurst, R. J. & Harrison, S. M. 1991. Triassic 'Gondwana' granites of the Gastre district, North Patagonian Massif. TRANS R SOC EDINBURGH EARTH SCI **83**, 291–304.

Rapela, C. W. & Shaw, D. M. 1979. Trace and major element models of granitoid genesis in the Pampean Ranges, Argentina. GEOCHEM COSMOCHIM ACTA **43**, 1117–29.

Sabatier, H. 1980. Vaugnérites et granites: une association particulière de roches grenues acides et basiques. BULL MINERAL **103**, 507–22.

Stein, H. J. & Crock, J. G. 1990. Late Cretaceous–Tertiary magmatism in the Colorado Mineral Belt; rare earth element and samarium–neodymium isotopic studies. *In* Anderson, J. L. (ed.) *The nature and origin of Cordilleran magmatism.* MEM GEOL SOC AM **174**, 195–223.

Storey, B. C., Alabaster, T., Hole, M. J., Pankhurst, R. J. & Wever, H. E. 1992. Role of subduction-plate boundary forces during the initial stages of Gondwana break-up: evidence from the proto-Pacific margin of Antarctica. *In* Storey, B. C., Alabaster, T. & Pankhurst, R. J. (eds) *Magmatism and the causes of continental break-up.* SPEC PUBL GEOL SOC LONDON **68**, 149–63.

Varela, R., Pezzuchi, H., Genini, A. & Zubia, M. 1991. Dataciones en el Jurásico inferior de rocas magmáticas del nordeste del Macizo del Deseado, Santa Cruz. REV ASOC GEOL ARGENTINA **46(3–4)**, 257–62.

Weaver, S. G. 1988. *The Patagonian batholith at 48°S latitude, Chile: implication for the petrochemical and geochemical evolution of calc-alkaline batholiths.* Ph.D. Thesis, Colorado School of Mines.

Winkler, H. G. F. & Lindemann, W. 1972. The system Qz–Or–An–H₂O within the granitic system Qz–Or–Ab–An–H₂O. Application to granitic magma formation. N JAHRB MINERAL MONATSCHR 49–61.

CARLOS W. RAPELA, Centro de Investigaciones Geológicas, Universidad de La Plata, Calle 1 No 644, 1900 La Plata, Argentina.
ROBERT J. PANKHURST, British Antarctic Survey, c/o NERC Isotope Geosciences Laboratory, Keyworth, Nottingham NG12 5GG, U.K.

Transactions of the Royal Society of Edinburgh: Earth Sciences, **87**, 205–215, 1996

Petrogenesis of slab-derived trondhjemite–tonalite–dacite/adakite magmas

M. S. Drummond, M. J. Defant and P. K. Kepezhinskas

ABSTRACT: The prospect of partial melting of the subducted oceanic crust to produce arc magmatism has been debated for over 30 years. Debate has centred on the physical conditions of slab melting and the lack of a definitive, unambiguous geochemical signature and petrogenetic process. Experimental partial melting data for basalt over a wide range of pressures (1–32 kbar) and temperatures (700–1150°C) have shown that melt compositions are primarily trondhjemite–tonalite–dacite (TTD). High-Al ($>15\%$ Al_2O_3 at the 70% SiO_2 level) TTD melts are produced by high-pressure ($\geqslant 15$ kbar) partial melting of basalt, leaving a restite assemblage of garnet + clinopyroxene $\pm$ hornblende. A specific Cenozoic high-Al TTD (adakite) contains lower Y, Yb and Sc and higher Sr, Sr/Y, La/Yb and Zr/Sm relative to other TTD types and is interpreted to represent a slab melt under garnet amphibolite to eclogite conditions. High-Al TTD with an adakite-like geochemical character is prevalent in the Archean as the result of a higher geotherm that facilitated slab melting. Cenozoic adakite localities are commonly associated with the subduction of young (<25 Ma), hot oceanic crust, which may provide a slab geotherm (≈ 9–$10°C$ km^{-1}) conducive for slab dehydration melting. Viable alternative or supporting tectonic effects that may enhance slab melting include highly oblique convergence and resultant high shear stresses and incipient subduction into a pristine hot mantle wedge. The minimum P–T conditions for slab melting are interpreted to be 22–26 kbar (75–85 km depth) and 750–800°C. This P–T regime is framed by the hornblende dehydration, $10°C/km$, and wet basalt melting curves and coincides with numerous potential slab dehydration reactions, such as tremolite, biotite + quartz, serpentine, talc, Mg-chloritoid, paragonite, clinohumite and talc + phengite. Involvement of overthickened (>50 km) lower continental crust either via direct partial melting or as a contaminant in typical mantle wedge-derived arc magmas has been presented as an alternative to slab melting. However, the intermediate to felsic volcanic and plutonic rocks that involve the lower crust are more highly potassic, enriched in large ion lithophile elements and elevated in Sr isotopic values relative to Cenozoic adakites. Slab-derived adakites, on the other hand, ascend into and react with the mantle wedge and become progressively enriched in MgO, Cr and Ni while retaining their slab melt geochemical signature. Our studies in northern Kamchatka, Russia provide an excellent case example for adakite–mantle interaction and a rare glimpse of trapped slab melt veinlets in Na-metasomatised mantle xenoliths.

KEY WORDS: adakite, trondhjemite, tonalite, dacite, slab melting, Archean, tectonics, mantle metasomatism, Kamchatka.

The focus of granitic rock petrology has leaned towards the study of high-K granites and rhyolites and their relevance to the generation and growth of continental crust through time. This pursuit has allowed us to find that a significant proportion of the granites represent recycling of continental crustal components and contribute sparingly to new net addition for continental crustal growth. To understand the birth of continental crust and its subsequent growth, we should refocus our studies towards the sodic end of the 'granite' family, the trondhjemite–tonalite–dacite group (TTD). Many continental growth models indicate that over 70% of the present day continental crust was produced by the end of the Archean (Armstrong 1981; Dewey & Windley 1981; McLennan & Taylor 1982; Reymer & Schubert 1984), with tonalite–trondhjemite suites constituting 60–70% (Condie 1981) of the preserved Archean crust. Wedepohl (1995) has suggested a tonalitic composition for the bulk continental crust. This may have resulted from ridge subduction, concomitant slab melting and associated TTD generation in the Archean (Martin 1986, 1993; Nelson & Forsythe 1989; Drummond & Defant 1990).

Numerous petrogenetic models, experimental studies and thermomechanical calculations have sought to explain the role of the subducted oceanic lithosphere in arc magma genesis. Early studies (Green & Ringwood 1968; Marsh & Carmichael 1974) postulated a role for slab melting in the genesis of arc magmatism. More recently it has been argued that, in general, the slab does not melt, but dehydrates, contributing fluids to and enhancing melting conditions in the overlying mantle wedge (e.g. Gill 1981; Mysen 1982; Arculus & Powell 1986; Tatsumi et al. 1986; Ellam & Hawkesworth 1988; Kushiro 1990). However, slab and mantle wedge sources need not be mutually exclusive or individually the sole source for arc magmatism.

This paper will evaluate the derivation of sodic, intermediate to felsic igneous rocks from a subducted slab (MORB) source. Although the process of slab melting has persisted through time (Defant & Drummond 1990; Drummond & Defant 1990), the rate of slab melting has slowed dramatically since the Archean. It is the goal of this study to further refine the geochemical characteristics of slab melts and to delimit the physical conditions of slab melting. The geochemical discrimination of slab melts (adakites) from mafic lower crustal melts is presented to more fully distinguish adakites as a recognisable lithology with important tectonic implications. A mantle geochemical component is recorded in numerous adakite localities due to reaction between adakite melt and sub-arc mantle on ascent. Mantle xenoliths from northern Kamchatka will be used for the analysis of slab melt–mantle interaction. Metasomatism of the mantle wedge by adakite melts presents fundamental information on source heterogeneity for mantle-derived arc magmas.

1. Methods of TTD generation

Models for TTD genesis have been summarised in previous studies (Arth 1979; Barker 1979; Drummond & Defant 1990) and commonly involve either the partial melting of a basaltic source or fractional crystallisation of basaltic parental magma. Partial melting processes are considered to be a more efficient mechanism for TTD production as 80–90% fractional crystallisation is required to produce residual TTD liquids from basaltic (Spulber & Rutherford 1983) or boninitic (Meijer 1983) mantle-derived parental magmas. Many TTD localities lack high proportions of mafic to intermediate igneous rocks, which argues against the fractional crystallisation process as the dominant TTD petrogenetic model.

Experimental studies have shown that partial melting of a basaltic source over a wide range of temperature (700–1150°C) and pressure (1–32 kbar) results in a TTD melt product. Figure 1 shows the distribution of the available basalt partial melt experimental data (170 samples) with respect to the granitic rock subdivision of Barker (1979). Those experimental glasses containing <56% SiO_2 (30 samples, not shown) or >50% normative An proportion (28 samples, Fig. 1) are unlike natural TTD compositions and will not be considered further. Most experimental melt compositions are tonalite (60%), trondhjemite (23%) and granodiorite (12%), with a minor (5%) granite melt representative. Thus the dominant product from partial melting of a low-K_2O basaltic source is represented by the Ab-rich end of the granitic rock spectrum. A first-order screen of the samples under consideration for low-K_2O basalt partial melt genesis should be that the Ab–An–Or data reside primarily in the tonalite–trondhjemite fields.

Drummond and Defant (1990) and Defant and Drummond (1990) defined a specific subtype of Cenozoic TTD (adakite) similar geochemically to Archean high-Al (>15% Al_2O_3 at 70% SiO_2 level) TTD gneisses. Table 1 summarises over 2000 published analyses of continental and island arc andesite–dacite–rhyolite (ADR) suites, low-Al TTD (<15% Al_2O_3 at 70% SiO_2 level), ophiolitic plagiogranites, boninites and the adakite/high-Al TTD group. The low-Al TTD group represents non-ophiolitic, low pressure partial melts of a basaltic source without accompanying mafic to intermediate fractionation products (Barker et al. 1976; Drummond & Defant 1990). Qualitative comparison of the Table 1 data indicates that the adakite/high-Al TTD group has elevated Sr, La/Sm, Sr/Y and Zr/Y and depleted Y, Sc and Yb values. This geochemical signature for the adakite/high-Al TTD group is interpreted to be due to high-pressure partial melting of a basaltic source leaving a garnet amphibolite to eclogite restite assemblage, as discussed in Section 1.2. Mesozoic, Palaeozoic and Proterozoic TTD samples are also included as part of the cumulative adakite/high-Al TTD database (394 total samples). Adakite/high-Al TTD rocks range in age from the 1980 Mount St Helens dacitic volcanics (Defant & Drummond 1993) to the 3·96 Ga Acasta gneiss (Bowring et al. 1990).

The 13 localities used to define the Cenozoic adakite lithology (Defant & Drummond 1990) are associated with the

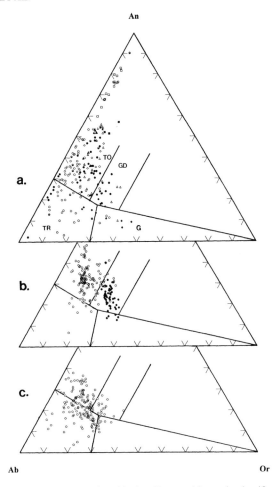

Figure 1 CIPW normative Ab–An–Or granitic rock classification after Barker (1979). TR, Trondhjemite; TO, tonalite; GD, granodiorite; and G, granite. (a) Experimental basalt partial melting data from: Holloway and Burnham (1972), closed triangle; Helz (1976), open triangle; Hacker (1990), open diamond; Rapp (1990) and Rapp et al. (1991), open circle; Beard & Lofgren (1991, dehydration melting only), closed circles; Rushmer (1991), closed square; Winther & Newton (1991), half-closed circle; Wolf & Wyllie (1991, 1994), open square; and Sen & Dunn (1994), closed diamond. (b) Cenozoic adakite (open circle) and CVZ/CBB (closed circle) data. (c) Archean high-Al TTD data.

subduction of oceanic lithosphere younger than 25 Ma. Parsons and Sclater (1977) found that oceanic crust younger than 25 Ma has a heat flow in the range 2·8–8·0 HFU (heat flow units), whereas older crust has a relatively constant heat flow of 1·0–2·5 HFU. We have proposed that in orogenic terranes where young, hot oceanic crust subducts, the slab has the ability to melt, producing adakite under garnet amphibolite to eclogite conditions. Numerous petrogenetic studies of Archean high-Al TTD have proposed that these magmas were also produced by high-pressure partial melting of the subducted oceanic crust (Jahn et al. 1981, 1984; Martin 1986, 1987, 1993; Nisbet 1987; Luais & Hawkesworth 1994). High heat flow, rapid convection and subduction of hotter, smaller oceanic plates are thought to be unique tectonic elements in the Archean which optimized conditions for the transformation of subducted oceanic crust into sial via partial melting (Abbott & Hoffman 1984; Martin 1986, 1993; Drummond & Defant 1990).

1.1. Pressure–temperature parameters for slab melting

The question of whether the subducted oceanic lithosphere directly contributes to magma production in volcanic arc terranes has provoked considerable debate since the advent of plate tectonics. One of the key issues is whether the

Table 1 Average compositional data on a variety of TTD types, arc ADR suites and boninites. ASI = alumina saturation index [(mol Al$_2$O$_3$)/(mol Na$_2$O + K$_2$O + CaO)]; Mg# = [100 ● mol MgO/(mol MgO + FeO)].

	Adakite and high-Al TTD (n=394)	Cenozoic adakite (n=140)	Archean high Al-TTD (n=174)	Continental arc ADR (n=815)	CVZ, Andes ADR (n=55)	Island arc ADR (n=473)	Low-Al TTD (n=121)	Plagiogranite (n=60)	Boninite (n=134)
SiO$_2$	67.91	63.89	70.20	63.87	63.30	67.72	73.47	68.83	58.74
TiO$_2$	0.42	0.61	0.33	0.68	0.89	0.49	0.38	0.50	0.23
Al$_2$O$_3$	16.58	17.40	15.74	17.02	16.39	15.44	13.27	14.61	11.22
FeO*	3.03	4.21	2.56	4.34	4.84	4.43	3.40	4.59	8.53
MnO	0.06	0.08	0.04	0.11	0.08	0.11	0.22	0.08	0.17
MgO	1.53	2.47	1.09	2.22	2.37	1.48	1.16	1.12	12.09
CaO	3.89	5.23	3.17	4.80	4.58	3.98	2.54	3.83	6.77
Na$_2$O	4.77	4.40	4.87	4.38	4.11	3.29	4.45	5.65	1.69
K$_2$O	1.67	1.52	1.88	2.37	3.14	2.95	1.02	0.66	0.47
P$_2$O$_5$	0.14	0.19	0.12	0.21	0.30	0.11	0.09	0.13	0.09
Sr	668	869	495	424	730	229	133	150	100
Ba	615	485	746	501	991	514	248	133	43
Rb	41	30	50	83	91	109	20	15	11
Cs	1.95	1.19	1.73	9.45	3.42	0.76	0.78	0.10	0.49
U	0.96	0.99	1.03	3.61	3.31	0.67	0.93	1.42	0.19
Th	4.50	3.52	5.98	13.2	11.9	4.96	3.15	4.51	0.40
Y	8.8	9.5	6.8	27.4	15.0	28.5	45	83	6.1
Zr	131	117	149	190	207	133	149	351	32
Hf	3.4	3.5	3.4	5.5	5.3	2.6	4.1	6.6	1.1
Nb	6.5	8.3	5.4	15.0	9.8	15.8	4.1	10.4	2.2
Ta	0.54	0.53	0.55	1.17	1.94	0.49	0.51	1.87	0.07
Sc	7.4	9.1	4.0	12.1	8.6	10.9	13.7	12.3	33.7
V	61	72	38	84	108	71	66	42	172
Cr	38	54	36	29	25	21	12	21	969
Co	15	13	15	12	14	19	12	11	38
Ni	26	39	19	19	18	9	18	6	223
Cu	23	24	26	30	48	27	15	59	19
Zn	51	57	46	59	95	58	52	41	70
La	22.62	17.55	29.84	29.94	36.05	32.01	12.07	16.00	2.07
Ce	41.03	34.65	51.63	60.15	79.02	55.27	30.14	43.17	4.58
Nd	18.53	20.14	19.92	28.27	35.77	17.85	15.54	28.62	2.67
Sm	2.88	3.15	2.79	5.28	6.56	6.40	4.26	7.85	0.75
Eu	0.91	0.97	0.91	1.18	1.59	3.60	0.96	1.65	0.28
Gd	2.05	2.25	2.04	4.26	4.33	6.99	3.60	11.52	1.06
Tb	0.30	0.37	0.25	0.72	0.62	0.76	0.86	1.33	0.15
Dy	1.22	1.43	1.16	3.57	2.65	6.88	4.07	13.79	1.35
Er	0.64	0.76	0.59	2.20		5.03	2.76	9.51	0.90
Yb	0.65	0.91	0.46	2.44	1.22	4.26	4.03	7.19	0.89
Lu	0.11	0.15	0.09	0.39	0.17	0.47	0.65	1.05	0.15
Mg#	43	48	41	41	46	34	33	31	69
ASI	1.00	0.96	1.00	0.92	0.89	1.00	1.02	0.89	0.73
Na/K	3.87	3.08	3.08	2.15	1.18	1.73	12.9	90.2	3.89
La/Sm	8.3	5.4	12.0	5.1	5.8	2.3	2.9	2.0	2.4
Sr/Y	104	121	110	18	51	9.3	5.4	2.2	19
Zr/Sm	65	47	86	40	32	24	34	44	48

thermomechanical conditions of the slab can allow melting to occur. A number of studies (Wyllie 1984; Peacock 1990a, 1990b; Cloos 1993; Peacock *et al.* 1994) have shown that slab melting is possible under a restricted set of conditions: (1) a young, hot subducting slab; (2) initiation of subduction providing a pristine, warm mantle wedge setting; or (3) the presence of high shear stresses in the subduction zone associated with highly oblique convergence and/or slow subduction. The original adakite localities of Defant and Drummond (1990) were attributed to melting young, hot subducting slab material. Three new adakite localities are associated with the subduction of young, hot oceanic crust: Late Miocene Cerro Pampa andesites–dacites, southern Argentina (Kay *et al.* 1993), Quaternary volcanic rocks of the Cordillera de Talamancas, Costa Rica (Drummond *et al.* 1995) and Quaternary volcanic rocks of SW Japan (Morris 1995). On the other hand, the initiation of subduction and high shear stress are interpreted causes for adakite generation from new localities in Mindanao, Philippines (Sajona *et al.* 1993) and the western Aleutians (Yogodzinski *et al.* 1995), respectively.

Figure 2 illustrates a slab geotherm of $10°C\,km^{-1}$ which would correspond to the special conditions required for slab melting. In addition, a $10°C\,km^{-1}$ geotherm correlates with many natural *P–T* estimates from exhumed subduction zone material, such as: (1) the top of the slab geotherm measured from the Zagros Mountains (Bird *et al.* 1975); (2) the *P–T* field recorded by subduction-related (type C) eclogites (Raheim & Green 1975); and (3) the *P–T* conditions from the Betic ophiolite, Spain (650–700°C, 20 kbar; Puga *et al.* 1995), Tauern window eclogites, Austria (625°C, 20 kbar; Selverstone *et al.* 1992), Monviso ophiolite, western Alps (500±50°C, 10–11 kbar, Philippot & Selverstone 1991) and the Adula nappe, central Alps (500–820°C, 17–27 kbar; Droop *et al.* 1990), among others.

Partial melting of the slab would require a source of water from dehydration reactions given the unrealistic thermal requirements for the dry melting of basalt (curve 11, Fig. 2). Thus the wet basalt solidus (curve 10; Green 1982) provides the minimum thermal conditions for slab melting. Under typical arc conditions the slab is thought to dehydrate, contributing fluids to the overlying mantle wedge that initiates melting of the mantle and BADR (basalt–andesite–dacite–rhyolite) arc volcanic production (see Tatsumi & Eggins 1995

and references cited therein). Slab dehydration is a continuous and progressive process that may be tracked by a series of dehydration reactions. Figure 2 indicates that numerous potential dehydration reactions correspond to the *P–T* conditions of the wet basalt solidus and the $10°C/km$ geotherm. In addition to these potential slab dehydration reactions, serpentinised subducted oceanic mantle could provide a major source of water to the subduction system. Ulmer and Trommsdorff (1995) have shown experimentally that antigorite breaks down to forsterite + enstatite + H_2O at 25 kbar and 730°C, having the capacity to release up to 13 wt% water. If these potential dehydration fluids experience significant residence time in the slab under super-solidus conditions, then partial melting is an obvious consequence.

In combination with the $10°C/km$ slab geotherm and wet basalt solidus, the hornblende dehydration curve (curve 1) envelops a *P–T* regime (700–800°C, 22–26 kbar) that would correspond to slab melting under garnet amphibolite to hornblende eclogite conditions. At *P–T* conditions below minimum basalt partial melting (650°C and 22 kbar), Poli (1993) experimentally confirmed the presence of hornblende eclogite as the stable metamorphic lithology. Drummond and Defant (1990) call on the retention of hornblende at the site of partial melting to explain many of the adakite geochemical features (e.g. high Al_2O_3, low K/Rb and Nb). Garnet is a key restite phase to generate the low HREE, Y and Sc contents of the adakites (Table 1); thus minimum *P–T* conditions must correspond to garnet stability during basalt partial melting. The present consensus from experimental work has shown that *P–T* conditions of $\geqslant 15$ kbar and $\geqslant 750°C$ are required to generate TTD with the requisite major and trace element character of adakite/high-Al TTD in the presence of a garnet amphibolite to eclogite refractory assemblage (Rapp *et al.* 1991; Winther & Newton 1991; Rushmer *et al.* 1994; Sen & Dunn 1994; Thompson & Ellis 1994). Thompson and Ellis (1994) favour a *P–T* regime of 800°C and 26 kbar for partial dehydration melting of a basaltic source following the reaction: zoisite + hornblende + quartz = clinopyroxene + pyrope + high-Al TTD melt.

The question remains, however, whether the dehydration fluids residence time in the slab is sufficient to initiate partial melting. Drummond and Defant (1990) speculated that dehydration fluids associated with slab melting would be largely retained in the slab and involve intraslab migration. This problem has been studied in exhumed subduction zone complexes from the Monviso ophiolite (Philippot & Selverstone 1991) and Tauern window (Selverstone *et al.* 1992). These studies have found that during eclogitisation of the subducted slab only localised, limited fluid flow and metasomatism occur under subsolidus conditions. Tauern window eclogites demonstrate that isolated fluid-filled pockets can exist at 20 kbar, giving rise to clinozoisite–zoisite–phengite–rutile segregations under transient $P_{fluid} = P_{total}$ conditions. Metamorphic fluid studies in the exhumed eclogitic slabs indicate that at depths of $>50 \pm 10$ km the slab retains its dehydration fluid (Selverstone *et al.* 1992; Philippot 1993), limiting fluid migration to intraslab fluid flow. In a preserved subduction complex that has experienced partial melting of the amphibolitic to eclogitic MORB, such as the Catalina Schist terrane, pervasive fluid flow involved metasomatism of the mantle wedge (Sørensen 1988; Sørensen & Grossman 1989). This suggests that transfer of fluids from the slab into the mantle wedge at depths $>50 \pm 10$ km is facilitated by slab melting. Philipott (1993) further suggests that all deep transfer of water into the mantle is by a water-rich melt phase.

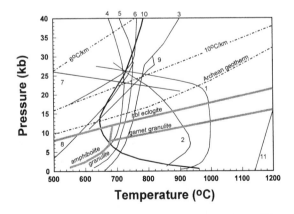

Figure 2 *P–T* diagram depicting potential slab melt conditions in concert with dehydration reactions. Pertinent curves include: (1) hornblende-out (Wyllie 1982); (2) tremolite-out (Obata & Thompson 1981); (3) biotite-out (Tatsumi & Eggins 1995); (4) serpentine-out (Tatsumi & Nakamura 1986); (5) talc-out (Wyllie 1982); (6) Mg–chloritoid-out (Schreyer 1988); (7) paragonite-out (Holland 1979); (8) clinohumite-out (Engi & Lindsley 1980); (9) talc + phengite-out (Massone & Schreyer 1989); (10) and (11) wet (Green 1982) and dry (Wyllie 1982) basalt solidus, respectively; Archean geotherm (Martin 1993); and metamorphic facies fields (Cloos 1993).

1.2. Geochemical parameters for slab melting

Figure 1b and 1c indicate that the Cenozoic adakites and Archean high-Al TTD data fall primarily within the trondhjemite, tonalite and granodiorite fields, corresponding to the compositional spread of experimental melts (Fig. 1a) from basaltic sources. Comparison between Cenozoic adakite and Archean high-Al TTD (Table 1, Fig. 1b, 1c) indicates that the Cenozoic adakites are relatively less felsic with a higher An proportion + ferromagnesian component + trace transition metal content and lower SiO_2. However, both groups exhibit low Y, Yb and Sc values indicative of a garnet-bearing residue at the site of basalt partial melting and are considered as a coherent petrogenetic group, i.e. high-pressure partial melting of a basaltic source.

Arth (1979) proposed a subdivision of TTD on the basis of Al_2O_3–Yb relationship (Fig. 3). Removal of subaluminous hornblende and garnet either as restite components or early crystallisation phases under high pressure conditions produces the low Yb–high Al_2O_3 character of adakite/high-Al TTD (Fig. 3). Trace element modelling (Kay 1978; Jahn et al. 1981; Martin 1986; Drummond & Defant 1990) has shown that the low Yb_N and high $(La/Yb)_N$ of adakite/high-Al TTD is due to a garnet amphibolite to eclogite restite assemblage. Low-Al TTD, island arc ADRs and plagiogranites exhibit a high Yb content (cumulatively referred to as the high-Yb TTD group) due to plagioclase + pyroxene extraction on partial melting and/or differentiation at low pressures (Drummond & Defant 1990; Beard 1995). The low-Al TTD has low Al_2O_3 and Sr values (Table 1) due to dominant plagioclase removal. Mantle-derived boninites define their own unique field with low Al_2O_3 and Yb. The overlap between adakite/high-Al TTD and specific continental ADR samples in Figure 3 is addressed in Section 2.

Y–Sr/Y relationships closely track the influence of garnet and plagioclase in TTD genesis (Drummond & Defant 1990; Defant & Drummond 1990). Strontium behaves incompatibly on high-pressure partial melting of basalt due to the absence or instability of plagioclase, whereas Y is dominantly controlled by the garnet fraction in the restite. Cenozoic adakites and Archean high-Al TTD samples are plotted on Figure 4 with respect to two partial melting models. The two models involve two basaltic source compositions, an Archean mafic composite (AMC; Drummond & Defant 1990) and an average MORB (Gill 1981), which experience 10–50% partial melting leaving either 10% garnet amphibolite and eclogite restite assemblages. The AMC and MORB partial melting models correspond well with the Archean high-Al TTD and Cenozoic adakite data, respectively. Independent analyses of Y–Sr/Y relationships with slab melting corroborate the model of adakite generation

by 10–50% partial melting of a MORB source under eclogitic to hornblende eclogitic conditions (Tsuchiya & Kanisawa 1994). The high-Yb TTD group (Fig. 3) display a high Y, low Sr/Y content (shaded area, Fig. 4) due to the incompatible and compatible behaviour of Y and Sr, respectively, associated with a dominant plagioclase–pyroxene extract assemblage.

A plot of La/Sm versus Zr/Sm (Fig. 5) subdivides the adakite/high-Al TTD group from the high-Yb TTD group and boninites at a La/Sm ratio of four. The high La/Sm reflects the strong affinity of garnet and hornblende for Sm relative to La. Partial melting (1–50%) curves of MORB leaving a 10% garnet amphibolite (upper curve) and eclogite (lower curve) are shown on Figure 5. Although the adakite/high-Al TTD exhibits a range of Zr/Sm, many (40% of the samples) display elevated values of Zr/Sm > 60. Pearce and co-workers (Pearce et al. 1992; Pearce & Peate 1995) have suggested that Zr behaves non-conservatively during slab melting due to the incompatibility of Zr in residual hornblende, which may explain the elevated Zr/Sm for some adakite/high-Al TTD under garnet amphibolitic melting conditions.

Optimum conditions for TTD melt segregation from a garnet- and hornblende-bearing basaltic protolith occur at 20–30% melting (Rapp 1995); however, above hornblende-out conditions the melt percentage increases to 50%, producing a

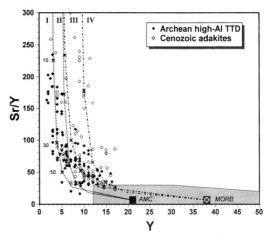

Figure 4 Y (ppm)–Sr/Y diagram with Archean high-Al TTD and Cenozoic adakites plotted relative to partial melt curves of MORB (values of Gill 1981) and Archean mafic composite (AMC values of Drummond & Defant 1990) sources leaving either eclogite (curves I and III) or 10% garnet amphibolite (curves II and IV) restite. Percentage partial melt values are labelled at the X on curve. High-Yb TTD group field is shaded.

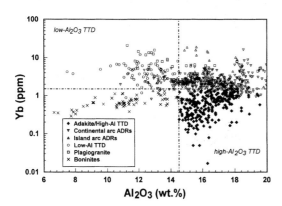

Figure 3 Al_2O_3 (wt%)–Yb (ppm) plot of various TTD types and boninites with high-Al_2O_3 and low-Al_2O_3 TTD subdivision of Arth (1979).

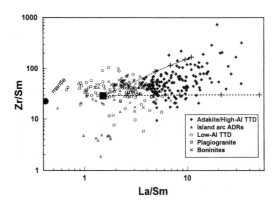

Figure 5 La/Sm–Zr/Sm plot of TTD types and boninites relative to partial melt curves of MORB (closed square) leaving either a 10% garnet amphibolite (upper curve) or eclogite (lower curve) restite. Percentage partial melt values are shown with a cross (+) and represent 1, 10, 30 and 50% melt with decreasing La/Sm. Mantle value from Hole et al. (1984).

more metaluminous, intermediate melt composition (Rapp & Watson 1995). Adakite/high-Al TTD samples that correspond to ≈50% partial melting on Figures 4 and 5 may be the result of hornblende decomposition or the addition of a minor mantle component (Section 3).

2. Slab melting versus lower crustal melting comparison

An alternative hypothesis to slab melting for high-Al TTD/adakite genesis involves the partial melting of overthickened continental arc crust (Kay & Kay 1991, 1993; Atherton & Petford 1993). In fact, some continental arc ADR localities, such as those in the central volcanic zone, Andes (CVZ), possess high La/Yb and Sr/Y ratios (Kay & Kay 1991, 1993; Feeley & Hacker 1995). To compare adakite compositions against ADR from overthickened continental arc terranes, 55 samples from the CVZ (Nevados de Payachata volcanic region—Davidson *et al.* 1990; volcan Tata Sabaya—de Silva *et al.* 1993; volcan Ollugue—Feeley & Davidson 1994; Feeley & Hacker 1995) and from the Cordillera Blanca batholith (CBB), Peru (Atherton & Sanderson 1987) are used (average shown in Table 1). In addition, partial K₂O–Rb–Sr isotopic data (Harmon *et al.* 1984) from the CVZ is also used for comparative purposes. The Quaternary volcanic rocks of the CVZ and the 5–7 Ma volcanic and plutonic rocks of the CBB correspond to a crustal thicknesses of 60–70 km and 50 km, respectively.

Continental arc ADR data from arcs of normal crustal thickness (e.g. 30–40 km) exhibit elevated Cs and U values relative to adakite/high-Al TTD due to the incorporation of or derivation from continental crust (Table 1). Figure 6 displays the high Y and Cs values of the continental arc ADRs indicative of a garnet-free crustal signature. However, the CVZ/CBB and Cenozoic adakite data compositionally overlap at low Y–Cs values (Fig. 6). The low Y and Yb values of the CVZ/CBB are attributed to one of the following models: (1) direct partial melting of a garnet-bearing mafic lower crustal source in the case of the CBB (Atherton & Petford 1993); (2) AFC of mantle-derived basaltic parental magmas with garnet-bearing lower crust for volcan Ollugue, CVZ (Feeley & Hacker 1995); or (3) mixing of mantle-derived normal arc magmas with lower crustal melts in the Nevados de Payachata volcanic region, CVZ (Davidson *et al.* 1990). Atherton and Petford (1993) and Feeley and Hacker (1995) further suggest that adakites should be re-evaluated relative to their petrogenetic models for the CVZ/CBB.

Adakites have a low Sr isotopic character (Kay 1978; Defant *et al.* 1992; Kay *et al.* 1993). The adakites of Panama and Costa Rica have some of the lowest radiogenic isotopic ratios of any Central American arc volcanic rocks (Defant *et al.* 1992). In the modern central and eastern Aleutians only the Adak Island adakites contain a MORB-like isotopic signature (Kay 1978; Yogodzinski *et al.* 1994). The CVZ volcanic rocks, on the other hand, are noted for their elevated Sr and O isotopic signature resulting from a lower continental crustal component (Harmon *et al.* 1984). A clean separation exists between adakite and CVZ/CBB Sr isotopic data at a Sr isotopic value of 0·7045 (Fig. 7a, 7b). Although some K₂O and Rb overlap occurs between the two data sets, the CVZ/CBB data are generally more K₂O- and Rb-enriched (Fig. 7a, 7b). Comparison of average K₂O, Rb and Na/K between adakite and CVZ/CBB (Table 1) indicates the more potassic character of the CVZ/CBB samples. As stated earlier, slab melts should reside primarily within the trondhjemite–tonalite fields (Fig. 1). The CVZ/CBB data fall consistently within the granodiorite–granite fields (Fig. 1c), which argues against a MORB source. Therefore, magmas derived from or mixed with a lower crustal component in overthickened (>50 km) arc terranes may exhibit some adakite-like compositional characteristics due to the influence of garnet and hornblende stability; however, close scrutiny of Sr isotopes in combination with the K₂O and Rb content should discriminate between slab melting and a lower crustal derivation.

3. Adakite–mantle interaction

Emplacement of adakitic magmas from the slab into the overlying continental or island arc crust requires ascent through the mantle wedge. The degree to which adakite magmas and the mantle interact is dependent on many factors, such as the ascent rate, subduction angle, slab geotherm and

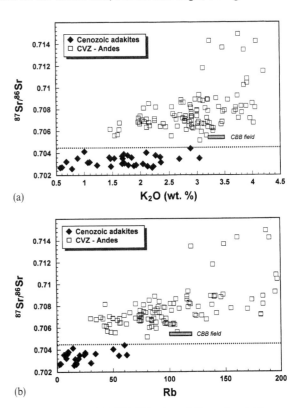

(a)

(b)

Figure 7 (a) K₂O (wt%)–^{87}Sr/^{86}Sr plot of Cenozoic adakites versus CVZ and CBB (shaded field), Andes data. CBB field represents range of values reported by Atherton & Sanderson (1987) and Atherton & Petford (1993). (b) Rb (ppm)–^{87}Sr/^{86}Sr plot of Cenozoic adakites versus CVZ and CBB (shaded field), Andes data.

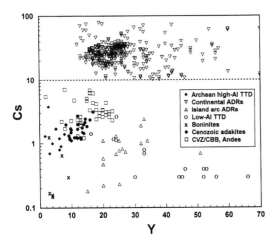

Figure 6 Y (ppm)–Cs (ppm) diagram of TTD types, boninites and CVZ/CBB Andes data.

depth of melting, thickness of the overriding lithosphere and the amount of prior adakite–mantle reaction within a ascent conduit. Beard *et al.* (1993) state 'that slab melts should show some evidence of equilibration (or, at least, reaction) with mantle peridotite during ascent' and chemical quantification of this process would aid in further defining adakites as a unique lithology.

Experimental studies of TTD melt–peridotite reactions at elevated pressures (15–30 kbar) have shown that the melt will react with peridotite producing orthopyroxene ± garnet ± hornblende ± clinopyroxene in the metasomatised peridotite (Carroll & Wyllie 1989; Johnston & Wyllie 1989; Sen & Dunn 1995). The dominant reaction involves melt + olivine ± clinopyroxene reacting to orthopyroxene ± hornblende. Sen and Dunn (1995) indicate that mantle metasomatism will initially produce hornblende-bearing harzburgite followed by hornblende-bearing orthopyroxenite production. The TTD melt would become more CaO- and MgO-enriched until it becomes saturated with respect to the metasomatic phases, whereupon further reaction/assimilation would simply cause additional reactant phase(s) precipitation and ultimate melt consumption (Carroll & Wyllie 1989; Johnston & Wyllie 1989).

Recent work by Yogodzinski and co-workers (1994, 1995) in the western Aleutians has evaluated geochemically the effects of slab melt–mantle interaction in a volcanic arc setting. They define two high-Mg andesite (HMA) types: Piip-type and Adak-type. Piip-type HMA is interpreted to represent volcanic rocks from a enriched peridotite source that was generated by mixing depleted MORB mantle with minor ($\approx 4\%$) adakite melts. The HMAs from the Setouchi belt, SW Japan (Tatsumi & Ishizaka 1982) were included into the Piip-type category (Yogodzinski *et al.* 1994). The Adak-type HMA, which we refer to as transitional adakites, represent CaO- and MgO-rich adakites from the western Aleutians, Adak Island–central Aleutians (Kay 1978), the Austral Andes (Stern *et al.* 1984) and Baja California (Saunders *et al.* 1987) and are produced as slab melts that incompletely interacted with the mantle wedge on ascent (Kay 1978; Yogodzinski *et al.* 1995).

A Cr–Ni plot (Fig. 8) demonstrates the relationship between adakites, transitional adakites, Piip-type HMAs and boninites. Partial melting of average mantle (Frey *et al.* 1978) leaving either a garnet peridotite or harzburgite residuum generates model curves that closely coincide with the boninite samples

and produces a Cr:Ni ratio of 4:1 to 5:1. Melting a MORB source under eclogitic to garnet amphibolitic conditions produces model curves with lower Cr:Ni ratios of 1.5:1 to 2:1, which corresponds to the low Cr:Ni ratios of adakite. A mixture of mantle and modelled slab melt compositions yields a mantle mix region (1–10% mantle component, Fig. 8) that coincides with many of the transitional adakite samples. The low Cr:Ni ratio of adakites and transitional adakites may be attributed to the strong affinity of restitic garnet–hornblende–clinopyroxene for Cr relative to Ni. Comparison between transitional adakites and boninites at comparable Ni values indicates that the transitional adakites are 100–200 ppm deficient in Cr, which is partially due to the relatively low Cr values of the initial slab melt. In addition, the reaction between adakite and mantle decomposes olivine, liberating Ni and Cr, with Cr being preferentially accepted in the metasomatic assemblage of orthopyroxene, hornblende and garnet. Most of the Piip-type HMAs display higher Cr values than the transitional adakites, which is consistent with a more dominant mantle source with minor slab melt component. Type-area boninites (Chichijima, Bonin Islands) may also have a minor (2%) slab melt component in the mantle source to explain their high Zr/Sm ratios (≈ 60) (Taylor *et al.* 1994). The minor overlap between Piip-type HMAs and some boninites may be due to a common petrogenesis.

Major element relationships, such as MgO–SiO$_2$ (Fig. 9), between adakites, transitional adakites, Piip-type HMAs and boninites supports minor mantle–slab melt interaction for the transitional adakites and adakite-contaminated mantle source for the Piip-type HMAs. Transitional adakites fall within the mantle mix region (1–10% mantle component), whereas the Piip-type HMAs are generally more enriched in MgO than the transitional adakites. Kelemen (1995), in a review of HMA geochemistry, concludes that HMA genesis is best explained by silicate melt–peridotite reaction. In this study, Kelemen states that a reaction between felsic slab melt and mantle peridotite will cause a sharp increase in the MgO and Ni content of the melt due to olivine dissolution and/or reaction-out to orthopyroxene. Our observations on Ni (Fig. 8) and MgO (Fig. 9) behaviour during adakite–mantle reaction agree with those of Kelemen (1995). An additional lithology shown on Figure 9 is commonly temporally and spatially related to adakites, Nb-enriched arc basalts (NEAB) (Defant *et al.* 1992), and delineates a continuous trend with the adakites and transitional adakites. The unusual composition of NEAB is apparently influenced by slab melting and adakite–mantle mixing (Section 3.1).

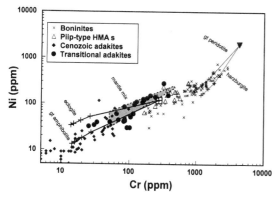

Figure 8 Cr (ppm)–Ni (ppm) diagram with MORB (square with cross) partial melting curves leaving 10% garnet amphibolite and eclogite restite. Mantle (Frey *et al.* 1978) partial melt curves on both garnet peridotite (olivine$_{60}$orthopyroxene$_{20}$clinopyroxene$_{10}$garnet$_{10}$ peridotite of Mysen 1982) and harzburgite (olivine$_{80}$orthopyroxene$_{20}$) stability fields. Percentage partial melt values are depicted by a cross (+) on curves and represent 1, 10, 30 and 50% partial melt with increasing Ni and Cr values. Shaded mantle mix box represents 1–10% mix of mantle component into 1–50% MORB partial melt component.

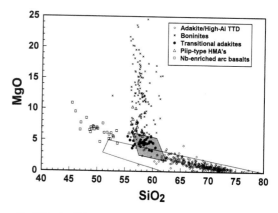

Figure 9 SiO$_2$ (wt%)–MgO (wt%) plot with experimental basalt partial melts (open field, see Fig. 1 for data sources) and 1–10% mantle (pyrolite of Ringwood 1975)–slab melt mix region (shaded field) shown relative to adakite-types, boninites and NEABs.

3.1. Case example of adakite–mantle interaction: northern Kamchatka, Russia

Our studies have focused on the northern Kamchatka adakites (Kepezhinskas 1989; Defant & Drummond 1990; Hochstaedter *et al.* 1994) and interaction of the slab melts with the sub-arc mantle. Adakites are found in two northern Kamchatka localities, the Tymlat volcanic field (Kepezhinskas 1989; Defant & Drummond 1990) and the Valovayam volcanic field (Hochstaedter *et al.* 1994). Adakite generation is interpreted to be associated with the subduction of young (<15 Ma), hot Komandorsky basin oceanic crust that apparently ceased about 2 Ma (Kepezhinskas 1989; Hochstaedter *et al.* 1994). Contemporaneous Nb-enriched arc basalts (7 Ma) occur with transitional adakites (6–8 Ma) in Valovayam as interlayered flows or as individual cinder cones or dykes in the vicinity of the transitional adakites (Hochstaedter *et al.* 1994; Kepezhinskas *et al.* in press).

The Valovayam NEABs host a suite of mantle xenoliths (dunites, websterites, spinel lherzolites, amphibole wehrlites, spinel pyroxenites, amphibole pyroxenites and garnet pyroxenites) that have experienced various degrees of Na metasomatism (Kepezhinskas *et al.* 1995). A metasomatic assemblage of Cr-poor, Al–Fe–Mg and Al–Ti–Fe spinel, Al-rich (up to 10 wt% Al_2O_3) augite, Cr-diopside, high-Al and Na pargasitic amphibole and almandine–grossularite garnet is found replacing mantle olivine and pyroxene (Kepezhinskas *et al.* 1995). In addition, andesine megacrysts occur in the host NEAB due to Na-metasomatic effects. The metasomatic assemblage is similar to that predicted from TTD melt–peridotite reactions from experimental studies (Carroll & Wyllie 1989; Sen & Dunn 1995), with the exception of lacking orthopyroxene.

Metasomatism of these mantle xenoliths requires the addition of Si, Al and Na, which may be introduced from a adakitic melt phase. Kepezhinskas *et al.* (1995) have suggested that trondhjemitic veinlets found in some of the Valovayam mantle xenoliths are the metasomatising agents. The trondhjemitic veinlets are comprised of felsic glass (SiO_2 = 63·88 wt%) with hornblende + plagioclase + magnetite. The glass has high Na/K (5·45), Sr (1736 ppm), La/Sm (8·0), La/Yb (22·8) and Sr/Y (289) and low Y (6 ppm), which is characteristic of adakite compositions (Kepezhinskas *et al.* 1996). However, the high Cr (349 ppm) and Yb (2.92 ppm) content in the trondhjemitic glass suggests that reaction and selective contamination with mantle components has occurred in the veinlets. The trace element chemistry of the mantle clinopyroxenes indicates an increase in Na and Al with increasing Sr, La/Yb and Zr/Sm, which is attributed to reaction with an adakite melt component (Kepezhinskas *et al.* 1996).

A model for adakite–transitional adakite–Na metasomatism of the sub-arc mantle in northern Kamchatka would include: (1) partial melting of the slab at 75–85 km depth; (2) ascent of adakite into overlying mantle; (3) adakite–peridotite interaction; and (4) melting of Na-metasomatised mantle to produce NEAB magmas that host the cognate mantle xenoliths (Kepezhinskas *et al.* 1995, 1996). The lithological association of adakite-transitional adakite–Piip-type HMA–NEAB is interpreted to represent the progressive interaction between slab melts and sub-arc mantle.

4. Conclusions

Trondhjemite–tonalite–dacite magmas (high-Al TTD/adakite) with a unique compositional character can be generated by high-pressure partial melting of a basaltic source under garnet amphibolite to eclogitic conditions. Under the elevated geotherms of the Archean, the production of high-Al TTD is considered to be a major continental crust forming mechanism (Martin 1993). In Cenozoic tectonic regimes where young, hot oceanic crust has subducted, melting of the subducted slab can occur. We consider the approximate upper age limit for slab melt generation to be 25 Ma, primarily on the basis of natural occurrences of adakite localities (Drummond & Defant 1990; Defant & Drummond 1990). However, melting of older oceanic lithosphere is found in tectonic settings associated with oblique subduction and attendant high shear stresses (Yogodzinski *et al.* 1995) and incipient subduction into warm, pristine mantle wedge material (Sajona *et al.* 1993). Magnetic lineation maps of the world's ocean basins (Cande *et al.* 1989) indicates that approximately 20% of the ocean floor is comprised of Miocene or younger crust, which implies that adakite generation may not be as rare as originally thought.

Intermediate to felsic arc volcanic and plutonic rocks from an overthickened continental arc (CVZ/CBB, Andes) setting have some compositional similarities with adakites (e.g. high Sr/Y and La/Yb) due to plagioclase instability and involvement of garnet in their petrogenesis. However, adakite compositions can be differentiated from CVZ/CBB data on the basis of lower Sr isotopic composition, lower large ion lithophile element concentrations (Fig. 10), positive Sr anomaly and common presence of a mantle component (MgO, Cr and Ni) in the slab melts (Fig. 10). Adakites and transitional adakites show progressive enrichment in Cr and Ni content due to mantle contamination. The positive Ni anomaly (Fig. 10) is interpreted to result from the affinity of Cr for both restite phases and metasomatic phases at the site of partial melting and melt–mantle interaction, respectively, relative to Ni.

In conclusion, progress has been made on three fronts: (1) categorisation of the physical, tectonic and geochemical parameters of slab melting has been more narrowly defined; (2) discrimination between lower crustal and slab melt geochemistries is possible; and (3) the geochemical signature imparted on a adakite magma by the mantle wedge (and vice versa) is recognised.

Acknowledgements

We acknowledge NSF grants EAR-9103062 and EAR-9401931 for support of our past and present research in Kamchatka. Alfred Hochstaedter has been a valuable member of our Kamchatka research team and we give thanks for his work on this project. Discussions with Jim Beard, George DeVore and the numerous participants in the 1992 Penrose Conference on 'Origin and emplacement of low-K silicic magmas in subduction settings' have contributed to our understanding of

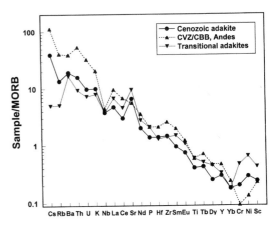

Figure 10 Multi-element variation diagram of average Cenozoic adakite, CVZ/CBB Andes (Table 1) and transitional adakite. MORB normalizing values of Taylor and McLennan (1985).

TTD magmas. Journal reviews of Mike Atherton, Jim Beard and Bob Kay are warmly appreciated.

6. References

Abbott, D. H. & Hoffman, S. E. 1984. Archaean plate tectonics revisited 1. Heat flow, spreading rate, and the age of subducting oceanic lithosphere and their effects on the origin and evolution of continents. TECTONICS 3, 429–48.

Arculus, R. J. & Powell, R. 1986. Source component mixing in the regions of arc magma generation. J GEOPHYS RES 91, 5913–26.

Armstrong, R. L. 1981. Radiogenic isotopes: the case for crustal recycling on a near-steady-state no-continental-growth Earth. PHIL TRANS R SOC LONDON A301, 443–72.

Arth, J. G. 1979. Some trace elements in trondhjemites—their implications to magma genesis and paleotectonic setting. In Barker, F. (ed.) Trondhjemites, dacites, and related rocks, 123–132. New York: Elsevier.

Atherton, M. P. & Petford, N. 1993. Generation of sodium-rich magmas from newly underplated basaltic crust. NATURE 362, 144–6.

Atherton, M. P. & Sanderson, L. M. 1987. The Cordillera Blanca Batholith: a study of granite intrusion and the relation of crustal thickening to peraluminosity. GEOL RUNDSCH 76, 213–32.

Barker, F. 1979. Trondhjemite: definition, environment and hypotheses or origin. In Barker, F. (ed.) Trondhjemites, dacites, and related rocks, 1–12. New York: Elsevier.

Barker, F., Arth, J. G., Peterman, Z. E. & Friedman, I. 1976. The 1.7- to 1.8-b.y.-old trondhjemites of southwestern Colorado and northern New Mexico. GEOL SOC AM BULL 87, 189–98.

Beard, J. S. 1995. Experimental, geological, and geochemical constraints on the origins of low-K silicic magmas in oceanic arcs. J GEOPHYS RES 100, 15 593–600.

Beard, J. S. & Lofgren, G. E. 1991. Dehydration melting and water-saturated melting of basaltic and andesitic greenstones and amphibolites at 1, 3, and 6.9 kb. J PETROL 32, 365–401.

Beard, J. S., Bergantz, G. W., Defant, M. J. & Drummond, M. S. 1993. Origin and emplacement of low-K silicic magmas in subduction setting: Penrose Conference Report. GSA TODAY 3, 38.

Bird, P., Toksov, M. N. & Sleep, N. H. 1975. Thermal and mechanical models of continent–continent convergence zones. J GEOPHYS RES 80, 4405–16.

Bowring, S. A., Housh, T. B. & Isachsen, C. E. 1990. The Acasta gneisses: remnant of earth's early crust. In Newsom, H. E. & Jones, J. H. (ed.) Origin of the earth, 319–43. New York: Oxford.

Cande, S. C., LaBreque, J. L., Larson, R. L., Pitman, W. C., Golovchenko, X. & Haxby, W. F. 1989. Magnetic lineations of the world's ocean basins. Tulsa: American Association of Petroleum Geologists.

Carroll, M. R. & Wyllie, P. J. 1989. Experimental phase relations in the system tonalite–peridotite–H$_2$O at 15 kb; implications for assimilation and differentiation processes near the crust–mantle boundary. J PETROL 30, 1351–82.

Cloos, M. 1993. Lithospheric buoyancy and collisional orogenesis: subduction of oceanic plateaus, continental margins, island arcs, spreading ridges, and seamounts. GEOL SOC AM BULL 105, 715–37.

Condie, K. C. 1981. Archean greenstone belts. New York: Elsevier.

Davidson, J. P., McMillan, N. J., Moorbath, S., Worner, G., Harmon, R. S. & Lopez-Escobar, L. 1990. The Nevados de Payachata volcanic region (18°S/69°W, N. Chile) II. Evidence for widespread crustal involvement in Andean magmatism. CONTRIB MINERAL PETROL 105, 412–32.

Defant, M. J. & Drummond, M. S. 1990. Derivation of some modern arc magmas by melting of young subducted lithosphere. NATURE 347, 662–5.

Defant, M. J. & Drummond, M. S. 1993. Mount St. Helens: potential example of the partial melting of the subducted lithosphere in a volcanic arc. GEOLOGY 21, 547–50.

Defant, M. J., Jackson, T. E., Drummond, M. S., de Boer, J. Z., Bellon, H., Feigenson, M. D., Maury, R. C. & Stewart, R. H. 1992. The geochemistry of young volcanism throughout western Panama and southeastern Costa Rica: an overview. J GEOL SOC LONDON 149, 569–79.

de Silva, S. L., Davidson, J. P., Croudace, I. W. & Escobar, A. 1993. Volcanological and petrological evolution of Volcan Tata Sabaya, SW Bolivia. J VOLCANOL GEOTHERM RES 55, 305–35.

Dewey, J. F. & Windley, B. F. 1981. Growth and differentiation of the continental crust. PHIL TRANS R SOC LONDON A301, 189–206.

Droop, G. T. R., Lombardo, B. & Pognante, U. 1990. Formation and distribution of eclogite facies rocks in the Alps. In Carswell, D. A. (ed.) Ecologite facies rocks, 225–59. New York: Chapman & Hall.

Drummond, M. S. & Defant, M. J. 1990. A model for trondhjemite-tonalite–dacite genesis and crustal growth via slab melting: Archean to modern comparisons. J GEOPHYS RES 95, 21 503–521.

Drummond, M. S., Bordelon, M., de Boer, J. Z., Defant, M. J., Bellon, H. & Feigenson, M. D. 1995. Igneous petrogenesis and tectonic setting of plutonic and volcanic rocks of the Cordillera de Talamanca, Costa Rica—Panama, Central American arc. AM J SCI 295, 875–919.

Ellam, R. M. & Hawkesworth, C. J. 1988. Elemental and isotopic variations in subduction related basalts: evidence for a three component model. CONTRIB MINERAL PETROL 98, 72–80.

Engi, M. & Lindsley, D. H. 1980. Stability of titanium clinohumite: experiments and thermodynamic analyses. CONTRIB MINERAL PETROL 72, 415–24.

Feeley, T. C. & Davidson, J. P. 1994. Petrology of calc-alkaline lavas at Volcan Ollague and the origin of compositional diversity at central Andean stratovolcanoes. J PETROL 35, 1295–340.

Feeley, T. C. & Hacker, M. D. 1995. Intracrustal derivation of Na-rich andesitic and dacitic magmas: an example from Volcan Ollague, Andean Central Volcanic zone. J GEOL 103, 213–25.

Frey, F. A., Green, D. H. & Roy, S. 1978. Integrated models of basalt petrogenesis: a study of quartz tholeiites to olivine melilites from southeastern Australia utilizing geochemical and experimental petrologic data. J PETROL 19, 463–513.

Gill, J. B. 1981. Orogenic andesites and plate tectonics. New York: Springer-Verlag.

Green, T. H. 1982. Anatexis of mafic crust and high pressure crystallization of andesite. In Thorpe, R. S. (ed.) Andesites—orogenic andesites and related rocks, 465–87. New York: Wiley.

Green, T. H. & Ringwood, A. E. 1968. Genesis of the calc-alkaline igneous rock suite. CONTRIB MINERAL PETROL 18, 105–62.

Hacker, B. R. 1990. Amphibolite-facies-to-granulite-facies reactions in experimentally deformed, unpowdered amphibolite. AM MINERAL 75, 1349–61.

Harmon, R. S., Barreiro, B. A., Moorbath, S., Hoefs, J., Francis, P. W., Thorpe, R. S., Deruelle, B., McHugh, J. & Viglino, J. A. 1984. Regional O-, Sr-, and Pb-isotope relationships in late Cenozoic calc-alkaline lavas of the Andean Cordillera. J GEOL SOC LONDON 141, 803–22.

Helz, R. T. 1976. Phase relations of basalts in their melting ranges at $P_{H_2O} = 5$ kb. Part II. Melt composition. J PETROL 17, 139–93.

Hochstaedter, A. G., Kepezhinskas, P. K., Defant, M. J., Drummond, M. S. & Bellon, H. 1994. On the tectonic significance of arc volcanism in northern Kamchatka. J GEOL 102, 639–54.

Hole, M. J., Saunders, A. D., Marriner, G. F. & Tarney, J. 1984. Subduction of pelagic sediments: implications for the origin of Ce-anomalous basalts from the Mariana Islands. J GEOL SOC LONDON 141, 453–72.

Holland, T. J. B. 1979. Experimental determination of the reaction paragonite = jadeite + kyanite + H$_2$O, and internally consistent thermodynamic data for part of the system Na$_2$O–Al$_2$O$_3$–SiO$_2$–H$_2$O, with applications to eclogites and blueschists. CONTRIB MINERAL PETROL 68, 293–301.

Holloway, J. R. & Burnham, C. W. 1972. Melting relations of basalt with equilibrium water pressure less than total pressure. J PETROL 13, 1–29.

Ishizaka, K. & Carlson, R. W. 1983. Nd–Sr systematics of the Setouchi volcanic belt, southwest Japan: a clue to the origin of orogenic andesite. EARTH PLANET SCI LETT 64, 327–40.

Jahn, B. M., Glikson, A. Y., Peucat, J. J. & Hickman, A. H. 1981. REE geochemistry and isotopic data of Archean silicic volcanics and granitoids from the Pilbara Block, Western Australia: implications for the early crustal evolution. GEOCHIM COSMOCHIM ACTA 45, 1633–52.

Jahn, B. M., Vidal, P. & Kroner, A. 1984. Multi-chronometric ages and origin of Archean tonalitic gneisses in Finnish Lapland: a case for long crustal residence time. CONTRIB MINERAL PETROL 86, 398–408.

Johnston, A. D. & Wyllie, P. J. 1989. The system tonalite–peridotite–H$_2$O at 30 kbar with applications to hybridization in subduction zone magmatism. CONTRIB MINERAL PETROL 102, 257–64.

Kay, R. W. 1978. Aleutian magnesian andesites: melts from subducted Pacific ocean crust. J VOLCANOL GEOTHERM RES 4, 117–32.

Kay, R. W. & Kay, S. M. 1991. Creation and destruction of lower continental crust. GEOL RUNDSCH **80**, 259–78.

Kay, R. W. & Kay, S. M. 1993. Delamination and delamination magmatism. TECTONOPHYSICS **219**, 177–89.

Kay, S. M., Mpodozis, C., Ramos, V. A. & Munizaga, F. 1991. Magma source variations for mid–late Tertiary magmatic rocks associated with a shallowing subduction zone and a thickening crust in the central Andes (28° to 33°S). GEOL SOC AM SPEC PAP **265**, 113–38.

Kay, S. M., Ramos, V. A. & Marquez, M. 1993. Evidence in Cerro Pampa volcanic rocks for slab melting prior to ridge–trench collision in southern South America. J GEOL **101**, 703–14.

Kelemen, P. B. 1995. Genesis of high Mg# andesites and the continental crust. CONTRIB MINERAL PETROL **120**, 1–19.

Kepezhinskas, P. K. 1989. Origin of the hornblende andesites of northern Kamchatka. INT GEOL REV **26**, 246–52.

Kepezhinskas, P. K., Defant, M. J. & Drummond, M. S. 1995. Na metasomatism in the island arc mantle by slab melt–peridotite interaction: evidence from mantle xenoliths in the north Kamchatka arc. J PETROL **36**, 1505–27.

Kepezhinskas, P. K., Defant, M. J. & Drummond, M. S. 1996. Progressive enrichment of island arc mantle by melt–peridotite interaction inferred from Kamchatka xenoliths. GEOCHIM COSMOCHIM ACTA **60**, 1217–29.

Kepezhinskas, P. K., McDermott, F., Defant, M. J., Hochstaedter, A., Drummond, M. S., Hawkesworth, C., Koloskov, A., Maury, R. C. & Bellon, H. Trace element and Sr–Nd–Pb isotope geochemistry of the Kamchatka volcanic arc, Russia. GEOCHIM COSMOCHIM ACTA, in press.

Kushiro, I. 1990. Partial melting of mantle wedge and evolution of island arc crust. J GEOPHYS RES **95**, 15929–39.

Luais, B. & Hawkesworth, C. J. 1994. The generation of continental crust: an integrated study of crust-forming processes in the Archaean of Zimbabwe. J PETROL **35**, 43–93.

Marsh, B. D. & Carmichael, I. S. E. 1974. Benioff zone magmatism. J GEOPHYS RES **79**, 1196–206.

Martin, H. 1986. Effect of steeper Archean geothermal gradient on geochemistry of subduction-zone magmas. GEOLOGY **14**, 753–56.

Martin, H. 1987. Petrogenesis of Archaean trondhjemites, tonalites, and granodiorites from eastern Finland: major and trace element geochemistry. J PETROL **28**, 921–53.

Martin, H. 1993. The mechanisms of petrogenesis of the Archaean continental crust—comparison with modern processes. LITHOS **30**, 373–88.

Massone, H. J. & Schreyer, W. 1987. Phengite geobarometry based on the limiting assemblage with K-feldspar, phlogopite, and quartz. CONTRIB MINERAL PETROL **96**, 212–24.

McLennan, S. M. & Taylor, S. R. 1982. Geochemical constraints on the growth of the continental crust. J GEOL **90**, 347–61.

Meijer, A. 1983. The origin of low-K rhyolites from the Mariana frontal arc. CONTRIB MINERAL PETROL **83**, 45–51.

Morris, P. A. 1995. Slab melting as an explanation of Quaternary volcanism and aseismicity in southwest Japan. GEOLOGY **23**, 395–8.

Mysen, B. O. 1982. The role of mantle anatexis. *In* Thorpe, R. S. (ed.) *Andesites–orogenic andesites and related rocks*; 489–522. New York: Wiley.

Nelson, E. P. & Forsythe, R. D. 1989. Ridge collision at convergent margins: implications for Archean and post-Archean crustal growth. TECTONOPHYSICS **161**, 307–15.

Nisbet, E. G. 1987. *The young earth.* Boston: Allen and Unwin.

Obata, M. & Thompson, A. B. 1981. Amphibole and chlorite in mafic and ultramafic rocks in the lower crust and upper mantle. CONTRIB MINERAL PETROL **77**, 74–81.

Parsons, B. A. & Sclater, J. G. 1977. An analysis of the variation of ocean floor bathymetry and heat flow with ages. J GEOPHYS RES **82**, 803–27.

Peacock, S. M. 1990a. Fluid processes in subduction zones. SCIENCE **248**, 329–37.

Peacock, S. M. 1990b. Numerical simulation of metamorphic pressure–temperature–time paths and fluid production in subducting slabs. TECTONICS **9**, 1197–211.

Peacock, S. M., Rushmer, T. & Thompson, A. B. 1994. Partial melting of subducting oceanic crust. EARTH PLANET SCI LETT **121**, 227–44.

Pearce, J. A. & Peate, D. W. 1995. Tectonic implications of the composition of volcanic arc magmas. *In* Wetherill, G. W., Albee, A. L. & Burke, K. C. (eds) ANNU REV EARTH PLANET SCI 251–85.

Pearce, J. A., van der Laan, S. R., Arculus, R. J., Murton, B. J. &

Ishii, T. 1992. Boninite and harzburgite from ODP Leg 125 (Bonin–Mariana forearc): a case study of magma genesis during the initial stages of subduction. *In* Fryer, P., Pearce, J. A. & Stokking, L. B. (eds) *Proceedings ODP Scientific Results, Leg 125*, 623–59.

Philippot, P. 1993. Fluid–melt–rock interaction in mafic ecologites and coesite-bearing metasediments: constraints on volatile recycling during subduction. CHEM GEOL **108**, 93–112.

Philippot, P. & Selverstone, J. 1991. Trace-element-rich brines in eclogitic veins: implications for fluid composition and transport during subduction. CONTRIB MINERAL PETROL **106**, 417–30.

Poli, S. 1993. The amphibolite–eclogite transformation: an experimental study on basalt. AM J SCI **293**, 1061–107.

Puga, E., Diaz de Federico, A. & Demant, A. 1995. The eclogitized pillows of the Betic Ophiolitic Association: relics of the Tethys Ocean floor incorporated in the Alpine chain after subduction. TERRA REV **7**, 31–43.

Raheim, A. & Green, D. H. 1975. P,T paths of natural eclogites during metamorphism—a record of subduction. LITHOS **8**, 317–28.

Rapp, R. P. 1990. *Vapor-absent partial melting of amphibolite/eclogite at 8–32 kbar: implications for the origin and growth of the continental crust, Troy.* Unpublished Ph.D. Thesis, Rensselaer Polytechnic Institute.

Rapp, R. P. 1995. Amphibole-out phase boundary in partially melted metabasalt, its control over liquid fraction and composition, and source permeability. J GEOPHYS RES **100**, 15601–18.

Rapp, R. P. & Watson, E. B. 1995. Dehydration melting of metabasalt at 8–32 kbar. Implications for continental growth and crust–mantle recycling. J PETROL **36**, 891–931.

Rapp, R. P., Watson, E. B. & Miller, C. F. 1991. Partial melting of amphibolite/eclogite and the origin of Archean trondhjemites and tonalites. PRECAMBRIAN RES **51**, 1–25.

Reymer, A. & Schubert, G. 1984. Phanerozoic addition rates to the continental crust and crustal growth. TECTONICS **3**, 63–77.

Ringwood, A. E. 1975. *Composition and petrology of the earth's mantle.* New York: McGraw-Hill.

Rushmer, T. 1991. Partial melting of two amphibolites: contrasting experimental results under fluid-absent conditions. CONTRIB MINERAL PETROL **107**, 41–59.

Rushmer, T., Pearce, J. A., Ottolini, L. & Bottazzi, P. 1994. Trace element behavior during slab melting: experimental evidence. EOS, TRANS AM GEOPHYS UNION **75**, 746.

Sajona, F. Z., Maury, R. C., Bellon, H., Cotten, J., Defant, M. J. & Pubellier, M. 1993. Initiation of subduction and the generation of slab melts in western and eastern Mindanao, Philippines. GEOLOGY **21**, 1007–10.

Saunders, A. D., Rogers, G., Marriner, G. F., Terrell, D. J. & Verma, S. P. 1987. Geochemistry of Cenozoic volcanic rocks, Baja California, Mexico: implications for the petrogenesis of post-subduction magmas. J VOLCANOL GEOTHERM RES **32**, 223–45.

Schreyer, W. 1988. Experimental studies on metamorphism of crustal rocks under mantle pressures. MINERAL MAG **52**, 1–26.

Selverstone, J., Franz, G., Thomas, S. & Getty, S. 1992. Fluid variability in 2 GPa eclogites as an indicator of fluid behavior during subduction. CONTRIB MINERAL PETROL **112**, 341–57.

Sen, C. & Dunn, T. 1994. Dehydration melting of a basaltic composition amphibolite at 1.5 and 2.0 GPa: implications for the origin of adakites. CONTRIB MINERAL PETROL **117**, 394–409.

Sen, C. & Dunn, T. 1995. Experimental modal metasomatism of a spinel lherzolite and the production of amphibole-bearing peridotite. CONTRIB MINERAL PETROL **119**, 422–32.

Sørensen, S. S. 1988. Petrology of amphibolite-facies mafic and ultramafic rocks from the Catalina Schist, southern California: metasomatism and migmatization in a subduction zone metamorphic setting. J METAMORPHIC GEOL **6**, 405–35.

Sørensen, S. S. & Grossman, J. N. 1989. Enrichment of trace elements in garnet amphibolites from a paleo-subduction zone: Catalina Schist, southern California. GEOCHIM COSMOCHIM ACTA **53**, 3155–77.

Spulber, S. D. & Rutherford, M. J. 1983. The origin of rhyolite and plagiogranite in oceanic crust: an experimental study. J PETROL **24**, 1–25.

Stern, C. R., Futa, K. & Muehlenbachs, K. 1984. Isotope and trace element data for orogenic andesites from the Austral Andes. *In* Harmon, R. S. & Barriero, B. A. (eds) *Andean magmatism—chemical and isotopic constraints.* 31–46. Cheshire: Shiva.

Tatsumi, Y. & Eggins, S. 1995. *Subduction zone magmatism.* Oxford: Blackwell.

Tatsumi, Y. & Ishizaka, K. 1982. Origin of high-magnesian andesites in the Setouchi volcanic belt, southwest Japan, I. Petrographic and chemical characteristics. EARTH PLANET SCI LETT **60**, 293–304.

Tatsumi, Y. & Nakamura, N. 1986. Composition of aqueous fluid from serpentine in the subducted lithosphere. GEOL J **20**, 191–96.

Tatsumi, Y., Hamilton, D. L. & Nesbitt, R. W. 1986. Chemical characteristics of fluid phase released from a subducted lithosphere and origin of arc magmas: evidence from high-pressure experiments and natural rocks. J VOLCANOL GEOTHERM RES **29**, 293–309.

Taylor, S. R. & McLennan, S. M. 1985. *The continental crust: its composition and evolution.* Oxford: Blackwell.

Taylor, R. N., Nesbitt, R. W., Vidal, P., Harmon, R. S., Auvray, B. & Croudace, I. W. 1994. Mineralogy, chemistry, and genesis of the boninite series volcanics, Chichijima, Bonin Islands, Japan. J PETROL **35**, 577–617.

Thompson, A. B. & Ellis, D. J. 1994. $CaO + MgO + Al_2O_3 + SiO_2 + H_2O$ to 35 kb: amphibole, talc, and zoisite dehydration and melting reactions in the silica-excess part of the system and their possible significance in subduction zones, amphibolite melting, and magma fractionation. AM J SCI **294**, 1229–89.

Tsuchiya, N. & Kanisawa, S. 1994. Early Cretaceous Sr-rich silicic magmatism by slab melting in the Kitakami Mountains, northeast Japan. J GEOPHYS RES **99**, 22 205–20.

Ulmer, P. & Trommsdorff, V. 1995. Serpentine stability to mantle depths and subduction-related magmatism. SCIENCE **268**, 858–61.

Wedepohl, K. H. 1995. The composition of the continental crust. GEOCHIM COSMOCHIM ACTA **59**, 1217–32.

Winther, K. T. & Newton, R. C. 1991. Experimental melting of hydrous low-K tholeiite: evidence on the origin of Archaean cratons. BULL GEOL SOC DENMARK **39**, 213–28.

Wolf, M. B. & Wyllie, P. J. 1991. Dehydration-melting of solid amphibolite at 10 kbar: textural development, liquid interconnectivity and applications to the segregation of magmas. MINERAL PETROL **44**, 151–79.

Wolf, M. B. & Wyllie, P. J. 1994. Dehydration-melting of amphibolite at 10 kbar: the effects of temperature and time. CONTRIB MINERAL PETROL **115**, 369–83.

Wyllie, P. J. 1982. Subduction products according to experimental prediction. GEOL SOC AM BULL **93**, 468–76.

Wyllie, P. J. 1984. Sources of granitoid magmas at convergent plate boundaries. PHYSICS EARTH PLANET INTER **35**, 12–8.

Yogodzinski, G. M., Volynets, O. N., Koloskov, A. V., Seliverstov, N. I. & Matuenkov, V. V. 1994. Magnesian andesites and the subduction component in a strongly calc-alkaline series at Piip volcano, far western Aleutians. J PETROL **35**, 163–204.

Yogodzinski, G. M., Kay, R. W., Volynets, O. N., Koloskov, A. V. & Kay, S. M. 1995. Magnesian andesite in the western Aleutian Komandorsky region: implications for slab melting and processes in the mantle wedge. GEOL SOC AM BULL **107**, 505–19.

MARK S. DRUMMOND, Department of Geology, University of Alabama at Birmingham, Birmingham, AL 35294-2160, U.S.A.

MARC J. DEFANT, Department of Geology, University of South Florida, Tampa, FL 33620-5200, U.S.A.

PAVEL K. KEPEZHINSKAS, Department of Geology, University of South Florida, Tampa, FL 33620-5200, U.S.A.

Transactions of the Royal Society of Edinburgh: Earth Sciences, **87**, 217–223, 1996

Non-linear dynamics, chaos, complexity and enclaves in granitoid magmas

James Flinders and John D. Clemens

ABSTRACT: Most natural systems display non-linear dynamic behaviour. This should be true for magma mingling and mixing processes, which may be chaotic. The equations that most nearly represent how a chaotic natural system behaves are insoluble, so modelling involves linearisation. The difference between the solution of the linearised and 'true' equation is assumed to be small because the discarded terms are assumed to be unimportant. This may be very misleading because the importance of such terms is both unknown and unknowable. Linearised equations are generally poor descriptors of nature and are incapable of either predicting or retrodicting the evolution of most natural systems. Viewed in two dimensions, the mixing of two or more visually contrasting fluids produces patterns by folding and stretching. This increases the interfacial area and reduces striation thickness. This provides visual analogues of the deterministic chaos within a dynamic magma system, in which an enclave magma is mingling and mixing with a host magma. Here, two initially adjacent enclave blobs may be driven arbitrarily and exponentially far apart, while undergoing independent (and possibly dissimilar) changes in their composition. Examples are given of the wildly different morphologies, chemical characteristics and Nd isotope systematics of microgranitoid enclaves within individual felsic magmas, and it is concluded that these contrasts represent different stages in the temporal evolution of a complex magma system driven by non-linear dynamics. If this is true, there are major implications for the interpretation of the parts played by enclaves in the genesis and evolution of granitoid magmas.

KEY WORDS: linearisation, non-linear dynamics, interfaces, striation, deterministic chaos, temporal evolution, fractals.

1. Why non-linear dynamics?

Physics attempts to quantify nature through the solution of equations that have usually been constructed on the basis that they can be solved. A dynamic system that can be treated in this classical manner may exist either in a steady state or exhibit periodic or quasiperiodic behaviour. In the last century, scientists were aware that deterministic systems can also behave in what appeared to be random ways. However, to learn anything about extremely complex systems, with many degrees of freedom, variables and parameters, they had to simplify their analysis.

In petrology and geochemistry it is commonplace to assume that outliers in data sets are either erroneous or represent 'noise'; they are commonly rejected when models are erected. The classical procedure is to linearise the non-linear by discarding those terms that make an equation insoluble. It is assumed that because the discarded terms in the insoluble higher order differential equation are small (which is true), then the difference between the solution of the linearised and the 'true' equation must also be small (which is not necessarily so). The importance of the discarded terms is both unknown and unknowable; the imposition of linearity is a psychological trap as the behaviour of linear equations is atypical of nature. Linear equations are incapable of retrodicting the future or the past evolution of most natural systems; they are descriptive but non-predictive.

Systems, such as a convecting fluid, have the capacity to surprise the investigator with the onset of previously hidden disorderly (chaotic) behaviour as, with deterministic chaos (sensitive dependence on initial conditions), prediction beyond a certain, usually short, time interval becomes impossible; the system is subject to fortuitous phenomena. The problem in dealing with many geological systems is that direct and/or prolonged observation is humanly impossible. Thus we must infer behaviour from the characteristics of 'frozen' products (rocks), the states of which represent only single instants in the time evolution of the system that produced them.

It is standard practice to analyse a real system by making a mathematical model, which is subsequently analysed by a computer program and compared with observations on the natural system. Many real systems can be described as deterministic, differentiable, dynamic systems with some added noise (Ruelle 1994). These are chaotic systems that have mostly predictable time evolutions, but with some 'noise' that cannot be accurately quantified beyond a certain limiting period. As a result there are numerous cases for which the mathematical models are quantitatively satisfactory. These are usually based on equations that are well known and, although chaos may be inherent in the system, the characteristic period for the onset of chaotic dynamics is long enough to enable them to give usable results. This characteristic time is a function of the interval between interactions of the particles comprising the system (e.g. planetary motion—several millions of years—and weather forecasting—several days). Such models, although intrinsically interesting, are unable to answer many questions posed by natural systems.

The onset of hydrodynamic turbulence and the mingling and mixing of high viscosity melts are areas where time evolution is chaotic, in the sense that small changes in the initial conditions greatly influence the future evolution of the system.

Chaos embraces both ordered periodic complexity and aperiodic disorder and is never truly random. Chaotic patterns, from any particular system, are recognisable on micro-, macro- and meso-scales and possess elements of self-similarity (i.e. the geometry of the pattern continues to be recognisable at different 'levels' of observation). Commonly, the complex geometries are not only fractal but multifractal, described by dimensional coefficients that are both fractal and changing in value, suggesting that, for some elements of nature, looking for a characteristic scale is non-essential. Petford et al. (1993) provide an example of the possible application of fractal analysis to igneous petrology.

Igneous-textured microgranitoid enclaves in granitoids are widely accepted as indicating the mingling and mixing of separate magma fractions, whatever the origin of those fractions. What we have discussed here has relevance to the topography of the interfaces between the host monzogranite and the monzonitic enclaves within the Glen Fyne intrusion of Scotland, where some of the quasi-two-dimensional thin sections exhibit interfaces with a fractal dimension >1 but <2; equivalent to >2 but <3 in three dimensions.

2. Magma mingling and mixing

In this paper, usage of the terms 'mingling' and 'mixing' assumes that they are different processes. *Mingling* occurs when two or more different magmas simultaneously occupy parts of a chosen volume and yet retain their chemical and physical identities. The boundaries between the individual magmas are marked by differences in colour (modal mineralogy) and texture. *Mixing* occurs when the physical and chemical identities of once separate magmas have been irrevocably changed, so that it is no longer possible to identify the initial, separate components. The resulting mixed magma exhibits new properties acquired through both chemical reaction/diffusion and mechanical mixing.

One of the possible signatures of chaos, mentioned earlier, is the sensitive dependence on initial conditions. In fluid flow, as in mingling and mixing magmas, this translates into the analysis of patterns produced by folding and stretching. Stewart (1993) illustrated the complexity of the stretching and folding process by deforming a two-dimensional cartoon of a duck. After only three iterations of this stretching and doubling, the form of the duck is totally unrecognisable. Moreover, Stewart pointed out that, although the formula for this transformation is simple, it cannot be solved to discover the actual result following a number of iterations. Such patterns provide a visual analogue of the deterministic chaos within the system, where two spatially juxtaposed points may be driven arbitrarily and exponentially far apart. It is impossible to predict or calculate the time evolution vectors of points within partially miscible fluids and hence to know the spatial and temporal history of any point. This has special relevance to the apparently chaotic spatial distribution of microgranitoid enclaves within the Glen Fyne granite, as evidenced by their chemistry and Sm–Nd isotope systematics (ε_{Nd} values).

In two-dimensional flow there are two types of periodic point, and it is the character of these points that determines how the complex flow evolves. A melt packet will move towards a hyperbolic point in one direction and away from it in another, whereas it will circulate around an elliptical point. In three-dimensional flow, material can neither enter nor leave the vicinity of a periodic *elliptical* point. In the present case, these regions would be islands of magma that hinder efficient mixing. Periodic *hyperbolic* points attract material from one direction and expel it in another, and *transverse homoclinic* points are established where outflows intersect inflows. Intersecting flows, from two different hyperbolic points, create *transverse heteroclinic* points. The identification of a single *transverse homoclinic* point is sufficient evidence to imply the existence of a horseshoe map and efficient mixing due to chaotic flow. However, this has not yet been observed in a granitoid, possibly because such things have not been looked for.

Mingling and mixing will form a time continuum within a given magma system. Two initially distinct magmas, with physical and chemical properties that will allow mixing, will first mingle and then mix. This occurs as a result of the simultaneous increase in interfacial area and decrease of striation (schlieren) thickness, causing disaggregation of the less viscous component (enclave magma) and subsequent hybridisation (mixing) through diffusion and chemical reaction. The amount of enclave break-up will be very much less within 'islands' (associated with periodic elliptical points) than within chaotic regions (associated with transverse homoclinic points). Within any chosen volume of magma, moving as a result of either mass flow or convection, parts will experience chaotic flow, with elongation of less viscous blobs (enclaves), and parts will experience non-chaotic flow, where the blobs are much less affected. Thus it is probable that the enclaves that we observe in the frozen system (granitic rock) are mostly those that have experienced only limited mechanical mingling.

The varying views that we get of the different shapes of an enclave magma within a more viscous host granite might be due to the temporal level of our viewpoint in relation to the evolution of the system. Early snapshots would show pillows or schlieren only. A view after continued chaotic mingling would first reveal disaggregation of the less viscous schlieren into numerous elongate enclaves and then into subspheroidal ones. A view following further mingling and/or mixing could introduce double (or multiple) enclaves formed by the coalescence or engulfment of earlier formed enclaves. It is highly unlikely that this entire sequence could be observed in a rock at any single locality. This is because the time evolution of a dynamic system involves chaotic motion, probably on many length scales. This will produce similarity between the disposition of inviscid elongated magma blobs within a more viscous host (observed at a scale of tens of cubic metres) and that of the individual early formed minerals within a still partially molten individual enclave (seen at a scale of a few cubic centimetres).

3. Images of natural systems

Figure 1 illustrates a transient, two-dimensional chaotic pattern of foam on a water surface, resulting from fluid flow within the underlying liquid volume. The factors that control the transfer of fluid motion from within the body of the fluid to the surface are unknown. It might be that motion transfer is due to a boundary layer effect. It may be independent of the water depth. It may operate over an optimum distance, or perhaps be controlled by the water velocity and resulting mass transfer rates. Regardless of the details of the mechanism responsible for their formation, such patterns are a two-dimensional illustration of the complexity produced in a three-dimensional system, such as a flowing magma. Computer modelling of the surface of a stirred fluid, and experiments on two-dimensional closed systems experiencing chaotic folding and stretching, produce similar patterns to those in Figure 1. (Ottino 1989; Ottino et al. 1988).

Although the fluid dynamics of a wide compositional range of magmas will produce chaotic folding and stretching, it is within those of very different ultimate colour index that the patterns of lobes (magma pillows) and striations (schlieren) are most obvious. Striations of dark- and light-coloured rock (schlieren) represent the first stages of mingling, with only slight chemical interaction at the interfaces. Magmas with this

Figure 1 Distribution of foam on a slow-moving water surface. The pattern forms by the two-dimensional movement of the water surface in response to the three-dimensional, non-linear dynamics of the water volume. This may provide an analogy with processes in igneous systems containing two partially miscible magmas. Those areas swept clear of foam are experiencing linear mingling, whereas those with complex striations, lobes and disaggregated blobs have been subject to chaotic dynamics.

type of pattern have not experienced any dynamic processes that would cause a significant degree of chemical or physical homogenisation (mixing).

In Figure 2 the dark microgranitoid enclaves stand out clearly from the more felsic host granite. Most commonly, the enclaves appear elongate, with pointed, cuspate, terminations parallel or subparallel to their major axes. This strongly suggests that the three-dimensional shapes of the enclaves will approximate to elongated lobes, rather than to disks. The aspect ratios of the enclaves vary from ≈ 2 to ≈ 20. Clearly, these enclaves were produced by the break-up of a far more voluminous mass of more mafic magma. The shear forces associated with the viscosity contrast and the relative velocity between the host and enclave were sufficient to cause elongation, extrusion and eventual disaggregation.

Figure 3 shows part of an exposure near the River Fyne at U.K. grid reference NN 21/31 228 160. The area of the exposure at this locality is ≈ 10 m $\times$ 6 m and the photograph shows an area of ≈ 4 m $\times$ 3 m. The host is a quartz monzodiorite.

The most important features in this figure are: the clusters of enclaves (lettered A to D); large individual enclaves (lettered F and G); eroded troughs with enclave remnants (lettered I and J); and sinuous trails of large orthoclase feldspar phenocrysts (megacrysts) between the arrows. The enclave clusters (A, B, C and D) and the eroded troughs (I and J) are within the megacryst trails. The individual, large enclaves (F and G) are outside the trails.

As this two-dimensional surface is subhorizontal, it immediately conveys the impression that the megacryst-rich trails had a predominantly horizontal motion vector. However, it is unlikely that horizontal fluid flow could assemble and maintain a vertically orientated enclave cluster, which experiences considerable frictional drag, as it moves within a viscous fluid. If the trails are a flow feature, which seems likely, it is more probable that the major motion vector was vertical (normal to the rock surface). The disposition of the enclave clusters suggests a three-dimensional matrix of vertical pipes and connecting sheets.

This concentration of the feldspar phenocrysts into sinuous trails, within which similar sized clusters of enclaves occur at almost regular distances, on a subhorizontal surface, is an abnormal feature within the Glen Fyne intrusion. However,

Figure 2 Swarm of microgranitoid enclaves within a granodioritic host from Yosemite National Park, California. The outcrop is approximately 2·5 m high. Spatially juxtaposed enclaves have dissimilar two-dimensional sections and wildly different three-dimensional morphologies. The spatial, chemical and temporal histories of each enclave are essentially unknowable, making it impossible for such an assemblage to yield useful data on the strain history at either the magmatic or post-magmatic stages. In any case, it is clear from Figure 1 that strain may vary, by orders of magnitude, over relatively short distances in systems undergoing chaotic mingling.

the intrusion as a whole appears to be unexceptional. Based on our experience with a large number of granitoid bodies, we would say that the Glen Fyne contains a normal abundance of apparently chaotically distributed, relatively small enclave remnants, mostly with maximum dimensions < 20 cm. As was clear from one of the reviews of this paper, enclave researchers would hardly look twice at such an unimpressive, ordinary granite. The extraordinarily spectacular occurrences of very enclave-rich areas which usually figure in granite enclave studies (e.g. Fig. 2) are probably atypical of granitic intrusions on the whole. In our experience, the Glen Fyne style of enclave abundance and distribution is typical. It seems possible that the features seen in places such as Glen Fyne, and the inferred chemical and dynamic processes responsible for their origin, are of more value in formulating a general model of host–enclave magma interaction.

Figure 4 shows a bowl-shaped pothole, with a diameter of ≈ 2 m and a depth of $\approx 1·5$ m, scoured into the bed of the River Fyne. Numerous enclave remnants, forming a loose vertical cluster, are exposed on the wall of the pothole. The near-side wall also contains enclave remnants, but fewer. The locality of Figure 4 lies ≈ 10 m to the right of the top right-hand corner of Figure 3, in the centre of the river. The rim of the depression in Figure 4 is ≈ 4 m lower than the rock surface shown in Figure 3. This photograph reveals the (usually unseen) third dimension in this unit.

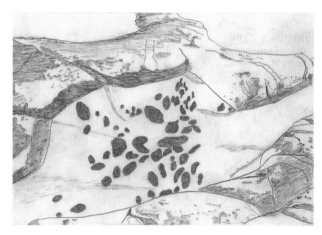

Figure 3 Photograph and sketch of several clumps of microgranitoid enclaves (A, B, C and D) within an exposure of the Glen Fyne quartz monzodiorite, Scotland. The clumps commonly occur within trails of very large orthoclase feldspar phenocrysts (arrowed). Separated enclaves within the body of the granitoid (F and G), are commonly larger than those within the phenocryst trails. Some enclave clumps have been partially (J) or totally weathered-out (I).

Figure 4 Photograph and sketch of a pothole within the bed of the river Fyne (width of field of view ≈ 2 m). As enclaves occur within each wall of the pothole, this suggests preferential weathering-out of a vertical, pipe-like, enclave cluster.

Throughout the Glen Fyne pluton, enclaves are preferentially weathered out. Trough I on Figure 3 is a good example of an enclave-rich locality that has undergone rapid selective weathering. The rock platform in Figure 3 is usually above the river level and so experiences minimal stream-driven erosion compared with the river bed locality in Figure 4, usually ≈ 2 m below the water surface.

To determine the volumes of the enclaves still remaining in the host rock would be extremely difficult. To determine the volumes of the enclaves eroded away is impossible. However, the hemispherical shape of the depression, with enclave remnants in each wall, the presence of enclaves around the rim, and the ease with which the enclaves are preferentially eroded, all suggest that a large number of enclaves once occupied this relatively small volume. Thus we interpret the enclave clumps visible on the rock platform as horizontal sections through vertical clusters of enclaves. This suggests that some form of constricted pipe flow concentrated the enclaves.

Serial sectioning of numerous, large, host–enclave samples has yielded photographic evidence showing the disaggregation of enclaves by the host magma and ingestion of host by the enclave magma. The polished slab in Figure 5 clearly shows a euhedral orthoclase phenocryst, from the host, in the process of being ingested by the enclave magma. Other serial sections of this sample show numerous xenocrysts of orthoclase and quartz in the enclave. The internal enclave–host interfaces are

Figure 5 Polished slab from a large serially sectioned sample showing ingestion of a host orthoclase phenocryst by the enclave magma.

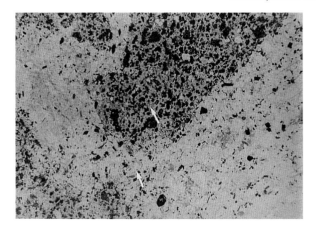

Figure 6 Plane polarized light macrophotograph of the central part of the face shown in Figure 5, with swirls of hornblende crystals in both the host and enclave (arrowed).

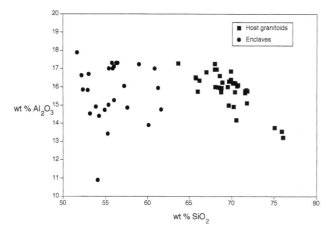

Figure 7 Example of the distribution of major oxide data showing a zero correlation between host and enclave analyses.

highly complex, commonly lobate and spatially variable. Several adjacent slabs, only ≈ 3 mm apart, show completely different morphologies. Figure 6 shows a large thin section of the central part of the face shown in Figure 5. Swirls of hornblende crystals, on a scale of millimetres to centimetres, occur within both the enclave and host (arrowed). We infer from this that the enclave magma was in a plastic state long enough to undergo complex stretching and kneading as the magma system flowed.

The external appearance or two-dimensional shape of an enclave gives no indication of the complexity of either its internal structure or its interface topography. Computer images constructed from serially sectioned enclaves clearly show multi-lobate (amoeboid), three-dimensional shapes that could yield an infinite number of ellipsoidal two-dimensional sections (Srogi & Lutz 1990). Thus measuring the aspect ratios of such sections provides no data useful either for determining the large-scale strain history within either the host magma or the solidified granite, or for correlating enclave area with chemical parameters. Beware of the blob!

Mobile host and enclave magmas continually modify their textures and compositions right up to the point of freezing, producing enclaves whose chemical compositions and shapes are unique to one time instant only, and probably unique to each enclave. If the compositions, shapes and locations of enclaves can indeed be modelled as resulting from chaotic processes, Stewart (1993:41) points out that '... the price we pay for this new knowledge is the realisation that... finding the right equations may not tell us much about their solutions.' Stewart comments further that extremely complex-looking patterns can be produced from very simple equations and so the information content of an observed pattern may be very low. This has obvious relevance for any models we may devise for enclave–host magma systems.

4. Host and enclave chemistry

Figure 7 shows a graph of SiO_2 versus Al_2O_3 for Glen Fyne host monzogranite and microgranitoid enclave analyses. The distribution of points is typical of many of the major oxide plots from these sample sets. It shows loose clustering of data among the hosts (possibly with a host lineage trend) and a wide, uncorrelated spread of the enclave analyses (which do not appear to belong to the host lineage; see also Roberts & Clemens 1995). None of the enclaves has cumulate textures; mixing between cumulates and differentiated host magma cannot explain the enclave chemistry. In any case, cumulate

or one- or two-magma differentiation and mixing models are ruled out by the Nd isotope systematics (see later).

Some of the major oxide plots exhibit approximately linear trends with small negative gradients. It is true that functional dependence, as a result of fractional crystallisation, for example, could produce this type of correlation. However, it is also true that random associations between the variables would produce exactly the same type of plot. Computed values for the slopes [from $E = 1/(1-M)$, where E is the slope and M is the number of chemical variables], for values of M from 5 to 8, lie between -0.25 and -0.14. Consequently, we would be entitled to dismiss such trends as petrogenetically equivocal, or even meaningless. This is because there are induced, enhanced correlations in all such major oxide bivariate plots as a result of the formation of closed arrays, and because very different processes can produce very similar trends (Chayes 1962; Wall et al. 1987; Rollinson 1993).

Figure 8 shows a graph of SiO_2 versus Zr for the same sample set. Almost all the trace element plots exhibit similar distributions—wide, uncorrelated spreads of data points from both hosts and enclaves. Only the data for vanadium show a loosely linear distribution. However, to assign significance to this one plot would be wildly over-optimistic. It is extremely unlikely that this type of plot, for this one element, would be capable of uniquely categorising the complexities associated with the temporal evolution of a magma system driven by non-linear dynamics.

Owing to the evident inadequacy of major and trace element data for deciphering the petrogenesis of the Glen Fyne rocks,

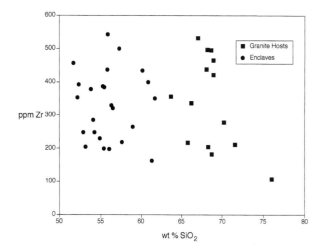

Figure 8 Example of the distribution of trace element data showing a zero correlation between host and enclave analyses.

Sm–Nd isotope systematics were investigated to 'see through' post-emplacement reactions and avoid the limitations of other chemical systems. There are no Rb–Sr isotope contrasts between hosts and enclaves. We interpret this to mean that diffusion has re-equilibrated the enclave initial $^{87}Sr/^{86}Sr$ to something very close to that of the original host magma. The Sm–Nd system is known to be more resistant to re-equilibration of this kind (e.g. Lesher 1990). Figure 9 compares the ε_{Nd} values of host and enclave pairs. Clearly, the enclaves have consistently higher ε_{Nd} than their hosts, indicating a less evolved source/s for the enclaves.

Figure 10 shows a plot of SiO_2 versus ε_{Nd} for three monzogranite hosts, each of which has multiple corresponding enclave analyses; hosts and enclaves are joined by tie lines. The negative gradients of the tie lines show that the enclaves are always less silicic than their hosts, that they have more positive ε_{Nd} and that the differences in ε_{Nd} are large enough to be real and are not artefacts of the analytical technique. More important, however, is the spread of enclave values for each host. The host with $\varepsilon_{Nd} = -0.7$ has enclaves with values of $+0.6$ and -0.2. The host with $\varepsilon_{Nd} = -1.2$ has enclaves with $+0.8$ and 0, whereas the host with a value of -0.9 has enclaves with $+0.9$, $+0.4$, 0, -0.5 and -0.6.

All the enclaves were collected from areas of $<1 m^2$, for those hosts with values of -0.7 and -1.2, and from within the cluster of enclaves labelled A (on Fig. 3) for the host with an ε_{Nd} value of -0.9. Spatially juxtaposed enclaves display a complex scattered distribution of their ε_{Nd}. It is difficult to see how this could be interpreted as being derived from any single magma batch, or even a chaotic mixing and mingling process involving only two magmas. Lesher (1990) showed that mixing of magmas could result in transient states in which isotopic

and elemental trends depart significantly from those expected for bulk mixing. However, these transient states appear to last only a few hours and produce curvilinear trends, not a broad scatter. The best interpretation (of our two-dimensional view of the problem) would seem to be that several distinct magma fractions coexisted and were modified by internal differentiation, diffusion and chaotic stretching and folding. This possibility, though highly complex, may be capable of detailed analysis. However, the results are unlikely to be gratifying if we expect enclaves to tell us something fundamental about the origin of the host granite. Many enclave suites may prove to be little more than accessory magmas derived from protolith heterogeneities and are thus unlikely to yield data that can adequately constrain their origin, original composition or their role (if any) in petrogenetic processes operating in a host granitoid magma. Isotope-based models are inherently non-unique in any case.

There are a number of other published enclave–host geochemical studies that have revealed complex, scattered variations in enclave chemistry, and a lack of correlation between host and enclave distributions on variation diagrams (e.g. Chen et al. 1990). As a counter example, we refer the reader to Metcalf et al. (1995), who describe the Mt Perkins pluton in Arizona, in which host and enclave magmas appear to form a single magmatic lineage. Cocirta and Orsini (1986) interpret their several northern Sardinian granite–enclave suites (which plot as a radial pattern on a total Fe–MgO diagram) as resulting from mixing between a single fractionating granitic parent and a continuously evolving single mafic magma. This is possible, but isotopic constraints are lacking. Such processes could lead to scatter in major and trace element variations. Similar issues have been comprehensively discussed by O'Hara and Mathews (1981: 237), who pointed out that '... these relationships cannot be inverted in order to deduce uniquely the magma chamber parameters or ... source compositions ...'

Only time and further careful studies will show whether the Glen Fyne type or the Mt Perkins type of situation is more common, and why. It seems possible that, for many granitic intrusions, enclave studies will prove to be an interesting, but time consuming, petrological cul de sac.

Acknowledgements

This paper benefited from the comments of an anonymous reviewer as well as editorial suggestions by Ed Stephens.

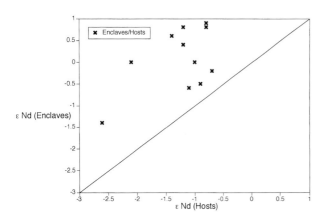

Figure 9 Graph of corresponding enclave–host ε_{Nd} values.

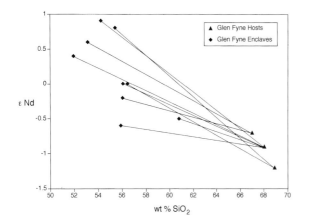

Figure 10 SiO_2 versus ε_{Nd} plot for Glen Fyne host granites and their enclaves. Tie lines join corresponding hosts and enclaves.

References

Chayes, F. 1962. Numerical correlation and petrographic variation. J GEOL **70**, 440–52.

Chen, Y., Price, R. C., White, A. J. R. & Chappell, B. W. 1990. Mafic inclusions from the Glenbog and Blue Gum granite suites, southeastern Australia. J GEOPHYS RES **B95**, 17 757–85.

Cocirta, C. & Orsini, J.-B. 1986. Signification de la diversité de composition des enclaves «microgrenues» sombre en contexte plutonique. L'example des plutons calco-alcalin de Bono et Budduso (Sardaigne septentrionale). C ACAD SCI PARIS SER (II) **302**, 331–6.

Lesher, C. E. 1990. Decoupling of chemical and isotopic exchange during magma mixing. NATURE **344**, 235–7.

Metcalf, R. V., Smith, E. I., Walker, J. D., Reed, R. C. & Gonzales, D. A. 1995. Isotopic disequilibrium among co-mingled hybrid magmas: evidence for a two-stage magma mixing–co-mingling process in the Mt Perkins pluton, Arizona. J GEOL **103**, 509–27.

O'Hara, M. J. & Mathews, R. E. 1981. Geochemical evolution in an advancing, periodically replenished, periodically tapped, continuously fractionated magma chamber. J GEOL SOC LONDON **138**, 237–77.

Ottino, J. M. 1989. The mixing of fluids. SCI AM **260**, 40–9.

Ottino, J. M., Leong, C. W., Rising, H. & Swanson, P. D. 1988.

Morphological structures produced by mixing in chaotic flows. NATURE **333**, 419–25.

Petford, N., Byron, D., Atherton, M. P. & Hunter, R. H. 1993. Fractal analysis in granitoid petrology: a means of quantifying irregular grain morphologies. EUR MINERAL **5**, 593–8.

Roberts, M. P. & Clemens, J. C. 1995. Feasibility of AFC models for the petrogenesis of calc-alkaline magma series. CONTRIB MINERAL PETROL **121**, 139–47.

Rollinson, H. R. 1993. *Using geochemical data: evaluation, presentation, interpretation.* Harlow: Longman Scientific & Technical.

Ruelle, D. 1994. Where can one hope to profitably apply the ideas of chaos? PHYS TODAY **47**, 24–30.

Srogi, L. & Lutz, T. M. 1990. Three-dimensional morphologies of metasedimentary and mafic enclaves from Ascutney Mountain, Vermont. J GEOPHYS RES **95**, (B11), 17 829–40.

Stewart, I. 1993. Chaos. *In*: Howe, L. & Wain, A. (eds) *Predicting the future*, 24–51. Cambridge: Cambridge University Press.

Wall, V. J., Clemens, J. D. & Clarke, D. B. 1987. Models for granitoid evolution and source compositions. J GEOL **95**, 731–50.

JAMES FLINDERS, Department of Earth Sciences, The University of Manchester, Oxford Road, Manchester M13 9PL, U.K.

JOHN D. CLEMENS, School of Geological Sciences, Kingston University, Penrhyn Road, Kingston-upon-Thames, Surrey KT1 2EE, U.K.

Transactions of the Royal Society of Edinburgh: Earth Sciences, **87**, 225–232, 1996

Trace element and isotopic exchange during acid–basic magma interaction processes

G. Poli, S. Tommasini and A. N. Halliday

ABSTRACT: Interaction processes between acid and basic magmas are widespread in the Sardinia–Corsica Batholith. The resulting hybrid magmas are extremely variable and can be broadly divided into: (i) microgranular mafic enclaves with geochemical characteristics of both magmatic liquids and cumulates; (ii) basic gabbroic complexes with internal parts mainly formed by cumulates and with interaction zones developing only in the marginal parts; and (iii) basic septa with the form of discrete, lenticular-like bodies often mechanically fragmented in the host rock. Different styles of interaction, ranging from mixing to mingling, have been related to variations in several physico-chemical parameters, such as: (i) the initial contrast in chemical composition, temperature and viscosity; (ii) the relative mass fractions and the physical state of interacting magmas; and (iii) the static versus dynamic environment of interaction.

A model is presented for the origin and history of interaction processes between basic and acid magmas based on the geochemical characteristics of hybrid magmas. Physico-chemical processes responsible for the formation of hybrid magmas can be attributed to: (i) fractional crystallisation of basic magma and contamination by acid magma; (ii) loss of the liquid phase from the evolving basic magma by filter pressing processes; (iii) mechanical mixing between basic and acid magmas; and (iv) liquid state isotopic diffusion during the attainment of thermal equilibrium.

KEY WORDS: Acid–basic interaction, Sardinia–Corsica Batholith, mixing, mingling, physico-chemical parameters.

The interaction between acid and basic magmas is increasingly considered to be one of the main mechanisms for the genesis of composite batholiths (e.g. Halliday *et al.* 1980; Reid *et al.* 1983; Frost & Mahood 1987; Holden *et al.* 1987; Poli *et al.* 1989; Didier & Barbarin 1991a).

The generation of felsic magmas mainly occurs in those portions of the continental crust to which hotter mafic magmas have access (Huppert & Sparks 1988). The coexistence of at least two magmas of different composition and different temperatures is inherent in these models. There is abundant field, petrographic and petrochemical evidence that co-mingling and mixing do occur in the plutonic environment, not only on a local scale, but to produce considerable batches of magma. It has been suggested that large volumes of calc-alkaline magma, now represented by granitic, granodioritic and tonalitic plutons, may be generated by mixing (e.g. Barbarin 1988; Poli *et al.* 1989).

Much remains to be understood regarding how and when this interaction takes place and what parameters are involved. To better understand the physical and chemical mechanisms that control interaction processes, we review the occurrence of hybrid magmas in the Hercynian Sardinia–Corsica Batholith (SCB). This is condensed into an outline, and from this a model is proposed regarding the possible origin and history of interaction processes between basic and acid magmas.

1. Associations of basic and acid rocks in the Sardinia–Corsica Batholith

The SCB was formed during the Hercynian orogeny and consists of metamorphic terranes, multiple coalescent granitoid plutons and subordinate gabbroic complexes. The magmatic activity related to the Hercynian orogeny developed during a time span of 70 million years, from 350 to 280 Ma, and has a

typical continent–continent collision, calc-alkaline signature (e.g. Poli *et al.* 1989; Rossi & Cocherie 1991; Carmignani *et al.* 1992). The sequence of intrusive events can be divided in three cycles on the basis of relationships with metamorphic rocks and structural features: (a) intrusions emplaced in a compressive syn-tectonic regime, representing 1–2% of the whole area of the batholith, and consisting of peraluminous granites *sensulato*; (b) intrusions emplaced in a late–post-tectonic regime, representing the main frame of the batholith (*c.* 70%), and consisting of metaluminous granites *sensulato*; and (c) intrusions emplaced in a distinctly post-tectonic regime, representing 20–25% of the whole area of the batholith and consisting of metaluminous leucogranites.

In general, the field occurrence and petrographic and geochemical characteristics of hybrid magmas are extremely variable (e.g. Didier & Barbarin 1991b). In the SCB (Fig. 1) they can be divided into three broad groups (e.g. Poli *et al.* 1989; Tommasini & Poli 1992; Tommasini 1993).

1. Microgranular mafic enclaves (MME) with geochemical characteristics of both magmatic liquids and cumulates.
2. Basic complexes (BC) in the form of discrete entities in acid plutons, with internal parts mainly formed by cumulates and with interaction zones developing only in the marginal parts.
3. Basic septa (BS) in the form of discrete, lenticular-like bodies with peninsular-like connections to the main wallrocks that do not have known floors (Best 1982); they are often mechanically fragmented in the host rock. These represent the link between groups (1) and (2) as they are similar in terms of petrography and geochemistry.

Mafic microgranular enclaves are ubiquitous in all plutons of the second group and are absent in the third group. Their abundance varies from *c.* 10% in tonalites and granodiorites to 1–2% in monzogranites.

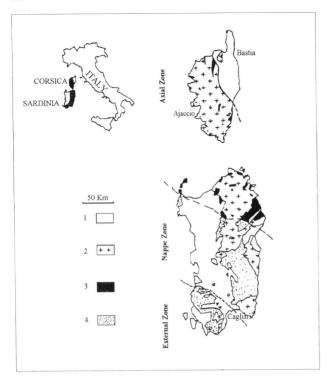

Figure 1 Geological sketch maps of the Sardinia–Corsica Batholith. (1) Post-Permian formations. (2) Granitic plutons and basic complexes belonging to the Hercynian orogenic cycle. (3) Hercynian metamorphic rocks of amphibolite facies and migmatites. (4) Hercynian metamorphic rocks of greenschist facies and anchimetamorphism.

Basic stratified complexes have been found in association with late tectonic intrusions. These outcrop over small areas and consist of gabbros and diorites with some cumulitic facies. Such complexes represent <5% of the whole area of the batholith and they are often found as septa or as large enclaves within granodioritic and monzogranitic plutons. Their emplacement is coeval with the host granites facies and mixing and mingling phenomena along the contacts indicate that such basic magmatism had an important role in the genesis of the SCB (Tommasini 1993).

Mafic microgranular enclaves vary in composition from gabbros and diorites to tonalites and generally have fine-grained hypidiomorphic equigranular textures, although ophitic textures are present in the less evolved samples. Mono- and polymineralic fragments of the host granite are ubiquitous in hand specimen. The main minerals are quartz + plagioclase ± K-feldspar + biotite ± amphibole ± clinopyroxene, whereas the accessory phases are oxides + apatite + zircon ± titanite ± allanite. Clinopyroxenes may occur as relics enclosed within hornblende cores and plagioclase displays complex zonation patterns with patchy textures. Other textures attributable to magma mixing—for example, acicular apatite, spike zones in plagioclase, poikilitic/oikocrystic K-feldspar and quartz (Hibbard 1991)—are often found.

Basic complexes and septa consist of medium-grained amphibole-bearing gabbros and quartz gabbros with autoallotriomorphic and cumulate textures. In the outer parts of the BC and BS, samples are medium- to fine-grained with hypidiomorphic porphyritic textures and glomerocrystals of plagioclase and amphibole. The main minerals are the same as for the MME, but mafic phases are modally more prevalent, together with amphibole–clinopyroxene textures, and patchy zoning in plagioclase. When present, the K-feldspar is found only as an interstitial phase.

Metaluminous granitoids consist of medium–coarse-grained rocks with hypidiomorphic heterogranular textures due to

K-feldspar megacrysts. The main minerals are quartz + plagioclase + K-feldspar + biotite ± amphibole, whereas the accessory phases are oxides + apatite + zircon ± titanite ± allanite. In the tonalitic rocks plagioclase commonly has a patchy zoning texture and within hornblende cores clinopyroxenes may occasionally be found.

Elements representative of different geochemical behaviour are plotted in Figure 2 to give an overview of the main trends followed by the calc-alkaline magmas in the SCB. The arrows indicate the trends caused by various processes. The white arrow indicates the evolutionary trend of mantle–derived magmas caused by simple fractional crystallisation or contamination plus fractional crystallisation processes (CFC; Poli & Tommasini 1991a). This process produces magmas rich in lithophile elements and poor in compatible elements (Fig. 2). Most MME and the outer zones of gabbroic masses and BS, in contact with the enclosing granitoids, are representative of these magmas. It is worth noting, however, that most inner zones of gabbroic masses and basic septa show high values of all the elements at low silica content (hatched arrow, Fig. 2). This can be due, in part, to having plotted in the diagrams all the data from the different areas of the batholith, though other processes are also superimposed. Samples plotting along the hatched arrow, in fact, have petrographic and geochemical characteristics (e.g. rare earth element pattern, Fig. 3A) suggesting that they have experienced crystal accumulation, mainly of phases such as amphibole and accessory minerals (Co, La, Th, Zr, Fig. 2). High contents of Rb and K are probably caused by replacement of amphibole by biotite (e.g. Johnston & Wyllie 1988). In addition, it is notable that samples belonging to MME have geochemical characteristics (Fig. 2 and 3B) of both evolved (white arrow trend) and cumulate (hatched arrow trend) rocks, though none has petrographic textures evidencing cumulus processes. Such a feature is fairly common and it is supposed to be the result of filter pressing processes (e.g. Hibbard 1995), which are likely to occur during the interaction between magmas of different viscosities. Thus MME can be representative of either evolved basic liquids or basic magmas with cumulus-like geochemical signatures caused by mechanical squeezing.

The other two arrows in Figure 2 point to the two main evolutionary processes of acid magmas. The grey arrow trend is caused by interaction processes with evolving basic magmas. Two distinct periods of interaction can be envisaged namely (i) during the attainment of thermal equilibrium mainly mingling processes occur (Poli & Tommasini 1991a) and (ii) after the attainment of thermal equilibrium mainly mixing processes occur between derivative products of basic magmas (trend followed by grey arrow) and acid magmas (e.g. Frost & Mahood 1987; Poli & Tommasini 1991a). The black arrow trend is mainly caused by crystal fractionation processes. It is notable that leucogranites follow only this trend as they do not have any field, petrographic and geochemical evidence of interaction processes with basic magmas, such as, for instance, the presence of microgranular enclaves and patchy zoning of plagioclase.

For the sake of clarity we assume that, from a geochemical point of view, evolutionary trends during basic acid magma interaction can be subdivided into three main types on the basis of three parameters: the compositions of acid (C_a) and basic (C_b) magmas and the bulk partition coefficient (D) of a given element in the basic magma (Poli & Tommasini 1991a). For elements with $D < 1$ and $C_a \approx C_b$, mafic microgranular enclaves and outer zones of gabbroic masses and basic septa (e.g. Zr and La, Fig. 2) follow an upward trend determined by simple fractional crystallisation or contamination plus fractional crystallisation processes acting during the attainment of

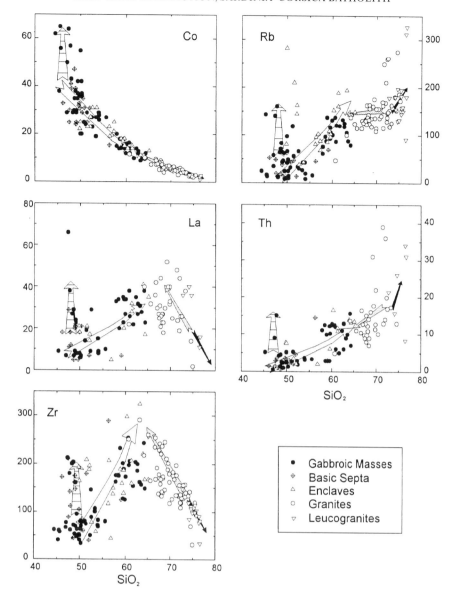

Figure 2 Harker diagrams illustrating the main trends exhibited by rock samples of the Sardinia–Corsica Batholith. General trends caused by different processes are indicated by arrows (see text for explanation): white arrows, evolution of basic magmas by CFC processes; hatched arrows, evolution of basic magmas by cumulus and/or filter pressing processes; grey arrows, evolution of acid magmas by interaction processes; and black arrows, evolution of acid magmas by crystal fractionation processes. Data from Tommasini (1993).

thermal equilibrium. As soon as this is attained the evolved magmas mix with the acid magma to give hybrid granitoids and in the geochemical space a bell-shaped pattern can be recognised (e.g. La and Zr, Fig. 2). For elements with $D < 1$ and $C_b \ll C_a$ in the geochemical space an upward trend for all the groups can be recognised (e.g. Th and Rb, Fig. 2). For elements with $D > 1$ and $C_b \gg C_a$ a downward trend for all the groups can be recognised (e.g. Co, Fig. 2).

2. Isotope geochemistry: case study of the Punta Falcone, northern Sardinia

From an isotopic point of view the problem seems more complex, as the isotopic characteristics of the products of interaction between acid and basic magmas cannot be explained by simple CFC processes. Microgranular mafic enclaves and granites are relatively uniform in isotopic terms, whereas the internal facies of gabbros and septa are slightly more distinct from granites. This behaviour is common in the Sardinia–Corsica Batholith (Tommasini 1993), but also in other associations of MME and BC and granitic plutons (e.g.

Halliday *et al.* 1980; Holden *et al.* 1987, 1991; Pin *et al.* 1990). This does not necessarily prove a genetic connection between mafic bodies and granites (e.g. Stephens *et al.* 1991), but nevertheless remains a particular characteristic that must be taken into account in petrogenetic modelling of the interaction between acid and basic magmas.

The Punta Falcone gabbroic complex was emplaced into a granitic magma during the Hercynian orogeny (Poli & Tommasini 1991b; Tommasini & Poli 1992). The complex has a stratified subvertical structure and consists of three zones developing from the bottom to the top of the magma chamber. An interaction zone (IZ), a maximum of 2 m wide, can be recognised along contacts with the surrounding granite and consists of finer grained and more evolved rocks than the interior of the gabbroic complex. Trace element geochemistry indicates that the IZ is the result of the processes of fractional crystallisation plus contamination by the acid magma (CFC process, Tommasini & Poli 1992).

Strontium isotopic analyses of representative samples of the Punta Falcone complex are reported in Table 1. Isotopic modelling indicates that the mantle source of the Punta

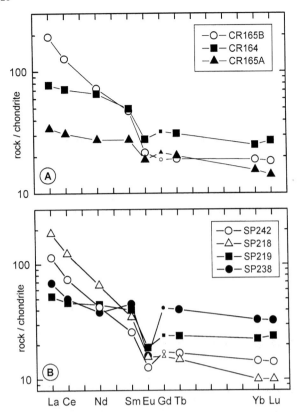

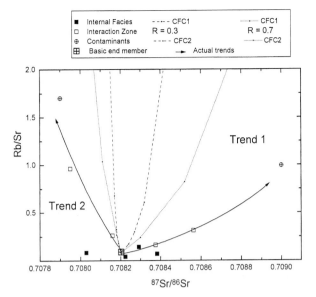

Figure 3 Chondrite-normalised (Haskin *et al.* 1966) rare earth element patterns of selected rocks from the Sardinia–Corsica Batholith. (A) Basic septa that experienced no (open symbol) or variable amounts (closed symbols) of amphibole accumulation; (B) mafic microgranular enclaves that experienced (closed symbols) or did not experience (open symbols) filter pressing processes. Normalised values for Gd are interpolated.

Figure 4 Evolution lines of the CFC process applied to the Sr isotopes of the Punta Falcone complex. Bulk partition coefficients (D) of the evolution lines for $r = 0.3$ (broken lines) are the same as those used for trace elements (Tommasini & Poli 1992; Tommasini 1993). The values of D are $^{Rb}D_{CFC1} = 0.1$, $^{Rb}D_{CFC2} = 0.7$, $^{Sr}D_{CFC1} = 1.05$ and $^{Sr}D_{CFC2} = 1.25$. Evolution lines for $r = 0.7$ (dotted lines) are calculated using the same bulk partition coefficients. Intervals (10%) of degree of fractionation (F) are reported as small closed circles. The actual lines are hand-drafted (solid lines). Contaminating magmas: for trend 1 SP58 and for trend 2 an hypothetical composition on the basis of SP146 sampled at the contact with IZ is assumed (Tommasini 1993).

Falcone basic magma was enriched in incompatible elements via sediment entrainment through subduction in the Ordovician (Tommasini *et al.* 1995). This explains the radiogenic Sr isotope characteristics of the inner parts of the complex which did not suffer low pressure contamination processes with the surrounding granites. Applying the CFC process to reproduce Sr isotopic ratios in IZ samples, there are significant discrepancies between the observed and calculated compositions, both using the value of r (contamination rate over fractionation rate; DePaolo 1981) estimated for trace elements by Tommasini and Poli (1992) ($r = 0.3$; broken lines in Fig. 4), as well as trying to use higher r values ($r = 0.7$; dotted lines in Fig. 4), which are thermodynamically unlikely and which are not consistent with the absolute trace element concentrations in the IZ samples. This suggests that isotopic variations in the IZ samples are controlled by other mechanisms in addition to a CFC process, which would have caused only slight variations in Sr isotopes (Fig. 4). Alternatively, Sr isotope variations in the IZ samples could have been determined by a liquid state diffusion in addition to contamination during the attainment of thermal equilibrium. In this case of liquid state chemical diffusion it is essential to take into account the values of chemical and thermal diffusion coefficients (Blake & Koyaguchi 1991). As thermal diffusion coefficients are many orders of magnitude (certainly more than 4–5; Hoffman 1980) greater than chemical diffusion coefficients, thermal equilibrium is reached much sooner than chemical equilibrium. This, along with the fact that diffusion coefficients are inversely correlated with temperature (e.g. Hofmann 1980), notably limits the effects of chemical interaction processes due to diffusion mechanisms during the attainment of thermal equilibrium, at least as far as the diffusion of elements is

Table 1 Rb–Sr isotopic data. All Sr isotopic analyses were normalised to $^{86/88}Sr = 0.1194$. Repeated analyses of NBS987 standard gave an average of $^{87}Sr/^{86}Sr$ of 0.71025 ± 1 (2σ, $v = 20$). Uncertainties in measured and initial isotopic ratios refer to least significant digits and represent $\pm 2\sigma$ run precision and $\pm 2\sigma$ error propagation. GIF, Gabbroic Internal Facies; IZ, Interaction Zone; GR IZ, Granite–Interaction Zone; and HGR, Host Granite.

Sample	Unit	Rb	Sr	$^{87}Rb/^{86}Sr$	$^{87}Sr/^{86}Sr$ 2σ	$^{87}Sr/^{86}Sr_i$ 2σ
SP96	GIF	18·44	223·1	0·2391	0·708995 ± 31	0·70803 ± 3*
SP39	GIF	12·08	299·3	0·1168	0·708695 ± 17	0·70822 ± 2
SP77	GIF	26·06	381·4	0·1978	0·709182 ± 21	0·70838 ± 2
SP81	GIF	42·16	299·9	0·4067	0·709935 ± 13	0·70829 ± 2
SP67	IZ	43·24	266·4	0·4697	0·710273 ± 14	0·70838 ± 2
SP102	IZ	78·89	252·5	0·9042	0·712217 ± 31	0·70856 ± 3
SP147	IZ	74·37	287·1	0·7496	0·711188 ± 21	0·70816 ± 2
SP32	IZ	136·6	142·0	2·785	0·719203 ± 19	0·70795 ± 2
SP146	GR IZ	118·2	203·3	1·683	0·714768 ± 16	0·70797 ± 2
SP55	HGR	131·1	200·8	1·891	0·716139 ± 13	0·70850 ± 2
SP58	HGR	155·2	156·6	2·872	0·720584 ± 15	0·70898 ± 2

*Initial Sr isotopes have been calculated at 284 Ma, i.e. the most probable age of PF complex based on Rb/Sr isochrone (MSWD = 1·05; age 284 ± 7 Ma (Tommasini 1993).

concerned. On the other hand, experimental studies (Baker 1989; Lesher 1990, 1994) pointed out that isotope diffusion coefficients are much (at least more than three orders of magnitude) higher than chemical diffusion coefficients, indicating that isotope re-equilibrium occurs more rapidly than chemical re-equilibrium.

Quantitatively, using Equation (1), an increase of 10^4 in the diffusion coefficient changes the order of magnitude of the effective area of diffusion by a factor of two, i.e. from centimetres to metres. Therefore the probability that chemical changes take place in a basic magma interacting with an acid magma while reaching thermal equilibrium is greater for isotopes than for trace elements. Thus it is reasonable to hypothesise that the CFC evolution of the IZ was associated with diffusion processes which substantially modified the Sr isotopes, but did not have any major influence on trace elements.

To assess the importance of diffusion processes, we assumed that the isotopic differences between the values modelled by the CFC process using trace elements (Fig. 4; CFC trends for $r = 0.3$, broken lines) and the values on the actual trends (Fig. 4; solid lines) have been caused by diffusion. It was further assumed that the initial values for the starting basic end member are the same, whereas the acid end-members are different. It is noteworthy that in trend 1 the basic end-member has lower Sr isotope values than the acid end-member, whereas the opposite holds for trend 2. Although CFC and diffusion processes act together, the net result can be modelled for each sample as the sum of isotopic ratios deriving from the two processes. This means that in modelling isotope diffusion, we considered the isotopic ratios and Sr values calculated using CFC process as initial composition of the basic end-member for each sample.

Strontium isotope diffusion in the IZ of the gabbroic complex has been calculated using the equation for a semi-infinite plane sheet medium with initial concentration C_0 (gabbroic mass) and constant surface concentration C_1 (host granite). The general solution [Equation (1)] for a diffusion process in such a geometry can be found in Crank (1975: eq. 3.13)

$$\frac{C - C_1}{C_0 - C_1} = \mathrm{erf}\, \frac{x}{2(Dt)^{1/2}} \tag{1}$$

where erf is the error function, C is the concentration at the distance x from the contact, D is the diffusion coefficient and t is the time elapsed from the beginning of the process.

To resolve such an equation as a function of x we need an estimation of the time scale of the duration of the interaction process. Furlong et al. (1991: 441) reported equations for such a problem and we have used their eq. 7

$$T = T_0 + (T_1 - T_0) * \left[\frac{1}{2} \left(\mathrm{erf}\, \frac{\zeta + 1}{2\tau^{1/2}} - \mathrm{erf}\, \frac{\zeta - 1}{2\tau^{1/2}} \right) \right];$$

$$\tau = \frac{kt}{a^2}; \qquad \zeta = \frac{x}{a} \tag{2}$$

where T is the temperature at the distance x from the contact, T_0 is the initial temperature of the host granite, T_1 is the initial temperature of the gabbroic mass, k is the thermal diffusivity, t is the time elapsed from the beginning of the process and a is half of the thickness of the gabbroic mass.

To take into account the latent heat of crystallisation, we considered the effective thermal diffusivity, i.e. the thermal diffusivity in a magma between its liquidus and solidus (Philpotts 1990), which is lower by about one order of magnitude with respect to thermal diffusivity. Using this approach we calculated an effective thermal diffusivity of

10^{-7} m^2 s^{-1}. Considering a system composed of a basic magma at 1200°C, an acid magma at 750°C, and the solidus of the basic magma at c. 900°C, we estimated a time span of c. 2700–4500 a for cooling to proceed 2 m from the contact.

Using such values of time in Equation (1) and using the diffusion coefficient measured by Lesher (1995) for Sr isotopes, we calculated the variation in Sr isotopes as a function of the distance from the contact. The results are reported in normalised form in Figure 5, along with the normalised values of samples from IZ. Three out of four samples fit the modelled curves fairly well, indicating that the modelled process is probably reasonable. Sample SP147 has isotope ratios very close to the basic end-member, but it was sampled close to the host. Such behaviour may be due to several causes that cannot be clearly defined in this instance, e.g. (i) the sample being very close to the host it solidified more quickly that the internal parts of the interaction zone; or (ii) as the gabbroic mass tilted (Tommasini & Poli 1992), such a sample could have been in contact with different isotopic compositions from the host.

In summary, although the method used is model-dependent, the goodness of fit of the result lends credibility to the idea that the dichotomy between trace element and isotopic evolution during basic–acid magma interaction is due to isotopic diffusion during the attainment of thermal equilibrium. Further work is, however, necessary on well constrained transects between basic and acid rocks to better constrain the topology and kinetics of the isotopic diffusion process.

3. Fluid dynamic and thermodynamic parameters

Some further general features have to be evaluated to derive a genetic model for basic–acid interaction processes. Fluid dynamic and thermodynamic parameters are also critical in controlling interaction processes.

Inertial and viscous forces associated with the movement of fluids and with the floating effect (the difference in density between the two magmas) are the main fluid dynamic parameters controlling interaction processes between magmas of different chemical compositions. Experimental studies (e.g.

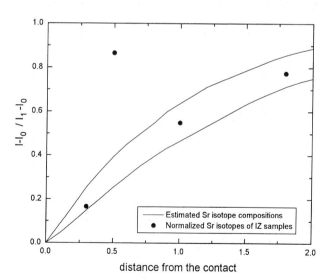

Figure 5 Calculated (solid lines) and observed Sr isotope compositions of IZ samples versus distance (m) from the granite contact. Data are presented normalised to the following equation: $(I - I_0)/(I_1 - I_0)$, where I is the Sr isotope composition of the IZ sample, I_0 the Sr isotope composition of the host granite and I_1 is the Sr isotope composition of the gabbroic complex. The estimated Sr isotope compositions have been calculated for the maximum and minimum span of time necessary for cooling to proceed 2 m inwards from the granite contact (2700–4500 a; see text).

Campbell & Turner 1985, 1986, 1989), indicate significant fragmentation and dispersion of basic magma batches into acid magma with the occurrence of extensive mechanical mixing (mingling) when the rheological behaviour of acid magma is Newtonian (e.g. Arzi 1978; Fernandez & Barbarin 1991).

When an acid magma is intruded by one or more injections of basic magma which is hotter and less viscous than the acid magma, the two magmas do not mix easily because of fluid dynamic barriers and hence remain as discrete entities while thermal equilibrium is being attained. Freezing of the basic magma and superheating of the acid magma does occur along their boundaries. The thermodynamic parameters which must be taken into account are the temperatures of the two magmas, the thermal capacities of the solid and liquid phases, the heat of fusion and the heat of crystallisation of the solid phases. These parameters, along with the relative mass fraction of basic and acid magmas, control the temperature at which the two magmas attain thermal equilibrium. At the equilibrium temperature, the final viscosities of the two magmas will be the main physical parameter controlling interaction mechanisms. Chemical mixing processes are possible when these viscosities are low and of the same order of magnitude.

Using the computer program developed by Frost and Lindsay (1988), it can be seen that, at the initial physicochemical conditions (temperature, pressure and composition) of magmas from the SCB (e.g. Tommasini & Poli 1992; Tommasini 1993), low final viscosities of the same order of magnitude in the two magmas are obtained only for high portions of basic magma in the system (*c.* 65–70%; Fig. 6). Therefore, the possibility of having chemical mixing processes between acid and basic magmas does not seem plausible, as it requires special conditions (very large amounts of basic magma) which cannot be considered appropriate for many interaction products. However, during the attainment of thermal equilibrium, the basic end–member is continuously evolving (Poli & Tommasini 1991a). To account for the evolution of the basic magma we used the program by Frost and Lindsay (1988) assuming, for the sake of simplicity, the radical view that during each step the interacting liquids do not contain crystals. Formally we used extreme values of the parameter σ (see Frost & Lindsay 1988) to reach this condition.

In addition, we divided the evolving period into many steps and used for each step a more evolved chemical composition and a different initial temperature to calculate the viscosity at thermal equilibrium. Given all these assumptions, only a first approximation model of the system can be derived. Nevertheless, the model is considered valid for acid members in view of its overheating along the contacts with the basic member (Huppert & Sparks 1985) and for the basic member in view of the fact that it is an evolved liquid and not a crystal mush, even if in actual situations it may contain some crystals.

In Figure 6 it can be seen that the amount of basic magma in the overall system can be as low as 30% for interacting magmas starting at 1200°C (basic) and 800°C (acid), respectively. Such values are much closer to those inferred for a variety of interaction products such as plutons containing MME (e.g. Didier & Barbarin 1991a).

4. Model for the association of basic and acid rocks

The geochemical and physical aspects of acid–basic magma interaction have been analysed by attempting to isolate different phenomena. The highly variable geochemical characteristics of interaction products are the result of several processes acting concomitantly following the intrusion of basic magmas into anatectic crustal environments and can be attributed to: (i) fractional crystallisation of basic magma and contamination by acid magma (CFC process, Poli & Tommasini 1991a); (ii) alkali and H_2O exchanges; (iii) loss of the liquid phase from the evolving basic magma by filter pressing processes; (iv) mechanical mixing between basic and acid magmas; and (v) isotopic diffusion processes during the attainment of thermal equilibrium. It is noteworthy, in addition, that the internal facies of the gabbroic complex and basic septa become cumulate by the settling of crystallising phases.

From the physical point of view many aspects have to be taken into account.

1. Magmas of different compositions and thermodynamic parameters do not mix because of fluid dynamic barriers and hence remain as discrete entities until thermal equilibrium is attained. As a corollary, freezing of the basic magma and superheating of the acid magma does occur along their boundaries.

2. The initial effective viscosity of acid magma acts as the main parameter for the evolution of the system. In fact, for high values, corresponding to solid–plastic rheology, mechanical mixing and interaction processes are possible along limited parts of the boundaries.

3. Chemical mixing, generating hybrid rocks, will be enhanced when the final viscosities of the two magmas are low and of the same order of magnitude, whereas mechanical mixing will take place when the viscosities are very different from each other.

4. Larger amounts of basic magma in the overall acid–basic system favours the chemical mixing process. Large initial differences in temperature, viscosity and chemical composition in the two magmas, however, tend to inhibit chemical mixing and limit the mechanical mixing interaction processes.

The MME, BS and BC are, hence, the result of the conjunction and superimposition of the physico-chemical processes suffered by basic magma after the intrusion in the crustal anatectic environment. Poli and Tommasini (1991) proposed a genetic model to explain the genesis of enclaves and it is also valid for all products with further reference to the rheology of interacting magmas. The model consists of three main stages concerning the evolution and mixing versus

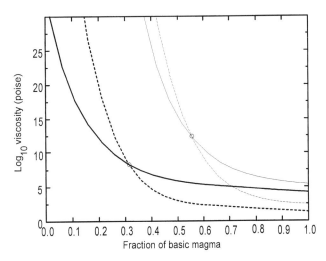

Figure 6 Variations of final viscosity in a basic (solid line) and acid (broken line) liquid at the thermal equilibrium temperature as a function of the fraction of basic magma in the system. Circles represent the points of equal viscosity for a basic magma starting at 1193°C and acid magma starting at 800°C, respectively. Light lines follow Frost & Lindsay (1988). Heavy lines assume that the interacting liquids do not contain crystals and use different evolved chemical compositions of basic magma (see text).

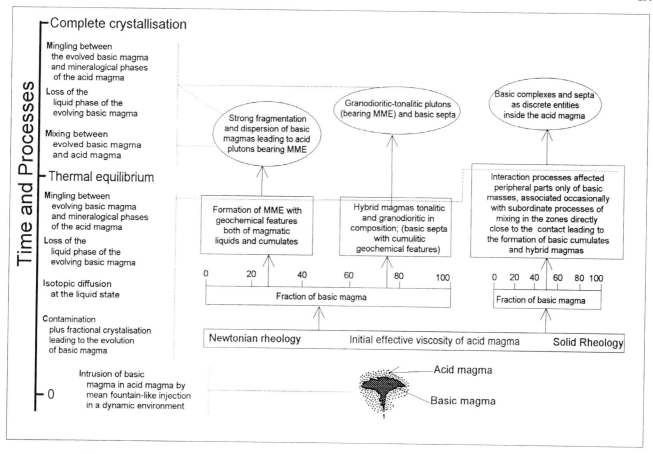

Figure 7 Schematic diagram summarising various environmental settings of interaction processes among basic and acid magmas.

mingling processes which basic magmas experienced once injected into anatectic crustal environments.

A schematic picture of the different environmental settings during basic acid magma interaction is shown in Fig. 7. A mechanism of fountain-like injection is considered, but processes of internal instability of the basic magma leading to fingering of the acid in the basic magma and vice versa must also be taken into account. During the attainment of thermal equilibrium the evolution of the basic magma is determined by (i) contamination plus fractional crystallisation along with (ii) liquid state isotopic diffusion, (iii) mechanical mixing of the evolving magma with mineralogical phases of the acid magma and (iv) loss of liquid phase of the evolving magma.

The different products of interaction depend on two physical parameters: the relative fraction of basic magma and the initial effective viscosity of acid magma, i.e. the viscosity of bulk magma taking into account the pressure, temperature, chemical composition and percentage and texture of suspended crystals. Two main cases can be envisaged: (i) granitic magmas with rheology intermediate between visco-plastic and solid; and (ii) granitic magmas with rheology intermediate between Newtonian and visco-plastic.

In the first case, the fraction of basic magma is relatively important as the rheological behavior of acidic magma does not allow the fragmentation of basic magmas which preserve their shape discrete from acid magmas and hence during the attainment of thermal equilibrium these processes occur only along the borders, and no significant effect of mixing between the acid and evolved basic magma along the contacts can be recognised after thermal equilibrium has been attained. Therefore in this setting only BC and BS can be found as discrete entities within acid magma. The internal parts of these

bodies behave as simple gabbroic complexes leading to cumulitic rocks.

In the second setting (Newtonian rheology of acid magma), the fraction of basic magma is important in understanding the interaction processes. For low values of basic magma during the attainment of thermal equilibrium, basic magma is subject to all the described processes, leading to the formation of an evolved magma that can undergo processes of mixing and mingling with the acid magma and the formation of dispersed MME in granitic plutons. The rheological behaviour of the host acidic magma and the relatively low amount of basic magma in the system gives rise to strong fragmentation of the evolved magma and the MME are representative of the evolutionary processes undergone by the basic magma after intrusion into the crustal anatectic environment.

When the fraction of basic magma is high, during the attainment of thermal equilibrium, basic magma is subject to all these processes, leading to the formation of large amounts of hybrid magma of tonalitic–granodioritic composition that can undergo, after thermal equilibrium has been attained, processes of mechanical mixing and loss of the liquid phase.

Such mechanical mixing can lead to the formation of BS with geochemical characteristics of both magmatic liquids and cumulates, but MME can also be found in this instance. Relatively less extreme values of the basic magma fraction and/or relatively low differences in temperature and chemistry between the acid and basic magmas allow fragmentation of the evolving magma. Therefore this environmental setting leads to the formation of hybrid magmas with tonalitic to granodioritic compositions and BS with geochemical characteristic of both magmatic liquids and cumulates.

5. Acknowledgements

The authors thank P. Manetti and A. Peccerillo for useful discussion and suggestions, and S. Conticelli and L. Francalanci for their invaluable help. We are also grateful to Ed Stephens and an anonymous referee for helpful suggestions and criticism in reviewing the manuscript. This work is based in part on a Ph.D. thesis carried out by one of author (S. T.) at the Department of Earth Sciences, University of Perugia. Research was financially supported by Italian MURST Grants (40%) and Italian CNR funds to G. P. The isotopic data were obtained at the University of Michigan and were partially supported by NSF grants to A. N. H.

6. References

Arzi, A. A. 1978. Critical phenomena in the rheology of partial melted rocks. TECTONOPHYSICS **44**, 173–84.

Baker, D. R. 1989. Tracer versus trace element diffusion: diffusional decoupling of Sr concentration from Sr isotope composition. GEOCHIM COSMOCHIM ACTA **53**, 3015–23.

Barbarin, B. 1988. Field evidence and mingling between the Piolard Diorite and the Saint-Julien-la-Vetre Monzogranite (Nord-Forez, Massif Central, France) CAN J EARTH SCI **25**, 49–59.

Best, M. G. 1982. *Igneous and Metamorphic Petrology*. San Francisco: Freeman.

Blake, S. & Koyaguchi, T. 1991. Insights on the magma mixing model from volcanic rocks. *In*: Didier, J. & Barbarin, B. (eds) *Enclaves and Granite Petrology*, 403–13. Amsterdam: Elsevier.

Campbell, I. H. & Turner, J. S. 1985. Turbulent mixing between fluids with different viscosities. NATURE **313**, 39–42.

Campbell, I. H. & Turner, J. S. 1986. The influence of viscosity on fountains in magma chamber. J PETROL **27**, 1–30.

Campbell, I. H. & Turner, J. S. 1989. Fountains in magma chambers. J PETROL **30**, 885–923.

Carmignani, L., Barca, S., Carosi, R., Di Pisa, A., Gattiglio, M., Musumeci, G., Oggiano, G. & Pertusati, P. C. 1992. Schema dell'evoluzione del Basamento Sardo. *In* Carmignani, L. (ed) *Struttura della catena ercinica in Sardegna. Informal Group of Structural Geology, Field-book*, 11–38. Siena: Department of Earth Sciences.

Crank, J. 1975. *The Mathematics of Diffusion*. Oxford: Clarendon.

DePaolo, D. J. 1981. Trace element and isotopic effects of combined wallrock assimilation and fractional crystallization. EARTH PLANET SCI LETT **53**, 189–202.

Didier, J. & Barbarin B. 1991a. *Enclaves and Granite Petrology*. Amsterdam: Elsevier.

Didier, J. & Barbarin B. 1991b. The different types of enclaves in granites–nomenclature. *In* Didier, J. & Barbarin, B. (eds) *Enclaves and Granite Petrology* 19–23. Amsterdam: Elsevier.

Fernandez, A. N. & Barbarin, B. 1991. Relative rheology of coeval mafic and felsic magmas: nature of resulting interaction processes and shape and mineral fabrics of mafic microgranular enclaves. *In*: Didier, J. & Barbarin, B., (eds) *Enclaves and Granite Petrology*, 263–75. Amsterdam: Elsevier.

Frost, T. P. & Lindsay, J. R. 1988. Magmix: a basic program to calculate viscosities of interacting magmas of differing composition, temperature, and water content. COMPUT GEOSCI **14**, 213–28.

Frost, T. P. & Mahood, G. A. 1987. Field, chemical and physical constraints on mafic felsic magma interaction in the Lamark Granodiorite, Sierra Nevada, California. GEOL SOC AM BULL **99**, 272–91.

Furlong, K. P., Hanson, R. B. & Bowers, J. R. 1991. Modelling thermal regimes. *In* Kerrick, D. M. (ed) *Contact Metamorphism*. REV MINERAL **26**, 437–505.

Halliday, A. N., Stephens, W. E. & Harmon, R. S. 1980. Rb/Sr and O isotopic relationships in three zoned Caledonianan granitic plutons, Southern Uplands, Scotland: evidence for varied sources

and hybridization of magmas. J GEOL SOC LONDON **137**, 329–48.

Haskin, L. A., Frey, F. A., Schmitt, R. A. & Smith, R. H. 1966. Meteoric, solar and terrestrial abundances of the rare earths. PHYS CHEM EARTH **7**, 167–321.

Hibbard, M. J. 1991. Textural anatomy of twelve magma-mixed granitoid systems. *In* Didier, J. & Barbarin, B. (eds) *Enclaves and Granite Petrology*, 431–44. Amsterdam: Elsevier.

Hibbard, M. J. 1995. *Petrography to Petrogenesis*. Englewood Cliffs: Prentice-Hall.

Hofmann, A. W. 1980. Diffusion in natural silicate melts: a critical review. *In* Hargraves, L. (ed.) *Physics of Magmatic Processes*, 385–417. Princeton: Princeton University Press.

Holden, P., Halliday, A. N. & Stephens, W. E. 1987. Microdiorite enclaves: Nd and Sr isotopes evidences for a mantle input to granitoid production. NATURE **330**, 53–6.

Holden, P., Halliday, A. N., Stephens, W. E. & Henney, P. J. 1991. Chemical and isotopic evidence for major mass transfer between mafic enclaves and felsic magma. CHEM GEOL **92**, 135–52.

Huppert, H. E. & Sparks R. S. J. 1985. Cooling and contamination of mafic and ultramafic magmas during ascent through continental crust. EARTH PLANET SCI LETT **74**, 371–86.

Huppert, H. E. & Sparks R. S. J. 1988. The generation of granitic magmas by intrusion of basalt into continental crust. J PETROL **29**, 599–624.

Johnston, A. D. & Wyllie, P. J. 1988. Interaction of granitic and basic magmas: experimental observations on contamination processes at 10 kbar with H_2O. CONTRIB MINERAL PETROL **98**, 352–62.

Lesher, C. E. 1990. Decoupling of chemical and isotopic exchange during magma mixing. NATURE **344**, 235–7.

Lesher, C. E. 1994. Kinetics of Sr and Nd exchange in silicate liquids: theory, experiments, and applications to uphill diffusion, isotopic equilibration, and irreversible mixing of magmas. J GEOPHYS RES **99**, (B5), 9585–604.

Philpotts, A. R. 1990. *Principles of Igneous and Metamorphic Petrology*. Englewood Cliffs. Prentice-Hall.

Pin, C., Binon, M., Belin, J. M., Barbarin, B. & Clemens, J. D. 1990. Origin of microgranular enclaves in granitoids: equivocal Sr–Nd isotopic evidence from Hercynian rocks in the Massif Central (France). J GEOPHYS RES **95**, 17821–8.

Poli, G. & Tommasini, S. 1991a. Model for the origin and significance of microgranular enclaves in calcalkaline granitoids. J PETROL **32**, 657–66.

Poli, G. & Tommasini, S. 1991b. A geochemical approach to the evolution of granitic plutons: a case study, the acid intrusions of Punta Falcone (northern Sardinia, Italy). CHEM GEOL **92**, 87–105.

Poli, G., Ghezzo, C. & Conticelli, S. 1989. Geochemistry of granitic rocks from the Hercynian Sardinia–Corsica Batholith: implication for magma genesis. LITHOS **23**, 247–66.

Reid, J. B., Evans, O. C. & Fates, D. G. 1983. Magma mixing in granitic rocks of the central Sierra Nevada, California. EARTH PLANET SCI LETT **66**, 243–61.

Rossi, P. & Cocherie, A. 1991. Genesis of a Variscan batholith: field, petrological and mineralogical evidence from the Corsica–Sardinia batholith. TECTONOPHYSICS **195**, 319–46.

Stephens, W. E., Holden, P. & Henney, P. J. 1991. Microdioritic enclaves within the Scottish Caledonian granitoids and their significance for crustal magmatism. *In* Didier, J. & Barbarin, B. (eds) *Enclaves and Granite Petrology*, 125–34. Amsterdam: Elsevier.

Tommasini, S. 1993. *Petrologia del Magmatismo Calcalcalino del Batolite Sardo-Corso: processi genetici ed evolutivi dei magmi in aree di collisione continentale e implicazioni geodinamiche*. Ph.D. Thesis, University of Perugia.

Tommasini, S. & Poli, G. 1992. Petrology of the late-Carboniferous Punta Falcone gabbroic complex, northern Sardinia, Italy. CONTRIB MINERAL PETROL **110**, 16–32.

Tommasini, S., Poli, G. & Halliday, A. N. 1995. The role of sediment subduction and crustal growth in Hercynian plutonism: isotopic and trace element evidence from the Sardinia–Corsica Batholith. J PETROL **36**, 1305–32.

GIAMPIERO POLI and SIMONE TOMMASINI*, Department of Earth Sciences, University of Perugia, Piazza Università, 06100 Perugia, Italy.
*Present address: Dipartimento di Scienza, Piazzale delle Cascine 16, 50144 Firenze.

ALEX N. HALLIDAY, Department of Geological Sciences, University of Michigan, 1006 C.C. Little Building, Ann Arbor, MI 48109–1063, U.S.A.

Transactions of the Royal Society of Edinburgh: Earth Sciences, **87**, 233–242, 1996

Mafic–silicic layered intrusions: the role of basaltic injections on magmatic processes and the evolution of silicic magma chambers

R. A. Wiebe

ABSTRACT: Plutonic complexes with interlayered mafic and silicic rocks commonly contain layers (1–50 m thick) with a chilled gabbroic base that grades upwards to dioritic or silicic cumulates. Each chilled base records the infusion of new basaltic magma into the chamber. Some layers preserve a record of double-diffusive convection with hotter, denser mafic magma beneath silicic magma. Processes of hybridisation include mechanical mixing of crystals and selective exchange of H_2O, alkalis and isotopes. These effects are convected away from the boundary into the interiors of both magmas. Fractional crystallisation and replenishment of the mafic magma can also generate intermediate magma layers highly enriched in incompatible elements.

Basaltic infusions into silicic magma chambers can significantly affect the thermal and chemical character of resident granitic magmas in shallow level chambers. In one Maine pluton, they converted resident I-type granitic magma into A-type granite and, in another, they produced a low-K (trondhjemitic) magma layer beneath normal granitic magma. If comparable interactions occur at deeper crustal levels, selective thermal, chemical and isotopic exchange should probably be even more effective. Because the mafic magmas crystallise first and relatively rapidly, silicic magmas that rise away from deep composite chambers may show little direct evidence (e.g. enclaves) of their prior involvement with mafic magma.

KEY WORDS: compositional gradients, crystallisation, diffusion, gabbro, granite, hybridisation, magma chambers.

Contemporaneous melting of both mantle and crustal sources is a common phenomenon, probably because mantle magmatism acts to heat the overlying crust and initiate partial melting there (Huppert & Sparks 1988; Bergantz 1989). Although there has been a growing appreciation that heat from basaltic magmatism commonly plays an important part in the generation of granitic magmas, it seems to be less widely recognised that basaltic magmatism might also play a significant part in the chemical and thermal evolution of the silicic magma. As these contemporaneous magmas rise through the crust, there are ample opportunities for both to occupy a single chamber. Once crustal melting begins, later injections of basalt will almost surely come into contact with migmatite or pockets of silicic magma, often ponding at the base of any temporarily stable chambers. At deep crustal levels blending between these two magmas may be possible in multiply replenished, dynamic magma chambers (Hildreth & Moorbath 1988). Where basaltic magma invades silicic magma chambers at mid-crustal levels, the rapid transfer of heat may provide essential energy to keep the silicic magma moving upwards. Here, because of rapid crystallisation of the basaltic magma, most mafic material would probably be left behind as hornblende-rich gabbro. Dynamic processes within these compositionally stratified chambers and subsequent upwards movement of the granite may determine whether or not mafic magmatic inclusions become entrained and are identifiable as enclaves at the final level of emplacement.

In upper crustal settings, basaltic magma that enters an existing chamber of silicic magma commonly ponds on the floor of the chamber (Wiebe 1993a and references cited therein). Depending on the relative volumes and temperatures of the two magmas, a compositionally stratified magma chamber may develop (Sparks & Marshall 1986). This possibility is greatly enhanced if multiple pulses of basaltic magma enter the chamber. There is widespread evidence from the study of silicic volcanic rocks (Hildreth 1981; Fridrich & Mahood 1987; Druitt & Bacon 1988; Nixon 1988; McGarvie et al. 1990; Hildreth et al. 1991) and from mafic–silicic layered intrusions (MASLI) (Wiebe 1993a) that these compositionally stratified composite chambers have commonly existed in a wide range of tectonic settings.

1. Mafic–silicic layered intrusions

Mafic–silicic layered intrusions are typically dominated by gently dipping, alternating layers of gabbroic and granitic rocks that may vary in thickness from less than a metre to several tens of metres (Wiebe 1993a). Most intrusions are basin-form with average diameters up to a few tens of kilometres. All appear to have been emplaced at shallow crustal levels. Field relations in these composite intrusions, strongly supported by petrographic and geochemical studies, indicate that most silicic layers are cumulates (Wiebe 1993a), commonly hybrid in origin, and the basal parts of many mafic layers are quenched basaltic liquids (Wiebe 1974, 1994). The criteria for the recognition of tops of layers and sequence of deposition are as strong as those used for evaluating sedimentary sequences (Wiebe 1993a). The term 'sill' is inappropriate to describe layers in MASLI for two reasons: (1) deposition has occurred at an interface between underlying highly viscous cumulates and overlying, lower viscosity magma and (2) the layers preserve a stratigraphic record of sequential deposition from the bottom upwards and have not been emplaced at random levels as might be implied by the use of 'sill'. Pipes and diapirs commonly project upwards from silicic layers into basally chilled gabbro and formed in response to gravitational instability caused by basaltic magma resting on silicic cumulate mush. These relations indicate that MASLI have formed by multiple injections of basaltic magma into floored silicic magma chambers rather than by multiple sill-like injections of silicic magma into pre-existing, partly solidified mafic rocks. MASLI therefore appear to be the plutonic expression of compositionally stratified magma chambers thought to exist beneath many long-lasting silicic volcanic systems.

MASLI typically contain macrorhythmic units, from less than one to several tens of metres thick, characterised by a chilled gabbroic base that grades upwards to medium–grained gabbro, diorite or highly evolved silicic cumulates (Wiebe 1993a). Each chilled base records the introduction of new basaltic magma into the magma chamber. The top of each unit is truncated by the chilled base of the overlying unit. Cumulates beneath the chilled base of a macrorhythmic unit vary widely in composition, suggesting that the level at which the basaltic magmas ponded in the magma chamber was controlled by a rapid inwards (upwards) drop in crystallinity and viscosity (the magma crossing the critical melt fraction boundary of Miller et al. 1988) rather than by neutral bouyancy.

Where the input of new mafic magma was small and the compositional contrast with resident magma great, gabbro typically forms either thin layers or discontinuous zones of pillow-like bodies that are chilled on all margins against enclosing granitic to dioritic material.

Macrorhythmic units that grade upwards from chilled gabbro to highly silicic cumulates preserve a record of double-diffusive convection with stably stratified, hotter, denser mafic magma beneath silicic magma. The lower gabbroic rocks are commonly hornblende-rich with large corroded biotite crystals; the upper silicic rocks of many units are relatively anhydrous with ternary feldspars and minor pyroxene. Intermediate rocks in the units commonly show evidence for mechanical mixing of crystals and selective exchange of H_2O, alkalis and isotopes. Crystallisation of biotite and hornblende in mafic magma at the boundary contributed to the exchange of H_2O and alkalis. Although diffusion and turbulent mixing were probably restricted to a thin boundary layer, convection away from that boundary transferred the effects of hybridisation into the interiors of both magmas.

MASLI can be recognised by the occurrence of a small number of distinctive field relations.

1. Basally chilled gabbroic layers with lobate bases similar to sedimentary load cast structures rest on silicic layers (Fig. 1).
2. Pipes, diapirs or veins of silicic material commonly extend upwards into basally chilled gabbroic layers (Figs 2, 4).
3. Silicic layers commonly display feldspar lamination and discontinuous faint modal layering, indicating that the layers are cumulates, not intrusive veins, dikes or sills. Silicic rocks may occur at the top of macrorhythmic units (Fig. 1) or occur intercalated with thin layers, lenses and pillows of chilled mafic material (Fig. 3).
4. Layers of intermediate composition are often texturally and compositionally heterogeneous with variably assimilated mafic enclaves (Fig. 4).
5. Where dioritic to granitic rocks are dominant in MASLI, they commonly show evidence of upwards fractional crystallisation that is only temporarily interrupted by the basaltic infusions (e.g. Wiebe 1974; Chapman & Rhodes 1992).

In addition to providing many valuable insights into magma chamber processes, the stratigraphy of a MASLI has the potential to provide a record of the compositions of basic magmas that invaded the chamber. Chambers of silicic magma act as traps for denser mafic magmas, and carefully selected samples from the chilled aphyric bases of many gabbroic layers appear to have compositions that are equivalent to mafic liquids. The sequence of basally chilled gabbroic layers therefore provides a temporal record of the compositions of mafic magma intruded into a restricted area of crust—a record comparable with that recorded by a pile of lava flows. In some MASLI, the compositions of chilled mafic layers vary widely. In the Pleasant Bay intrusion (Wiebe 1993b), for example, the cation ratio $Mg/(Mg + Fe)$ of different mafic chill zones varies from about 0·60 to 0·06.

2. Occurrence

Plutons that display these characteristic relations occur in many different tectonic settings. They appear to be widespread in calc-alkaline plutonic suites. Calc-alkaline intrusions that

Figure 1 Chilled gabbroic base of one macrorhythmic unit rests on the silicic top of the underlying unit. Layers dip approximately 30° to the left.

Figure 2 Subhorizontal outcrop surface cuts roughly perpendicular to steeply plunging, irregularly shaped silicic pipes that have risen approximately 8 m from an underlying silicic layer into the gabbroic base of the overlying unit. Circular pipes average about 12 cm in diameter.

Figure 3 Silicic cumulate layers intercalated with chilled gabbroic sheets and pillows. Gabbroic layers average about 40 cm thick. Each mafic layer represents a single pulse of magma that spread across the gradually rising floor of the silicic magma magma chamber.

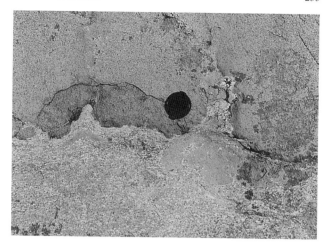

Figure 4 Small diapirs of hybrid silicic to intermediate material have formed between load–casts along the chilled gabbroic base of a macrorhythmic unit. Note the deformation of the fabric and fine-grained mafic enclaves in the underlying cumulate.

have previously been interpreted in this manner include small gabbroic and dioritic intrusions near Ingonish, Nova Scotia (Wiebe 1974). There, silicic cumulates are dominant and fractionated from mafic hornblende diorite at the base to biotite tonalite at the top. These cumulate sequences were interrupted repeatedly by the emplacement of three different types of mafic magmas, each of which formed basally chilled layers on the cumulate floor and beneath silicic magma. In the Massif Central, France, Barbarin (1988) described a large sheet of basally chilled diorite within the Saint–Julien-la-Vetre Monzogranite and interpreted it as the record of the emplacement and ponding of mafic liquid on a floor of a granitic magma chamber. Michael (1991) described evidence for the emplacement and ponding of mafic magmas in the Cordillera del Paine pluton in southern Chile.

MASLI have also been identified in large Proterozoic anorthosite complexes. In Labrador, the Hybrid Series of the Newark Island layered intrusion records the emplacement of a wide compositional range of Fe-rich basic liquids into the chamber when it was floored by highly silicic magma (Wiebe 1988). In the Laramie complex (Wyoming) the Maloin Ranch pluton contains coarse-grained cumulate monzosyenites to granites intercalated with chilled layers of Fe-rich monzonite and biotite gabbro. These chilled layers are thought to record the entry and ponding of two different types of mafic liquids into a silicic magma chamber (Kolker & Lindsley 1989).

In the Coastal Maine Magmatic Province (Hogan & Sinha 1989), many gabbro–diorite plutons contain interlayered mafic and silicic rocks with features characteristic of MASLI. The basin-form Pleasant Bay intrusion, which is roughly 12 by 20 km in area with a maximum thickness of 3 km, provides a stratigraphic record of hundreds of basaltic injections into a chamber floored by silicic magma (Wiebe 1993b). Chapman and Rhodes (1992) showed that the interlayered diorites and gabbros on Isle au Haut record upwards fractionation of dioritic cumulates interrupted by periodic injection and ponding of basaltic magma on the chamber floor. Gabbro–diorites that form gently dipping layers in the lower part of the Cadillac Mountain granite represent numerous basaltic infusions into a floored chamber of granitic magma (Wiebe 1994).

Several classic areas of magma co-mingling and mixing should probably be reinterpreted as MASLI. The Cadomian layered gabbro–diorites of Guernsey and Jersey consist of complexly interlayered mafic and silicic rocks that display in great abundance the basally chilled mafic layers and granitic pipe structures that are diagnostic of MASLI (Elwell et al. 1962; Topley et al. 1982). Field relations and petrographic descriptions strongly suggest that the silicic layers are feldspar-rich cumulates that were trapped by multiple infusions of basic magma onto the floor of an existing chamber of silicic magma. Comparable field relations are well displayed in layered mafic and granitic rocks in the Tertiary Slieve Gullion Complex (Elwell 1958).

3. Interactions between stably stratified basaltic and silicic magmas

The extent of hybridisation that can result during the initial emplacement of basaltic magma into a low temperature silicic magma chamber should generally be limited because of the large temperature, density and viscosity contrasts between the magmas (Campbell & Turner 1985). Because heat can rapidly convect away from the underlying basaltic layer, a single small infusion might solidify so rapidly that its upper boundary remains solid. On the other hand, large infusions of basaltic magma into previously heated granitic magma (due to earlier replenishments), could, at least for some time, lead to the development of a stable liquid–liquid boundary between them (Sparks & Marshall 1986). Double-diffusive convection should occur along this boundary and a finite hybrid layer could develop (Oldenburg et al. 1989). A variety of processes might occur within this zone: partial mixing due to turbulent stirring (Huppert et al. 1984); selective diffusion of alkalis (Watson & Jurewicz 1984); nucleation and growth of new phases within the basaltic magma (Wiebe 1973; Johnston & Wyllie 1988; van der Laan & Wyllie 1993); and decoupling of chemical and isotopic exchange (Lesher 1990, 1994). The cap of silicic magma should also act to trap any evolved magma formed by fractional crystallisation of the underlying mafic magma.

Macrorhythmic units in MASLI provide cumulate records of the interaction between hotter, denser mafic magma with overlying cooler, less dense silicic magma (Wiebe 1993a). Detailed studies of these units in several Silurian plutons of coastal Maine indicate that the following processes play important parts in these composite magma chambers.

3.1. Selective diffusion

It is likely that a double-diffusive boundary is commonly established between the two magmas (Huppert *et al.* 1984) and this should provide an effective setting for the rapid exchange of heat and selective chemical and isotopic diffusion. H_2O moves into the mafic magma, destabilising plagioclase and leading to the crystallisation of hornblende and biotite (Coulon *et al.* 1984). Selective diffusion of alkalis (Watson & Jurewicz 1984) is an important process, with movement of K into the mafic magma (generally also promoted by the growth of biotite) and Na into the silicic magma.

Differential exchange of isotopes (Lesher 1990, 1994) may also have been effective. In macrorhythmic units of the Pleasant Bay intrusion, the initial $^{87}Sr/^{86}Sr$ ratio gradually increases upwards, whereas little or no variation is seen in the Nd isotopes (Wiebe *et al.*, unpublished data). This behaviour resembles that seen in the Muskox intrusion (Stewart & DePaolo 1992) and is consistent with the relative diffusivities of Sr and Nd isotopes (Lesher 1994). In the time available for the crystallisation of these individual units, diffusion should have occurred over distances of only a few centimetres. The cumulates, however, show isotopic gradients over distances of up to 50 m (Wiebe *et al.*, unpublished data). If diffusive exchange within the boundary layer was the main process, then these effects must have been transported away from the boundary by convection.

3.2. Effects of crystallisation at the boundary

Crystallisation at the boundary (within the mafic magma) is an important factor in the subsequent evolution of both magmas. Initial temperatures at the boundary and subsequent temperatures of equilibration generally appear to fall between 900 and 1000°C. Biotite and hornblende commonly appear to have nucleated and grown there in abundance. Biotite, particularly, can play a major part in promoting the exchange of alkalis, with K concentrating in the mafic magma and Na in the silicic magma (Johnston & Wyllie 1988). The growth of hornblende and biotite also tends to dehydrate the base of the overlying silicic magma.

These effects are probably transferred away from the boundary by convection in both magmas. Some macrorhythmic units clearly increase upwards from hornblende-rich gabbroic rocks to essentially anhydrous silicic rocks with two pyroxenes (Wiebe 1993b). Gabbroic cumulates in the lower parts of these units commonly contain large crystals of biotite that are now partly corroded and enclosed within hornblende and/or pyroxene. At some boundaries it appears that other phases (including Fe–Ti oxides, zircon and apatite) have also crystallised and then been convected downwards, where they now occur sparsely in gabbroic cumulates as corroded crystals armoured by pyroxene, hornblende and biotite.

Under the right circumstances multiple injections of basalt might be capable of generating low-K, relatively anhydrous silicic magma at the base of normal high-K silicic magma. These two magmas might not mix if the density of the low-K magma is increased more by the loss of water than it is decreased by thermal expansion. If this happened, a convecting layer of low-K silicic magma could develop beneath normal high-K magma (see section 4.2).

3.3. Mixing

Exchange of crystals (and, presumably, some melt) occurs between the two magma layers. Evidence for this is most easily seen in some hybrid tops of macrorhythmic units (e.g. Fig. 4). More subtle evidence includes sparse calcic plagioclase xenocrysts in some silicic parts of macrorhythmic units and sodic plagioclase and alkali feldspar in some of the mafic rocks, either as xenocrysts or as cores to more calcic plagioclase. Some of the isotopic variation in macrorhythmic units may have been due to the variable mixing of liquid and crystals between mafic and silicic magmas.

3.4. Fractional crystallisation of the mafic magma

If crystallisation does not proceed too rapidly, the mafic magma undergoes fractional crystallisation and produces cumulates that are typical of the lower parts of many macrorhythmic units. These rocks consist of a touching framework of subhedral plagioclase ± subhedral olivine, augite and Fe–Ti oxides, with a high percentage of interstitial material. Because heat can be transferred convectively to the overlying silicic layer, most of the nucleation and growth of crystals in the mafic liquid probably occurred along the horizontal boundary between the two magmas. When this relatively dense, crystal–rich boundary layer reached some minimum thickness, it probably convected downwards to the chamber floor. Thicker units (e.g. 50 m) commonly preserve a record of the incoming of cumulus phases, starting with plagioclase + olivine + augite + Fe–Ti oxides + apatite. After Fe–Ti oxides begin to crystallise, residual liquid should tend to collect upwards and remain trapped beneath overlying silicic melt (Sparks & Huppert 1984).

In many macrorhythmic units, some normally incompatible elements (e.g. Ba and Zr) increase rapidly upsection to very high abundances (Fig. 5). These changes appear to reflect strong compositional gradients in the basal liquid rather than the incoming of new cumulus phase for three reasons: (1) the abundances of the elements do not reach a peak and then decline sharply (Fig. 5); (2) the steep increase in Zr is marked not by large 'cumulus' zircons, but by gradually increasing amounts of extremely fine-grained and disseminated crystals; and (3) the steep increase in Ba may occur even though plagioclase remains the only feldspar. This decoupling of major and incompatible trace elements can be explained if the residual liquid that produced the upper parts of these layers developed through multiple replenishments of basaltic magmas that mixed and underwent fractional crystallisation in a manner proposed by O'Hara and Mathews (1981) for mid-

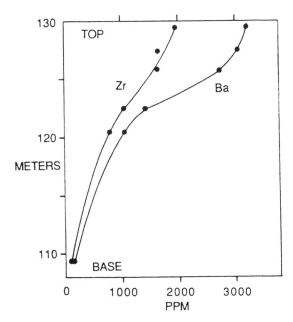

Figure 5 Plot of Ba and Zr versus stratigraphic height in a macrorhythmic unit in the Pleasant Bay intrusion (Reproduced from Wiebe 1993b, *Journal of Petrology* **34**, 482, by permission of Oxford University Press.).

ocean ridge magma chambers. The stratigraphies of the MASLI strongly support this model because they commonly record hundreds of mafic replenishments. The process of decoupling can be very efficient in MASLI because the basaltic magmas cannot erupt through the overlying silicic magma that traps them.

Many studies of silicic volcanic rocks have proposed that the fractional crystallisation of underlying mafic magmas played a major part in establishing strong compositional gradients of the source magma chambers (e.g. Ferriz & Mahood 1987; Druitt & Bacon 1988). Cumulates in MASLI appear to provide a record of these gradients. Considerations

of fractionation density (Sparks & Huppert 1984) suggest that multiple replenishments could lead to the generation of more than one discrete compositional layer near the base of a silicic chamber.

4. Effects of basaltic infusions on the evolution of granitic magma

The compositional and thermal effects of repeated infusions of basaltic magma may greatly affect the character of granite that crystallises from overlying silicic magma. In the Silurian plutons of coastal Maine, processes that operated in MASLI

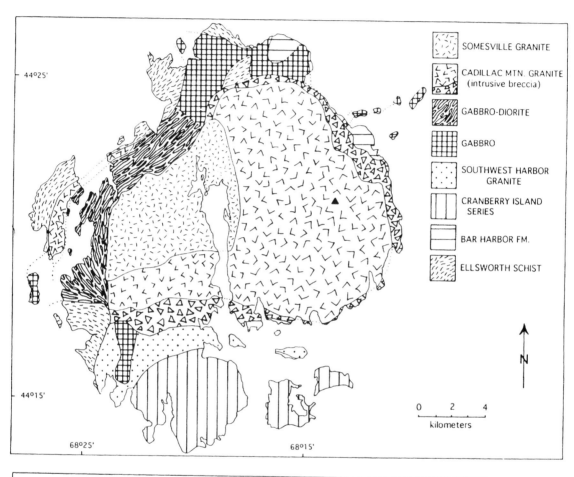
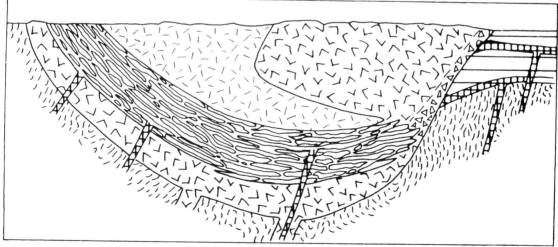

Figure 6 Geological map and cross-section of the Cadillac Mountain intrusive complex. The E–W cross-section is drawn about 1 km south of the summit of Cadillac Mountain (closed triangle). Gabbroic feeders are schematic; there are actually hundreds of dykes with thicknesses ranging from 50 cm to 3 m. (This figure first appeared in Wiebe 1994, *Journal of Geology* **102**, 425–6, and is reproduced with the permission of The University of Chicago Press).

appear to have had profound and varying effects on the evolution of overlying granites and their enclaves. In the Cadillac Mountain granite (CMG; Fig. 6), basaltic infusions have apparently converted an I-type granite into a hypersolvus granite with A-type characteristics. In the Gouldsboro granite repeated injections of basaltic magma appear to have mainly promoted the exchange of alkalis and led to the development of a low-K (trondhjemitic) magma layer beneath normal granitic magma (Wiebe & Adams, in press). The different responses of the silicic systems are probably related to the rate of influx of basaltic magma and the relative proportions of mafic and silicic magma within the chamber. These two intrusions appear to represent highly divergent paths of evolution and may provide insights into a broad continuum of possible interactions between resident silicic magma and basaltic injections in composite magma chambers.

4.1. Evolution of silicic magma in the Cadillac Mountain chamber

Field, petrographic and geochemical compositions strongly suggest that basaltic infusions into the Cadillac Mountain magma chamber have transformed silicic magma that produced a subsolvus I-type granite (lower CMG) into a hypersolvus granitic magma with A-type characteristics (upper CMG) (Wiebe *et al.*, unpublished data). The lower granite is characterised by two feldspars with early crystallising biotite and hornblende, whereas the upper granite is characterised by hypersolvus feldspar (averaging about $An_3Or_{32}Ab_{65}$) with late interstial hornblende. The upper CMG has much higher Ga/Al, higher Fe/Mg and is greatly enriched in REEs, Y and other incompatible elements (Fig. 7). Neither these chemical nor thermal characteristics can be generated by fractional crystallisation of the granite (Wiebe *et al.*, unpublished data). Instead, these changes can best be explained by the 'mixing in' of hotter, evolved mafic to intermediate magmas that formed at the base of the chamber due to fractional crystallisation and mixing of the basaltic replenishments (see section 4.2). The development of these A-type characteristics by processes operating in a magma chamber is compatible with the commonly gradational character of A-type granites with respect to nearby I- and S-types (Whalen *et al.* 1987). It

suggests that the thermal, chemical and isotopic character of granites with A-type characteristics may not accurately reflect their sources or original conditions of melting.

4.2. Origin of enclaves in the Cadillac Mountain granite

Granite beneath the gabbro–diorite unit (lower CMG) has only very scarce enclaves, in contrast with the upper CMG in which enclaves are relatively abundant and widely scattered. These relations suggest that most of the enclaves originated from mafic to intermediate magma that evolved near the base of the chamber during the deposition of the gabbro–diorite unit. Enclaves range in SiO_2 from about 55 to 75%. Relative to the chilled basalts from the gabbro–diorite unit, the most mafic enclaves are greatly enriched in Mn, Ba, Zr, Rb, Y, Yb, Ce, Nb, Zn, Be and Ga and strongly depleted in Mg, Sr, Cu, Co and V (Fig. 8). Concentrations of the incompatible trace elements are higher than in the host granite and are greatest in enclaves with the lowest SiO_2, decreasing strongly as SiO_2 increases (Fig. 9). These relations suggest that the chemical characteristics of the enclaves are largely controlled by the replenishment and fractional crystallisation of mafic magmas at the base of the chamber. There are, in addition, several

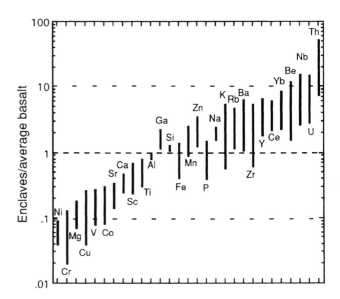

Figure 8 Compositions of all analysed mafic enclaves in the Cadillac Mountain granite (CMG) normalised to the average composition of chilled basaltic layers in the gabbro–diorite unit. These layers record the composition of basaltic magma that was injected into the CMG magma chamber.

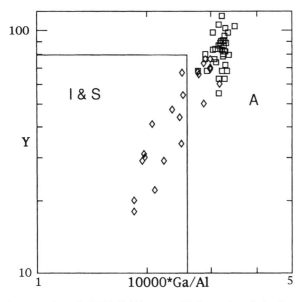

Figure 7 Plot of 10 000 Ga/Al versus Y for rocks of the lower Cadillac Mountain granite (CMG) (diamonds) and upper CMG (squares). The lower CMG lies mainly within the field of I- and S-type granites, whereas the upper CMG lies within the field of A-type granites. Field boundaries as in Whalen *et al.* (1987).

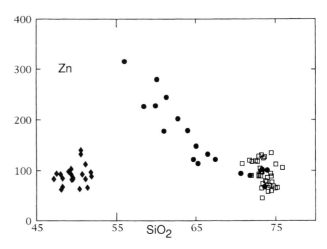

Figure 9 Zn versus SiO_2 for Cadillac Mountain granite (CMG) enclaves (circles), chilled basaltic rocks from the gabbro–diorite unit (diamonds) and the upper CMG (squares).

reasons why it appears that the enclaves were not significantly modified by chemical exchange with the host granite after they were disrupted from basal mafic to intermediate magma layers (Wiebe *et al.*, unpublished data): (1) enclaves with very different compositions occur in granite of constant composition; (2) enclaves appear to lack internal mineral and compositional gradations; and (3) hornblende-dominated and clinopyroxene-dominated enclaves occur in the same outcrop. Although exchange between the enclave and host granite may be important in most other settings (e.g. Christiansen & Venchuarutti 1990; Blundy & Sparks 1992), there are others where exchange has apparently been limited (Metcalf *et al.* 1995). The relatively anhydrous character of the CMG may have reduced the potential for exchange between the upper CMG magma and its enclaves.

Most mafic enclaves appear to be quenched samples of magma that developed by the mixing and fractional crystallisation of multiple infusions of basaltic magma at the base of the CMG magma chamber (Wiebe *et al.*, unpublished data). Intermediate to silicic enclaves in the CMG appear to represent quenched samples of magmas that developed at the base of the chamber due to hybridisation between the highly evolved mafic liquids and overlying silicic magmas. Some groups of enclaves in silicic volcanic rocks are similarly enriched in incompatible trace elements (Grunder 1994; Montanini *et al.*). Grunder (1994) has proposed a similar model to explain the comparable enrichment of incompatible trace elements in magmatic enclaves.

4.3. Nature and origin of magma stratification as preserved by large enclave-rich zones in the Gouldsboro granite

Based on gravity modelling, Hodge *et al.* (1982) suggested that the Gouldsboro granite (Fig. 10) may be as little as 800 m thick and rests on top of a thicker mass of gabbroic material. The granite is a fine- to medium-grained, leucocratic biotite hornblende granite with common miarolitic cavities. A well-exposed basal transition from gabbro to granite occurs north of the area shown in Figure 10. It is marked by mainly chilled mafic rocks that occur in gently dipping sheets and pillows

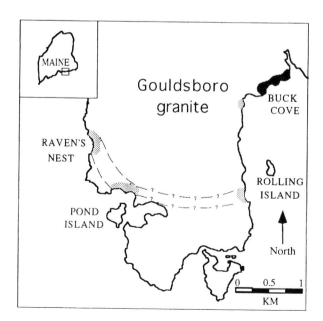

Figure 10 Map of Schoodic peninsula (all Gouldsboro granite) showing the locations of enclave-rich zones. Light grey areas have large enclaves with SiO$_2$ between 65 and 70%, whereas dark area near Buck Cove contains mainly enclaves with SiO$_2$ between 54 and 59%.

(typically from 30 to 100 cm thick) of chilled gabbro in granite (Fig. 3). This transition is locally up to 1 km thick. The granite is cut by many steeply dipping basaltic, silicic and composite dykes that trend roughly N–S.

In the highest level exposures of the Gouldsboro granite there are areas up to 300 m in width and possibly up to 1 km in length that contain roughly 30–50% of chilled intermediate to silicic globular enclaves (ranging in diameter from tens of centimetres to several metres) in a matrix of granite (Fig. 10) (Wiebe & Adams, in press). In the Raven's Nest, Pond Island, and Rolling Island areas, there are two different types of enclaves (high-K$_2$O and low-K$_2$O) with identical SiO$_2$ (Fig. 11). For most major and trace elements (e.g. FeO$_T$) plotted against SiO$_2$ both enclaves and host granite produce coherent trends, with SiO$_2$ ranging from 65 and 71% in the globules and from 73 to 78% in the granite (Fig. 11). Because the host granites crystallised at fairly shallow depths and because these zones occur only in the structurally highest parts of the plutons, it is likely that the chilled silicic globules represent samples of magma from the chamber interior that were trapped during eruption through the mushy roof of the chamber.

The compositions of these enclaves and the host granite strongly suggest that they evolved within the same chamber. The low-K enclaves are concentrated in the central portion of the enclave-rich areas, whereas the high-K enclaves are dominant in the margins. Some enclaves are composite, consisting of both high-K and low-K types separated by a sharp, but highly irregular, crenulate boundary. Scarce chilled pillows and angular fragments of basalt occur only in the low-K enclaves and some of these basaltic enclaves are unusually rich in K$_2$O. A shallow level of emplacement is indicated by the presence of some amygdales filled with feldspar and quartz in most silicic enclaves and vesicles in the basaltic inclusions.

The enclaves probably preserve a record of magma compositions within the chamber at the time of eruption. On the basis of models for magma withdrawal (e.g. Trial *et al.* 1992), the distribution of the two types of enclaves suggests that the chamber was compositionally stratified with low-K magma below high-K magma. The restriction of chilled basalt inclusions to the low-K type suggests that basaltic magma entered the base of the chamber shortly before the eruption and may have triggered it.

The only significant compositional differences between these two types of silicic magmas appear to be in their K, Na, Rb and Ba contents. Although the high-K magma appears to be normal granite, the low-K magma has the composition of a trondhjemite. Locally in the vicinity of the enclaves, the host granite is also a high-SiO$_2$, very low-K trondhjemite (Fig. 11). The fact that some basaltic inclusions in the low-K globules are unusually K-rich suggests that the low-K magma could have been derived from the high-K magma by the selective exchange of alkalis between silicic and mafic magmas (Watson & Jurewicz 1984).

More mafic, chilled enclaves dominate the enclave-rich zone in the Buck Cove area of the granite. These have SiO$_2$ that varies from about 54 to 59%; their major and trace element compositions form tight trends against SiO$_2$, except for the alkalis, which vary widely at a constant value of SiO$_2$ (Fig. 11). In terms of most elements (except mainly the alkalis) these mafic enclaves have compositions which suggest they evolved largely by fractionation from a basaltic parent. The host granite near these enclaves has highly variable K$_2$O and ranges from normal granite to trondhjemite with very low K$_2$O. The compositional variation of alkalis in both enclaves

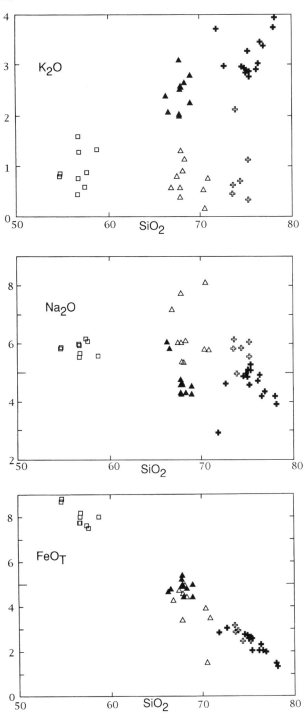

Figure 11 Plots of FeO$_T$, K$_2$O and Na$_2$O versus SiO$_2$ for Gouldsboro granitic rocks (crosses) and their chilled inclusions. Buck Cove area, mafic enclaves (open squares); Raven's Nest, Pond Island, and Rolling Island areas, low-K chilled enclaves (open triangles); and high-K chilled enclaves (solid triangles). See Figure 10 for locations.

and granitic host appears to be best explained by selective exchange during co-mingling.

In the Gouldsboro magma chamber it appears that the low-K (trondhjemitic) silicic magma developed from the high-K magma through the repeated exchange of alkalis with basaltic infusions into the base of the chamber. On a small scale, a comparable exchange of alkalis is commonly observed at the margins of chilled pillows in granite (Wiebe 1973). Exchange between the resident silicic magma and the many replenishments of basaltic magma may have generated a hotter, drier (and more dense) layer of low-K magma between the basalt and normal high-K magma. These relationships

point to an intriguing new model for the origin of trondhjemitic magma—one that neither requires a special source for melting nor the fractional crystallisation of an unusually low-K mafic magma. Unlike existing models it appears capable of explaining why K$_2$O in some associations keeps decreasing as SiO$_2$ increases (e.g. Barnes *et al.* 1992).

4.4. Causes of different paths of magma evolution

How and why have basaltic injections into two different silicic magma chambers produced such different results? The answer may be related to the rates of influx of basaltic magma and the relative proportions of mafic and silicic magma in the chamber. If basaltic magma comes in rapidly enough and in large enough volumes (relative to the overlying silicic magma), it should have a substantial thermal effect on the silicic magma. Moreover, with a large volume of resident basaltic magma there is a much greater likelihood that a double-diffusive boundary with overlying silicic magma would commonly be present, allowing some turbulent mixing in addition to the extensive exchange of heat and selective diffusion of chemicals and isotopes. The larger volume of basaltic magma could undergo fractional crystallisation and this fractional crystallisation, in combination with multiple replenishment, would lead to the production of an intermediate layer rich in incompatible elements immediately beneath the silicic magma (see section 3.4). Periodic disturbances to the layer boundaries (due to new injections and eruptions) could generate enclaves from this layer and allow the gradual stirring in of the enriched intermediate magma. The gabbro–diorite unit in the CMIC contains many thick macrorhythmic units which preserve evidence of these processes. The enriched enclaves and A-type character of the upper CMG appear to be the result of these thermal, chemical and physical interactions between mafic and silicic magmas.

On the other hand, if basaltic injections are of smaller volume and occur in a larger reservoir of silicic magma, the basaltic magma will tend to chill rapidly against the silicic magma. This rapid quenching would only allow the most rapid processes (e.g. exchange of alkalis) to have a significant effect. The transition between gabbro and granite in the Gouldsboro complex is, in fact, characterised by many thin sheets and pillows of thoroughly chilled basaltic magma in granite, whereas thick macrorhythmic units with hybrid and silicic tops are lacking. This is consistent with the low-K trondhjemitic magma in the Gouldsboro pluton having been formed largely by alkali exchange.

5. Further implications for granitic magmatism

If silicic plutons commonly trap basaltic magma at some stage in their movement upwards from their source, processes such as selective diffusion (chemical, isotopic and thermal), fractional crystallisation of the trapped basalt and mixing may be important in establishing the thermal and chemical character of the final granite. As these processes can significantly affect relatively simple upper crustal systems, they are probably capable of even more profoundly affecting magmas that have interacted at depth and for potentially longer periods of time.

It may be difficult to determine whether or not a particular granitic magma has been affected by these processes at some point during its movement upwards. Because ponded mafic magmas are relatively dense and, moreover, crystallise first and relatively rapidly, silicic magmas that rise away from deep composite chambers may show little direct evidence of their prior involvement with mafic magma. Although the occurrence of mafic magmatic enclaves in a granite could indicate prior interactions with mafic magma, the absence of visible enclaves does not rule out earlier interactions between basaltic and granitic magmas. As a result, the chemical and isotopic compositions of many granites may not necessarily provide, as is often assumed, a reliable indication of their source.

Basaltic injections into silicic magma chambers may also play a significant part in enabling silicic magma to rise to shallow crustal levels. Basaltic magma that ponds in silicic chambers will transfer heat efficiently through double-diffusive convection to the overlying silicic magma. This heat should reduce the percentage of restite or phenocrysts and make the silicic magma less viscous. The growth of hornblende and biotite along the boundary should tend to dehydrate the silicic magma, thereby increasing its potential to rise without crystallising.

6. Acknowledgements

This research was supported by National Science Foundation Grants EAR-9003712 and 9204475. Reviews by Richard d'Lemos and Sheila Seaman were most helpful.

References

Barbarin, B. 1988. Field evidence for successive mixing and mingling between the Piolard Diorite and the Saint-Julien-la-Vetre Monzogranite (Nord-Forez, Massif Central, France). CAN J EARTH SCI 25, 49–59.

Barnes, C. G., Barnes, M. A. & Kistler, R. W. 1992. Petrology of the Caribou Mountain pluton. J PETROL 33, 95–124.

Bergantz, G. W. 1989. Underplating and partial melting: implications for melt generation and extraction. SCIENCE 245, 1093–95.

Blundy, J. D. & Sparks, R. S. J. 1992. Petrogenesis of mafic inclusions in granitoids of the Adamello Massif, Italy. J PETROL 33, 1039–104.

Campbell, I. H. & Turner, J. S. 1985. Turbulent mixing between fluids with different viscosities. NATURE 313, 39–42.

Chapman, M. & Rhodes, J. M. 1992. Composite layering in the Isle au Haut Igneous Complex, Maine: evidence for periodic invasion of a mafic magma into an evolving magma reservoir. J VOLCANOL GEOTHERM RES 51, 41–60.

Christiansen, E. H. & Venchiarutti, D. A. 1990. Magmatic inclusions in rhyolites of the Spor Mountain Formation, western Utah: limitations on compositional inferences from inclusions in granitic rocks. J GEOPHYS RES 95, 17,717–28.

Coulon, C., Clocchiati, R., Maury, R. C. & Westercamp, D. 1984. Petrology of basaltic xenoliths in andesitic to dacitic host lavas from Martinique (Lesser Antilles): evidence for magma mixing. BULL VOLCANOL 47, 705–34.

Druitt, T. H. & Bacon, C. R. 1988. Compositional zonation and cumulus processes in the Mount Mazama magma chamber, Crater Lake, Oregon. TRANS R SOC EDINBURGH: EARTH SCI 79, 289–97.

Elwell, R. W. D. 1958. Granophyre and hybrid pipes in a dolerite layer of Slieve Gullion. J GEOL 66, 57–71.

Elwell, R. W. D., Skelhorn, R. R. & Drysdall, A. R. 1962. Net-veining in the diorite of northeast Guernsey, Channel Islands. J GEOL 70, 215–26.

Ferriz, H. & Mahood, G. A. 1987. Strong compositional zonation in a silicic magmatic system: Los Humeros, Mexican Neovolcanic Belt. J PETROL 28, 171–209.

Fridrich, C. J. & Mahood, G. A. 1987. Compositional layers in the zoned magma chamber of the Grizzly Peak Tuff. GEOLOGY 15, 299–303.

Grunder, A. L. 1994. Interaction of basalt and rhyolite in a bimodal suite. GEOL SOC AM ABSTR PROGRAM 26, A476–7.

Hildreth, W. 1981. Gradients in silicic magma chambers: implications for lithospheric magmatism. J GEOPHYS RES 86, 10, 153–93.

Hildreth, W. & Moorbath, S. 1988. Crustal contributions to arc magmatism in the Andes of central Chile. CONTRIB MINERAL PETROL 98, 455–89.

Hildreth, W., Halliday, A. N. & Christiansen, R. L. 1991. Isotopic and chemical evidence concerning the genesis and contamination of basaltic and rhyolitic magma beneath the Yellowstone Plateau volcanic field. J PETROL 32, 63–138.

Hodge, D. S., Abbey, D. A., Harbin, M. A., Patterson, J. L., Ring, M. J. & Sweeney, J. F. 1982. Gravity studies of subsurface mass distributions of granitic rocks in Maine and New Hampshire. AM J SCI 282, 1289–324.

Hogan, J. P. & Sinha, A. K. 1989. Compositional variation of plutonism in the coastal Maine magmatic province: mode of origin and tectonic setting. In Tucker, R. D. & Marvinney, R. G. (eds) Studies of Maine geology: igneous and metamorphic geology, vol. 4, 1–33. Augusta, Maine: Maine Geological Survey, Department of Conservation.

Huppert, H. E. & Sparks, R. S. J. 1988. The generation of granitic magmas by intrusion of basalt into continental crust. J PETROL 29, 599–624.

Huppert, H. E., Sparks, R. S. J. & Turner, J. S. 1984. Some effects of viscosity on the dynamics of replenished magma chambers. J GEOPHYS RES 89, 6857–77.

Johnston, A. D. & Wyllie, P. J. 1988. Interaction of granitic and basic magmas: experimental observations on contamination processes at 10 kbar with H_2O. CONTRIB MINERAL PETROL, 98, 352–62.

Kolker, A. & Lindsley, D. H. 1989. Geochemical evolution of the Maloin Ranch pluton, Laramie anorthosite complex, Wyoming: petrology and mixing relations. AM MINERAL, 74, 307–24.

Lesher, C. E. 1990. Decoupling of chemical and isotopic exchange during magma mixing. NATURE 344, 235–7.

Lesher, C. E. 1994. Kinetics of Sr and Nd exchange in silicate liquids: theory, experiments, and applications to uphill diffusion, isotopic equilibration, and irreversible mixing of magmas. J GEOPHYS RES 99, 9585–604.

McGarvie, D. W., Macdonald, R., Pinkerton, H. & Smith, R. L. 1990. Petrogenetic evolution of the Torfajokull volcanic complex, Iceland II. The role of magma mixing. J PETROL 31, 461–81.

Metcalf, R. V., Smith, E. I., Walker, J. D., Reed, R. C. & Gonzales, E. A. 1995. Isotopic disequilibrium among commingled hybrid magmas: evidence for a two-stage magma mixing–commingling process in the Mt. Perkins pluton, Arizona. J GEOL 103, 509–27.

Michael, P. J. 1991. Intrusion of basaltic magma into a crystallizing granitic magma chamber: the Cordillera del Paine pluton in southern Chile. CONTRIB MINERAL PETROL 108, 396–418.

Miller, C. F., Watson, E. B. & Harrison, T. M. 1988. Perspectives on the source, segregation and transport of granitoid magmas. TRANS R SOC EDINBURGH: EARTH SCI 79, 135–56.

Montanini, A., Barbieri, M. & Castorina, F. 1994. The role of fractional crystallization, crustal melting and magma mixing in the petrogenesis of hyolites and mafic inclusion-bearing dacites from the Monte Arci volcanic complex (Sardinia, Italy). J VOLCANOL GEOTHERM RES 61, 95–120.

Nixon, G. T. 1988. Petrology of the younger andesites and dacites of Iztaccihuatl volcano, Mexico: II. Chemical stratigraphy, magma mixing, and the composition of basaltic magma influx. J PETROL 29, 265–303.

O'Hara, M. J. & Mathews, R. E. 1981. Geochemical evolution in an advancing, periodically replenished, periodically tapped, continuously fractionating magma chamber. J GEOL SOC LONDON 138, 237–77.

Oldenburg, C. M., Spera, F. J., Yuen, D. A. & Granville, S. 1989. Dynamic mixing in magma bodies: theory, simulations, and implications. J GEOPHYS RES 94, 9215–36.

Sparks, R. S. J. & Huppert, H. E. 1984. Density changes during fractional crystallization of basaltic magmas: fluid dynamic implications. CONTRIB MINERAL PETROL 85, 300–9.

Sparks, R. S. J. & Marshall, L. A. 1986. Thermal and mechanical constraints on mixing between mafic and silicic magmas. J VOLCANOL GEOTHERM RES 29, 99–124.

Stewart, B. W. & DePaolo, D. J. 1992. Diffusive isotopic contamination of mafic magma by coexisting silicic liquid in the Muskox intrusion. SCIENCE 255, 708–11.

Topley, C. G., Brown, M. & Power, G. M. 1982. Interpretation of field relationships of diorites and associated rocks with particular reference to northwest Guernsey, Channel Islands. GEOL J 17, 323–43.

Trial, A. F., Spera, F. J., Greer, J. & Yuen, D. A. 1992. Simulations of magma withdrawal from compositionally zoned bodies. J GEOPHYS RES 97, 6713–33.

van der Laan, S. R. & Wyllie, P. J. 1993. Experimental interaction of granitic and basaltic magmas and implications for mafic enclaves. J PETROL 34, 491–517.

Watson, E. B. & Jurewicz, S. R. 1984. Behavior of alkalis during diffusive interaction of granitic xenoliths with basaltic magma. J GEOL 92, 121–31.

Whalen, J. B., Currie, K. L. & Chappell, B. W. 1987. A-type granites: geochemical characteristics, discrimination, and petrogenesis. CONTRIB MINERAL PETROL 95, 407–19.

Wiebe, R. A. 1973. Relations between coexisting basaltic and granitic magmas in a composite dike. AM J SCI **273**, 130–51.

Wiebe, R. A. 1974. Coexisting intermediate and basic magmas, Ingonish, Cape Breton Island. J GEOL **82**, 74–87.

Wiebe, R. A. 1988. Structural and magmatic evolution of a magma chamber: the Newark Island layered intrusion, Nain, Labrador. J PETROL **29**, 383–411.

Wiebe, R. A. 1993a. Basaltic injections into floored silicic magma chambers. EOS, TRANS AM GEOPHYS UNION **74**, 1, 3.

Wiebe, R. A. 1993b. The Pleasant Bay layered gabbro–diorite, coastal Maine: ponding and crystallization of basaltic injections into a silicic magma chamber. J PETROL **34**, 461–89.

Wiebe, R. A. 1994. Silicic magma chambers as traps for basaltic magmas: the Cadillac Mountain Intrusive Complex, Mount Desert Island, Maine. J GEOL **102**, 423–37.

Wiebe, R. A. & Adams, S. D. Enclave-rich zones in the Gouldsboro granite, Maine: a record of eruption and compositional stratification in a silicic magma chamber. *In* Brown, M. & Piccoli, P. M. (eds) *The Origin of Granites and Related Rocks.* US GEOL SURV CIRC **1129**, 165.

R. A. WIEBE, Department of Geosciences, Franklin and Marshall College, Lancaster, PA 17604, U.S.A.

Transactions of the Royal Society of Edinburgh: Earth Sciences, **87,** 243–250, 1996

Crystallisation of anhydrite-bearing magmas

Leslie L. Baker and Malcolm J. Rutherford

ABSTRACT: Anhydrite has been identified as a phenocrystic phase in some silicic volcanic magmas, but it is not commonly described in plutonic rocks. Anhydrite-bearing magmas tend to form in arc environments and to contain hydrous, low-temperature, oxidised mineral assemblages. Phenocrystic anhydrite coexists with sulphur-enriched apatite and sometimes with pyrrhotite, in silicate melt that contains from 50 ppm to 1 wt% S, depending on temperature and f_{O_2} conditions. Vapour coexisting with anhydrite- and water-saturated magma may contain from a few tenths of a mole per cent to a few mole per cent sulphur gases (SO_2 and H_2S), with the exact composition and gas speciation depending on temperature and oxygen fugacity. Samples of one anhydrite-bearing magma, the 1991 Pinatubo dacite, have been experimentally crystallised to determine whether the magma retains its characteristic sulphur-rich mineral phases during solidification. Results show that anhydrite and sulphur-rich apatite are retained throughout crystallisation and vapour phase evolution. This suggests that anhydrite-bearing intrusive equivalents of the Pinatubo dacite should be present in arc plutonic complexes.

KEY WORDS: Pinatubo, El Chichón, sulphur, volatiles, arc magmas.

The eruption of anhydrite-bearing magmas by Pinatubo, El Chichón and other volcanoes has drawn the attention of volcanologists to these apparent petrological oddities. It is thought that such sulphur-rich explosive eruptions may under some circumstances have a significant effect on the earth's climate. In addition, the origin and evolution of anhydrite-bearing magmas pose many interesting and still unanswered petrological questions. It may be that such magmas occur more commonly than has been realized to date (Luhr *et al.* 1984), yet plutonic equivalents of these volcanic rocks have not been identified. One example is known of an intrusive rock containing phenocrystic anhydrite, but it is a dyke, not a pluton: the Bulolo dyke in the Julcani district of Peru is composed of anhydrite-bearing dacite (Drexler & Munoz 1985). If more anhydrite-bearing intrusive rocks could be identified and studied, they could provide additional insight into the development of relatively sulphur-rich, oxidised, silicic magma systems.

The sulphur concentration in high-silica melts saturated with anhydrite or pyrrhotite at near-liquidus conditions is of the order of tens of ppm to 1 wt% S, depending on temperature and melt composition (Carroll & Rutherford 1987; Luhr 1990). Recently erupted anhydrite-bearing magmas contained from several tenths of a weight per cent to several weight per cent sulphur trioxide before eruption, significantly more sulphur than was soluble in the melt phase of these magmas. In the following discussion, the term 'sulphur-rich magma' refers to a magma which is saturated with a sulphide or sulphate phase.

Sigurdsson (1990) has reviewed the history of climate-altering volcanic eruptions such as Tambora (1815), Krakatau (1888) and Laki (1783). The potential of a given volcano for producing climate-altering eruptions appears to be directly related to the amount of sulphur that could be degassed from its magma (Devine *et al.* 1984). Some recent eruptions have produced one to two orders of magnitude more sulphur dioxide gas than could have been dissolved in the melt phase of the erupted magma (Westrich & Gerlach 1992; Gerlach *et al.* 1994, 1996; Gerlach & McGee 1994). This has been referred to as the 'excess sulphur problem' and it has been most pronounced in eruptions of magmas that contain phenocrystic anhydrite.

The origin and pre-eruptive state of excess sulphur in volcanic gas emissions is an important question with regard to the evolution of such magmatic systems. Another important question is whether anhydrite-bearing magmas can be reliably identified in the geological record. Other questions include whether the sulphur-enriched character of the magmas is the product of sulphur addition processes late in the magma's history and, if so, how these processes operate. These questions are important in the study of plutonic as well as volcanic igneous rocks.

The aim of this study is to define the characteristics that are likely to identify a plutonic rock as having crystallised from a magma that contained anhydrite and perhaps other sulphur-enriched mineral phases, both through examination of erupted anhydrite-bearing magmas and through experimental crystallisation studies of these magmas. This paper reviews the concentration and speciation of sulphur dissolved in silicic melts and the common characteristics of several anhydrite-bearing magmas. We then describe crystallisation experiments on a sulphur-rich magma, focusing on the stability and composition of sulphur-bearing phases throughout crystallisation.

1. Sulphur speciation and concentration in silicate melts

The sulphur concentration in silicic magmas was experimentally examined by Katsura and Nagashima (1974). More recently, Carroll and Rutherford (1985, 1987, 1988) and Luhr (1990) examined the variation in the solubility of sulphur phases with temperature, pressure and oxygen fugacity. Magmatic dissolved sulphur takes one of two possible valence states, depending on the oxygen fugacity of the system. At reducing conditions, it dissolves as sulphide ion, S^{2-}, whereas under oxidising conditions it has a valence of 6^+ and takes the form of sulphate ion, SO_4^{2-} (Katsura & Nagashima 1974; Carroll & Rutherford 1987). Figure 1 shows the variation of sulphur oxidation state with f_{O_2} relative to the Ni–NiO (NNO) buffer. The sulphur minerals that will crystallise from a sulphur-saturated melt tend to show a similar variation, with reduced melts crystallising sulphides (typically pyrrhotite) and oxidised magmas crystallising sulphates (generally anhydrite), although the addition of components such as copper will stabilise sulphide minerals above NNO+2 (McKenzie & Helgeson 1985). This study is primarily concerned with magmas that are at oxygen fugacities above the NNO buffer, i.e. magmas that are oxidised enough to contain significant sulphate (Fig. 1).

The solubility of sulphate in oxidised melts is strongly dependent on both oxygen fugacity and temperature, as shown in Figure 2. Data from Luhr (1990) for El Chichón trachyandesite and from Carroll and Rutherford (1988) for El Chichón trachyandesite and Mount St Helens dacite show the combined effect of changing temperature and melt composition on sulphur concentrations along crystallisation trends in these magmas. Reduction in the iron content of the melt with increasing crystallinity probably accounts in part for the drop in sulphur content with temperature, because sulphur solubility is dependent on the melt iron content (Haughton et al. 1974; Carroll & Rutherford 1985). Superliquidus data on sulphate solubility in a water-saturated rhyolite melt at 2 kbar and the $MnO-Mn_3O_4$(MNO) buffer (Fig. 2) show a weaker dependence on temperature (Baker & Rutherford 1996). This difference may be due in part to the constant melt composition in these above-liquidus experiments.

Despite the low solubility of sulphide and sulphate in silicic melt, abundant sulphur-bearing minerals are not a common feature of silicic magma systems. The Pinatubo dacite and El Chichón trachyandesite are unusual in that their bulk sulphur contents are high enough to have stabilised phenocrystic anhydrite. Determining the origin of these high sulphur contents would be an important addition to our understanding of magmatic volatile budgets.

The sulphur in anhydrite phenocrysts in these two magmas is isotopically light; values of +7 to +9 in El Chichón trachyandesite (Rye et al. 1984) and +5 to +11 in the Pinatubo dacite (McKibben & Eldridge 1993) indicate that these crystals are of magmatic origin and did not result from contamination of the magma with isotopically heavy evaporitic sulphur from the country rocks. This implies that the sulphur in the anhydrite crystals was ultimately transported by magmas, probably oxidised and of basaltic composition,

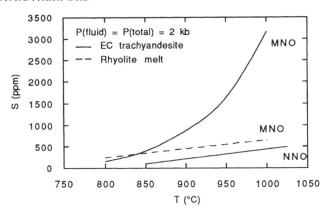

Figure 2 Solubility of anhydrite in El Chichón trachyandesite at NNO and MNO buffers (Carroll & Rutherford 1987; Luhr 1990) and rhyolite melt at MNO (Baker & Rutherford 1996). Data for trachyandesite are along crystallization trends and reflect the effect of changing melt composition as well as temperature. Data for rhyolite are above-liquidus at 2 kb $P(H_2O)$.

ascending from depth (Luhr 1990). There is still some disagreement, however, over whether anhydrite-saturated magmas require the open system addition of sulphur from other magmas to reach their eruptive sulphate contents.

Luhr (1990) has argued that the El Chichón trachyandesite did not require the addition of sulphur from an external source, based on the experimentally measured dependence of anhydrite solubility on pressure and temperature. He concludes that retention of the original sulphur complement of a fairly oxidised (>NNO) arc basalt throughout fractionation to a trachyandesitic composition would be sufficient to produce the bulk sulphate content of the El Chichón trachyandesite.

In contrast, Hattori (1993) asserts that the Pinatubo dacite displays evidence of fluid addition from underlying basaltic magma and argues that the dacite's sulphur content was increased by this fluid addition process. Although there is clear evidence for interaction of the Pinatubo dacite with a basaltic magma before its eruption, and although that basalt may have contributed sulphur-rich vapour to the eruption (Pallister 1992, 1996), there is no direct evidence to suggest that an external fluid source is required to account for the high bulk sulphate content of the Pinatubo dacite. The observation that ancient magmas erupted at Pinatubo were also saturated with anhydrite (Pallister et al. 1995) suggests that sulphate saturation is in fact a persistent feature of this magmatic system.

2. Examples of sulphur-rich oxidised magmas

The known examples of anhydrite-saturated magmas share a number of common characteristics that are typical of the arc environments in which they formed. They are oxidised, low temperature (800–900°C), amphibole-bearing, highly crystalline and are interpreted to be vapour-saturated. The small number of known anhydrite-bearing magmas precludes further generalisation. Some magmas have been erupted that had all these characteristics and yet were apparently not anhydrite-saturated, e.g. the Yn tephra of Mount St Helens (Geschwind & Rutherford 1992), indicating that other factors are necessary for the production of a sulphate-saturated magma. As discussed earlier, this could mean either evolution from a sulphur-rich parent, or the late-stage addition of sulphur to the evolved magma.

The following discussion will use the Pinatubo dacite and the El Chichón trachyandesite as primary examples, as they have been closely studied in this context. The 1951 Mt Lamington andesite is another example of an anhydrite-

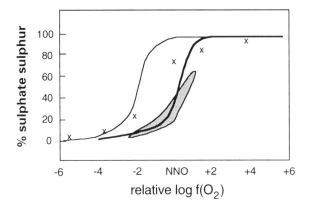

Figure 1 Sulphur speciation in silicate melts versus log oxygen fugacity relative to the nickel–nickel oxide (NNO) buffer. Heavy curve is from experimental data of Carroll and Rutherford (1988) on andesite and dacite melts. Light curve is from experimental data of Nagashima and Katsura (1973) on $Na_2O-3SiO_2$ melts; the relatively high sulphate contents of these melts are probably due to their high alkali contents. Crosses show data from Katsura and Nagashima (1974) on tholeiite basalt melt. Stippled field shows location of natural back-arc basalt glasses (Nilsson & Peach 1993) and mid-ocean ridge basalt glasses (Wallace & Carmichael 1991).

bearing magma (Arculus *et al.* 1983), but will not be discussed in detail. A single intrusive anhydrite-bearing rock, the Bulolo dyke of the Julcani district, Peru (Drexler & Munoz 1985), will also be briefly described.

The dacite erupted by Mt Pinatubo in 1991 was a highly evolved magma of 35–50 vol% crystallinity (Rutherford & Devine 1996; Pallister *et al.* 1996). It contained phenocrysts of plagioclase, quartz, hornblende (commonly rimmed with cummingtonite), iron–titanium oxides, biotite, anhydrite and apatite (Pallister *et al.* 1992; 1996). Compositions of the rhyolite matrix glass and mineral phases are given in Pallister *et al.* (1995a). Phase equilibrium studies and analyses of iron–titanium oxide compositions and the aluminium content of hornblende (Rutherford & Devine 1996) define the pre-eruptive conditions in the magma chamber to be $780 \pm 20°C$, $2·2 \pm 0·5$ kbar and an oxygen fugacity of three log units above the NNO oxygen buffer, as shown in Figures 3 and 4. Melt inclusions in plagioclase phenocrysts contain up to 6·4 wt% H_2O and 80 ppm S, indicating saturation of the dacite with anhydrite and a mixed $H_2O–SO_2$ fluid at magma chamber conditions (Rutherford & Devine 1996).

The trachyandesite pumice erupted by El Chichón in 1982 contained 58 wt% (24 vol%) phenocrysts, including plagioclase, hornblende, augite, titanomagnetite, anhydrite, apatite, sphene and biotite (Luhr *et al.* 1984). Before eruption, it was at a temperature of 800°C, a pressure of 2 kbar and an oxygen fugacity of one log unit above NNO (Luhr 1990). As shown in Figure 1, NNO + 1 is the reaction boundary between sulphur dissolved in silicic melt as sulphide (S^{2-}) and as sulphate (SO_4^{2-}); this is borne out by the coexistence of anhydrite and pyrrhotite in the El Chichón magma. The andesite was probably saturated with a water-rich vapour phase at the time of eruption (Luhr 1990).

The 1951 eruption of Mt Lamington, Papua, New Guinea produced a suite of high-K, anhydrite-bearing andesite lavas. Phenocryst phases in these rocks include plagioclase, amphibole, biotite and magnetite; accessory minerals included pyroxene, apatite, zircon and pyrite (Arculus *et al.* 1983).

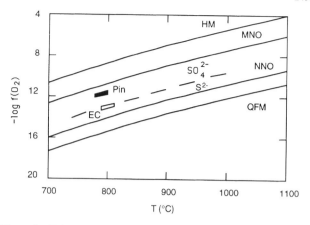

Figure 4 $T–f_{O_2}$ diagram showing Pinatubo dacite (Pin) and El Chichón trachyandesite (EC), modified after Rutherford and Devine (1996). Broken line is sulphide–sulphate reaction boundary, as in Figure 1.

Iron–titanium oxides from one sample gave a temperature of 650°C and a log oxygen fugacity of −16 (almost two log units above the NNO buffer at this temperature), indicating re-equilibration of the oxide compositions at subsolidus temperatures (Arculus *et al.* 1983). Two sulphur isotope analyses of coexisting pyrite and anhydrite give temperatures of 700 and 800°C (Arculus *et al.* 1983). The coexistence of pyrite and anhydrite phenocrysts suggests an oxygen fugacity for this magma near NNO + 1. Although melt inclusion compositions are not available, the common amphibole and biotite phenocrysts imply that the pre-eruptive water content of the melt was at least 4 wt% (Merzbacher & Eggler 1984).

The Bulolo dyke of the Julcani Volcanic district, Peru was emplaced in the late stages of activity in this region. The dacitic magma was probably water-saturated and contained 20–25% phenocrysts. In addition to anhydrite, the dacite contains phenocrysts of plagioclase, biotite, amphibole, apatite and quartz, and small amounts of sphene, pyroxene, iron–titanium oxides and pyrrhotite (Drexler & Munoz 1985). General magmatic conditions for the Julcani magmatic suite are temperatures of 830–908°C and oxygen fugacities of $10^{-10·1}$ to $10^{-11·3}$; no specific conditions are given for the Bulolo dacite (Drexler & Munoz 1985).

3. Sulphur-bearing phases in the Pinatubo and El Chichón magmas

Estimates of the bulk sulphur content of the Pinatubo dacite pumice range from 0·12 to 0·42 wt% sulphur as SO_3 (Bernard *et al.* 1991, Pallister *et al.* 1996). This sulphur is primarily contained in anhydrite, with trace amounts in apatite and the silicate melt. The 20 MT of SO_2 that was degassed during eruption of the dacite is equivalent to an additional 0·1–0·2 wt% SO_3 in the pre-eruptive magma. This results in a total range of possible bulk sulphur contents for the pre-eruptive magma of 0·22–0·62 wt% SO_3. This estimate is a minimum, as the measurement of 20 MT of sulphur degassed includes only SO_2 injected into the stratosphere (Bluth *et al.* 1992). The 1982 El Chichón pumices contained 1·2 wt% SO_3 in anhydrite and <0·1 wt% SO_3 in pyrrhotite; the eruption degassed an additional 1·3 wt% SO_3, resulting in a total pre-eruptive bulk sulphur content of 2·6 wt% SO_3 (Luhr *et al.* 1984; Varekamp *et al.* 1984).

3.1. Anhydrite

The bulk of the sulphur in the Pinatubo dacite is stored in anhydrite phenocrysts. These are fairly small and comprise one volume per cent or less of the pumice (Pallister *et al.*

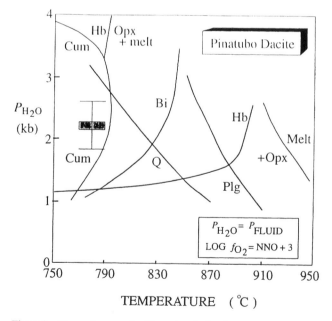

Figure 3 Phase diagram for Pinatubo dacite, from Rutherford and Devine (1996). Shaded block shows pressure from Al-in-Hb geobarometry and temperature from Fe–Ti oxide compositions. These estimated conditions are consistent with the phenocryst assemblage, particularly the stable cummingtonite rims on amphibole in the dacite. Oxygen fugacity for the phase equilibrium experiments (NNO + 3) is that given by Fe–Ti oxide compositions in the dacite pumice. Bi = biotite, Cum = cummingtonite, Hb = hornblende, Opx = orthopyroxene, Plg = plagioclase, and Q = quartz.

1996). Pinatubo phenocrystic anhydrite is generally equidimensional to rectangular in thin-section (Fig. 5); some crystals are embayed or rounded. It is found in contact with matrix glasses and sometimes in aggregates with apatite, as well as in inclusions in other phenocrysts. Because of the three good cleavage directions of anhydrite, and because of the prevalence of shattered crystals in the pumices, however, many of the anhydrite phenocrysts may no longer be in their original form. Lath-shaped anhydrite, for example, may be in its original crystal form or may be a fragment of a larger phenocryst.

Anhydrite also stores most of the sulphur in the El Chichón trachyandesite. This magma contained considerably more bulk sulphur and more phenocrystic anhydrite than the Pinatubo dacite. Anhydrite phenocrysts in the trachyandesite are subhedral to euhedral and do not display reaction coronas (Luhr et al. 1984). Anhydrite accounted for at least 1·2 wt% of the 2·6 wt% total SO_3 in the trachyandesite (Luhr 1990).

3.2. Apatite

Sulphur is known to replace phosphate groups in apatite via a coupled substitution with silica (Rouse & Dunn 1982). Apatite phenocrysts may therefore reflect the sulphur contents of their host magmas to some extent. Apatite in the 1991 Pinatubo pumice contains 0·1 to 1% SO_3, with an average sulphur content of 0·3%; apatites from ancient Pinatubo eruption products have similar average S contents (Pallister et al. 1995). Apatite phenocrysts within a single thin section may have sulphur contents that span the entire range from 0·1 to 1% (Pallister, pers. comm.). These variations in apatite sulphur content may be due to kinetic effects during growth, or they may reflect heterogeneities in the melt.

Sulphur-rich apatite is also present in the El Chichón andesite, where it contains 0·26–0·33% SO_3 (Luhr 1990; Pallister et al. 1996). This is comparable with the average sulphur content of apatites in the Pinatubo dacite, but the El Chichón apatites display far less variability in sulphur content.

3.3. Silicate melt

The melt phase of the Pinatubo magma was not particularly sulphur-rich. Its composition is essentially rhyolitic (Rutherford & Devine 1996; Pallister et al. 1996) and the solubility of sulphate in rhyolite is low at 780°C, as shown in Figure 2. At magma chamber conditions, experiments indicate that the melt contained approximately 80 ppm S, dissolved entirely as sulphate due to the magma's high oxygen fugacity (Rutherford & Devine 1996).

The melt phase of the El Chichón trachyandesite was also

sulphur-poor, containing 0·05 wt% SO_3 (Luhr 1990). Once again, this is consistent with studies of sulphur concentrations in low-temperature, oxidised magmas saturated with pyrrhotite or anhydrite (Carroll & Rutherford 1985, 1987; Luhr 1990). Despite its lower oxygen fugacity, this melt is more sulphur-rich than the Pinatubo dacite because of its higher iron and alkali contents.

3.4. Vapour

It is likely that the Pinatubo dacite contained an excess vapour phase before its eruption and it has been suggested that this vapour phase contained significant sulphur (Westrich & Gerlach 1992; Gerlach et al. 1996). Experiments to determine the sulphur content of a water-rich vapour phase in equilibrium with anhydrite and rhyolite melt at 2 kbar, 800°C and the MNO buffer give values of 0·1–0·6 mol% SO_2 (Baker et al. unpublished data). These results are from multiple mass spectrometric analyses of vapour bubbles trapped in quenched rhyolite glasses and include forward and reversal experiments. This sulphur content is not large, but it indicates nonetheless that the vapour is capable of transporting considerably more sulphur than the melt at these conditions.

Luhr et al. (1984) and Luhr (1990) have suggested that much of the sulphur released during the eruption of El Chichón was stored in a free vapour phase. At 800°C, this magma was only slightly hotter than the Pinatubo dacite, but was several log units more reduced. To determine the effect of changing oxygen fugacity on vapour composition, we have modelled the composition of a vapour in equilibrium with anhydrite in the Ca–H–O–S system, after the method described by Holloway (1987). Details of this calculation are given in Appendix 2. Figure 6 shows that the model vapour composition is strongly dependent on oxygen fugacity. This is not surprising, as the reaction controlling the sulphur content of the vapour in equilibrium with anhydrite is thought to be the following

$$CaSO_4(xtl) \rightarrow Ca(melt) + SO_2(vapour) + O_2 \qquad (1)$$

This reaction will progress to the right with lower f_{O_2} at constant T, P and a_{CaO}^{melt}. In addition, as described earlier, the solubility of anhydrite in the melt decreases with lower oxygen fugacity. Therefore, more reducing conditions should favour transfer of sulphur from the melt to the vapour, as sulphur in the vapour is in a less oxidised state. The stable melt species is SO_4^{2-}, with a sulphur valence of $+6$, whereas the stable sulphur species in the vapour are SO_2, with a sulphur valence of $+4$, and H_2S, with a sulphur valence of -2. Thus it is possible that

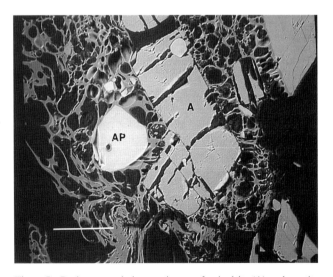

Figure 5　Back-scattered electron image of anhydrite (A) and apatite (AP) crystals in Pinatubo dacite pumice. Scale bar = 100 μm.

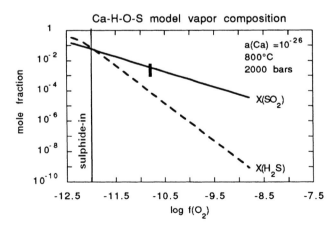

Figure 6　Model vapour composition in equilibrium with anhydrite at 800°C, plotted against oxygen fugacity. Solid line shows mole fraction of SO_2 in vapour; broken line shows mole fraction of H_2S. Bar shows range of experimental vapour SO_2 contents at 800°C, MNO buffer and 2 kb (Baker et al. unpublished data). Vertical line is sulphide–sulphate boundary.

a vapour phase coexisting with the El Chichón trachyandesite did contain significant sulphur, perhaps on the order of several mole per cent. H_2S would dominate in the vapour at these conditions: at NNO + 1 and 800°C, calculations show that the $H_2S:SO_2$ ratio would be approximately 24:1. This ratio is strongly dependent on oxygen fugacity, and approaches 1:1 at NNO + 1·75. These values are in agreement with those calculated by Luhr (1990) for El Chichón trachyandesite.

3.5. Sulphides

A sulphide phase was not stable in the very oxidised Pinatubo dacite, although sulphides have been found in the more reduced mixed magmas that were produced early in the Pinatubo eruption (Pallister *et al.* 1996). Pyrrhotite is not stable in magmas at more than two log units above NNO (Fig. 1; Carroll & Rutherford 1988; Luhr 1990). The trachyandesite erupted by El Chichón in 1982 was on the sulphide–sulphate boundary (Fig. 4) and contained both anhydrite phenocrysts and pyrrhotite blebs (Luhr *et al.* 1984; Luhr 1990). As discussed earlier, anhydrite was the primary sulphur-rich phase in the El Chichón magma, with only a small fraction of the magmatic sulphur stored in the pyrrhotite.

4. Experimental crystallisation of the Pinatubo dacite

We have experimentally examined the crystallisation of Pinatubo dacite to determine how the phases present in sulphur-rich magma systems evolve during cooling, crystallisation and vapour evolution. Details of these experiments are given in Appendix 1 and Table 1. Our focus was on whether any indications remain of the magma's high sulphur content on total crystallisation. Under natural conditions, volatile loss on cooling and crystallisation may result in the loss of much of the sulphur to the surrounding hydrothermal system. An important question addressed by these experiments is whether sulphur-rich phases such as anhydrite and apatite survive during crystallisation of a granitic magma and exsolution of a vapour phase, as indicators of that magma's enrichment in sulphur.

All experimental run products contained rhyolite glass, plagioclase, quartz, hornblende, cummingtonite, iron–titanium oxides, biotite, anhydrite and apatite, except for runs H4 and H7, which were reduced and contained pyrrhotite (replacing anhydrite). Crystallisation proceeded to an estimated 90% in the runs at 700°C, the lowest equilibration temperature in this study, with cummingtonite becoming relatively more abundant at lower temperatures. All the experimental magmas also contained a vapour phase produced by exsolution of volatiles from the melt as crystallisation progressed.

4.1. Anhydrite

Anhydrite was generally preserved in the experimental magmas throughout crystallisation as tabular grains of up to 100 µm in length (Fig. 7). The crystals were commonly partly surrounded by vapour bubbles, however, and were ragged or embayed from dissolution at the bubble contacts. This is inferred to have taken place during crystallisation of the magma: the proportion of SO_2 to H_2O exsolved from the melt would not have been high enough to maintain the equilibrium SO_2 content of the fluid, so dilution of the fluid with H_2O was balanced by anhydrite dissolution. Evidence of dissolution did not occur where anhydrite crystals contacted glass or other crystals, suggesting that anhydrite was still stable in the magma. No other sulphate phase appeared in the experiments, either independently or as a replacement mineral for anhydrite.

These results suggest that natural anhydrite-bearing magmas should retain their anhydrite phenocrysts throughout crystallisation. Preservation of anhydrite after solidification may be less likely, however. Interaction with groundwater would probably dissolve many anhydrite grains. Anhydrite inclusions in other minerals would be less subject to such attack, and would therefore be most likely to be preserved in rocks exposed at the earth's surface. Plutonic rocks sampled by drill

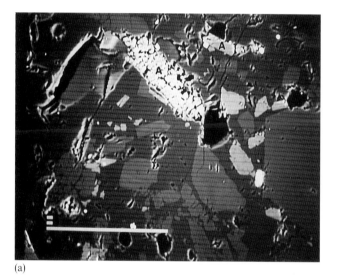

(a)

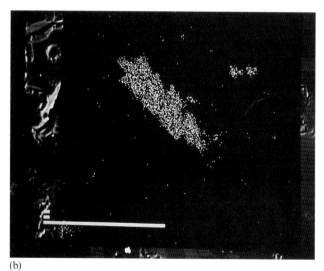

(b)

Figure 7 Anhydrite (labelled A) in 700°C experimental dacite. Scale bars are 100 µm. (a) Back-scattered image; (b) Sulphur X-ray map.

Table 1 Experimental temperatures, run times and glass and apatite sulphur contents.

Run	T_{final} (°C)	Dwell time (h)[1]	S in glass (ppm)	S in apatite (ppm)
Pin[2]	780		77	1000–10 000[3]
H2	800	23[4]	93 ± 20	—
H9/H10	800	3[4]	—	—
H3	780	71·5	75 ± 15	—
H4	760	94·5	—	—
H12	760	111·5	52 ± 30	473 ± 35
H5	740	120	[5]	—
H8	740	118	[5]	—
H6	720	141	[5]	2206
H7	700	157·5	[5]	987 ± 275
H11	700	115	[5]	—

Notes: (1) Time annealed at final temperature; (2) 1991 natural Pinatubo dacite pumice; (3) from Pallister *et al.* (1996); (4) fusion of starting material for experiments; (5) below detection limit of 30 ppm; and not analysed.

core would be more likely to preserve igneous-textured anhydrite that was not armoured with some other mineral. Even if igneous anhydrite has been preserved in a rock, it may be difficult to identify in thin sections that have been ground in a water slurry.

These results suggest that many sulphate-saturated magmas may be going unidentified because their anhydrite phenocrysts have dissolved away. Even when anhydrite is present in a thin section, it may be easily overlooked or ascribed to alteration processes because of its rarity as an igneous mineral. Because of these drawbacks, the presence of anhydrite as an inclusion in other phenocrysts is probably the most reliable indicator of anhydrite saturation of the rock.

4.2. Apatite

When analysing apatite grains in the experiments, we restricted our analyses to those that appeared texturally to have crystallised during the experiments (i.e. small needle-like crystals in contact with glass, or included in the rims of larger crystals). Sulphur concentrations in these crystals are given in Table 1 and show no clear correlation with temperature, although they are generally low compared with the sulphur contents of apatites in the Pinatubo dacite pumice (Pallister *et al.* 1995). The lack of correlation with temperature may result partly from accidental inclusion of analyses of original apatite from the dacite. However, given the variability in apatite sulphur content in the Pinatubo dacite reported by Pallister *et al.* (1995), the lack of a clear temperature trend is not surprising.

4.3. Silicate melt

The sulphur contents of the rhyolitic matrix glasses in the new experiments are shown in Fig. 8 and Table 1. The sulphur content at 780°C is 75 ± 15 ppm, in good agreement with the average melt sulphur content of 77 ± 29 ppm measured by Rutherford and Devine (1996) in the Pinatubo melt inclusions. The sulphur concentration in the experimental glasses decreases steadily with temperature and drops below 30 ppm, the detection limit of the electron microprobe, by 740°C. This is in accordance with extrapolation of the sulphate solubility trends found by Carroll & Rutherford (1987), Luhr (1990) and Baker & Rutherford (1996) at the MNO buffer.

4.4. Vapour phase

The vapour compositions in these experiments can be calculated using the thermodynamic model described in Appendix 2. A reduction in temperature should cause reaction 1 to progress to the left, lowering the sulphur content of the

vapour. Figure 9 shows that changing temperature at constant oxygen fugacity has a strong effect on the sulphur content of the H_2O-rich vapour; lowering the temperature from 800 to 700°C results in a reduction of almost three orders of magnitude in the total sulphur content of the vapour. The speciation of sulphur in the vapour is also temperature-dependent; at 800°C, SO_2 dominates strongly, whereas H_2S becomes increasingly important as the temperature drops and begins to dominate at roughly 600°C.

The strong dependence of the model vapour composition on temperature suggests that the capacity of the exsolved vapour to carry sulphur will drop as the magma crystallises. Thus, although the magma will release 6·4 wt% water as it crystallises, the vapour will require progressively less dissolution of anhydrite to achieve its equilibrium sulphur content. This is in accordance with the observation that anhydrite is preserved during experimental crystallisation of the Pinatubo dacite.

4.5. Sulphides

Because of the strong dependence of sulphur speciation on oxygen fugacity, anhydrite is easily reduced to a sulphide phase through reaction with the melt if the oxygen fugacity drops even to NNO+1. Partial reduction, leaving the anhydrite rimmed with sulphides, can occur in only a few hours at magmatic conditions (Baker & Rutherford unpublished data). In experiment H4 for this study (Table 1), all anhydrite was completely reduced to sulphides in less than 95 hours; anhydrite in run H7 was similarly reduced in less than 160 hours. A rock that originally contained sulphates but was reduced during or after cooling might contain anhydrite only as small inclusions in phenocrysts or as armoured remnants in sulphide or oxide phases. The compositions of other mineral phases that re-equilibrate more slowly at new oxidation states (e.g. amphibole or iron–titanium oxides) could provide a clue that this process has occurred.

5. Anhydrite-bearing intrusive rocks

The Bulolo dyke, a single example of an intrusive anhydrite-bearing dacite, was studied by Drexler & Munoz (1985) and has been described earlier. Because the Julcani magmatic system has been extensively studied, the formation of this anhydrite-bearing magma can be placed in a more complete geological context than recently erupted magmas.

The Bulolo dyke was emplaced during the fourth stage of activity at the Julcani volcanic field (Drexler & Munoz 1985). Stages I and II produced pyroclastic and dome eruptions of rhyolitic to dacitic rocks. During Stage III, mineralisation produced the economically important ore deposits in the

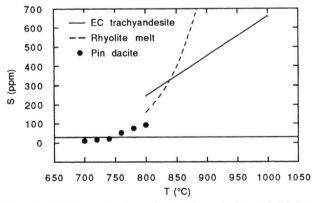

Figure 8 Solubility of sulphur in experimental glasses. Solid line shows data on rhyolite melt from Fig. 2 (Baker & Rutherford, 1996). Broken line shows trend from Luhr (1990) for El Chichón trachyandesite, MNO buffer. Points show data from experiments in this study; sulphur solubility was below the microprobe detection limit (horizontal line at 30 ppm) at temperatures of 740°C and below.

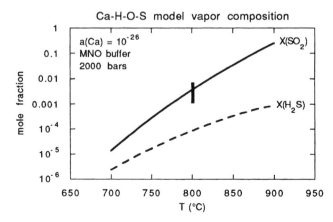

Figure 9 Model vapour composition in equilibrium with anhydrite, plotted against temperature. Solid line shows mole fraction of SO_2 in vapour; broken line shows mole fraction of H_2S. Bar as in Fig. 6.

region. Stage IV, and the eruption of the anhydrite-bearing dacite, only occurred after the end of ore-forming activity. This timing suggests that the eruption of the anhydrite-bearing magma was related to the ore formation in some way.

Drexler & Munoz (1985) suggest that the Stage IV magma of the Bulolo dyke was not related to the Stage III magmas and that it probably instead represented degassed magmas from deeper levels of the volcanic system. The degassed character of the magma, as inferred by Drexler & Munoz (1985) from F/Cl ratios in apatite, hornblende and biotite, implies that the parent magma of the Stage IV dacite was also the source of volatiles during earlier episodes of mineralisation. This proposed origin for the Bulolo dacite may be compared with the models discussed in section 1 for the formation of recently erupted anhydrite-saturated magmas. The models which invoke the addition of fluid from a basaltic magma underlying the magmatic system, as suggested for Pinatubo (Hattori 1993; Pallister et al. 1992, 1996) are not applicable to the situation at Julcani, where the anhydrite-saturated Bulolo dacite itself is thought to have come from the evolution of a more primitive magma body underlying the volcanic system. Instead, the origin of the Stage IV dacite is probably similar to that of the El Chichón trachyandesite as proposed by Luhr (1990): simple fractional crystallisation of a parental arc basalt.

6. Conclusions

Our experimental results suggest that magmas that are sulphur-rich enough to contain anhydrite and sulphur-rich apatite should retain these mineral indicators of their high sulphur content throughout crystallisation and vapour evolution. Despite this, the plutonic equivalent of a Pinatubo dacite does not appear to have been described in geological publications. It may be that the subtle mineral indicators of sulphur enrichment have been overlooked in plutonic rocks, or ascribed to later alteration processes. However, the results of this study suggest that such indications should exist in the rock record.

If the plutonic equivalents of anhydrite-bearing volcanic magmas exist, they will most likely be found in calc-alkaline arc systems. They are likely to contain a large percentage of hydrous minerals such as amphibole and to contain oxidised mineral assemblages. If the rocks contain apatite, it will be enriched in sulphur to some degree, perhaps containing several tenths of a weight per cent SO_3. Minor anhydrite should remain as a phenocryst phase, but it may only be retained as inclusions in other minerals such as amphibole. The discovery of such plutonic rocks would complement the ongoing studies of their volcanic counterparts. Their study would provide useful new insights into the provenance, evolution and frequency of occurrence of sulphur-rich magmas.

7. Acknowledgements

This paper has benefited from helpful discussions with Paul Hess, Phil Candela and John Pallister, who also kindly provided additional unpublished data on ancient eruptions of Mt Pinatubo and Mt St Helens. Joe Devine ably assisted with the microprobe analyses. The manuscript was improved by thoughtful comments from Aleta Finnila, Dina Venezky and Chuck Halfen. Jim Luhr, Calvin Miller and Phil Candela also provided thorough and helpful reviews. Support for this study was provided by grant #EAR 94-18157 and by a NASA Graduate Researchers Fellowship to the first author.

Appendix 1: experimental and analytical methods

The starting material for our crystallisation experiments was a natural Pinatubo pumice, which was ground to a powder under ethanol in an agate mortar. The powder was fused in a sealed 5 mm gold tube with water in excess of that required for saturation, at 800°C, 2400 bar and the intrinsic oxygen fugacity of the Renee metal pressure vessel and Ni filler rod (near $NNO+1$; Rutherford & Hill 1993). This ensured that at the run starting pressure of 2200 bar, the capsule would already contain a small amount of excess vapour. The fused glass-crystal mixture was re-ground under ethanol in an agate mortar. Because of the low oxygen fugacity of the fusion experiment, some anhydrite was probably reduced to sulphide. A small amount (≈ 1 wt%) of ground anhydrite powder was therefore added to the starting material to ensure that anhydrite was present in all experimental runs.

All experiments were performed at 2200 bar and buffered with the assemblage $MnO+Mn_3O_4+H_2O$. The run configuration was that of standard double-capsule hydrothermal experiments. The charge was contained in a sealed 2 mm AgPd tube, which was nested inside a sealed 5 mm AgPd tube containing the buffer materials. The experiments ran in Renee metal vessels pressurised with water, in horizontal resistance furnaces.

Each experiment was initially taken to a temperature of 800°C and pressure of at least 2200 bar. After the temperature had stabilised, it was programmed to begin ramping down to final temperature at a rate of 2°C per hour. Once the experiment had reached the final temperature, it was allowed to equilibrate for several days, then quenched in a jet of compressed air. Specific run data are listed in Table 1.

Thin sectioning of the run products was complicated by the solubility of anhydrite in water. We attempted to compensate for this by grinding and cleaning the sections in ethanol rather than water. However, it was necessary to finish the final polishing in a water slurry to obtain an acceptable polish for microprobe analysis. Although each thin section was only exposed to water for one to two minutes during the final polishing, this may have been enough time for some dissolution of anhydrite to take place. In addition, vapour bubbles were commonly attached to the anhydrite crystal faces, allowing more surface area over which dissolution could potentially take place. Ordinary thin sectioning and ultrasonic cleaning under water would be likely to obscure the presence of very small anhydrite crystals in ordinary rock samples.

The run products were thin sectioned, examined under the optical microscope and analysed for mineral and glass compositions on a Cameca Camebax electron microprobe. For mineral analyses, the accelerating voltage was 15 kV and the beam current 15 nA. For analyses of sulphur in glasses, we used an accelerating voltage of 15 kV, a beam current of 30 nA and long count times of 100 s, to compensate for the extremely low sulphur concentrations. For major elements in glasses, we used a beam current of 10 nA and corrected for sodium loss from the hydrous glass (Nielsen & Sigurdsson 1981; Devine et al. 1995).

Appendix 2: vapour composition model

Holloway (1987) described a method of calculating vapour compositions using equilibrium constants and mass balance. This method may be modified to apply to anhydrite-bearing systems. The simplest system is Ca–H–O–S, which completely describes the anhydrite–water vapour equilibrium if the activity of calcium can be fixed. Because of the low solubility of sulphur in all phases except anhydrite in the Pinatubo magma, this system is a reasonable approximation of the actual Pinatubo vapour phase. The controlling reactions for the phase compositions in this system are those for the formation of $CaSO_4$, H_2O, SO_2 and H_2S from the elements. The presence of pure anhydrite fixes the activity of $CaSO_4$ at 1, and T, P and f_{O_2} are varied as wished. For this study, data for the equilibrium constants of all components in this system were taken from Robie et al. (1978). Fugacity coefficients were calculated from the critical data after the method of Ryzhenko and Volkov (1971). We assumed ideal mixing of the gases, which should be reasonable for the water-rich vapour phase coexisting with the Pinatubo dacite.

The remaining unknown quantity is the activity of calcium in the system. There is unfortunately no easy way to determine a_{Ca} in the Pinatubo magma from the mineral compositions. The El Chichón trachyandesite, however, contains clinopyroxene, titanomagnetite, pyrrhotite and anhydrite, making it possible to calculate the activity of Ca in this magma. Luhr (1990) used the compositions of these phases to calculate a vapour phase composition for the trachyandesite. We have used this system both to calculate the activity of Ca in the magma and to study its dependence on parameters such as temperature, oxygen fugacity and melt composition. This suggested a reasonable value for the Pinatubo magma of $a_{Ca} = 10^{-26}$. In the model calculations, the activity of calcium is fixed at this value, making it possible to examine the variation in the sulphur concentration and speciation in the vapour with changing T and f_{O_2}.

7. References

Arculus, R. J., Johnson, R. W., Chappell B. W., McKee, C. O. & Sakai, H. 1983. Ophiolite-contaminated andesites, trachybasalts, and cognate inclusions of Mount Lamington, Papua New Guinea: anhydrite-amphibole-bearing lavas and the 1951 cumulodome. J VOLCANOL GEOTHERM RES 18, 215–47.

Baker, L. & Rutherford, M. J. 1992. Anhydrite breakdown as a possible source of excess sulfur in the 1991 Mount Pinatubo eruption. EOS, TRANS AM GEOPHYS UNION 73, 625.

Baker, L. & Rutherford, M. J. Sulfur diffusion in rhyolite melts. CONTRIB MINERAL PETROL 123, 335–44.

Bernard, A., Demaiffe, D., Mattielli, N. & Punongbayan, R. S. 1991. Anhydrite-bearing pumices from Mount Pinatubo: further evidence for the existence of sulphur-rich silicic magmas. NATURE 354, 139–40.

Bluth, G. J., Doiron, S. D., Schnetzler, C. C., Krueger, A. J. & Walter, L. S. 1992. Global tracking of the SO_2 clouds from the June, 1991 Mount Pinatubo eruptions. GEOPHYS RES LETT 19, 151–54.

Carroll, M. & Rutherford, M. J. 1985. Sulfide and sulfate saturation in hydrous silicate melts. Proceedings of 15th Lunar and Planetary Science Conference. J GEOPHYS RES 90, C601–12.

Carroll, M. & Rutherford, M. J. 1987. The stability of igneous anhydrite: experimental results and implications for sulfur behavior in the 1982 El Chichón trachyandesite and other evolved magmas. J PETROL 28, 781–801.

Carroll, M. & Rutherford, M. J. 1988. Sulfur speciation in hydrous experimental glasses of varying oxidation state: results from measured wavelength shifts of sulfur x-rays. AM MINERAL 73, 845–9.

Devine, J. D., Sigurdsson, H., Davis, A. N. & Self, S. 1984. Estimates of sulfur and chlorine yield to the atmosphere from volcanic eruptions and potential climate effects. J GEOPHYS RES 89, 6309–25.

Devine, J. D., Gardner, J. E., Brach, H. P., Layne, G. D. & Rutherford, M. J. 1995. Comparison of microanalytical methods for estimation of H_2O contents of silicic volcanic glasses. AM MINERAL 80, 319–28.

Drexler, J. W. & Munoz, J. L. 1985. Highly oxidized, pyrrhotite-anhydrite-bearing silicic magmas from the Julcani Ag–Cu–Bi–Pb–Au–W District, Peru: physiochemical conditions of a productive magma. In: Canadian Institute of Mining Conference on Granite-related Mineral Deposits, Extended Abstracts, 87–100.

Gerlach, T. M. & McGee, K. A. 1994. Total sulfur dioxide emissions and pre-eruption vapor-saturated magma at Mount St. Helens, 1980–1988. GEOPHYS RES LETT 21, 2833–6.

Gerlach, T. M., Westrich, H. R., Casavedall, T. J. & Finnegan, D. L. 1994. Vapor saturation and accumulation in magmas of the 1989–1990 eruption of Redoubt volcano, Alaska. J VOLCANOL GEOTHERM RES 62, 317–37.

Gerlach, T. M., Westrich, H. R. & Symonds, R. B. 1996. Pre-eruption vapor in magma of the climactic Mount Pinatubo eruption: source of the giant stratospheric sulfur dioxide cloud. In: Punongbayan, R. S. & Newhall, C. G. (eds) The 1991 eruptions of Pinatubo Volcano, Philippines, 415–33. The Philippine Institute of Volcanology & Seismology and The Unversity of Washington Press.

Geschwind, C. -H. & Rutherford, M. J. 1992. Cummingtonite and the evolution of the Mount St. Helens magma system: an experimental study. GEOLOGY 20, 1011–4.

Hattori, K. 1993. High-sulfur magma, a product of fluid discharge from underlying mafic magma: evidence from Mount Pinatubo, Philippines. GEOLOGY 21, 1083–6.

Haughton, D., Roeder, P. L. & Skinner, B. J. 1974. Solubility of sulfur in mafic magmas. ECON GEOL 69, 451–67.

Holloway, J. R. 1987. Igneous fluids. In: Carmichael, I. S. E. & Eugster, H. P. (eds) Thermodynamic modeling of geological materials: minerals, fluids and melts. REV MIN 17, 211–33.

Katsura, T. & Nagashima, S. 1974. Solubility of sulfur in some magmas at 1 atm pressure. GEOCHIM COSMOCHIM ACTA 38, 517–31.

Luhr, J. F. 1990. Experimental phase relations of water- and sulfur-saturated arc magmas and the 1982 eruptions of El Chichón volcano. J PETROL 31, 1071–114.

Luhr, J. F., Carmichael, I. S. E. & Varekamp, J. C. 1984. The 1982 eruptions of El Chichón volcano, Chiapas, Mexico: mineralogy and petrology of the anhydrite-bearing pumices. J VOLCANOL GEOTHERM RES 23, 69–108.

McKenzie, W. F. & Helgeson, H. C. 1985. Phase relations among silicates, copper iron sulfides, and aqueous solutions at magmatic temperatures. ECON GEOL 80, 1965–73.

McKibben, M. A. & Eldridge, C. S. 1995. Sulfur isotopic systematics of the June 1991 eruptions of Mount Pinatubo: a SHRIMP ion microprobe study. EOS, TRANS AM GEOPHYS UNION 74, 668.

Merzbacher, C. & Eggler, D. H. 1984. A magmatic geohydrometer: application to Mount St. Helens and other dacitic magmas. GEOLOGY 12, 587–90.

Nagashima, S. & Katsura, T. 1973. The solubility of sulfur in $Na_2O–SiO_2$ melts under various oxygen partial pressures at 1100, 1250 and 1300°C. BULL CHEM SOC JPN 46, 3099–103.

Nielsen, C. H. & Sigurdsson, H. 1981. Quantitative methods of electron microprobe analysis of sodium in natural and synthetic glasses. AM MINERAL 66, 547–52.

Nilsson, K. & Peach, C. L. 1993. Sulfur speciation, oxidation state, and sulfur concentration in backarc magmas. GEOCHIM COSMOCHIM ACTA 57, 3807–13.

Pallister, J. S., Hoblitt, R. P. & Reyes, A. G. 1992. A basalt trigger for the 1991 eruptions of Pinatubo volcano? NATURE 356, 426–8.

Pallister, J. S., Meeker, G. P. & Luhr, J. F. 1995. Recognizing ancient sulfur-rich eruptions: lessons from Pinatubo, El Chichón, and Mount St. Helens. IUGG XXI GENERAL ASSEMBLY, ABSTR A279.

Pallister, J. S., Hoblitt, R. P., Meeker, G. P., Knight, R. J. & Sierns, D. F. 1996. Magma mixing at Mount Pinatubo: petrographic and chemical evidence from the 1991 deposits. In: Punongbayan, R. S. & Newhall, C. G. (eds) The 1991 eruptions of Pinatubo Volcano, Philippines, 687–731. The Philippine Institute of Volcanology & Seismology and The University of Washington Press.

Robie, R. A., Hemingway, B. S. & Fisher, J. R. 1978. Thermodynamic properties of minerals and related substances at 298·15 K and 1 bar (10⁵ Pascals) pressure and at higher temperatures. US GEOL SURV SPEC PAP 1452.

Rouse, R. C. & Dunn, P. J. 1982. A contribution to the crystal chemistry of ellestadite and the silicate sulfate apatites. AM MINERAL 67, 90–6.

Rutherford, M. J. & Devine, J. D. 1996. Pre-eruption pressure–temperature conditions and volatiles in the 1991 dacitic magma of Mount Pinatubo. In: Punongbayan, R. S. & Newhall, C. G. (eds) The 1991 eruptions of Pinatubo Volcano, Philippines, 751–61. The Philippine Institute of Volcanology & Seismology and The University of Washington Press.

Rutherford, M. J. & Hill, P. 1993. Magma ascent rates from amphibole breakdown: an experimental study applied to the 1980–1986 Mount St. Helens eruptions. J GEOPHYS RES 98, 19,667–85.

Rye, R. O., Luhr, J. F. & Wasserman, M. D. 1984. Sulfur and oxygen isotopic systematics of the 1982 eruptions of El Chichón volcano, Chiapas, Mexico. J VOLCANOC GEOTHERM RES 23, 109–23.

Ryzhenko, B. N. & Volkov, V. P. 1971. Fugacity coefficients of some gases in a broad range of temperatures and pressures. GEOCHEM INT 8, 468–81.

Sigurdsson, H. 1990. Assessment of the atmospheric impact of volcanic eruptions. GEOL SOC AM SPEC PAP 247, 99–110.

Varekamp, J. C., Luhr, J. F. & Prestegaard, K. L. 1984. The 1982 eruptions of El Chichón volcano (Chiapas, Mexico): character of the eruptions, ash-fall deposits and gasphase. J VOLCANOL GEOTHERM RES 23, 39–68.

Wallace, P. J. & Carmichael, I. S. E. 1992. Sulfur in basaltic magmas. GEOCHIM COSMOCHIM ACTA 56, 1863–74.

Westrich, H. R. & Gerlach, T. M. 1992. Magmatic gas source for the stratospheric SO_2 cloud from the June 15, 1991 eruption of Pinatubo volcano. GEOLOGY 20, 867–70.

LESLIE L. BAKER and MALCOLM J. RUTHERFORD, Department of Geological Sciences, Brown University, Providence, RI 02912, U.S.A.

Transactions of the Royal Society of Edinburgh: Earth Sciences, **87,** 251–259, 1996

Hydrodynamics of magmatic and meteoric fluids in the vicinity of granitic intrusions

R. Brooks Hanson

ABSTRACT: Numerical models that account for fluid flow, magmatic and metamorphic fluid production, topography and thermal expansion of the fluid following emplacement of a granitic magma in the upper crust reveal controls on the distribution of magmatic fluids during the evolution of a hydrothermal system. Initially, fluid pressures are close to lithostatic in and near an intrusion, and internally generated magmatic and metamorphic fluids are expelled. Later, fluid pressures drop to hydrostatic values and meteoric fluids circulate throughout the system. High permeabilities and low rates of fluid production accelerate this transition. Fluid production in the magma and wallrocks is the dominant mechanism elevating fluid pressures to lithostatic values. For granitic intrusions, about three to five times as much magmatic fluid is produced as metamorphic fluid. Continuous fluid release from a granitic magma with a vertical dimensions of 10 km produces a dynamic permeability of up to several tens of microdarcies.

Near the surface, topography associated with a typical volcano acts to maintain a shallow meteoric flow system and drive fluids laterally. The exponential decay with depth of the influence of topography on fluid pressures results in a persistent zone of mixing at a depth of 1–2 km between these meteoric fluids and magmatic fluids despite variations in the strength of the magmatic hydrothermal system. However, in shallow systems where fluid release is episodic, dramatic changes in the region of mixing are still possible because fluid pressure is sensitive to variations in the rates of fluid production. At depth, high rates of metamorphic fluid production in the wallrocks and low permeabilities (<1 μD) produce elevated fluid pressures, which hinder the lateral flow of magmatic fluids. Together, these patterns are consistent with the distribution and evolution of skarns and hydrothermal ore deposits around granitic magmas.

KEY WORDS: hydrodynamics, magmatic fluids, meteoric fluids, granitic intrusions, numerical models, skarns, hydrothermal ore deposits.

Observations of ore deposits, particularly porphyry copper deposits, and active and ancient magmatic hydrothermal systems have led to a general heuristic model of hydrothermal flow in and around granitic magmas (Burnham 1979; Henley & McNabb 1978; Taylor 1990; Hedenquist & Lowenstern 1994). In this model, a cupola forms at the boundary between regimes where magmatic and meteoric fluids dominate. Inside the cupola, fluids are magmatic and fluid pressures are high, at near-lithostatic values. Outside of it, meteoric fluids circulate at lower temperatures and fluid pressures are near hydrostatic values. The cupola collapses inwards as the magma cools. Episodic release of fluids from the magma produces overlapping vein relations. Hydrofracturing is concentrated in the receding cupola. Mixing of the magmatic and meteoric fluids and fluid–rock reactions produce ore deposits and extensive hydrothermal alteration (Hedenquist & Lowenstern 1994). Although this general model is representative, a wide variation in vein styles and abundances and extent and type of mineralisation and alteration are observed among ancient magmatic hydrothermal systems and hydrothermal ore deposits, and the factors leading to this diversity are not clear.

Numerical models of magmatic hydrothermal systems are useful for examining the sensitivity of this general pattern to the variation of many parameters. However, most numerical models have not mimicked even this general evolution of the magmatic hydrothermal system well. One reason is that most models have not included magmatic fluids, a key component of the system, a main carrier of many ore metals and a critical factor in generating lithostatic fluid pressures and producing hydrofractures. Many other processes important in hydrother-

mal systems have also often been neglected, such as deformation, heterogeneous permeability and the evolution of permeability with time.

In this paper, a series of numerical models is presented illustrating controls on the variation and evolution of granitic hydrothermal systems that specifically include the generation of magmatic fluids and the effects of topography above shallow intrusions (Sammel *et al.* 1988; Birch 1989; Forster & Smith, 1990). Earlier work has shown that permeability, which varies over many orders of magnitude in crustal rocks, exerts a first-order control on flow patterns (see Norton 1982 and Johnson & Norton 1985 for dramatic examples). However, the evolution and distribution of permeability in wallrocks and crystallising magma are uncertain, and it is difficult to model permeability in a non-*ad hoc* way (rather than, say, proscribing it). Here, the focus is on understanding the effects of the main processes driving fluid flow following the emplacement of a granitic magma; these effects will be superimposed on those due to heterogeneous or variable permeability. The models are similar to those used earlier to examine hydrothermal flow during contact metamorphism generally (Hanson 1995). General aspects of the duration of magmatic hydrothermal systems have been given by a number of workers (Norton & Cathles 1979; Cathles 1981; Furlong *et al.* 1991).

1. Magmatic fluids in granitic hydrothermal systems

A granitic magma has enough energy to drive roughly its mass in meteoric fluid circulation (Norton & Cathles 1979;

Cathles 1981). Meteoric fluids would thus seem to dominate the magmatic fluid component of even up to 10 wt% or so for some granitic magmas (Burnham 1979; Gaetani *et al.* 1993; Johnson *et al.* 1994) except when the permeability is so low that the meteoric hydrothermal flow is limited. However, several factors act to focus magmatic fluids in parts of the hydrothermal system and during certain times in the evolution of the system, and as a result magmatic fluids can locally dominate over meteoric fluids and should not be ignored. Magmatic fluids are released only while the intrusion is crystallising, approximately one-fifth of the total duration of the hydrothermal system (cf. Cathles 1981; Furlong *et al.* 1991; Hanson 1995). In some young active systems, magmatic fluids are seen escaping regularly at the surface (e.g. Hedenquist & Lowenstern 1994) and thus are the locally dominating fluid. Fractionation of magmatic fluids to the upper part of the magma chamber can focus the release of fluids in a small region of the crust compared with the full extent of the hydrothermal system. The presence of significant topography of several kilometres above a shallow intrusion may further focus the discharge of magmatic fluids (Forster & Smith 1990).

One of the main uncertainties in modelling the hydrothermal evolution of granitic systems is our incomplete knowledge about how, when and where magmatic fluids are released from a crystallising magma (see, for example, Candela 1991). In many systems, a large amount of fluid was evidently released in a few distinct events presumably towards the top of the magma chamber (e.g. Burnham 1979). Such behaviour is difficult to include in numerical models in a non-*ad hoc* way. Monitoring of active systems has shown, on the other hand, that at least over years to decades some fluids are continuously being released from cooling magmas (Symonds *et al.* 1994). The permeability of magmas and its effect on the lateral flow of fluids out of the magma as margins of the magma crystallise are also not known.

2. Models

Hydrothermal systems associated with the emplacement of a granitic magma were simulated using an explicit finite-difference model that solves for the conservation of mass, momentum and energy of the rock–fluid system (see Hanson 1995 for details). During each time step the numerical model first solves the transient conservation of energy equation. Cooling or heating rates are then used to infer the rates of magmatic or metamorphic fluid production or rates of expansion or contraction of pore fluids and to obtain initial estimates of fluid density and viscosity. The combined conservation of mass and momentum equations are then solved by iteration, using these values, to obtain the fluid pressure at each grid point. Fluid pressure and viscosity are recalculated during the iteration. The fluid fluxes obtained are then used to account for advective heat flow in the next time step. Porosity, permeability and thermal properties (specific heat, thermal conductivity and so forth) are specified for each cell. When fluid pressure locally exceeds the lithostatic value, permeability (and porosity) is increased to the degree that is necessary to return fluid pressure to the local lithostatic value (see Hanson 1995).

All fluids in the model are assumed to be pure H_2O. This assumption has the effect of maximising the effect of thermal buoyancy in driving fluid flow because the increase in mineral solubilities with pressure and temperature will tend to increase the fluid density and thus decrease fluid buoyancy. The model also does not account for boiling or critical behaviour (the equation of state of H_2O is smoothed through the critical region). Together, these assumptions preclude the examination of the effect of brine or phase separation on the hydrothermal system. Phase separation is thought to occur at shallow depths above an intrusion (Hedenquist & Lowenstern 1994), but modelling of boiling and phase separation still faces considerable challenges (Lowell 1991; Ingebritsen & Hayba 1994). The critical region is reached in some models at shallow depths, but the presence of separate phases is assumed not to affect flow in these models. Further work is needed to evaluate this effect.

Magmatic fluids in the model are released between the emplacement temperature and the solidus at each cell within the intrusion. The rate of release at each cell is in proportion to the amount of cooling. This approach is different from that in Hanson (1995), where the fluids in any interior column of magma were released at the cell above the magma where fracture permeability was developed, rather than at each cell throughout the column of magma. Thus it was assumed that all of the fluids fractionated upwards within the magma. The two approaches essentially represent different notions of how fluids are released from a magma. The assumption here is essentially of no fractionation or vertical pooling of fluid-rich magma, such that the entire magma is saturated in fluid. The assumption in Hanson (1995), where the focus was on the overall interaction of different sources in affecting fluid flow, is essentially that except along the original side of the intrusion, all of the fluid pools and is released at the top of the magma. A series of *ad hoc* models exists between these end-members. It is assumed here that fluids are released smoothly as the magma cools, not in discrete bursts. The models are useful for predicting the overall controls on the distribution of fluids and areas of mixing of fluids; specific applications are not intended.

2.1. Reference model

To focus on the effects of key parameters on flow patterns–namely, permeability distribution, topography, amount of magmatic fluid production and the presence of metamorphic devolatilisation—some parameters are kept constant in all models. All models began with the emplacement of a granitic magma at a temperature of 800°C. The magma was assumed to crystallise to 650°C, releasing $1 \cdot 7 \times 10^5$ J kg^{-1} of latent heat (Hanson & Barton 1989). The thermal diffusivity of rocks (wallrocks and crystallised magma) was $0 \cdot 92 \times 10^{-6}$ m^2 s^{-1}. The thermal diffusivity of the magma or wallrocks was increased to account for the release of latent heat during crystallisation or heat consumed by metamorphic reactions, respectively (see Hanson & Barton 1989).

In a first series of models, a system with a granitic intrusion having a half-width of 4 km and emplaced at a depth of 6 km was considered. The magma released 5% H_2O by weight during crystallisation. Magmatic fluids were released linearly with cooling in each cell of the intrusion (see earlier) and there was no metamorphic fluid. The upper 6 km of the crust was assigned a permeability of $0 \cdot 1$ mD. Numerical models of several porphyry copper systems have suggested that permeabilities of about 1 mD (1 D is 10^{-18} m^2) are appropriate as average values, although the evolution of permeability with time is not clear (e.g. Villas & Norton 1977; Norton 1982; Johnson & Norton 1985). At a permeability of $0 \cdot 1$ mD, the total amount of magmatic fluids released represents a significant component of the fluid budget compared with meteoric fluids that can be driven by the energy of the intrusion (see Norton 1982). The permeability of the deeper crust was 1 μD, which allows the brief generation of lithostatic fluid pressures. The intrusion crystallised in $\approx 200\,000$ a and cooled to near-background in about 1 Ma.

The simulations focus on the evolution of the system while

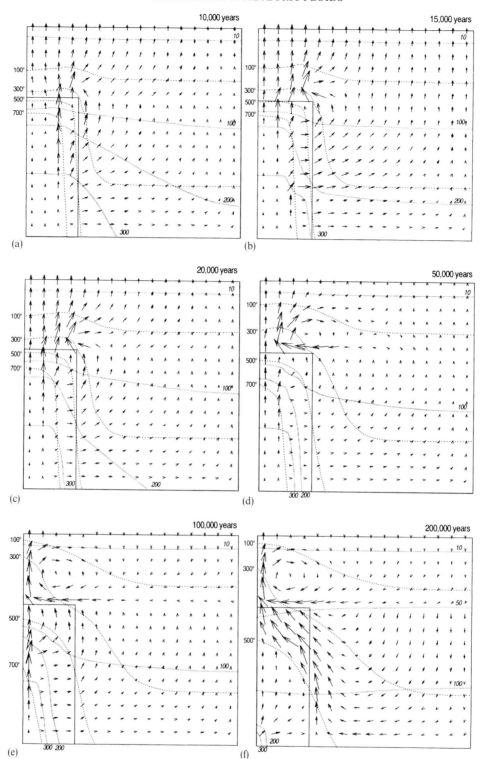

Figure 1 Numerical model of intrusion of a granitic magma into anhydrous wallrocks. Grid spacing is 1 km. Intrusion at 800°C is emplaced at a depth of 5·5 km into wallrocks with a permeability of 1 µD. Crust above the intrusion has a permeability of 100 µD. The base of the model has a constant heat flux and is impermeable; the right-hand side is impermeable and a natural boundary for temperature; the left-hand side is a symmetrical boundary. The top is permeable and held at 0°C. The initial temperature gradient in the wallrocks is 25°C km^{-1}. The water content of the intrusion is 5% by weight; fluid is released linearly with temperature change to 650°C. The intrusion is assumed to fracture and have wallrock permeability at temperatures less than 500°C and to have a permeability of 0·1 µD between 650 and 500°C. The permeability of the unfractured intrusion is 0·1 µD. (A) Fluid flux vectors, temperatures (dotted lines, °C) and fluid pressures (heavy lines, MPa) 10 000 a after emplacement. Maximum flux is 6×10^{-7} kg m^{-2} s^{-1}. (B) After 15 000 a; maximum flux is 5×10^{-7} kg m^{-2} s^{-1}. (C) After 20 000 a; maximum flux is 5×10^{-7} kg m^{-2} s^{-1}. (D) After 50 000 a; maximum flux is 6×10^{-7} kg m^{-2} s^{-1}. (E) After 100 000 a; maximum flux is 1×10^{-6} kg m^{-2} s^{-1}. (F) After 200 000 a; maximum flux is 7×10^{-7} kg m^{-2} s^{-1} in the upper crust and 3×10^{-8} kg m^{-2} s^{-1} in the lower crust (re-scaled for visibility). The intrusion has essentially crystallised. Thermal diffusivity 0.92×10^{-6} m^2 s^{-1}.

magmatic volatiles are being produced. This is the time when many high-temperature ore deposits and skarns are produced. Once a magma has crystallised, the flow is controlled by metamorphic fluid production and thermal buoyancy, and these flow patterns have been well described and are reasonably well predicted (Norton & Knight 1977; Furlong *et al.* 1991; Hanson 1992).

Initially, the release of magmatic fluids coupled with low permeability increased fluid pressures noticeably near the intrusion and slightly near the surface (Fig. 1). Even this slight increase in fluid pressures and influx of fluids from below was sufficient to overwhelm and inhibit any tendency for fluid circulation driven by thermal buoyancy. As a result, all fluids were expelled. The large proportion of the magmatic fluids flowing out from the sides of the intrusion also inhibited the development of a shallow circulation system by preventing the formation of a focused, hot hydrothermal plume. By 100 000 a, a shallow meteoric circulation system developed and meteoric fluids began to mix with magmatic fluids above the intrusion. The timing of this transition depended primarily on permeability, but was also sensitive to the width of the model domain. At permeabilities of about 1 mD or higher, the impact of the magmatic fluids on the shallow hydrothermal system became insignificant, regardless of the flow pattern at depth, and a shallow circulation system developed essentially immediately. Advective transport of heat dominated at shallow depths; conduction dominated at depth. Fluid pressures remained elevated to near-lithostatic values at depth near and within the intrusion, but fell rapidly to near hydrostatic values at shallow depths and in the wallrocks.

For this simple model, the time-integrated flux of magmatic volatiles above the intrusion was of the order of 10^5 kg m^{-3} after the first 50 000 a, and higher, 10^6 kg m^{-3} after 200 000 a (Fig. 2), when the intrusion had just crystallised. Rocks above the intrusion as well as nearby wall rocks interacted with only magmatic fluids for the first 100 000 years or so (Fig. 3). The

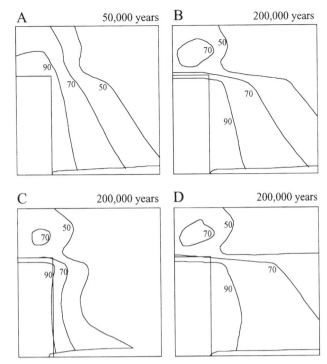

Figure 3 Proportion of magmatic fluids in per cent of time-integrated fluid flux (Fig. 3) in models in Figure 1 (A and B) and Figure 4 (C). Model in Figure 4 differs from the model in Figure 1 by including the metamorphic devolatilisation in wallrocks with a permeability of 0·1 µD. Note that lateral extent of interaction of magmatic fluids with the wallrocks is limited by metamorphic fluid production and low permeability. As a result, magmatic fluids are focused upwards. (D) Model as in Figure 1 except most (70%) of the fluids are fractionated into residual magma until the fluid content is 8% by weight, then released normally. Note that there is less interaction with meteoric waters in the lower crust. This occurs because the enhanced production of magmatic fluids later during cooling inhibits the full development of the meteoric hydrothermal system.

continued flow of meteoric fluids throughout the lifetime of the system, however, resulted in a dominance of meteoric fluids in the time-integrated fluid budget at shallow depths (see Hanson 1995).

Comparison of the results of models in which a few processes were varied shows that several mechanisms control the distribution of magmatic fluids, both at depth in the wallrocks and near the surface, and that a wide variety of flow patterns is possible even with homogeneous permeability. The fractionation of fluids into the residual magma in this model did not reduce the interaction of magmatic fluids with the wallrocks significantly (Fig. 3). This result differs from that in Hanson (1995) because of the way magmatic fluids are released (see earlier). In the models here, fluids are produced in the interior of the intrusion rather than concentrated near the top and can thus flow uninhibited out the intrusion sides.

The importance of metamorphic devolatilisation in affecting the distribution of magmatic fluids depends greatly on permeability (Hanson 1992). In models where wallrock permeabilities were less than about 1 µD, even a small amount (1% by weight total released from 300 to 700°C) of metamorphic fluid production raised the fluid pressures in wallrocks sufficiently to reduce the lateral flow of magmatic fluids at depth (Fig. 4). At permeabilities of 0·1 µD, the production of metamorphic fluids in proportion to the heating rate of rocks elevated the fluid pressures briefly to lithostatic values and fluid pressures were increased above hydrostatic values for several hundred thousand years. The effect of the added fluid source in the wallrocks was to focus magmatic fluids upwards. Thus metamorphic fluid production in turn

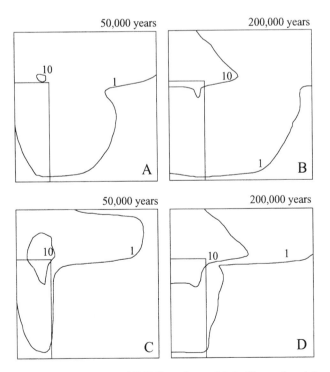

Figure 2 Time-integrated fluid fluxes for models in Figures 1 and 4. Values are in units of 10^6 kg m^{-2}. (A) Model in Figure 1 after 50 000 a. (B) Model in Figure 1 after 200 000 a. (C) Model in Fig. 4 after 50 000 a. (D) Model in Figure 4 after 200 000 a. Note that the lateral flow of magmatic fluids is much reduced in the model with metamorphic fluid production and lower permeability.

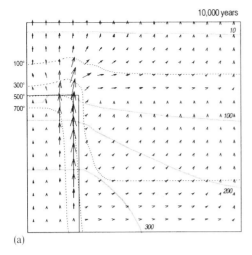

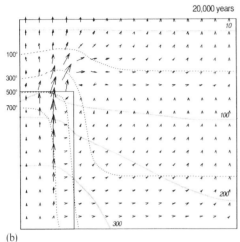

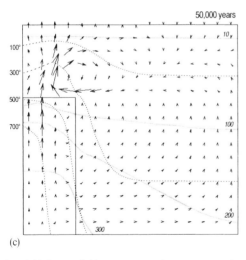

Figure 4 Fluid fluxes, fluid pressures and temperatures in a model as in Figure 1, except that here the production of metamorphic fluids is included and the wallrock permeability in the model is 0·1 so that fluid pressures are elevated in the wallrocks. (A) Flux vectors, temperatures (dotted lines, °C) and fluid pressures (heavy lines, MPa) 10 000 a after emplacement. Maximum flux is 1×10^{-6} kg m^{-2} s^{-1}. (B) After 15 000 a; maximum flux is 1×10^{-6} kg m^{-2} s^{-1}. (C) After 20 000 a; maximum flux is 1×10^{-6} kg m^{-2} s^{-1}. (D) After 50 000 a; maximum flux is 6×10^{-7} kg m^{-2} s^{-1}. (E) After 100 000 a; maximum flux is 1×10^{-6} kg m^{-2} s^{-1}. (F) After 200 000 a; maximum flux is 7×10^{-7} kg m^{-2} s^{-1} in the upper crust and 3×10^{-8} kg m^{-2} s^{-1} in the lower crust (re-scaled for visibility).

accelerated the development of the shallow hydrothermal circulation and enhanced lateral temperature gradients near the surface by promoting the formation of a coherent plume of magmatic fluids.

Permeabilities within the intrusion required to account for the equilibrium rate of magmatic fluid production (that is, produced by the release of fluids proportional to the cooling rate of the magma) were of the order of 1 to 10 µD (Fig. 5). Such permeabilities could be produced by relatively few narrow veins (e.g. 0·01 mm wide veins 1 m apart; Norton & Knapp 1977). This value is clearly a minimum because any sudden release of fluids would require larger, more closely spaced veins and because it represents the instantaneous vein width, not the total amount or thickness of veins that might be produced during degassing of the magma. This permeability is, however, about an order of magnitude higher than that produced by metamorphic dehydration in the aureole (Hanson 1995).

2.2. Epithermal intrusions and topography

In these models, and those in Hanson (1995), large intrusions were emplaced at mesozonal depths, below highly permeable crust. Many ore deposits seem to be associated with smaller epithermal stocks, emplaced in permeable crust. Some of these stocks may be intruded within a volcano. The effect of topography on flow must be considered in these systems (for example, Sammel et al. 1988; Forster & Smith 1990). The potential effects and other aspects of flow associated with shallow intrusions are illustrated in two models (Figs 6, 7), one with and one without topography. The granite intrusion in these models had a half-width of 1·5 km and was emplaced at a depth of 3·25 km below the surface. The grid spacing was 0·5 km. In the model with topography, the boundary condition at the surface of the model reflected a hydraulic head associated with an elevation rise of 1·5 km over 2·5 km (30° slope). The crust had a permeability of 100 µD. No metamorphic fluid production was included, as would be the case for the common situation of the intrusion of a magma into coeval, essentially anhydrous volcanic rocks. This model is similar to that described by Forster and Smith (1990) except that it includes magmatic fluid production and the effects of transient thermal expansion of the fluid, and the intrusion is not permeable at high temperatures.

In the model without topography (Fig. 6), as for the mesozonal intrusion in Figure 1, the initial production of magmatic fluids drives fluids away from the intrusion. The flow of magmatic fluids is dispersed over a wide area. There is no one area, over time, where meteoric and magmatic fluids consistently mix (compare Figs 3 and 8). Fluid pressures are high within the intrusion because of the production of magmatic fluids, but near hydrostatic values elsewhere. With time, hydrothermal circulation develops and flow is concentrated along the margin of the intrusion.

In the model with topography (Fig. 7), initially the production of fluids from the granite overwhelms the head associated with the topography and magmatic fluids are expelled everywhere, including from areas of high topography. Within a few thousand years, however, the expulsion of magmatic fluids is focused near the base of the topography by the topographically driven flow of meteoric fluids. Furthermore, this pattern persists throughout the time of crystallisation of the intrusion; such a pattern is also evident in models that did not consider magmatic fluid production (Birch 1989; Forster & Smith 1990). Thus there is a persistent region where magmatic and meteoric fluids mix (Fig. 8), even with uniform permeability. This region is stable, despite the variation in the strength of the magmatic fluid flow as the magma crystallises and cools, because the topographic forcing decays exponentially with depth and has a minimal effect below the wavelength of the topography (see, for example, Fehn et al. 1978; Forster & Smith 1990). Thus it takes great

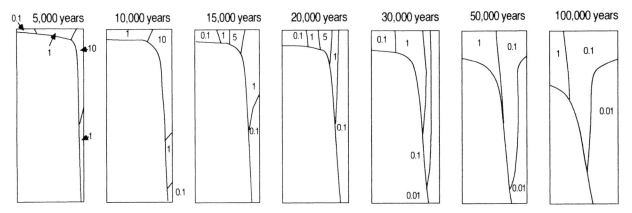

Figure 5 Permeabilities in the crystallised roof and walls of the intrusion in Figure 4 produced by hydrofracturing of the magma to allow the release of magmatic fluids while maintaining an lithostatic fluid pressure gradient. Values are in microdarcies. The 650°C isotherm (solidus) is shown. Maximum permeability shifts to above the core of the intrusion with time and the maximum value decreases in this model because the rate of cooling and thus generation of fluids decreases. Permeabilities would increase with time if most of the fluid was fractionated into the residual magma.

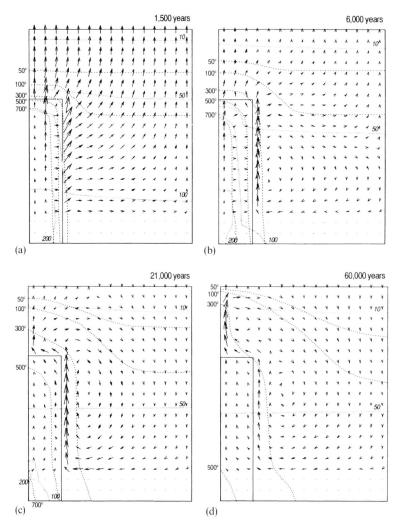

Figure 6 Model of an epithermal intrusion emplaced at a depth of 3·25 km. Intrusion width is 1·5 km; grid spacing is 0·5 km. No metamorphic fluid production. Crust to a depth of 8 km has a permeability of 100 μD; crust is impermeable at deeper levels. Intrusion, as in Figure 1, is emplaced at 800°C and has 5% water by weight, which is released linearly in each cell to 650°C. (A) After 1500 a. Maximum fluid flux is 2×10^{-6} kg m^{-2} s^{-1}. (B) After 6000 a; maximum flux is 2×10^{-6} kg m^{-2} s^{-1}. (C) After 21 000 a; maximum fluid flux is 2×10^{-6} kg m^{-2} s^{-1}; intrusion has just crystallised. (D) After 60 000 a; maximum fluid flux is 1×10^{-6} kg m^{-2} s^{-1}.

changes in the buoyancy or vigour of the flow at depth to alter the position of the interface with the topographically driven flow. As pointed out by these earlier workers, the precipitation of minerals as a result of the mixing of

compositionally distinct fluids might further stabilise this region. Other features of this model, such as the pronounced flow along the intrusion margin, are consistent with the model without topography.

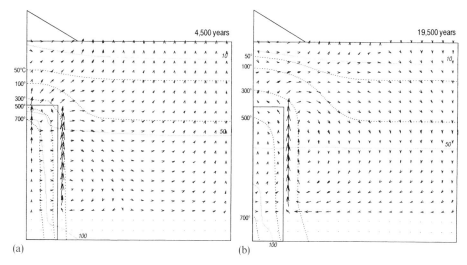

Figure 7 Model as in Figure 6, but with inclusion of topography equivalent to that for a small volcano extending 1.5 km above mean elevation above the intrusion with a $30°$ slope. Boundary condition is adjusted in the region of topography to account for increased head. (A) After 4500 a. Maximum fluid flux is $2 \times 10^{-6} \mathrm{~kg~m}^{-2} \mathrm{~s}^{-1}$. (B) After 19 500 a; maximum fluid flux is $2 \times 10^{-6} \mathrm{~kg~m}^{-2} \mathrm{~s}^{-1}$. (C) After 19 500 a; maximum fluid flux is $2 \times 10^{-6} \mathrm{~kg~m}^{-2} \mathrm{~s}^{-1}$; intrusion has just crystallised.

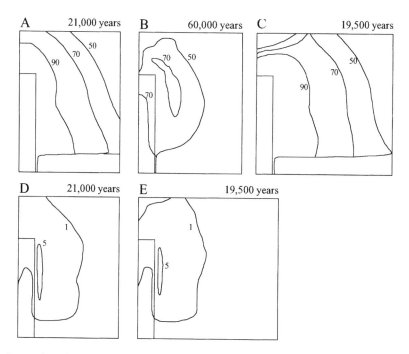

Figure 8 Proportion of magmatic fluid in time-integrated fluid flux (A to C; in per cent) and time-integrated fluid flux (D and E) for models in Figures 6 (A, B, and D) and 7 (for C and E). Time-integrated fluid flux is in units of $10^6 \mathrm{~kg~m}^{-2}$. Note strong gradient in contribution of magmatic fluids to time-integrated fluid flux in (C) beneath the topography.

3. Implications for controls on ore genesis and metasomatism

The overall evolution of the hydrothermal flow in these models is indeed analogous to the general heuristic models of hydrothermal ore deposits (for example, Burnham 1979), but the models show that the observed variability might be systematically related to several causes. Variability in permeability, which was not considered in depth here, could account for significant diversity. Fluids will obviously be focused into permeable rocks, such as along fault zones (see Norton & Knight 1977 for one example of such a model). Similarly, the temporal evolution of permeability, resulting from the precipitation or solution of minerals or thermal cracking, will affect the types of interactions. Some of these effects were illustrated in similar models in Hanson (1995).

Obviously, increasing permeability by an order of magnitude or more will correspondingly decrease both the overall importance of magmatic fluids in the hydrothermal system and the time over which magmatic fluids may dominate.

Despite the significance of permeability, the results in these models show that much variability can be produced solely by reasonable interactions of the mechanisms driving fluid flow. These models imply that significant lateral interaction of magmatic fluids with wallrocks at depth is promoted by a wallrock permeability sufficiently high to limit the importance of metamorphic fluid production in generating anomalous fluid pressures ($>1 \mu D$), but sufficiently low to prevent the rapid development of a meteoric hydrothermal system ($<100 \mu D$). Because conditions of low rates of metamorphic devolatilisation and high permeability are unusual below depths of several kilometres, the outward flow of magmatic fluids probably decreases with depth. Interestingly, several prominent contact metamorphic systems where the large lateral transport of magmatic fluids is evident or has been inferred, Alta and Notch Peak (Cook & Bowman 1994; Nabelek & Labotka 1993), were at depths of 5–8 km during metamorphism and wallrock permeabilities there are inferred to have been between 1 and 100 μD. In these systems, the presence of horizontal bedding or faults probably also favoured the lateral flow of fluids. Comparison of the results with the models in Hanson (1995) suggest that the early release of fluids from the magma (that is, not all of the fluids produced by crystallisation on the margins are fractionated into the magma) might also be important in promoting the lateral flow of fluids (Fig. 3).

At shallow depths, the extent of lateral flow will roughly mimic the scale of the topography. Buoyancy-driven circulation of meteoric fluids can also produce a plume of magmatic fluids outwards from the top of the intrusion where there is an impermeable horizon at shallow depths (Parmentier & Schedl 1981).

The geometry of the hydrothermal circulation, superposed on the pattern of expulsion of magmatic fluids, can produce large variations in the time-integrated fluxes of magmatic fluids, of an order of magnitude or more, even with a uniform permeability (Figs 2, 8). Areas of relict interaction with magmatic fluids are preserved in the centres of fluid circulation systems because the relative fluxes of meteoric fluids are less there, even though the distribution of magmatic fluids was more uniform initially. Thus observations of areas of interaction of rocks with magmatic fluids in ore deposits may not necessarily indicate where most of the magmatic fluids flowed, only the areas of fortunate preservation.

Shallow systems in particular will experience abrupt changes in hydrological regime. One reason is that metamorphic devolatilisation is likely to be minimal at shallow depths. Compared with magmatic devolatilisation, metamorphic devolatilisation probably proceeds more or less continuously at the scale of the aureole, because of the wide range of temperatures and rock compositions in the aureole, and because many metamorphic reactions are continuous reactions. Thus the relatively continuous devolatilisation tends to stabilise fluid pressures in the aureole. At shallow depths near granites, however, the main mechanism elevating fluid pressures is magmatic devolatilisation. Episodic degassing, hydrofracturing and cracking could therefore lead to abrupt changes in flow regime. One factor stabilising the flow would seem to be topography, however, as similar patterns are observed in models with topography, but with or without magmatic fluid production.

Hydrothermal flow occurs throughout the duration of cooling of the system and thus while temperatures are changing greatly in the wallrocks. In general, magmatic fluids are interacting with rocks on the prograde path and much of the meteoric hydrothermal circulation is occurring on the retrograde path. Only a fraction of the flow is associated with peak temperatures at any one place (see Hanson 1995). As pointed out by Podladchikov and Wickham (1994) the nature of interaction of magmatic fluids with wallrocks depends critically on the timing of release of the fluids and distance from the contact because of the time-transgressive nature of heating in the wallrocks. Magmatic fluids released early are likely to interact with all wallrocks on the prograde heating path; fluids released later may interact with rocks in the inner aureole that are cooling from peak temperatures.

4. Acknowledgements

I thank B. Dutrow and L. Baumgartner for reviews, and T. Drake for assistance with programming.

5. References

Barton, M. D., Staude, J.-M., Snow, E. A. & Johnson, D. A. 1991. Aureole systematics. In Kerrick, D. M. (ed.) Contact Metamorphism. MINERAL SOC AM. REV MINERAL 26, 723–847.
Birch, M. U. 1989. Groundwater flow systems and thermal regimes near cooling igneous plutons: influence of surface topography. M.S. Thesis. Utah State University, Logan.
Burnham, C. W. 1979. Magmas and hydrothermal fluids. In Barnes, H. L. (ed.) Geochemistry of hydrothermal ore deposits, 2nd edn, 71–136. New York: Wiley.
Candela, A. 1991. Physics of aqueous phase evolution in plutonic environments. AM MINERAL 76, 1081–91.
Cathles, L. M. 1981. Fluid flow and genesis of hydrothermal ore deposits. ECON GEOL 75th ANNIV VOL 424–57.
Cook, S. J. & Bowman, J. R. 1994. Contact metamorphism surrounding the Alta stock: thermal constraints and evidence of advective heat transport from calcite + dolomite geothermometry. AM MINERAL 79, 513–25.
Fehn, U., Cathles, L. M. & Holland, H. D. 1978. Hydrothermal convection and uranium deposits in abnormally radioactive plutons. ECON GEOL 73, 1556–66.
Forster, C. & Smith, L. 1990. Fluid flow in tectonic regimes. In Nesbitt, B. E. (ed.) Crustal fluids. MINERAL ASSOC CAN SHORT COURSE HANDB 18, 1–47.
Furlong, K. P., Hanson, R. B. & Bowers, J. R. 1991. Modeling thermal regimes. In Kerrick, D. M. (ed.) Contact metamorphism. MINERAL SOC AM REV MINERAL 26, 437–506.
Gaetani, G. A., Grove, T. L. & Bryan, W. B. 1993. The influence of water on the petrogenesis of subduction-related igneous rocks. NATURE 365, 332–4.
Hanson, R. B. 1992. Effects of fluid production on fluid flow during regional and contact metamorphism. J METAMORPH GEOL 10, 87–97.
Hanson, R. B. 1995. The hydrodynamics of contact metamorphism. GEOL SOC AM BULL 107, 595–611.
Hanson, R. B. & Barton, M. D. 1989. Thermal development of low-pressure metamorphic belts: results from two-dimensional numerical models. J GEOPHYS RES 94, 10,363–77.
Hedenquist, J. W. & Lowenstern, J. B. 1994. The role of magmas in the formation of hydrothermal ore deposits. NATURE 370, 519–27.
Henley, R. W. & McNabb, A. 1978. Magmatic vapor plumes and ground water interaction in porphyry copper emplacement. ECON GEOL 74, 1600–12.
Ingebritsen, S. E. & Hayba, D. O. 1994. Fluid flow and heat transport near the critical point of H_2O. GEOPHYS RES LETT 21, 2199–292.
Johnson, J. W. & Norton, D. 1985. Theoretical prediction of hydrothermal conditions and chemical equilibria during skarn formation in porphyry copper systems. ECON GEOL 80, 1797–823.
Johnson, M. C., Anderson, A. T. Jr & Rutherford, M. J. 1994. Pre-eruptive volatile contents of magmas. In Carroll, M. R. & Holloway, J. R. (eds) Volatiles in Magmas. MINERAL SOC AM REV MINERAL 30, 281–330.
Lowell, R. P. 1991. Modeling continental and submarine hydrothermal systems. REV GEOPHYS 29, 457–76.
Nabelek, P. I. & Labotka, T. C. 1993. Implication of geochemical fronts in the Notch Peak contact-metamorphic aureole, Utah, USA. EARTH PLANET SCI LETT 119, 539–59.
Norton, D. L. 1982. Fluid and heat transport phenomena typical of copper-bearing pluton environments. In Titley, S. R. (ed.) Geology of the porphyry copper deposits, southwestern North America, 59–72. Tucson: University of Arizona Press.
Norton, D. & Cathles, L. M. 1979. Thermal aspects of ore deposition. In Barnes, H. L. (ed.) Geochemistry of hydrothermal ore deposits, vol. 2, 611–31. New York: Wiley.
Norton, D. & Knapp, J. 1977. Transport phenomena in hydrothermal systems. The nature of porosity. AM J SCI 277, 913–36.
Norton, D. & Knight, J. 1977. Transport phenomena in hydrothermal systems. Cooling plutons. AM J SCI 277, 937–81.
Parmentier, E. M. & Schedl, A. 1981. Thermal aureoles of igneous intrusions: some possible indications of hydrothermal convective cooling. J GEOL 89, 1–22.
Podladchikov, Y. Y. & Wickham, S. M. 1994. Crystallization of hydrous magmas: calculation of associated thermal effects, volatile fluxes, and isotopic alteration. J GEOL 102, 25–45.
Sammel, E. A., Ingebritsen, S. E. & Mariner, R. H. 1988. The

hydrothermal system at Newberry Volcano, Oregon. J GEOPHYS RES **93,** 10,149–62.

Symonds, R. B., Rose, W. I., Bluth, G. J. S. & Gerlach, T. M. 1994. Volcanic-gas studies: methods, results, and applications. *In* Carroll, M. R. & Holloway, J. R. (eds) *Volatiles in magmas.* MINERAL SOC AM **30,** 1–66.

Taylor, H. P. Jr 1990. Oxygen and hydrogen isotope constraints on the deep circulation of surface waters into zones of hydrothermal metamorphism and melting. *In* Bredehoeft, J. D. & Norton, D. L. (eds) *The role of fluids in crustal processes,* 72–95. Washington, DC: National Academy Press.

Villas, R. N. & Norton, D. 1977. Irreversible mass transfer between circulating hydrothermal fluids and the Mayflower stock. ECON GEOL **72,** 1471–504.

R. BROOKS HANSON, *Science*, 1200 New York Avenue, NW, Washington, DC 20005, U.S.A.
E-mail: bhanson@aaas.org.

Transactions of the Royal Society of Edinburgh: Earth Sciences, **87**, 261–280, 1996

Granitic magmatism and metallogeny of southwestern North America

Mark D. Barton

ABSTRACT: In southwestern North America, late Palaeozoic through Cenozoic granitoids and their related mineral deposits show consistent patterns that can be interpreted in terms of combined provincial, exposure and process controls. Voluminous Cordilleran magmatism began in the Permian and continued with few major interruptions through the Mesozoic and Cenozoic, reaching maximum fluxes in the mid-Jurassic, Late Cretaceous and Oligocene. Two distinctive types of broad-scale igneous suites formed. The first type consists of calc-alkaline to alkaline suites that vary regularly with time from early intermediate–mafic centres to late felsic centres over intervals lasting 20–50 Ma. These suites formed during periods of stable convergence and compressional tectonics, most notably in the late Mesozoic and early–mid-Cenozoic. The second type is compositionally varied, but shows no obvious secular variation in composition. This type formed during neutral to extensional tectonics in the mid-Mesozoic and the mid- to late Cenozoic. Regional (west to east) and secular (old to young) changes from calcic to alkalic compositions do not correspond to basement types; they point to tectonic rather than crustal controls on magmatic evolution, although basement signatures are clearly transmitted in isotopic systematics. Contrasting types of intrusive centres formed in the same lithospheric columns, suggesting that variability reflects thermal and stress regimes, subcrustal magma flux and crustal thickness. Simple thermal and mechanical models of limits on assimilation and magma uprise are broadly consistent with these patterns.

Igneous-related mineralisation is ubiquitous where epizonal environments are preserved, thus preservation (and exposure) form the first-order filter on metallogeny. Mineralisation includes porphyry, skarn, epithermal, replacement and syngenetic deposits of widely varying styles, metal contents and links to magmatic heat and materials. Metal contents and alteration styles correlate closely with igneous compositions and are broadly independent of setting, although systematic regional variations in metal ratios are documented. Ore element suites vary from Cu–Au–Fe associated with (quartz) dioritic to monzonitic intrusive centres through Cu–Zn–Mo–Pb–Ag–W–Au associated with broadly granodioritic centres, and finally to F–Mo–Zn–W–Ag–Be associated with metaluminous to strongly peraluminous granitic centres. A model that includes both composition and process controls rationalises this igneous correlation and the lack of strong regional control. Key features are (1) mineralogical controls on fluid compositions and (2) the efficacy of magmatic processes in producing voluminous ore-forming aqueous fluids. This interpretation is supported by field relationships, igneous petrographic and isotopic data, and theoretical considerations.

KEY WORDS: granite, ore deposits, isotopes, Cordillera, hydrothermal processes.

Phanerozoic magmatism in the Cordillera of southwestern North America is associated with diverse types of world-class mineral deposits. Patterns in magmatism and mineralisation were recognised and emphasised in the first half of the 20th century by Waldemar Lindgren and other geologists of the US Geological Survey (e.g. Lindgren 1915, 1933: 889–94). Although our knowledge has increased tremendously since then, as, for example, with the recognition of seafloor massive sulphide deposits and the development of plate tectonics, these early observations formed the foundation for much of the subsequent work on regional geology and ore formation. Diverse topical studies have grown from this heritage, such as those in the Sierra Nevada batholith which originated with Paul Bateman's work during the Second World War on the tungsten resources of that region (Bateman 1965, 1992). In the last few decades, advances in the study of both igneous rocks and mineral deposits has led to largely separate treatments of granitic magmatism and mineralisation. Igneous petrologists have emphasised petrogenesis and broader tectonic controls, whereas economic geologists have emphasised ore-forming processes at the scale of individual districts (e.g. Kirkham et al. 1993).

In contrast with the abundance of topical studies, relatively few studies have developed the broader links between ore-forming environments and magmatism. From the petrotectonic perspective in southwestern North America, exceptions include plate tectonic approaches (Sillitoe 1976; Titley 1982; Clark et al. 1982; Damon et al. 1983; Sawkins 1990) and studies linking regional metallogenic patterns and magmatic patterns (Westra & Keith 1982; Farmer & DePaolo 1984; Keith 1986; Barton et al. 1988, 1995; Ague & Brimhall 1988; Barton 1990; Bookstrom 1990). From the deposit or system perspective, advances in the understanding of chemical and physical links of late magmatic and hydrothermal processes and chemistry have been important (e.g. Burnham 1979; Burnham & Ohmoto 1980; Norton 1982; Candela 1989, 1991; Blevin & Chappell 1992).

This paper takes a two-fold approach to an integrated view of granitoids and metallogeny: firstly, it reviews the broad characteristics and distribution of granitoids and associated mineral deposits and, secondly, it examines likely provincial, process and preservational controls that govern the distribution and characteristics of magmatism and mineralisation.

Table 1 Hydrothermal mineralisation associated with magmatism, southwestern North America.

Deposit types	Associated magmatism
Magmatic fluids and heat essential*	
Porphyry Cu–Au; skarn/replacement Cu–Fe–Au	(Qz) monzonite–(qz) diorite
Porphyry Cu–Mo; skarn Cu–Zn–Ag–Pb	Qz monzodiorite–granodiorite
Porphyry Mo–Cu–W; skarn/replacement W–Zn–Cu–Mo–Ag–Pb	Granodiorite–monzogranite
Porphyry Mo–F–W; skarn/replacement F–Zn–Mo–Ag–Pb–Be	Metaluminous granite
Greisen W–Be–F–Mo; skarn/replacement F–Zn–Zn	Peraluminous monzogranite
Complex rare element pegmatites	Granodiorite–granite
Magmatic heat essential, magmatic fluids problematic*	
Epithermal (vein) deposits (dilute to moderately saline, meteoric ± magmatic water)	Andesitic to rhyolitic volcanic and hypabyssal intrusive centres; minor correlation with composition
Volcanogenic massive sulphides (moderately saline, seawater)	Basaltic to rhyolitic volcanism and hypayssal intrusions; minor correlation with composition
Fe oxide (–Cu–REE) systems (highly saline, evaporitic?)	Basaltic to rhyolitic volcanic and hypabyssal intrusive centres, minor correlation with composition
Magmatic link problematic*	
Au–quartz systems (Mother Lode and Carlin types; dilute meteoric or metamorphic water)	No close link demonstrated; alkaline to peraluminous, mafic to felsic intrusions have been postulated

Note: * There is not universal agreement on this division. Also, more than one process can commonly be documented.

Key issues include: what are the characteristics of these systems; what petrogenetic links exist; and how can one sort out the relative importance of many different influences? Studies in southwestern North America offer an insight into these questions through the consideration of metallogeny and magmatism across contrasting times and terranes.

1. Background

1.1. Types of igneous-related mineralisation
Magmatism contributes to the generation of mineral deposits by providing materials, energy and suitable hosts. Multiple types of mineralisation are linked to magmatism (Table 1), although the nature of such links remains controversial in some instances. Metal and alteration suites correlate with magmatic compositions as indicated in Figure 1. A direct link with magmatic fluids is indicated for porphyry, skarn and many replacement deposits (e.g. Burnham 1979; Einaudi *et al.* 1981; Sawkins 1990). In these deposits, aqueous fluids provide the essential medium for mass transfer. Subsequent interaction with external fluids (meteoric or otherwise) during hypogene or supergene processes may profoundly modify the characteristics of mineralisation. The correlation of igneous compositions, metal suites and the nature of hydrothermal alteration indicates a basic linkage (Fig. 1; Barton *et al.* 1995). Although

some rare metal deposits form primarily by igneous fractionation (Pollard & London 1995), such systems appear to be rare in the Cordillera.

In contrast with those deposits clearly linked with magmatic fluids, a magmatic heat source with variable contributions of magmatic materials is inferred for epithermal, volcanogenic massive sulphide and some types of iron oxide deposits. In these deposits, geochemical and geological evidence point to a predominance of external fluids. Metal contents vary systematically with salinity and sulphur content from metal-poor (precious metal dominated) epithermal systems formed from meteoric waters (Field & Fifarek 1985; Hedenquist & Lowenstern 1994), through base metal rich volcanogenic massive sulphide deposits formed from seawater (Franklin *et al.* 1981), to iron oxide dominated deposits formed from non-magmatic brines (Barton & Johnson, 1996).

Temporal links with magmatism exist for gold-bearing quartz-rich deposits, including those of the Mother Lode and Carlin types, but a process link is problematic. Some workers have proposed that magmatism directly contributes heat and perhaps materials (e.g. Sillitoe & Bonham 1990), whereas others have argued that ore-forming hydrothermal systems are primarily driven by tectonism independent of magmatism, even though igneous activity is often contemporaneous (Nesbitt *et al.* 1989; Ilchik & Barton, in press).

1.2. Geological framework
The North American Cordillera is divisible into several broad types of terranes across which younger magmatism and mineralisation are superimposed (Fig. 2; Coney 1989). The region thus allows the examination of temporal evolution within the same type of crust and the comparison of coeval magmatism developed in contrasting crustal segments. Distinctions among terranes reflect differing ages of formation and degrees of cratonisation. The broadest divisions are: (1) cratonal areas of Proterozoic and Archean age with at most thin Phanerozoic cover; (2) transitional areas of miogeoclinal and platformal sequences, commonly with cratonal basement; and (3) oceanic/accretionary/arc terranes, some of which may be exotic to North America. Superimposed on these terranes are magmatic belts of various ages; the best mineralised range from Jurassic to Miocene in age.

The oldest crust, the Archean of the Wyoming province, occurs in the northeastern part of this region. Proterozoic arc-related provinces young southeastwards from Wyoming

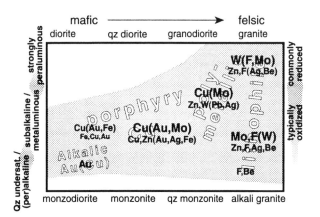

Figure 1 Generalised variation of ore deposits with magmatic composition as described in this paper. Compare Table 1. Metal suites are in bold; generalised deposit types are in outline font. Shading indicates ranges of interest and transitions. For further documentation see Blevin and Chappell (1992) Barton *et al.* (1995).

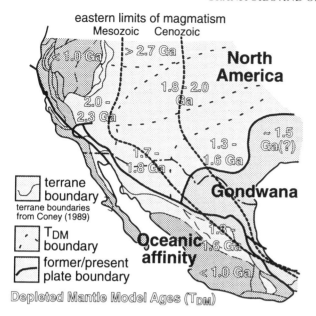

Figure 2 Geological framework of southwestern North America showing basement ages, terrane boundaries and limits of Mesozoic and Tertiary plutonism. Terrane boundaries from Coney (1989), basement boundaries from Bennett and DePaolo (1987), Nelson and DePaolo (1985) and Patchett and Ruiz (1989), and plutonic limits from Ortega-Guitérrez et al. (1992) and King and Beikman (1974).

(Nelson & DePaolo 1985; Bennett & DePaolo 1987). Slivers of North American basement make up the core of northern and central Mexico. These are abutted by Gondwanan basement that surrounds the western margin of the Gulf of Mexico. These older terranes rocks are cut by extensive mid-Proterozoic 'anorogenic' granitoids at ≈ 1·4 Ga (Anderson & Morrison 1992) and in the eastern areas by late Proterozoic, Grenvillian magmatism at ≈ 1·1 Ga (Patchett & Ruiz 1989).

Phanerozoic crust consists of miogeoclinal and eugeoclinal sequences (Fig. 2; Coney 1989). The former were generated following rifting during the late Proterozoic (Pacific margin) and early Mesozoic (Gulf of Mexico margin). The eugeoclinal (mainly arc) terranes were added to the western margin by a series of accretionary events primarily in the Palaeozoic and Mesozoic. Although older components are present, much of this material was added to the crust in the Phanerozoic (DePaolo 1981; Centeno et al. 1993). Notable segments include the Palaeozoic terranes of northern California and Nevada and the early Mesozoic Guerrero terrane of western Mexico (Coney 1989; Miller et al. 1992; Centeno et al. 1993).

2. Time–space distribution of magmatism and mineralisation

In southwestern North America, magmatic fluxes, compositions and styles vary systematically with time and space and correlate closely with styles of mineralisation. These patterns are consistent with a primary igneous control on mineralisation. The regular patterns in magmatism correlate with times and styles of regional tectonism, notably between times of compression and extension. Contrasts between igneous suites associated with extensional and compressional regimes point to contrasting physical controls on the generation of the magmas themselves, the formation of related mineralisation and the subsequent probability of preserving ore deposits. Igneous rocks occur sporadically throughout the early and middle Palaeozoic, but it is not until the Permian that broadly continuous magmatism is recognisable along the western margin of North America. Likewise, even though there is rare

Palaeozoic igneous-related mineralisation, nearly all ore deposits are Jurassic or younger.

2.1. Magmatism and mineralisation through time
Magmatism and mineralisation have been largely continuous along the western margin of North America for the last 250 Ma. During this time, two distinctive styles correlate with tectonic environment (Figs 3, 4). Periods of broadly uniform convergence and compressional tectonics developed extensive batholithic belts that evolved progressively with time. The Cretaceous coastal batholiths and early Tertiary (Laramide) interior batholiths provide type examples (Fig. 3). Periods of variable convergence and neutral to extensional tectonics produced distributed magmatic activity which was generally more variable in composition than compressional epochs. Mid-Tertiary volcanism in the Great Basin and Jurassic volcanism along much of the margin provide examples (Fig. 3).

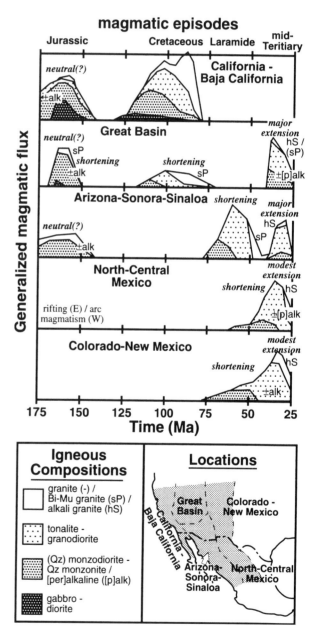

Figure 3 Magmatic fluxes (reflecting areas of exposed rock of a given age per unit time) for selected areas, based on compilations of King and Beikman (1974), Barton et al. (1988), Ortega-Guiterrez et al. (1992) and Barton et al. (1995). Broad compositional ranges are indicated; note the restriction of alkaline to peralkaline rocks to extensional events. Central and southern Mexico are not included due to the scarcity of compositional data on igneous rocks.

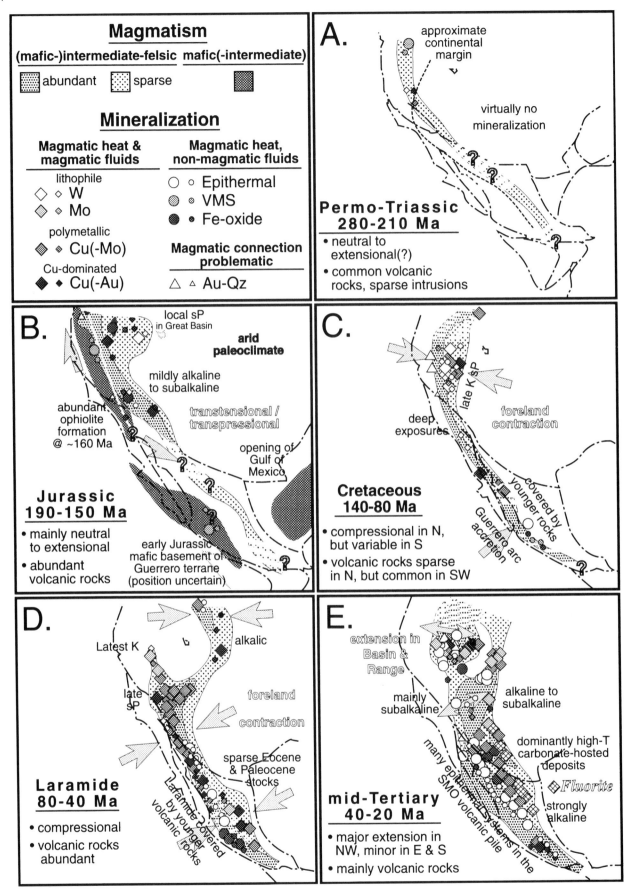

Figure 4 Time panels for magmatism and mineral deposits in southwestern North America. Compiled from various sources including Guild *et al.* (1981) and Barton *et al.* (1988, 1995). See text for discussion. Plate positions are from Scotese and Denham (1988).

Some mineralisation occurs with each of the igneous suites; it is prominent where epizonal levels are preserved. Parallel with variations in magmatism, the alteration and metal suites of porphyry, skarn and replacement systems vary regularly in time during compressional episodes, but are diverse during extensional episodes. Deposit types that lack a strong magmatic fluid connection (Table 1; epithermal, volcanic massive sulphide, Fe oxide, Au–quartz) are prominent during neutral to extensional episodes. These patterns of magmatism and mineralisation are supported by a wealth of data (e.g. summaries in Clark *et al.* 1982; Barton *et al.* 1988, 1995; Anderson 1990; Bookstrom 1990; Miller & Barton 1990; Bateman 1992; Lipman 1992; Miller *et al.* 1992; Saleeby & Busby-Spera 1992).

Permo-Triassic (260–210 Ma). Prominent magmatism along the western margin of North America is first recorded in a Permian arc that is irregularly exposed from northern California and northwestern Nevada through southern Mexico (Fig. 4A, Miller *et al.* 1992; Sedlock *et al.* 1993). In northern California and Nevada, this arc is oceanic in affinity and has been interpreted by some to be exotic to North America. Others have suggested that it represents the continuation of contemporaneous magmatism that is rooted in cratonal materials from central and south California. Permo-Triassic intrusive rocks are relatively sparse, but appear to be predominantly calc-alkaline intermediate to felsic composition granitoids (Torres-Vargas 1993). Volcanic assemblages in the northern part of the arc are mafic to intermediate in composition (LaPierre *et al.* 1994). Described mineralisation is sparse, although sodic hydrothermal alteration is apparently widespread (Fig. 4A; Battles & Barton 1995). Known mineral deposits are limited to volcanic massive sulphide deposits in the north (e.g. East Shasta district, California; Albers 1981) and rare porphyry-related mineralisation in the southern Great Basin (e.g. Royston district, Nevada; Seedorff 1991b).

Jurassic (190–150 Ma). Magmatism resumed in the early Jurassic following a lull at the end of the Triassic. Jurassic magmatism was compositionally diverse, notably in the Middle and Late Jurassic. Intrusive centres and volcanic rocks are abundant from northwestern California east to Utah and south through the Basin and Range into southern Mexico (Fig. 4B). Mineralisation is widespread and volcanic rocks and hypabyssal intrusions are typically hydrothermally altered, commonly to sodic compositions (Barton *et al.* 1988; Battles & Barton 1995).

Magmatism in the Early Jurassic is localised along the western margin of the Jurassic arc. Dioritic to granodioritic plutons are prominent parts of multiple arc terranes recognised in the eugeoclinal basement of northern California (Barnes *et al.* 1992), whereas monzonitic compositions characterise the Jurassic arc over continental basement in the USA and northern Mexico (Miller 1978; Tosdal *et al.* 1989; Saleeby & Busby-Spera 1992). Mineralisation associated with Early to Middle Jurassic magmatism includes porphyry and epithermal Cu(±Au) mineralisation in southern Arizona (e.g. Bisbee; Titley 1982), western Nevada (e.g. Yerington; Dilles 1987) and eastern California (Fig. 4B). These deposits have some similarities with the early Mesozoic alkalic porphyry copper suites of British Columbia (Lang *et al.* 1995).

In the Middle and Late Jurassic (170–155 Ma) compositionally expanded suites developed in the western USA as magmatism broadened eastwards from central California across the northern Great Basin. Magmatic fluxes, as estimated from modern outcrop areas, were fairly high during this episode (Barton *et al.* 1988). Most evidence indicates that neutral to extensional characteristics predominated within and behind the arc, although there was some Late Jurassic

shortening in northern Nevada and California (Saleeby & Busby-Spera 1992). Gabbroic and dioritic intrusions formed abundantly in California and the Great Basin at this time. The most prominent of these are the forearc/intra-arc ophiolitic sequences of central and western California (Saleeby & Busby-Spera 1992). Volcanic massive sulphide deposits are prominent with these rocks in the west where seawater was available (Foothills belt, California; Albers 1981), whereas iron oxide rich hydrothermal systems formed in the interior, perhaps sourced from evaporitic brines (Humboldt complex, Nevada; Barton & Johnson 1996). These latter deposits and widespread sodic alteration probably reflect the unusually arid palaeoclimate of the Jurassic (Battles & Barton 1995). Intermediate to felsic members of the Middle–Late Jurassic arc are alkali-calcic to calc-alkaline granitoids and include locally strongly peraluminous phases (e.g. in eastern Nevada, Lee & Christensen 1983). These are associated with widespread intrusion-hosted and skarn tungsten and copper mineralisation in Nevada and southeastern California (Fig. 4B). Extremely aluminous gold-bearing metavolcanic rocks of the lower Colorado river area may represent Jurassic advanced argillic epithermal systems caught up and metamorphosed in the Cretaceous foreland metamorphic belt.

Jurassic plutonism is less well documented in central and southern Mexico, but apparently extends far south both inland on older basement and in the forerunner of the predominantly Cretaceous Guerrero terrane (Ortega-Guitérrez *et al.* 1992; Sedlock *et al.* 1993). Along the eastern margin of Mexico, basaltic magmatism accompanied the opening of the Gulf of Mexico. Volcanic massive sulphides and some iron oxide rich occurrences in Baja California may belong to this suite (Fig. 4B). Porphyry, replacement and epithermal mineralisation of Jurassic age have not been described in Mexico.

Cretaceous (130–80 Ma). In contrast with compositionally chaotic magmatism during the Jurassic, Cretaceous magmatism evolved progressively over almost 50 Ma. The tectonic style was predominantly compressional (Burchfiel *et al.* 1992). Porphyry, skarn and replacement deposits are widespread in epizonal environments; styles vary in parallel with magmatic changes (Barton 1990).

In the southwestern U.S.A. and northwestern Mexico plutonism changed from primarily quartz dioritic rocks in the Early Cretaceous to felsic metaluminous and peraluminous suites in the Late Cretacecous (Fig. 3). Contemporaneous volcanic rocks are scarce. Magmatism during this period was localised in the coastal batholithic belt with minor though continuous magmatism in the back-arc regions (Fig. 4C; Miller & Barton 1990). Further south in central and southern Mexico, Cretaceous intrusions and a thick volcano-sedimentary package make up much of the Guerrero terrane (Campa & Coney 1983; Centeno *et al.* 1993). Relatively little is known about these rocks, but at their northern end they grade into the southern continuation of the coastal batholiths (Henry 1975). Strongly peraluminous rocks are most obvious inboard of the main batholithic belts in areas of more mature crust as in the Great Basin and southeastern California (Miller & Bradfish 1980). Strongly peraluminous granitoids make up much of the Idaho batholith to the north (Foster & Hyndman 1990) and are locally important in the younger plutons of the Peninsular Ranges batholith (Todd & Shaw 1985; Walawender *et al.* 1990).

Slow migration of the batholithic axes eastwards over time (Silver *et al.* 1979; Chen & Moore 1981) may have led to the observed more felsic compositions as the result of movement onto more mature crust (Section 3.1). Alternatively, the secular variation in composition may be due to changes in the physical characteristics of the lithosphere over time, analogous to changes in the back-arc in the Great Basin (Section 3.2; Fig. 5).

Mineralisation is widespread in the Cretaceous (Fig. 4C). The most common types are skarn and porphyry deposits in the U.S.A. and northern Mexico, whereas volcanic massive sulphide and iron oxide rich deposits are more common in southern Mexico. The major gold–quartz veins of the Mother Lode and related districts formed along the western margin of the arc in the Early Cretaceous (Bohlke & Kistler 1986), perhaps associated with transpressional tectonics of the Early Cretaceous (Saleeby & Busby-Spera 1992). Epithermal systems are uncommon, but are locally present in the Great Basin where Cretaceous volcanic rocks are preserved (e.g. Rochester, Nevada; Vikre 1981). They may also be present in western Mexico.

Cretaceous porphyry, replacement and greisen style systems are widespread in the Great Basin where many shallow intrusive centres are preserved (Barton 1990; Albino 1995). The most abundant deposits are polymetallic systems with Mo(Cu) porphyry style mineralisation (e.g. Eureka district; Nolan 1962) associated with mid-Cretaceous biotite $\pm$ hornblende granodiorites. Cu(–Au–Mo) porphyry mineralisation is associated with Early Cretaceous hornblende–biotite quartz monzonites in eastern Nevada (Robinson district; Westra 1982). Subeconomic lithophile element systems are

common with Late Cretaceous two-mica granites (Barton 1987). Porphyry copper systems are rare in the Cretaceous batholithic belts of California (Lights Creek), Baja California (El Arco) and Sinaloa, but most mineralisation in these regions consists of widespread small to moderate sized polymetallic skarns dominated by tungsten (Newberry and Einaudi 1981; e.g. Bishop District, California; Bateman 1965). Further south, in Jalisco, Michoacan and adjacent states of south-central Mexico iron- and copper-bearing skarns and some porphyry style mineralisation occur with poorly dated, but nominally Cretaceous, intermediate composition intrusive centres (Barton et al. 1995).

Gold–quartz veins of the Mother Lode and related districts formed on the west of the Cretaceous batholith apparently coeval with its early stages ($\approx$ 120–130 Ma; Bohlke & Kistler 1986) and perhaps in a transpressional environment (Saleeby & Busby-Spera 1992). A close connection with magmatism is problematic; most models invoke metamorphic or meteoric fluids, but perhaps driven by batholith emplacement (e.g. Bohlke & Kistler 1986; Nesbitt et al. 1989). Volcanogenic massive sulphide deposits are apparently absent in the north, consistent with the subaerial character of the Cretaceous arc, but there are a number of occurrences in the Cretaceous marine rocks of the Guerrero terrane in southwestern Mexico (Miranda-Gasca 1994).

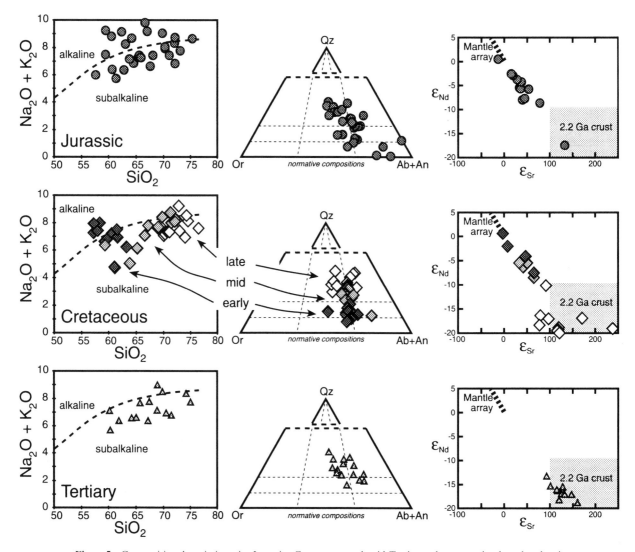

Figure 5 Compositional variations in Jurassic, Cretaceous and mid-Tertiary plutons and selected volcanic rocks along the miogeocline in the Great Basin. This region has fairly uniform lithosphere, thus the secular changes must reflect physical controls. See text for discussion. Data compiled from various sources cited in Barton (1990) and new analyses (Barton, unpublished data).

Laramide (80–40 Ma). As during the Cretaceous, Laramide igneous activity accompanied compressive deformation and produced a prominent magmatic belt with a pronounced secular progression in compositions (Dickinson & Snyder 1978; Fig. 3). From central Mexico northwards this belt is displaced markedly eastwards of the older coastal batholiths (Fig. 4D). The Laramide magmatic transgression ultimately generated intrusive centres up to 1500 km from the Pacific margin. Simultaneously, magmatism ceased near the margin in California, Nevada and Baja California. Further south, Laramide age magmatism is abundant and distributed across a broad area, but it did not move as prominently inland as it did in areas to the north. The eastwards migration has been interpreted in terms of a flattening slab, presumably driving the locus of subcrustal magmatic sources east with time (e.g. Coney & Reynolds 1977; Damon et al. 1983). Although the most voluminous activity shifted eastwards with time, magmatism spread over an unusually wide region during the Laramide and reached its eastern limit in the Late Cretaceous. Magmatic activity continued with little interruption into the middle Tertiary in the eastern and southern areas, whereas there is a gap of varying duration from northwestern Mexico into the Great Basin (Fig. 3).

Compositions of Laramide granitoids vary regionally and with time. In all areas, older magmatism is relatively mafic and silica-poor (diorites to monzonites) and becomes more felsic with time, ultimately to (variably peraluminous) monzogranites and metaluminous alkali granites (Miller & Barton 1990). Compositional suites become progressively more alkalic going from southern Mexico to the north. Calcic diorite–quartz diorite–tonalite suites in central and southern Mexico give way to calc-alkaline quartz diorite–granodiorite suites in northern Mexico. Calc-alkaline to alkali–calcic suites in Arizona and New Mexico (quartz monzodiorite–monzogranite; Keith 1986) are superseded by alkalic suites (syenite to alkali granite) in Colorado, Wyoming and Montana (Bookstrom 1990). Two-mica granitoids comprise the prominent local last centres in Laramide in Arizona and Sonora (55–35 Ma; Haxel et al. 1984; Roldan Quintana 1991).

Hydrothermally altered volcanic rocks are common in many parts of the Laramide arc, in contrast with their scarcity in the Cretaceous arc. Only the more westerly and northerly portions (e.g. Sinaloa and Sonora batholiths, Colorado Mineral Belt) lack significant volcanic superstructure. World-class porphyry Cu(–Mo) mineralisation (e.g. Morenci, Arizona; Cananea, Sonora) formed with many parts of the Laramide arc, notably within 400 km of the USA–Mexico border (Fig. 4D). Epithermal, carbonate-replacement and skarn systems are also widespread. In the border region, more than 100 porphyry-related Cu(–Mo–Ag) deposits are associated with variably exposed Laramide volcano-plutonic complexes (Titley 1982). These complexes range in size from a few square kilometres to >1000 km^2 and commonly have prolonged magmatic histories, beginning with mafic volcanism and ending with granite porphyries. Strikingly, major porphyry style mineralisation in Arizona and Sonora appears to have formed in two discrete episodes (≈70 and ≈60 Ma) during the 30 Ma evolution of the Laramide arc, indicating a regional tectonic trigger (75–45 Ma; McCandless & Ruiz 1993). To the north in Colorado, the earliest magmatism along the Colorado mineral belt produced alkaline, silica-poor intrusive centres associated with Au(–Cu–Mo) mineralisation (e.g. Central City, Rice et al. 1985; Bookstrom 1990). To the south, in central and southern Mexico, numerous Laramide intermediate to felsic intrusive centres are associated with Fe(Cu) skarn (e.g. Peña Colorado, Jalisco), Cu(Mo) porphyry and skarn (e.g. Santo Tomas, Sinaloa) and Pb–Zn–Ag skarn (e.g. San Martin,

Zacatecas), carbonate replacement and vein mineralisation (e.g. Taxco, Guerrero; see Fig. 4D; Clark et al. 1982; Barton et al. 1995). In northern Mexico and Arizona, scattered W–Mo(–Cu)-bearing greisens and skarns associated with felsic, variably peraluminous intrusions constitute a distinct lithophile element rich late phase of Laramide metallogeny.

Mid-Tertiary (40–20 Ma). Voluminous magmatism during the late Eocene to middle Miocene occurred through much of the Cordillera (Fig. 4E). The compositions of this magmatism vary with both time and space. Volcanic rocks predominate; many were erupted from centres that presently expose few or no intrusive rocks. Where intrusive centres are common, porphyry, skarn and replacement deposits are widespread (e.g. in the northern Great Basin, in eastern and southern Mexico, Fig. 4E). Epithermal deposits are widespread in volcanic rocks and Carlin-type gold–quartz deposits are common in miogeoclinal rocks. Volcanic-hosted iron oxide systems occur locally in northern Mexico.

In the western U.S.A. and northern Mexico magmatism reached an eastern high water mark during the Eocene and then retreated over a broad region towards the Pacific margin (Coney & Reynolds 1977; Damon et al. 1983). In central and southern Mexico migration was subdued and the locus remained largely near the coast. The large areas of exposed mid-Tertiary igneous rocks indicate that magmatic fluxes typically exceeded those of earlier times. Although felsic rocks dominate, intermediate and some mafic rocks are also widespread; the transition to bimodal volcanism characterising Basin and Range tectonism is generally younger than 20 Ma (Fig. 3; Lipman 1992). Most rock suites are metaluminous or weakly peraluminous, but some peralkaline and strongly peraluminous centres also occur. In the interior, particularly in central Colorado, central New Mexico and northeastern Mexico, mid-Tertiary magmatism has a strongly alkalic character, whereas it is mainly subalkaline further to the west. In most areas, magmatism accompanied crustal extension, which was minor in the east, but profound in much of the Basin and Range from Nevada through northern Mexico (Gans et al. 1989; Dickinson 1991; Henry & Aranda–Gomez 1992). Where extension was modest or absent, as in the eastern regions, magmatic compositions continue the secular progression begun in the Laramide (Fig. 3). In contrast, in the highly extended areas to the west, diverse compositions were produced resembling the Jurassic, but unlike the Cretaceous and Laramide. The main secular variation was the westward sweep of the magmatic front during the Oligocene.

As with magmatism, the associated mineralisation is diverse; it varies regularly with time only in the east (Fig. 4E). Many hundreds, probably thousands, of porphyry, skarn and replacement deposits occur with intrusive centres; many thousands of volcanic and sedimentary hosted epithermal occurrences formed. Iron oxide rich systems are present in northern Mexico, commonly with evidence for evaporitic materials (J. Ruiz, pers. comm.). Porphyry style mineralisation in the eastern portion of the mid-Tertiary magmatic zone is dominated by lithophile element systems such as the porphyry molybdenum deposits of Colorado to west Texas and the fluorine-rich lithophile element systems of west Texas into eastern Mexico (Fig. 4E). To the west, porphyry style mineralisation is prominent in the Great Basin (Seedorff 1991a) and may underlie parts of the Sierra Madre Occidental in western Mexico (Staude 1995). It is striking that in the northern Great Basin intrusive centres of approximately the same age (40–35 Ma) produced Cu(–Au–Mo) systems associated with quartz monzonites (Bingham, Utah; Copper Canyon, Nevada), Mo(–Cu) systems associated with monzongranites (Ward, Nevada), Climax-type Mo systems with alkali granites (Mt

Hope, Nevada; Pine Grove, Utah) and W(–Be) systems with biotite–muscovite monzogranites (Harrison Pass, Nevada). This variety, all generated from the miogeocline within a span of 5 Ma, attests to diverse petrogenetic and ore-forming processes taking place concurrently in the same region. The other mid-Tertiary porphyry deposit province is in southern Mexico, where erosion levels have exposed abundant intrusive rocks of the Sierra Madre del Sur. Quartz diorite to granodioritic stocks are associated with Cu–Au–Fe porphyry and skarn mineralisation in numerous centres in Michoacan, Guerrero and adjacent states (e.g. La Verde, Michoacan; Barton *et al.* 1995).

Volcanic terranes contain hundreds of epithermal Au–Ag deposits and many thousands of occurrences. They may or may not connect with deeper mineralisation, although alteration and metals appear to correlate with igneous compositions. Acid-rich systems with fairly high Au/Ag and modest Cu appear to be related to intermediate composition centres (e.g. Mulatos, Sonora; Summitville, Colorado), whereas large Ag-rich systems more commonly occur with felsic compositions (e.g. Guanajuato, Guanajuato; Bolaños, Jalisco; Barton *et al.* 1995; Staude 1995). Meteoric fluids, some perhaps of saline composition, formed the majority of the sources, yet were circulated at magmatic centres (O'Neil & Silberman 1974; Criss & Taylor 1986). Sparse iron oxide rich systems in northern Mexico are also associated with volcanic centres; they may represent the circulation of evaporitic fluids derived from local enclosed basins in the arid palaeoclimate (Barton & Johnson 1996; see Lyons 1988, for an alternative interpretation). Sediment-hosted gold–quartz deposits of the Carlin type are widespread in the northern Great Basin and possibly in Sonora. They have been interpreted as related to magmatic hydrothermal systems of various ages (Sillitoe & Bonham 1990), but the lack of close spatial or demonstrated temporal links. Moreover, the alternative links to crustal extension (Seedorff 1991a) support the notion that they may represent extension-driven fluid flow with at most a tenuous connection to magmatism (Ilchik & Barton, in press).

2.2. Synthesis of temporal and regional patterns

Temporal patterns. Temporal variations in magmatism and associated mineralisation are generally more prominent than regional variations. Temporal patterns reflect tectonic regimes, whereas regional patterns can be attributed in part to differences in crustal compositions, but most appear to be a function of the changing locus of magmatism with time and tectonics. The principal type of secular variation, seen in many areas during compressional events and independent of crust and age, is progression from mafic to felsic systems over the course of compressional episodes that last several tens of millions of years (cf. Brown *et al.* 1984).

Figure 5 illustrates temporal variations along the Cordilleran miogeocline in the Great Basin, an area of fairly uniform lithosphere in the back-arc of the Mesozoic batholiths (Barton 1990). These patterns illustrate the diverse secular changes in magmatism in southwestern North America (Fig. 3) and are reinforced by changes in Nd isotopes with time (Fig. 6). Times of mainly neutral or extensional tectonics exhibit coeval magmatic centres of widely differing compositions (Jurassic and mid-Tertiary, both ≤15 Ma duration). In contrast, compositions are more restricted at any given time in arcs that formed during compressional tectonics (Cretaceous, Laramide, both ≥25 Ma duration), even though the overall range of compositions was comparable when integrated over complete episodes. Compresion-related magmatism also appears to have been more spatially restricted (during any given interval) than extension-related magmatism.

Mineralisation in both instances follows magmatic compositions (Fig. 1), although certain types of mineralisation such as volcanic massive sulphide, iron oxide and gold–quartz seem more closely associated with extensional or translational tectono-magmatic settings. Within overall igneous epochs, mineralisation may be restricted in time, as suggested by Re–Os dating of molybdenites from Arizona and Sonora (McCandless & Ruiz 1993). These multiple variations probably

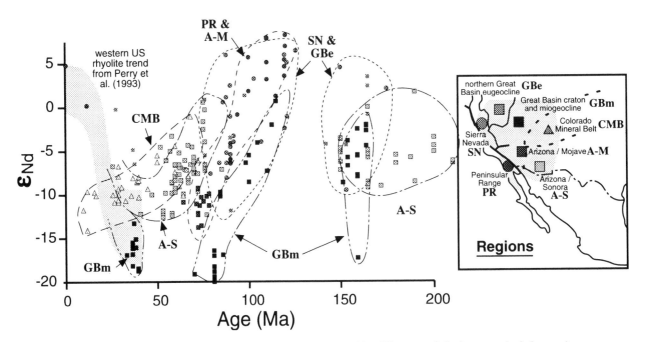

Figure 6 ε_{Nd} variations with time for different regions (denoted by different symbols shown on the index map) in the western U.S.A. and northern Mexico. Compare Figures 3 and 8. The paucity of U–Pb dates on older suites prevents a detailed assessment of these patterns. Data compiled from DePaolo (1980, 1981), Farmer and DePaolo (1983, 1984), Anthony and Titley (1989), Barton (1990, unpublished data), Miller *et al.* (1990), Stein and Crock (1990), Asmerom *et al.* (1991) and Wodzicki (1995).

reflect both direct igneous compositional controls and broader tectonic controls.

Mineral deposit and igneous environments also change on a longer time-scale. Most epithermal deposits are Tertiary in age, known porphyry deposits are primarily of early Tertiary or late Mesozoic age and the older arc rocks of late Palaeozoic to middle Mesozoic age are sparsely mineralised with the exception of seafloor hydrothermal systems. These changes, and a corresponding variation in the abundance of volcanic rocks with time, probably reflect the preservation potential of the uppermost crust (Section 4.1).

Regional patterns. At the broadest scale, igneous and ore suites do not correlate strongly with crustal type (Fig. 3). All regions have igneous suites that range from mafic to felsic. Important second-order characteristics such alkalinity, aluminosity and specific metal suites do vary in space, perhaps reflecting varying crust types as well as temporal changes. Once the effect of secular variations are removed, regional variations are most obvious in terms of the variability in the nature of the basement going from cratonal materials in the north and east to eugeoclinal materials in the south and west (Fig. 6). Magmatism becomes more felsic and alkaline moving from young crust to regions of older crust (Lipman 1992), but this effect is superimposed on the broader and larger secular variations. Local crust clearly governs the overall range of isotopic values that may be obtained, although particular intervals may be fairly restricted in composition (Fig. 6). Comparison of magmatism on a regional basis demonstrates that many types occur within areas of rather uniform crust (Barton 1990; Fig. 5). To what extent these variations reflect the nature of the crust itself is unclear. In many instances, regional variations appear to be related to the changing nature of subduction, the most obvious example being the Tertiary eastwards sweep of magmatism and its correlative change from calc-alkaline to alkaline magmatism (Lipman *et al.* 1971; Westra & Keith 1982).

Magmatic belts cut across contrasting basement terranes, but maintain similar compositional systematics through much of the Mesozoic and Cenozoic. The Tertiary alkaline province of the Cordilleran interior extends from the Archean Wyoming province in the north, through various Proterozoic terranes and into Phanerozoic terranes bordering the western margin of the Gulf of Mexico (compare Fig. 2 with Fig. 4E). Likewise, the Cretaceous batholiths of the coastal region extend across the late Proterozoic margin with relatively minor difference from north to south, but with major differences as a function of time. Some of the latter may be due to the general eastwards transgression of the Cretaceous batholiths onto more mature crust (e.g. Silver *et al.* 1979; Bateman 1992), but the results could also be interpreted in terms of increasing ease of assimilation of crust in the warming arc (Barton & Hanson 1989; see Section 3). Local variability in the Sierra Nevada batholith correlates with adjacent framework rocks and has been interpreted to reflect local compositional controls (Ague & Brimhall 1988). There may be a substantial crustal control of the generation of strongly peraluminous granitoids, which are most abundant in the inner part of the Mesozoic arcs, but even two-mica granites are widespread across regions of contrasting age, composition and position relative to the main arcs (Miller & Barton 1990).

Regional differences in mineralisation are also clear. Porphyry copper deposits are scarce in the deep interior and porphyry molybdenum deposits are absent west of the Proterozoic margin. Phanerozoic volcanic massive sulphide and gold–quartz systems are apparently restricted to regions of Phanerozoic or transitional crust. The scarcity of epithermal and porphyry deposits in the coastal batholiths surely reflects

the removal of hypabyssal environments. Titley (1987) and others document crustal control of metal ratios and lead isotopes of ore deposits in the southern Basin and Range. Many other regional patterns have been described (e.g. Sillitoe 1976; Westra & Keith 1982). Despite this spatial variability, single regions may exhibit nearly the full spectrum of mineralisation types. For example, the Great Basin has world-class Cu(–Au) and gold–quartz deposits, large lithophile element deposits of both alkalic (porphyry Mo) and strongly peraluminous (W–F–Be) types with a variety of intermediate types (Barton 1990; Seedorff 1991a). These deposit suites are either diachronous, as in the Cretaceous, or broadly synchronous, as in the mid-Tertiary or Jurassic.

2.3. Linkages

Although the descriptive framework for magmatism and mineralisation in time and space is reasonably well established, controls on these patterns are controversial. The systematic patterns among tectonics, magmatism and mineralisation must reflect the interplay among various processes, a number of which are schematically illustrated in Figure 7. These linkages are explored in the following sections, first looking at possible causes of magmatic systematics, then at possible controls on igneous-related mineralisation. Because granitoids reflect the nature of the melt sources and the physical processes that govern their evolution, their characteristics are the product of tectonics and the local lithosphere (Fig. 7, area I). Among many factors, the state of stress, the thermal structure and crustal thickness may be most significant (Section 3). Crustal compositions are also important, but the geological evidence indicates that they are subordinate to the physical factors. Magmas evolve by mixing and assimilation–fractional crystallisation (AFC) processes in response to these various factors, which thus ultimately govern what is emplaced in the upper crust and available to generate ore-forming systems.

In the upper crust, magmas evolve by fractional crystallisation, volatile loss by second boiling and eruption, and various mixing processes including recharge and assimilation (Fig. 7, area II). The scale and efficacy of these processes will govern the volume and composition of magmatic–hydrothermal fluids and thus the nature of fluid–rock interactions (Section 4). Subsequently, granitoids and their related alteration will be affect by post-depositional processes as a function of tectonic history (Fig. 7, area III; Section 4.1).

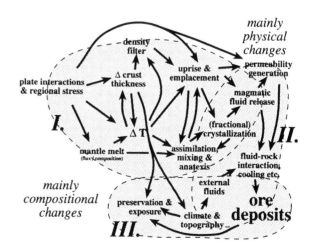

Figure 7 Schematic links among processes in the generation of granitoid-related ore deposits. Magmatic evolution is in group I, hydrothermal evolution is in group II and post-mineralisation processes are in group III. This scheme neglects most of the dynamics of the systems, as well as the many magmas and processes that can occur. See text for discussion.

3. Interpretation of magmatic patterns

The systematics of granitoids and associated volcanic rocks in southwestern North America reflect physical, mainly tectonic, controls influenced by provincial factors. These issues have been widely discussed (e.g. see review by Lipman 1992); they are briefly treated here. Provincial controls are clearly indicated by isotopic data and other compositional variations along strike in contemporaneous magmatic belts (Fig. 6). Physical controls, particularly the stress and thermal regimes, govern mixing processes and the rise of magmas (Fig. 7). The observed temporal variations within particular regions and the broadly parallel changes across regions of contrasting lithosphere indicate that physical controls dominate magmatic and metallogenic patterns in southwestern North America. Simple thermal and mechanical models of limits on assimilation and magma uprise, when constrained by regional tectonics, are broadly consistent with the general petrogenetic patterns and predict the observed temporal variations in composition.

3.1. Provincial effects

Variations in crustal compositions have long been used to explain differences in both granitoids and regional metallogeny. As summarised in Section 2.2, the overall variation in igneous characteristics and mineralisation between regions in southwestern North America bears a general correspondence to regional crustal differences. Likely examples illustrating compositional maturity were noted earlier. Metamorphic maturity may also be significant. Granulite facies sources have been suggested to be a significant factor in the generation of dry fluorine-rich Cenozoic magmas such as topaz rhyolites and their mineralised intrusive equivalents (Christensen & Lee 1986; Ruiz 1985). Crust undergoing first-time high-grade regional metamorphism could contribute to the abundance of two-mica granitoids in the Great Basin and elsewhere (Barton 1990; Patino Douce et al. 1990).

Overall, provincial effects must be second order, as evidenced by the compositional heterogeneity within uniform domains and the many similarities between areas with distinct crustal histories. It is apparent that the broad chemical characteristics are not strongly dependent on crustal age or composition. In contrast with isotopic correlations, this lack of strong regional chemical correlations indicates that factors such as bulk compositional factors, crustal maturity or evolutionary history are not the dominant factors controlling composition.

3.2. Physical controls

Feedbacks among thermal regimes, state of stress, crustal thickness and subcrustal magma flux (Fig. 7) are helpful in rationalising the temporal distribution and compositions of Cordilleran granitoids (Figs 3, 8). Warm crust enhances assimilation or anatexis by reducing the thermal input needed for melting; it also eases deformation. Extensional tectonics and cool crust favour magma rise due to a favourable minimum stress orientation for dyking and a less effective crustal density filter. Conversely, thick crust may impede the rise of mafic and intermediate magmas due to the lower density of the crust and the presence of intervening felsic melts. Thick crust and elevated subcrustal magma flux lead to higher temperatures and larger degrees of melting.

Some likely consequences of these factors for open system magmatic evolution are illustrated in Figure 9. Distinct evolutionary paths will result from contrasting thermal, thickness, and stress regimes. Cool, thin crust, particularly if under tension, should contribute only modestly to mantle-derived magmas. Isotopically primitive magmas would result, with their compositional evolution dominated by fractional crystallisation (Fig. 9A), 'cool or tensile' path). This situation

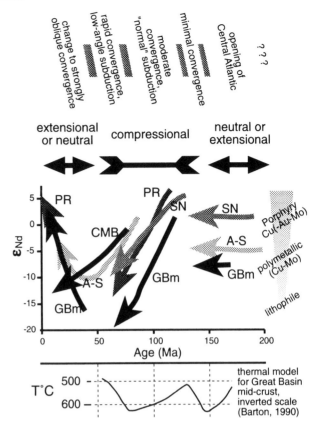

Figure 8 Tectonism, Nd isotopic trends and mineral deposit types for granitoids in southwestern North America. Systematic changes in magmatism reflect tectonics and the temperature of the crust. Locations, abbreviations and data are given in Figure 6.

appears to be appropriate for much of the Jurassic and perhaps for the Late Cenozoic. Extension-related mafic–intermediate volcano-plutonic complexes emplaced in island arc (thin?) crust in western Nevada and California have positive ε_{Nd} (Saleeby et al. 1989); likewise much of the mildly alkaline Jurassic arc inboard and south of these areas appears to have a dominantly mantle heritage (Miller 1978). Late Cenozoic felsic magmatism post-dates major crustal extension and also has only a small crustal component (Perry et al. 1993). The reduced volumes of these two magmatic episodes compared to Cretaceous through mid-Tertiary episodes may reflect cooler ambient crust and thus reduced crustal incorporation into those magmatic centers. DePaolo et al. (1992) proposed similar thermal effects to rationalise the contrasting characteristics of large and small felsic volcanic centres in the Tertiary.

In contrast with suites developed in cool crust, magmas generated in tensional regimes with thicker or warmer crust would probably be more varied. The lower crust could be heated by the local influx of basaltic magma or regional asthenospheric uprise during an extensional episode. Alternatively, the base of thick crust at the beginning of extension would be warm in its own right. These high temperatures allow the assimilation of moderate to large amounts of lower crust (Fig. 9A, 'warm and tensile' path). In the most favourable thermal cases, some magmas could be dominated by lower crust. The mid-crust may contribute less to magmatism during extension than it does during compression because either magma rise is more easily accommodated by extension or the relatively short-lived extensional events did not significantly heat the mid-crust. These processes, coupled with a leaky density filter of extending hot crust, could be responsible for much of the magmatic (and thus

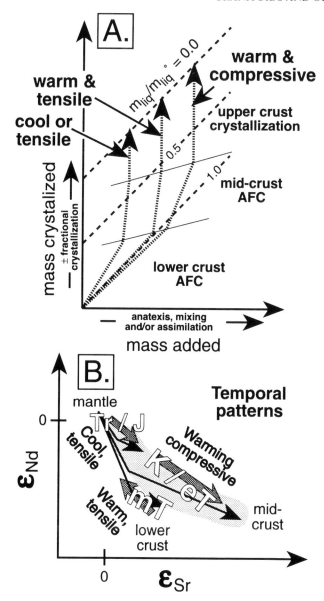

mantle-rich to more felsic, crust-rich compositions (Figs 3, 8). These episodes correspond to long periods of stable convergence. Although the overall trends are clear, their interpretation is open. The preferred model here is: firstly, that the increasing crustal content of granitoids primarily reflect warming of the crust over time (Fig. 9, 'warm and compressive path') and, secondly, that warm thick crust under compression retards mafic and even intermediate magmas from rising into the upper crust. Considerable evidence points to crustal warming (e.g. Haxel *et al.* 1984; Miller *et al.* 1987) as a response to combinations of crustal thickening (Farmer & DePaolo 1983; Patino Douce *et al.* 1990), increased mantle heat flux (Armstrong 1982; Barton 1990) or higher subcrustal magmatic input (Speed *et al.* 1988; Barton & Hanson 1989; DePaolo *et al.* 1992; Leventhal *et al.* 1995). Alone or in combination, these factors rationalise the secular trends inboard of the coastal batholiths where variations in crustal composition are subordinate. In the Sierra Nevada and Peninsular Range batholiths, the progression in time and space from quartz diorites to granites may reflect progressive heating of the crust or migration eastwards from Mesozoic arc lithosphere onto cratonal lithosphere (Chen & Moore 1981; Silver *et al.* 1979). Although isotopes clearly indicate changing lithosphere from west to east, progressive heating may better explain bulk composition shifts given (1) the magmatic diversity in any particular region and (2) the similarity in plutons of similar age as the batholiths cut north and south across lithospheric boundaries.

3.3. Summary of magmatic controls

In summary, the physical processes involved in magma generation dominate the patterns seen in granitoids of southwestern North America (Fig. 9B). The composition of the local lithosphere is important in isotopic systematics and may impact, but does not appear to govern, bulk compositions. This creates difficulties for models that call on source-dependent melting either in the crust (e.g. Roberts & Clemens 1993) or lithospheric mantle (e.g. Coleman *et al.* 1992). Control by thermal and mechanical regimes, as outlined earlier, better rationalises patterns of magmatism (Fig. 7; e.g. Barton 1990; Sandiford *et al.* 1992).

Early to middle Mesozoic arc systems exhibit heterogeneous compositions without pronounced secular variation. Mid-Tertiary compositional heterogeneity is pronounced in extensional domains, but is subdued in the less extended arcs of western and southern Mexico and Colorado–New Mexico. Magmatic flare-ups during times of transtension suggest that this variability is due primarily to the variable interaction of mantle-derived melts with transiently heated lower crust. Extension accommodated the rise of multiple types of magmas ranging from basaltic to strongly peraluminous to rare peralkaline. Larger centres evolved substantial volumes of felsic calc-alkaline melts, whereas smaller centres were varied. Differences in crustal contributions to Jurassic and mid-Tertiary extension-related granitoids may be due to the presence of thicker and warmer(?) crust in the mid-Tertiary. Uniform evolution from mantle-dominated to crust-dominated magmas in the Cretaceous and early Tertiary reflects warming and thickening crust, the consequence of stable convergence and compressional tectonics.

4. Controls on mineralisation

Hydrothermal activity is a natural consequence of epizonal magmatism; most shallow intrusive centres show some evidence of mineralisation (Barton *et al.* 1991). The scale and efficiency of igneous-related hydrothermal systems in concentrating

Figure 9 Schematic paths for open-system magmatic behaviour in contrasting tectonic environments. (A) Mixing, assimilation and melting compared with the removal of crystals and volatiles for different parts of the crust and different physical regimes. m_{liq}/m_{liq}° is the mass of melt divided by the original mass of melt. (B) Schematic trends observed in Nd and Sr isotopic systematics in southwestern North America (cf. Fig. 8) as a function of physical regime. The dark arrows represent overall changes, the light arrows indicate paths from (A).

metallogenic) diversity of the mid-Tertiary and Jurassic. Many Jurassic and mid-Tertiary volcano-intrusive suites exhibit a broad range in bulk compositions and a dominant crustal component (e.g. Asmerom *et al.* 1991; Fig. 5). For example, compositionally diverse Jurassic intrusions (160 ± 10 Ma) in the central Great Basin range from two-mica granites with ε_{Nd} as low as -16 to monzonites with ε_{Nd} near 0 (Barton, unpublished data; Farmer & DePaolo 1983; Wright & Wooden 1991). Likewise, voluminous crust-dominated mid-Tertiary magmatism throughout the Basin and Range province is consistent with highly variable mixing of hot lower crustal materials with moderate amounts of mantle-derived melt. Chemically primitive, isotopically evolved magmas that are commonly observed in these settings are consistent with this type of heat balance (Reiners *et al.* 1995).

The compressive, warming regimes of the Cretaceous and Laramide generated a systematic change from intermediate, materials are governed by multiple factors, of which the two most important are probably compositional and fractionation controls (Fig. 7, group II). Before the importance of these direct controls is considered, it is appropriate to remove the filter of preservation and exposure (Fig. 7, group III). Differences in level of exposure give profoundly different views of the same igneous system (e.g. the Jurassic volcano-plutonic complex at Yerington, Nevada; Dilles 1987) and thus can strongly bias comparisons of regions with different erosional histories.

4.1. The preservation/exposure filter

It has been long recognised that most ore deposits form in the upper few kilometres of the crust (Lindgren 1933; Sawkins 1990). This has been made explicit in models of some igneous-related hydrothermal systems such as porphyry copper deposits (e.g. Sillitoe 1973). In southwestern North America, the distribution of porphyry type deposits reflects exposure level (Fig. 10; Barton *et al.* 1988, 1995). Areas with extensive exposure of hypabyssal intrusive regimes (1–5 km depth) commonly host intrusion-related mineralisation. In the uppermost 1–2 km, epithermal systems predominate. Below 4–6 km palaeodepth, stockwork style mineralisation becomes rare, although skarns are common to 8–10 km depth. This depth variation in mineralisation has complementary variations in the nature of contact metamorphic aureoles and can be rationalised in terms of the energetics of the magmatic systems (Barton *et al.* 1991).

Erosion and tectonic denudation first expose and subsequently destroy the ore-bearing intervals. Burial by younger rocks under favourable circumstances may preserve deposits for longer periods. The time–space patterns described in Section 2 are consistent with these notions—most types of epizonal mineral deposits are sparse in Mesozoic rocks, whereas barren batholiths are common. Porphyry deposits are common in the Late Cretaceous and early Tertiary, and epithermal deposits are most abundant in the mid-Tertiary (Fig. 4).

A simple predictive model that rationalises these variations

is shown in Figure 11. This approach predicts the fraction of a particular crustal interval that is preserved as a function of time across an orogenic belt. The model is based on a diffusional model for the removal of topographic relief and local isotostatic compensation (e.g. cf. Koons 1989). In the example shown, thick crust (typical of the Cretaceous to middle Tertiary crust of much of southwestern North America) is eroded across a symmetrical orogen. The parameters are chosen to yield plausible results across a variety of examples; they are not well constrained by independent data. The test of the model is the internal consistency among the various geological examples. An additional geological constraint, not shown in Figure 11, is the timing of the end of major erosion either by change of base level (e.g. crustal extension) or by burial.

The model results are consistent with the notion that most

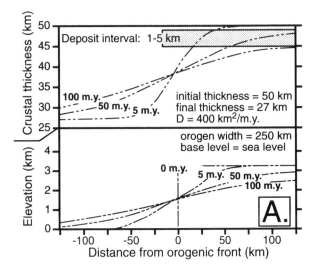

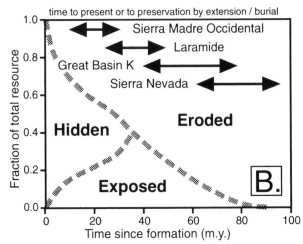

Figure 11 Results of a diffusional model for exposure and removal of intrusion-related mineral deposits. These results simulate conditions relevant to the Andean type margin of southwestern North America in the late Mesozoic and early Tertiary. In this example, the orogen is considered to be 250 km wide and symmetrical about its centre (thus 125 km half-width). Erosion of the key crustal interval begins at time 0 with the initial topographic profile and the isostatically compensated crustal thickness taken as a step function. (A) Crustal thickness (top) and elevation above sea level (bottom) for three time intervals after inception of erosion. Model parameters are given in the figure. (B) Fractional amount of mineralised interval (here taken as 1–5 km original depth) that is hidden, exposed and removed for the model in (A). The arrows show approximate times for which each of the labelled regions underwent erosion before the present. In some instances, erosion was effectively terminated by extension before the present (e.g. in the Great Basin, extension took place at about 35 Ma, 40–80 Ma after the formation of the Cretaceous granitoids and related ore deposits).

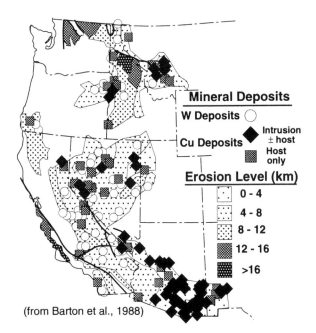

Figure 10 Distribution of Mesozoic and Laramide intrusion-related copper and tungsten deposits as a function of average depth of exposure of granitoids in the western U.S.A. Note the concentration of porphyry copper deposits with the shallowest intrusions and tungsten and copper-bearing skarns with intermediate depths. Similar relationships hold in Mexico (Barton *et al.* 1995).

igneous centres contain some type of high to low temperature, vertically zoned hydrothermal systems (Barton *et al.* 1991). A large fraction of mid-Tertiary porphyry systems, if present, would still be hidden because they are young, except in southern Mexico where the mid-Tertiary arc has undergone rapid erosion. Cretaceous porphyry and epithermal systems would be removed, except in the Great Basin where they may be preserved by a combination of mid-Tertiary extension (40–80 Ma after formation) and their central position in the relatively wide Cretaceous orogen (Fig. 4C). Igneous systems of comparable age in the coastal batholiths are eroded to 5–15 km and expose mainly tungsten skarn systems (Fig. 10). Laramide systems are well exposed in the central and eastern parts of the Laramide arc due to their inboard position and due to major extension following mineralisation by only 25–40 Ma. In the Sierra Madre Occidental of western Mexico, the southern limit of numerous Laramide porphyry copper occurrences is probably caused by burial beneath mid-Tertiary ignimbrites rather than changes in the nature of the arc (Barton *et al.* 1995). Other deposit types will have different preservation potentials (Sawkins 1990). Gold–quartz deposits form at relatively great depths (4–15 km), thus their chance for preservation is higher in older rocks. Volcanogenic massive sulphide deposits form where sedimentary and volcanic rocks are accumulating, thus they will be preserved until tectonically uplifted.

4.2. Igneous controls

It is generally recognised that magmas provide the energy needed to move fluids and to create permeability (e.g. Burnham 1979; Cathles 1981; Norton 1982; Hanson 1995). Less certain are the roles of igneous compositions and magmatic evolution in governing the generation, composition and style of mineralisation. Although provincial effects undoubtedly occur (Section 4.3), most evidence in the southern Cordillera points to a predominance of magmatic compositional and process controls.

Compositional controls. Compositional controls on mineralisation are implicated in porphyry and skarn mineralisation by the involvement of magmatic fluids and the correlations between igneous compositions and metal suites (Fig. 1; Section 2). In contrast, metal suites correlate poorly with the compositions of associated igneous rocks in deposit types that lack evidence for prominent magmatic fluid involvement (Table 1). Two compositional controls are important: firstly, the concentration (abundance or scarcity) of ore-forming components in the magma; and, secondly, equilibria imposed by the minerals ($\pm$melt) on aqueous fluids. Both reflect the petrogenetic history, but could be produced in different settings.

Relatively little has been systematically published on the concentrations of ore metals in granitoids in southwestern North America, in contrast, for example, to the data sets for eastern Australia (Blevin & Chappell 1992; Blevin *et al.*, this issue). Studies of copper mineralisation show that associated intermediate composition intrusions have rather ordinary concentrations of copper. Some data indicate that copper concentrations reach maxima in the intrusive suites immediately before the ore-forming ('productive') intrusions are emplaced (Dilles 1987; Hendry *et al.* 1985; Wodzicki 1995). Subsequent depletion presumably reflects the loss of metal from the magma to ore-forming fluids. The scarcity of economic granite-related uranium and tin deposits in the southern Cordillera reflects the low concentrations of those elements in granitoids that might otherwise be appropriate sources (Shaw & Guilbert 1990). Perhaps the best general evidence for a bulk igneous control is the broad correlation of chalcophile and siderophile elements with intermediate

composition intrusive suites and of lithophile elements with felsic intrusive suites, which contain, respectively, higher concentrations of their respective ore-forming elements.

In contrast with the modest role inferred for bulk enrichments or depletions, consideration of aqueous fluid–mineral $\pm$ melt equilibria systematises many aspects of intrusion-related mineralisation in Mexico and the western U.S.A. (Barton *et al.* 1995). Perhaps best established is the role of oxidation state, which strongly affects the mobility of metals and the nature of related alteration (e.g. Burnham & Ohmoto 1980; Ishihara 1981; Candela 1992). Oxidised systems (magnetite $\pm$ sphene-bearing) that are rich in sulphur produce voluminous acid alteration by hydrolysis of sulphur hydroxide, mobilise abundant chalcophile elements and exhibit oxidised and pyrite-rich alteration assemblages. Virtually all copper-dominated porphyry and skarn deposits belong to this group (Titley & Beane 1981; Einaudi 1982). Oxidised, but less sulphur-rich systems are broadly similar in character, but they have less abundant acid alteration and commonly less pyrite. In the southern Cordillera, these systems include iron oxide-rich skarns associated with dioritic stocks (Einaudi *et al.* 1981) and porphyry molybdenum deposits associated with felsic metaluminous to weakly peraluminous stocks (Mutschler *et al.* 1981). Reduced systems (magnetite-poor) have more common lithophile elements, such as tungsten, tin and beryllium, and lack abundant pyrite and acid alteration. Cordilleran types are mainly skarn and greisen lithophile element deposits that are associated with weakly to strongly peraluminous granites (Barton 1990; van Middelaar & Keith 1990), but may also include some tin-bearing rhyolites.

Beyond redox state, igneous mineralogy helps to rationalise other facets of metallogenic systematics. Figure 12 shows a simplified petrogenetic grid for quartz-saturated granitoids in terms of the activities of lime (CaO) and alumina (Al_2O_3). Limiting and continuous mineral buffers (e.g. reactions 1a–d; Fig. 12A)

$$KAlSi_3O_8 + [Al_2O_3] + H_2O = KAl_2[AlSi_3O_{10}](OH)_2 \quad (1a)$$

$$[CaO] + MgSiO_3 + SiO_2 = CaMgSi_2O_6 \quad (1b)$$

$$[CaO] + [Al_2O_3] + 2SiO_2 = CaAl_2Si_2O_8 \quad (1c)$$

$$3[CaO] + 3FeTiO_3 + 3SiO_2 + 0 \cdot 5O_2 = 3CaTiSiO_5 + Fe_3O_4 \quad (1d)$$

allow common rocks to be placed on this diagram (Fig. 12B). Peralkaline rocks occur on the left (sodic pyroxene buffer), strongly peraluminous rocks on the right (cordierite and andalusite buffers). Plagioclase becomes more sodic to the lower left. This approach is similar to that pioneered by Carmichael *et al.* (1974) for volcanic rocks; some aspects have been applied to ore deposits by others including Burnham (1979), Candela and Picoli (1995) and Urabe (1985).

The utility of this approach to fluid–rock interaction comes from consideration of reactions such as Reactions (2a) and (2b).

$$NaAlSi_3O_8 + H^+ = Na^+ + 0 \cdot 5[Al_2O_3] + 3SiO_2 + 0 \cdot 5H_2O \quad (2a)$$

$$[CaO] + 2HF = CaF_2 + H_2O \quad (2b)$$

Reaction (2a), when evaluated for appropriate plagioclase compositions [Reaction (1c)], shows that a_{Na^+}/a_{H^+} increases to the right on Figure 12. This helps explain the distribution of metals and the abundance of acid alteration with rock type (Fig. 12C). Because the transport of base metals is strongly pH dependent (e.g. $MS + 2H^+ = M^{+2} + H_2S$), they will move in metaluminous and peraluminous systems preferentially over alkaline systems. A similar rationale applies to the development of acid alteration and supergene enrichment. Granodiorites

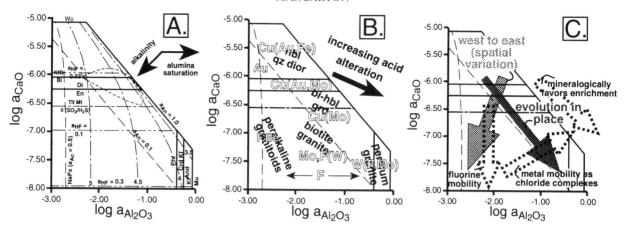

Figure 12 Lime (CaO)–alumina (Al_2O_3) activities in quartz-saturated igneous rocks, illustrative reactions and possible relationships to associated mineralisation. Calculated at 600°C and $P_{H_2O} = 1$ kbar using data mainly from Helgeson *et al.* (1978) assuming ideal site mixing for solid solutions. (A) Mineral–mineral and mineral–fluid reactions. (B) Illustrative reactions are given in the text [Equations (1a–1d), (2a–2b)]. Igneous rock and metal types (cf. Fig. 1). (C) General trends in time (dark arrow) and space (grey arrow) for southwestern North America and trends in fluid–rock interaction (dotted arrows). See text for discussion.

and monzogranites that lie to the right on the diagram offer relatively weak buffers against hypogene or supergene acid alteration. Acid alteration is virtually absent in peralkaline systems and, although the most voluminous acid alteration occurs with sulphur-rich metaluminous rocks, greisen-type (i.e. muscovite-stable) alteration occurs to the highest temperatures in Cordilleran peraluminous systems (e.g. Barton 1987). For similar reasons, supergene enrichment is commonly well developed in pyrite-rich porphyry copper deposits formed in granodiorites and monzogranites, but poorly developed in diorite-related systems. These systematics [Reaction (2b), Fig. 12C] also rationalise the abundance of fluorine-rich deposits across a wide spectrum of calcium-poor rock types from peralkaline (as in the mid-Tertiary of eastern Mexico, Fig. 4E) through metaluminous (Climax-type porphyry molybdenum deposits) to strongly peraluminous (as in the Great Basin; Barton 1987).

Magmatic evolution. Enrichment of metals and volatiles by fractional crystallisation have long been advocated to contribute to the formation of metal deposits (e.g. Lehmann 1987; Blevin & Chappell, 1992; Pollard & London 1995). Conversely, a lack of fractionation, mixing with unenriched materials and episodic devolatilisation (e.g. by eruption) would all probably be unfavourable for mineralisation. Mass balance and energy balance require that metals, volatiles and heat are extracted from tens to hundreds of cubic kilometres of magma to form ore deposits (e.g. Hedenquist & Lowenstern 1994). Thus the small exposed areas of many ore-related intrusions in the western U.S.A. are only the tops of those systems (Barton *et al.* 1991: 811–5) making it difficult in most instances to evaluate the magmatic history, especially where alteration is extensive.

Evidence for the late-stage magmatic enrichment of metals or other components is equivocal in the southern Cordillera. Most studies have found little systematic difference between mineralised and barren centres or within single centres (e.g. Anthony & Titley 1988; Lang 1991). The best tested hypothesis, noted earlier, is that metal loss to hydrothermal fluids causes the depletions observed in 'productive' intrusive phases. Although metal and volatile concentrations have not been systematically documented, other geochemical evidence is consistent with the best mineralised intrusions being magmatically well mixed, but closed to significant contamination by mixing or assimilation in their late stages. This is illustrated in Figure 13, which schematically summarises [Cl], [Sr] and

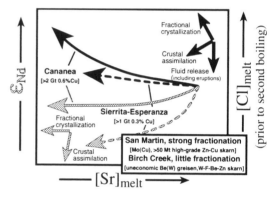

Figure 13 Schematic compositional changes and links to mineralisation for several Cretaceous and early Tertiary intrusive centres. Light arrows indicate Nd isotopic changes as a function of magmatic evolution (monitored by Sr concentration), whereas the dark arrows indicate changes in Cl concentration in magma or the coexisting aqueous fluid. Generation of evolved magmatic compositions, which should be most favourable for mineralisation (e.g. high Cl), is promoted by fractional crystallisation. Late-stage assimilation or mixing (changes in ε_{Nd}), fluid release (decrease in [Cl]) or lack of fractionation (no change in [Sr]) would be unfavourable for mineralisation. Cananea, solid lines; Sierrita, broken lines. See text for data sources and discussion.

ε_{Nd} evolution for several well-studied intrusive suites associated with hydrothermal activity. Two world-class deposits, the granodiorite-associated porphyry Cu(Mo) system at Cananea, Sonora and the granite-associated Zn(–Cu) skarn at San Martin, Zacatecas show considerable crystal fractionation, but little Nd isotopic change (Graf & Barton 1995; Wodzicki 1995). High-salinity fluids released in the later stages of both systems formed high-grade base metal mineral deposits. The Sierrita, Arizona porphyry Cu(Mo) system, which is similar in age and size to Cananea, shows larger ε_{Nd} variations and has significantly lower hypogene copper grades (Anthony & Titley 1988). Contamination of a fractionating magma by the continuing incorporation of new crustal material (as indicated by changes in ε_{Nd}) would dilute previously fractionated magma, hindering generation of an highly evolved, Cl-rich fluid and related high-grade mineralisation. The Late Cretaceous Birch Creek pluton formed a large barren magmatic hydrothermal system through the more or less continuous evolution of Cl-poor fluid from a strongly peraluminous magma (Barton & Trim 1991). Isotopic and other geochemical data for Birch

Creek show that multiple magma batches were emplaced, but underwent little mixing and large-scale crystal fractionation (Barton *et al.* 1994). Consequently, fluids remained Cl-poor and metals were not concentrated.

Similar reasoning can be applied to more deeply exposed batholiths. Suites in the Sierra Nevada such as the Tuolumne intrusive series, which have very large volumes of uniform magma with fairly uniform isotopic compositions (although there is variation between major phases; Kistler *et al.* 1986), could have had large mineralised systems at shallower levels. Higher levels might have looked much like the current distribution of the Tertiary porphyry copper deposits of Chile and Peru. At present exposure depths (10 ± 5 km), the presence of large W(–Mo) skarns with some of these plutons is consistent with the notion that they were productive overall. For example, the largest tungsten deposit in the U.S.A., the Pine Creek deposit in the central Sierra Nevada, is related to the Mono Crest pluton, which is contemporaneous with the Tuolumne suite (Bateman 1965). On the other hand, bodies composed of multiple pulses of material that did not homogenise, such as the San Jacinto pluton of the northern Peninsular Ranges (Hill & Silver 1988), may not have produced large hydrothermal systems, consistent with the scarcity of deep skarn mineralisation in that region (or, alternatively, with greater overall depth; Barton *et al.* 1988).

Figure 14 compares Nd concentration and isotopic data for pairs of primitive and evolved rocks from 23 dioritic to granitic intrusive centres in the southwestern U.S.A. and Mexico. Extreme compositions and volumetrically minor phases (e.g. aplites and mafic dykes) were excluded. Well mineralised systems exhibit only small isotopic variations ($\leqslant \pm 1.5$ epsilon units), which is consistent with the lack of late-stage assimilation or magma injection in the later stages. In contrast, late-stage assimilation and magma mixing appear to be common in barren and many poorly mineralised systems. Changes in Nd concentrations do not correlate well with mineralisation, although the most evolved compositions are those associated with some ore deposits. Other elements such as Rb or Sr would be superior monitors of crystal fractionation if fresh rocks can be found, but these are difficult to find in many mineral districts. This model should be tested in other regions, such as the central Andes, for which appropriate data might readily be acquired.

Lipman and Sawyer (1985) suggested that caldera-forming

eruptions vent the volatile content of large intrusive centres, precluding the formation of magmatic–hydrothermal ore deposits. Although sensible, the evidence for this hypothesis remains equivocal. It is consistent with the paucity of porphyry type systems in much of the mid-Tertiary and perhaps in the Jurassic (Fig. 4). Nevertheless, some caldera systems do have large porphyry deposits as at Questa, New Mexico (the Latir volcanic complex; Johnson *et al.* 1990; Lipman 1983; see Elston 1994). Contrasting levels of exposure of caldera complexes and porphyry type systems may explain some of the discrepancy (Section 4.1), as also might recharge of chambers over prolonged magmatic histories, for example as indicated by geochronometry in the Laramide (Titley 1982; McCandless & Ruiz 1993). Although the loss of volatiles during eruption may reduce the ore-forming potential of some systems in the western U.S.A., lack of eruption does not necessarily enhance mineralisation. Many intrusive centres did not vent and yet show only minor economic mineralisation at their tops (e.g. two-mica granites of the Cordilleran interior; Barton 1987; Shaw & Guilbert 1990).

A final magmatic control, as yet poorly understood, is the efficiency of physical and chemical focusing of ore-forming fluids, both at the late magmatic stages and post-magmatic stages (e.g. Norton 1982; Candela 1991; Hanson 1995). Although current models are successful in rationalising many observations in particular systems, it remains difficult to predict the broader patterns observed in southwestern North America.

4.3. Other controls

Each of the regions in Figure 2 shows evidence for many types of intrusion-related ore deposits. The details differ—dioritic Cu(–Au–Fe) to granitic W(–Mo) in the coastal batholiths to monzonite Au(–Cu) to alkali granite Mo(–F–W) in Colorado—but the parallels are obvious. Although most variation in metal contents appears to be a function of magma composition and evolution, some variations appear to be clearly linked to local crust independent of magmatism. Titley (1987) has shown that the systematic differences in gold to silver ratios between Proterozoic terranes in Arizona are independent of deposit type or age. Other regional metal suites have been interpreted in this framework, such as the restriction of lead-rich deposits to cratonic crust in the Cordilleran interior (e.g. Clark *et al.* 1982; Titley 1993), nominally because of the increased content of lead in mature crust. Such an interpretation is supported by abundant lead isotope data, which show that ore lead is usually representative of local crust (e.g. Zartman 1974). Some regional variation is also consistent with the physical characteristics of contrasting crust, primarily through the influence of physical factors on magmatism. This can be seen, for example, in the contrast between mainly 'cratonal' Cu(Mo) porphyry deposits in Arizona and Sonora which are associated with granodioritic intrusive suites and the mainly 'island arc' Cu(Au) porphyry deposits of southwestern Mexico, which are associated with intermediate (dioritic) intrusive centers, a pattern recognised globally (cf. Titley 1982). Even though similar overall igneous compositional ranges occur in both settings, the large, well-mixed, efficiently fractionating intrusions that should be most favourable for mineralisation ought to be more felsic in crust that is either more fertile or a better density filter (i.e. 'cratonal' settings).

Lastly, igneous-related deposits that form primarily from non-magmatic fluids (most epithermal, volcanic massive sulphide, iron oxide; Table 1) reflect the availability of meteoric, marine and evaporitic waters, respectively. This is fundamentally a palaeogeographical control which is readily apparent

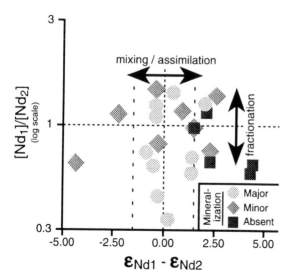

Figure 14 Nd concentration ratios versus maximum difference in ε_{Nd} for 23 intrusive suites from southwestern North America. Degree of mineralisation is based on present level of exposure. See text for discussion. Sources are given in Figure 6.

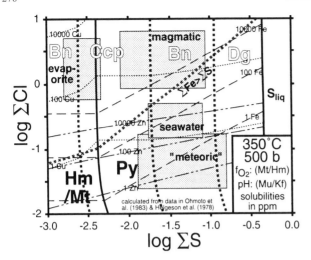

Figure 15 Generalised ore fluid compositions (boxes), Cu–Fe–O–S mineral stabilities (heavy lines) and Cu, Zn, Fe solubilities (in ppm; thin lines) as a function of total chlorine and total sulphur. These relationships rationalise some of the differences among various types of igneous-related ore deposits as a function of plausible magmatic or external fluid sources.

in the time–space distribution of these deposits (Fig. 4). The gold–quartz deposits (Mother Lode, Carlin), which have only a weak correlation with magmatism, have an even stronger correlation with region (forearc and extensional systems, respectively) and host/source rocks (greenschist facies metamorphism of reduced sedimentary or volcanic rocks).

Magmatic energy may drive these systems and magmatic constituents may dominate the materials transported and thus influence the metal ratios and some aspects of the alteration. Nevertheless, their overall metallogenic characteristics are governed by the fluids that are involved (Fig. 15). Dilute fluids (e.g. meteoric waters) are relatively ineffective at transporting copper and iron at moderate and low temperatures, but they can be fairly effective at transporting gold and variable amounts of silver, lead and zinc. These suites characterise epithermal and gold–quartz deposits. Modestly saline fluids (seawater or contaminated meteoric water) can transport a fair amount of base metal, leading to either base metal rich epithermal systems or volcanogenic massive sulphide deposits. High-temperature deposits formed from such fluids are commonly fairly poor in iron and base metals, but rich in lithophile elements such as tungsten and molybdenum that do not require chloride complexing. High-salinity fluids generated either by second boiling or by the involvement of evaporitic sources will transport abundant metals (Fig. 15). If sulphur is limited, only the most chalcophile elements (primarily copper) will precipitate as sulphides; other precipitates will have to form oxides (iron, rare earth elements) or native elements (e.g. gold). This suite is typical of the diorite type porphyry deposits and igneous-related iron oxide deposits of the Kiruna or Olympic Dam types. The latter are commonly attributed to a direct magmatic origin, but may form from hydrothermal fluids with evaporitic components as indicated by palaeogeography and lack of correlation with magma type (Barton & Johnson 1996).

5. Conclusions

Granitic magmatism and mineralisation have been studied extensively for a century in the southwestern U.S.A. and Mexico. The bulk compositions of granitoids systematically reflect their tectonic environments; local crust is reflected most strongly in isotopic signatures. In turn, the broad characteristics of igneous-related mineralisation reflect the type of

magmatism, the efficacy of late-stage fractionation, the nature of external fluids and hosts, and exposure level. All of these factors must be considered together to fully rationalise the distribution of granitoid-related ore deposits.

Two modes of granitic magmatism stand out: (1) progressive compositional evolution over tens of millions of years that is associated with compressional tectonics; and (2) heterogeneous magmatism of variable duration that is associated with neutral or extensional regimes. These modes are seen, respectively, in the Cretaceous and Laramide, and the Jurassic and mid-Tertiary of the Basin and Range province. The heterogeneity of magmatism in local regions of uniform lithosphere and the homogeneity of coeval belts across lithospheric boundaries argues against a primarily lithospheric control. The temporal, spatial and compositional patterns in the southern Cordillera can be rationalised semi-quantitatively in terms of the combined duration of magmatism and the effects of thermal and stress regimes. Warm crust is necessary for abundant generation of felsic (crust-dominated) magmas by assimilation or direct melting. Warm, thick crust under compression will provide an effective density filter impeding magma rise, particularly for relatively mafic compositions. Observed compositional progressions in the Cretaceous and in the Laramide have approximately the right time-scales for lithospheric heating. Extending crust will not provide as effective a density filter and thus could more easily generate the spectrum of magmatic compositions that occur synchronously in the Jurassic and mid-Tertiary. Magmatic episodes associated with major extension have highly variable duration and, at best, show modest secular variation. Geological evidence and the application of a simple erosional model explain the overall temporal and spatial distributions of hydrothermal, metamorphic and igneous rocks.

In the southern Cordillera, systematic regional variations in metal ratios suggest an inherited (basement) control in some regions, but the overall variability within single provinces points to the importance of igneous and external fluid controls. Igneous enrichments and mineralogical (major element) controls offer promising explanations for many aspects of the metallogeny of the southern Cordillera. Unusual primary enrichments are not documented, although the overall correspondence of metal suites to igneous suites with similar trace element compositions supports a broad control. Fluid compositions governed by igneous oxidation state, alkalinity (a_{CaO}) and aluminousity ($a_{Al_2O_3}$) help rationalise metal transport and the volumes and types of hydrothermal alteration. Efficient evolution and a lack of catastrophic volatile loss appear to be keys for generating productive systems. Geochemical and geological evidence indicates that intrusive centres which underwent late mixing or did not mix well are less likely to form economic deposits. The formation of large deposits during the late stages of the evolution of major intrusive centres may be the consequence of advanced evolution or a requirement for preservation against subsequent magmatic events. The chlorine and sulphur content of hydrothermal fluids, magmatic or external, rationalise the distribution of deposit types in different Cordilleran environments.

6. Acknowledgements

I thank Jane Hammerstrom and Phil Candela for helpful reviews. This work has been supported by the National Science Foundation (EAR 91-17372, EAR 90-96294) and the University of Arizona Mexico Mineral Resource Consortium. Thanks also to the convenors of Hutton III for their invitation to present this work and for their patience during completion of the manuscript.

7. References

Ague, J. J. & Brimhall, G. H. 1988. Regional variations in bulk chemistry, mineralogy, and the compositions of mafic and accessory minerals in the batholiths of California. GEOL SOC AM BULL **100**, 891–911.

Albers, J. P. 1981. A lithologic–tectonic framework for the metallogenic provinces of California. ECON GEOL **76**, 765–90.

Albino, G. V. 1995. Porphyry copper deposits of the Great Basin—Nevada, Utah, and adjacent California. *In* Pierce, F. W. & Bolm, J. G. (eds) *Porphyry copper deposits of the American Cordillera.* ARIZONA GEOL SOC DIG **20**, 267–96.

Anderson, J. L. 1990. The nature and origin of Cordilleran magmatism. GEOL SOC AM MEM **174**.

Anderson, J. L. & Morrison, J. 1992. The role of anorogenic granites in the Proterozoic crustal development of North America. *In* Condie, K. C. (ed.) *Proterozoic crustal evolution.* DEV PRECAMBRIAN GEOL **10**, 263–99.

Anthony, E. Y. & Titley, S. R. 1988. Progressive mixing of isotopic reservoirs during magma genesis at the Sierrita porphyry copper deposit, Arizona; inverse solutions. GEOCHIM COSMOCHIM ACTA **52**, 2235–50.

Armstrong, R. L. 1982. Cordilleran metamorphic core complexes—from Arizona to southern Canada. ANNU REV EARTH PLANET SCI **10**, 119–54.

Asmerom, Y., Patchett, P. J. & Damon, P. E. 1991. Crust–mantle interaction in continental arcs: inferences from the Mesozoic arc in the southwestern US. CONTRIB MINERAL PETROL **107**, 124–34.

Barnes, C. G., Petersen, S. W., Kistler, R. W., Prestvik, T. & Sundvoll, B. 1992. Tectonic implications of isotopic variation among Jurassic and Early Cretaceous plutons, Klamath Mountains. GEOL SOC AM BULL **104**, 117–26.

Barton, M. D. 1987. Lithophile element mineralization associated with Late Cretaceous two-mica granites in the Great Basin. GEOLOGY **15**, 337–40.

Barton, M. D. 1990. Cretaceous magmatism, mineralization and metamorphism in the east-central Great Basin. *In* Anderson, J. L. (ed.) *The nature and origin of Cordilleran magmatism.* GEOL SOC AM MEM **174**, 283–302.

Barton, M. D. & Hanson, R. B. 1989. Magmatism and the development of low-pressure metamorphic belts: implications from the western United States and thermal modeling. GEOL SOC AM BULL **101**, 1051–65.

Barton, M. D. & Johnson, D. A. 1996. An evaporitic-source model for igneous-related Fe-oxide(–REE–Cu–Au–U) mineralization. GEOLOGY **24**, 259–62.

Barton, M. D. & Trim, H. E. 1991. Late Cretaceous two-mica granites and lithophile-element mineralization in the Great Basin. *In* Schafer, R. W. & Wilkinson, W. H. (eds) *Geology and ore deposits of the Great Basin, Geological Society of Nevada Symposium Proceedings,* 529–38. Reno: Geological Society of Nevada.

Barton, M. D., Battles, D. A., Bebout, G. E., Capo, R. C., Christensen, J. N., Davis, S. R., Hanson, R. B., Michelsen, C. J. & Trim, H. 1988. Mesozoic contact metamorphism in the Western United States. *In* Ernst, W. G. (ed.) *Metamorphism and crustal evolution, western conterminous United States, Rubey Volume VII,* 110–78. Englewood Cliffs: Prentice Hall.

Barton, M. D., Staude, J-M., Snow, E. A. & Johnson, D. A. 1991. Aureole systematics. *In* Kerrick, D. M. (ed.) *Contact metamorphism.* REV MINERAL **26**, 723–847.

Barton, M. D., Ghidotti, G. A., Holden, P. & Grossman, J. N. 1994. Petrochemical characteristics of the strongly peraluminous end-member of Cretaceous magmatism in the Great Basin: the Birch Creek pluton. GEOL SOC AM ABSTR PROGRAMS **26**, 369.

Barton, M. D., Staude, J-M. G., Zürcher, L. & Megaw, P. K. M. 1995. Porphyry copper and other intrusion-related mineralization in Mexico. *In* Pierce, F. W. & Bolm, J. G. (eds) *Porphyry copper deposits of the American Cordillera.* ARIZONA GEOL SOC DIG **20**, 487–524.

Bateman, P. C. 1965. Geology and tungsten mineralization of the Bishop district, California. US GEOL SURV PROF PAP **470**.

Bateman, P. C. 1992. Plutonism in the central part of the Sierra Nevada batholith, California. US GEOL SURV PROF PAP **1483**.

Battles, D. A. & Barton, M. D. 1995. Arc-related sodic hydrothermal alteration in the western US. GEOLOGY **23**, 913–6.

Bennett, V. C. & DePaolo, D. J. 1987. Proterozoic crustal history of the western United States as determined by neodymium isotopic mapping. GEOL SOC AM BULL **99**, 674–85.

Blevin, P. L. & Chappell, B. W. 1992. The role of magma sources, oxidation states and fractionation in determining the granite metallogeny of eastern Australia. TRANS R SOC EDINBURGH EARTH SCI **83**, 305–17.

Bohlke, J. K. & Kistler, R. W. 1986. Rb–Sr, K–Ar, and stable isotope evidence for the ages and sources of fluid components in the northern Sierra Nevada foothills metamorphic belt, California. ECON GEOL **81**, 296–322.

Bookstrom, A. A. 1990. Igneous rocks and carbonate-hosted ore deposits of the central Colorado Mineral Belt. ECON GEOL MONOGR **7**, 45–65.

Brown, G. C., Thorpe, R. S. & Webb, P. C. 1984. The geochemical characteristics of granitoids in contrasting arcs and comments on magma sources. J GEOL SOC LONDON **141**, 369–77.

Burchfiel, B. C., Cowan, D. S. & Davis, G. A. 1992. Tectonic overview of the Cordilleran orogen in the western United States. *In* Burchfield, B. C., Lipman, P. W. & Zoback, M. L. (eds) *The Cordilleran orogen, conterminous US. The Geology of North America,* Vol. G-3, 407–79. Boulder: Geological Society of America.

Burnham, C. W. 1979. Magmas and hydrothermal fluids. *In* Barnes, H. L. (ed.) *Geochemistry of hydrothermal ore deposits,* 71–136. New York: Holt, Rinehart and Winston.

Burnham, C. W. & Ohmoto, H. 1980. Late-stage processes of felsic magmatism. *In* Ishihara, S. & Takenouchi, S. (eds) *Granitic magmatism and related mineralization.* MIN GEOL SPEC ISSUE **8**, 1–11.

Campa, M. F. & Coney, P. J. 1983. Tectono-stratigraphic terranes and mineral resource distributions in Mexico. CAN J EARTH SCI **20**, 1040–51.

Candela, P. A. 1989. Magmatic ore-forming fluids: thermodynamic and mass-transfer calculations of metal concentrations. REV ECON GEOL **4**, 203–21.

Candela, P. A. 1991. Physics of aqueous phase evolution in plutonic environments. AM MINERAL **76**, 1081–91.

Candela, P. A. 1992. Controls on ore metal ratios in granite-related ore deposits: an experimental and computational approach. TRANS R SOC EDINBURGH EARTH SCI **83**, 317–26.

Candela, P. A. & Piccoli, P. M. 1995. Model ore-metal partitioning from melts into vapor and vapor/brine mixtures. *In* Thompson, J. F. H. (ed.) *Granites, fluids, and ore deposits.* MINERAL ASSOC CAN **23**, 101–28.

Carmichael, I. S., Turner, F. J. & Verhoogen, J. 1974. *Igneous petrology.* New York: McGraw-Hill.

Cathles, L. M. 1981. Fluid flow and genesis of hydrothermal ore deposits. *In* Skinner, B. J. (ed.) *75th Anniversary Volume.* ECON GEOL 424–57.

Centeno, G. E., Ruiz, J., Coney, P. J., Patchett, P. J. & Ortega, G. F. 1993. Guerrero Terrane of Mexico; its role in the Southern Cordillera from new geochemical data. GEOLOGY **21**, 419–22.

Chen, J. & Moore, J. 1981. Uranium–lead isotopic ages from the Sierra Nevada batholith, California. J GEOPHYS RES **87**, 4761–84.

Christensen, E. & Lee, D. E. 1986. Fluorine and chlorine in granitoids from the Basin and Range province, western United States. ECON GEOL **81**, 1484–94.

Clark, K. F., Foster, C. T. & Damon, P. E. 1982. Cenozoic mineral deposits and subduction-related magmatic arcs in Mexico. GEOL SOC AM BULL **93**, 533–44.

Coleman, D. S., Frost, T. P. & Glazner, A. F. 1992. Evidence from the Lamarck granodiorite for rapid Late Cretaceous crust formation in California. SCIENCE **258**, 1924–6.

Coney, P. J. 1989. Structural aspects of suspect terranes and accretionary tectonics in western North America. J STRUCT GEOL **11**, 107–25.

Coney, P. J. & Reynolds, S. J. 1977. Cordilleran Benioff Zones. NATURE **270**, 403–6.

Criss, R. E. & Taylor, H. P. Jr 1986. Meteoric–hydrothermal systems. *In* Valley, J. W., Taylor, H. P. Jr. & O'Neil, J. R. (eds) *Stable isotopes in high temperature geological processes.* REV MINERAL **16**, 373–424.

Damon, P. E., Shafiqullah, M. & Clark, K. F. 1983. Geochronology of the porphyry copper deposits and related mineralization of Mexico. *In* Dawson, K. M. (ed.) *Symposium; metallogeny and tectonics of the North American Cordillera.* CAN J EARTH SCI **20**, 1052–71.

DePaolo, D. J. 1980. Sources of continental crust: neodymium isotope evidence from the Sierra Nevada and Peninsular Ranges. SCIENCE **209**, 684–7.

DePaolo, D. J. 1981. A neodymium and strontium isotopic study of the Mesozoic calc-alkaline granitic batholiths of the Sierra Nevada and Peninsular Ranges, California. J GEOPHYS RES **86**, 10 470–88.

DePaolo, D. J., Perry, F. V. & Baldridge, W. S. 1992. Crustal versus mantle sources of granitic magmas: a two-parameter model based on neodymium isotope studies. TRANS SOC EDINBURGH EARTH SCI **83**, 439–46.

Dickinson, W. R. 1991. Tectonic setting of faulted Tertiary strata associated with the Catalina core complex in southern Arizona. GEOL SOC AM SPEC PAP **264**.

Dickinson, W. R. & Snyder, W. S. 1978. Plate tectonics of the Laramide orogeny. GEOL SOC AM MEM **151**, 355–66.

Dilles, J. H. 1987. The petrology and geochemistry of the Yerington batholith, Nevada: evidence for the evolution of porphyry copper ore fluids. ECON GEOL **82**, 1750–89.

Einaudi, M. T. 1982. General features and origin of skarns associated with porphyry copper plutons. *In* Titley, S. R. (ed.) *Advances in the geology of the porphyry copper deposits, Southwestern North America*, 185–209. Tucson: The University of Arizona Press.

Einaudi, M. T., Meinert, L. D. & Newberry, R. J. 1981. Skarn deposits. *In* Skinner, B. J. (ed.) *75th Anniversary Volume*. ECON GEOL 327–91.

Elston, W. E. 1994. Siliceous volcanic centers as guides to mineral exploration: review and summary. ECON GEOL **89**, 1662–86.

Farmer, G. L. & DePaolo, D. J. 1983. Mesozoic and Tertiary granite in the western United States and implications for pre-Mesozoic crustal structure—1. Nd and Sr isotopic studies in the geocline of the northern Great Basin. J GEOPHYS RES **88**, 3379–401.

Farmer, G. L. & DePaolo, D. J. 1984. Origin of Mesozoic and Tertiary granite in the western United States and implications for pre-Mesozoic crustal structure. 2. Nd and Sr isotopic studies of unmineralized and Cu– and Mo–mineralized granite in the Precambrian craton. J GEOPHYS RES **89**, 10 141–60.

Field, C. W. & Fifarek, R. H. 1985. Light stable-isotope systematics in the epithermal environment. *In* Berger, B. R. & Bethke, P. M. (eds) *Geology and geochemistry of epithermal systems*. REV ECON GEOL **2**, 99–128.

Foster, D. A. & Hyndman, D. W. 1990. Magma mixing and mingling between synplutonic mafic dikes and granite in the Idaho–Bitterroot Batholith. *In* Anderson, J. L. (ed.) *The nature and origin of Cordilleran magmatism*. GEOL SOC AM MEM **174**, 347–58.

Franklin, J. M., Lydon, J. W. & Sangster, D. F. 1981. Volcanic-associated massive sulfide deposits. *In* Skinner, B. J. (ed.) *75th Anniversary Volume*. ECON GEOL 485–627.

Gans, P. B., Mahood, G. A. & Schermer, E. 1989. Synextensional magmatism in the Basin and Range province; a case study from the eastern Great Basin. GEOL SOC AM SPEC PAP **233**.

Graf, A. & Barton, M. D. 1995. Geology and porphyry-style mineralization of the Cerro de la Gloria stock associated with high-T carbonate-hosted Zn–Cu–Ag(–Pb) mineralization, San Martin district, Zacatecas, Mexico. *In International field conference on carbonate-hosted lead–zinc deposits, St Louis Missouri*, 115–6.

Guild, P. W. 1981. *Metallogenic map of North America: scale 1 : 5,000,000*. Reston: US Geological Survey.

Hanson, R. B. 1995. Hydrodynamics of contact metamorphism. GEOL SOC AM BULL **107**, 595–611.

Haxel, G. B., Tosdal, R. M., May, D. J. & Wright, J. E. 1984. Latest Cretaceous and Early Tertiary orogenesis in south-central Arizona: thrust faulting, regional metamorphism and granitic plutonism. GEOL SOC AM BULL **95**, 631–53.

Hedenquist, J. W. & Lowenstern, J. B. 1994. The role of magmas in the formation of hydrothermal ore deposits. NATURE **370**, 519–27.

Helgeson, H. C., Delaney, J. M., Nesbitt, H. W. & Bird, D. K. 1978. Summary and critique of the thermodynamic properties of rock forming minerals. AM J SCI **278-A**, 1–229.

Hendry, D. A. F., Chivas, A. R., Long, J. V. P. & Reed, S. J. B. 1985. Chemical differences between minerals from mineralizing and barren intrusions from some North American porphyry copper deposits. CONTRIB MINERAL PETROL **89**, 317–29.

Henry, C. D. 1975. *Geology and geochronology of the grantic batholithic complex, Sinaloa, Mexico*. Ph.D. Dissertation, University of Texas.

Henry, C. D. & Aranda-Gomez, J. J. 1992. The real southern Basin and Range, mid- to late Cenozoic extension in Mexico. GEOLOGY **20**, 701–4.

Hill, R. I. & Silver, L. T. 1988. San Jacinto intrusive complex; 3, constraints on crustal magma chamber hypotheses from strontium isotope heterogeneity. J GEOPHYS RES **93**, 10 373–8.

Ilchik, R. P. & Barton, M. D. Evaluation of an amagmatic origin for Carlin-type gold deposits: ECON GEOLOGY, in press.

Ishihara, S. 1981. The granitoid series and mineralization. *In* Skinner, B. J. (ed.) *75th Anniversary Volume*. ECON GEOL 458–84.

Johnson, C. M., Lipman, P. W. & Czmanske, G. K. 1990. H, O, Sr, Nd, and Pb isotope geochemistry of the Latir volcanic field and cogenetic intrusions, New Mexico, and relations between evolution of a continental magmatic center and modifications of the lithosphere. CONTRIB MINERAL PETROL **104**, 99–124.

Keith, S. B. 1986. Petrochemical variations in Laramide magmatism and their relationship to Laramide tectonic and metallogenic evolution in Arizona and adjacent regions. *In* Beatty, B. & Wilkinson, P. A. K. (eds) *Frontiers in geology and ore deposits of Arizona and the southwest*. ARIZONA GEOL SOC DIG **16**, 89–101.

King, P. B. & Beikman, H. M. 1974. *Geologic map of the United States and Alaska: scale 1 : 2,500,000*. Reston: US Geological Survey.

Kirkham, R. V., Sinclair, W. D. Thorpe, R. I. & Duke, J. M. (eds) 1993. Mineral deposit modeling. GEOL ASSOC CAN SPEC PAP **40**.

Kistler, R. W., Chappell, B. W., Peck, D. L. & Bateman, P. C. 1986. Isotopic variation in the Tuolumne Intrusive Suite, central Sierra Nevada, California. CONTRIB MINERAL PETROL **94**, 205–20.

Koons, P. O. 1989. The topographic evolution of collisional mountain belts: a numerical look at the southern Alps, New Zealand. AM J SCI **289**, 1041–69.

Lang, J. R. 1991. *Isotopic and geochemical characteristics of Laramide igneous rocks in Arizona*. Ph.D. Dissertation, University of Arizona.

Lang, J. R., Stanley, C. R. & Thompson, J. F. H. 1995. Porphyry copper–gold deposits related to alkalic igneous rocks in the Triassic–Jurassic arc terranes of British Columbia. *In* Pierce, F. W. & Bolm, J. G. (eds) *Porphyry copper deposits of the American Cordillera*. ARIZONA GEOL SOC DIG **20**, 219–36.

LaPierre, H., Charvet, J. Blein, O. & Rouer, O. 1994. Les séquences d'arcs insulaires permo-tiasiques du Nevada nord-occidental (Etats Unis): éléments clés dans l'évolution géodynamiques des Cordilleras nord-américaines. BULL SOC GEOL FR **165**, 541–57.

Lee, D. E. & Christiansen, E. H. 1983. The granite problem as exposed in the southern Snake Range, Nevada. CONTRIB MINERAL PETROL **83**, 99–116.

Lehmann, B. 1987. Tin granites, geochemical heritage, magmatic differentiation. GEOL RUNDSCH **76**, 177–85.

Leventhal, J. A., Reid, M. R., Montana, A. & Holden, P. 1995. Mesozoic invasion of crust by MORB-source asthenospheric magmas, US Cordilleran interior. GEOLOGY **23**, 399–402.

Lindgren, W. 1915. The igneous geology of the Cordilleras and its problems. *In Problems of American geology*, 234–86. New Haven: Yale University Press.

Lindgren, W. 1933. *Mineral deposits*, 4th edn. New York: McGraw-Hill.

Lipman, P. W. 1983. The Miocene Questa Caldera, northern New Mexico; relation to batholith emplacement and associated molybdenum mineralization. *In The genesis of Rocky Mountain ore deposits: changes with time and tectonics. Proceedings, Denver Regional Geological Society Symposium*, 133–47. Denver: Denver Regional Geological Soc.

Lipman, P. 1992. Magmatism in the cordilleran United States; progress and problems. *In* Burchfield, B. C., Lipman, P. W. & Zoback, M. L. (eds) *The Cordilleran orogen, conterminous US. The geology of North America*, Vol. G-3, 481–514. Boulder: Geological Society of America.

Lipman, P. & Sawyer, D. 1985. Mesozoic ash-flow caldera fragments in southeast Arizona and their relation to porphyry copper deposits. *Geology* **13**, 652–6.

Lipman, P. W., Christiansen, R. L. & Postka, H. J. 1971. Evolving subduction zones in the western United States, as interpreted from igneous rocks. *Science* **148**, 821–5.

Lyons, J. I. 1988. Volcanogenic iron oxide deposits, Cerro de Mercado and vicinity, Durango, Mexico. ECON GEOL **83**, 1886–906.

McCandless, T. E. & Ruiz, J. 1993. Rhenium–osmium evidence for regional mineralization in southwestern North America. *Science* **261**, 1282–6.

Miller, C. F. 1978. Monzonitic plutons, California, and a model for generation of alkali-rich, near silica-saturated magmas. CONTRIB MINERAL PETROL **67**, 349–55.

Miller, C. F. & Barton, M. D. 1990. Phanerozoic granitoids of the inner Cordillera of the western United States. *In* Kay, S. M. & Rapela, C. W. (eds) *Plutonism from Antarctica to Alaska*. GEOL SOC AM SPEC PAP **241**, 213–32.

Miller, C. F. & Bradfish, L. J. 1980. An inner Cordilleran belt of muscovite-bearing plutons. *Geology* **8**, 412–6.

Miller, D. M., Hillhouse, W. C., Zartman, R. E. & Lanphere, M. A. 1987. Geochronology of intrusive and metamorphic rocks in the Pilot Range, Utah and Nevada, and comparison with regional patterns. GEOL SOC AM BULL **99**, 866–79.

Miller, E. L., Miller, M. M., Stevens, C. H., Wright, J. E. & Madrid, R.

1992. Plate Paleozoic paleogeographic and tectonic evolution of the western US cordillera. *In* Burchfiel, B. C., Lipman, P. W. & Zoback, M. L. (eds) *The Cordilleran orogoen: conterminous US. The geology of North America*, Vol. G-3, 57–106. Boulder: Geological Society of America.

Miranda Gasca, M. A. 1994. The VMS and SEDEX of the Guerrero Terrane, Mexico. *In Society of Mining and Exploration Annual Meeting and Exhibit, Albuquerque, New Mexico, abstracts*, 49.

Mutschler, F. E., Wright, E. G., Ludington, S. & Abbott, J. T. 1981. Granite molybdenite systems. ECON GEOL **76**, 874–97.

Nelson, B. K. & DePaolo, D. J. 1985. Rapid production of continental crust 1·7 to 1·9 b.y. ago: Nd isotopic evidence from the basement of the North American mid-continent. GEOL SOC AM BULL **96**, 746–54.

Nesbitt, B. E., Muehlenbachs, K. & Murowchick, J. B. 1989. Genetic implications of stable isotope characteristics of mesothermal Au deposits and related Sb and Hg deposits in the Canadian Cordillera. ECON GEOL **84**, 1489–506.

Newberry, R. J. & Einaudi, M. T. 1981. Tectonic and geochemical setting of tungsten skarn mineralization in the Cordillera. *In* Dickinson, W. R. & Payne, W. D. (eds) *Relations of tectonics to ore deposits in the southern Cordillera*. ARIZONA GEOL SOC DIGEST **14**, 99–112.

Nolan, T. B. 1962. The Eureka mining district, Nevada. US GEOL SURV PROF PAP **406**.

Norton, D. 1982. Fluid and heat transport phenomena typical of copper-bearing pluton environments; southeastern Arizona. *In* Titley, S. R. (ed.) *Advances in geology of porphyry copper deposits; southwestern North America*, 59–72. Tucson: University of Arizona Press.

Ohmoto, H., Drummond, S. E., Eldridge, C. S., Pisutha-Arnond, V. & Barton, P. B. Jr. 1983. Chemical processes of Kuroko formation. *In* Ohmoto, H. & Skinner, B. J. (eds) *The Kuroko and related volcanogenic massive sulfide deposits*. ECON GEOL MONGR **5**, 570–604.

O'Neil, J. R. & Silberman, M. L. 1974. Stable isotope relations in epithermal Au–Ag deposits. ECON GEOL **69**, 902–909.

Ortega-Gutiérrez, F., Mitre Salazar, L. M., Roldán Quintana, J., Aranda Gomez, J., Morán Zenteno, D., Alaniz Alvarez, S. & Nieto Samaniego, A. 1992. *Carta geológica de la República Mexicana, scale 1 : 2,000,000*, 5th edn. Mexico City: Editorial Americana.

Patchett, P. J. & Ruiz, J. 1989. Nd isotopes and the origin of Grenville-age rocks in Texas: implications for Proterozoic evolution of the United States Mid-continent region. J GEOL **97**, 685–95.

Patino Douce, A. E., Humphreys, E. D. & Johnston, A. D. 1990. Anatexis and metamorphism in tectonically thickened continental crust exemplified by the Sevier hinterland, western North America. EARTH PLANET SCI LETT **97**, 290–315.

Perry, F. V., DePaolo, D. J. & Baldridge, W. S. 1993. Isotopic evidence for a decline in crustal contributions to caldera-forming rhyolites of the western United States during the middle to late Cenozoic. GEOL SOC AM BULL **105**, 872–82.

Pollard, P. J. & London, D. 1995. A special issue devoted to the geology of rare metal deposits: an introduction and overview. ECON GEOL **90**, 489–94.

Reiners, P. W., Nelson, B. K. & Ghiorso, M. S. 1995. Assimilation of felsic crust by basaltic magmas: thermal limits and extents of crustal contamination of mantle-derived magmas. GEOLOGY **23**, 563–6.

Rice, C. M., Harmon, R. S. & Shepard, C. J. 1985. Central City, Colorado: the upper part of an alkaline porphyry molybdenum system. ECON GEOL **80**, 1769–96.

Roberts, M. P. & Clemens, J. D. 1993. Origin of high-potassium, calc-alkaline, I-type granitoids. GEOLOGY **21**, 825–8.

Roldan Quintana, J. 1991. Geology and chemical composition of the Jaralito and Aconchi batholiths in east-central Sonora, Mexico. *In* Perez Segura, E. & Jaques Ayala, C. (eds) *Studies of Sonoran geology*. GEOL SOC AM SPEC PAP **254**, 19–36.

Ruiz, J. 1985. Petrology, distribution and origin of rhyolites associated with tin mineralization in the Sierra Madre Occidental, Mexico. *In* Taylor, R. P. & Strong, D. F. (eds) *Granite-related mineral deposits; geology, petrogenesis, and tectonic setting. Conference on granite related mineral deposits*, 322–30. Ottawa: Canadian Institute of Mining.

Saleeby, J. & Busby-Spera, C. J. 1992. Early Mesozoic tectonic evolution of the western United States cordillera. *In* Burchfield, B. C., Lipman, P. W. & Zoback, M. L. (eds) *The Cordilleran orogen, conterminous US. The geology of North America*, Vol. G-3, 107–68. Boulder: Geological Society of America.

Saleeby, J. B., Shaw, H. F., Niemeyer, S., Moores, E. M. & Edelman,

S. H. 1989. U/Pb, Sm/Nd and Rb/Sr geochronological and isotopic study of northern Sierra Nevada ophiolitic assemblages, California. CONTRIB MINERAL PETROL **102**, 205–20.

Sandiford, M., Marin, N., Zhou, S. & Turner, S. 1992. Granite genesis and the mechanism of convergent orogenic belts with application to the southern Adelaide fold belt. TRANS R SOC EDINBURGH EARTH SCI **83**, 83–93.

Sawkins, F. J. 1990. *Metal deposits in relation to plate tectonics*, 2nd edn. New York: Springer-Verlag.

Scotese, C. R. & Denham, C. R. 1988. *Terra mobilus*. Austin: *Earth in Motion Technologies*.

Sedlock, R. L., Ortega-Gutiérrez, F. & Speed, R. C. 1993. Tectonostratigraphic terranes and tectonic evolution of Mexico. GEOL SOC AM SPEC PAP **278**.

Seedorff, E. 1991a. Magmatism, extension, and ore deposits of Eocene to Holocene age in the Great Basin—mutual effects and preliminary proposed genetic relationships. *In* Raines, G. L., Lisle, R. E., Schafer, R. W. & Wilkinson, W. H. (eds) *Geology and ore deposits of the Great Basin, symposium proceedings*, 113–78. Reno: Geological Society of Nevada.

Seedorff, E. 1991b. Royston District, western Nevada; a Mesozoic porphyry copper system that was tilted and dismembered by Tertiary normal faults. *In* Raines, G. L., Lisle, R. E., Schafer, R. W. & Wilkinson, W. H. (eds) *Geology and ore deposits of the Great Basin, symposium proceedings*, 359–92. Reno: Geological Society of Nevada.

Shaw, A. L. & Guilbert, J. M. 1990. Geochemistry and metallogeny of Arizona peraluminous granitoids with reference to Appalachian and European occurrences. *In* Stein, H. J. & Hannah, J. L. (eds) *Ore-bearing granitic systems: petrogenesis and mineralizing processes*. GEOL SOC AM SPEC PAP **246**, 317–56.

Sillitoe, R. H. 1973. The tops and bottoms of porphyry copper deposits. ECON GEOL **68**, 799–815.

Sillitoe, R. H. 1976. Andean mineralization: a model for the metallogeny of convergent plate boundaries. GEOL ASSOC CAN SPEC PAP **14**, 59–100.

Sillitoe, R. H. & Bonham, H. F. Jr 1990. Sediment-hosted gold deposits; distal products of magmatic–hydrothermal systems. GEOLOGY **18**, 157–61.

Silver, L. T., Taylor, H. P. Jr & Chappell, B. W. 1979. Some petrologic and geochronologic observations of the Peninsular Ranges batholith near the international border of the United States of America and Mexico. *In* Abott, P. L. & Todd, V. R. (eds) *Mesozoic crystalline rocks: Peninsular Ranges batholith and pegmatites, Point Sal Ophiolite*. GEOL SOC AM FIELD TRIP GUIDE, 83–116.

Speed, R. C., Elison, M. W. & Heck, F. R. 1988. Phanerozoic tectonic evolution of the Great Basin. *In* Ernst, W. G. (ed.) *Metamorphism and crustal evolution, western conterminous United States. Rubey Volume VII*, 572–605. Englewood Cliffs: Prentice-Hall.

Staude, J.-M. 1995. *Epithermal mineralization in the Sierra Madrea Occidental and the metallogeny of northwestern Mexico*. Unpublished Ph.D. Thesis, University of Arizona.

Stein, H. J. & Crock, J. G. 1990. Late Cretaceous–Tertiary magmatism in the Colorado mineral belt; rare earth element and samarium–neodymium isotopic studies. *In* Anderson, J. L. (ed.) *The nature and origin of Cordilleran magmatism*. GEOL SOC AM MEM **174**, 195–223.

Titley, S. R. 1982. Geologic setting of porphyry copper deposits; southeastern Arizona. *In* Titley, S. R. (ed.) *Advances in geology of the porphyry copper deposits southwestern North America*, 37–58. Tucson, University of Arizona Press.

Titley, S. R. 1987. The crustal heritage of silver and gold ratios in Arizona ores. GEOL SOC AM BULL **99**, 814–26.

Titley, S. R. 1993. Characteristics of high-temperature, carbonate-hosted massive sulphide ores in the United States, Mexico and Peru. GEOL ASSOC CAN SPEC PAP **40**, 585–614.

Titley, S. R. & Beane, R. E. 1981. Porphyry copper deposits; part I, geologic settings, petrology and tectogenesis. *In* Skinner, B. J. (ed.) *75th Anniversary Volume*. ECON GEOL 214–35.

Todd, V. R. & Shaw, S. E. 1985. S-type granitoids and an I–S line in the Peninsular Ranges Batholith, southern California. GEOLOGY **13**, 231–3.

Torres-Vargas, R. 1993. *A Permo-Triassic granitic belt in Mexico: its origin and tectonic implications*. Unpublished M.S. Thesis, University of Arizona.

Tosdal, R. M., Haxel, G. B. & Wright, J. E. 1989. Jurassic geology of the Sonoran Desert region, southern Arizona, southeastern California, and northernmost Sonora; construction of a continental-margin magmatic arc. *In* Reynolds, S. J. & Jenny, J. P. (eds)

Summary of Arizona geology. ARIZONA GEOL SOC DIG **17**, 397–434.

Urabe, T. 1985. Aluminous granite as a source of hydrothermal ore-deposits: an experimental study. ECON GEOL **80**, 148–57.

van Middelaar, W. T. & Keith, J. D. 1990. Mica chemistry as an indicator of oxygen and halogen fugacities in the CanTung and other W-related granitoids in the North America Cordillera. *In* Stein, H. J. & Hannah, J. L. (eds) *Ore-bearing granitic systems: petrogenesis and mineralizing processes*. GEOL SOC AM SPEC PAP **246**, 21–34.

Vikre, P. G. 1981. Silver mineralization in the Rochester Mining District, Pershing County, Nevada. ECON GEOL **76**, 580–609.

Walawender, M. J., Gastil, R. G., Clinkenbeard, J. P., McCormick, W. V., Eastman, B. G., Wernicke, R. S., Wardlaw, M. S., Gunn, S. H. & Smith, B. M. 1990. Origin and evolution of the zoned La Posta-type plutons, eastern Peninsular Ranges Batholith, Southern and Baja California. *In* Anderson, J. L. (ed.) *The nature and origin of Cordilleran magmatism*. GEOL SOC AM MEM **174**, 1–18.

Westra, G. 1982. Alteration and mineralization in the Ruth porphyry copper deposit near Ely, Nevada. ECON GEOL **77**, 950–70.

Westra, G. & Keith, S. B. 1982. Classification and genesis of stockwork molybdenum deposits. ECON GEOL **76**, 844–73.

Wodzicki, W. A. 1995. *Relationships between magmatism and mineralization, Cananea district, Sonora, Mexico*. Unpublished Ph.D. Thesis, University of Arizona.

Wright, J. E. & Wooden, J. L. 1991. New Sr, Nd, Pb isotopic data from plutons in the northern Great Basin: implications for crustal structure and granite petrogenesis in the hinterland of the Sevier thrust belt. GEOLOGY **19**, 457–60.

Zartman, R. E. 1974. Lead isotope provinces in the Cordillera of the western United States and their geologic significance. ECON GEOL **69**, 792–805.

MARK D. BARTON, Department of Geosciences, University of Arizona, Tucson, AZ 85721, USA. E-mail: barton@geo.arizona.edu

Transactions of the Royal Society of Edinburgh: Earth Sciences, **87**, 281–290, 1996

Intrusive metallogenic provinces in eastern Australia based on granite source and composition

Phillip L. Blevin, Bruce W. Chappell and Charlotte M. Allen

ABSTRACT: Ore element ratios in intrusion-related mineralisation are in part a function of the relative oxidation state and degree of fractionation of the associated granite suite. A continuum from Cu–Au through W to Mo dominated mineralisation related to progressively more fractionated, oxidised I-type magmas can be traced within single suites and supersuites. Such systematic relationships provide strong evidence for the magmatic source of ore elements in granite-related mineral deposits and for the production of the observed ore element ratios dominantly through magmatic processes. The distribution of mineralised intrusive suites can be used to define a series of igneous metallogenic provinces in eastern Australia. In general, there is a correlated evolution in the observed metallogeny (as modelled based on the compatibility of ore elements during fractionation) with increasing degree of chemical evolution of the associated magmatic suite. This is from Cu–Au associated with chemically relatively unevolved magmas, through to Sn and Mo-rich mineralisation associated with highly evolved magmas that had undergone fractional crystallisation. Provinces recognised in that way do not necessarily correlate with the tectonostratigraphic boundaries defined by the near-surface geology, indicating that the areal distribution of some granite source regions in the deep crust is unrelated to upper crustal geology.

KEY WORDS: Australia, granites, metallogenic provinces, mineral deposits, fractionation.

Mineralisation in eastern Australia that can be related to intrusive rocks is heterogeneously distributed, both spatially and in terms of the dispersal of ore elements. Similarly, no apparent correlation exists in eastern Australia between the volume of magmatism now exposed at the surface and the distribution of related mineralisation, not a unique circumstance. In magmatic–hydrothermal systems, the evolution of the magma and its source materials is also the history of their contained ore elements. An understanding of why such distribution patterns exist, and their degree of correlation with magmatic processes and source materials, is critical if the petrography and composition of intrusive rocks are to be used as a meaningful exploration guide.

Ore elements, like the source materials of granites, were ultimately derived from the mantle. Granites themselves can be the product of the magmatic reworking of those materials during periods of new crust formation, or the result of the remagmatisation and vertical redistribution within the crust of older igneous or metasedimentary materials (Chappell & Stephens 1988). Chappell (1979) suggested that granites compositionally image their source rocks, which for most granites are located in the mid- to lower crust. The regional petrographic and compositional similarities among granites in the Lachlan Fold Belt (LFB) that define nine distinctive petrographic provinces were correlated by Chappell *et al.* (1988) with basement terranes, the boundaries between which generally correlate poorly with any tectonostratigraphic boundaries based on geological features exposed at the surface. The distribution of those granite basement terranes was subsequently shown by Blevin and Chappell (1992) to be strongly correlated with the distribution of granite-related mineralisation within the LFB, suggesting that strong genetic links exist between granite source compositions and the ore element ratios of associated mineralisation. If magmatic compositions and processes determine ore element ratios in

granite-related mineralisation, then the distribution of those source regions will be one of several important controls on the distribution of such mineralisation.

Other 'process' type controls such as the level of emplacement of intrusions, physical and chemical factors at the site of mineralisation, and the role of tectonic setting are critical factors in determining whether the ore-forming potential of fertile magma systems is realised. In this study, however, we have concentrated on granite source compositions to recognise broad-scale intrusive metallogenic provinces in eastern Australia, defined in the first instance on the inferred source character and compositions of mineralised granite suites, using a general intrusive (granite) metallogenic scheme.

1. Magmatic framework of eastern Australia with special reference to mineralised intrusive suites

Granites comprise a significant proportion of the Palaeozoic fold belts of eastern Australia and are associated with a wide range of types of mineralisation (Fig. 1). Several distinct phases of ore-related granitic and related magmatism are present, with peaks in activity occurring during Silurian–Devonian, Carboniferous–Permian, Permian–Triassic and Cretaceous times. Those related to significant mineralisation are described in the following.

1.1. Lachlan Fold Belt granites

1.1.1. Ordovician intrusive rocks. Mafic to felsic extrusive and minor intrusive igneous rocks of Ordovician age are exposed in N–S trending belts in the central zone of the LFB (Fig. 1). These igneous rocks are typically shoshonitic in character, being high in K but low in Ti (Owen & Wyborn 1979; Wyborn 1992). Some calcalkaline and tholeiitic suites are also present. Mineralisation is dominantly, but not exclusively, associated with the shoshonitic series and com-

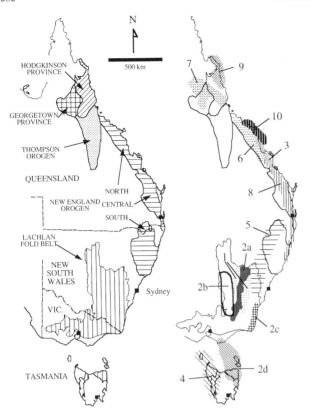

Figure 1 Map of eastern Australia showing distribution of fold belts mentioned in the text (left) and the distribution of intrusive metallogenic provinces (numbered on right). Numbers refer to provinces as listed in Table 1.

prises Cu–Au porphyry and skarn types, spatially related to monzonitic to granodioritic intrusions. Wyborn (1992) proposed that the Ordovician magmatism resulted from delayed melting of subcontinental lithosphere previously enriched during an earlier (Cambrian) subduction event. Pb isotope studies (Carr *et al.* 1995) define very precise mantle arrays with no crustal Pb present in these suites or their related ore deposits, consistent with the postulated mantle origin for these magmas.

1.1.2. Silurian–Devonian and Carboniferous granites. The Silurian–Devonian and Carboniferous granites of the LFB have been the focus of considerable study (reviewed in Chappell & White 1992). Granites of these ages comprise 20% (61 000 km²) of the total exposed area of the LFB, whereas the remnants of associated volcanic rocks cover an additional 5% of the area. Granite magmatism mostly falls into a relatively restricted age range of 430–390 Ma (Chappell *et al.* 1988; I. S. Williams pers. comm.). Slightly younger granites and volcanic rocks occur in the central southern mainland portion of the LFB and western Tasmania, whereas Carboniferous granites (≈330–315 Ma, Shaw & Flood 1993) are present in the NE LFB (Fig. 1). World-class mineralisation associated with LFB granites is restricted to Sn in the western portions of the LFB in New South Wales and Sn–W in both the eastern and western portions of the LFB in Tasmania. Numerous minor deposits of Sn, W, Mo, Au and Cu have also been exploited (Blevin & Chappell 1992).

Both the S- and I-type granites of the LFB are dominantly potassic and relatively mafic (gabbroic and dioritic igneous rocks are rare). S-type granites are inferred to have been derived by the partial melting of large volumes of metasedimentary rocks in the middle crust. Chemical and isotopic criteria (Chappell & White 1992) are consistent with the derivation of both S- and I-type granites of the LFB by partial melting of

older continental crust with little or no interaction, assimilation or mixing with materials of direct mantle origin (Chappell this issue). The Carboniferous granites of the NE LFB have chemical compositions distinct from those of other granites of the LFB, being much more similar to those of the Moonbi Supersuite of the southern New England Orogen (NEO; Chappell *et al.* 1988; Shaw & Flood 1993; Chappell 1994).

1.2. Mineralised intrusive rocks of the New England Orogen

Contrasts in the magmatic and tectonic evolution of the NEO and LFB were examined by Chappell (1994). Notably, the NEO is more easily accommodated into a continental margin model. Granites in the NEO are isotopically less evolved than those of the Silurian–Devonian granites of the LFB (Fig. 2) and attest to the strong difference in the history of those two belts (Chappell 1994) and the nature and age of the source materials for their contained granites.

1.2.1. Devonian magmatism. Mid- to late Devonian magmatism in the NEO is widely distributed in the form of volcanic rocks and volcaniclastic sediments. Known significant intrusive magmatism of any scale is restricted to the Mount Morgan Tonalite Complex, a member of the Calliope Volcanic Assemblage in the northern portion of the NEO (Fig. 1); it is closely spatially and temporally related to the major Mount Morgan Au (250 t) and Cu (300 000 t) deposit (Arnold & Sillitoe 1989). The chemical composition of that complex, a low alumina gabbro–tonalite–trondjhemite series with high Yb/Al_2O_3 and very low LILE and REE abundances (Blevin unpublished data) is oceanic in character and similar to the M-type granites of New Britain (Fig. 3; Whalen 1985). The tectonic setting and petrogenesis of the Devonian magmatism in the NEO is unclear. Chappell (1968) showed that the Upper Devonian Baldwin Formation of the southern NEO contains volcanic greywackes of andesite composition. Leitch and Willis (1982) concluded that the igneous clasts in those rocks are of island arc derivation or were derived from an arc built on thin continental crust. Morand (1993) concluded that the Calliope Volcanic Assemblage in the northern NEO was derived from an arc built either on thin continental crust or comprised a mature island arc close to a continent. The first appearance of LFB-derived detritus in the southern NEO

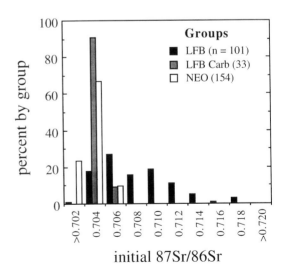

Figure 2 Initial $^{87}Sr/^{86}Sr$ ratios of granites from the Lachlan Fold Belt (LFB) and the New England Orogen (NEO). Note that the Carboniferous granites of the NE Lachlan Fold Belt are similar to the granites of the New England Orogen. Data sources: McCulloch and Chappell (1982 and sources cited therein); Hensel *et al.* (1985); Gray (1990); Shaw & Flood (1993); C. M. Allen unpublished data; Bryant *et al.* unpublished data.

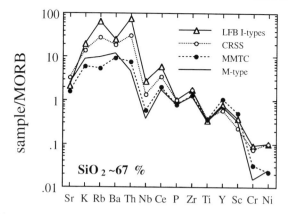

Figure 3 Representative compositions of granites (at 67% SiO₂) from the LFB, CRSS (Clarence River Supersuite of the NEO), MMTC (Mount Morgan Tonalite Complex) and M-type (New Britain) normalised to MORB (Pearce 1983).

during the Late Devonian Famennian stage (Flood & Aitchison 1992) indicates that the NEO, if it was originally located in an intra-oceanic setting, had been or was accreting to the margin of Gondwana (the LFB) by that time.

1.2.2. New England Batholith. Permian to Triassic I-type magmatism in the New England Batholith of the southern NEO is associated with Sn, W, Mo and minor Cu mineralisation. Mineralisation is mainly associated with the high-K Moonbi Supersuite and with two highly fractionated leuco-monzogranites (Gilgai and Mole granites, Blevin & Chappell 1993). Granites of the Moonbi Supersuite are K-rich and have distinctive trace element compositions, most notably high Sr, Ba, Pb, Th and Cs and relatively low Y, indicating derivation from source rocks that were shoshonitic in character (Chappell 1994). Both I- and S-type granites within the southern NEO are isotopically primitive despite their generally felsic nature (Hensel *et al.* 1985), indicating that they were derived from only slightly older source rocks.

1.2.3. Magmatism in the central and northern NEO. The central and northern portions of the NEO were the sites of extensive plutonism in the late Carboniferous to early Permian (Urannah Batholith in the northern NEO; Allen & Chappell 1993) and the early Triassic, extending down along the central NEO (Webb & McDougall 1968) into the southern NEO. These granites are typically low- to medium-K diorites, tonalites and granodiorites, with chemical and isotopic signatures indicative of continental margin affinity (Allen & Chappell 1993; Bryant *et al.* unpublished data). Early Cretaceous magmatism also developed along the eastern exposed portion of the northern NEO (Ewart *et al.* 1992; Allen & Chappell 1993). Although the earlier magmatic–plutonic episodes in the central NEO and northern NEO were probably subduction related, early Cretaceous magmatism was related to rifting and opening of the Tasman Sea (Ewart *et al.* 1992). All these magmatic stages in the central northern NEO are associated with numerous subeconomic porphyry style Cu–Mo–Au systems (Horton 1978).

1.3. Mineralised intrusive suites of North Queensland
The mineralised granites of North Queensland fall into two groups: the 'Carboniferous I-type granite' group (CIG; Champion & Chappell 1992), and the Permian I- and S-type granites of the northern Hodgkinson Province (Champion & Bultitude 1994). The three Carboniferous I-type supersuites that are mineralised (Sn, W–Mo–Bi and Cu–Pb–Ag–Au) are potassic, felsic and show evidence of strong fractional crystallisation. The CIG were probably derived from the partial melting of an enriched andesitic to dacitic source underplated in the

Proterozoic (Champion & Chappell 1992). Permian granites of the northern Hodgkinson Province are associated with Sn and W mineralisation and are isotopically and chemically less mature than those of the CIG (Champion & Bultitude 1994).

1.4. Summary of the evolutionary and source history of mineralised eastern Australian granites
Granites in the fold belts of eastern Australia have a range of granite compositions as a result of their derivation from a variety of source materials and varying degrees of fractionation. They may be grouped into three broad categories, those produced directly from the mantle (LFB Ordovician), those derived from the melting of early Palaeozoic to Proterozoic crustal materials (Silurian–Devonian magmatism in the LFB, Permian–Carboniferous magmatism in northern Queensland), and those resulting from partial melting of Palaeozoic crustal materials recently added to the crust where fusion was associated temporally and spatially with an active subduction or rifting process (Fig. 4). This evolutionary trend is also apparent in the chemical composition of I-type granites from different regions (Fig. 3), where both old material in the crust and derived granites are generally more felsic and chemically evolved (elevated K, high relative LILEs, REEs except Eu, and incompatible/compatible element ratios).

2. A working model for granite metallogeny

Blevin and Chappell (1991, 1992, 1995) studied relationships between granite chemical compositions, source and intensive variables (most importantly oxidation state) and the ore element assemblages of associated mineralisation within the LFB. A trend was apparent, from Cu (–Au) associated with the more mafic end of the granite spectrum, W with the intermediate compositional range and Mo (±W) and Sn (±W) related to the felsic, fractionated granites. A strong oxidation control was also present, with the Cu–Au–Mo series associated with oxidised rocks and Sn with reduced rocks. W is associated with both series, but was best developed in granite suites with intermediate to relatively reduced oxidation states. Thus a combination of oxidation state and fractionation

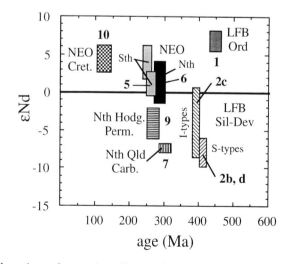

Figure 4 ε_Nd for east Australian granites plotted against general age range of intrusion. The ranges represent all available data grouped according to geographical and temporal association (e.g. New England Orogen Cretaceous granites). Numbers refer to associated intrusive metallogenic provinces named in Table 1. Note the general division between granites derived from juvenile sources (ε_Nd > 0) and from older, more evolved sources (ε_Nd < 0). Data sources: McCulloch and Chappell (1982); Hensel *et al.* 1985; Champion and Bultitude (1994); C. M. Allen unpublished data; P. L. Blevin unpublished data; Bryant *et al.* unpublished data; D. Wyborn pers. comm.

adequately explains the metallogeny of granite suites within the LFB (Blevin & Chappell 1992, 1995), the southern NEO (Blevin & Chappell 1993) and of granites associated with significant mineralisation globally (Blevin & Chappell 1995).

2.1. The importance of magma processes

That there is a relationship between certain ore element associations and the oxidation state of related granitic rocks has long been known (Ishihara 1981). However, the importance of the role of fractionation and the continuity of ore element ratio variations in mineralisation associated with fractionating granite suites has not previously been emphasised. These features may be effectively modelled on the known and inferred compatible/incompatible behaviour of ore elements under different oxidation states (summarised in Blevin & Chappell 1992). The relative abundances of ore elements change in granite suites during fractional crystallisation under different oxidation states. For example, in oxidised metaluminous magma suites Mo is relatively incompatible (Lowenstern *et al.* 1993), Cu is removed by fractional crystallisation as a compatible element, whereas Mo slowly increases in abundance. Mineralisation with high Cu/Mo ratios is more typically associated with less felsic and fractionated oxidised granites than is the case for mineralisation with low Cu/Mo ratios. Alternatively, significant fractionation of reduced granites removes both Cu and Mo and allows the build up of Sn (Blevin & Chappell 1992). In suites that evolve to very felsic compositions, such that Fe and Ti contents fall to such low levels that Fe–Ti oxides, sphene and biotite only crystallise late, if at all, then Sn may be concentrated in the melt fraction even if the magma is relatively oxidised.

These relationships between ore element behaviour during fractionation can be followed using ore element abundances in magmas (see Blevin & Chappell 1992, 1995) and also by comparing where mineralised granites belonging to certain ore element associations (e.g. Sn–W, W–Mo, Cu–Au, etc.) fall within the fractionation trends of individual suites, supersuites or even groups of supersuites (Fig. 5). These supersuites each have a range of mineral deposits associated with them, although within each there is a systematic order of ore element ratio variations in associated mineralisation as they become more evolved. Such a systematic relationship between ore element ratios in mineralisation and the chemical composition of ore-related granites provides good evidence for the control of such ratios by magmatic processes acting on a magma whose metal content, volatile inventory and oxidation states are themselves functions of the magma source.

Fractionation of volatiles and halogens may also occur along with other elements in granitic magma systems. For example, F enrichments in bulk rocks, and in biotites from eastern Australia (when corrected for reciprocal site effects; Munoz 1992) are strongly correlated with the degree of fractionation undergone by the host granite, as monitored, for example, by Rb (Fig. 6). The ratio of Cl to H_2O in biotites is also strongly source-controlled, with S-type biotites having significantly lower Cl/H_2O ratios than those from I-type granites (Blevin 1993; Blevin & Chappell 1992). The timing of volatile saturation with respect to the progress of fractional crystallisation within granite plutons is very important. Within a single magma body undergoing such crystallisation, exsolution of the fluid phase early rather than late may sample a different ratio of incompatible to compatible ore elements, as the relative concentrations of these elements change during crystallisation. These factors are not considered here; however, this scenario has been explored for Cu–Mo mineralising plutons by Candela and Holland (1986).

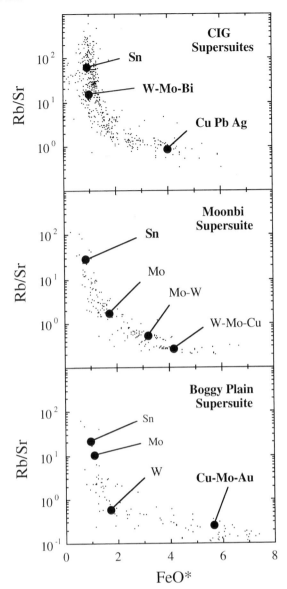

Figure 5   Rb/Sr versus FeO* (all Fe as FeO) for three groups of I-type granites from eastern Australia: the Carboniferous supersuites of north Queensland (CIG Supersuites), the Moonbi Supersuite of the southern New England Orogen and the Boggy Plain Supersuite of the Lachlan Fold Belt. Also plotted are the average compositions of mineralised granites within each group by commodity association.

2.2. A magmatic template for interpreting intrusive metallogeny

The composition of granite source materials is important in understanding related metallogeny as they are a primary control on the composition of derived magmas, and of parameters such as f_{O_2} and the S, H_2O and halogen contents of the magma. I-type granites can be divided conceptually into three broad groups: M-type, I-tonalite and I-granodiorite (Chappell & Stephens 1988). M-types are derived dominantly from juvenile mantle or mantle wedge materials. Tonalitic I-types are derived dominantly from the fusion of these M-type igneous materials of basaltic to andesitic composition. Granodioritic I-types, typical of the LFB, represent the remagmatised products of igneous rocks similar in composition to those of I-tonalitic composition in the deep crust. Progressively more felsic and chemically evolved I-type granites are thus generated by the remelting of older more mafic and chemically primitive igneous rocks. S-type granites were derived by the partial melting of large volumes of metasedimentary rocks in the middle crust. Such source rocks

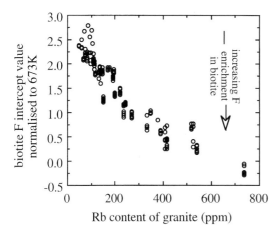

Figure 6 Intercept values of F enrichment (corrected for Fe–Mg avoidance) in biotites versus the Rb content of the host granite for the I-type granites of eastern Australia. Fluorine intercept values have been calculated using the equations of Munoz (1992) and normalised to 673 K for comparative purposes. Decreasing intercept values correspond to increasing F enrichment (i.e. F/OH) ratio in the biotite.

were fractionated during weathering, transport and sedimentation with the loss of Na, Ca, Sr (Chappell & White 1992) and Cl (Blevin 1993).

The relationships between the evolution of granite compositions and their source, and processes observed within fractionating granite magmas, can be viewed as a continuum, the ore element ratio in the melt fraction of the magma at volatile saturation representing the end product of a series of successive fractionation events. Ore element ratios within the hydrothermal fluid reflect (but do not directly image) the ore element ratios in the melt fraction of the magma at the time of volatile exsolution. For relatively oxidised I-type ore element associations, a continuum exists from Cu-dominant, through Cu–Mo to Mo-dominant mineralisation with progressive evolution of the ore-related granite. In reduced and fractionated S-type granites, dominant Sn ± W mineralisation develops. Sn mineralisation without Mo may be present in relatively reduced I-type granites, but in granites of intermediate oxidation state, both Sn and Mo may occur together (e.g. felsic members of the Moonbi Supersuite). Coupled with this is the role of the degree of chemical evolution of the granite magma itself. Each subsequent generation of granite magma is a step further down this metallogenic evolutionary trend than its predecessor.

The magmatic chemical aspects of this model can be demonstrated on a plot showing the conceptualised compositional and relative oxidation states of mineralised intrusive rocks relative to each other for the main ore element assemblages associated with such magmatism in eastern Australia (Fig. 7). Here, Rb/Sr acts as a monitor both of evolving source rocks and within magmatic suites as that ratio increases with fractional crystallisation. Also shown is a field enclosing 90% of chemical analyses of around 3500 granites from eastern Australia (unpublished data). The typical range of Rb/Sr and Fe_2O_3/FeO (representing relative oxidation state) of granite provinces in eastern Australia is shown in Figure 8. The relationship between Rb/Sr and SiO_2 for granites in eastern Australia (Fig. 9) demonstrates that SiO_2 (or any major element) is relatively poor indicator of the overall degree of fractionation undergone by evolved granitic rocks (see also Blevin & Chappell 1995).

3. Recognition and general aspects of intrusive metallogenic provinces

An important consequence of the described model is that a single magmatic suite which has fractionated over a large

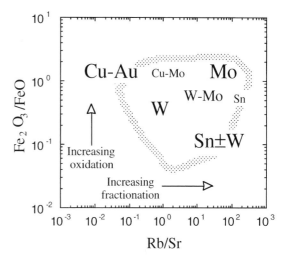

Figure 7 Interpretative plot showing the relationship between whole rock Fe_2O_3/FeO and Rb/Sr of granites and the dominant ore element associations in related mineralisation. The field shown encloses ≈ 3500 analyses of granite (s.l.) from eastern Australia (B. W. Chappell, unpublished data). Larger text size corresponds to the dominant 'end member' commodity associations.

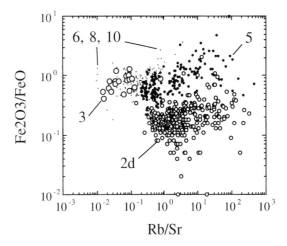

Figure 8 Examples of Fe_2O_3/FeO–Rb/Sr relationships for granites from some intrusive metallogenic provinces as numbered (cf. Table 1 and Fig. 7). Note that each province has its own characteristic range of oxidation state and relative degrees of fractionation, each being consistent with the observed range of mineral deposits associated with them.

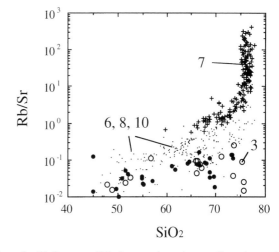

Figure 9 Rb/Sr versus SiO_2 for some intrusive metallogenic provinces (numbered as in Table 1). Also included is data for M-types (closed circles; data of Whalen 1985). Note that despite often overlapping SiO_2 ranges, Rb/Sr ranges are distinctive, particularly at high SiO_2.

compositional range may be associated with a range of types of mineralisation. In Figure 8 some supersuites (e.g. Moonbi) range compositionally over fields that Figure 7 shows are associated with W, Cu–Mo, W–Mo, Mo and even minor Sn mineralisation. Thus rather than grouping intrusive-related mineralisation together on the basis of a specific ore type (e.g. Cu deposits), the source character of the ore-related intrusive suites is here used as the primary division for classification purposes. Provinces so recognised consequently have a range of deposit types whose ore element assemblages are determined by magmatic factors, in concert with the degree of evolution of the magmas present and the extent of fractional crystallisation that they have experienced.

The main intrusive metallogenic provinces recognised in eastern Australia are shown in Figure 1 and their chemical and metallogenic character described in Table 1. Provinces may contain granites or other intrusive rocks of more than one type. They may also overlap with each other if they are defining magmatic source regions at different levels within the crust or igneous source materials that were emplaced into or beneath the crust at different times.

Some intrusive metallogenic provinces contain a relatively limited set of ore element associations (Cu–Au in the case of the central New South Wales Ordovician province), whereas others, such as the North Queensland Carboniferous I-type and the Moonbi-Bathurst provinces, are polymetallic (Sn, W, Mo, Cu, Au). In general, the larger the range of intrusive rock compositions present, particularly when that variation results from fractional crystallisation, the more diverse are the associated ore element associations (Fig. 7). Some provinces are defined on a single supersuite of distinctive character, such as the Yeoval–Boggy Plain Province, which is solely related to the Boggy Plain Supersuite (Wyborn *et al.* 1987). This supersuite is anomalously mineralised relative to other I-type supersuites in the LFB and was derived from a distinctive source that was both more mafic and younger than those of other granites of Silurian–Devonian age in the LFB.

Other provinces comprise groups of granite suites of more diverse origin and type, but which can be grouped together because their associated metallogenic character is very similar. For example, Sn mineralisation in the 'Wagga Tin Belt' is associated with both S- and I-type granites. Sn is also associated in this province with dyke swarms of unknown affinity and Pb model ages from sulphides in the Dorradilla Sn skarn deposits of the northern part of the belt indicate a Carboniferous age for this mineralisation (Carr *et al.* 1995), as opposed to Silurian–Devonian ages on Sn mineralisation elsewhere in the province. All Sn mineralised granites are, however, fractionated and relatively reduced. In the western Tasmanian Province, Sn and W are also associated with both I- and S-type granites, both of which are highly fractionated. In the Bega Province, Mo and Au mineralisation occurs as small deposits associated with I-type granites belonging to different suites with different chemical and isotopic characteristic ranging from the Moruya Supersuite to the more typically I-granodioritic and isotopically evolved Bemboka Supersuite (Chappell *et al.* 1990).

The age range of magmatism may vary within individual provinces. The Moonbi–Bathurst Province contains a range of magmatic–mineralisation ages, being Carboniferous where it occurs in the LFB, Late Permian in the southern NEO (250–265 Ma; Hensel *et al.* 1985) and ranging into the Triassic further to the NE. Potassic I-type supersuites within this province are similar both chemically and isotopically, whereas variations in magmatic ages probably reflect variations in the time of application of heat to broadly similar source materials.

A strong asymmetry in the distribution of ore elements in related mineralisation is also shown in that province, with Mo, Au and W mineralisation concentrated mainly in the portion that outcrops within the LFB and Sn restricted to the northern part of the southern NEO (Fig. 1).

3.1. Correlation of mineralised intrusive compositions and source with metallogenic character

Blevin and Chappell (1995) noted that in a general way mineralisation dominated by chalcophile elements is associated with less evolved granitic magmas, whereas that involving lithophile elements is related to more chemically evolved (i.e. fractionated) granitic magmas. However, this relationship is not necessarily mirrored in the isotopic composition of these granites. This is because the isotopic composition is also a function of the age of the source materials of the granites. Thus felsic, fractionated granites in eastern Australia are not necessarily more isotopically evolved than relatively unfractionated granites.

Copper and Mo mineralised granites in eastern Australia have relatively low initial $^{87}Sr/^{86}Sr$ ratios (Fig. 10). This is partly a result of the alliance of those elements with I-type granites, but also of the clear association of Cu with chemically (and for the most part isotopically) less evolved granites. Sn is related to granites with a broad range of initial $^{87}Sr/^{86}Sr$ ratios, reflecting the association of such mineralisation with a wide range of granite types, varying from S-type granites derived from old source rocks at one extreme (e.g. in the LFB) to I-type granites in the southern NEO derived from young source materials. It should not be assumed therefore that the derivation of magmas from evolved continental crustal materials is an *a priori* requirement for the production of lithophile mineralisation. What is required is a granite magma of appropriate composition and oxidation state that has arrived at that composition through extended fractional crystallisation. However, it is granites derived from an already more mature source that are able to attain the more highly evolved compositions required for the production of Sn mineralisation.

3.2. Correlation of intrusive metallogenic provinces with Palaeozoic surface geology

The boundaries of the intrusive metallogenic provinces recognised in this study tend to be rather distinct, except where buried under cover sequences, suggesting that discontinuities between differing source regions in the lower crust are relatively sharp. These boundaries, however, like the basement terranes defined by Chappell *et al.* (1988), generally do not correspond with major structural discontinuities or terrane boundaries proposed on the basis of surface geology. This is particularly the case for provinces defined on the basis of mineralised I-type granites. For example, the NEO comprises two zones of distinctly different metallogenic character (Blevin & Chappell 1993). The northern and central NEO are Cu–Au–Mo provinces, whereas the southern NEO is dominated by Sn–W–Mo. The Moonbi–Bathurst province extends from the southern NEO southwards into the LFB (Fig. 1). I-type granites of the North Queensland CIG province intrude either side of the Palmerville Fault which separates the Palaeozoic Hodgkinson Province from the Proterozoic Georgetown Province.

Provinces based on the distribution of S-type granites have boundaries that generally coincide with tectonostratigraphic boundaries defined by near-surface geology. S-type granites are derived from source rocks in the mid-crust whose distribution is more likely to reflect the geology of the upper

Table 1 Character of some intrusive metallogenic provinces of eastern Australia. Abbreviations: lK, mK, hK = low, medium, high K (potassic); Sh = shoshonitic. M = M-type, I-t = I-tonalitic, I-g = I-granodioritic, S = S-type. Ox, Ox-Rd, Rd = oxidised, intermediate and reduced relative to QFM. CF, Granite suite undergoing strong fractional crystallisation.

Province No.	Intrusive metallogenic province	Age of igneous activity	Mineralised intrusive types	Character of mineralised intrusive suites	$({}^{87}Sr/{}^{86}Sr)_i$*	Dominant ore element associations	Deposit examples
1	Central NSW Ordovician	Ordovician	hK-Sh	Ox	<0·704	Cu-Au	Goonumbla
2	LFB Siluro-Devonian Granites		lK	Ox		Cu-Au	Copper Hill
2a	Yeoval–Boggy Plain	Devonian	hK, I-g	Ox, CF	<0·705	Cu-Au, Mo, W	Yeoval
2b	Wagga	Devonian	hK, S	Rd, CF, Fr	>0·710	Sn	Ardlethan
						Sn	Beechworth
						Au	Braidwood
2c	Bega	Devonian	hK, I-g	Ox	<0·710	Mo	Whipstick
2d	Eastern Tasmanian	Devonian	hK, S	Rd, CF, Fr	>0·710	Sn, W	Lottah
3	Mount Morgan–Calliope	Devonian	lK, I-t/M	Ox	0·704	Cu Au	Mount Morgan
4	Western Tasmania	Devonian	hK, I-g	Ox-Rd, UnFr	0·712–0·714	W	King Island
			hK, I-g	Rd, CF, Fr	>0·74	Sn	Renison
5	Moonbi–Bathurst	Carboniferous–Triassic	hK, I-g	Ox, Fr	0·704–0·706	Mo, Cu, W, Au	Yetholme, Attunga
				Ox-R; Fr		Sn, Mo, W	
6	Northern New England Orogen	Permo-Carboniferous	l-mK, I-t	Ox, UnFr	0·704–0·7065	Cu, Mo, Au	Stanthorpe/Ruby Ck
7	North Queensland Carboniferous I-type	Permo-Carboniferous	hK, I-g	Ox, Ox-Rd	0·71	W-Mo, Cu, Au	Bamford Hill
				Rd	0·71	Sn W	Herberton
8	Central New England Orogen	Permo-Triassic	l-mK, I-t	Ox, UnFr	—	Cu Mo Au	Moonmera, Coulstoun
9	North Hodgkinson	Permian	hK, S	Rd, CF, Fr	0·710–0·716	Sn,	Collingwood
				Bd, CF, Fr		W	
10	Central Queensland Cretaceous	Cretaceous	mK, I-g	Ox	<0·704	Minor Cu-Au	Mount Carbine Dittmar

Notes: *See Figure 2 caption for data sources.

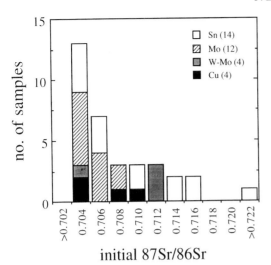

Figure 10 Initial $^{87}Sr/^{86}Sr$ ratios of granites associated with Sn, Mo, W–Mo and Cu mineralisation in eastern Australia. Note that although Mo and Cu mineralisation is associated with granites having relatively low initial $^{87}Sr/^{86}Sr$ ratios, a broad range is apparent for Sn. See Fig. 2 for data sources

crust. I-type granites are interpreted as having been sourced from regions nearer the base of the crust (Chappell & White 1992). Contrasts in relationships between the distribution of these intrusive metallogenic provinces and upper crustal geology suggest that rock relationships in the lower crust are very different to those of the upper crust in many places in eastern Australia, as recognised by Chappell et al. (1988). These strong discontinuities between the upper and lower crustal geological features for I-type metallogenic provinces are interpreted to be a reflection of I-type source materials having been underplated beneath the crust, often subsequent to the assembly of crustal terranes which provide the substrate to the middle Palaeozoic fold belts of eastern Australia (Chappell et al. 1988).

3.3. Upper and lower crustal relationships: the vertical dimension

The relationship between the lower to mid-crust (comprising the inferred source regions for most granite magmas) and the upper crust, where accessible granites and associated mineralisation reside, also exert important controls on the nature and distribution of intrusive-related mineralisation. Another important aspect in this regard is the present relative level of exposure of the provinces, and the depth of exhumation, on the distribution of mineralisation.

Following partial fusion, granite magmas rise towards the upper crust. Those being emplaced to very shallow levels cross from the ductile into the brittle structural regime of the uppermost crust. The level of emplacement attained by a granite magma on ascent is important because of the significant effect of pressure on volatile solubility and consequent behaviour in magmas (e.g. Candela 1991). Volatile solubility in magmas decreases strongly with pressure. A magma of a given composition and volatile content will undergo volatile exsolution much earlier at shallow (low pressure) levels in the crust than a more deeply emplaced equivalent. Timing of volatile exsolution relative to the degree of crystallisation in magmas is important, as significant changes in ore element ratios within the remaining melt fraction may occur within the crystallisation interval (see earlier).

Strong structural controls at all scales also affect the

distribution of mineralisation and related intrusive rocks. This is evident in the often elongate nature of many batholiths and igneous suites in eastern Australia and the alignment of mineralisation along 'corridors' and 'lineaments' that may or may not have a physical surface expression (O'Driscoll 1990). Such structures are dominantly developed in the upper (brittle) crust and theoretically provide 'least lines of resistance' to the emplacement of magmas to high levels.

4. Some implication of intrusive metallogenic provinces

Ore element assemblages in intrusive-related mineralisation are functions of relative oxidation state as well as fractionation. Many fractionated and mineralised intrusive suites are chemically indistinguishable at mafic compositions, compared with other intrusive suites which are neither strongly fractionated nor mineralised. In the LFB for example, the Koetong Supersuite of the Wagga Basement Terrane became associated with Sn mineralisation as a result of significant fractional crystallisation, whereas the initially compositionally similar Bullenbalong Supersuite of the Kosciusko Basement Terrane, did not (Chappell & White 1990; Chappell this issue). Both initially had similar Sn contents; the reasons why the Koetong Supersuite became Sn mineralised therefore did not lie in different initial metal contents, but in the reason, not yet understood, for its undergoing prolonged fractional crystallisation.

The question should therefore be asked: what are the mechanisms and controls that influence the distribution of granite source materials, the nature and timing of their melting, and the oxidation state and whether or not fractionation will take place in the subsequent magma? The distribution of granite source materials of specific character in the lower crust is closely related to processes by which the crust was constructed over time. Granite source materials may have been resident in the crust for extended periods before fusion, the age of granite magmatism recording the time when sufficient heat was applied to the crust to cause melting, rather than the timing of metal and/or melt extraction from the mantle per se. 'Metallogenic epochs', if defined on the ages of granite intrusion and crystallisation, are not particularly useful as an exploration guide if used in this way.

The role of tectonics, and the tectonic setting, are important aspects of metallogenic studies. Metallotectonic provinces as applied to granite-related mineralisation, however, have the disadvantage that because granites may inherit a chemical signature of their source, the tectonic setting during their formation may be radically different to that at the time of granite formation (Chappell & Stephens 1988). For example, although some of the Devonian LFB I-type granites have trace element signatures consistent with their source materials having been through an arc-forming event at some stage in their history, the tectonic setting of the LFB at the time of actual granite formation and emplacement was very different (Coney 1992; Chappell 1994).

Oxidation states of granites in eastern Australia are consistent within magmatic suites, and differ between suites, often over large areas, implying that source controls are paramount in this regard. The reasons why certain suites undergo extended fractional crystallisation while others of similar initial compositions do not are unclear, but may involve a range of factors such as contrasting thermal regimes, H_2O contents, degree of partial melting, fertility of source rocks, rate of ascent and crystallisation, etc. Marked contrasts must have existed in the thermal regimes during the time of granite formation between the different regions of eastern Australia depending on whether granite formation was driven by subduction, or rifting, or occurred in intracratonic settings perhaps driven by increasing thermal gradients as a result of crustal thinning. In intracratonic environments in which widespread partial melting of the crust occurred, such as the LFB, a problem may be that insufficient heat becomes available to achieve high degrees of partial melting, or that more fertile source regions allowed relatively high degrees of partial melting at relatively low temperatures. Fractionation may have been inhibited in these magmas by their restite-rich nature. Understanding these controls is important if the relationship between intrusive metallogenic provinces and unmineralised, but chemically similar, granite provinces, as well as the localisation of potential mineralised intrusive rocks within provinces, is to be better known.

Acknowledgements

P. L. B and B. W. C acknowledge the Australian Mineral Industry Research Association (AMIRA), and the Australian Research Council (ARC) for their support and ongoing funding of granite-related research in eastern Australia. This is Publication 50 of the Key Centre for Geochemical Evolution and Metallogeny of Continents.

References

Allen, C. M. & Chappell, B. W. 1993. Contrasting Carboniferous–Permian and Cretaceous plutonism in the Urannah Batholith, northern New England Fold Belt. In Flood, P. G. & Aitchison, J. C. (eds) New England Orogen, Eastern Australia, 573–9. Armidale: University of New England.

Arnold, G. O. & Sillitoe, R. H. 1989. Mount Morgan gold–copper deposit, Queensland, Australia: evidence for an intrusion-related replacement origin. ECON GEOL 84, 1805–16.

Blevin, P. L. 1993. Halogen (F, Cl) contents of biotites from I- and S-type granites of eastern Australia: possible metallogenic implications. GEOL SOC AUST ABSTR 34, 83–4.

Blevin, P. L. & Chappell, B. W. 1991. Relationships between granites and mineral deposits in the Lachlan Fold Belt. GEOL SOC AUST ABSTR 29, 4.

Blevin, P. L. & Chappell, B. W. 1992. The role of magma sources, oxidation states and fractionation in determining the granite metallogeny of eastern Australia. TRANS R SOC EDINBURGH EARTH SCI 83, 305–16.

Blevin, P. L. & Chappell, B. W. 1993. The influence of fractionation and magma redox on the distribution of mineralisation associated with the New England Batholith. In Flood, P. G. & Aitchison, J. C. (eds) New England Orogen, Eastern Australia, 423–29. Armidale: University of New England.

Blevin, P. L. & Chappell, B. W. 1995. Chemistry, origin and evolution of mineralized granites in the Lachlan Fold Belt, Australia; the metallogeny of I- and S-type granites. ECON GEOL 90, 1604–19.

Candela, P. A. 1991. Physics of aqueous phase evolution in plutonic environments. AM MINERAL 76, 1081–91.

Candela, P. A. & Holland, H. D. 1986. A mass transfer model for copper and molybdenum in magmatic hydrothermal systems: the origin of porphyry-type ore deposits. ECON GEOL 81, 1–19.

Carr, G. R., Dean, J. A., Suppel, D. W. & Heithersay, P. S. 1995. Precise lead isotope fingerprinting of hydrothermal activity associated with Ordovician to Carboniferous metallogenic events in the Lachlan Fold Belt of New South Wales. ECON GEOL 90, 1467–505.

Champion, D. C. & Bultitude, R. J. 1994. Granites of the Eastern Hodgkinson Province. II. Their geochemical and Nd–Sr isotopic characteristics and implications for petrogenesis and crustal structure in north Queensland. GEOL SURV QUEENSLAND REC 1994/1.

Champion, D. C. & Chappell, B. W. 1992. Petrogenesis of felsic I-type granites: an example from northern Queensland. TRANS R SOC EDINBURGH EARTH SCI 83, 115–26.

Chappell, B. W. 1968. Volcanic greywackes from the Upper Devonian Baldwin Formation, Tamworth–Barraba district, New South Wales. J GEOL SOC AUST 15, 87–102.

Chappell, B. W. 1979. Granites as images of their source rocks. GEOL SOC AM PROGRAM ABST 11, 400.

Chappell, B. W. 1994. Lachlan and New England: fold belts of contrasting magmatic and tectonic development. J PROC R SOC NSW 127, 47–59.

Chappell, B. W. & Stephens, W. E. 1988. Origin of infracrustal (I-type) granite magmas. TRANS R SOC EDINBURGH EARTH SCI 79, 71–86.

Chappell, B. W. & White, A. J. R. 1990. Tin granites: their evolution from fertile sediments by partial melting and fractional crystallisation. 10TH AUST GEOL CONVENTION, R. L. STANTON SYMP ABST, 17–8.

Chappell, B. W. & White, A. J. R. 1992. I- and S-type granites in the Lachlan Fold Belt. TRANS R SOC EDINBURGH EARTH SCI 83, 1–26.

Chappell, B. W., White, A. J. R. & Hine, R. 1988. Granite provinces and basement terranes in the Lachlan Fold Belt, southeastern Australia. AUST J EARTH SCI 35, 505–521.

Chappell, B. W., Williams, I. S., White, A. J. R. & McCulloch, M. T. 1990. Granites of the Lachlan Fold Belt. ICOG 7 Field Excursion A-2. BUR MINERAL RESOUR CANBERRA REC 1990/48.

Coney, P. J. 1992. The Lachlan belt of eastern Australia and Circum-Pacific tectonic evolution. TECTONOPHYSICS 214, 1–25.

Ewart, A., Schon, R. W. & Chappell, B. W. 1992. The Cretaceous volcanic–plutonic province of the central Queensland (Australia) coast—a rift related 'calc-alkaline' province. TRANS R SOC EDINBURGH EARTH SCI 83, 327–45.

Flood, P. G. & Aitchison, J. C. 1992. Late Devonian accretion of the Gamilaroi Terrane to eastern Gondwana: province linkage suggested by the first appearance of Lachlan Fold Belt-derived quartzarenite. AUST J EARTH SCI 39, 539–44.

Gray, C. M. 1990. A strontium isotope traverse across the granitic rocks of southeastern Australia: petrogenetic and tectonic implications. J GEOL SOC AUST 37, 331–49.

Hensel, H.-D., McCulloch, M. T. & Chappell, B. W. 1985. The New England Batholith: constraints on its derivation from Nd and Sr isotopic studies of granitoids and country rocks. GEOCHIM COSMOCHIM ACTA 49, 369–84.

Horton, D. J. 1978. Porphyry-type copper–molybdenum mineralization belts in eastern Queensland. ECON GEOL 73, 904–21.

Ishihara, S. 1981. The granitoid series and mineralization. ECON GEOL 75TH ANNIV VOL, 458–84.

Leitch, E. C. & Willis, S. G. A. 1982. Nature and significance of plutonic clasts in Devonian conglomerates of the New England Fold Belt. J GEOL SOC AUST 29, 83–9.

Lowenstern, J. B., Mahood, G. A., Hervig, R. L. & Sparks, J. 1993. The occurrence and distribution of Mo and molybdenite in unaltered peralkaline rhyolites from Pantelleria, Italy. CONTRIB MINERAL PETROL 114, 119–29.

McCulloch, I. & Chappell, B. W. 1982. Nd isotopic characteristics of S- and I-type granites. EARTH PLANET SCI LETT 58, 51–64.

Morand, V. J. 1993. Stratigraphy and tectonic setting of the Calliope Volcanic Assemblage, Rockhampton area, Queensland. AUST J EARTH SCI 40, 15–30.

Munoz, J. L. 1992. Calculation of HF and HCl fugacities from biotite compositions: revised equations. GEOL SOC AM ABSTR PROGRAM 24, A221.

O'Driscoll, E. S. T. 1990. Lineament tectonics of Australian ore deposits. In Hughes, F. E. (ed) Geology of the mineral deposits of Australia and Papua New Guinea, 33–44. Melbourne: The Australasian Institute of Mining and Metallurgy.

Owen, M. & Wyborn, D. 1979. Geology and geochemistry of the Tantangara and Brindabella 1:100 000 sheet areas. BULL BMR GEOL GEOPHYS 204, 52 pp.

Pearce, J. A. 1983. Role of the sub-continental lithosphere in magma genesis at active continental margins. In Hawkesworth, C. J. & Norry, M. J. (eds) Continental basalts and mantle xenoliths, 231–49. Nantwich: Shiva.

Shaw, S. E. & Flood, R. H. 1993. Carboniferous magmatic activity in the Lachlan and New England Fold Belts. In Flood, P. G. & Aitchison, J. C. (eds) New England Orogen, Eastern Australia, 113–21. Armidale: University of New England.

Webb, A. W. & McDougall, I. 1968. The geochronology of the igneous rocks of eastern Queensland. J GEOL SOC AUST **15,** 313–46.

Whalen, J. B. 1985. Geochemistry of an island-arc plutonic suite: the Uasilau-Yau Yau Complex, New Britain, P.N.G. J PETROL **26,** 603–32.

Wyborn, D. 1992. The tectonic significance of Ordovician magmatism in the eastern Lachlan Fold Belt. TECTONOPHYSICS **214,** 177–92.

Wyborn, D., Turner, B. S. & Chappell, B. W. 1987. The Boggy Plain Supersuite: a distinctive belt of I-type igneous rocks of potential economic significance in the Lachlan Fold Belt. AUST J EARTH SCI **34,** 21–43.

PHILLIP L. BLEVIN, BRUCE W. CHAPPELL and CHARLOTTE M. ALLEN, Key Centre for Geochemical Evolution and Metallogeny of Continents (GEMOC), Department of Geology, Australian National University, Canberra, ACT 0200, Australia.

Transactions of the Royal Society of Edinburgh: Earth Sciences, **87**, 291–303, 1996

Exsolved magmatic fluid and its role in the formation of comb-layered quartz at the Cretaceous Logtung W–Mo deposit, Yukon Territory, Canada

Jacob B. Lowenstern and W. David Sinclair

ABSTRACT: Comb-layered quartz is a type of unidirectional solidification texture found at the roofs of shallow silicic intrusions that are often associated spatially with Mo and W mineralisation. The texture consists of multiple layers of euhedral, prismatic quartz crystals (Type I) that have grown on subplanar aplite substrates. The layers are separated by porphyritic aplite containing equant phenocrysts of quartz (Type II), which resemble quartz typical of volcanic rocks and porphyry intrusions. At Logtung, Type I quartz within comb layers is zoned with respect to a number of trace elements, including Al and K. Concentrations of these elements as well as Mn, Ti, Ge, Rb and H are anomalous and much higher than found in Type II quartz from Logtung or in igneous quartz reported elsewhere. The two populations appear to have formed under different conditions. The Type II quartz phenocrysts almost certainly grew from a high-silica melt between 600 and 800°C (as β-quartz); in contrast, the morphology of Type I quartz is consistent with precipitation from a hydrothermal solution, possibly as α-quartz grown below 600°C. The bulk compositions of comb-layered rocks, as well as the aplite interlayers, are consistent with the hypothesis that these textures did not precipitate solely from a crystallising silicate melt. Instead, Type I quartz may have grown from pockets of exsolved magmatic fluid located between the magma and its crystallised border. The Type II quartz represents pre-existing phenocrysts in the underlying magma; this magma was quenched to aplite during fracturing/degassing events. Renewed and repeated formation and disruption of the pockets of exsolved aqueous fluid accounts for the rhythmic banding of the rocks.

KEY WORDS: volatiles, magma, degassing, comb quartz, comb layering, unidirectional solidification textures, Mo, W, porphyry, infra-red, FTIR, trace elements.

Evidence from volcanic systems suggests that magmatic vapours and brines commonly exsolve and exist as separate phases within many shallow magma chambers (Johnson *et al.* 1994; Lowenstern 1995, Gerlach *et al.* In press). In intrusive systems, exsolving magmatic fluids result in hydrothermal alteration and contribute to a host of magmatic–hydrothermal ore deposits (Burnham 1979; Hedenquist & Lowenstern 1994). Though workers have presented models to explore the movement of magmatic volatiles through crystallising intrusions (Candela 1994; Shinohara & Kazahaya 1995), the igneous textures that result from degassing have received relatively little attention. Miarolitic cavities provide unambiguous evidence for a magmatic aqueous phase, though they represent aqueous fluid that remained within the intrusion, rather than leaving the system. As such, it would be useful to identify rock textures that provide information on the means of egress of magmatic volatiles from intrusions and the physical conditions that prevail at the magma–wallrock interface.

One such texture, comb-layered quartz, is a distinctive feature of felsic dykes at the Cretaceous Logtung W–Mo deposit on the Yukon–British Columbia border in NW Canada. We describe here the petrography and mineralogy of the Logtung comb-layered quartz and explore various possible mechanisms for its origin. We conclude that comb-layered quartz forms at the top of shallow, apical intrusions from exsolved aqueous fluid that accumulates at the (inwardly migrating) contact between wallrock and magma.

1. Comb-layered quartz

1.1. Characteristics

Comb-layered quartz (Fig. 1) is a relatively common, though volumetrically minor, texture of silicic intrusions associated with porphyry ore deposits, especially those associated with Mo and W mineralisation. It is typically found at the roofs of small, shallow, apical intrusions (Kirkham & Sinclair 1988). Also referred to as brain rock, ribbon rock, crenulate quartz, unidirectional solidification texture and other terms, the rock type is characterised by repeated bands of coarse-grained euhedral quartz separated by bands of fine-grained, equigranular granite (i.e. aplite).

Shannon *et al.* (1982) used the metallurgical term 'unidirectional solidification textures' (USTs) to describe several textures in granite porphyries that form by growth inward from, and perpendicular to, the outer edge of the intrusion. They described three principal types of USTs from the Henderson Mo deposit of Colorado, U.S.A: (I) monomineralic crenulate layers consisting of terminated crystals of quartz or alkali feldspar that extend from a planar substrate, where individual layers represent time lines of crystallisation; (II) dendritic layers containing branching dendrites of quartz or alkali feldspar set in an aplitic matrix; and (III) intergrowth layers comprising quartz intergrown with alkali feldspar and including micrographic and granophyric textures. Shannon *et al.* (1982) showed that all three types of USTs are useful

Figure 1 Photographs of slabs of comb-layered textures from the Logtung deposit. (a) Sample SYA85-13C contains coarse layers of comb-textured quartz separated by thin aplite interlayers that contain equant Type II quartz phenocrysts. Growth direction towards bottom right. (b) Sample SYA85-15B displays multiple thin layers of comb-textured quartz and aplite interlayers. Growth direction towards bottom left. (c) Detail of sample SYA85-15B. Growth direction towards bottom left. (d) Sample SYA85-14B. Crenulate comb layers are cut by a later quartz + molybdenite vein. Growth direction towards bottom left.

for identifying the relative ages of multiple intrusions composed of similar material, as well as the borders of individual intrusions.

Kirkham and Sinclair (1988) reviewed more than 20 occurrences of comb-layered quartz at a number of Mo, W and Cu porphyry deposits. They discussed previous investigations of comb layering and various hypotheses for its origin and also provided detailed descriptions of individual localities, primarily in the U.S.A. and Canada. Principally, they focused on the crenulate layers (type I) of Shannon *et al.* (1982), which they called comb-layered quartz.

The term 'comb layering' (crescumulate) is also used to describe needle-like laths of plagioclase, pyroxene and amphibole found in dioritic and gabbroic intrusions (Lofgren & Donaldson 1975; Moore & Lockwood 1973). Moreover, it is

used to describe textures in low-temperature hydrothermal veins. This paper is restricted to textures consisting of comb-layered quartz with aplite interlayers. These textures are 'pegmatoidal' in the sense that they commonly contain large (> 1 cm) crystals, but should not be confused with large, zoned pegmatites and pod-like bodies with high concentrations of rare lithophile elements (e.g. London 1992), nor with stockscheiders in rare element granites.

1.2. Genetic models

Several models for the origin of comb-layered quartz have been suggested. One scenario involves crystallisation of both comb quartz and associated aplite from a silicate melt. White *et al.* (1981) postulated that the textures form during inward crystallisation of a static melt. Rhythmic variations in P_{H_2O}

and P_{HF} act to move the position of the ternary (qz–or–ab) minimum, causing repeated enlargement and shrinkage of the stability field of quartz. Decreases in P_{H_2O} and P_{HF} would be caused by degassing events associated with fracturing or possibly by diffusion. However, exsolved fluid would be absent during the crystallisation of the comb layers themselves. Carten et al. (1988b) took a similar stance, stating: 'individual layers are interpreted to have formed by inward growth from successively solidified margins of a magma body.'

Shaver (1988) found gashes within dendrite-bearing aplites at the Hall deposit in Nevada. The gashes were filled with comb-layered quartz and Shaver (1988) concluded that the quartz had precipitated from a hydrothermal fluid. He hypothesised that the gashes, as well as the layers of quartz parallel to the intrusion periphery, had formed during cooling-related shrinkage of the intrusions. He considered these tear fractures to have been filled with exsolved aqueous fluid and to have formed sequentially as the intrusion cooled inwards.

Stewart (1983) and Kirkham and Sinclair (1988) put forward the idea that the quartz layers precipitated from an aqueous fluid that had separated from the magma by resurgent boiling (crystallisation-induced volatile saturation) and had accumulated at the apex of the intrusion. The aplite layers crystallised from silicate melt that was quenched during explosive release of the accumulated aqueous fluid into the surrounding wallrock.

This paper presents new data that corroborates the model of Kirkham and Sinclair (1988) by showing that there are two very different populations of quartz within the comb-layered rocks, and that one population was probably precipitated from an exsolved magmatic fluid (vapour). This hydrothermal quartz is rich in trace elements. Moreover, the composition of the Logtung comb-layered rocks are far too silica-rich to have formed from a silicate melt.

2. The Logtung Deposit

The Logtung deposit lies on the Yukon Territory–British Columbia border, Canada (Fig. 2), about 130 km E of Whitehorse, Yukon Territory. The deposit is low grade and remains unmined, but contains 162 million tons of 0·103% W and 0·03% Mo (Noble et al. 1984). It consists mainly of a stockwork of mineralised fractures and quartz veinlets associated with a Cretaceous monzogranite stock and related dykes of quartz–feldspar porphyry and aplite that have intruded Palaeozoic metasedimentary rocks (Fig. 2). Noble et al. (1984) recognised several different cross-cutting vein types in the stockwork representing multiple stages of mineralisation. The earliest stage, consisting of quartz–molybdoscheelite veinlets, is associated with the monzogranite stock. Later quartz–scheelite and quartz–molybdenite veinlets are related to irregular, branching felsic dykes of quartz–feldspar porphyry and aplite. The final stage of mineralisation is represented by widespread, polymetallic quartz veins. In the vicinity of the W–Mo deposit, these veins contain scheelite, molybdenite, bismuthinite and beryl; to the NE, they contain Pb and Zn sulphides with Ag and minor amounts of Sn.

The monzogranite stock at Logtung averages 35% modal quartz, 32% plagioclase, 30% orthoclase perthite and 3% biotite (Stewart 1983). The primary texture is seriate, although aplitic and pegmatoidal textures are also present. Minor accessory phases include Fe oxides and sulphides, apatite and zircon, muscovite, epidote and fluorite. Stewart (1983) inter-

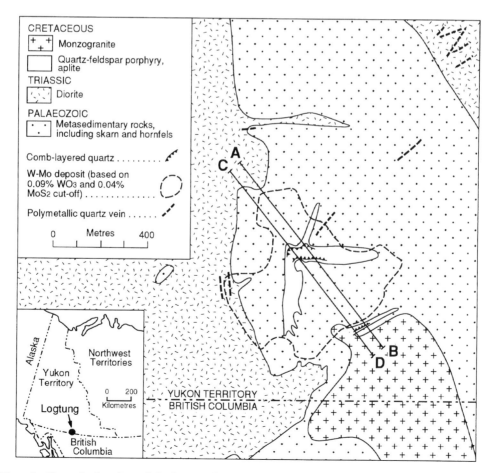

Figure 2 Generalised geology of the Logtung W–Mo deposit, Yukon Territory (after Harris & van der Poll, unpublished data; Noble et al. 1984).

preted some of these phases to be introduced subsolidus as secondary, deuteric alteration minerals.

No clear cross-cutting relationships have been observed, but the felsic dykes probably post-date the monzogranite (Noble *et al.* 1984). Although textures differ, the compositions of the dykes are broadly similar and they probably emanated from the same magma. The dykes average 30% phenocrysts, consisting of 31% modal quartz, 40% perthitic alkali feldspar, 28% plagioclase and 1% other minerals. The matrix itself is similar, containing 37% quartz, 41% orthoclase and 22% plagioclase (Stewart 1983).

Comb-layered quartz is a prominent feature of the felsic dykes, which are composed of high-silica granite with 75–77% SiO_2. Macroscopic descriptions were provided first by Stewart (1983), who mapped these textures in a currently inaccessible decline. The following information is based on his work, along with our own observations on outcrops, drill core and samples of dump material from the decline.

The comb-layered quartz is best developed in the NE arm of the largest of the felsic dykes, which contains layers of comb quartz that range in thickness from less than 1 mm to more than 1 m. We estimate the number of individual layers as between 100 and 1000. The layering is clearly contact-related and occurs primarily at the top and sides of the intrusions (Fig. 3). The layering is best developed on the hanging wall of the dyke, particularly in regions with shallow dips. In the smaller dyke to the SE, comb layers have grown inwards from both walls of the dyke towards a quartz core (comparable features at the Climax and Henderson Mo deposits, Colorado, have been referred to as 'vein dykes' by White *et al.* 1981). Other parts of the dykes are characterised by parting veins (Kirkham & Sinclair 1988) or 'streaky quartz veins' (Stewart 1983), which consist of parallel quartz veins separated by laminar aplites. These veins appear to be closely related to the comb-layered quartz. They cross-cut comb-quartz layers in places and probably post-date the early quartz–molybdoscheelite veinlets, but preceded the quartz–molybdenite stage of mineralisation (Stewart 1983).

No significant regional deformation has occurred subsequent to emplacement of the dykes; however, significant faults are present in the area and may have caused some gentle tilting.

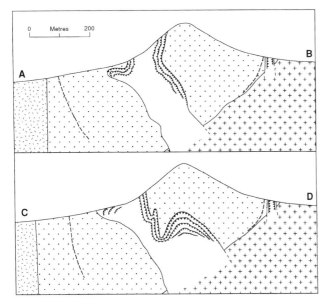

Figure 3 Cross-sections A–B and C–D (locations indicated in Fig. 2) depicting the distribution of comb-layered quartz in felsic intrusions at Logtung, Yukon Territory. Legend as in Fig. 2.

3. Petrographic and chemical characteristics of comb-layered rocks of the Logtung deposit

3.1. Macroscopic characteristics of comb-layered quartz from Logtung

Typically, the comb-quartz layers range from 1 to 4 mm in thickness. Individual layers can be planar or crenulate, the latter presumably due to ductile deformation of the hot, plastic quartz subsequent to its precipitation. In general, thin layers are less continuous and less planar than thick layers. Some layers wrap around large crystals protruding from underlying layers (Fig. 4a, 4b, 4e). The base of each comb layer, which appears sharp in hand specimen, is irregular on a microscopic scale and conforms to grain boundaries in the underlying aplite. Terminations of the comb layers, however, are euhedral. Prismatic quartz grains are oriented with their c-axes perpendicular to the layering (Stewart 1983). The space between prisms is filled by aplite. The quartz morphology within the comb layers (i.e. prisms and elongate facets; Fig. 4) is characteristic of that for α-quartz, the low-temperature variety of quartz found in hydrothermal assemblages and most pegmatites. This may imply that the quartz grew at a temperature <600°C (at 100 MPa). Comparison with photographs from Kirkham and Sinclair (1988), Shannon *et al.* (1982) and other papers indicates that comb layers from most, if not all, localities consist of similar quartz. Conceivably, this morphology could have grown within the field of stability of the high-temperature β-quartz if the crystal grew from a substrate, particularly if growth was rapid. Therefore, the prismatic quartz found in comb layers is given the generic term Type I quartz.

The aplite interlayers typically consist of small, 10–20 μm sized grains of quartz, sodic plagioclase and alkali feldspar. Minor muscovite, apatite and garnet are also present. Granophyric intergrowths of quartz and alkali feldspar are common, especially along the edges of crystals in the bordering comb layers and around phenocrysts within porphyritic aplite. Such intergrowths are often interpreted to form during episodes of rapid growth (Barker 1970; Lentz & Fowler 1992). Phenocrysts of both feldspars and quartz (0·5–2 mm in diameter) make up 5–20% of the porphyritic aplite interlayers. The quartz grains have a different morphology from quartz grains in the comb layers and are called Type II quartz. Prismatic crystals, for example, are absent in the aplite; instead, quartz shows equant hexagonal bipyramidal growth forms (Fig. 5). Re-entrants are common and may represent episodes of cellular growth (Swanson & Fenn 1986; MacLellan & Trembath 1991), or in some instances, embayments due to resorption. This morphology, with equant bipyramids and common reentrants, is typical of quartz in rhyolites and granite porphyries and represents the normal form of volcanic quartz (β-quartz). Because the β/α polymorphic transition (573°C at atmospheric pressure) is non-quenchable, all Logtung quartz would have an α-quartz structure at room temperature, although the relict growth morphologies would remain.

3.2. Secondary alteration

Most phenocrysts within the aplite layers are relatively unaltered. Plagioclase compositions are highly albitic (An_{10}) and perhaps reflect re-equilibration under subsolidus conditions. Alkali feldspars are perthitic and sericitised. Compositions of 20 μm sized plagioclase (An_{10}) and alkali feldspar (Or_{95}) crystals in the aplitic groundmass are consistent with final equilibration at temperatures below 400°C (Brown & Parsons 1981), presumably under subsolidus conditions. Quartz, both as phenocrysts within aplite and in the comb

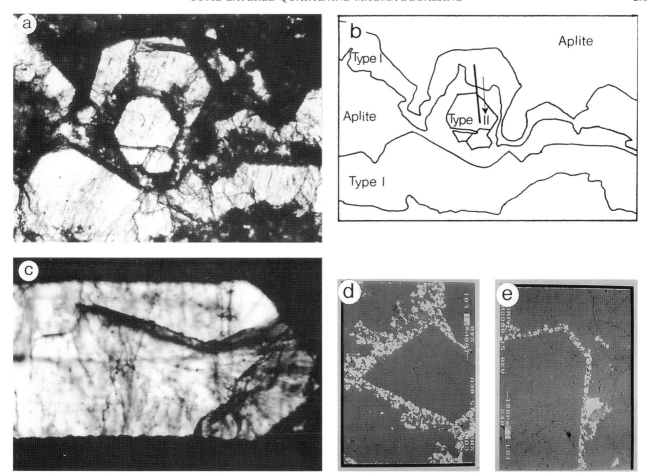

Figure 4 Quartz crystals from comb-layered rocks. (a) Two successive comb layers from sample SYA 85-15B. Growth direction towards top. Field of view is approximately 5 mm across. Photomicrograph taken in plane polarised light. (b) Simplified diagram illustrating features in (a). Top comb layer has overgrown a Type II phenocryst from the aplite interlayer. Bold line and arrow indicate the location and direction of trace element scans shown in Fig. 8 and discussed in text. (c) Elongate, prismatic Type I quartz from sample SYA85-15B. Crystal is 1 mm from top to bottom. Photomicrograph taken in plane polarised transmitted light. (d) Backscattered electron image of prismatic Type I quartz from sample SYA85-15B. Field of view in photo is ≈3 mm high. Whitish grains are feldspars from aplite interlayers between successive layers of Type I quartz. (e) Backscattered electron image of two comb layers separated by a 50 μm thick band of aplite from SYA85-15B. Field of view in photo is ≈3 mm high.

layers, is generally optically continuous. However, some grains, though seemingly single crystals, appear in cross-polars to be composites of numerous, small anhedral crystals with unrelated orientations and extinctions. Evidently, some large continuous crystals recrystallised subsequent to their initial growth. The recrystallised regions, and some of the single, optically continuous grains, can contain sprays of acicular apatite.

3.3. Fluid inclusions and growth layering

Fluid inclusions also show that these rocks have undergone some secondary alteration. Crystals in the comb quartz layers and aplites contain numerous trains of fluid inclusions oriented along planes oriented oblique to the direction of growth (Fig. 6). All inclusions appear to be secondary, with the exception of some submicrometre sized inclusions that will be discussed later. The trains of inclusions are interpreted to be located along healed fracture planes formed after the initial growth of the comb-layered textures. More than 90% of these secondary inclusions are two- and three-phase inclusions, containing liquid + vapour and two liquids + vapour, respectively. The latter are CO_2-bearing, as distinguished by infrared spectroscopy and their behaviour during heating and freezing experiments (J.B. Lowenstern, unpublished data). Other less abundant inclusion types include vapour-rich and

halite-bearing inclusions. No secondary inclusions of any kind are greater than 30 μm in greatest diameter and most inclusions are less than 10 μm in size.

Microscopic examination revealed that large, optically continuous quartz grains within the comb layers contain bands of brownish, submicrometre sized inclusions. As shown in Fig. 6, the bands are in nearly all instances parallel to crystal faces and show no correlation with fracture-controlled planes that contain the much larger secondary fluid inclusions. Often, the abundance of the brownish inclusions is greatest near the outer edges of comb layers and adjacent to the 'overlying' (younger) aplite. The identity of the individual inclusions making up the bands could not be determined with an optical microscope; however, secondary electron images of polished quartz grains showed a correlation of micrometre sized pits in the sample surface with discoloration due to the presence of the brownish bands. The pits, as well as the infra-red data discussed later, indicate that the brown bands consist partially of submicrometre sized fluid inclusions.

The planar nature of these bands, their location parallel to crystal faces and their abundance near the edges of crystals is consistent with their formation during the growth of the quartz. Variations in the density of the growth bands are interpreted to be due to changes in the crystallisation environment during growth episodes. The bands are always

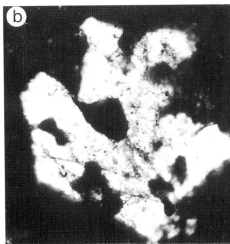

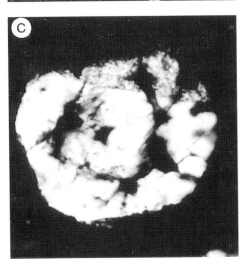

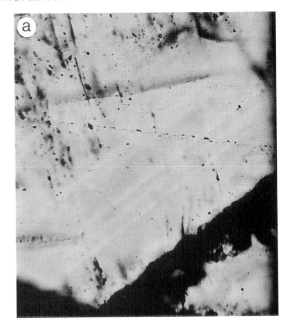

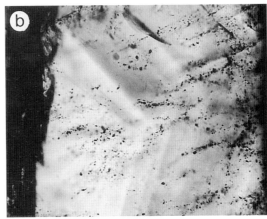

Figure 6 Transmitted light photomicrographs of typical 'growth' banding of comb-layered quartz from Logtung samples. (a) Banding runs from lower left to upper right and trends parallel to crystal face. Greater magnification allows discrimination of countless individual submicrometre sized inclusions within the band. Healed fractures that contain secondary fluid inclusions, generally 5–20 μm in size, have orientations unrelated to those of the crystal faces. One such band trends roughly E–W across the centre of the image. Field of view is 2 mm across. (b) Growth banding trends from upper left to centre of field, turns and trends down to the centre of the bottom of the image. The bands run parallel to crystal faces (not shown here). Dark, discontinuous bands of secondary inclusions trend primarily E–W and have no correlation with the growth bands. Field of view is 2 mm across.

Figure 5 Transmitted light photomicrographs of aplite-hosted quartz (Type II). Crystals have embayments, are equant and show typical morphology of volcanic (β-form) quartz. (a) Crystal (1.7 mm across) from aplite interlayer in SYA85-14B. (b) Crystal (1.5 mm across) from aplite interlayer in SYA85-15B. (c) Crystal (1.0 mm across) from aplite interlayer in SYA85-15B.

absent in regions where Type I quartz has recrystallised. No band is present in the aplite-hosted, Type II quartz.

3.4. Trace element concentrations in Logtung quartz

Comb-layered quartz that contains growth banding also displays anomalous concentrations of many trace elements. As shown by electron- and X-ray microprobe analyses (see Appendix), the growth banding corresponds with higher

amounts of Na, Al, K, Ca, Mn, Ti, Fe, Ge and Rb. Even grains where banding is not visible may have anomalous concentrations of these metals, particularly if the grains are large and optically continuous. As shown in Table 1, one quartz crystal contained high concentrations of Al (3690 ppm by weight; ppmw), K (930 ppmw), Ti (110 ppmw), Mn (30 ppmw), Ge (8 ppmw) and Rb (36 ppmw). Most grains contained lower concentrations of these metals, although 400–800 ppmw Al and 300–600 ppmw K were common. Figure 7 shows the results of maps made with the electron microprobe: the grey level corresponds to count rates for Al Kα X-rays determined by a wavelength-dispersive detector. The large Type I quartz grains within comb layers are often zoned in Al and contain higher Al concentrations than Type II quartz within aplite interlayers, as well as domains within the Type I quartz that have recrystallised (arrow).

The Type II quartz phenocrysts within porphyritic aplite

Table 1 Representative analyses of quartz.

	(1) Type I	(2) Type I	(3) Type II	(4) β-qtz
Al (ppmw)	3690	940	110	100
K (ppmw)	930	500	230	180
Ti (ppmw)	110	50	20	6
Ge (ppmw)	8	3	1	1
Rb (ppmw)	36	4	0·7	1
H_2O (ppmw)	1500	690	190	—
Al (ppma)	8330	2110	250	220
K (ppma)	1450	780	360	290
Ti (ppma)	135	64	25	8
Ge (ppma)	7	3	1	1
Rb (ppma)	26	3	0·5	1
H_2O (ppma)	10 050	4630	1310	—

Notes: ppmw is parts per million by weight. ppma is parts per million atomic or $10^6 \times$ moles element divided by moles Si. Samples: (1) trace element rich Logtung Type I quartz from comb layering; (2) Type I quartz from comb layering (Fig. 4a, 4b); (3) Type II quartz phenocryst (β-quartz) from aplite interlayer (Fig. 4a, 4b); and (4) β-quartz from volcanic fallout, tuff of Pine Grove (Lowenstern 1994). Analytical techniques: Al and K by electron microprobe, estimated analytical uncertainty is 5% relative; Ti, Ge and Rb by X-ray microprobe, estimated accuracy within 50% relative; H by infra-red spectroscopy, estimated accuracy within 25% relative. See Appendix for analytical details

layers are unzoned in Al and other elements and contain low concentrations of all elements except Si and O (e.g. Al < 200 ppmw, K < 300 ppmw). Quartz grains making up part of the groundmass of the aplite (30–70 μm in size) contained similar concentrations of trace elements as the Type II quartz. Figure 8 shows the results of line scans with both the electron microprobe and X-ray microprobe. The beam traversed a crystal of Type I quartz, continuous with comb layering, that grew over a Type II quartz phenocryst from the aplite interlayer (Fig. 4a, 4b). The overgrowth layer has much higher concentrations of all elements analysed, including Al, K, Ti, Mn, Ge and Rb.

The low trace element concentrations within the Type II/β-quartz is consistent with the low trace element abundances found in igneous quartz (Smith & Steele 1984). As a comparison with the Logtung samples, 30 quartz grains from five different volcanic systems were analysed by electron microprobe for their Al, Fe, Ti, Na and K contents (Fig. 9). Quartz was extracted from rhyolites from the Pine Grove (Table 1), Pantelleria, Katmai, Bishop Tuff and Lassen Peak volcanic systems, covering a range from peralkaline to peraluminous magma types and from continental to arc tectonic settings. In all instances, only Al and K were reliably measured above the detection limits and both elements had consistently low concentrations relative to the Type I quartz from comb layers of the Logtung deposit (Fig. 9).

3.5. Infra-red spectroscopy of Logtung quartz

Quartz grains rich in trace elements have unique spectra when studied with Fourier transform infra-red spectroscopy. Spectrum 2 in Figure 10 shows data from a large, single crystal of Type I quartz from one of the comb layers. It shows two prominent features, a sharp peak at 3630 cm^{-1} superimposed on a broad, symmetrical peak from 3700 to 3000 cm^{-1}. The height of this broad peak correlates strongly with the density of the brown bands of submicrometre sized inclusions described earlier. The broad peak is similar in shape and width to that expected from liquid water, as in either fluid inclusions or the nanometre sized domains of water molecules that are often found in quartz (Aines *et al.* 1984; Kronenberg 1994). Aines and Rossman (1984) found that the broad symmetrical peak from 3700 to 3000 cm^{-1} in both synthetic and natural α-

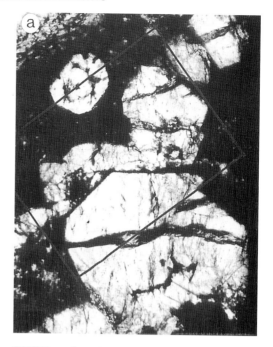

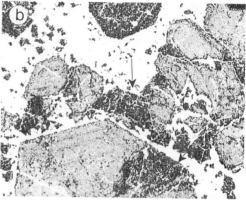

Figure 7 Photomicrograph and trace element maps of comb-layered texture from sample SYA85-15B. (a) Photomicrograph of 300 μm thick section taken with plane polarised light. Box displays area of scan shown in (b). (b) Map of Al distribution (light colours are high Al concentrations) within area shown in box from (a). Round grain at top centre (only bottom half shown) is Type II quartz hosted by aplite. The comb layer below has much higher Al concentrations, as does the Type I crystal at bottom which displays Al zoning parallel to crystal faces. Field of view is 5.1 mm across. Arrows show region of recrystallisation of primary quartz into a mosaic of anhedral grains. The matrix is bright due to the presence of feldspar. (c) Map of Al distribution in a Type I quartz crystal (bottom right) and three Type II crystals (top left). The Type I grain is zoned in Al with Al-rich and Al-poor layers parallel to crystal faces. These same layers are visible optically as brown bands of submicrometre sized inclusions as shown in Fig. 6. Field of view is 6·5 mm across.

quartz was due to molecular water trapped in the quartz structure. Individual water droplets were too small to freeze at liquid nitrogen temperatures and were presumed to be due to domains of water containing less than 200 molecules. They also found that synthetic quartz and natural, low-temperature (α) quartz have identical signatures in the infra-red, with

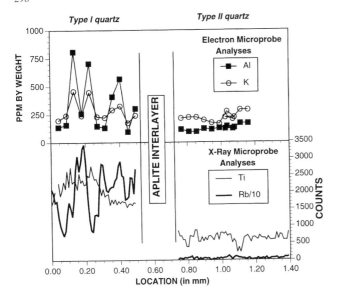

Figure 8 Line scans through Type II quartz phenocryst and Type I overgrowth shown in Figure 4a, 4b. Top panel displays concentration data for Al and K acquired with the electron microprobe. The scan went from the overgrowth layer, shown on the left, passed through the interlayer aplite (data not shown) and into the Type II quartz. Bottom panel displays count rates for Ti and Rb $K\alpha$ X-rays acquired with the X-ray microprobe at Brookhaven National Laboratory. See Table 2 for Ti and Rb concentrations in similar samples.

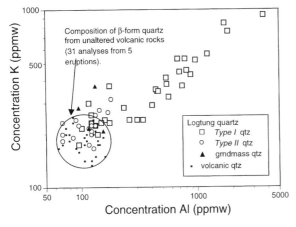

Figure 9 Log–log plot of Al versus K in ppm by weight. Data include Type I quartz from comb layers, Type II quartz from phenocrysts within aplite interlayers, quartz from the aplite groundmass (grndmass qtz) and volcanic β-quartz from the Katmai, Pantelleria, Lassen, Bishop tuff and Pine Grove volcanic systems. All volcanic quartz as well as Logtung quartz from aplite interlayers contains low concentrations of both Al and K. In contrast, Type I quartz from comb layers show highly variable Al and K, and concentrations as high as 3700 and 930 ppm, respectively. The Al- and K-rich areas correspond to the banding shown in Figures 6 and 7.

characteristic broad-band peaks that increase in size as a function of growth rate and the consequent incorporation of water into the quartz structure.

The sharp peak at 3630 cm^{-1} shown in Figure 10 corresponds with that of the peak due to structural OH$^-$ in muscovite (see also Mainprice & Paterson 1984). A small peak at 4550 cm^{-1} could also be due to muscovite inclusions, though we cannot rule out the presence of some OH$^-$ within the quartz structure. It thus appears that water is located within the Type I quartz as submicrometre sized water droplets and muscovite flecks and possibly as minor structurally bound hydroxyl.

In contrast, the infra-red data for Type II quartz from Logtung indicates that it, like all β-form quartz (e.g. spectrum

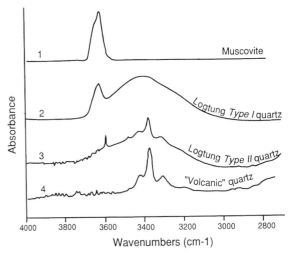

Figure 10 Infra-red spectra of quartz and muscovite, shown in plots of wavenumbers versus absorbance. Scale for the Y-axis is different for each spectrum. All spectra collected with an unpolarised beam. Peaks from 3700 to 3200 cm^{-1} correspond to the O–H stretch and provide information on the abundance and speciation of water in the host quartz. Logtung Type II quartz (spectrum 3) has low H concentrations and several sharp bands, similar to those in volcanic quartz, such as spectrum 4 of β-quartz from the Pine Grove volcanic system (Lowenstern 1994). Such sharp peaks are due to hydroxyl within the quartz structure. The Type I quartz (spectrum 2) has more water, primarily as submicrometre sized liquid H$_2$O domains, though a sharp peak at 3630 cm^{-1} may be due to inclusions of muscovite (spectrum 1). The two Logtung spectra shown are from an overgrowth of Type I on Type II quartz so that the two grains have identical orientation (Fig. 4a). Other Type I and II quartz grains showed similar spectral characteristics to these two end-members, regardless of crystal orientation.

4 of Fig. 10), lacks significant H$_2$O. Spectra 2 and 3 in Figure 10 compare Type II quartz with an overgrowth of Type I quartz (Figs 4a, 4b, 8). The two grains are optically continuous and, therefore, the crystal orientation relative to the infra-red beam is identical. The broad-band peak for the Type II quartz is much smaller than for the Type I quartz.

The IR data show that the two populations of quartz from the Logtung comb-layered rocks are dissimilar. The Type II quartz crystals contain little H$_2$O and most of it is probably secondary fluid inclusions. The Type I quartz has up to ten times more water as (1) abundant submicrometre sized droplets or inclusions of molecular water found in growth layering *and* (2) significant hydroxyl as either structural OH$^-$ or inclusions of muscovite (spectrum 1). The height of both sharp and broad-band peaks correlates with the density of submicrometre sized inclusions (the brownish bands).

4. Significance of impurity-rich quartz

The compositions of the Type I quartz grains hint that they may not have formed from a silicate melt. A survey of relevant publications shows that most igneous quartz (from unaltered volcanic rocks and intrusions) contains extremely low concentrations of all trace elements (Scotford 1975; Dennen 1966, 1967; Smith & Steele 1984; Ghiorso *et al.* 1979; Fig. 9). Only at Macusani Peru, where the β-quartz–spodumene solid solution (virgilite) is found, are high concentrations of Al reported for volcanic quartz (along with Li, which, in contrast, is <10 ppmw in both Logtung quartz types; SIMS analyses by R.L. Hervig of Arizona State University).

In contrast with β-quartz, some hydrothermal α-quartz can have relatively high concentrations of Al, K, H$_2$O and other constituents (Brice 1985; Perny *et al.* 1992; Heany 1994). Perny *et al.* (1992) reported concentrations of Al in inclusion-free hydrothermal quartz from the Swiss Alps that are similar to the values found in the Logtung samples.

Given that natural quartz has grown at unknown rates, and generally poorly constrained temperatures and pressures, it is worth considering the relevant publications on synthetic quartz grown in the laboratory. Brice (1985) provided a detailed review of the growth of synthetic quartz for piezo-electric resonators. Typically, these crystals are grown at temperatures close to 350°C in large autoclaves placed within a temperature gradient. Natural quartz crystals are dissolved at the hot end of the autoclave and precipitate at the cool end on suitably oriented seed crystals. The quartz grows from an aqueous solution, often a 0·5–1·0 M solution of NaOH or Na_2CO_3, though other 'mineralisers' are sometimes used. Considerable effort has been placed on growing quartz free from trace elements, which is best suited for industrial purposes. Brice (1985) showed that H and other impurities increase as a function of the growth rate. Crystals grown at 2–3 mm/day can contain more than 2000 ppma H (parts per million atomic; i.e. molar $H/Si \times 10^6$); the H correlates directly with dislocation densities. Aines et al. (1984) found more than 3000 ppma H in some crystals (grown at similar rates). These concentrations are similar to those found in Logtung Type I quartz (Table 1). There are a number of ways in which impurities can be incorporated into the growing quartz: (1) as inclusions of fluid or mineral phases; (2) as interstitial atoms within the quartz structure; and (3) as substitutions for Si or O. According to Brice (1985), most Li, Na, K and Al is interstitial, although Al may substitute for Si along some crystal faces. Aines et al. (1984) showed that hydrogen is incorporated primarily as liquid H_2O, in small, non-freezable aggregates. Whether or not other elements may accompany the H_2O within these aggregates is unknown.

Several factors may be responsible for the high trace element concentrations in the Logtung Type I quartz. Although H, Li, Na and K could reasonably be ascribed to the presence of fluid inclusions, Al and Ge are more likely to reside within the quartz structure. Submicrometre sized inclusions of muscovite would result in high Al, H, K and Rb, but should not result in the large broad peak in the middle of the infra-red, which indicates liquid H_2O either as inclusions or microaggregates within the quartz (Aines & Rossman 1984). The trace constituents are thus most likely to be a combination of elements within the quartz structure plus submicrometre sized inclusions of liquid H_2O. Tiny inclusions of muscovite may also be present and are consistent with the relatively high Al/K in the quartz.

In general, high dislocation and inclusion densities, trace element and H_2O concentrations all correlate with fast growth rates in synthetic crystals (Brice 1985). Such growth rates (up to >3 mm/day) are 50 times faster than the highest known growth rates for magmatic β-quartz (skeletal crystals grown by Swanson & Fenn 1986). Given that temperature and pressure alone do not appear to have a significant affect on Al substitution in quartz (Scotford 1975; Pavlishin et al. 1978; Stavrov et al. 1979), the trace element variations may tentatively be assigned to variations in growth rate. Therefore, if the Logtung Type I quartz crystals grew from the same solvent as the Type II quartz and groundmass quartz within the aplite (i.e. they grew from a silicate melt), this implies that they grew more quickly than the aplite-hosted quartz pheno-crysts (Type II). Rapid growth textures are also exhibited in the aplite layers, but as dendrites and granophyric in-tergrowths, rather than euhedral, prismatic quartz. Like the Type II quartz, the granophyric quartz has low trace element concentrations. The similarity of the Type I quartz with synthetic α-quartz grown from hydrothermal solutions hints

at another possibility—that the Type I quartz rich in trace elements grew from a different medium than the Type II quartz and aplite.

5. Genesis of comb-layering at Logtung and in other silicic porphyries

5.1. Composition of comb-layered rock

We believe that the Type II quartz are simply pre-existing phenocrysts that were present within the magma before the quench events. The aplites represent silicate melt quenched during fracturing/fluid loss events. In contrast, we contend that the comb layers of nearly monomineralic quartz grew from thin pockets of exsolved aqueous fluid that had accumulated at or near the top of the magma reservoir. To bolster this argument, it is important to show why the bulk comb-layered rocks are unlikely to have formed solely due to crystallisation of a silicate melt.

Bulk-rock analyses and point counts of the comb-layered rocks show that they do not represent magmatic compositions. Because of the abundance of the comb layers themselves, which contain little other than quartz, these rocks contain well over 60% normative (and modal) quartz. Figure 11 displays the haplogranite ternary, including the location of minimum melt compositions for H_2O-saturated magmas from 0·5 to 3 kbar (50–300 MPa). Though aplite dykes and porphyry intrusions from Logtung have normal magma compositions, the comb-layered rocks are very quartz rich and correspond to magmas with unreasonable melting temperatures over 1000°C, regardless of the amount of H_2O present (Tuttle & Bowen 1958). The composition of H_2O-saturated minima may be shifted by the addition of fluxing agents to the magma (e.g. Li, B, F, P; Manning & Pichavant 1988), but all such components move the minima toward the albite apex, contrary to what is observed in the comb-layered rocks.

One possible mechanism for the formation of comb-layered rocks by purely magmatic crystallisation would be under-cooling, poisoning of nucleation sites for feldspar and non-equilibrium growth of quartz. London (1992) has shown that this occurs in systems rich in Li, B and F. Swanson and Fenn (1986) conducted experiments on simplified rhyolitic melts, but found that strongly undercooled growth forms were dendritic, rather than as large euhedral crystals. Even if euhedral quartz growth was possible, however, an effect should be seen in the composition of the neighbouring melt. Aplites bordering quartz layers should be shifted in composition away from the quartz apex in Figure 11. Point counts and chemical analyses of aplites, however, give normal magma compositions for the aplite interlayers (Fig. 11) and are similar to the whole-rock compositions of Stewart (1983). Indeed, there do not appear to be any albite- or feldspar-rich zones that reflect the subtraction of quartz from the silicate melt during crystalli-sation of the comb layers, nor is there evidence for Na-rich mineralisation associated with hydrothermal alteration. Instead, as in other W–Mo porphyry systems, high-temperature mineral assemblages are typically Si and K-rich (Carten et al. 1988b). It is therefore difficult to envision how closed system crystallisation of a normal silicate melt could result in these quartz-rich textures.

5.2. Implications of the presence of prismatic quartz

The morphology of the quartz may also yield information, both that the quartz grew from an aqueous fluid and that it crystallised at temperatures below 600°C. The Type I quartz from Logtung has α-form morphology, as does that from all other examples of comb-layered quartz described previously

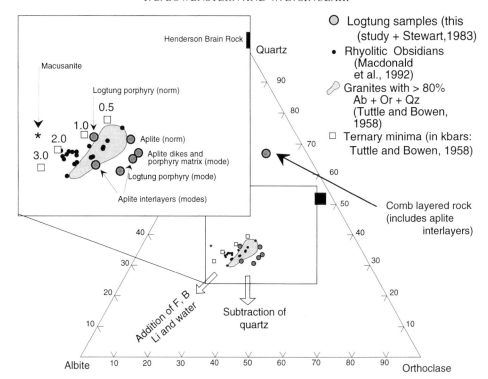

Figure 11 Ternary diagram of the haplogranite system. Macusanite and rhyolitic obsidians saturated with two feldspars + quartz from Macdonald *et al.* (1992). Logtung aplite interlayers from this study (point counts; $n \geq 1000$). Logtung aplites and porphyries (from Stewart 1983) plot near water-saturated minima and natural silicic granites and obsidians. Brain rock from Henderson (Carten *et al.* 1988a) and comb-layered rock from Logtung (Stewart 1983) are shifted towards the quartz apex. The shift is inconsistent both with closed system subtraction of quartz from a minimum melt and with the addition of fluxing agents such as Li, B and F.

(e.g. Shannon *et al.* 1982; Carten *et al.* 1988a; Kirkham & Sinclair 1988). If this quartz indeed grew within the stability field of α-quartz, this requires that it grew at temperatures less than 600°C at 100 MPa (Martin 1982; Heany 1994). For a silicate melt to reach such low temperatures, particularly at the shallow depths (2–4 km) at which porphyries are emplaced (White *et al.* 1981; Carten *et al.* 1988b; Keith & Shanks 1988), fluxing agents must be added to the magma, otherwise equilibrium crystallisation would occur at temperatures of 670–725°C (Tuttle & Bowen 1958). The presence of F in many W–Mo systems should help to decrease the magma temperature, although, as noted earlier, the compositions of the comb-layered rocks are inconsistent with the addition of large percentages of fluxing agents to the magma. There is no strong evidence for their former presence either, notwithstanding degassing, as the Logtung rocks do not contain more than 6000 ppmw F or 100 ppmw Li or B (Stewart 1983; Kirkham & Sinclair 1988). Non-equilibrium crystallisation of quartz from an undercooled silicate melt would also be possible, but is inconsistent with the observation that the aplite layers show no evidence of the subtraction of quartz and consequent enrichment in albite/feldspar (as discussed earlier).

If the Type I quartz grew as α-quartz from aqueous fluid, this implies that the roof of the magma chamber contained supercooled magma that was undergoing little, if any, crystallisation. Both conduction through the top of the magma reservoir and adiabatic cooling during pressure release events would have contributed to the low temperature.

Plausibly, the quartz is an unusual form of β-quartz and offers no information about temperature. Externally nucleated quartz, grown either from an aqueous fluid or silicate melt, could conceivably show an elongate morphology (P. Fenn, Corning Corp., pers. comm.). Although no-one has yet produced such prismatic quartz in experiments on rock melts or aqueous fluids in the β-quartz stability field, it may be

premature to make firm conclusions based on quartz morphologies.

5.3. Growth of comb-layered quartz from a pocket of exsolved fluid
Figure 12 and Table 2 outline a model for the growth of comb layering in porphyry and other shallow intrusions. The model is similar to that of Stewart (1983) and Kirkham & Sinclair (1988), but provides a more detailed explanation of degassing and crystallisation events.

Layers of Type I quartz crystallise from exsolved fluid that has accumulated at the top or side of a shallow stock of

Table 2 Sequence of events during growth of comb-layered textures

I Bubble coalescence and collapse of vesiculated magma produces a pocket of exsolved magmatic fluid adjacent to rhyolitic magma that contains phenocrysts of two feldspars and quartz (β-form) + minor mafic and accessory minerals

II Prismatic quartz grows from pocket of exsolved magmatic fluid to produce comb-layered bands. Little to no growth of phenocrysts occurs in subjacent undercooled magma. Continued degassing of melt and accumulation of fluid results in significant overpressures

III Overpressures cause fracturing of previously deposited comb layers. The pocket of fluid is drained as fluid flows out towards cooler country rocks, creating quartz-rich mineralisation, often accompanied by molybdenite. Precipitation of quartz and feldspar results in system sealing. Melt/aplite often enters the veins, forming dyke-like bodies (vein dykes; White *et al.* 1981)

IV Initial reduction in pressure, associated with fracturing, causes vesiculation of magma, even at great distances from the comb layer–magma interface. A thin layer of magma is quenched to phenocryst-bearing aplite along the comb-layered quartz. Aplite crystallisation may be caused by loss of H_2O from the melt, the negative heat of vaporisation, conduction of heat into the comb quartz and adiabatic depressurisation, all of which should promote crystallisation of the melt. Growth of dendrites within the aplite layers also occurs during undercooling associated with this stage

V Return to stage I for repetition of events

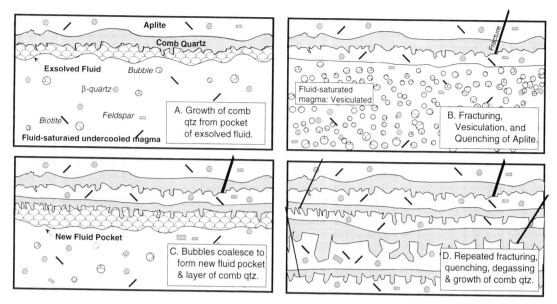

Figure 12 Schematic model for the formation of comb-layered textures in porphyry intrusions. See text and Table 2 for details.

magma (Fig. 12A). The exsolved magmatic fluid is plausibly supplied by convective degassing, where large volumes of fluid/exsolved vapour are fed to the cupola by the degassing of a large, subjacent magma body (Shinohara *et al.* 1995). Such open-system degassing does not require that all silica and metals are supplied by the cupola itself, but by the larger magmatic system, for which the cupola serves as a conduit. The exsolved aqueous fluid would be silica-rich and quartz-saturated (Burnham 1967) and would also serve as a medium for rapid diffusion and transport of silica from the magma to the growing quartz (Jahns 1982). Fluorine is thought to be an important volatile species in porphyry Mo systems (Carten *et al.* 1988b) and its presence would aid the hydrothermal transport of silica by increasing the diffusion rates (Dingwell 1988) and quartz solubility in any exsolved aqueous fluid (Haselton 1984).

The exsolved volatiles are envisioned to form films or thin pockets (centimetre to metre sized) that accumulate below the impermeable, crystallised top of the intrusion or along its upper sides. Elongate quartz nucleates on the crystalline aplite from the previously formed band. The crystals grow perpendicular to these bands and thus the intrusion wall, even if the exsolved aqueous fluid has accumulated on the side of the intrusion. The growth of quartz stops when pressure builds up to cause fracturing of the overlying wallrock, including the previously crystallised comb layers. The aqueous fluid then escapes, causing alteration of the overlying rocks and, often, W and Mo mineralisation. Newly formed fractures seal as the temperature drops and quartz precipitates out of the aqueous fluid; this sealing causes the system to repressurise. Repeated fracturing can result in the formation of numerous parallel quartz veins causing regions of streaky quartz (Stewart 1983) or parting veins (Kirkham & Sinclair 1988).

The decompression events force magma vesiculation and contact between the newly grown Type I quartz and the melt (Fig. 12B). A thin layer of aplite (typically 10 μm to 20 cm) forms as a quenched rind on the layer of comb quartz. Quenching is caused by the loss of water from the melt, as well as adiabatic (isentropic) cooling of the aqueous fluid and melt due to rapid depressurisation. Indeed, granophyric and dendritic textures indicate rapid crystallisation of the aplite layers. Aplite formation would occur only in the region directly in contact with the comb-layered quartz. Below, the magma

would remain molten and able to flow. The phenocryst abundance in the aplite layers and thus the magma (represented by Type II quartz as well as feldspars) was about 20–40%, similar to that in the porphyries. Rhyolitic magmas with similar phenocryst contents, and H_2O pressures of 50 MPa or more, should remain fluid enough to convect and degas (Shinohara *et al.* 1995), thereby supplying more aqueous fluid to the intrusion roof.

Continued magma degassing and bubble coalescence results in the formation of a new pocket of exsolved fluid below the new layer of quenched aplite (Fig. 12C). Repetition of this process results in the formation of numerous comb layers (Fig. 12D). Folding and crenulation of the comb layers occurs during and subsequent to their formation and is probably caused by deformation related to magmatic intrusion and pressure release events. At temperatures above 500°C, the quartz would be exceedingly ductile and likely to flow when deformed at low strain rates. Catastrophic depressurisation, however, results in the brittle fracturing associated with the loss of fluid pressure and the formation of quartz veins in the superjacent wallrock.

5.4. Advantages of the present model

The following statements summarise the arguments for formation of comb layering in silicic intrusions by the magmatic–hydrothermal model outlined here.

1. The quartz-rich nature of the comb-layered rocks (70 normative per cent) is consistent with the open-system addition of quartz-saturated aqueous fluid to the top of the intrusion. It is not consistent with the crystallisation of silicate melt, even if fluxing agents (Li, P, F, B) were present, as there is no evidence for albitic pods or sodic alteration at comb quartz localities.

2. High trace element concentrations in the comb layering is consistent with rapid crystal growth from a hydrothermal solution. Enrichment in Al and H commonly occurs in synthetic quartz, when rapid growth causes the entrapment of inclusions and high defect densities (Brice 1985).

3. The quartz-dominated compositions of the comb layers are consistent with those expected for aqueous fluid in equilibrium with a minimum melt granite. Burnham (1967) showed that at 100 MPa and 500–700°C the aqueous fluid in equilibrium with Spruce Pine pegmatite contained

several wt% dissolved silicates, over 80% of which were normative silica.

4. In this model, individual comb layers begin growing at a definite site within the magma—at the interface between exsolved fluid and crystallised intrusion/wallrock. They nucleate heterogeneously from crystals at the bottom of the previous layer of quenched aplite. In contrast, if the quartz layers grew from a silicate melt, it is unclear where and why they nucleate.

5. The presence of molybdenite flakes intergrown with comb quartz from many localities (Kirkham & Sinclair 1988) is consistent with precipitation from aqueous fluid, particularly given the trifling solubility of Mo in silicate melts (Lowenstern *et al.* 1993). Carten *et al.* (1988a, b) and Kirkham and Sinclair (1988) discuss the close spatial relationship between comb-layered quartz and hydrothermal alteration and mineralisation. In some instances, hydrothermal veins can be traced into individual comb-layered rocks (Carten *et al.* 1988a, b).

6. The close association of two forms of quartz is consistent with crystallisation of the Type I quartz from exsolved aqueous fluid, at $T < 600°C$, in contact with a supercooled magma containing quartz phenocrysts (Type II, β-quartz). If the Type I quartz is actually a previously unrecognised form of elongate β-quartz, the model outlined here would hold, although without any thermal constraints. The comb layers would grow instead from exsolved fluid at temperatures above that of the β–α inversion (600°C at 100 MPa).

6. Comb layering and porphyry ore deposits

Models that explain the growth of comb-layered rock from initially vapour-undersaturated magma have the additional problem that they must explain how Mo and W porphyry magmas can reach shallow crustal levels without first becoming volatile-saturated. Carten *et al.* (1988a, b) point out the tremendous mass of altered and mineralised rock associated with these small, cigar-shaped intrusions. Keith *et al.* (1993) argue convincingly that such deposits must be created by the streaming of volatiles through partially crystallised magma, rather than closed-system crystallisation (second boiling) of a unique, volatile-enriched melt. Volatile exsolution may begin at depths as great as 10–15 km. For example, Lowenstern (1994) showed that magmas from the Pine Grove Mo porphyry would have been saturated with a CO_2–H_2O vapour at pressures of >400 MPa, corresponding to depths of 16 km. Shinohara *et al.* (1995) presented a model whereby degassing within a column of convecting silicic magma supplies exsolved fluid to the top of a shallow stock. They calculate that convective degassing is rapid enough to form an Mo ore deposit within 100 a. Such a steady supply of exsolved magmatic fluid appears consistent with the repetitive comb layering at the roofs of many stocks. According to the model presented herein, such textures require the continuous supply of exsolved fluid from subjacent magma and may provide important clues to the processes that operate to create some granite-hosted ore bodies.

7. Acknowledgements

We thank R. Kirkham for providing samples from the Logtung and other deposits as well as encouragement and helpful reviews. Sasâ Bajt of the University of Chicago helped quantify the X-ray microprobe spectra, L. Calk of the USGS helped with data collection on the electron microprobe and R. Hervig of Ariz. State University performed the SIMS analyses. We appreciate helpful reviews and/or comments by C. Bacon, A. Campbell, P. Candela, R. Fournier, B. Hanson, D. Lentz, D. London, S. Ludington, J. Stimac and B. Taylor. The work was funded through the Geothermal Studies and Mineral Resource Surveys Programs of the Geologic Division of the USGS. Contribution of the Geological Survey of Canada to this paper is acknowledged as GSC Contribution No. 35095.

Appendix: analytical methods

Electron microprobe
Maps of Al and other elements were made by using a 50 nA beam focused to a 10 µm spot while rastering the stage in 10 µm increments in grids up to 3 mm on a side. The five-spectrometer JEOL 8900 electron microanalyser was set to count each spot for 0·3 s. Most maps took 10–16 h to collect. Individual points were analysed with a 50 nA beam focused to a 20 µm spot with count times of 40 s. Standards included Tiburon albite (for Na), OR1 (for K), An 100 glass (for Ca and Al), quartz (for Si) and Fe_2O_3 (for Fe). Estimated uncertainties are $\pm 5\%$ relative.

X-ray microprobe
Synchrotron radiation, collimated to a 10×8 µm spot, was used to induce fluorescence within Logtung quartz. The resulting spectra were collected with a Si(Li) detector set up on beamline X26a at the National Synchrotron Light Source at Brookhaven National Laboratory, in Upton, NY, U.S.A. The incident angle is 45° so that the beam passes obliquely through the sample. Because of this geometry, and the fact that many samples are zoned, a single, unzoned sample rich in trace elements was used as a standard. The concentration of Ca within this quartz grain, as determined by electron microprobe, was used as an internal standard (Lu *et al.* 1989). Other samples were then quantified by accounting for the change in incident beam flux during subsequent analyses, as well as sample thickness. Because this technique is less rigorous than normal methods, and due to the zoning within the crystals, the calculated concentrations are estimated to be valid only within a factor of two. The reproducibility of the technique is within 2–3% relative. Line scans were performed by moving the sample with a motorised stage through the beam in 10 µm steps, while counting 100 s at each step. The raw number of counts for seven regions of interest were recorded (Ca, K, Fe, Mn, Ti, Ge, Rb). No attempt was made to calculate concentrations from the line scan data.

Infra-red spectroscopy
Spectra were collected on a Nicolet Magna 550 Fourier transform infra-red spectrometer with an attached SpectraTech Analytical-IR microscope that utilises a liquid N_2 cooled MCT-A detector. The entire system is purged with dry air and is located at the U.S. Geological Survey in Menlo Park, CA, U.S.A. The microscope can acquire usable spectra of spots down to 15–20 µm in size. Between 512 and 4096 scans were collected at 4 cm^{-1} resolution from 400 to 6000 cm^{-1}. All spectra were collected with unpolarised light. Quantification was carried out according to the method of Aines and Rossman (1984), which should be accurate to within 20%, given that most of the quartz-hosted water was present in molecular form.

8. References

Aines, R. D. & Rossman, G. R. 1984. Water in minerals? A peak in the infrared. J GEOPHYS RES **89**, 4059–71.
Aines, R. D., Kirby, S. H. & Rossman, G. R. 1984. Hydrogen speciation in synthetic quartz. PHYS CHEM MINERALS **11**, 204–12.
Barker, D. S. 1970. Compositions of granophyre, myrmekite and graphic granite. GEOL SOC AM BULL **81**, 3339–50.
Brice, J. C. 1985. Crystals for quartz resonators. REV MOD PHYS **57**, 105–46.
Brown, L. & Parsons, I. 1981. Toward a more practical two feldspar geothermometer. CONTRIB MINERAL PETROL **76**, 369–77.
Burnham, C. W. 1967. Hydrothermal fluids at the magmatic stage. *In* Barnes, H. L. (ed.) *Geochemistry of hydrothermal ore deposits*, 37–76. New York: Holt, Rinehart & Winston.
Burnham, C. W. 1979. Magmas and hydrothermal fluids. *In* Barnes, H. L. (ed.) *Geochemistry of hydrothermal ore deposits*, 71–136. New York: Wiley.
Candela, P. A. 1994. Combined chemical and physical model for pluton devolatilization: a non-Rayleigh fractionation algorithm. GEOCHIM COSMOCHIM ACTA **58**, 2157–67.
Carten, R. B., Geraghty, E. P., Walker, B. M. & Shannon, J. R. 1988a. Cyclic development of igneous features and their relationship to

high-temperature hydrothermal features in the Henderson porphyry molybdenum deposit, Colorado. ECON GEOL **83**, 266–96.

Carten, R. B., Walker, B. M., Geraghty, E. P. & Gunow, A. J. 1988b. Comparison of field-based studies of the Henderson porphyry molybdenum deposit, Colorado, with experimental and theoretical models of porphyry systems. *In* Taylor, R. P. & Strong, D. F. (eds) *Recent advances in the geology of granite-related mineral deposits*. CIM SPEC VOL **31**, 351–65.

Dennen, W. H. 1966. Stoichiometric substitution in natural quartz. GEOCHIM COSMOCHIM ACTA **30**, 1235–41.

Dennen, W. H. 1967. Trace elements in quartz as indicators of provenance. GEOL SOC AM BULL **78**, 125–30.

Dingwell, D. B. 1988. The structures and properties of fluorine-rich magmas: a review of experimental studies. *In* Taylor, R. P. & Strong, D. F. (eds) *Recent advances in the geology of granite-related mineral deposits*. CIM SPEC VOL **31**, 1–12.

Gerlach, T., Westrich, H. R. & Symonds, R. B. Pre-eruption vapor saturation in magma of the climactic Mount Pinatubo eruption: source of the giant stratospheric sulfur dioxide cloud. *In* Newhall C. (ed.). U.S. GEOL. SURV. PROF. PAP., in press.

Ghiorso, M. S., Carmichael, I. S. E. & Moret, L. K. 1979. Inverted high-temperature quartz: unit cell parameters and properties of the α–β inversion. CONTRIB MINERAL PETROL **68**, 307–23.

Haselton, H. T. Jr 1984. The solubility of quartz in dilute HF solutions at 600°C and 1 kbar. EOS, TRANS AM GEOPHYS UNION **65**, 308.

Heaney, P. J. 1994. Structure and chemistry of the low-pressure silica polymorphs. *In* Heaney, P. J., Prewitt, C. T. & Gibbs, G. V. (eds) *Silica*. MINERAL SOC AM REV MINERAL **29**, 1–40.

Hedenquist, J. W. & Lowenstern, J. B. 1994. The role of magmas in the formation of hydrothermal ore deposits. NATURE **370**, 519–27.

Jahns, R. H. 1982. Internal evolution of pegmatite bodies. *In* Cerny, P. (ed.) *Granitic pegmatites in science and industry*. MIN SOC CAN SHORT COURSE VOL **8**, 293–327.

Johnson, M. C., Anderson, A. T. Jr & Rutherford, M. J. 1994. Pre-eruptive volatile contents of magmas. *In* Carrol, M. R. & Holloway, J. R. (eds) *Volatiles in magmas*. MINERAL SOC AM REV MINERAL **30**, 281–330.

Keith, J. D. & Shanks, W. C. III 1988. Chemical evolution and volatile fugacities of the Pine Grove porphyry molybdenum and ash-flow tuff system, southwestern Utah. *In* Taylor, R. P. & Strong, D. F. (eds) *Recent advances in the geology of granite-related mineral deposits*. CIM SPEC VOL **31**, 402–22.

Keith, J. D., Christiansen, E. H. & Carten, R. B. 1993. The genesis of porphyry molybdenum deposits. *In* Whiting, B. H., Mason, R. & Hodgson, C. J. (eds) *Giant ore deposits*. SOC ECON GEOL SPEC PUBL **2**, 285–317.

Kirkham, R. V. & Sinclair, W. D. 1988. Comb quartz layers in felsic intrusions and their relationship to porphyry deposits. *In* Taylor, R. P. & Strong, D. F. (eds) *Recent advances in the geology of granite-related mineral deposits*. CIM SPEC VOL **31**, 50–71.

Kronenberg, A. K. 1994. Hydrogen speciation and chemical weakening of quartz. *In* Heaney, P. J., Prewitt, C. T. & Gibbs, G. V. (eds) *Silica*. MIN SOC AM REV MINERAL **29**, 123–76.

Lentz, D. R. & Fowler, A. D. 1992. A dynamic model for graphic quartz-feldspar intergrowths in granitic pegmatites in the southwestern Grenville Province. CAN MINERAL **30**, 571–85.

Lofgren, G. E. & Donaldson, C. H. 1975. Curved branching crystals and differentiation in comb-layered rocks. CONTRIB MINERAL PETROL **49**, 309–19.

London, D. 1992. The application of experimental petrology to the genesis and crystallization of granitic pegmatites. CAN MINERAL **30**, 499–540.

Lowenstern, J. B. 1994. Dissolved volatile concentrations in an ore-forming magma. GEOLOGY **22**, 893–6.

Lowenstern, J. B. 1995. Applications of silicate melt inclusions to the study of magmatic volatiles. *In* Thompson, J. F. H. (ed.) *Magmas fluids and ore deposits*. MIN SOC CAN SHORT COURSE VOL **23**, 71–100.

Lowenstern, J. B., Mahood, G. A., Hervig, R. L. & Sparks, J. 1993. The occurrence and distribution of Mo and molybdenite in unaltered peralkaline rhyolites from Pantelleria, Italy. CONTRIB MINERAL PETROL **114**, 119–29.

Lu, F. Q., Smith, J. V., Sutton, S. R., Rivers, M. L. & Davis, A. M. 1989. Synchrotron X-ray fluorescence analysis of rock-forming minerals: (1) comparison with other techniques (2) White-beam energy-dispersive procedure for feldspars. CHEM GEOL **75**, 123–43.

Macdonald, R., Smith, R. L. & Thomas, J. E. 1992. Chemistry of the subalkalic silicic obsidians. U.S. GEOL SURV PROF PAP **1523**, 214pp.

MacLellan, H. E. & Trembath, L. T. 1991. The role of quartz crystallization in the development and preservation of igneous texture in granitic rocks: experimental evidence at 1 kbar. AM MINERAL **76**, 1291–305.

Mainprice, D. H. & Patterson, M. S. 1984. Experimental studies of the role of water in the plasticity of quartzites. J GEOPHYS RES **89**, 4257–69.

Manning, D. A. C. & Pichavant, M. 1988. Volatiles and their bearing on the behaviour of metals in granitic systems. *In* Taylor, R. P. & Strong, D. F. (eds) *Recent advances in the geology of granite-related mineral deposits*. CIM SPEC VOL **31**, 13–24.

Martin, R. F. 1982. Quartz and the feldspars. *In* Cerny P. (ed.) *Granitic pegmatites in science and industry*. MIN SOC CAN SHORT COURSE VOL **8**, 41–62.

Moore, J. G. & Lockwood, J. P. 1973. Origin of comb layering and orbicular structure, Sierra Nevada Batholith, California. GEOL SOC AM BULL **84**, 1–20.

Noble, S. R., Spooner, E. T. C. & Harris, F. R. 1984. The Logtung large tonnage, low-grade W (scheelite)-Mo porphyry deposit, south-central Yukon Territory. ECON GEOL **79**, 848–68.

Pavlishin, V. I., Mazykin, V. V., Matyash, I. V. & Voznyak, D. K. 1978. Variations in the proportion of substitutional aluminum during growth of a quartz crystal. GEOCHEM INT **15**, 158–65.

Perny, B., Eberhardt, P., Ramseyer, K., Mullis, J. & Pankrath, R. 1992. Microdistribution of Al, Li and Na in α quartz: possible causes and correlation with short-lived cathodoluminescence. AM MINERAL **77**, 534–44.

Scotford, D. M. 1975. A test of aluminum in quartz as a geothermometer. AM MINERAL **60**, 139–42.

Shannon, J. R., Walker, B. M., Carten, R. B. & Geraghty, E. P. 1982. Unidirectional solidification textures and their significance in determining relative ages of intrusions at the Henderson Mine, Colorado. GEOLOGY **10**, 293–7.

Shaver, S. A. 1988. Petrology, petrography, and crystallization history of intrusive phases related to the Hall (Nevada Moly) molybdenum deposit, Nye County, Nevada. CAN J EARTH SCI **7**, 1000–19.

Shinohara, H. & Kazahaya, K. 1995. Degassing processes related to magma chamber crystallization. *In* Thompson, J. F. H. (ed.) *Magmas fluids and ore deposits*. MIN SOC CAN SHORT COURSE VOL **23**, 47–70.

Shinohara, H., Kazahaya, K. & Lowenstern, J. B. 1995. Volatile transport in a convecting magma column: implications for porphyry Mo mineralization. GEOLOGY **23**, 1091–4.

Smith, J. V. & Steele, I. M. 1984. Chemical substitution in silica polymorphs. N JAHRB MINERAL MON **3**, 137–44.

Stavrov, O. D., Moiseyev, B. M. & Rakov, L. T. 1979. Relation between content of alkali metals and concentration of aluminum centers in quartz. GEOCHEM INT **15**, 5–10.

Stewart, J. P. 1983. *Petrology and geochemistry of the intrusives spatially associated with the Logtung W–Mo prospect, south-central Yukon Territory*. M.Sc. Dissertation. University of Toronto.

Swanson, S. E. & Fenn, P. M. 1986. Quartz crystallization in igneous rocks. AM MINERAL **71**, 331–42.

Tuttle, O. F. & Bowen, N. L. 1958. Origin of granite in the light of experimental studies in the system $NaAlSi_3O_8$–$KAlSi_3O_8$–SiO_2–H_2O. GEOL SOC AM MEM **74**.

White, W. H., Bookstrom, A. A., Kamilli, R. J., Gansta, M. W., Smith, R. P., Ranta, D. E. & Steininger, R. C. 1981. Character and origin of Climax-type molybdenum deposits. *In* Skinner, B. J. (ed.) ECON GEOL 75TH ANNIV VOL 270–316.

JACOB B. LOWENSTERN, U.S. Geological Survey, Mail Stop 910, 345 Middlefield Road, Menlo Park, CA 94025 U.S.A.
W. DAVID SINCLAIR, Geological Survey of Canada, 601 Booth Street, Ottawa, Ontario, K1A 0E8 Canada

Transactions of the Royal Society of Edinburgh: Earth Sciences, **87**, 305–319, 1996

Granitic pegmatites

David London

ABSTRACT: The primary focus of this review is on $P–T$ conditions, mineralogical indicators of melt or fluid composition and textural evolution; lesser treatment is given to pegmatite sources or to pegmatite–wallrock interactions. Investigations of stable and radiogenic isotopes have revealed that the source materials for pegmatites are likely to be more heterogeneous or varied than previously thought, especially for peraluminous pegmatites, but that overall pegmatites bear a clear intrusive relationship with their hosts, as opposed to an origin *in situ*. The $P–T$ conditions of crystallisation of some lithium-rich pegmatites have been constrained by lithium aluminosilicate stability relations in combination with stable isotope or fluid inclusion methods. Experimental studies have elucidated the effects of components such as Li, B, P and F, which are common in some classes of pegmatites, to liquidus relations in the hydrous haplogranite system. Experimentation has also provided corroboration of an old concept of pegmatite crystallisation—that pegmatites owe their distinctive textures and mineral/chemical zonation to relatively rapid crystallisation of melt from the margins inwards at conditions far from the equilibrium (i.e. from supercooled liquids). The origin of aplites, whether alone, layered, or paired with pegmatites, remains an active area of research. Studies of fluid inclusions, crystal–vapour equilibria and wallrock alteration have helped to define the timing and compositions of vapour phases in pegmatites and to aid in the economic evaluation of deposits.

KEY WORDS: granitic pegmatites, $P–T$ conditions, mineralogical indicators, fluid composition, textural evolution, pegmatite sources, pegmatite–wallrock interactions.

Granitic pegmatites are characterised by at least some mineral phases of very coarse grain size, but through usage the grain size that distinguishes pegmatite from granite is more relative than quantitative. Pegmatites are also distinguished from other intrusive equivalents (granite, diorite, etc.) by internal hetero-geneity in pegmatites, including abrupt variations in grain size or in mineralogy, anisotropy in fabric and a wide range of crystal morphologies (e.g. skeletal, graphic, euhedral). Magmas of all compositions generate pegmatites. Granitic pegmatites are so prevalent, however, that the term pegmatite carries this implied composition. There is a good reason why pegmatites of basic, intermediate or alkaline composition are uncommon: the crystallisation response on undercooling of these magmas is different from that of granitic melts (discussed in the following sections).

The purpose of this paper is to present a synopsis of results and interpretations stemming from recent research on a variety of pegmatite-related topics, but focusing mainly on the aspects of internal evolution that produce granitic pegmatites (i.e. distinguish them from granites). Recent is arbitrarily defined as the quarter century since 1970, but with emphasis on the past decade. Selected examples are just that: reference to individual studies (predominantly from North America) are mostly to be taken as indicative of similar efforts elsewhere.

Recent special publications dedicated solely to pegmatites include issues of periodicals (e.g. Brown & Ewing 1986; Martin & Cerny, 1992) short course notes (e.g. Cerny 1982a), and other symposia (e.g. Modreski 1986; Novak & Cerny 1992a). Topical and summary papers abound in other periodicals and special publications. A tally of the reference list to this paper shows a large proportion of recent pegmatite-related research published in *The American Mineralogist* and *The Canadian Mineralogist*. Several field trip guides to pegmatites have appeared in connection with geological meetings (e.g. Foord

et al. 1991; Novak & Cerny 1992b; Francis *et al.* 1993). Since 1991, the Pegmatite Interest Group (PIG) has co-ordinated activities and disseminated news about pegmatite studies. Readers who would like to be informed of PIG activities, or have announcements to post, should contact the group's representatives, Petr Cerny (University of Manitoba, for the Mineralogical Association of Canada) or David London (for the Mineralogical Society of America).

1. Background

In *The Study of Pegmatites*, Jahns (1955) compiled a remarkably thorough history of pegmatite investigation, beginning well into the 19th century, in which virtually every conceivable explanation for the origins of pegmatites had been espoused. Jahns (1955) agreed that no one hypothesis was fully satisfactory in all instances, but in an earlier paper, Jahns (1953a) championed the prevailing view at his time:

> 'Many investigators have concluded that pegmatite zones were formed by fractional crystallization of magma, with incomplete reaction between successive crops of crystals and rest-liquid. The giant crystals in pegmatites are here considered to be primary constituents of those zones, mainly on the basis of their systematic distribution, age relations, textural and structural relations, and variations in composition. They are thought to have crystallized directly from pegmatite liquid that was rich in hyperfusible components, probably under restricted-system conditions involving a rather delicate thermal and chemical balance ... All known features of pegmatite zones seem reasonably explainable on the basis of crystallization from a melt of low viscosity, with or without end-stage deuteric or hydrothermal activity. Many of these features also suggest that masses of pegmatite crystallized inward from the walls of an original chamber ... rather than in some sort of channelway or thoroughfare under more open-system

conditions ... In this writer's opinion, all the observed features of numerous pegmatite bodies, especially those of complex mineralogy, are most readily explained by a combination of two processes: (1) progressive crystallization of segregated or injected rest-magma under restricted-system conditions, accompanied by some reaction between the crystalline phase and the residual liquid, and (2) resurgent boiling of the residual liquid, condensation of the vapor at points with the pegmatite system, and partial (commonly selective) replacement of the crystalline phase by the condensate. The latter process need not have been operative within pegmatite bodies in which there is little evidence of replacement, and neither process would necessarily apply to pegmatite bodies formed instead by metasomatism of earlier non-pegmatitic rocks.'

With this summary, Jahns (1953a) concluded that the features of most pegmatites are consistent with a magmatic origin, with or without the involvement of hydrothermal solutions (referred to as vapour in this paper) in addition to magma.

In the 1960s, Jahns and Burnham (1969) introduced an important change to the pegmatite paradigm that Jahns had expounded in the 1950s. In what has come to be recognised as the Jahns–Burnham (1969) model, they proposed that what distinguishes pegmatites and their distinctive characteristics from typical granites is the presence of a vapour phase (a single and dominantly aqueous fluid) in addition to magma in pegmatites, i.e. that the transition from granitic to pegmatitic textures marks the point of aqueous vapour saturation of melt; by implication, all 'granite' was the crystalline product of H_2O-undersaturated magma. In the Jahns–Burnham model, conditions of crystallisation close to equilibrium were thought to prevail; this contrasted with the previous interpretations wherein crystal–liquid equilibrium was generally not attained.

2. Pegmatite compositions and regional zonation

Jahns and Burnham (1969) further promoted the hypothesis that granitic pegmatites were derived from larger batches of granite by crystal fractionation. The bulk compositions of pegmatites are difficult to establish accurately (e.g. Jahns 1953b). Where this has been done, however, it is important that the bulk compositions of most pegmatites lie close to the granite minimum (e.g. Jahns & Tuttle 1963; Norton & Redden 1990). This apparent fact strengthens the connection between granite magmatism and pegmatites.

Rare-element pegmatites whose compositions deviate substantially from those of simple granites constitute less than about 2% of pegmatites in well-exposed regions (Cerny 1991). Nonetheless, as in granite petrology, the minor and trace element signatures of pegmatites have been studied for the basis of classification, tectonic setting, discrimination of sources and processes of fractionation (e.g. Cerny et al. 1985).

Cerny (1991) classified pegmatite compositions into two broad families based on distinctive trace element signatures. The NYF (Nb–Y–F) family is associated mostly with subalkaline to metaluminous A-type granites (e.g. Whalen et al. 1987), though peraluminous NYF pegmatite fields are known (e.g. Smeds 1994). The pegmatites are markedly enriched in LREEs ($100\times$ to $350\times$ chondritic: Cerny 1991), with monazite, samarskite, euxenite and yttrofluorite as some common accessory minerals. The Pikes Peak batholith (Colorado, U.S.A.) and associated pegmatites of the South Platte and Lake George districts are illustrative of the NYF family (e.g. Wobus & Anderson 1978; Desborough et al. 1980; Simmons & Heinrich 1980).

The LCT family (Li–Cs–Ta; Cerny 1991) is more varied in apparent source materials (supracrustal metasedimentary sequences plus lower crustal granulites; e.g. Cerny 1991), though

most LCT granite–pegmatite systems are peraluminous and are enriched in elements (e.g. B and Cs) that are (thought to be) liberated from rock-forming minerals at the onset of initial anatexis in metasedimentary rocks. The LCT family of pegmatites is therefore mostly associated with the tectonic regimes and source materials of S-type granites (e.g. Chappell & White 1992). The Harney Peak granite (South Dakota, U.S.A.) and its associated pegmatite aureole (e.g. Shearer et al. 1992) are chemically representative of the LCT family.

Trace element and isotopic signatures have been utilised to explore potential genetic links between pegmatites, granites and host rocks. Much of the recent work has been conducted on rocks of the Superior Province of Manitoba, Canada (e.g. Cerny et al. 1981; Goad & Cerny 1981; Cerny & Brisbin 1982) and in the Black Hills of South Dakota, U.S.A. (e.g. Shearer et al. 1987, 1992; Walker et al. 1986b, 1992; Jolliff et al. 1992; Nabelek et al. 1992a; Krogstad et al. 1993; Nabelek & Glascock 1995). In some instances, smooth elemental variations from granite to pegmatite are suggestive of a genetic link, as by derivation of pegmatite from granite (e.g. phosphorus; Tomascak & Cerny 1992). In detail, however, the processes that might link pegmatites with granites or source rocks with granite–pegmatite systems are more difficult to confirm (e.g. Cerny 1995; Krogstad & Walker, this issue). One problem in trace element modelling of leucogranite systems has been a dearth of reliable partition coefficients for relevant minerals and melt compositions. A relevant and mostly complete data set is now available (e.g. Icenhower et al. 1994; Icenhower 1995; Icenhower & London 1995, 1986a).

Cerny (1992) cites a general sequence of zonation of subtypes within a group of LCT pegmatites outward from their source granite as: (1) barren; (2) beryl–columbite; (3) beryl–columbite–phosphate; (4) spodumene or petalite $\pm$ amblygonite; (5) lepidolite; (6) albite–spodumene; and (7) albite. Beryl–REE pegmatites are locally important between (2) and (3) (Cerny 1992). The zoning sequence is taken to reflect a greater mobility of the most chemically fractionated magmas down the steep thermal gradient that surrounds the central source of heat for granites and pegmatites. The origins of the zonation, however, are not known. London (1992a) noted that most pegmatite groups emanate (where exposed) from the cupolas of their source granites, which themselves are pegmatitic (pegmatite–aplite) at the roof, but otherwise grade downwards into typical granitic facies (e.g. Lawler Peak granite, Bagdad, Arizona and White Rocks, Middletown, Connecticut, both U.S.A.). Local structures dominate the distribution of dykes, but most appear to have been emplaced in a single episode with only limited cross-cutting relations, rather than by repeated, successive episodes of intrusion (cf. the Harney Peak Granite, e.g. Shearer et al. 1987, 1992; Duke et al. 1988). The pegmatites lack entrained phenocrysts and individual conduits appear to have sealed shortly after emplacement. London (1992a) suggested that, in general, much of the chemical zonation found in pegmatite groups is inherited from similar vertical zonation within the source pluton and that this is augmented by crystal fractionation that takes place along a P–T gradient among pegmatite bodies, as long as these remain connected; once isolated, individual pegmatites fractionate internally as recognised by Jahns (1953a) and numerous others. In the case of the Harney Peak Granite, the individual granite–pegmatite layers, each representing an injection of magma, are isotopically heterogeneous (Krogstad & Walker, this volume). Repeated dyking of increasingly fractionated melts from a single crystallising magma body is not viable for this locality. Though data are sparse, the trace element signatures of pegmatites commonly cannot be linked with those of exposed granites by models that entail fractional crystallisation of large batches of melt; in such instances, the pegmatites have been interpreted as the products of small volumes of anatectic melt, rather than derivatives of the granites (e.g. Shearer *et al.* 1992; Simmons *et al.* 1995).

3. Internal zonation

The view of Jahns (1953a) and others that individual pegmatite dykes (or isolated bodies) crystallise as largely closed or restricted magmatic systems remains well entrenched in contemporary thought (but cf. Roedder 1992), though there is chemical evidence from major and trace elements and isotopic systems that limited communication between pegmatites and their enclosing host rocks exists in the early stages, if not throughout crystallisation, and again at the end of solidification (e.g. Taylor & Friedrichsen 1983; Walker *et al.* 1986b; London 1990). The principal arguments that have been cited against an open system origin, i.e. streaming of vapour or melt through open fractures, include: (1) the lack of exposed connectivity among pegmatite dykes; (2) a general absence of flow structure within pegmatites; (3) the concentric nature of sequential zoning within individual pegmatites; (4) the magmatic (granitic) composition of pegmatites; (5) variations in mineral compositions that are consistent with closed system fractionation; (6) the low solubility of alkali aluminosilicate components in alternative fluids such as aqueous–carbonic vapour; and (7) except for local metasomatic effects, the general lack of wallrock alteration, all indicative of small vapour/rock ratios, within enclosing metamorphic rocks.

A scheme of internal mineralogical zoning that was formalised by Cameron *et al.* (1949) among others and revised by Norton (1983) recognised the apparently sequential nature of pegmatite crystallisation from magma. Internal mineralogical zoning is common and was recognised as such even in simple pegmatites whose compositions are close to the granite minimum. Exclusive of Li minerals, the general zoning sequence has been expressed (Cameron *et al.* 1949), from first formed to last formed, as: (1) plagioclase–quartz–muscovite; (2) plagioclase–quartz; (3) quartz–perthite–plagioclase(–muscovite); (4) quartz–perthite; and (5) quartz. The important concept is that magmas of near-eutectic composition were thought to have undergone non-eutectic crystallisation— sequential, not simultaneous crystallisation of all phases at the eutectic, as advocated by Jahns (1953a).

In developing the Jahns–Burnham (1969) model, Jahns rejected the sequential nature of pegmatite crystallisation in favour of eutectic crystallisation, which required near-equilibrium and simultaneous crystallisation of quartz, plagioclase and alkali feldspar in eutectic proportions, but in spatially separated zones. The essence of the model entailed the simultaneous crystallisation of sodic aplites (very fine-grained albite-rich rocks) in the lower or footwall portions of pegmatites, giant microcline crystals (commonly graphically intergrown with quartz) in the apical or hanging wall portions, and pure quartz as 'discoidal masses in the interior parts of the dike' (Jahns 1982). Selective fractionation of K from Na into ascending aqueous vapour was proposed as the driving mechanism for both the fine-grained albitic rocks—formed by 'compositional quench' caused by a shift of melt composition away from the eutectic (Jahns & Tuttle 1963)—and for the growth of giant microcline crystals in the apical portions. The mechanistic details of how this vapour phase transfer process operated around the growing and neutrally buoyant masses of pure quartz in the centre of a pegmatite were not explained.

The Jahns–Burnham (1969) model was extended by Burnham and Nekvasil (1986) to include a peraluminous component, Li, F and B. The experimental data, however, were the same as those of the earlier model, in which two pegmatite bulk compositions were used (Spruce Pine, North Carolina and Harding pegmatite, New Mexico, both U.S.A.). The sequences of crystallisation in the H_2O-saturated melts (neither was eutectic in bulk composition) matched those observed at the pegmatites from which the samples were collected. The simultaneous crystallisation of quartz + plagioclase + orthoclase $\pm$ spodumene $\pm$ mica at the eutectic, however, presented the same conundrum as faced Jahns (1982), wherein nearly pure masses of minerals (e.g. quartz cores) were required to crystallise simultaneously, but in spatially segregated parts of a pegmatite dyke. The model presented by Burnham and Nekvasil (1986) differed little from that of Jahns and Burnham (1969), though Burnham and Nekvasil (1986) acknowledged that the K* (K* = molar 100 K/[K + Na]) of aqueous vapour with and without chloride was actually equal to that of coexisting granitic melt under most conditions, rather than far from unity as hypothesised in and required by the models of Jahns and Tuttle (1963) and Jahns and Burnham (1969). In addition, $\approx 60\%$ of the massive quartz in the central portions of pegmatites could not be generated by the equilibrium crystallization of melt, leaving Burnham and Nekvasil (1986) to conclude that this quartz must be hydrothermal in origin 'although the actual mechanism of transport and deposition remain to be determined.'

An alternative explanation for the internal zoning of pegmatites was presented by London *et al.* (1989), also on the basis of experimentation. Their explanation was that the mineral zoning patterns, sharp changes in grain size, a multitude of mineral textures and oriented fabrics that typify pegmatites all resulted from the slow crystallisation response of volatile-bearing granitic melts to cooling. The further the melt was cooled below its equilibrium liquidus before crystallisation commenced, the greater was the tendency to form pegmatitic as opposed to granitic fabrics. In turn, the crystallisation response of the melt to undercooling was a complex function of the bulk composition of melt and the degree of melt undercooling. Virtually all of the textural and mineralogical features of pegmatites, including fine-grained plagioclase–quartz borders, graphic alkali feldspar + quartz intergrowths, central zones of very coarse-grained alkali feldspar + quartz $\pm$ petalite (virgilite) $\pm$ mica, even monomineralic segregations of quartz and later fine-grained lepidolite + albite deposits, were replicated by experiments that entailed cooling crystal-free melts along various T–t paths below their liquidus temperatures. In addition, London (1992a) illustrated that pivotal experiments cited by Jahns (1982) and depicted by Wyllie (1963) (Fig. 1) were produced along run paths similar to those employed by London *et al.* (1989). London (1992a), therefore, further diminished the inferred role of a separate aqueous phase as a causative agent in pegmatite formation, noting that tangible evidence for an aqueous phase during the primary crystallisation of pegmatites was mostly weak or absent, and that experiments in which crystallisation occurred from melt through an aqueous vapour phase tended to produce results that more resembled miarolitic granite than pegmatite (London *et al.* 1988, 1989). London (1992a) suggested that the cause and effect in the experiments cited by Jahns (1982) and Wyllie (1963) had been misinterpreted—that the product resulted from undercooling granitic melt (and consequent crystal growth far from the equilibrium with melt), whether H_2O-saturated or not, rather than uniquely from a transition from H_2O-undersaturated to saturated conditions (which was never documented in the experiments). This interpretation applied to the experimental simulation of pegmatites; the obvious scaling problem of size (but not necessarily time) remains an important unknown in linking

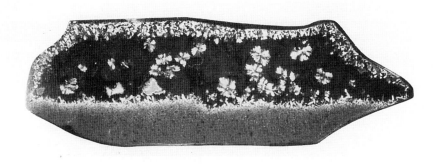

Figure 1 Photomicrograph (under crossed nicols) reproduced from figure 39 of Wyllie (1963) showing a reverse-direction experiment that was cited by Jahns (1982) as a pivotal experiment in his investigations (with C. W. Burnham) of pegmatite-forming processes. See Jahns (1982) for his interpretation of this experiment. The caption in Wyllie (1963) notes that the section measures 0·5 inches across and that a 'fine-grained crystalline border encloses glass (isotropic) studded with radial groups of larger crystals … The development of a wide border facies at the bottom of the sample is due to a vertical temperature gradient.' Reprinted from Wyllie (1963) by permission of the publisher.

experiments with natural rocks. London (1992a) nevertheless noted that this new explanation for pegmatite textures and zonation constituted an experimentally based reaffirmation of the original magmatic concepts espoused by Jahns (1953a). This shift of emphasis back to magmatic crystallisation does not necessarily mean that pegmatite magmas are vapour-undersaturated, but that the role of a vapour phase, if present, in generating pegmatite is secondary.

3.1. Pegmatites of composition other than granitic
Pegmatites are rare in igneous rock compositions other than granitic, and an obvious question is why that should be. The model proposed by London *et al.* (1989; London 1992a) contains an inherent explanation. High-silica melts tend to persist metastably as liquids or glasses, so that crystallisation, if it occurs, proceeds from melts that are undercooled far below their equilibrium liquidi (see under Pressure–temperature conditions). In experiments, the greater the undercooling before the beginning of crystallisation, the more the results resemble pegmatites (and eventually at very high undercooling, spherulitic rhyolites). Less siliceous melts of basic, intermediate and especially alkalic compositions intrinsically nucleate crystals more readily and hence their crystallisation commences and tracks closer to equilibrium than for granitic melts. The result, which is well known and described in published work on petrology and metallurgy, is that the crystallisation textures of lower-silica melts tend towards the homogeneity and uniform grain size typical of non-pegmatitic plutonic rocks.

4. Pegmatites, aplites and composite dykes

The Jahns–Burnham (1969) model had as its central objective an explanation for dykes that contained paired pegmatite and aplite units. That model, as restated without change by Jahns (1982) and Burnham and Nekvasil (1986), invoked compositional quenching of magma along the footwall of dykes by loss of K to vapour, which moved the magma along the footwall from the eutectic composition into the field of albite + quartz, and hence to a condition of greater undercooling that promoted abundant crystal nucleation and consequently fine-grained sodic rocks. Layering in sodic aplites, represented primarily by alternations in quartz and albite, was attributed to cyclic shifts in the quartz–feldspar field boundary as internal vapour overpressures were repeatedly bled off. The addition of K to the upper portions of pegmatite promoted the growth of giant microcline + quartz.

Though the upper and lower domains of pegmatite bodies were displaced in equivalent but opposite compositional directions away from the eutectic, and both were vapour-saturated (i.e. bubble-rich), the process was envisioned to explain the formation of saccharoidal texture in one region and giant pegmatitic texture in the other.

London (1992a) outlined several objections to this model, claiming that most of the consequences of vapour saturation in pegmatites would promote effects (if any) that were opposite to those envisioned in the Jahns–Burnham model. In addition to the diminished role of aqueous fluid in fractionating K from Na (Burnham & Nekvasil 1986), London (1992a) proposed that the thermal profile that would emerge from the exsolution and ascent of aqueous vapour would actually make the upper portions of pegmatites hotter than the lower regions, and that in such a thermal profile the migration of Na and K would be opposite to that observed in pegmatites (cf. Orville 1963). In the experiments of London *et al.* (1989), the initial products of crystallisation of H_2O-undersaturated melts in experiments with Macusani glass invariably yielded fine-grained sodic albite–quartz intergrowths that evolved sharply into very coarse-grained Or-rich alkali feldspar (e.g. see figures 3a and 7a of London *et al.* 1989 and figure 8a of London 1992a).

The Jahns–Tuttle–Burnham model has been invoked in recent studies of aplite–pegmatite or quartz–feldspar layered rocks (e.g. Lowenstern, this issue), but other explanations have appeared recently that deviate from the existing paradigm. For example, Rockhold *et al.* (1987) explain layering in paired pegmatite-layered aplites from the Harney Peak Granite in connection with cyclic increases in the activity of H_2O in melt (i.e. repeatedly saturated and undersaturated in H_2O at constant pressure). Duke *et al.* (1988) and particularly London (1992a) suggested that double-diffusive fractional crystallisation or oscillating self-organisational models (e.g. McBirney & Noyes 1979; Ortoleva 1994) as a result of rapid crystallisation represent viable methods of producing alternate banding. Importantly, the banding in aplites is not always in terms of quartz and feldspar, but may be defined by concentrations of garnet, tourmaline, muscovite and gahnite. Citing bedding and cross-cutting structures from layered aplites dykes of San Diego County, California (U.S.A.), Kleck (1991) proposed sedimentary process as the cause of layering. Jahns (1955; Jahns & Tuttle 1963) discounted crystal settling as a viable hypothesis for layering in aplites on the grounds that the viscosity of the melt would be too high to allow settling.

Webber *et al.* (1995) estimated that the viscosities of melt would indeed be too high to allow settling of fine-grained quartz or feldspar in a reasonable time frame. Based on analogies with diffusional bands in gels, Webber *et al.* (1995) also suggested oscillating self-organisation of monominerallic bands as the probable mechanism from layering in aplites. As a final consideration, layer-parallel foliation and lineation of tourmaline or mica in some aplites bears evidence of viscous flow (e.g. London 1984, 1985b), but such textures are sufficiently uncommon that flow does not appear as a reasonable general mechanism for layering in aplitic rocks.

The reality of the geology, as noted by Jahns and Tuttle (1963), includes: (1) pegmatites without any associated aplites; (2) pegmatites with aplites (layered and non-layered); and (3) aplites without pegmatites. Cross-cutting relations of pegmatite alone cutting aplite or vice versa are common enough (e.g. Harney Peak Granite, South Dakota; Maromas Granite, Connecticut) as to require the injection of distinct melts for each aplite or pegmatite. Indeed, Jahns and Tuttle (1963) noted paired pegmatite–aplite dykes that bifurcate into separate pegmatite and aplite intrusions. Jahns and Tuttle (1963) also illustrated dykes that had pegmatite margins and aplitic centres. There is no unique temporal relationship between aplite and pegmatite. Where paired, aplite appears to mostly precede the injection or generation of pegmatite, and aplite serves as a substrate for oriented pegmatitic crystal growth; however, aplites tend to be present throughout the zoning sequences in pegmatites and are included among the latest and most fractionated units (e.g. the Harding pegmatite, New Mexico, U.S.A.—Chakoumakos & Lumpkin 1990; the Tanco pegmatite, Manitoba—Cerny 1982b). Jahns and Tuttle (1963) note that aplites of different compositions may arise from different processes—pressure quench and compositional quench—in the temporal evolution of a single pegmatite body. To date, however, there are no thorough studies of multiple aplite occurrences in a single pegmatite.

One important conclusion from these observations is that any paradigm that entails the gravitational separation of vapour from melt as its key precept must always work, and in a predictable way that produces the same results. The integral role of vapour ascent in the Jahns–Burnham model means that regardless of texture, all pegmatites should have complementary potassic pegmatite and sodic aplite domains, which they do not. The model also requires that potassic pegmatite overlies sodic aplite, which is not always the case (e.g. the Hale pegmatite in Portland, Connecticut, U.S.A.; London 1985b). As products of compositional quench and its implied and sudden increase in viscosity due to rapid crystallisation, aplites would not be mobile as fluids, but they clearly are as discordant fracture-fillings in the least evolved to most evolved pegmatite systems. In summary, then, several alternatives to the Jahns–Burnham model for aplite–pegmatite pairs and of the causes of layering in aplites have appeared in recent years. Though there are significant problems with the Jahns–Burnham model, the new explanations currently lack either the generality or the factual confirmation to be acceptable as a new paradigm.

5. Mineralogy

The study of pegmatite mineralogy has burgeoned in recent years, so much so that a thorough summary is beyond this review. The focus of this section is on developments in the mineralogy of granitic pegmatites that have aided particularly in understanding processes. It is inevitable that some good work and important mineral groups will be omitted here, including studies whose relevance to pegmatites have not yet been (fully) explored or understood.

5.1. Quartz and feldspars

Recent investigations of quartz have entailed studies of its nucleation and growth characteristics (Martin 1982; Swanson & Fenn 1986; MacLellan & Trembath 1991), trace element chemistry (e.g. Dennen *et al.* 1970; Hervig & Peacock 1989; London 1984; London *et al.* 1988; Perny *et al.* 1992), and indirectly its role as a host for fluid inclusions (e.g. Barker & Sommer 1974; Aines *et al.* 1984; Cordier *et al.* 1988; Yurimoto *et al.* 1989; Bakker & Jansen 1994). Among these topics, the one most readily connected with pegmatite-forming processes is the interpretation of quartz habits. Experimental studies of the kinetics of nucleation and growth of quartz from granitic melts were intended to relate quartz morphology to conditions of growth (degree of melt undercooling). Together with Fenn (1986), these works reproduced the skeletal and graphic textures of quartz so commonly found in pegmatites. The principal conclusion reached from the correlation of experimental with natural textures of quartz is that graphic or skeletal quartz morphologies reflect rapid crystal growth from highly undercooled (i.e. metastable) silicate melts (e.g. Martin 1982). This experimental work, and its replication by London *et al.* (1989), figures prominently in the pegmatite model presented by London (1992a).

In the chemistry of feldspars, variations in trace element abundance, primarily the K/Rb and K/Cs ratios of alkali feldspars, have been developed by Cerny (e.g. Cerny & Macek 1972; Cerny *et al.* 1981) and widely used by others to monitor fractionation trends in pegmatites. These variations become more useful in the light of recent experimental determinations of trace element partitioning between feldspars and melt (e.g. Icenhower & London 1996a).

Feldspar chemistry has also been utilised and calibrated as a tool for monitoring the phosphorus content of granitic melts including pegmatites. London (1992b; London *et al.* 1990, 1993) showed that alkali feldspars from peraluminous pegmatites and chemically similar granites and rhyolites contain elevated P_2O_5, with averages near $0.5\,wt\%$ P_2O_5 for K-feldspar. Substitution via the exchange $AlPSi_{-2}$ in alkali feldspars was demonstrated by experimental synthesis (Simpson 1977). The $AlPSi_{-2}$ substitution in feldspars was correlated experimentally with the aluminum saturation index (ASI = molar $Al/\sum M^+ + \sum 2M^{2+}$, M^+ = alkali, M^{2+} = alkaline earth) of melt (London *et al.* 1993) and the data fit a linear equation of the form

$$D_p^{Af/m} = 2.05(ASI) - 1.75$$

Using this experimental data, London (1992b) illustrated that, to various extents, granitic rocks generally do not conserve the phosphorus originally present in the magma.

A small number of experimental programmes have explored how undercooling and bulk composition affect the crystallisation of feldspar from granitic magmas (e.g. Fenn 1977; Petersen & Lofgren 1986; London *et al.* 1989). In relation to pegmatites, London *et al.* (1989) demonstrated that the sequential crystallisation of plagioclase–quartz ⇒ K-feldspar–quartz (graphic) ⇒ K-feldspar–quartz (monophase) ⇒ quartz is a normal consequence of undercooling melt, and the more undercooled, the more sharply zoned is the crystalline product. In a more systematic study, Fenn (1977) illustrated the combined relations of crystal nucleation numbers and growth rate to undercooling, bulk composition and, importantly, to the H_2O content of undersaturated melt. An important result of Fenn's (1977) work was that in the undercooling regions where crystal growth occurred, the nucleation numbers of feldspar fell sharply with increasing H_2O, with the result that fewer, larger crystals grew with increasing H_2O in the melt.

5.2. Lithium aluminosilicates

Stability relations among the lithium aluminosilicates in quartz-saturated portions of the system $LiAlSiO_4–SiO_2–H_2O$ were thoroughly outlined by London and Burt (1982a, b, c) and determined experimentally by London (1984). Reactions of spodumene or petalite to virgilite or β-spodumene limit lithium aluminosilicate crystallisation in pegmatites to temperatures below about 700°C at any pressure. The isochemical reaction of petalite to spodumene + quartz is a consequence of cooling (Stewart 1963; London 1984). The stability field of eucryptite + quartz, predicted by London and Burt (1982c), was verified by London (1984) as a low-temperature (subsolidus with respect to pegmatite bulk compositions) replacement of either spodumene or petalite. The quartz-saturated phase relations among the lithium aluminosilicates represent the most reliable P–T indicator for pegmatites with appropriate assemblages (London 1984) because of the lack of complication from solid solutions away from the end-member compositions.

5.3. Phosphates

The phosphate mineralogy of pegmatites is at least as varied as that of silicates, but far less is known about the phosphates. The names and formulae of the large number of primary and secondary phosphates can be daunting to the non-specialist (e.g. see Moore 1973, 1982), but some recent studies of phosphate minerals are especially relevant to pegmatites.

Various phosphate–silicate assemblages now constrain the abundance of phosphorus in granitic melt. The P_2O_5 content of melt at apatite saturation in peraluminous melts is $\leqslant 0.7$ wt% when apatite is the sole source of Ca, P and F in the melt (Wolf & London 1994). Reactions among biotite and triplite buffer melt composition near 0.8 wt% P_2O_5 (London et al. 1995b). The equilibria among lithium aluminosilicates, amblygonite–montebrasite and quartz (e.g. London & Burt 1982a) regulate P_2O_5 near 1–2 wt% P_2O_5 in melt (London et al. 1995a) and garnet–phosphate equilibria (e.g. spessartine-beusite, spessartine–sarcopside) occur between ~ 0.5 and 2.0 wt% P_2O_5 in melt, with the higher values associated with sarcopside equilibria. Thus there is a variety of reactions among silicates (biotite, garnet, Li–Al silicates) and phosphates that will maintain the P_2O_5 content of melt between about 0.5 and 2.5 wt%.

Keller and co-workers (1988; Keller et al. 1994) focused attention on the common occurrence of skeletal or dendritic phosphate crystals in pegmatites branching inwards from the margin. Keller (1988) interpreted these large, inwardly branching skeletal growth forms as indicative of rapid crystal growth in a fluid (magmatic) medium. The implied disequilibrium is evident in chemical zonation of some phosphates—for example, the distributions of the REEs in apatite from pegmatites (e.g. Jolliff et al. 1989). An important corollary of Keller's work is that the skeletal phosphates are interpreted as primary. The matrix for these phosphate crystals is usually rock of nearly pure albite (coarse-grained or aplitic) $\pm$ mica; such rocks have previously been regarded as replacive by most workers.

5.4. Tourmaline

The chemistry of tourmaline is complex enough, with a large number of real and theoretical end-member compositions (e.g. Burt 1989), that variations in its solid solutions can elucidate fractionation trends and zonation sequences in pegmatites. Zoning of tourmaline compositions within individual pegmatite bodies yields consistent trends among the following elements: Mg and Ti decrease, whereas Fe, Al ($\pm$ Li) and Mn

increase (e.g. Staatz et al. 1955; Foord 1976; Manning 1982; Jolliff et al. 1986) with location from margin to core, which is usually interpreted to represent the direction of increasing fractionation. Other element variations may be systematic within an individual granite or pegmatite, but different between locations so that no generality can be made (e.g. typical of Na, F and Ca).

Recent experimental studies of tourmaline-forming equilibria in hydrous granite–pegmatite melts have provided constraints on the B_2O_3 contents of tourmaline-saturated melts (e.g. Benard et al. 1985; Holtz & Johannes 1991; Harvey 1991; London et al. 1994; London & Manning, in press). Among these studies, only the work of London et al. (1994) actually reversed the tourmaline–silicate equilibria, and London et al. (1994) reported that in excess of 2 wt% B_2O_3 in melt was required to stabilise tourmaline at 750°C and 200 MPa (H_2O).

London (1990; London & Manning 1995; London et al. in press) suggested that tourmaline at the margins of pegmatites arises by mixing between two reservoirs—pegmatites as sources of B and wallrocks as sources of Fe and Mg—and that the deposition of tourmaline marks the site of mixing (within pegmatites or within adjacent wallrocks). If correct, the location of tourmaline concentrations denote the timing and extent of chemical communication between pegmatite and wallrocks.

5.5. Other phases

Oxides. A virtual explosion of new data on complex oxide solid solutions in the system containing Fe–Mn–Nb–Ta–Ti–Sn and other components (e.g. REEs, Y, U, Th, Sb, Al, Cs and Na) is summarised in, e.g. Cerny and Ercit (1989), Ercit et al. (1992a, b, c, 1995) and Wise and Cerny (1996), as an outgrowth of many detailed examinations of crystal structure, composition, order–disorder and zonation from individual crystals to pegmatite fields (e.g. Spilde & Shearer 1992). Our understanding of the causes of chemical variation, and especially order–disorder in these oxide phases, is not yet sufficient to draw clear petrogenetic conclusions.

Pollucite. Pollucite, ideally $CsAlSi_2O_6$ but actually pollucite–analcime solid solutions in natural phases, is found only in the most fractionated of LCT pegmatites. Most studies of pollucite have examined variations in composition and the limits to pollucite–analcime solid solution (e.g. Beger 1969; Cerny 1974; Cerny & Simpson 1978; Smeds & Cerny 1989; Teertstra et al. 1992, 1993; Drysdale 1992; Lagache 1995). Among the existing experimental investigations of pollucite (Henderson & Manning 1984; Beall & Rittler 1982; Icenhower & London 1996b), Icenhower and London established that hydrous granitic systems become saturated in pollucite-analcime solid solutions at > 4 wt% Cs_2O in melt. The immediate petrological significance of this work is an appreciation of how extreme fractionation in pegmatites can be from the likely starting abundance of Cs in the range of a few ppm.

Beryl. Variations in the composition of beryl correlate well with other chemical fractionation indices by showing increased contents of Cs and other alkalis (Cerny & Simpson 1977; Northrup et al. 1989; Sherriff et al. 1991) from crystal cores to rims, and with zonal position in the inferred crystallisation sequence. Volatile constituents, including H_2O and CO_2, may occupy channel voids in beryl (e.g. Aines & Rossman 1984) and the abundances of these components are effectively trapped by the additional presence of alkalis (e.g. Brown & Mills 1986). When liberated by heating, however, the δD values of the H_2O released vary smoothly and logically with the degree of chemical fractionation as determined by other monitors, e.g. K/Rb in alkali feldspars (Taylor et al. 1992).

Garnet. Garnet zoning trends from layered aplites and pegmatites show an expected increase in $Mn/(Mn+Fe+Mg)$, but are punctuated by non-variation and reversals of trend (e.g. Foord 1976). Summaries of larger databases for garnet compositions show smooth increases in $Mn/(Mn+Fe+Mg)$ of garnet from parental granites to fractionated pegmatites (e.g. Cerny et al. 1985; Solokov & Khlestov 1990; Whitworth 1992), though the explanations for this chemical variation are mostly conjectural and not immediately obvious. At moderate magmatic temperatures and pressures (650–750°C at 200 MPa), the equilibrium constant for the reaction

$$Fe_3Al_2Si_3O_{12} \text{ (garnet)} + 3 \text{ Mn (melt)}$$

$$= Mn_3Al_2Si_33O_{12} \text{ (garnet)} + 3 \text{ Fe (melt)}$$

is slightly >1, i.e. the crystallisation of garnet reduces the $Mn/(Mn+Fe+Mg)$ of coexisting melt (Icenhower et al. 1994; Icenhower 1995; Icenhower & London 1995), so that the almandine component of garnet (rim) increases with cooling (Mukhopadhyay & Holdaway 1994). Thus the overall increase in $Mn/(Mn+Fe+Mg)$ in garnet from granite to pegmatite (e.g. Cerny et al. 1985) and within individual pegmatites is dictated by the crystallisation of biotite or tourmaline, not by garnet itself (cf. Haralampiev & Grover 1993).

6. Pressure–temperature conditions

6.1. Liquidus relations

Experimentally determined liquidus–solidus boundaries are widely used to delimit pegmatite emplacement conditions. These most commonly utilise the published H_2O-saturated liquidi and solidi for: (1) the haplogranite (Ab–Or–Qtz) minimum (e.g. Luth et al. 1964); (2) the Harding pegmatite rock composite (e.g. Fenn 1986); and (3) the Macusani rhyolitic glass (London et al. 1989). Two widely held assumptions are that pegmatite-forming magmas begin to crystallise at the temperatures of their liquidi and that they are H_2O-saturated at the liquidus.

Solidus–liquidus diagrams are points of reference for systems that are close to equilibrium, but process is obviously important in the evolution of geological materials. Because the slopes $(\partial P/\partial T)$ of crystal–liquid reactions are steep, the most obvious factor in determining reaction processes in magmas is undercooling relative to the equilibrium of the system. Relevant experiments of quartz–feldspar nucleation and growth in hydrous granitic melts (e.g. Fenn 1977, 1986; Swanson & Fenn 1986) amply demonstrate the importance of undercooling on the crystallisation of granitic liquids.

The question of cooling rate and maintenance of equilibrium are critical to understanding pegmatite formation and different opinions on these two topics have figured prominently in models for pegmatite genesis (discussed earlier). It is widely assumed and stated in most petrology texts that the giant crystal size in pegmatites requires very long periods of undisturbed crystal growth, i.e. that pegmatite magmas cool slowly. London (1992a), however, noted that constant crystal growth rates of $\approx 10^{-6} \text{ cm s}^{-1}$ could produce quartz and feldspar crystals of pegmatic dimension in a few years. In a model of the cooling history of the large Harding pegmatite dyke, New Mexico, Chakoumakos and Lumpkin (1990) applied conservative boundary conditions (e.g. heat loss by conduction only) with a magma–wallrock temperature difference of 300°C at emplacement and calculated that the centre of the pegmatite dyke would have cooled below its equilibrium solidus in ≈ 1–2 years. This model raises the important question of how closely crystallisation within pegmatites can track the crystal–liquid equilibria. A cooling history such as

Chakoumakos and Lumpkin (1990) have proposed approaches the time-scale of experiments. London et al. (1989) reported that for chemically evolved peraluminous melts, the crystal nucleation delay was significant even for experiments that were slowly cooled (relative to typical experiments) within the liquidus field for over six months. If pegmatite magmas experience large undercooling before crystallisation commences (e.g. 200–300°C below the equilibrium liquidus), then the equilibrium boundaries are of little value for establishing P–T conditions attending crystallisation.

For lithium aluminosilicate-bearing pegmatites, phase relations in the quartz-saturated portions of the system $LiAlSiO_4$–SiO_2 (London & Burt 1982c; London 1984) have been utilised to establish broad P–T conditions of actual crystallisation, and in a few instances multiple and sequential lithium aluminosilicate assemblages have defined well-constrained cooling curves, e.g. for the Tanco, Harding and Bikita pegmatites (London 1984). The established cooling curves contain two segments: essentially isobaric cooling starting from ≈ 650–675°C down to ≈ 350–450°C, with an inflection and steeper P–T cooling trajectory down to ≈ 200°C, below which lithium aluminosilicate assemblages provide no further constraint (e.g. London 1986a; Chakoumakos & Lumpkin 1990). Pressures at the level of emplacement for some of the larger rare-element pegmatites (Tanco, Harding, Bikita) are constrained to the range of ≈ 250–350 MPa. The isobaric portion of the cooling curves is construed to represent the cooling of magma by contact with wallrocks and provides one of the few indications that wallrock temperatures are ≈ 300–350°C below those of Li-rich pegmatite magmas at emplacement. The steeper portions of the cooling curves are comparable with some proposed Archean and Proterozoic continental geotherms and hence may track the uplift of the pegmatite bodies.

Fluid inclusion analysis and isotopic data have been combined to estimate P–T conditions as follows: isochores based on average estimated densities at inclusion homogenisation are extrapolated to temperatures estimated from ^2H and ^{18}O isotopic distributions between coexisting phases. By this method, Taylor et al. (1979) estimated emplacement conditions near 700–750°C at 200 MPa for miarolitic pegmatites of San Diego County, California. London (1986b) noted that these P–T conditions lie well within the petalite field, but that at least some of the pegmatites studied contain primary spodumene + quartz assemblages; thus London (1986b) suggested that the actual P–T conditions must lie closer to the spodumene field at higher pressure and/or lower temperature. Using an intersection of fluid inclusion isochores with lithium aluminosilicate stability relations, London (1986b) estimated that geologically similar miarolitic pegmatites in Afghanistan crystallised close to 475–500°C at 240–280 MPa. The lower temperatures proposed by London (1986b) could be a measure of how far the melts were undercooled before crystallisation proceeded. In anticipation of equilibrium crystallisation on the liquidus, there is a general tendency to disregard the lower temperatures obtained from fluid inclusion or stable isotopic measurements as reset by subsolidus re-equilibration. That is a possibility, but not the only possible interpretation of the data. Conversely, intersections of isochores from H_2O–CO_2 inclusions have been utilised to suggest a magmatic solidus for the Tanco pegmatite to approximately <265°C (Thomas et al. 1988); using inclusions in spodumene and quartz, London (1986a) proposed a solidus of between 450 and 475°C for the Tanco pegmatite, which turned out to be very close to the experimental solidus of 450°C for the similarly evolved Macusani glass (London et al. 1989).

7. Isotope systematics

The stable (O, H) and radiogenic (K–Ar, Rb–Sr, U–Pb, Sm–Nd) isotope systematics of rare-element pegmatite fields and their associated granites have been comprehensively compiled and assessed by Cerny (1991) so that only some pertinent conclusions with a few additional examples are given here. There are important examples of pegmatite–granite suites where high initial $^{87}Sr/^{86}Sr$ values (typically between 0.71 and 0.74) and high $\delta^{18}O$ are consistent with a derivation of magma from previously unmelted (i.e. mica-rich) metapelites (e.g. Brookins 1986) at comparatively low T. Cerny (1991) notes that there are more documented instances, however, where initial $^{87}Sr/^{86}Sr$ and εNd values are consistent with mixed supracrustal (mica schist) and lower crustal (paragneiss and granulite) sources for the LCT family of deposits (e.g. the Li-rich and peraluminous Leinster Granite and pegmatites of SE Ireland with an initial $^{87}Sr/^{86}Sr$ of 0·705; O'Connor et al. 1991). Multiple source rocks for cogenetic magmas of peraluminous granite–pegmatite systems (e.g. the Harney Peak granite, South Dakota; e.g. Walker et al. 1986a, 1992; Nabelek et al. 1992b; Krogstad et al. 1993, Krogstad & Walker, this issue) appear to be the rule rather than the exception. The data are less variable for the NYF pegmatite family, whose stable and radiogenic isotope systematics are consistent with combined lower crustal and mantle sources (Cerny 1991).

An additional observation derived from both stable (O, H) and radiogenic (Rb–Sr) systematics is that pegmatite bodies usually exhibit marked isotopic disequilibrium with their surrounding hosts (e.g. Longstaffe et al. 1981; Taylor & Friedrichsen 1983; Walker et al. 1986a; Nabelek et al. 1992b), and this disparity is the strongest argument against the derivation of pegmatites in situ by local anatectic (e.g. Stewart 1978) or metamorphic processes (e.g. Gresens 1969).

Brookins (1986) and Clark and Cerny (1987) detail some of the difficulties of conducting K–Ar and Rb–Sr geochronological studies on pegmatites, wherein different host minerals exhibit excess or loss of ^{40}Ar and ^{87}Sr. Nevertheless, Brookins (1986) compared apparent K–Ar ages with Rb–Sr emplacement ages and proposed that pegmatites of the Middletown district, Connecticut spent 10–40 Ma above the closure temperature for ^{40}Ar loss from feldspars and micas (cf. the estimates of cooling rate proposed by Chakoumakos & Lumpkin 1990 and discussed earlier).

Other recent developments of isotope systems for pegmatite studies include applications of existing systems to new mineral phases (e.g. U–Pb of columbite and apatite; Romer & Wright 1992; Krogstad & Walker 1994) and increased interest in other light isotope systems, particularly of $^{11}B/^{10}B$ (e.g. Spivack et al. 1987; Swihart & Moore 1989; Palmer 1991; Palmer et al. 1992). Boron is enriched as tourmaline in peraluminous granites and pegmatites and the natural range of $\delta^{11}B$ is large. Although the fractionation of boron isotopes between tourmaline and fluid phases appears to be a complex function of P, T and perhaps fluid chemistry (Palmer et al. 1992), it may still be possible to unambiguously distinguish the isotopic source of boron in pegmatitic tourmaline and to use $^{11}B/^{10}B$ as other isotope systems to study mixing phenomena between isotopically distinct rock–fluid reservoirs. Tomascak et al. (1995) have initiated investigations of the $^{6}Li/^{7}Li$ system and its application to granitic pegmatites.

8. Magma chemistry and element partitioning

8.1. Liquidus effects of non-haplogranite components

Additional components of pegmatites (granites and rhyolites) that may achieve concentrations capable of modifying melt properties include Li, Be, B, F and P. The liquidus effects of these components, as far as they are known, have been summarised by London (1992a), and viscosity–density changes have been presented by Dingwell et al. (this issue); readers should consult this and related papers for additional details and references.

Among these components, Be is essentially unstudied; its effects on melt compositions or properties and the Be content needed to saturate melt in beryl are unknown. Lithium aluminosilicate saturation of melts occurs at ≈ 2 wt% Li_2O (Stewart 1978). The addition of Li lowers the eutectic in hydrous granitic melt by $\approx 50°C$, with no displacement from the (projection of the) Li-absent haplogranite minimum (Martin & Henderson 1984). Icenhower and London (1996b) have demonstrated that the Cs mineral pollucite is stable on the liquidus of hydrous haplogranite with >4 wt% Cs_2O in melt. Unlike Li, however, the addition of Cs expands the liquidus fields of orthoclase (Icenhower & London, 1996b) and possibly quartz (Henderson & Manning 1984), causing the melt to shift markedly towards more sodic compositions.

Individually and together, B, F and P lower solidus temperatures, decrease viscosities, promote an increase in the activity coefficient of SiO_2 in melt and consequent expansion of the liquidus field of quartz, and to various extents appear to increase the solubility of H_2O in melts. Probably the most important effect of these components is to extend magmatic crystallisation down to low temperatures. The solidi of haplogranitic systems with added B, F and P are not as well established as is commonly assumed, but the solidi at 100–200 MPa (H_2O) probably lie below 500°C (e.g. London et al. 1993). These elements may further lead to crystallisation of melts at low T by inhibiting the nucleation of feldspar phases (e.g. London et al. 1989), as all three elements tend to form stable complexes with alkalis and Al (see London 1992 for review). Though B, F and P are thought to promote melt fractionation towards sodic compositions (e.g. towards melts that generate albite–spodumene pegmatites), the experimental results actually show little or no change in the K* of melt with up to several wt% of these components added to melt by crystal fractionation (London et al. 1993). Investigations of the effects of these components on the solubilities of ore-forming minerals such as cassiterite, mangantantalite and REE phosphates are incomplete, but preliminary results indicate that F and P tend to increase the solubilities of high field strength elements (HFSEs) in melts (all other variables equal), whereas B does not and may actually decrease the solubility of HFSEs in melt by competition for alkalis (Keppler 1993; Wolf & London 1993, 1995; Wolf et al. 1994).

In melt–vapour systems, B is moderately volatile ($D_B^{melt/vapour}$ from ≈ 1 to 0·3; Pichavant 1987; London et al. 1988) and F and P are distinctly non-volatile ($D_F^{melt/vapour} \approx 3$ and $D_P^{melt/vapour} \approx 10$; e.g. Manning 1981; London et al. 1993). Enrichment of these elements in hydrothermal vein or contact metasomatic aureoles around pegmatites results more from their incompatibility in crystalline phases than their volatility in melt–vapour systems.

Estimating abundances of non-haplogranite components. Though the potential effects of Li, Cs, B, F and P in pegmatites (and granites and rhyolites) have received considerable attention, determining their actual concentrations in original melts (as opposed to actual rock) remains difficult. Like H_2O, the components of Li, Cs, B, P and F are mostly not conserved by pegmatite whole rocks. In the absence of sufficient crystallisation to fully conserve these components in minerals (e.g. especially B and F), they are eventually transferred from pegmatite to wallrocks via exsolved vapour; subsolidus alteration of earlier pegmatite minerals (e.g. Li from lithium aluminosilicates, Rb from feldspars, P from feldspars) also contributes to the flux of solutes from pegmatite to wallrocks.

From experimental studies (e.g. Stewart 1978; London *et al.* 1994) it appears that melts must contain about $\geqslant 2$ wt% Li_2O or B_2O_3 to attain saturation in lithium aluminosilicates and tourmaline, respectively. Whereas saturation in lithium aluminosilicates is likely to conserve most Li, various factors preclude sufficient deposition of tourmaline to effectively control or conserve boron (e.g. London, 1986a; London *et al.* 1995a; London & Manning 1995). Phosphorus concentrations in melts are buffered by a number of mineral–melt equilibria such that pegmatitic melts probably contain between 0·5 and 2·5 wt% P_2O_5 (London *et al.* 1995a); whole-rock P_2O_5 contents tend to be lower than magmatic values because sericitic and argillic alteration of feldspars liberates P (London 1992b). Estimates of concentrations of F in pegmatitic melts are undetermined, but pegmatites of the NYF family appear to be richer in F than those of the LCT family (e.g. Cerny 1991). Using amblygonite–montebrasite solid solutions as a guide, the F content of LCT pegmatites may be around 1 wt% of less (London *et al.* 1995a).

8.2. Crystal–melt partitioning of LILEs

Most attempts to model the derivation of pegmatites from granites (e.g. Shearer *et al.* 1992) have utilised the fractionation trends of the large ion lithophile elements (LILEs) of Rb, Cs, Sr and Ba. These elements reside principally in the major minerals of feldspar and mica groups. Until recently, the partitioning of these elements between minerals and granitic melts was not well established. There is now an adequate and mostly complete body of recent experimental work (e.g. Long 1978; Icenhower *et al.* 1994; Icenhower 1995; Icenhower & London 1995, 1996a) that details the partitioning of these elements between micas, feldspars, peraluminous melts and other phases such as cordierite, garnet and tourmaline. Important general observations from these studies are that the bulk melt composition tends to be less important than mineral chemistry and structure in controlling partition coefficients and that some crystal–melt partition coefficients are *T*-dependent whereas others are not.

For pegmatites that may be derived from metapelite sources, the experimental partitioning of Li and LILEs among feldspars, micas and peraluminous melts helps to explain how these magmas acquire their distinctly Li- and Cs-enriched (LCT) signatures (London 1995). Muscovite-rich (albite-poor) protoliths are capable of generating only small quantities of low-*T* melt. Caesium is incompatible in feldspars and micas and hence is partitioned into melt. Though Li is moderately compatible in white and dark micas (Icenhower *et al.* 1994), the terminal reaction of muscovite to Or-rich alkali feldspar + corundum at *T* slightly above the melting reaction of $Ab + Qtz + Ms + H_2O = L + Als$ can liberate a large quantity of Li into a small volume of melt. Rubidium and Ba are sufficiently compatible in micas and especially the alkali feldspar produced that may be relatively depleted from melt. The resulting LILE signature of melt is rich in Li and Cs and poor in Ba and possibly Rb.

9. Compositions of vapour phase

9.1. Fluid inclusions

During the decade 1965–75, Soviet petrologists collected a wealth of data on the properties of fluid inclusions found in pegmatite minerals (see Roedder 1984 and London 1986a for references). This research consisted mostly of the indirect identification of fluid compositions (from microthermometry) and lesser analysis of actual fluids or solid phases contained by inclusions. Two generalities can be deduced from this body of work: inclusions in minerals from Li–Cs-rich LCT pegmatites contain mostly low-salinity aqueous liquid (complex solutions of alkali and alkaline earth chloride, borate and phosphate components) and vapour with a minor carbonic or methane component, and sometimes are silicate crystal-rich; inclusions from F-rich NYF pegmatites (miarolitic fluorite–topaz type) contain aqueous hypersaline fluids with diverse assemblages of daughter salt minerals. From inclusions, therefore, the salinities of gases (= fluids) associated with NYF pegmatites appear to be high, and much higher than the LCT family of peraluminous pegmatites. In pegmatitic rocks associated with alkaline syenite–carbonatite systems, the inclusion contents also tend to be saline brines with abundant daughter minerals, but with the common addition of methane and higher alkanes (see also Roedder 1984 for a literature review).

Recent studies within and outside the large Tanco pegmatite, Manitoba provide the most detailed assessment yet of fluid inclusion contents in a peraluminous rare-element pegmatite (London 1986a; Thomas *et al.* 1988; Thomas & Spooner 1988; Channer & Spooner 1994), though the results are similar to those from less complete studies. Among these different investigators, a generally consistent observation is that fluid inclusions in the border zone contain methane in addition to aqueous and carbonic components and these differ from inclusions within the interior portions of the pegmatite where only aqueous–carbonic fluids are found (London 1986a; Thomas *et al.* 1988). The CO_2 content of the fluid before phase separation was estimated at $\leqslant 5$ mol% CO_2 (London 1986a; Thomas *et al.* 1988); methane in the border zone is derived from host amphibolites (Thomas *et al.* 1988), which reflects the open system nature noted from isotopic and mineralogical studies cited earlier. The salinity of the aqueous phase starts low (3–5 wt% equivalent $CaCl_2$, but actually a more complex solution saturated in borates) and increases with decreasing *T*, possibly by the unmixing of aqueous and carbonic components and by the influx of externally derived components during the subsolidus fluid exchange with host rocks (London 1986a). Thomas *et al.* (1988) interpreted the aqueous–carbonic vapour as a fluid phase present throughout the consolidation of the Tanco body, but London (1986a) suggested that this fluid phase appeared near the solidus on crystallisation of the small remaining fraction of melt, which was unusually hydrous, peralkaline and carbonate–borate-rich; London based this view in part on cross-cutting relations of the inclusion populations in spodumene (silicate-rich inclusions have a 'primary' distribution, whereas aqueous–carbonic inclusions occur only along crossing fractures) and from experiments in which complete miscibility between H_2O and a simple analogue melt existed at pressures as low as 200 MPa. Thomas *et al.* (1988) also recognised crystal-rich inclusions and suggested that these might represent melts that were present down to $\approx 262\,°C$; London (1986a) combined melting experiments with the crystal-rich inclusions and with synthetic melts to propose a solidus near 420–470°C for the pegmatite.

Despite differences of interpretation, these and other studies of inclusions in peraluminous LCT pegmatites consistently reveal that the vapour phase (whenever it is exsolved from melt) is dominantly aqueous with minor CO_2, and possesses low total salinity, especially in miarolitic pegmatites (e.g. London 1986b). Salt species are normally recast as equivalent NaCl, but actual solutes are much more complex. London and co-workers (1986b; Foord *et al.* 1991) noted also that crystal-rich inclusions representing trapped silicate melts are consistently present in lithium aluminosilicates, feldspars and garnet, sometimes in topaz and tourmaline, but almost never in quartz, the host phase that is most studied for its inclusions. London (1985a) has also suggested that the general paucity of melt inclusions in quartz, compared with other silicates, results from the tendency for quartz to recrystallise so pervasively that any early inclusions are obliterated.

9.2. Experimental studies

The composition of vapour in equilibrium with crystals or melt has been investigated for simple and complex systems that include many relevant to pegmatites. Candela and Picolli (1995) review ore–metal partitioning between melt and vapour (multiple gas and solute components). The partitioning of F and Cl between melt and vapour has been assessed in simple and complex melts (e.g. London et al. 1989; Webster 1992a, b; Webster & Holloway 1990), as have B and P. Experimental and theoretical investigations of mineral–vapour equilibria that are relevant to pegmatites include systems containing combinations of albite, beryl, lithium aluminosilicates, pollucite and quartz (e.g. Manier-Glavinaz et al. 1989; Sebastian & Lagache 1990, 1991; Lagache & Sebastian 1991; Dujon et al. 1991; Wood & Williams-Jones 1993; Lagache 1995).

Building on the important observations of Burnham (1967) and Kilinc (1969), recent experiments in melt–vapour systems have shown a high degree of congruency in the dissolution of major melt components into vapour, especially with regard to the elemental ratios of the alkali and alkaline earth elements. Though Cl is well known to suppress the solubility of Al in vapour (e.g. Anderson & Burnham 1981), melts that are low in Cl but high in B, P and F (e.g. the Macusani glass) dissolve essentially congruently with respect to all major components (London et al. 1988). Experiments in crystal–vapour systems also show that alkali and alkaline earth elemental ratios vary little when buffered by multiphase assemblages found in pegmatites; values of $Na/(Li + Na)$ and $Cs/(Cs + Na)$ are ≈ 0.5 and $0.1–0.2$, respectively, for appropriate buffer assemblages (Sebastian & Lagache 1990, 1991). The experimental evidence indicates that an aqueous vapour phase does not promote a high degree of element fractionation among crystalline, melt and vapour phases, and hence the deposition of essentially monominerallic zones cannot be attributed to the selective fractionation of elements between crystalline or melt phases and vapour (cf. Stewart 1978).

The HFSEs tend to partition strongly in favour of melt over vapour, even in the presence of B, P and F (e.g. London et al. 1988; Keppler & Wyllie 1991), except in Cl-rich systems. The solubilities of certain metals in vapour, such as Cu and Sn, increase rapidly with Cl content, but the solubilities of Mo and W are unchanged or decrease with Cl in vapour. Overall the lithophile ore metals found in pegmatites will not be extensively remobilised by the separation of an aqueous vapour phase except for Sn in Cl-rich systems. It is an important and commonly overlooked fact that ore-grade deposits of Sn, Nb or Ta in pegmatites are coarse-grained, primary and, importantly, endogenic. The subsolidus alteration that removes alkalis, alkaline earths and fluxes from the pegmatite to the wallrock does not substantially redistribute the HFSEs.

10. Wallrock alteration

Wallrock alteration around pegmatites is mostly notable for its absence. When found, it is usually sporadically distributed around only the most evolved pegmatite types, and then mostly along fractures or pocket blow-outs from the pegmatite interior. In only a few instances (e.g. the huge Tanco pegmatite) is wallrock alteration known to pervasively surround an entire pegmatite body.

Alteration entails metasomatism by the introduction of pegmatite-derived components and recrystallisation, with or without metasomatism, to assemblages indicative of the greenschist facies (350–550°C) of metamorphism (cf. Thomas & Spooner 1988, who proposed temperatures of $\approx 720–780°C$ for wallrock alteration around the Tanco pegmatite based on fluid inclusion studies). In mafic–ultramafic hosts, typical propylitic assemblages include chlorite + epidote + actinolite + titanite + calcite; in metapelite hosts, recrystallised white and dark micas plus spessartine garnet are common. Metasomatic phases present with these assemblages include tourmaline, holmquistite (London 1986c; Shearer & Papike 1988), and rare-alkali-enriched biotite in mafic hosts; and tourmaline and rare-alkali-enriched biotite and muscovite in metapelitic hosts. Rarer metasomatic phases include axinite in metacarbonates and bityite in some unusual circumstances (e.g. the Harding pegmatite; Chakoumakos & Lumpkin 1990).

Substantial quantities of the alkali, B and F content of pegmatite magma can be lost to wallrocks (Shearer et al. 1984, 1986; Morgan & London 1987, 1989; Shearer & Papike 1988). This is especially true of B; in the Tanco pegmatite, an estimated 60–80% of the B originally contained by the pegmatite magma was lost to wallrocks (Morgan & London 1987, 1989). The difficulties of making such mass balance calculations are obvious: a lack of three-dimensional control on most pegmatite bodies and, particularly, the localised or sporadic nature of wallrock alteration. The latter point is made well by pegmatite exposures in the miarolitic gem pegmatites of San Diego County, California (e.g. Foord et al. 1991). In the San Diego County pegmatites, norite–granodiorite hosts show virtually no alteration adjacent to pegmatite contacts exposed over metres of strike length. Where miarolitic pockets have ruptured, however, the adjacent wallrocks are converted completely to tourmaline. Clearly the wallrocks were very reactive with pegmatite-derived fluids where those fluids gained access to the wallrocks. Therefore, the general lack of wallrock alteration even within millimetres of the pegmatite contacts indicates a lack of mass transfer (via vapour) or communication from pegmatite to host throughout most of the history of pegmatite consolidation. Evidence derived from stable isotope systematics (e.g. Taylor et al. 1979), fluid inclusions (e.g. Thomas et al. 1988) and mineral chemistry (e.g. of tourmaline; London 1990; London & Manning 1995; London et al. 1995a) reveal, however, that mass transfer from wallrocks to pegmatite does occur in the early stages of magma consolidation.

A second important point derived from these occurrences is that wallrock alteration reflects the chemistry of the immediately adjacent pegmatite, including the innermost units. At Tanco, for example, tourmaline metasomatism is most intense near a late-stage albitic body that was a likely crystallisation product of the B-rich silicate melt identified from fluid inclusions by London (1986a). Holmquistite alteration is most intense adjacent to the large spodumene bodies within Tanco and elsewhere (London 1986c). At Tanco, dark micas, loosely referred to as biotite, are particularly Li-rich near spodumene units, Rb-rich near large masses of altered microcline and, importantly, Cs-rich over the large pollucite bodies at Tanco (Morgan & London 1987). These observations signify that all of the major primary units of pegmatites, down to the latest pollucite masses and albitic rocks, are in place before the development of wallrock alteration. This and other evidence cited here necessitate that fluid loss from pegmatites is a near-solidus event.

10.1. Wallrock alteration as a tool for exploration

Virtually all types of host rocks are reactive with the chemically evolved aqueous fluids derived from pegmatites and the chemistry of wallrock metasomatism mirrors the mineralogy of (i.e. fluids are derived from) the innermost fractionated zones of pegmatites. These two features make wallrock alteration a potentially valuable guide to pegmatite exploration and assessment, especially as the simple mineralogy of the exposed outer zones of pegmatites may conceal more fractionated interiors. The principal limitations of wallrock alteration, however, include: (1) its absence around pegmatites that may be economically viable sources of beryl or ceramic-grade feldspars, micas and quartz; (2) the narrowness and discontinuous nature of the aureoles; and (3) the lack of correlation between the chemical signature of wallrock alteration and the actual tonnage or grade of mineralisation within (e.g. Cerny 1989). Some manifestations of wallrock alteration are fairly reliable; for example, holmquistite is found only around large pegmatites in which spodumene or petalite are early crystallising and volumetrically abundant (London 1986c), and tourmalinisation around known miarolitic pegmatites is a good indicator of pocket development in adjacent pegmatite. The Nb and Ta contents of metasomatic micas have been utilised to discriminate barren and mineralised deposits of these elements (e.g. Morteani & Gaupp 1989). Where developed, macroscopic wallrock alteration is confined mostly to within a few metres of the pegmatite contact and geochemical anomalies rarely extend much further (e.g. Trueman & Cerny 1982).

11. Concluding remarks

Incomplete as it is, this review demonstrates that the study of pegmatites has flourished in recent years. The review has not adequately conveyed the fact that many field-based studies of pegmatites are in progress throughout the world. Petrologists are just on the verge of using the now extensive knowledge of mineral chemistry to link field and experimental data towards a better understanding of process.

The origins of most pegmatites are widely, though not universally, ascribed to the fractionation of larger plutons of granite. The recent investigations of granite–pegmatite relations in the Black Hills of South Dakota illustrate, however, that we are a long way from understanding the details of when and how pegmatite groups are derived from granite, and whether the accumulation of trace elements in pegmatitic melts occurs by low degrees of anatexis or by extended fractionation of large magma batches (or both).

The textural features of pegmatites originate from an interplay of magma composition with conditions of emplacement (cooling history). Though the chemical and physical properties of the magmas are now much better defined, that is not yet true of the P–T regimes into which pegmatite magmas are injected. The numerical simulation by Chakoumakos and Lumpkin (1990) has important implications for pegmatite crystallisation if the wallrock temperatures (and hence the calculated cooling history of the magma) are even approximately correct. There are few, if any, well constrained estimates of wallrock temperatures in specific cases of pegmatite emplacement.

The origins of aplites, pegmatites and paired aplite–pegmatite intrusives are receiving renewed scrutiny. The vapour transport model of Jahns and Tuttle (1963; Jahns & Burnham 1969) as it is currently defined is not consistent with recent experimentation, and exceptions are too numerous in nature for a gravitational model based (as it is) on buoyancy. It has been demonstrated that there is no causal relation between aqueous vapour saturation and the development of pegmatitic textures in experiments; it has also been demonstrated experimentally that the efficacy of an aqueous vapour phase in promoting chemical segregation (fractionation) of alkalis, 'volatiles' such as B, P and F and HFSEs is far less than was postulated originally (i.e. essentially zero in the case of alkalis). In this context of pegmatites, aplites and pegmatite–aplite composite dykes, the question of whether a magma is vapour-saturated or not is a moot point.

12. Acknowledgements

Financial support for this research has been provided by an NSF-EPSCoR contract EHR-9108771 for collaborative research with the University of Tulsa, grant DE-FG22-87FE1146 from the U.S. Department of Energy, and grants EAR-8516753, EAR-8720498 and EAR-8821950 from the National Science Foundation. Thanks to Brad Jolliff for his constructive review of the manuscript.

13. References

Aines, R. D., Kirby, S. H. & Rossman, G. R. 1984. Hydrogen speciation in synthetic quartz. PHYS CHEM MINER 11, 204–12.

Aines, R. D. & Rossman, G. R. 1984. The high temperature behavior of water and carbon dioxide in cordierite and beryl. AM MINER 69, 319–27.

Anderson, G. M. & Burnham, C. W. 1981. Feldspar solubility and the transport of aluminum under metamorphic conditions. AM J SCI 283-A, 283–97.

Bakker, R. J. & Jansen, J. B. H. 1994. A mechanism for preferential H_2O leakage from fluid inclusions in quartz, based on TEM observations. CONTRIB MINERAL PETROL 116, 7–20.

Barker, C. & Sommer, M. A. 1974. Potential method of geobarometry using quartz. NATURE 250, 402–4.

Beall, G. H. & Rittler, H. L. 1982. Glass-ceramics based on pollucite. In Simmons, J. H., Uhlmann, D. R. & Beall, G. H. (eds) Advances in ceramics, Vol. 4. Nucleation and crystallization of glasses, 301–12. Columbus: The American Ceramic Society.

Beger, R. M. 1969. The crystal structure and chemical composition of pollucite. Z KRISTALLOGR 129, 280–302.

Benard, F., Moutou, P. & Pichavant, M. 1985. Phase relations of tourmaline leucogranites and the significance of tourmaline in silicic magmas. J GEOL 93, 271–91.

Brookins, D. G. 1986. Rubidium–strontium geochronologic studies of large granitic pegmatites. N JAHRB MINERAL ABH 156, 81–97.

Brown, G. E. Jr & Ewing, R. C. (eds) 1986. R. H. Jahns memorial issue: the mineralogy, petrology, and geochemistry of granitic pegmatites and related granitic rocks. AM MINERAL 71, 233–654.

Brown, G. E. Jr & Mills, B. A. 1986. High-temperature structure and crystal chemistry of hydrous alkali-rich beryl from the Harding pegmatite, Taos County, New Mexico. AM MINERAL 71, 547–56.

Burnham, C. W. 1967. Hydrothermal fluids at the magmatic stage. In Barnes, H. L. (ed.) Geochemistry of hydrothermal ore deposits, 1st edn, 34–76. New York: Holt, Reinhart, and Winston.

Burnham, C. W. & Nekvasil, H. 1986. Equilibrium properties of granite pegmatite magmas. AM MINERAL 71, 239–63.

Burt, D. M. 1989. Vector representation of tourmaline compositions. AM MINERAL 74, 826–39.

Cameron, E. N., Jahns, R. H., McNair, A. H. & Page, L. R. 1949. Internal structure of granitic pegmatites. ECON GEOL MONOGR 2, 115 pp.

Candela, P. A. & Picolli, P. M. 1995. Model ore-metal partitioning from melts into vapor and vapor/brine mixtures. In Thompson, J. F. H. (ed.) Magmas, fluids, and ore deposits. MINERAL ASSOC CAN SHORT COURSE HANDB 23, 101–27.

Cerny, P. 1974. The present status of the analcime–pollucite series. CAN MINERAL 12, 334–41.

Cerny, P. (ed.) 1982a. Granitic pegmatites in science and industry. MINERAL ASSOC CAN SHORT COURSE HANDB 8, 555 pp.

Cerny, P. 1982b. The Tanco pegmatite at Bernic Lake, Manitoba. In Cerny, P. (ed.) Granitic pegmatites in science and industry. MINERAL ASSOC CAN SHORT COURSE HANDB 8, 527–43.

Cerny, P. 1989. Exploration strategy and methods for pegmatite deposits of tantalum. In Möller, P., Cerny, P. & Saupé, P. (eds)

Lanthanides, tantalum and niobium, 274–302. Berlin: Springer-Verlag.

Cerny, P. 1991. Fertile granites of Precambrian rare-element pegmatite fields: is geochemistry controlled by tectonic setting or source lithologies? PRECAMBRIAN RES **51**, 429–68.

Cerny, P. 1992. Rare-element granitic pegmatites. Part II: regional to global environments and petrogenesis. GEOSCI CAN **18**, 68–81.

Cerny, P. 1995. Regional zonation of rare-element pegmatite groups [abstract]. GAC-MAC PROG ABSTR **20**, A14.

Cerny, P. & Brisbin, W. C. 1982. The Osis Lake pegmatitic granite, Winnipeg River district, southeastern Manitoba. *In* Cerny, P. (ed.) *Granitic pegmatites in science and industry*. MINERAL ASSOC CAN SHORT COURSE HANDB **8**, 545–55.

Cerny, P. & Ercit, T. S. 1989. Mineralogy of niobium and tantalum: crystal chemical relationships, paragenetic aspects and their economic implications. *In* Möller, P., Cerny, P. & Saupé, F. (eds) *Lanthanides, tantalum, and niobium*, 27–79. Berlin: Springer-Verlag.

Cerny, P. & Macek, J. 1972. The Tanco pegmatite at Bernic Lake, Manitoba. V. Colored potassium feldspars. CAN MINERAL **11**, 679–89.

Cerny, P. & Simpson, F. M. 1977. The Tanco pegmatite at Bernic Lake, Manitoba. IX. Beryl. CAN MINERAL **15**, 489–99.

Cerny, P. & Simpson, F. M. 1978. The Tanco pegmatite at Bernic Lake, Manitoba. X. Pollucite. CAN MINERAL **16**, 325–33.

Cerny, P., Trueman, D. L., Ziehlke, D. V., Goad, B. E. & Paul, B. J. 1981. *The Cat Lake–Winnipeg River and the Wekusko Lake pegmatite fields, Manitoba*. Manitoba Department of Energy and Mines, Mineral Resources Division. ECON GEOL REP **ER80-1**, 234 pp.

Cerny, P., Meintzer, R. E. & Anderson, A. J. 1985. Extreme fractionation in rare-element granitic pegmatites: selected examples of data and mechanisms. CAN MINERAL **23**, 381–421.

Chakoumakos, B. C. & Lumpkin, G. R. 1990. Pressure–temperature constraints on the crystallization of the Harding pegmatite, Taos County, New Mexico. CAN MINERAL **28**, 287–98.

Channer, D. M., Der & Spooner, E. T. C. 1994. Combined gas and ion chromatographic analysis of fluid inclusions: applications to Archean granite pegmatite and gold–quartz vein fluids. GEOCHIM COSMOCHIM ACTA **58**, 1101–18.

Chappell, B. W. & White, A. J. R. 1992. I- and S-type granites in the Lachlan Fold Belt. TRANS ROY SOC EDINBURGH EARTH SCI **83**, 1–26.

Clark, G. S. & Cerny, P. 1987. Radiogenic ^{87}Sr, its mobility, and the interpretation of Rb–Sr fractionation trends in rare-element granitic pegmatites. GEOCHIM COSMOCHIM ACTA **51**, 1011–8.

Cordier, P., Boulogne, B. & Doukhan, J.-C. 1988. Water precipitation and diffusion in wet quartz and wet berlinite AlPO$_4$. BULL MINERAL **111**, 113–37.

Dennen, W. H., Blackburn, W. H. & Quesada, A. 1970. Aluminum in quartz as a geothermometer. CONTRIB MINERAL PETROL **27**, 332–42.

Desborough, G. A., Ludington, S. D. & Sharp, W. N. 1980. The Redskin Granite: a rare-metal-rich Precambrian pluton, Colorado, U.S.A. MINERAL MAG **43**, 959–66.

Drysdale, D. J. 1992. The pollucite–analcime series, 1974–1990. PAPERS DEP EARTH SCI UNIV QUEENSLAND **12**, 317–24.

Dujon, S. C., Lagache, M. & Sebastian, A. 1991. Experimental study of Li-rich granitic pegmatites: part III. Thermodynamic implications of the experiments in the Na–Li–Cs system: consequences for the properties of solutes. AM MINERAL **76**, 1614–19.

Duke, E. F., Redden, J. A. & Papike, J. J. 1988. Calamity Peak layered granite–pegmatite complex, Black Hills, South Dakota. 1. Structure and emplacement. GEOL SOC AM BULL **100**, 825–40.

Ercit, T. S., Cerny, P. & Hawthorne, F. C. 1992a. The wodginite group. III. Classifications and new species. CAN MINERAL **30**, 633–8.

Ercit, T. S., Cerny, P., Hawthorne, F. C. & McCammon, C. A. 1992b. The wodginite group. II. Crystal chemistry. CAN MINERAL **30**, 613–31.

Ercit, T. S., Hawthorne, F. C. & Cerny, P. 1992c. The wodginite group. I. Structural crystallography. CAN MINERAL **30**, 597–611.

Ercit, T. S., Wise, M. A. & Cerny, P. 1995. Compositional and structural systematics of the columbite group. AM MINERAL **80**, 613–9.

Fenn, P. M. 1977. The nucleation and growth of alkali feldspars from hydrous melts. CAN MINERAL **15**, 135–61.

Fenn, P. M. 1986. On the origin of graphic granite. AM MINERAL **71**, 325–30.

Foord, E. E. 1976. *Mineralogy and petrogenesis of layered pegmatite-aplite dikes in the Mesa Grande district, San Diego County, California*. Ph.D. Thesis, Stanford University.

Foord, E. E., London, D., Kampf, A. F., Shigley, J. E. & Snee, L. W. 1991. Gem-bearing pegmatites of San Diego County, California. *In* Walawender, M. J. & Hanan, B. B. (eds) *Geological excursions in southern California and Mexico. Guidebook for the 1991 meeting, Geological Society of America*. San Diego: Department of Geological Sciences, San Diego State University.

Francis, C. A., Wise, M. A., Kampf, A. R., Brown, C. D. & Whitmore, R. W. 1993. Granitic pegmatites in northern New England. *In* Cheney, J. T. & Hepburn, J. C. (eds) *Field trip guidebook for the northeastern United States: 1993 Boston GSA*, Vol. 1. DEP GEOL GEOGR UNIV MASSACHUSETTS CONTRIB **67**, E1–24.

Goad, B. E. & Cerny, P. 1981. Peraluminous pegmatitic granites and their pegmatite aureoles in the Winnipeg River district, southeastern Manitoba. CAN MINERAL **19**, 177–94.

Gresens, R. H. 1969. Tectonic–hydrothermal pegmatites. I. The model. CONTRIB MINERAL PETROL **15**, 345–55.

Haralampiev, A. & Grover, J. 1993. Synthesis experiments in the binary system tsilaisite–dravite, Na[Mn$_x$Mg$_{1-x}$]$_3$Al$_6$(BO$_3$)$_3$-Si$_6$O$_{18}$(OH)$_4$, at T = 375–700°C and P = 2000 bars: does garnet control the occurrence of tourmaline? [abstract] GEOL SOC AM ABSTR PROGR **25**, A94–5.

Harvey, M. J. 1991. *An experimental study of greywacke partial melting in the presence of water, fluorine, and boron*. Ph.D. Thesis, University of Manchester.

Henderson, C. M. B. & Manning, D. A. C. 1984. The effect of Cs on phase relations in the granite system: stability of pollucite. *In Progress in experimental petrology, sixth progress report of research supported by N.E.R.C., 1981–1984*. NAT ENVIRON RES COUN (UK) PUBL SER D **25**, 41–2.

Hervig, R. L. & Peacock, S. M. 1989. Implications of trace element zoning in deformed quartz from the Santa Catalina mylonite zone. J GEOL **89**, 343–50.

Holtz, F. & Johannes, W. 1991. Effect of tourmaline on melt fraction and composition of first melts in quartzofeldspathic gneiss. EUR J MINERAL **3**, 527–36.

Icenhower, J. P. 1995. *Experimental determination of element behavior in silicic systems during hydrous partial fusion*. Ph.D. Thesis, University of Oklahoma.

Icenhower, J. P. & London, D. 1995. An experimental study of element partitioning between biotite, muscovite, and coexisting peraluminous granitic melts at 200 MPa (H$_2$O). AM MINERAL **80**, 1229–51.

Icenhower, J. P. & London, D. 1996a. Experimental partitioning of Rb, Cs, Sr, and Ba between alkali feldspar and peraluminous melt. AM MINERAL **81**, 719–34.

Icenhower, J. P. & London, D. 1996b. Pollucite as a liquidus phase in granite pegmatite melts, in press.

Icenhower, J. P., London, D. & Layne, G. D. 1994. Element partitioning among biotite, muscovite, garnet, cordierite, and peraluminous melt: behavior of Li and Mn [abstract]. GEOL SOC AM ABSTR PROGR **26**, A-290.

Jahns, R. H. 1953a. The genesis of pegmatites. I. Occurrence and origin of giant crystals. AM MINERAL **38**, 563–98.

Jahns, R. H. 1953b. The genesis of pegmatites (II). Quantitative analysis of lithium bearing pegmatite, Mora County, New Mexico. AM MINERAL **38**, 1078–112.

Jahns, R. H. 1955. The study of pegmatites. ECON GEOL FIFTIETH ANNIV VOL, 1025–130.

Jahns, R. H. 1982. Internal evolution of pegmatite bodies. *In* Cerny, P. (ed.) *Granitic pegmatites in science and industry*. MINERAL ASSOC CAN SHORT COURSE HANDB **8**, 293–327.

Jahns, R. H. & Burnham, C. W. 1969. Experimental studies of pegmatite genesis: I. A model for the derivation and crystallization of granitic pegmatites. ECON GEOL **64**, 843–64.

Jahns, R. H. & Tuttle, O. F. 1963. Layered pegmatite–aplite intrusives. MINERAL SOC AM SPEC PAP **1**, 78–92.

Jolliff, B. L., Papike, J. J. & Shearer, C. K. 1986. Tourmaline as a recorder of pegmatite evolution: Bob Ingersoll pegmatites, Black Hills, South Dakota. AM MINERAL **71**, 472–500.

Jolliff, B. L., Papike, J. J. & Shearer, C. K. 1989. Inter- and intra-crystal REE variations in apatite from the Bob Ingersoll pegmatites, Black Hills, South Dakota. GEOCHIM COSMOCHIM ACTA **53**, 429–41.

Jolliff, B. L., Papike, J. J. & Shearer, C. K. 1992. Petrogenetic relationships between pegmatite and granite based on geochemistry of muscovite in pegmatitic wallzones, Black Hills, South Dakota. GEOCHIM COSMOCHIM ACTA **56**, 1915–39.

Keller, P. 1988. Dendritic phosphate minerals and their paragenetic relation to the silicate minerals of pegmatites from Namibia and from the Black Hills, South Dakota, U.S.A. N. JAHRB MINERAL ABH **159**, 249–81.

Keller, P., Fonatan, F. & Fransolet, A.-M. 1994. Intercrystalline cation partitioning between minerals of the triplite–zweiselite–magniotriple and triphylite–lithiophilite series in granitic pegmatites. CONTRIB MINERAL PETROL **118**, 239–48.

Keppler, H. 1993. Influence of fluorine on the enrichment of high field strength trace elements in granitic rocks. CONTRIB MINERAL PETROL **114**, 479–88.

Keppler, H. & Wyllie, P. J. 1991. Partitioning of Cu, Sn, Mo, W, U, and Th between melt and aqueous fluid in the systems haplogranite–H_2O–HCl and haplogranite–H_2O–HF. CONTRIB MINERAL PETROL **109**, 139–50.

Kilinc, I. A. 1969. *Experimental metamorphism and anatexis of shales and graywackes.* Ph.D. Thesis, Pennsylvania State University.

Kleck, W. D. 1991. The origin of line rock and a model for the formation of Pala-type pegmatite bodies [abstract]. GEOL SOC AM ABSTR PROGR **23**, A-329.

Krogstad, E. J. & Walker, R. J. 1994. High closure temperatures of the U-Pb system in large apatites from the Tin Mountain pegmatite, Black Hills, South Dakota, U.S.A. GEOCHIM COSMOCHIM ACTA **58**, 3845–53.

Krogstad, E. J., Walker, R. J., Nabelek, P. I. & Russ-Nabelek, C. 1993. Lead isotope evidence for mixed sources of Proterozoic granites and pegmatites, Black Hills, South Dakota, U.S.A. GEOCHIM COSMOCHIM ACTA **57**, 4677–85.

Lagache, M. 1995. New experimental data on the stability of the pollucite–analcime series: application to natural assemblages. EUR J MINERAL **7**, 319–23.

Lagache, M. & Sebastian, A. 1991. Experimental study of Li-rich granitic pegmatites: part II. Spodumene + albite + quartz equilibrium. AM MINERAL **76**, 611–6.

London, D. 1984. Experimental phase equilibria in the system $LiAlSiO_4$–Si_2O–H_2O: a petrogenetic grid for lithium-rich pegmatites. AM MINERAL **69**, 995–1004.

London, D. 1985a. Origin and significance of inclusions in quartz: a cautionary example from the Tanco pegmatite, Manitoba. ECON GEOL **80**, 1988–95.

London, D. 1985b. Pegmatites of the Middletown district, Connecticut. *In 77th Annual Meeting, New England Intercollegiate Geological Conference, Yale University; Connecticut Geological and Natural History Survey Guidebook*, Vol. 6. 509–33.

London, D. 1986a. The magmatic–hydrothermal transition in the Tanco rare-element pegmatite: evidence from fluid inclusions and phase equilibrium experiments. AM MINERAL **71**, 376–95.

London, D. 1986b. Formation of tourmaline-rich pockets in miarolitic pegmatites. AM MINERAL **71**, 396–405.

London, D. 1986c. Holmquistite as a guide to pegmatitic rare metal deposits. ECON GEOL **81**, 704–12.

London, D. 1990. Internal differentiation of rare-element pegmatites: a synthesis of recent research. *In* Stein, H. J. & Hannah, J. L. (eds) *Ore-bearing granite systems.* GEOL SOC AM SPEC PAP **246**, 35–50.

London, D. 1992a. The application of experimental petrology to the genesis and crystallization of granitic pegmatites. *In* Cerny, P. & Martin, R. F. (eds) *Granitic pegmatites* CAN MINERAL **30**, 499–540.

London, D. 1992b. Phosphorus in S-type magmas: the P_2O_5 content of feldspars from granites, pegmatites, and rhyolites. AM MINERAL **77**, 126–45.

London, D. 1995. Geochemical features of peraluminous granites, pegmatites, and rhyolites as sources of lithophile metal deposits. *In* Thompson, J. F. H. (ed.) *Magmas, fluids, and ore deposits* MINERAL ASSOC CAN SHORT COURSE HANDB **23**, 175–202.

London, D. & Burt, D. M. 1982a. Chemical models for lithium aluminosilicate stabilities in pegmatites and granites. AM MINERAL **67**, 494–509.

London, D. & Burt, D. M. 1982b. Lithium minerals in pegmatites. *In* Cerny, P. (ed.) *Granitic pegmatites in science and industry.* MINERAL ASSOC CAN SHORT COURSE HANDB **8**, 97–133.

London, D. & Burt, D. M. 1982c. Lithium aluminosilicate occurrences in pegmatites and the lithium aluminosilicate phase diagram. AM MINERAL **67**, 483–93.

London, D. & Manning, D. A. C. 1995. Chemical variation and significance of tourmaline from southwest England. ECON GEOL **90**, 495–519.

London, D., Hervig, R. L. & Morgan, G. B. VI 1988. Melt-vapor solubilities and elemental partitioning in peraluminous granite–pegmatite systems: experimental results with Macusani glass at 200 MPa. CONTRIB MINERAL PETROL **99**, 360–73.

London, D., Morgan, G. B. VI & Hervig, R. L. 1989. Vapor-undersaturated experiments in the system macusanite–H_2O, at 200 MPa, and the internal differentiation of granitic pegmatites. CONTRIB MINERAL PETROL **102**, 1–17.

London, D., Cerny, P., Loomis, J. L. & Pan, J. J. 1990. Phosphorus in alkali feldspars of rare-element granitic pegmatites. CAN MINERAL **28**, 771–86.

London, D., Morgan, G. B. VI, Babb, H. A. & Loomis, J. L. 1993. Behavior and effects of phosphorus in the system Na_2O–K_2O–Al_2O_3–SiO_2–P_2O_5–H_2O at 200 MPa (H_2O). CONTRIB MINERAL PETROL **113**, 450–65.

London, D., Wolf, M. B. & Morgan, G. B. VI 1994. Boron saturation in granitic magmas: tourmaline–biotite–cordierite equilibria [abstract]. GEOL SOC AM ABSTR PROGR **26**, A-516.

London, D., Wolf, M. B. & Morgan, G. B. VI, 1995a. Silicate–phosphate equilibria in peraluminous granites and pegmatites [abstract]. GEOL SOC AM PROGR ABSTR **27**, 411.

London, D., Wolf, M. B., Morgan, G. B. VI & Gallego, M. 1995b. The phosphorus cycle in peraluminous granitic magmas [abstract]. *In* Brown, M. & Picolli, P. M. (eds) *The origin of granites and related rocks, Third Hutton Symposium.* US GEOL SURV CIRC **1129**, 90–1.

London, D., Morgan, G. B. VI & Wolf, M. B. Boron in granitic rocks and their contact aureoles. *In* Grew, E. S. & Anovitz, L. (eds) *Boron: mineralogy, petrology, and geochemistry in the earth's crust.* MINERAL SOC AM REV MINERAL **33**, in press.

Long, P. E. 1978. Experimental determination of partition coefficients for Rb, Sr, and Ba between alkali feldspar and silicate liquid. GEOCHIM COSMOCHIM ACTA **42**, 833–46.

Longstaffe, F. J., Cerny, P. & Muehlenbachs, K. 1981. Oxygen-isotope geochemistry of the granitoid rocks in the Winnipeg River pegmatite district, southeastern Manitoba. CAN MINERAL **19**, 195–204.

Luth, W. C., Jahns, R. H. & Tuttle, O. F. 1964. The granite system at pressures of 4 to 10 kilobars. J GEOPHYS RES **69**, 759–73.

MacLellan, H. E. & Trembath, L. L. 1991. The role of quartz crystallization in the development and preservation of igneous texture in granitic rocks: experimental evidence at 1 kbar. AM MINERAL **76**, 1291–305.

Manier-Glavinaz, V., D'Arco, P. & Lagache, M. 1989. Alkali partitioning between beryl and hydrothermal fluids: an experimental study at 600°C and 1·5 kbar. EUR J MINERAL **1**, 645–55.

Manning, D. A. C. 1981. The effect of fluorine on liquidus phase relationships in the system Qz–Ab–Or with excess water at 1 kb. CONTRIB MINERAL PETROL **76**, 206–15.

Manning, D. A. C. 1982. Chemical and morphological variation tourmalines from the Hub Kapong batholith of peninsular Thailand. MINERAL MAG **45**, 139–47.

Martin, J. S. & Henderson, C. M. B. 1984. An experimental study of the effects of small amounts of lithium on the granite system. *In Progress in experimental petrology, sixth progress report of research supported by N.E.R.C., 1981–1984.* NAT ENVIRON RES COUN (UK) PUBL SER D **25**, 30–5.

Martin, R. F. 1982. Quartz and the feldspars. *In* Cerny, P. (ed.) *Granitic pegmatites in science and industry,* MINERAL ASSOC CAN SHORT COURSE HANDB **8**, 97–133.

Martin, R. F. & Cerny, P. (eds) 1992. Granitic pegmatites. CAN MINERAL **30**, 497–954.

McBirney, A. R. & Noyes, R. M. 1979. Crystallization and layering of the Skaergaard intrusion. J PETROL **20**, 487–554.

Modreski, P. J. (ed.) 1986. *Colorado pegmatites.* Denver: Colorado Chapter, Friends of Mineralogy, Denver Museum of Natural History, 160 pp.

Moore, P. B. 1973. Pegmatite phosphates: descriptive mineralogy and crystal chemistry. MINERAL REC **1973**, 103–30.

Moore, P. B. 1982. Pegmatite minerals of P(V) and B(III). *In* Cerny, P. (ed.) *Granitic pegmatites in science and industry.* MINERAL ASSOC CAN SHORT COURSE HANDB **8**, 267–91.

Morgan, G. B. VI & London, D. 1987. Alteration of amphibolitic wallrocks around the Tanco rare-element pegmatite, Bernic Lake, Manitoba. AM MINERAL **72**, 1097–121.

Morgan, G. B. VI & London, D. 1989. Experimental reactions of amphibolite with boron-bearing aqueous fluids at 200 MPa: implications for tourmaline stability and partial melting in mafic rocks. CONTRIB MINERAL PETROL **102**, 1–17.

Morteani, G. & Gaupp, R. 1989. Geochemical evaluation of the tantalum potential of pegmatites. *In* Möller, P., Cerny, P. & Saupé, P. (eds) *Lanthanides, tantalum and niobium.* Berlin: Springer-Verlag, 303–10.

Mukhopadhyay, B. & Holdaway, M. J. 1994. Cordierite–garnet–sillimanite–quartz equilibrium: I. New experimental calibration in the system $FeO–Al_2O_3–SiO_2–H_2O$ and certain $P–T–X_{H2O}$ relations. CONTRIB MINERAL PETROL **116**, 462–72.

Nabelek, P. I. & Glascock, M. D. 1995. REE-depleted leucogranites, Black Hills, South Dakota: a consequence of disequilibrium melting of monazite-bearing schists. J PETROL **36**, 1055–71.

Nabelek, P. I., Russ-Nabelek, C. & Denison, J. R. 1992a. The generation and crystallization conditions of the Proterozoic Harney Peak leucogranite, Black Hills, South Dakota, U.S.A.: petrologic and geochemical constraints. CONTRIB MINERAL PETROL **110**, 173–91.

Nabelek, P. I., Russ-Nabelek, C. & Haeussler, G. T. 1992b. Stable isotope evidence for the petrogenesis and fluid evolution in the Proterozoic Harney Peak leucogranite, Black Hills, South Dakota. GEOCHIM COSMOCHIM ACTA **56**, 403–17.

Northrup, C. J., Lumpkin, G. R. & Ewing, R. C. 1989. Variation in the composition of beryl and apatite from the Harding pegmatite, Taos county, New Mexico [abstract]. GEOL SOC AM ABSTR PROGR **21**, A198–9.

Norton, J. J. 1983. Sequence of mineral assemblages in differentiated granitic pegmatites. ECON GEOL **78**, 854–74.

Norton, J. J. & Redden, J. A. 1990. Relations of zoned pegmatites to other pegmatites, granite, and metamorphic rocks, in the southern Black Hills, South Dakota. AM MINERAL **75**, 631–55.

Novak, M. & Cerny, P. (eds) 1992a. *Lepidolite 200, abstracts of papers.* Brno: Masaryk University and Moravian Museum, 82 pp.

Novak, M. & Cerny, P. (eds) 1992b. *Lepidolite 200, field trip guide.* Brno: Masaryk University and Moravian Museum, 83 pp.

O'Connor, P. J., Gallagher, V. & Kennan, P. S. 1991. Genesis of lithium pegmatites from the Leinster Granite margin, southeastern Ireland: geochemical constraints. GEOL J **26**, 295–305.

Ortoleva, P. J. 1994. *Geochemical self-organization.* Oxford: Oxford University Press, 411 pp.

Orville, P. M. 1963. Alkali ion exchange between vapor and feldspar phases. AM J SCI **261**, 201–37.

Palmer, M. R. 1991. Boron isotope systematics of hydrothermal fluids and tourmalines: a synthesis. CHEM GEOL **94**, 111–21.

Palmer, M. R., London, D., Morgan, G. B. VI & Babb, H. A. 1992. Experimental determination of fractionation of $^{11}B/^{10}B$ between tourmaline and aqueous vapor: a temperature- and pressure-dependent isotopic system. CHEM GEOL **101**, 123–9.

Petersen, J. S. & Lofgren, G. E. 1986. Lamellar and patchy intergrowths in feldspars: experimental crystallization of eutectic silicates. AM MINERAL **71**, 343–55.

Perny, B., Eberhardt, P., Rasmeyre, K., Mullis, J. & Pankrath, R. 1992. Microdistribution of Al, Li, and Na in α quartz: possible causes and correlation with short-lived cathodoluminescence. AM MINERAL **77**, 534–44.

Pichavant, M. 1987. Effects of B and H_2O on liquidus phase relations in the haplogranite system at 1 kbar. AM MINERAL **72**, 1056–70.

Rockhold, J. R., Nabelek, P. I. & Glascock, M. D. 1987. Origin of rhythmic layering in the Calamity Peak satellite pluton of the Harney Peak Granite, South Dakota: the role of boron. GEOCHIM COSMOCHIM ACTA **51**, 487–96.

Roedder, E. 1984. Fluid inclusions. MINERAL SOC AM REV MINERAL **12**, 644 pp.

Roedder, E. 1992. Fluid inclusion evidence for immiscibility in magmatic differentiation. GEOCHIM COSMOCHIM ACTA **56**, 5–20.

Romer, R. L. & Wright, J. E. 1992. U–Pb dating of columbites: a geochronologic tool to date magmatism and ore deposits. GEOCHIM COSMOCHIM ACTA **56**, 2137–42.

Sebastian, A. & Lagache, M. 1990. Experimental study of the equilibrium between polluate, albite, and hydrothermal fluid in pegmatitic systems. MINERAL MAG **54**, 447–54.

Sebastian, A. & Lagache, M. 1991. Experimental study of lithium-rich granitic pegmatites: part I. Petalite + albite + quartz equilibrium. AM MINERAL **76**, 205–10.

Shearer, C. K. & Papike, J. J. 1988. Pegmatite–wallrock interaction: holmquistite-bearing amphibolite, Edison pegmatites, Black Hills, South Dakota. AM MINERAL **73**, 324–37.

Shearer, C. K., Papike, J. J. & Simon, S. B. 1984. Pegmatite/wallrock interactions, Black Hills, South Dakota: progressive boron metasomatism adjacent to the Tip Top pegmatite. GEOCHIM COSMOCHIM ACTA **48**, 2563–79.

Shearer, C. K., Papike, J. J., Simon, S. B. & Laul, J. C. 1986. Pegmatite–wallrock interactions, Black Hills, South Dakota: interaction between pegmatite-derived fluids and quartz-mica schist wallrock. AM MINERAL **71**, 518–39.

Shearer, C. K., Papike, J. J. & Laul, J. C. 1987. Mineralogical and chemical evolution of a rare-element granite–pegmatite system: the Harney Peak granite, Black Hills, South Dakota. GEOCHIM COSMOCHIM ACTA **51**, 473–86.

Shearer, C. K., Papike, J. J. & Jolliff, B. L. 1992. Petrogenetic links among granites and pegmatites in the Harney Peak rare-element granite–pegmatite system, Black Hills, South Dakota. CAN MINERAL **30**, 785–809.

Sherriff, B. L., Grundy, H. D., Hartman, J. S., Hawthorne, F. C. & Cerny, P. 1991. The incorporation of alkalis in beryl: multi-nuclear MAS NMR and crystal-structure study. CAN MINERAL **29**, 271–85.

Simmons, W. B. & Heinrich, E. W. 1980. Rare-earth pegmatites of the South Platte district, Colorado. COLORADO GEOL SURV RESOUR SER **11**, 131 pp.

Simmons, W. B., Foord, E. E., Falster, A. U. & King, V. T. 1995. Evidence for an anatectic origin of granitic pegmatites, western Maine, U.S.A. [abstract]. GEOL SOC AM ABSTR PROGR **27**, A-411.

Simpson, D. R. 1977. Aluminum phosphate variants of feldspar. AM MINERAL **62**, 351–5.

Smeds, S.-A. 1994. Zoning and fractionation trends of a peraluminous NYF granitic pegmatite field at Falun, south-central Sweden. GEOL FOREN STOCHOLMS FORH **116**, 175–84.

Smeds, S.-A. & Cerny, P. 1989. Pollucite from Proterozoic petalite-bearing pegmatites of Utö, Stockholm archipelago, Sweden. GEOL FOREN STOCHOLMS FORH **111**, 361–72.

Sokolov, Yu. M. & Khlestov, V. V. 1990. Garnets as indicators of the physicochemical conditions of pegmatite formation. INT GEOL REV **32**, 1095–107.

Spilde, M. N. & Shearer, C. K. 1992. A comparison of tantalum–niobium oxide assemblages in two mineralogically distinct rare-element granitic pegmatites, Black Hills, South Dakota. CAN MINERAL **30**, 719–37.

Spivack, A. J., Palmer, M. R. & Edmond, J. M. 1987. The sedimentary cycle of the boron isotopes. GEOCHIM COSMOCHIM ACTA **51**, 1939–49.

Staatz, M. H., Murata, K. J. & Glass, J. J. 1955. Variation of composition and physical properties of tourmaline with its position in the pegmatite. AM MINERAL **40**, 789–804.

Stewart, D. B. 1963. Petrogenesis and mineral assemblages of lithium-rich pegmatites [abstract]. GEOL SOC AM SPEC PAP **76**, 159.

Stewart, D. B. 1978. Petrogenesis of lithium-rich pegmatites. AM MINERAL **63**, 970–80.

Swanson, S. E. & Fenn, P. M. 1986. Quartz crystallization in igneous rocks. AM MINERAL **71**, 331–42.

Swihart, G. H. & Moore, P. B. 1989. A reconnaissance of the boron isotopic composition of tourmaline. GEOCHIM COSMOCHIM ACTA **53**, 911–6.

Taylor, B. E. & Friedrichsen, H. 1983. Light stable isotope systematics of granitic pegmatites from North America and Norway. ISOTOPE GEOSCI **1**, 127–67.

Taylor, B. E., Foord, E. E. & Friedrichsen, H. 1979. Stable isotope and fluid inclusion studies of gem-bearing granitic pegmatite-aplite dikes, San Diego Co., California. CONTRIB MINERAL PETROL **68**, 187–205.

Taylor, R. P., Fallick, A. E. & Breaks, F. W. 1992. Volatile evolution in Archean rare-element granitic pegmatites: evidence from the hydrogen isotopic compositions of channel H_2O in beryl. CAN MINERAL **30**, 877–93.

Teertstra, D. K., Cerny, P. & Chapman, R. 1992. Compositional heterogeneity of pollucite from High Grade dyke, Maskwa Lake, southeastern Manitoba. CAN MINERAL **30**, 687–97.

Teertstra, D. K., Sherriff, B. L., Xu, Z. & Cerny, P. 1993. MAS and DOR NMR study of Al–Si order in the analcime–pollucite series. CAN MINERAL **32**, 69–80.

Thomas, A. V. & Spooner, E. T. C. 1988. Fluid inclusions in the system $H_2O–CH_4–NaCl–CO_2$ from metasomatic tourmaline within the border unit of the Tanco zoned granitic pegmatite, S.E. Manitoba. GEOCHIM COSMOCHIM ACTA **52**, 1065–75.

Thomas, A. V., Bray, C. J. & Spooner, E. T. C. 1988. A discussion of the Jahns–Burnham proposal for the formation of zoned granitic pegmatites using solid–liquid–vapour inclusions from the Tanco pegmatite, S.E. Manitoba, Canada. TRANS ROY SOC EDINBURGH EARTH SCI **79**, 299–315.

Tomascak, P. B. & Cerny, P. 1992. Mineralogy and geochemistry of phosphorus in the Aylmer Lake pegmatite field, N.W.T. GAC–MAC PROGR ABSTR **17**, A110.

Tomascak, P. B., Lynton, S. J., Walker, R. J. & Krogstad, E. J. 1995. Li isotope geochemistry of the Tin Mountain pegmatite, Black Hills, South Dakota [abstract]. *In* Brown, M. & Picolli, P. M.

(eds) *The origin of granites and related rocks, Third Hutton Symposium.* US GEOL SURV CIRC **1129**, 151–2.

Trueman, D. L. & Cerny, P. 1982. Exploration for rare-element granitic pegmatites. *In* Cerny, P. (ed.) *Granitic pegmatites in science and industry.* MINERAL ASSOC CAN SHORT COURSE HANDB **8**, 463–93.

Walker, R. J., Hanson, G. N., Papike, J. J. & O'Neil, J. R. 1986a. Nd, O, and Sr isotopic constraints on the origin of Precambrian rocks, southern Black Hills, South Dakota. GEOCHIM COSMOCHIM ACTA **50**, 2833–46.

Walker, R. J., Hanson, G. N., Papike, J. J., O'Neil, J. R. & Laul, J. C. 1986b. Internal evolution of the Tin Mountain pegmatite, Black Hills, South Dakota. AM MINERAL **71**, 440–59.

Walker, R. J., Hanson, G. N. & Papike, J. J. 1992. Trace element constraints on pegmatite genesis: Tin Mountain pegmatite, Black Hills, South Dakota. CONTRIB MINERAL PETROL **101**, 290–300.

Webber, K. L., Falster, A. U., Simmons, Wm. B. & Foord, E. E. 1995. Origin of 'line rock' in composite aplite–pegmatite dikes (abstr.). GEOL SOC AM ABSTR PROGR **27**, A-411.

Webster, J. D. 1992a. Fluid–melt interactions involving Cl-rich granites: experimental study from 2 to 8 kbar. GEOCHIM COSMOCHIM ACTA **56**, 659–78.

Webster, J. D. 1992b. Water solubility and Cl partitioning in Cl-rich granitic systems: effects of melt composition at 2 kbar and 800°C. GEOCHIM COSMOCHIM ACTA **56**, 679–87.

Webster, J. D. & Holloway, J. R. 1990. Partitioning of F and Cl between magmatic fluids and highly evolved granitic magmas. *In* Stein, H. J. & Hannah, J. L. (eds) *Ore-bearing granite systems.* GEOL SOC AM SPEC PAP **246**, 21–34.

Whalen, J. B., Currie, K. L. & Chappell, B. W. 1987. A-type granites: geochemical characteristics, discrimination, and petrogenesis. CONTRIB MINERAL PETROL **95**, 407–19.

Whitworth, M. P. 1992. Petrogenetic implications of garnets associated with lithium pegmatites from SE Ireland. MINERAL MAG **56**, 75–83.

Wise, M. A. & Cerny, P. 1996. The crystal chemistry of the tapiolite series. CAN MINERAL **34**, 631–47.

Wobus, R. A. & Anderson, R. S. 1978. Petrology of the intrusive center at Lake George, southern Front Range, Colorado. J RES US GEOL SURV **6**, 81–94.

Wolf, M. B. & London, D. 1993. Preliminary results of HFS and RE element solubility experiments in 'granites' as a function of B and P [abstract]. EOS TRANS AM GEOPHYS UNION **74**, 343.

Wolf, M. B. & London, D. 1994. Apatite dissolution into peraluminous haplogranitic melts: an experimental study of solubilities and mechanisms. GEOCHIM COSMOCHIM ACTA **58**, 4127–45.

Wolf, M. B. & London, D. 1995. Incongruent dissolution of REE- and Sr-rich apatite in peraluminous granitic liquids: differential apatite, monazite, and xenotime solubilities during anatexis. AM MINERAL **80**, 765–75.

Wolf, M. B., London, D. & Morgan, G. B. VI 1994. Effects of boron on the solubility of cassiterite and tantalite in granitic liquids [abstract]. GEOL SOC AM ABSTR PROGR **26**, A-450.

Wood, S. A. & William-Jones, A. E. 1993. Theoretical studies of the alteration of spodumene, petalite, eucryptite, and pollucite in granitic pegmatites: exchange reactions with alkali feldspars. CONTRIB MINERAL PETROL **1114**, 255–63.

Wyllie, P. J. 1963. Applications of high pressure studies to the earth sciences. *In* Bradley, R. S. (ed.) *High pressure physics and chemistry,* Vol. 2, 1–89. New York: Academic Press.

Yurimoto, H., Kurosawa, M. & Sueno, S. 1989. Hydrogen analysis in quartz crystals and quartz glasses by secondary ion mass spectrometry. GEOCHIM COSMOCHIM ACTA **53**, 751–5.

School of Geology and Geophysics, University of Oklahoma, 100 East Boyd Street, SEC 810, Norman, OK 73019, U.S.A. E-mail: dlondon@uoknor.edu

Transactions of the Royal Society of Edinburgh: Earth Sciences, **88**, 321–329, 1996

Control of redox state and Sr isotopic composition of granitic magmas: a critical evaluation of the role of source rocks

Michel Pichavant, Tahar Hammouda and Bruno Scaillet

ABSTRACT: The current underlying assumption in most geochemical studies of granitic rocks is that granitic magmas reflect their source regions. However, the mechanisms by which source rocks control the intensive and compositional parameters of the magmas remain poorly known. Recent experimental data are used to evaluate the 'source rock model' and to discuss controls of (1) redox states and (2) the Sr isotopic compositions of granitic magmas.

Experimental studies have been performed in parallel on biotite–muscovite and tourmaline–muscovite leucogranites from the High Himalayas. Results under reducing conditions ($\log f_{O_2} =$ FMQ-0.5) at 4 kbar and variable f_{H_2O} suggest that the tourmaline–muscovite granite evolved under progressively more oxidising conditions during crystallisation, up to f_{O_2} values more than four log units above the FMQ buffer. Leucogranite magmas thus provide an example of the control of redox conditions by post-segregation rather than by partial melting processes.

Other experiments designed to test the mechanisms of isotopic equilibration of Sr during partial melting of a model crustal assemblage show that kinetic factors can dominate the isotopic signature in the case of source rocks not previously homogenised during an earlier metamorphic event. The possibility is therefore raised that partial melts may not necessarily reflect the Sr isotopic composition of their sources, weakening in a fundamental way the source rock model.

KEY WORDS: mechanisms of partial melting, redox state, Sr isotopes, experiments.

One underlying assumption in recent studies of granites and rhyolites is that source rocks exert a dominant control on the geochemical characteristics of the derived magmas. The 'source rock model' has been elaborated and championed mostly by some Australian scientists since about 1975 (see Chappell & White 1992 for a review). It has proved useful for rationalising many different aspects of granitic rocks such as whole-rock geochemistry, petrographic, mineralogical, trace element and isotopic characteristics. This model has also served to make first-order classifications of granites on the basis of differences between their source rocks and to use either granites or rhyolites as probes of deep crustal structures and lithologies. Some aspects of the source rock model have been vigorously debated (e.g. Chappell *et al.* 1987; Wall *et al.* 1987), but the general conclusion that 'granites image their sources' (Chappell 1979) has become widely accepted.

There is little doubt that granites (at least those produced by crustal anatexis without mixing with mafic magmas) inherit some of their characteristics from their source lithologies. However, little is known about how source rocks exert controls on the geochemical characteristics of the magmas formed. The first fundamental question concerns the control of the intensive parameters (chiefly P–T–f_{H_2O}–f_{O_2}) during partial melting. Are values of the intensive parameters buffered *internally* by source rocks or *externally* by some fluid? If intensive parameters such as f_{H_2O} and f_{O_2} are internally buffered, then source rocks can control the characteristics of the derived magmas. Alternatively, if f_{H_2O} and f_{O_2} are externally imposed (as if the melting event occurs, for example, as a result of the influx of external fluids), then a range of magma compositions can be generated, depending on values of the intensive parameters, and partial melting would be dominantly controlled by intensive param-

eters rather than by source rock lithologies. For example, Pichavant *et al.* (1988a, b) concluded that f_{H_2O} was internally controlled in the source region of the felsic peraluminous Macusani magmas. The mineralogical and geochemical contrasts between these magmas and S-type granites from the Lachlan Fold Belt was attributed to a fundamental difference in the respective source rock lithologies (metapelites versus metagreywackes). However, it could also be argued (see White *et al.* 1986) that these two types of magmas were produced by the melting of similar S-type sources under different f_{H_2O} regimes (externally versus internally buffered). In the same way, the Mid-Proterozoic anorogenic granites from North America have mafic silicate compositions and Fe–Ti mineralogy systematically changing across the continent, indicating low f_{O_2} in Labrador to high f_{O_2} in the SW U.S.A. (Anderson & Smith 1995). This could be explained either by different source rocks imposing different intrinsic redox states or by externally imposed f_{O_2} if the source rocks were identical.

The second important question concerns the attainment of equilibrium during partial melting. Several studies have stressed the importance of disequilibrium melting processes (e.g. Pichavant *et al.* 1988a; Rubie & Brearley 1990; Sawyer 1991; Harris & Inger 1992; Barbero *et al.* 1995), implying that the chemical composition of magmas can also be controlled by kinetic factors and reaction progress. It is worth noting that the source rock model makes no assumption about reactions and equilibrium in the source region and, therefore, is not inconsistent with disequilibrium melting. Nevertheless, the link between the geochemical characteristics of a melt and its source would be seriously weakened if crustal melting occurs under disequilibrium conditions.

The third question addresses the respective importance of

the multiple magmatic processes involved in granite (or rhyolite) genesis. Following anatexis and melt segregation, granitic magmas evolve by a variety of processes that take place during ascent and emplacement, such as crystal fractionation, wall rock assimilation, volatile degassing. These processes would probably modify the characteristics inherited at the source level. Therefore the interpretation of the geochemistry of granitic rocks in terms of source rock lithologies requires that post-segregation magmatic processes are identified and their importance carefully evaluated.

Here we focus on two parameters that play an important part in the geochemical characterisation of granitic magmas: redox state and Sr isotopic composition. Firstly, recent experimental data are used to discuss control of f_{O_2} in leucogranitic magmas. Two points are considered: the acquisition of the specific redox signature in the partial melting region and the influence of post-segregation magmatic processes. We conclude that leucogranites can record the redox state of the source mineralogy, but also that important f_{O_2} changes can result from post-segregation processes. Secondly, new experiments designed to test the kinetics of the isotopic equilibration of Sr during crustal melting are summarised. We emphasise the importance of reaction mechanisms in controlling the Sr isotopic composition of the melts and demonstrate the possibility that, under certain conditions, the Sr isotopic compositions of melts may differ from the bulk Sr isotopic composition of the source rock.

1. Redox state of leucogranitic magmas

1.1. Redox state of peraluminous leucogranites

The redox states of granite magmas are increasingly considered to reflect those of their source regions (Carmichael 1991; Blevin & Chappell 1992). Silicic magmas commonly show trends of decreasing f_{O_2} with temperature, parallel or nearly parallel to oxygen buffer curves (Carmichael 1991). In addition, there is a systematic relation between f_{O_2} and the type of ferromagnesian silicates present, suggesting that the magmas are little affected by post-segregation processes and, consequently, that their redox states reflect the conditions of generation (Carmichael 1991). However, detailed case studies have documented large f_{O_2} changes and trends of magmatic oxidation during differentiation (i.e. increasing f_{O_2} with decreasing temperature, trends nearly orthogonal to oxygen buffer curves, e.g. Czamanske & Wones 1973). In this paper, we use our experimental work on peraluminous leucogranites to test these two models.

Leucogranites (and their few volcanic equivalents) crystallise from primary magmas derived from dominantly crustal, metapelitic lithologies (Le Fort *et al.* 1987; Pichavant *et al.* 1988b; Inger & Harris 1993). These granites are widespread in the Hercynian belt of Europe, in the Himalayas (Le Fort 1981; Pichavant & Manning 1984; Le Fort *et al.* 1987; Scaillet *et al.* 1990; Inger & Harris 1993) and also in North America (Nabelek *et al.* 1992). Chemically, they are both felsic and strongly peraluminous, which distinguishes them from typical S-type granites from the Lachlan Fold Belt (Pichavant *et al.* 1988a, b). Their mineralogy is characterised by several specific features: (1) the presence of aluminium silicates (sillimanite, andalusite) and Al-rich phases (muscovite, biotite, tourmaline and rare cordierite, garnet and hercynite) in the crystallisation sequence; (2) a low abundance (and absence in some instances) of ilmenite and a lack of primary magnetite and sulphides. Stable and radiogenic isotope data typically yield crustal signatures with no indication of mixing with mafic magmas. The undoubtably crustal nature of the source rocks and the absence of a mantle component make peraluminous leucogranites among the best igneous rocks to test the influence of crustal sources on the geochemistry of derived magmas.

Nevertheless, determination of the redox states of these granites is difficult. Mineral assemblages are not suited for direct estimations of f_{O_2}, using, for example, ilmenite–magnetite equilibria. On the basis of the presence of ilmenite alone and of the low contents of Fe^{3+} of ferromagnesian phases (biotite, hercynite, ilmenite), it is commonly believed that these granites equilibrated at f_{O_2} equal or below that of the FMQ buffer (e.g. Pichavant *et al.* 1988a; see also Powers & Bohlen 1985). Zen (1985) formulated an oxybarometer for two-mica granites from the coexistence of biotite–garnet–muscovite–quartz–magnetite and found that this assemblage typically equilibrates under f_{O_2} intermediate between the FMQ and NNO buffers. However, such an assemblage is uncommon (it has been encountered neither in the Hercynian nor in the Himalayan examples). Therefore it is of little use in determining f_{O_2} in these leucogranites. Calculations of f_{O_2} based on the homogeneous equilibrium between the FeO and Fe_2O_3 melt components require samples of quenched melts from the peraluminous magmatic series, which are exceedingly rare. Such calculations have been performed on the Macusani obsidian glasses JV1 and JV2 using the regression from Kilinc *et al.* (1983) and compositional data from Pichavant *et al.* (1987). At 650°C, calculated $\log f_{O_2}$ values range between -19.45 (FMQ-1) and -19.94 (FMQ-1.5; FMQ buffer calculated at 1 bar from Chou 1987). These f_{O_2} values are consistent with estimates using phenocryst phases from the parental Macusani ash flow tuffs (f_{O_2} intermediate between the FMQ and WM buffers; Pichavant *et al.* 1988a).

In summary, although the redox states of peraluminous leucogranites and related volcanic rocks are not accurately known, the available information suggests moderately reducing f_{O_2} at or below the FMQ buffer.

1.2. Experimental studies

On the basis of this conclusion, the crystallisation sequence of natural leucogranites from the High Himalayas has been experimentally reproduced under moderately reducing conditions (FMQ-0.5), at 4 kbar (i.e. at the pressure of emplacement of the magmas), for variable f_{H_2O} and at temperatures ranging between 663 and 803°C (Scaillet *et al.* 1995a). The experiments were performed in parallel on two leucogranite samples, one from the Manaslu pluton (biotite–muscovite) and the other from the Gangotri massif (muscovite–tourmaline). These two samples have very similar whole-rock compositions, yet they differ in their ferromagnesian silicate phases. Biotite is nearly absent in the Gangotri sample: only scarce, small biotites are found included in tourmaline. Other samples show textural evidence of a reaction relationship between biotite and tourmaline (Fig. 1). The T–H_2O in melt phase diagram obtained for the Gangotri leucogranite is shown in Figure 2. It is characterised by a wide stability field for biotite, which is the liquidus phase for melt H_2O contents above 7.5 wt% (Fig. 2). The presence of biotite in the experiments contrasts with its absence in the starting granite sample. Moreover, the phase diagram implies that biotite is stable down to solidus conditions and should be a major mineral phase of the granite. This result is inconsistent with the mineralogy of the Gangotri granite sample, suggesting that biotite reacted out at some stage of the magmatic evolution, as indicated by the textural observations (Fig. 1). Hence the phase diagram (Fig. 2) only models an early, near-liquidus biotite-present stage and not the final (biotite-absent) conditions of solidification of the granite. Given our current knowledge about biotite and tourmaline stabilities in peralumi-

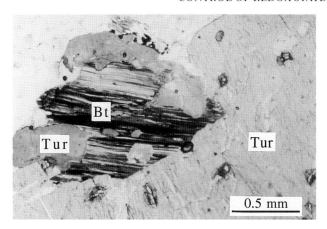

Figure 1 Textures showing replacement of biotite by tourmaline in the Gangotri leucogranite (peritectic reaction biotite + melt = tourmaline).

Table 1 Composition of the Manaslu and Gangotri leucogranites used in the experiments (Scaillet *et al.* 1995a).

	Manaslu DK 89	Gangotri GB 4
SiO_2	73·04	72·94
Al_2O_3	15·32	15·57
Fe_2O_3	0·17	0·52
FeO	0·74	0·27
MnO	0·01	0·01
MgO	0·20	0·14
CaO	0·85	0·57
Na_2O	3·85	4·56
K_2O	4·96	4·14
TiO_2	0·13	0·06
P_2O_5	0·14	0·19
CO_2	0·05	0·06
H_2O	0·65	0·64
F	0·09	0·11
Total	100·2	99·73
F = 0	0·04	0·05
Total	100·16	99·68
Li (ppm)	163	367
B	14	890

nous silicic liquids (see Scaillet *et al.* 1995a), the most likely explanation is that crystallisation of the Gangotri granite occurred under progressively increasing f_{O_2}. In particular, it is unlikely that biotite disappearance can be explained by increasing boron concentrations in the melt during crystallisation, because biotite and tourmaline are co-stable (with melt) in our experiments (Fig. 2) and also because such a mechanism would leave unexplained the Fe_2O_3/FeO data (see later). We conclude that an oxidation process affected the Gangotri granite magma. In contrast, the mineral assemblages and compositions of the Manaslu granite are adequately reproduced by our experiments at $\log f_{O_2} = FMQ - 0.5$ (Fig. 2).

Whole-rock analyses provide confirmation that an oxidation process took place during crystallisation of the Gangotri magma. The Fe_2O_3/FeO ratio of the Gangotri granite sample is elevated (1·92, Table 1) and sharply contrasts with that of the Manaslu sample (0·22, Table 1), although the two rocks have nearly identical total Fe content (FeO$_t$ 0·74 and 0·89 wt%, Table 1). T–$\log f_{O_2}$ paths of the two leucogranite magmas have been calculated (Fig. 3). The two paths are markedly different

for the two granites considered. In the case of the Manaslu sample, the T–$\log f_{O_2}$ path remains approximately parallel to the FMQ buffer, illustrating an evolution characterised by 'self-buffering' of the redox state during crystallisation (Carmichael & Ghiorso 1990). In contrast, the trend for the Gangotri magma cross-cuts the reference oxygen buffer curves at high angles toward more oxidising conditions. It also cuts the biotite–magnetite–sanidine equilibria, thus illustrating how an oxidation trend could lead to the progressive disappearance of biotite, as noted in the Gangotri sample. The Gangotri path corresponds to a second type of evolution, characterised by an 'unbuffered' redox state of the magma during crystallisation, with f_{O_2} more than four log units above the FMQ buffer attained at the end of the magmatic evolution. The lack of magnetite in the granite, despite conditions of crystallisation above the FMQ buffer, is explained by the incorporation of Fe^{3+} in muscovite and tourmaline.

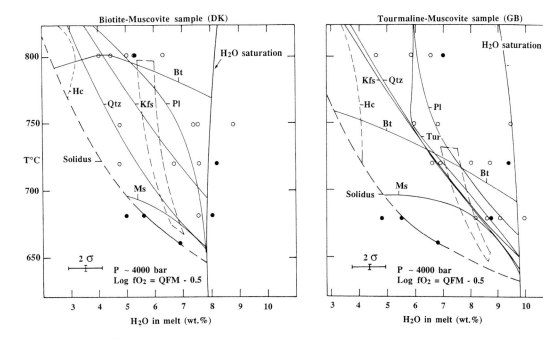

Figure 2 T–H_2O in melt phase diagrams for the biotite–muscovite (Manaslu) and muscovite–tourmaline (Gangotri) leucogranites at 4 kbar and $\log f_{O_2} = FMQ - 0.5$ (from Scaillet *et al.* 1995a). Note the presence of biotite and muscovite in the experimental crystallisation sequence of both granites. Tourmaline possesses a stability field in the Gangotri sample only, consistent with the boron contents of the respective starting samples (Table 1).

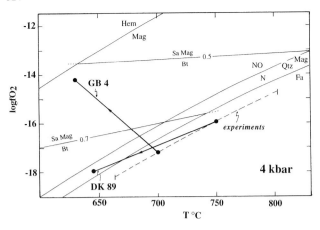

Figure 3 T–logf_{O_2} paths of the Manaslu (DK 89) and Gangotri (GB 4) magmas during crystallisation, compared with reference oxygen buffer curves and selected biotite equilibria (labelled with Fe/(Fe + Mg) biotite composition). For conditions of emplacement and crystallisation of the magmas, temperatures are obtained from plagioclase cores (750 and 700°C, respectively; see Scaillet *et al.* 1995a) and f_{O_2} values are assumed to be identical to our experimental values (FMQ – 0·5; see text). For conditions at the end of the magmatic evolution, temperatures are the H_2O-saturated solidus temperatures (645 and 630°C respectively; see Fig. 2) and logf_{O_2} values are calculated from Kilinc *et al.* (1983) using bulk granite analyses and Fe_2O_3/FeO ratios (see text, Table 1 and Scaillet *et al.* 1995a).

1.3. Discussion

The combined experimental, mineralogical and whole-rock analytical data are consistent with an evolution of the Manaslu granite under relatively reducing conditions, at or below FMQ during the entire magmatic stage. In contrast, for the Gangotri granite, the data demonstrate an early near-liquidus, relatively reducing biotite stage, followed by progressively more oxidising conditions during which biotite reacted out. Below, we explore the possibility that the relatively reducing stage recorded in both granites could reflect initial f_{O_2} conditions imposed by source rocks.

The similarity of the peraluminous leucogranites cropping out along more than 2000 km in the High Himalayas has been emphasised previously (Le Fort *et al.* 1987). One important implication is that these granites all broadly have the same origin and evolution, which in turn makes it unlikely that they were generated from source rocks with markedly different lithologies and mineralogical compositions. Previous studies have identified protoliths of the Himalayan leucogranites as mica schists and pelitic gneisses (Le Fort *et al.* 1987; Inger & Harris 1993). Because of the likely presence of carbonaceous material, pelitic rocks are on average reduced (e.g. Miyashiro 1964; Ohmoto & Kerrick 1977). The presence of graphite would buffer f_{O_2} in the source rocks through equilibrium with CO_2 (gas or melt). Note that, if graphite is present in sufficiently large amounts to buffer f_{O_2} through equilibrium with CO_2, then melting is likely to occur in the presence of a vapour phase because of the low solubility of CO_2 (and also of H_2, CO, CH_4) in granitic magmas. The Himalayan mica schists and gneisses are commonly graphite-bearing (France-Lanord 1987), implying that the postulated source rocks of the Himalayan leucogranites may have low intrinsic f_{O_2}. However, it is worth noting that there are examples of high-grade metapelitic rocks equilibrated at f_{O_2} below the FMQ buffer and which contain no graphite (e.g. Bhattacharya & Sen 1986). This stresses the need for more detailed studies on how reducing conditions are established in high-grade metamorphic rocks, particularly in graphite-free assemblages. In addition, the Himalayan cross-section does not directly expose rocks from the source zone of the granites,

which would be located down-dip and presumably equilibrated at higher metamorphic grades. Some uncertainty thus persists about mineral assemblages and the presence of graphite in the source region. Nevertheless, the Gangotri leucogranite contains elevated amounts of CO_2 (between 400 and 800 ppm; Scaillet *et al.* 1990, 1995c), suggesting that C-bearing sources are involved in magma generation. These CO_2 contents (if assumed magmatic), together with H_2O contents deduced from phase equilibria (Scaillet *et al.* 1995a) can be used to calculate f_{O_2} from homogeneous equilibria in the C–O–H system at fixed P and T. Results in the 700–800°C range and for 4 kbar are close to FMQ – 0·5, i.e. are consistent with other f_{O_2} estimates and with our experimental redox conditions (Fig. 3). Therefore, although the role of graphite in buffering f_{O_2} to low values in the source region of the leucogranite magmas is difficult to ascertain, the early moderately reducing biotite stage marked in both the Manaslu and Gangotri leucogranites is consistent with a control of their redox states by their metapelitic source rocks. Leucogranite magmas thus show permissive evidence for a control of their initial redox states by source rocks. In contrast, the situation is less clear for other granitoid magmas, such as the anorogenic granites associated with anorthosite complexes, which equilibrated at f_{O_2} 1–3 log units below the FMQ buffer, and whose strongly reduced nature is not attributed to the presence of graphite in their lower crustal source region (Emslie & Stirling 1993; Emslie 1995, pers. comm.). This raises the question of knowing how such reduced conditions are reached in their source region if graphite is not the reducing agent. Reduced fluids of mantle origin may be involved (Emslie 1995, pers. comm.). If confirmed, then redox states would be externally controlled in the source region of these granites.

The oxidising evolution recorded in the Gangotri granite demonstrates that the redox states of granitic magmas can be controlled by magmatic processes other than those occurring in the source region (Fig. 3; see also Czamanske & Wones 1973). Initial crystallisation of the High Himalayas leucogranite magmas may have started during ascent, although the available evidence favours the emplacement of the magmas as crystal-poor bodies at temperatures close to their liquidus (Scaillet *et al.* 1990; Inger & Harris 1993; Scaillet *et al.* 1995a, b). As the oxidation mechanisms occurred contemporaneously with crystallisation, it must have operated within the magma chambers at pressures of ~4 kbar.

Oxidising trends comparable with that described for the Gangotri granite are not exceptional and probably fairly common in felsic peraluminous magmas. Crystallisation of magmatic cassiterite has been described in highly fractionated leucogranites (Beauvoir granite, Cuney *et al.* 1992), pegmatites from Thailand (Linnen *et al.* 1992) and leucogranitic pegmatites from the High Himalaya (Visonà & Zantedeschi 1994) and constitutes a good mineralogical indicator of high oxidation conditions. SnO_2 solubility is very high in granitic melts under reducing conditions (0·89 wt% at FMQ, 850°C and 2 kbar for a haplogranite melt close to H_2O- saturation; Linnen *et al.* 1995) and is negatively correlated with increasing f_{O_2}. For cassiterite saturation to occur in natural tin-rich magmas, redox conditions have to be more oxidising than the NNO buffer (Linnen *et al.* 1995), thus suggesting a magmatic oxidation as in the Gangotri granite. Such oxidising evolutions have been documented in other types of granites, such as in the pyroxene–amphibole–biotite monzonite–granodiorite–granite complex of Finmarka (Czamanske & Wones 1973) and are probably of general importance.

Finally, mechanisms of oxidation and, more generally, controls of redox states in magma chambers remain poorly understood and still subject to debate (Carmichael & Ghiorso

1990). For hydrous magmas such as leucogranitic magmas, typical molar concentrations of H exceed by about 50 times the molar concentrations of Fe. Thus f_{O_2} will not be controlled by the Fe_2O_3/FeO equilibrium between melt components and mechanisms of oxidation such as fractionation of Fe^{2+}-rich phases (Carmichael 1991) cannot explain the observed oxidising evolutions. We suggest that the redox state of leucogranitic magmas is buffered by equilibria between C–O–H species, either dissolved in the melt or present in a coexisting fluid phase. Two possible mechanisms can account for a magmatic oxidation such as in the Gangotri granite: (1) the diffusive loss of H_2 (Sato 1978) and (2) open system fractionation of a C–O–H fluid phase containing H_2 (Candela 1986) and/or other reduced species (CO, CH_4).

2. Sr isotopic equilibration during partial melting

The fundamental assumption in the application of Sr isotopes to determine the compositions and crustal residence times of granite source rocks is that the equilibrium distribution of ^{86}Sr and ^{87}Sr between melt and crystals is attained during partial melting, i.e. that the $^{87}Sr/^{86}Sr$ ratio of the melt is identical to that of the source rocks at the time of the partial melting event. Such use of Sr isotopes is based on the assumption that diffusion rates of Sr are fast enough, compared with commonly accepted time-scales of partial melting events, for the homogenisation of $^{87}Sr/^{86}Sr$ between the minerals in a source rock undergoing high temperature metamorphism and anatexis, as discussed by Hofmann and Hart (1978) for the mantle. To provide detailed information on how a crustal anatectic melt acquires its Sr isotopic signature, we have experimentally tested the behaviour of Sr isotopes during the initial stages of partial melting of an isotopically heterogeneous model source rock (Hammouda 1995; Hammouda et al., unpublished data). We focused on the mechanisms and kinetics of equilibration of Sr isotopes between the source and the melt.

2.1. Experiments and results

Melting experiments were performed at 1 atm between 1100 and 1250°C and for run durations between 1 and 168 h on assembled pairs of single crystals of plagioclase and fluorphlogopite. The plagioclase used is a natural crystal of gem quality ($Ab_{31-33} An_{66-68}Or_{0-1}$, $^{87}Sr/^{86}Sr = 0.70$, Sr = 632 ppm). Fluorphlogopite, a mica phase stable at 1 atm up to its melting point (about 1400°C), was synthesised as single crystals using a batch melting technique allowing the addition of Sr of known $^{87}Sr/^{86}Sr$ composition (Hammouda et al. 1995). The Sr concentration in the mica (90 ppm) was chosen to match the $K_{d_{Sr}}$ between mica and plagioclase and the Sr isotopic composition of the mica ($^{87}Sr/^{86}Sr = 4.0$) selected to be in sharp contrast with that of plagioclase. The experimental charges were examined by scanning electron microscopy and analysed by electron (major elements) and ion (Sr concentrations and isotopic compositions) microprobes. Complete experimental and analytical details are given in Hammouda (1995) and Hammouda et al. (unpublished data).

Partial melting of the plagioclase–fluorphlogopite assemblage occurred for temperatures above 1150°C. In addition to melt, the reaction produced a calcic plagioclase and spinel (Fig. 4). The observed change in the kinetics of melting between 1175 and 1200°C is consistent with overstepping of a critical temperature (Tsuchiyama 1985) where the kinetics of melting is controlled by atomic diffusion within the melt. Below 1200°C, the kinetics of melting is much slower and probably controlled by a complex process involving dissolution of the starting plagioclase within the melt and crystallisation of the new plagioclase (Hammouda et al. unpublished data).

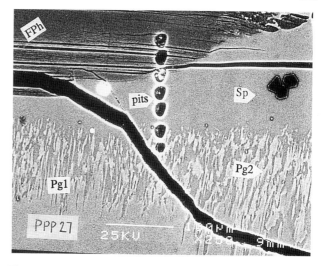

Figure 4 Scanning electron photomicrograph of partial melting run products and textures (1200°C, 10 h, from Hammouda 1995). L, Glass; FPh, fluorphlogopite; Pg1, starting plagioclase; Pg2, newly formed plagioclase; and Sp, spinel. The picture shows the initial stage of the melting reaction with glass located between residual mica and plagioclase. Note the presence of spinel crystals within the glass and without any preferential spatial relation with either mica or plagioclase. In contrast, the newly formed plagioclase grows inwards by lamellae and patch development controlled by the crystallography and orientation of the starting plagioclase. Also shown are the ion microprobe pits. Scale bar is 100 μm.

Mass fractions of products and reactants of the partial melting reaction (estimated from the major element data) show that fluorphlogopite is preferentially consumed over plagioclase during the early stages of the reaction.

The quenched melts have homogeneous major element compositions at 1175°C. Owing to the low amounts of liquid produced at that temperature, it was not possible to obtain reliable Sr analyses of these melts. At 1200°C and above, glasses are chemically zoned from the mica to the plagioclase interface. This is true not only of the major elements (Fig. 5), but also of Sr concentrations and isotopic compositions (Fig. 6). The newly formed plagioclase has variable Sr contents and $^{87}Sr/^{86}Sr$ ratios that differ from the starting plagioclase. Both textural (the presence of a sharp interface with the starting plagioclase) and isotopic data (abrupt change in $^{87}Sr/^{86}Sr$ across the interface with the starting plagioclase) suggest that the new plagioclase grew from the melt rather than by diffusion from the starting plagioclase. When fluorphlogopite had totally disappeared, the zonation in the liquid became less pronounced (Fig. 7) and $^{87}Sr/^{86}Sr$ of the glass decreased with time at a given temperature. However, the $^{87}Sr/^{86}Sr$ of the quenched melts remained significantly higher (up to twice) than that of the bulk 'source' (from 0.729 to 0.824 depending on the mass fractions of fluorphlogopite and plagioclase in the starting pair) even for the longest experimental runs. Note that our experiments were designed to investigate the initial steps of the melting reaction and consequently remained out of equilibrium, even for the longest experimental durations (96 h at 1200 and 1250°C) because some initial plagioclase did not react. These experiments provide examples of melts whose Sr isotopic compositions differ from that of their bulk source assemblage, thus indicating that homogenisation of Sr isotopes did not occur during melting. What was the process that controlled the Sr isotopic composition of these experimental melts?

Several lines of evidence suggest that the Sr isotopic composition of the melts at 1200 and 1250°C was controlled by the initial isotopic composition of the reacting minerals and by the proportion of minerals consumed in the melting

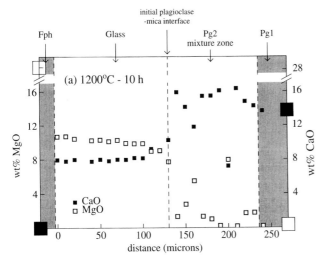

Figure 5 Electron microprobe traverses for CaO and MgO across the experimental charge illustrating a case where fluorphlogopite is still present at the end of the experiment (1200°C, 10 h, from Hammouda 1995). Glass, quenched liquid; Fphlog, residual fluorphlogopite; Pg1, residual plagioclase; and Pg2, newly formed plagioclase. Large symbols refer to the concentrations in the starting minerals and smaller symbols to the concentrations measured after the experiment. Note (1) the chemical zonation in the glass layer and (2) the agreement between the composition of the residual and starting plagioclase. The scatter in both CaO and MgO concentrations in the Pg2 mixture zone corresponds to mixtures either between the two plagioclases or between plagioclase and glass (see the textural relations in this zone in Fig. 4).

reaction. The $^{87}Sr/^{86}Sr$ ratios in the residual phases did not change within analytical uncertainty from values in the starting products (Figs 6, 7). This rules out processes of isotopic homogenisation by the exchange of isotopes between residual minerals and newly formed phases, including melt, as factors in controlling the Sr isotopic composition of the melt. Mass balance calculations closely reproduce the composition of the analysed melts (Hammouda 1995). During the initial stages of melting, the Sr isotopic composition of the melt was dominated by the dissolution of fluorphlogopite, which has an elevated $^{87}Sr/^{86}Sr$ ratio. With reaction progress, the mica was eventually consumed and the contribution of the dissolving plagioclase became dominant, yielding to progressively lower values of $^{87}Sr/^{86}Sr$ ratio in the melt.

These experiments stress the importance of the melting reaction in initially controlling the Sr isotopic composition of the melt formed in a given source assemblage that is not isotopically homogeneous. In a general way, it is critical to evaluate the kinetics of melting and to compare them with the kinetics of Sr homogenisation by diffusion in crystals. If melting proceeds faster than the Sr diffusion rate in crystals, then liquids with Sr isotopic composition that is different from the bulk 'source' can be produced, as in the experiments. This situation also leaves a residual phase assemblage isotopically heterogeneous when melting stops. In contrast, if the melting rate is slower than Sr diffusion in crystals, then isotopic homogenisation between the different mineral phases of the source becomes possible during melting. Then, melts that reflect the Sr isotopic composition of their bulk source can be produced, although this situation has not yet been investigated experimentally (Hammouda 1995; Hammouda *et al.* unpublished data).

It is necessary to evaluate the time-scales of isotopic equilibration between the melt and the residual phases in a situation analogous to the experiments at 1200 and 1250°C described earlier. If melting stops, then isotopic exchange between residual minerals and melt must take place to

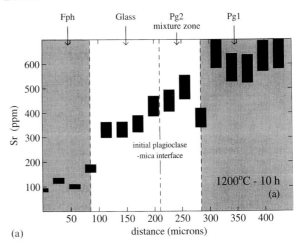

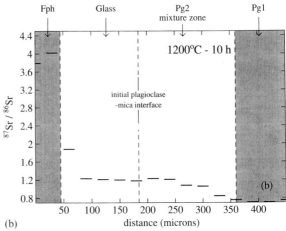

Figure 6 Ion microprobe traverses for (a) Sr concentrations and (b) Sr isotopic compositions illustrating a case where fluorphlogopite is still present at the end of the experiment (1200°C, 10 h, from Hammouda 1995). Same abbreviations as in Figure 5. Note the zonation of the glass layer in both Sr concentration and Sr isotopic composition. In contrast, the residual phases retain their initial Sr concentration and isotopic composition.

eliminate the isotopic gradients resulting from the disequilibrium melting. Isotopic equilibration will proceed by the exchange of Sr isotopes between the melt and the residual phases. If we assume a diffusion controlled process, plagioclases of 0·1 cm at 1200–1250°C (as in the experiments) would equilibrate in ≈ 10² a (diffusion data for Sr in plagioclase taken from Giletti & Casserly 1994; see also Hammouda *et al.* unpublished data). For temperatures of 800–1000°C, 10⁴–10⁶ a are necessary to equilibrate Sr isotopes by diffusion between the melt and residual plagioclases 0·1–1 cm in size.

2.2. Implications

These experimental results show that the Sr isotopic composition of a melt may differ from that of the bulk source when the source rock has not been homogenised for Sr isotopes before melting (either by a metamorphic or a first partial melting episode). This situation is plausible geologically, judging from the examples of preserved evidence of non-homogenisation of Sr isotopes, even after having experienced high-grade metamorphism (above 700°C) or partial melting (e.g. Peucat 1986; Deniel *et al.* 1987; Inger & Harris 1993; Barbero *et al.* 1995). Sr isotopic heterogeneities on the centimetre (hand specimen) scale have been documented for migmatites (Peucat & Martin 1985; Barbero *et al.* 1995). Because isotopic homogenisation is dramatically assisted by interstitial melt (Hofmann & Hart 1978; Peucat 1986), we now compare the time-scales of partial melting and melt

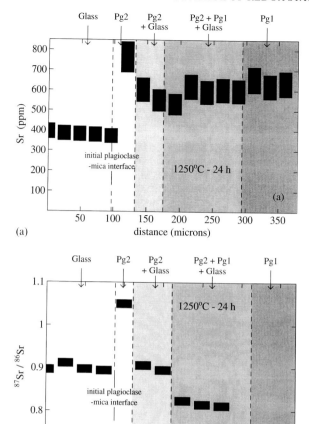

Figure 7 Ion microprobe traverses for (a) Sr concentrations and (b) Sr isotopic compositions illustrating a case where fluorphlogopite has totally reacted out (1250°C, 24 h, from Hammouda 1995). Same abbreviations as in Figure 5. Note the homogeneity of the glass layer in both Sr concentration and Sr isotopic composition. The residual plagioclase retains its initial Sr concentration and isotopic composition.

extraction with the time-scales of Sr isotopic homogenisation calculated earlier to assess the practical implications of our experimental results.

Recently proposed models of crustal melting by the intrusion and crystallisation of hot mafic magmas (Huppert & Sparks 1988) have become widely used to explain the generation of granitic magmas, in particular those formed at high temperatures. The basalt-induced crustal melting model is characterised by short time-scales (10^2–10^3 a) and high temperatures (well above the water-saturated solidus), and hence by high heat fluxes in the source region. Under these conditions, the stable melting equilibria are likely to be overstepped (Pichavant et al. 1988a), corresponding to conditions of melting above the critical temperature as in the experiments described earlier, and yielding a regime of partial melting that is kinetically controlled by atomic diffusion in the melt. Such a regime would make possible the production of anatectic melts with Sr isotopic signatures that are different from that of the bulk source rocks. Examples demonstrating that such mechanisms indeed occur in nature are provided by partially fused xenoliths in mafic magmas (Maury & Bizouard 1974; Pushkar & Stoeser 1975). There are also examples of batholitic-scale magmas with Sr isotopic disequilibria inherited from partial melting processes—such as, for example, the Macusani Volcanics from SE Peru (Pichavant et al. 1988b), the San Vicenzo rhyolites from Tuscany, Italy (Ferrara et al. 1989; Feldstein et al. 1994) and the anatectic granitoids from Toledo, Spain (Barbero et al.

1995). In the San Vicenzo rhyolites, the Sr isotopic heterogeneities have been explained by magma mixing (Feldstein et al. 1994), but could also be interpreted as resulting from disequilibrium partial melting as was first suggested by Ferrara et al. (1989).

Current views on the ascent of granitic magmas from their sources also tend to favour rapid transfer through dykes (e.g. Clemens & Mawer 1992; Scaillet et al. 1995b). Although such a mechanism may apply only to certain types of granite magmas (e.g. leucogranites), time-scales of magma ascent through dykes could be as short as $\approx 10^{2-3}$ a (Scaillet et al. 1995b). These time-scales are of the same order of magnitude as (or even shorter than) those calculated here for the equilibration of Sr isotopes between the melt and residual phases. This shows that magma segregation and extraction may occur before isotopic equilibrium is attained between the melt and the residual mineral assemblage (Sawyer 1991). However, clearly such fast ascent rates would only preserve isotopic disequilibria and not cause them. The key factor for producing Sr isotopic disequilibria resides in the kinetics of melting, which have to be faster than the kinetics of Sr exchange or equilibration between phases.

3. Discussion and conclusions

Examination of the redox states of leucogranitic magmas shows, using the example of the Himalayan leucogranites, that source rocks appear to initially control the redox states. However, additional work on how f_{O_2} is controlled during progressive metamorphism and is imposed on the melts during anatexis is needed, particularly to explain moderately to strongly reduced conditions (<FMQ) in graphite-free crustal sources. We have also illustrated, using a sample of the Gangotri granite, how the imprint of the source can be strongly modified by post-segregation processes occurring at the level of crystallisation. The idea that redox states of granite magmas reflect those of their source region (Carmichael 1991) should be viewed only as a general framework, whose validity must be established on a case by case basis.

Experiments have also provided some insight on how the Sr isotopic signature of a granite magma is acquired in the source region. $^{87}Sr/^{86}Sr$ ratios of the magmas can be controlled by the kinetics of partial melting, and by the individual melting rates of minerals, during the early stages of partial melting. The results for Sr equilibration clearly demonstrate that source rocks do not impose the sole control on the geochemical properties of granitic magmas. The need to consider kinetic factors weakens in a fundamental way the source rock model as a range of isotopic compositions of magmas can be produced from a given source rock undergoing anatexis under a given set of P–T–f_{H_2O}–f_{O_2} conditions. Magmas formed under disequilibrium conditions should carry information about melting mechanisms and rates. Provided that the kinetic information is recognised, granites could be used as indicators of heat transfer and kinetics of melting processes as well as images of their source rocks.

Acknowledgements

The ideas developed in this paper have benefited from discussions over several years with F. Albarède, P. Barbey, A. Brearley, B. Charoy, C. France-Lanord, F. Holtz, P. Le Fort and J.-M. Montel. J. Roux and M. Chaussidon also contributed to the experiments and analyses. The experimental study on the behaviour of Sr isotopes during partial melting was supported by the DBT Program of CNRS-INSU.

P Nabelek is gratefully acknowledged for his particularly thorough review.

References

Anderson, J. L. & Smith, D. R. 1995. The effects of temperature and fO$_2$ on the Al-in-hornblende barometer AM MINERAL **80**, 549–59.

Barbero, L., Villaseca, C., Rogers, G. & Brown, P. E. 1995. Geochemical and isotopic disequilibrium during crustal melting: an insight from the anatectic granitoids from Toledo, Spain. J GEOPHYS RES **100**, 15 745–65.

Bhattacharya, A. & Sen, S. K. 1986. Granulite metamorphism, fluid buffering and dehydration melting in the Madras charnockites and metajelites. J PETROL **27**, 1119–1141.

Blevin, P. L. & Chappell, B. W. 1992. The role of magma sources, oxidation states and fractionation in determining the granite metallogeny of eastern Australia. TRANS R SOC EDINBURGH EARTH SCI **83**, 305–16.

Candela, P. 1986. The evolution of aqueous vapour from silicate melts: effects on oxygen fugacity. GEOCHIM COSMOCHIM ACTA **50**, 1205–11.

Carmichael, I. S. E. 1991. The redox state of basic and silicic magmas: a reflection of their source regions? CONTRIB MINERAL PETROL **106**, 129–41.

Carmichael, I. S. E. & Ghiorso, M. 1990. The effect of oxygen fugacity on the redox state of natural liquids and their crystallizing phases. In Nicholls, J. & Russel, J. K. (eds) Modern methods of igneous petrology: understanding magmatic processes. REV MINERAL **24**, 191–212.

Chappell, B. W. 1979. Granites as images of their source rocks. GEOL SOC AM ABSTR PROGRAM **11**, 400.

Chappell, B. W. & White, A. J. R. 1992. I- and S-type granites in the Lachlan Fold Belt. TRANS R SOC EDINBURGH EARTH SCI **83**, 1–26.

Chappell, B. W., White, A. J. R. & Wyborn, D. 1987. The importance of residual source material (restite) in granite petrogenesis. J PETROL **28**, 1111–38.

Chou, I.-M. 1987. Oxygen buffer and hydrogen sensor techniques at elevated pressures and temperatures. In Ulmer, G. C. & Barnes, H. L. (eds) Hydrothermal experimental techniques, 61–99. New York: Wiley.

Clemens, J. D. & Mawer, C. K. 1992. Granitic magma transport by fracture propagation. TECTONOPHYSICS **204**, 339–60.

Cuney, M., Marignac, C. & Weisbrod, A. 1992. The Beauvoir topaz-lepidolite albite granite (Massif Central, France): the disseminated magmatic Sn–Li–Ta–Nb–Be mineralization. ECON GEOL **87**, 1766–94.

Czamanske, G. K. & Wones, D. R. 1973. Oxidation during magmatic differentiation, Finnmarka complex, Oslo area, Norway. Part 2, the mafic silicates. J PETROL **14**, 349–80.

Deniel, C., Vidal, P., Fernandez, A., Le Fort, P. and Peucat, J-J. 1987. Isotopic study of the Manaslu granite (Himalaya, Nepal): inferences on the age and source of Himalayan leucogranites. CONTRIB MINERAL PETROL **96**, 78–92.

Emslie, R. F. & Stirling, J. A. R. 1993. Rapakivi and related granitoids of the Nain Plutonic Suite: geochemistry, mineral assemblages and fluid equilibria. CAN MINERAL **31**, 821–47.

Feldstein, S. N., Halliday, A. N., Davies, G. R. & Hall, C. M. 1994. Isotope and chemical microsampling: constraints on the history of an S-type rhyolite, San Vicenzo, Tuscany, Italy. GEOCHIM COSMOCHIM ACTA **58**, 943–58.

Ferrara, G., Pietrini, R., Serri, G. & Tonarini, S. 1989. Petrology and isotope geochemistry of San Vicenzo rhyolites. BULL VOLCANOL **151**, 379–88.

France-Lanord, C. 1987. Chevauchement, métamorphisme et magmatisme en Himalaya du Népal central. Etude isotopique H,C,O. Thesis, University of Nancy.

Giletti, B. J. & Casserly, J. E. D. 1994. Strontium diffusion kinetics in plagioclase feldspars. GEOCHIM COSMOCHIM ACTA **58**, 3785–93.

Hammouda, T. 1995. Mécanismes et cinétique de fusion partielle d'assemblages à fluor-phlogopite: modélisation, expérimentation et implications géochimiques. Thesis, University of Orléans.

Hammouda, T., Pichavant, M., Barbey, P. & Brearley, A. J. 1995. Synthesis of fluorphlogopite single crystals. Application to experimental studies. EUR J MINERAL **7**, 1381–7.

Harris, N. B. W. & Inger, S. 1992. Trace element modelling of pelite-derived granites. CONTRIB MINERAL PETROL **110**, 46–56.

Hofmann, A. W. & Hart, S. R. 1978. An assessment of local and regional isotopic equilibrium in the mantle. EARTH PLANET SCI LETT **38**, 44–62.

Huppert, H. E. & Sparks, R. S. J. 1988. The generation of granitic magmas by intrusion of basalt into the continental crust. J PETROL **29**, 599–624.

Inger, S. & Harris, N. 1993. Geochemical constraints on leucogranite magmatism in the Langtang valley, Nepal Himalaya. J PETROL **34**, 345–68.

Kilinc, A., Carmichael, I. S. E., Rivers, M. L. & Sack, R. O. 1983. The ferric–ferrous ratio of natural silicate liquids equilibrated in air. CONTRIB MINERAL PETROL **83**, 136–40.

Le Fort, P. 1981. Manaslu leucogranite: a collision signature of the Himalayas. A model for its genesis and emplacement. J GEOPHYS RES **86**, 10 545–68.

Le Fort, P., Cuney, M., Deniel, C. France-Lanord, C., Sheppard, S. M. F., Upreti, B. N. & Vidal, P. 1987. Crustal generation of the Himalayan leucogranites. TECTONOPHYSICS **134**, 39–57.

Linnen, R. L., Williams-Jones, A. E. & Martin, R. F. 1992. Evidence of magmatic cassiterite mineralization at the Nong Sua aplite–pegmatite complex, Thailand. CAN MINERAL **30**, 739–61.

Linnen, R. L., Pichavant, M., Holtz, F. & Burgess, S. 1995. The effect of fO$_2$ on the solubility, diffusion, and speciation of tin in haplogranitic melt at 850°C and 2 kbar. GEOCHIM COSMOCHIM ACTA **59**, 1579–88.

Maury, R. C. & Bizouard, H. 1974. Melting of acid xenolith into a basanite: an approach to the possible mechanisms of crustal contamination. CONTRIB MINERAL PETROL **48**, 275–86.

Miyashiro, A. 1964. Oxidation and reduction in the Earth's crust with special reference to the role of graphite. GEOCHIM COSMOCHIM ACTA **28**, 717–29.

Nabelek, P., Russ-Nabelek, C. & Denison, J. R. 1992. The generation and crystallization conditions of the Proterozoic Harney Peak leucogranite, Black Hills, South Dakota, USA: petrologic and geochemical constraints. CONTRIB MINERAL PETROL **110**, 173–91.

Ohmoto, H. & Kerrick, D. 1977. Devolatilization equilibria in graphitic systems. AM J SCI **277**, 1013–44.

Peucat, J. J. 1986. Behaviour of Rb–Sr whole rock and U–Pb zircon systems during partial melting as shown in migmatitic gneisses from the St Malo Massif, NE Brittany, France. J GEOL SOC LONDON **143**, 875–85.

Peucat, J. J. & Martin, H. 1985. Are Rb–Sr thin slabs migmatite ages always meaningful? N JARHB MINERAL ABH **152**, 1–21.

Pichavant, M. & Manning, D. A. C. 1984. Petrogenesis of tourmaline granites and topaz granites: the contribution of experimental data. PHYS EARTH PLANET INTER **35**, 31–50.

Pichavant, M., Valencia Herrera, J., Boulmier, S., Briqueu, L., Joron, J.-L., Juteau, M., Marin, L., Michard, A., Sheppard, S. M. F., Treuil, M. & Vernet, M. 1987. The Macusani glasses, SE Peru: evidence of chemical fractionation in peraluminous magmas. In Mysen, B. O. (ed.) Magmatic processes: physicochemical principles. GEOCHEM SOC SPEC PUBL **1**, 359–73.

Pichavant, M., Kontak, D. J., Valencia Herrera, J. & Clark, A. H. 1988a. The Miocene–Pliocene Macusani Volcanics, SE Peru. I. Mineralogy and magmatic evolution of a two-mica aluminosilicate-bearing ignimbrite suite. CONTRIB MINERAL PETROL **100**, 300–24.

Pichavant, M., Kontak, D. J., Briqueu, L., Valencia Herrera, J. & Clark, A. H. 1988b. The Miocene–Pliocene Macusani Volcanics, SE Peru. II. Geochemistry and origin of a felsic peraluminous magma. CONTRIB MINERAL PETROL **100**, 325–38.

Powers, R. E. & Bohlen, S. R. 1985. The role of synmetamorphic igneous intrusions in the metamorphism and partial melting of metasediments. CONTRIB MINERAL PETROL **90**, 401–9.

Pushkar, P. & Stoeser, D. B. 1975. ^{87}Sr/^{86}Sr ratios in some volcanic rocks and some semifused inclusions of the San Francisco volcanic field. GEOLOGY **3**, 669–701.

Rubie, D. C. & Brearley, A. J. 1990. A model for rates of disequilibrium melting during metamorphism. In Ashworth, J. R. & Brown, M. (eds) High temperature metamorphism and crustal anatexis, 57–86. London: Unwin Hyman.

Sato, M. 1978. Oxygen fugacity of basaltic magmas and the role of gas-forming elements. GEOPHYS RES LETT **5**, 447–9.

Sawyer, E. W. 1991. Disequilibrium melting and the rate of melt–residuum separation during migmatisation of mafic rocks from the Grenville Front, Quebec. J PETROL **32**, 701–38.

Scaillet, B., France-Lanord C. & Le Fort P. 1990. Badrinath–Gangotri plutons (Garhwal, India): petrological and geochemical evidence for fractionation processes in a high Himalayan leucogranite. J VOLCANOL GEOTHERM RES **44**, 163–88.

Scaillet, B., Pichavant, M. & Roux, J. 1995a. Experimental crystalliz-
ation of leucogranite magmas. J PETROL **36**, 663–705.
Scaillet, B., Pêcher, A., Rochette, P. & Champenois, M. 1995b. The
Gangotri granite (Garhwal Himalaya): laccolithic emplacement
in an extending collisional belt. J GEOPHYS RES **100**, 585–607.
Scaillet, B., Pichavant, M. & Holtz, F. 1995c. Experimental constraints
on the petrogenesis of the high himalayan leucogranites. *In 10th
Himalaya Karakoram Tibet Workshop.* MITT GEOL INST ETH
UNIV ZURICH **298**.
Tsuchiyama, A. 1985. Partial melting of plagioclase–diopside pairs.
CONTRIB MINERAL PETROL **91**, 12–23.

Visonà, D. & Zantedeschi, C. 1994. Spodumene, petalite and cassiterite,
new occurrence in Himalayan leucogranite pegmatites: petrological implications. *In 16th IMA General Meeting, Pisa,* 429.
Wall, V. J., Clemens, J. D. & Clarke, D. B. 1987. Models for granitoid
evolution and source composition. J GEOL **95**, 731–49.
White, A. J. R., Clemens, J. D., Holloway, J. R., Silver, L. T., Chappell,
B. W. & Wall, V. J. 1986. S-type granites and their probable
absence in southwestern North America. GEOLOGY **15**, 115–8.
Zen, E-an 1985. An oxygen buffer for some peraluminous granites
and metamorphic rocks. AM MINERAL **70**, 65–73.

M. PICHAVANT & B. SCAILLET, Centre de Recherche sur la Synthèse et la Chimie des Minéraux, CNRS, 1A
rue de la Férollerie, 45071, Orléans, France.
T. HAMMOUDA, Geophysical Laboratory, 5251 Broad Branch Road, N. W., Washington, DC 20015–1305,
U.S.A.

Transactions of the Royal Society of Edinburgh: Earth Sciences, **87**, 331–337, 1996

Evidence of heterogeneous crustal sources: the Harney Peak Granite, South Dakota, U.S.A.

Eirik J. Krogstad and Richard J. Walker

ABSTRACT: The Early Proterozoic (1715 Ma) Harney Peak Granite (Black Hills, SD, U.S.A.) is a complex of hundreds of dykes and sills. Earlier studies of Nd, O and Pb isotope variations demonstrated that the complex was not derived from a single source, or even different sources of a single age. Instead, the granites can be divided into a group with sources probably dominated by Early Proterozoic sediments and a group with sources probably dominated by Archean sediments. New results on the Nd isotopic variations of many additional samples indicate that there is considerable overlap between Nd isotopic compositions within the complex. Values of ε_{Nd} (1715 Ma) of the Harney Peak Granite suite ($n = 20$) range from $-2{\cdot}0$, indicating an Early Proterozoic (2300–2200 Ma) crustal source, to $-13{\cdot}4$, indicating a Middle to Late Archean (3200–3100 Ma) protolith. These results suggest that the Early Proterozoic source may have included rocks such as the *c.* 2200–1900 Ma metasedimentary rocks that occur in the southern Black Hills. The Archean sources might have included rocks such as those exposed on the periphery of the Black Hills. The range in Nd model ages negates the usefulness of the concept of the 'average' age of the crust in this part of the craton. Because such heterogeneity is present in the magmatic compositions of the Harney Peak Granite, it can be inferred that at least as much heterogeneity was present in the sources. In this granite system, melts were evidently derived from isolated, heterogeneous zones and did not have the opportunity to coalesce into large magma bodies. In systems where coalescence does occur, the evidence for such highly heterogeneous sources may be lost. These results emphasise that inferences drawn from a few samples of plutonic rocks in which magma mixing and homogenisation occurred can lead to erroneous conclusions about the age and nature of protoliths and, consequently, the development of continental crust.

KEY WORDS: heterogeneous crustal sources, isotope studies, granitic melts, magma mixing.

Petrogenetic studies of granite sources have used their compositional particulars as probes of the deeper crust. Granite compositions are often used as recorders of the chemical characteristics of *particular* source compositions (Chappell & White 1974). Granite compositions have also been interpreted to record the *average* crustal composition in a particular region. Thus the initial isotopic compositions of granites have been used in models to determine the average age of the crust in a particular area (e.g. Nelson & DePaolo 1985; Bennett & DePaolo 1987). Any geological map of surface exposures of the continental crust and studies of the deepest exposed sections of the continental crust (e.g. Fountain & Salisbury 1981), however, show that the continental crust is heterogeneous on a fine scale. We know, in addition, that granites are commonly derived in small batches from a wide variety of source materials. In many instances the only demonstrated link between granites in a particular granite field is their age and potentially the depth of melting. In other instances, batches of magma from diverse sources have apparently been mixed together in a mid-crustal magma chamber. This process tends to obscure the original differences in elemental and isotopic compositions that would be indicative of multiple sources.

The Early Proterozoic (*c.* 1715 Ma) Harney Peak Granite of the Black Hills is well known for a variety of reasons. The granite and associated pegmatites were the objects of some of the first radiogenic age determinations (e.g. Tomlinson & Das Gupta 1953; Davis & Aldrich 1956; Wetherill *et al.* 1956; Goldich *et al.* 1966). The granite has also been used as a

model for S-type granites in experimental studies (Huang & Wylie 1981). In addition, the Harney Peak Granite, at Mount Rushmore, is among the most familiar granite exposures in North America. Various field studies have shown that the Harney Peak Granite is a compositionally and isotopically heterogeneous complex of dykes and sills. Because none of these dykes or sills could be expected to have sampled the entire crustal column, this complex is an ideal setting in which to observe the range of sources sampled by the 1715 Ma melting event.

1. Geological setting of the Harney Peak Granite

The Black Hills of western South Dakota and eastern Wyoming are a Laramide (*c.* 60 Ma) domal structure (Lisenbee 1978) (Fig. 1). Erosion of the dome has exposed a complex Precambrian terrane. This terrane consists of two Late Archean granites, Early Proterozoic metasedimentary and metavolcanic rocks and the *c.* 1715 Ma Harney Peak Granite, a complex of numerous sills and dykes (Fig. 2). The Harney Peak Granite was emplaced near the peak of metamorphism at depths equivalent to 3–4 kbar (Nabelek *et al.* 1992b), with a geothermal gradient in the crust of the area at that time about 40–45°C/km (Helms & Labotka 1991). Thus the country rocks in the area were fairly hot (600–650°C) at the time of peak metamorphism. The Harney Peak Granite complex is surrounded by a pegmatite field which consists of thousands of granitic pegmatites (e.g. Norton & Redden 1990). The internal structures and mineralogies of these pegmatites range from simple to complex.

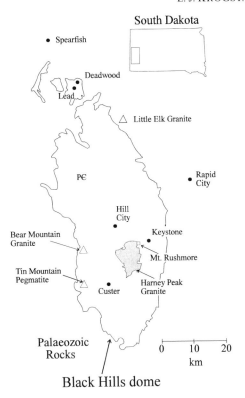

Figure 1 Location of the Harney Peak Granite complex, Little Elk Granite and Bear Mountain Granite.

The Little Elk Granite and the Bear Mountain Granite are the oldest rocks dated in the region. Gosselin *et al.* (1988) reported a U–Pb zircon upper intercept age of 2549 ± 11 Ma for the Little Elk Granite, in good agreement with the previously published U–Pb zircon age of 2560 Ma (Zartman & Stern 1967). Gosselin and co-workers also reported a highly discordant zircon upper intercept age of 2393 ± 230 Ma for

the Bear Mountain Granite. The latter age is probably a minimum age because the lower intercept age is approximately present day. Thus this granite probably had a late Archean crystallisation age. Zircons from metamorphosed tuffs and granophyric differentiates from the northern and central Black Hills have ages of *c.* 2200–1900 Ma (Redden *et al.* 1991). Walker *et al.* (1986) reported Nd T_{DM} (depleted mantle of DePaolo 1981) model ages of 2400–2200 Ma for similar metasedimentary rocks from the southern Black Hills. The combination of the U–Pb zircon ages of the tuffs and the juvenile T_{DM} model ages of the interbedded metasedimentary rocks can be interpreted to mean that these rocks are new crust formed in the Early Proterozoic. New crust of this age is uncommon in North America.

The age of the Harney Peak Granite complex has been determined using the Rb/Sr whole rock isochron technique by Riley (1970) and Walker *et al.* (1986). The combined data from both studies yield an isochron age of 1711 ± 21 Ma (MSWD = 6·4), with an $^{87}Sr/^{86}Sr_i = 0·7145 \pm 16$ (Krogstad *et al.* 1993). U–Pb data for concordant monazites taken from a single sill in the Harney Peak Granite complex give a similar, but more precise, age of 1715 ± 3 Ma (Redden *et al.* 1991). U–Pb analyses of several apatite samples separated from the Tin Mountain pegmatite yield concordant U–Pb ages of 1704–1700, establishing a temporal relationship between the Harney Peak Granite and at least one of the peripheral pegmatites (Krogstad & Walker 1994).

2. Previous work on granite magma sources

Earlier studies of the Harney Peak Granite suggested that the chemical diversity within the complex probably resulted from either the derivation of granite from compositionally diverse pelitic rocks (e.g. Redden *et al.* 1982) or extensive crystal–liquid fractionation of the parent magmas (e.g. Shearer *et al.* 1987). Because of the absence of isotopic data, these studies

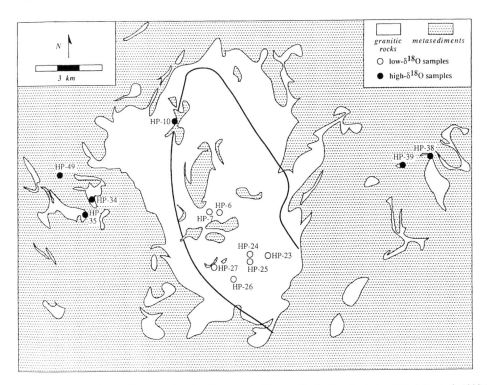

Figure 2 Location of sampling sites for the Proterozoic Harney Peak Granite (after Nabelek *et al.* 1992a). Note that the Harney Peak Granite is a dyke and sill complex, not a single intrusive body. The unornamented area shown on the map is where the volume of the dykes and sills of the Harney Peak Granite exceeds that of the country rock. The heavy line separates the low $\delta^{18}O$ samples in the middle of the granite from the high $\delta^{18}O$ samples at the periphery of the main pluton and in the satellite plutons.

were not able to constrain further the nature of the protoliths involved.

More recent radiogenic and stable isotopic studies (Walker *et al.* 1986; Nabelek *et al.* 1992a; Krogstad *et al.* 1993) showed that the granites were derived from at least two sources and that the Early Proterozoic pelites exposed near the granite complex cannot represent the sole source component of any of the granite and granitic pegmatite magmas. Nabelek *et al.* (1992a, b) divided the Harney Peak Granite into two groups based on their principal ferromagnesian minerals and their $\delta^{18}O$ values. These workers proposed that the granite group that is generally found in the core of the complex (biotite granites with $\delta^{18}O$ values of $+11\cdot5\pm0\cdot6\%$) was derived from a source distinct from the granites of the periphery of the complex (tourmaline granites with $\delta^{18}O$ values of $+13\cdot2\pm0\cdot8\%$).

Nabelek and Glascock (1995) have proposed that the two suites of granite in the Harney Peak Granite were derived from two distinct sources and under different melting conditions. They showed evidence that the high $\delta^{18}O$ granites were derived from low-temperature non-modal melting of pelites, leaving monazite armoured in biotites. This could result in the characteristic low REE abundances and low LREE/HREE ratio in this suite. In contrast, the granites with relatively lower $\delta^{18}O$ were proposed to be produced by melting of a similar pelitic rock, but accompanied by the breakdown of biotite and, thus, monazite also. The result was a suite of comparatively low $\delta^{18}O$ granites with generally steeper, negatively sloped REE patterns and higher REE abundances. Because of the higher temperatures required for biotite breakdown than for muscovite breakdown, it was suggested that this melting event was deeper and hotter than that required to produce the high $\delta^{18}O$ granite magmas.

The Nd data also show evidence for source heterogeneity. Walker *et al.* (1986), in a study of a limited suite of granite samples, showed that ε_{Nd} (1715 Ma) values for the Harney Peak Granite range from $-4\cdot0$ to $-13\cdot3$. This was interpreted to mean that two sources, one Early Proterozoic and the other Archean, were required. The Early Proterozoic metasedimentary rocks from the southern Black Hills examined in that study have ε_{Nd} (1715 Ma) values which range from $-1\cdot7$ to $-3\cdot8$ (Walker *et al.* 1986). Therefore, the Early Proterozoic source for the granites may have included metasedimentary rocks such as those seen in the southern Black Hills. Walker and co-workers proposed that an older, probably late Archean, source was required as one component of a source mixture which possibly included the Proterozoic pelites.

We began to investigate in greater detail the petrogenesis of the Harney Peak Granite, using first the Pb isotopic system (Krogstad *et al.* 1993) and subsequently the Nd isotope system (this work). In our study of Pb isotopic variations, we showed that the high $\delta^{18}O$ group of Nabelek *et al.* (1992a, b) have lower $^{207}Pb/^{204}Pb$ for their $^{206}Pb/^{204}Pb$ than do the low $\delta^{18}O$ group. We interpreted these data to mean that the high $\delta^{18}O$ group had sources dominated by Early Proterozoic sediments whose provenance materials resided in the upper crust for a maximum of 100–300 Ma before the *c.* 1715 Ma crystallisation age of the granite complex. These results indicated, as had the earlier results of Walker *et al.* (1986), that the Early Proterozoic source may have included metasedimentary rocks such as those seen in the southern Black Hills.

In contrast, the low $\delta^{18}O$ group had sources dominated by presumably Late Archean sediments. One of the source components of the Harney Peak granites with low $\delta^{18}O$ was an upper crustal (high U/Pb, low Th/U) lithology that was separated from a mantle source before the latest Archean. The

Late Archean Little Elk Granite, or its protoliths, may have been this source component.

Thus the Pb, Nd and O isotopic data for the mid-Proterozoic granites and granitic pegmatites of the Black Hills all confirm that the magmas were derived from at least two sources and preclude the interpretation that the compositional variations among the granitic rocks are exclusively the result of crystal–liquid fractionation of a single parent magma.

3. Samples and analytical techniques

The Harney Peak granite is a leucocratic rock essentially composed of quartz, plagioclase and K-feldspar. Minor amounts of biotite, muscovite, tourmaline, apatite and garnet are present in various samples. Sampling locations for all rocks are shown in Figures 1 and 2. Samples were collected from numerous dykes and sills of the Harney Peak Granite complex and most are the same as given in Nabelek *et al.* (1992b) and Krogstad *et al.* (1993).

The Pb and Nd analytical procedures were modified from those of Walker *et al.* (1994). Samples were completely dissolved for Nd analysis using the Li metaborate flux fusion technique. Samples were fused at 1050°C and the melt was quenched in dilute HNO_3 to which a multi-REE spike was added. All the REEs were initially separated from matrix by way of co-precipitation with Fe hydroxide and centrifugation (Evans 1987). The precipitate was redissolved in HCl and the REEs were subsequently separated as a group using a 9·5 ml HCl cation-exchange column, followed by a 2·4 ml 2-methyllactic acid cation-exchange column to separate individual REEs. Blanks for the Nd separation averaged 1·5 ng. Concentrations of Nd and Sm were determined by isotope dilution. Uncertainties in $^{147}Sm/^{144}Nd$ are about $\pm0\cdot2\%$, based on numerous analyses of the BCR-1 standard powder (Tomascak 1995). Aliquots of sample powders weighing approximately 200 mg were processed for Pb by Krogh-type bomb dissolution in HF and HNO_3. The solution was then evaporated to dryness and redissolved in 2 M HCl. Lead in this solution was separated in two passes on a 2 ml, HBr-form, AG1 × 8 anion-exchange column. The total blank for this procedure was 75–100 pg. Neodymium and Pb isotopic data were collected using the Bobcat II mass spectrometer, a VG Sector 54. A multidynamic programme was used for Nd and a static programme for Pb. Repeated analyses of the La Jolla Nd standard during the period of this study gave a mean $^{143}Nd/^{144}Nd$ of 0·511849 with a 2σ (population) of 20 ppm ($n=19$). Repeated measurements of SRM-982 were used to correct sample Pb ratios for fractionation. This correction averaged $+0·11\%$/a.m.u. The reproducibility of Pb standards was $\pm0·04\%$/a.m.u. (2σ of the population).

4. Results

4.1. Pb data

The interpretation of common Pb data on granites is hindered by difficulties in the estimation of initial ratios of magmas and potential source rocks. Unlike the Sm–Nd system, and to a lesser extent the Rb–Sr system, the U–Pb system in a rock is fairly easily and commonly disrupted from closed system behaviour during weathering and metamorphism. The initial Pb isotopic composition of a rock is not reliably estimated by simply measuring its Pb isotopic composition, and U and Pb concentrations. The isotopic composition of low μ ($^{238}U/^{204}Pb$) phases, such as K-feldspars, can be used to estimate the initial Pb isotopic compositions of older granites (e.g. Ludwig & Silver 1977). However, the isotopic compositions of potential source rocks at the time of magmatism are more difficult to establish. This is especially true for sediments whose low μ phases, feldspars, may be detrital and not equilibrated with the bulk rock during diagenesis and metamorphism. The Pb isotopic composition of the feldspars in a metasediment may not, therefore, be representative of the bulk Pb isotopic composition of the whole rock at the time it was melted. Instead, an estimation of the Pb isotopic composition of the metasediment at some time requires a knowledge of its present day Pb isotopic composition, a reference age (such as the age of the magmatic event) and the Pb isotopic composition of a reference point.

A useful reference point among the Pb data for the 1715 Ma igneous rocks of the southern Black Hills is $^{207}Pb/^{204}Pb = 15.47$ for $^{206}Pb/^{204}Pb = 16.0$. This point lies at the least radiogenic end of the array of K-feldspar Pb isotopic compositions (Krogstad et al. 1993), with the feldspars from the high $\delta^{18}O$ granites lying below (at lower $^{207}Pb/^{204}Pb$) than these coordinates, and the feldspars from the low $\delta^{18}O$ group lying above these coordinates. A rock with this initial Pb isotopic ratio, with a non-zero μ after 1715 Ma, will have developed more radiogenic Pb so that it now will lie to the right and above this reference point. The slope of the line connecting these two points (initial and present day $^{206}Pb/^{204}Pb$ and $^{207}Pb/^{204}Pb$) will have a slope (m) controlled exclusively by the beginning and ending ages of radiogenic Pb growth. A 1715 Ma rock with a Pb isotopic composition lying above this line must have had an initial $^{207}Pb/^{204}Pb$ higher than the reference value. A rock with a composition lying below this line must have had an initial $^{207}Pb/^{204}$ Pb less than the reference value.

In this work we represent differences in $^{207}Pb/^{204}Pb$ for a given $^{206}Pb/^{204}Pb$ by the notation $\delta7/4$. This is the part per thousand deviation of the $^{207}Pb/^{204}Pb$ of a sample, versus the $^{207}Pb/^{204}Pb$ of a reference point, for a given $^{206}Pb/^{204}Pb$. The equation for the value $\delta7/4$, which can only be used to compare rocks at a similar age, is

$$\delta7/4 = 10^3 * (^{207}Pb/^{204}Pb_s - m*(^{206}Pb/^{204}Pb_s - ^{206}Pb/^{204}Pb_r)$$
$$- ^{207}Pb/^{204}Pb_r)/^{207}Pb/^{204}Pb_s$$

where the subscript 's' refers to the sample, the subscript 'r' refers to the pertinent reference point and 'm' is the age-controlled growth slope. For the 1715 Ma granites and pegmatites of the southern Black Hills, $m = 0.105$ (for 1715 Ma to present), $^{207}Pb/^{204}Pb_r = 15.47$ and $^{206}Pb/^{204}Pb_r = 16.0$.

Because the $^{207}Pb/^{204}Pb - ^{206}Pb/^{204}Pb$ growth slope is uniquely dependent on the time over which uranogenic Pb growth occurs, this approach allows a comparison of the likely $^{207}Pb/^{204}Pb$ of the sources of rock samples whose U/Pb ratios may be different. Samples with high $\delta7/4$ are probably derived from sources with early, high U/Pb histories. This contrasts with samples with low $\delta7/4$, which probably had no early, high U/Pb episodes. Thus the distinction can be made between low $^{207}Pb/^{204}Pb$ granites derived from Early Proterozoic (high U/Pb) sources and high $^{207}Pb/^{204}Pb$ granites derived from Late Archean (high U/Pb) sources. New whole-rock Pb data and the Pb data from Krogstad et al. (1993) have been recalculated in the terms of the notation $\delta7/4$ (Table 1; Fig. 3).

The values of $\delta7/4$ for the K-feldspars from the Little Elk Granite and the Bear Mountain Granite are high (+48 and +58 at 1715 Ma; Table 1). These reflect the Archean ages of the rocks, the high U/Pb of the granites and possibly also some re-equilibration of the whole-rock Pb with the K-feldspars at about 1900–1800 Ma, as discussed by Krogstad et al. (1993).

The Proterozoic sediment samples are fairly radiogenic, but

they have, by contrast, low values of $\delta7/4(1715 Ma)$ (−6.3 to −7.0), less even than those of the least radiogenic granites (Krogstad et al. 1993) (Fig. 3). The $^{208}Pb/^{204}Pb - ^{206}Pb/^{204}Pb$ growth slope (not shown) requires that the Th/U ratio of the samples has been slightly higher than that of the Stacey–Kramers model (Th/U = 3.78; Stacey & Kramers 1975). This can be interpreted to mean that the rocks were not subjected to depletion of U with respect to Th, as might occur in the lower crust (Heier 1973). The low $\delta7/4$ of the metasedimentary rocks means that the provenance for these sediments included only a minor component, if any, of Archean crust. This conclusion is consistent with the interpretation of the Nd data for the same samples (see later). A four-point Pb–Pb isochron age for the four metasediment samples gives an age of 1683 ± 48 Ma (MSWD = 0.43). This age is younger than the 2200–1900 Ma U–Pb zircon ages of tuffs interbedded with the Early Proterozoic sediments (Redden et al. 1991), but does include, within its uncertainty, the established emplacement age for the Harney Peak Granite (1715 ± 3 Ma). Consequently, the Pb–Pb isochron may indicate that U and Pb were redistributed within the metasedimentary rocks when they were intruded by the granites. This may not be surprising. Large-scale fluid circulation in these rocks has been proposed (J.A. Redden, pers. comm.) based on Li concentration data. Such fluid circulation might have been capable of redistributing Pb and U within the granites and surrounding country rocks.

4.2. Nd data

Neodymium data for three additional samples of metasedimentary rocks (Table 1) are consistent with the data of Walker et al. (1986). The combined results indicate that these metasedimentary rocks are homogeneous in $^{147}Sm/^{144}Nd$, have $\varepsilon_{Nd}(1715)$ ranging only from −1.7 to −3.8, and have T_{DM} model ages of 2450–2200 Ma (Table 1, Fig. 3).

New results on the Nd isotopic variations of the granite samples indicate that there is considerable overlap between the Nd isotopic compositions (at 1715 Ma) of the high $\delta^{18}O$ and low $\delta^{18}O$ groups defined by Nabelek et al. (1992a, b). The $\varepsilon_{Nd}(1715 Ma)$ of all of the samples ($n = 20$) range from −2.0, indicating an Early Proterozoic (2300–2200 Ma) crustal source, to −13.4, indicating the incorporation of a Middle Archean (3200–3100 Ma) or older protolith (Figs 3, 4). The model age of this latter protolith is similar to that required for the magmatic sources of the 2500 Ma Little Elk Granite [ε_{Nd} (1715 Ma) = −14.2, Fig. 4]. The concentration-weighted average of these samples is $\varepsilon_{Nd}(1715 Ma) = −7.7$, with a T_{DM} model age (assuming a crustal source with an $^{147}Sm/^{144}Nd = 0.112$, typical of the metasediments) of 2700 Ma.

The average $\varepsilon_{Nd}(1715 Ma)$ of the high $\delta^{18}O$ group and low $\delta^{18}O$ group do not differ strongly. The Nd concentration-weighted average for the high $\delta^{18}O$ group is $\varepsilon_{Nd}(1715 Ma) = −7.2$ (−7.1 if sample 39A is omitted), and for the low $\delta^{18}O$ group the average is $\varepsilon_{Nd}(1715 Ma) = −7.9$. This represents a difference of just 70 Ma in the model crustal ages for these two groups. However, the high $\delta^{18}O$ samples with the lowest Nd concentrations (less than 2.5 ppm) have $\varepsilon_{Nd}(1715 Ma)$ higher than −6.0. All samples that have more than 2.5 ppm Nd have $\varepsilon_{Nd}(1715 Ma)$ lower than −6.0. The distinction in sources between those for the high and low $\delta^{18}O$ samples may, therefore, not be as distinctive as was seen in the smaller Pb data set.

5. Discussion

The weighted average for the Nd data shows that the average $^{147}Sm/^{144}Nd$ of the granites is 0.202 and the average ε_{Nd} (1715 Ma) is −7.7. This gives an average T_{DM} model age of

Table 1 Nd and Pb isotopic data for Proterozoic granites and metasedimentary rocks, and Archean granites from the southern Black Hills.

Sample	[Nd]	[Sm]	147/144	143/144	$\varepsilon_{Nd}(1715)$	$^{206}Pb/^{204}Pb$	$^{207}Pb/^{204}Pb$	$^{208}Pb/^{204}Pb$	$\delta7/4$
High $\delta^{18}O$ granites									
1–1	**3·95**	**1·28**	**0·1954**	**0·512347**	−5·4				
2–1	**0·76**	**0·244**	**0·1940**	**0·512310**	−5·7				
3–1B	**0·911**	**0·361**	**0·2396**	**0·512849**	−5·1				
4–1	**8·41**	**2·17**	**0·1561**	**0·511799**	−7·4				
3B	**4·88**	**1·99**	**0·2469**	**0·513084**	−2·0				
10D	1·272	0·697	0·3314	0·513869	−5·6	16·232	15·435	35·430	−3·9
34A	1·265	0·484	0·2312	0·512853	−3·4	16·301	15·493	35·458	−0·6
35A	2·422	0·776	0·1938	0·512340	−5·2	16·190	15·417	35·489	−4·7
35B						16·192	15·390	35·391	−6·5
38A	1·064	0·394	0·2238	0·512727	−4·2	16·103	15·418	35·472	−4·1
39A	2·645	1·569	0·3589	0·513780	−13·4				
39Arep	2·961	1·752	0·3580	0·513854	−11·8				
39C + D						16·869	15·577	35·738	1·0
49A	11·08	3·39	0·1851	0·512112	−7·7				
49B						16·312	15·440	35·508	−4·1
Low $\delta^{18}O$ granites									
8 4L	**2·93**	**1·63**	**0·3368**	**0·513868**	−6·4				
8 8U	**4·68**	**2·20**	**0·2841**	**0·513103**	−9·9				
6B	10·16	2·45	0·1456	0·511734	−6·4	16·234	15·510	35·547	1·0
7B	2·043	0·924	0·2736	0·513132	−7·3	16·379	15·566	35·395	3·6
20	**7·28**	**2·14**	**0·1780**	**0·512077**	−6·7				
23A	3·167	1·53	0·2913	0·513259	−8·7	16·245	15·533	35·597	2·4
23B						16·338	15·516	35·538	0·7
24B						16·384	15·588	35·681	5·0
25	14·01	4·34	0·1871	0·512091	−8·6	16·165	15·492	35·523	0·3
26	24·38	7·31	0·1813	0·512040	−8·3	16·384	15·543	35·547	2·1
27	7·141	2·57	0·2179	0·512483	−7·7	16·558	15·604	35·639	4·8
30A	13·26	3·97	0·1810	0·512083	−7·4				
Archean granites									
LElk	**67·0**	**10·7**	**0·0965**	**0·510776**	−14·2	21·289	16·837	40·060	48·2
BMtn	**1·46**	**0·726**	**0·3008**	**0·513959**	3·0	21·491	17·028	33·989	57·6
Proterozoic metapelites									
WC-4	**36·7**	**6·68**	**0·1101**	**0·511498**	−3·0	24·251	16·232	46·459	−6·4
23–1	**34·6**	**6·69**	**0·1168**	**0·511633**	−1·9				
23–2	**34**	**6·3**	**0·1122**	**0·511586**	−1·7	23·438	16·139	46·098	−7·0
26–2	**44·2**	**7·83**	**0·1072**	**0·511424**	−3·8	18·194	15·603	38·482	−6·3
9–4	27·7	5·14	0·1121	0·511511	−3·3	19·580	15·746	41·588	−6·4
9–3	28·1	5·22	0·1122	0·511563	−2·3				
9–1	36·4	7·54	0·1251	0·511726	−2·0	24·152	16·167	39·806	−9·9
40–1A	**45·9**	**8·26**	**0·1087**	**0·511519**	−2·3				

Note: Data in **bold** face previously published: Nd, Walker *et al.* (1986); Pb, Krogstad *et al.* (1993). [Sm] and [Nd] are in ppm. $\delta7/4$ is defined in the text. LElk = Little Elk Granite; BMtn = Bear Mountain Granite.

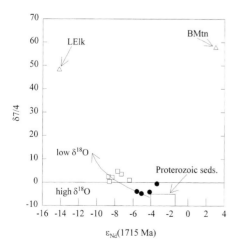

Figure 3 ε_{Nd} versus $\delta7/4$ diagram. The Pb data for the low $\delta^{18}O$ granites lie at higher $\delta7/4$ than do the Pb data for the high $\delta^{18}O$ granites of Nabelek *et al.* (1992a), although there is nearly overlap. In this plot of model initial ratios, a single hyperbolic array could indicate that the granites were derived from variable proportions of the two end-members shown: Little Elk Granite and Proterozoic sediments. The data do not scatter about any single curve, indicating that either more than two sources were involved, or some processes affected the Nd/Pb in resulting melts.

2700 Ma. The LREE depleted nature of this average, as well as those of the two subgroups (0·209 and 0·199), is not an expected result. Overall, the 1715 Ma melting event produced LREE-depleted granites and relatively LREE-enriched residues, compared with average continental crust. Thus conditions were such that crustal differentiation that produces the most evolved types of granites may, in some instances, lead to LREE enrichment in residual rocks.

The Pb and Nd isotopic compositions of all granite samples are bracketed by the isotopic compositions of the Late Archean granites and the Early Proterozoic metasediments. Thus no additional source is required for the granitic magmas. The Nd and Pb isotopic compositions of the local metasedimentary rocks at 1715 Ma lie in a restricted field, without evidence of significant involvement of older crust in their provenance. Such compositions allow the metasedimentary rocks to be the relatively juvenile crustal component that was melted to generate granites with the lowest $\delta7/4$ and highest ε_{Nd} (1715 Ma). On various plots in which mixing of two components would be apparent [i.e. Fig. 5, 1/Nd versus ε_{Nd} (1715 Ma)] the data do not define an array with a linear nature, such as would be produced by two-component mixing, subsequently modified by crystal–liquid fractionation. Nor do the Pb and Nd data for the granites, metasedimentary rocks and Archean granites define a single hyperbola on a plot of

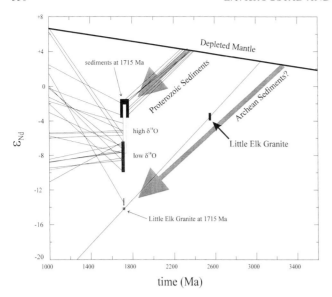

Figure 4 ε_{Nd} versus time diagram. The Proterozoic metasedimentary rocks have a limited spread of ε_{Nd} (1715 Ma), indicating that the provenance materials were, on average, of Early Proterozoic age. These metasedimentary rocks lie at the positive ε_{Nd} end of the granite array. This supports the hypotheses of Walker *et al.* (1986) and Krogstad *et al.* (1993), from limited Nd data and no Pb data on these rocks, that the metasedimentary rocks could be one of the juvenile crustal sources of the Harney Peak Granite magmas. The negative ε_{Nd} end of the granite array is bracketed by the ε_{Nd} (1715 Ma) of the Little Elk Granite, confirming the hypotheses of these workers that this unit, or one of similar age and history, may represent the older crustal end-member of the granite source rocks.

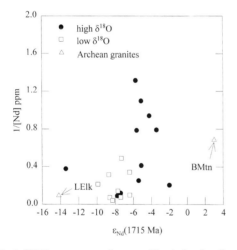

Figure 5 1 [Nd] versus ε_{Nd} diagram. The lack of a linear array shows, as did Figure 3, that simple two-component source mixing cannot explain the variations in the Nd concentrations and isotopic compositions of the Harney Peak Granite samples.

isotopic compositions such $\delta 7/4$ versus ε_{Nd} (1715 Ma) (Fig. 3) or $\delta^{18}O$ versus $^{207}Pb/^{204}Pb$ (Fig. 2a of Nabelek & Glascock 1995). Consequently, the data are inconsistent with simple two-component mixing and probably indicate the derivation of granites from a variety of isotopically and compositionally distinct protoliths. The curvature of the array in Figure 3 does indicate, however, that the melts of the Archean protoliths probably had higher Nd/Pb than did the melts of the Proterozoic metasedimentary rocks.

It has been argued (e.g. Hogan & Sinha 1991) that strongly non-modal melting can cause large changes in the Pb isotope ratios of resultant magmas. In such a model, high U/Pb phases, such as monazite and zircon, might not participate in early, low percentage melting events. Such magmas would, as a result, have low values of uranogenic Pb ($^{207}Pb/^{204}Pb$ and

$^{206}Pb/^{204}Pb$). Later magmas, derived by higher temperature melting, in which monazite and zircon were broken down, would have comparatively high $^{207}Pb/^{204}Pb$ and $^{206}Pb/^{204}Pb$. However, such melting relations cannot account for the relation between the high $\delta 7/4$ values for some of the Harney Peak Granite samples and the low $\delta 7/4$ values for others. In the Hogan and Sinha model, the $\delta 7/4$ for all melt products of a single source, early or late, would have essentially the same values of $\delta 7/4$. Consequently, the Pb and Nd data indicate that the Harney Peak Granite complex was derived from a heterogeneous source with a range of crustal ages spanning a minimum of about 1000 Ma, from 3200 to 2200 Ma. The apparent involvement of more than two distinct sources may indicate that magmas were derived from relatively small volumes of protolith. The isotopic heterogeneity of the granites suggests that their sources were at least as heterogeneous as the granites.

We conclude that the Harney Peak Granite complex was derived from isolated, heterogeneous zones within the crust. The depth zonation of these melting events is, at present, unknown. However, the results of the study by Nabelek *et al.* (1992b) suggest that melting occurred at two or more depths, with the magmas for the biotite granites being produced at higher temperatures, and greater depths, than the magmas for the tourmaline granites. The continuous variation in ε_{Nd} (1715 Ma) for these granites suggests that their simple model of two discrete sources and two discrete depths may not apply. Instead, the protoliths were probably various mixtures of sediments with different major and trace element characteristics. The melts were derived from these sources and did not subsequently have the opportunity to coalesce into large magma bodies in which the evidence for such source differences would have been lost. This study shows that inferences drawn from a few samples of plutonic rocks in which mixing occurred may lead to erroneous conclusions regarding the absolute age of protoliths and the nature of the melting processes. For example, trace element modelling within the Harney Peak Granite system, as a means of defining degree of fractionation, is probably futile because of the near certainty of trace element variations in the protoliths of the various sills and dykes.

The range of model ages brings into question any model for the 'average' age of the crust in this part of the craton. The average crustal model age of 2700 Ma may not be represented by any rock, or detrital material, in the Black Hills region. Instead, crustal formation ages of 3200 and 2200 Ma have been cryptically mixed to form this average. The Nd data further suggest a Wyoming Province component, rather than a Superior Province component, for the granites. This is because the older (*c.* 3300 Ma) source is a more common component of the former than of the latter. In the Superior Province, the typical depleted mantle model age would be, except for isolated settings such as the Cross Lake subprovince, 2900–2700 Ma (Shirey & Carlson 1989). By contrast, model ages, as well as zircon crystallisation ages of 3500–3000 Ma, are relatively common in the Wyoming Province (Wooden & Mueller 1988).

6. Summary

The Early Proterozoic Harney Peak Granite was not derived from a single source, or even different sources of a single age. Instead, the complex was evidently derived from mixtures of both Early Proterozoic and Middle to Late Archean metasedimentary protolith. Each granitic sill or dyke probably represents the melting of an isotopically and compositionally unique metasedimentary package. Consequently, the concept of average crust in this portion of the craton is not realistic.

These results highlight the complexity of petrogenetic modelling in an evolved granitic system.

7. Acknowledgements

Partial funding for this study was provided to EJK by a Summer Research Grant from the General Research Board, University of Maryland at College Park. Some samples were provided by P. I. Nabelek. Comments on an earlier version by P. I. Nabelek and C. J. Johnson substantially improved this work. Analytical facilities were supported by instrumentation and facilities grants from the NSF to EJK and RJW (EAR-91-05222, EAR-93-17136).

8. References

Bennett, V. C. & DePaolo, D. J. 1987. Proterozoic crustal history of the western United States as determined by neodymium isotopic mapping. GEOL SOC AM BULL **99**, 674–85.

Chappell, B. W. & White, A. J. R. 1974. Two contrasting granite types. PACIFIC GEOL **8**, 173–4.

Davis, G. L. & Aldrich, L. T. 1956. Determination of the age of lepidolites by the method of isotope dilution. GEOL SOC AM BULL **64**, 379–80.

DePaolo, D. J. 1981. A neodymium and strontium isotopic study of the Mesozoic calc-alkaline granitic batholiths of the Sierra Nevada and Peninsular ranges, California. J GEOPHYS RES **86**, 10,470–88.

Evans, O. C. 1987. *The petrogenesis of the Saganaga Tonalite revisited.* M.S. Thesis, State University of New York at Stony Brook.

Fountain, D. M. & Salisbury, M. 1981. Exposed cross-sections through the continental crust: implications for crustal structure, petrology, and evolution. EARTH PLANET SCI LETT **56**, 263–77.

Goldich, S. S., Lidiak, E. G., Hedge, C. E. & Walthall, F. G. 1966. Geochronology of the midcontinent region, United States, 2: northern area. J GEOPHYS RES **71**, 5389–408.

Gosselin, D. C., Papike, J. J., Zartman, R. E., Peterman, Z. E. & Laul, J. C. 1988. Archean rocks of the Black Hills, South Dakota: reworked basement from the southern extension of the Trans-Hudson orogen. GEOL SOC AM BULL **100**, 1244–59.

Heier, K. 1973. Geochemistry of granulite facies rocks and problems of their origin. PHIL TRANS R SOC LONDON A **273**, 429–42.

Helms, T. S. & Labotka, T. C. 1991. Petrogenesis of early Proterozoic pelitic schists of the southern Black Hills, South Dakota: constraints on regional low-pressure metamorphism. GEOL SOC AM BULL **103**, 1324–34.

Hogan, J. P. & Sinha, A. K. 1991. The effect of accessory minerals on the redistribution of lead isotopes during crustal anatexis: a model. GEOCHIM COSMOCHIM ACTA **55**, 335–48.

Huang, W. L. & Wyllie, P. J. 1981. Phase relationships of S-type granite with H_2O to 35 kbar: muscovite granite from Harney Peak, South Dakota. J GEOPHYS RES **86**, 515–29.

Krogstad, E. J. & Walker, R. J. 1994. High closure temperatures of the U–Pb system in large apatites. GEOCHIM COSMOCHIM ACTA **58**, 3845–53.

Krogstad, E. J. Walker, R. J., Nabelek, P. I. & Russ-Nabelek, C. 1993. Pb isotopic evidence for mixed sources for Proterozoic granites and pegmatites, Black Hills, South Dakota. GEOCHIM COSMOCHIM ACTA **57**, 4677–85.

Lisenbee, A. L. 1978. Laramide structure of the Black Hills uplift, South Dakota–Wyoming–Montana. GEOL SOC AM BULL **151**, 165–96.

Ludwig, K. R. & Silver, L. T. 1977. Lead-isotope inhomogeneity in Precambrian igneous K-feldspars. GEOCHIM COSMOCHIM ACTA **41**, 1457–71.

Nabelek, P. I. & Glascock, M. D. 1995. REE-depleted leucogranites, Black Hills, South Dakota: a consequence of disequilibrium melting of monazite-bearing schists. J PETROL **36**, 1055–71.

Nabelek, P. I., Russ-Nabelek, C. & Haeussler, G. T. 1992a. Stable isotope evidence for the petrogenesis and fluid evolution in the Proterozoic Harney Peak leucogranite, Black Hills, South Dakota. GEOCHIM COSMOCHIM ACTA **56**, 403–17.

Nabelek, P. I., Russ-Nabelek, C. & Denison, J. R. 1992b. The generation and crystallization conditions of the Proterozoic Harney Peak leucogranite, Black Hills, South Dakota: petrologic and geochemical constraints. CONTRIB MINERAL PETROL **110**, 173–91.

Nelson, B. K. & DePaolo, D. J. 1985. Rapid production of continental crust 1·7 to 1·9 b.y. ago: Nd isotopic evidence from the basement of the North American mid-continent. GEOL SOC AM BULL **96**, 746–54.

Norton, J. J. & Redden, J. A. 1990. Relations of zoned pegmatites to other pegmatites, granite, and metamorphic rocks in the southern Black Hills, South Dakota. AM MINERAL **75**, 631–55.

Redden, J. A. Norton, J. J. & McLaughlin, R. J. 1982. Geology of the Harney Peak Granite, Black Hills, South Dakota. OPEN FILE REP US GEOL SURV **82–481**.

Redden, J. A., Peterman, Z. E., Zartman, R. E. & DeWitt, E. 1991. U–Th–Pb geochronology and preliminary interpretation of Precambrian tectonic events in the Black Hills, South Dakota. *In* Lewry, J. F. & Stauffer, M. R. (eds) *The Early Proterozoic Trans-Hudson Orogeny of North America*, 229–51. SPEC PAP GEOL ASSOC CAN **37**.

Riley, G. H. 1970. Isotopic discrepancies in zoned pegmatites, Black Hills, South Dakota. GEOCHIM COSMOCHIM ACTA **34**, 713–25.

Shearer, C. K., Papike, J. J. & Laul, J. C. 1987. Mineralogical and chemical evolution of a rare-element granite–pegmatite system: Harney Peak Granite, Black Hills, South Dakota. GEOCHIM COSMOCHIM ACTA **51**, 473–86.

Shirey, S. B. & Carlson, R. W. 1989. The Pb and Nd isotopic evolution of the Archean mantle. *In* Ashwal, L. D. (ed.) *Workshop on the Archean mantle*, 82–4. LUNAR PLANET INST TECH REP **89–05**.

Stacey, J. S. & Kramers, J. D. 1975. Approximation of terrestrial lead isotope evolution by a two-stage model. EARTH PLANET SCI LETT **26**, 207–21.

Tomascak, P. B. 1995. *The petrogenesis of granitic rocks in southwest Maine.* Ph.D. Dissertation, University of Maryland at College Park.

Tomlinson, R. H. & Das Gupta, A. K. 1953. The use of isotope dilution in determination of geologic ages of minerals. CAN J CHEM **31**, 909–14.

Walker, R. J., Hanson, G. N., Papike, J. J. & O'Neil, J. R. 1986. Nd, O, and Sr isotopic constraints on the origin of Precambrian rocks, southern Black Hills, South Dakota. GEOCHIM COSMOCHIM ACTA **50**, 2833–46.

Walker, R. J. Morgan, J. W., Horan, M. F., Czamanske, G. K., Krogstad, E. J., Fedorenko, V. A. & Kunlov, V. E. 1994. Re–Os isotopic evidence for an enriched-mantle source for the Noril'sk-type, ore-bearing intrusions, Siberia. GEOCHIM COSMOCHIM ACTA **58**, 4179–97.

Wetherill, G. W., Tilton, G. R., Davis, G. L. & Aldrich, L. T. 1956. New determinations of the age of the Bob Ingersoll pegmatite, Keystone, S. Dakota. GEOCHIM COSMOCHIM ACTA **9**, 292–7.

Wooden, J. L. & Mueller, P. A. 1988. Pb, Sr, and Nd isotopic composition of a suite of late Archean, igneous rocks, eastern Beartooth Mountains; implications for crust/mantle evolution. EARTH PLANET SCI LETT **87**, 59–72.

Zartman, R. E. & Stern, T. W. 1967. Isotopic age and geologic relationships of the Little Elk Granite, northern Black Hills, South Dakota, 157–63. PROF PAP US GEOL SURV **575–D**.

EIRIK J. KROGSTAD and RICHARD J. WALKER, Isotope Geochemistry Laboratory, Department of Geology, University of Maryland at College Park, College Park, MD 20742, U.S.A.

Transactions of the Royal Society of Edinburgh: Earth Sciences, **87**, 339–352, 1996

New approaches to crustal evolution studies and the origin of granitic rocks: what can the Lu–Hf and Re–Os isotope systems tell us?

Clark M. Johnson, Steven B. Shirey and Karin M. Barovich

ABSTRACT: The Lu–Hf and Re–Os isotope systems have been applied sparsely to elucidate the origin of granites, intracrustal processes and the evolution of the continental crust. The presence or absence of garnet as a residual phase during partial melting will strongly influence Lu/Hf partitioning, making the Lu–Hf isotope system exceptionally sensitive to evaluating the role of garnet during intracrustal differentiation processes. Mid-Proterozoic (1·1–1·5 Ga) 'anorogenic' granites from the western U.S.A. appear to have anomalously high ε_{Hf} values, relative to their ε_{Nd} values, compared with Precambrian orogenic granites from several continents. The Hf–Nd isotope variations for Precambrian orogenic granites are well explained by melting processes that are ultimately tied to garnet-bearing sources in the mantle or crust. Residual, garnet-bearing lower and middle crust will evolve to anomalously high ε_{Hf} values over time and may be the most likely source for later 'anorogenic' magmas. When crustal and mantle rocks are viewed together in terms of Hf and Nd isotope compositions, a remarkable mass balance is apparent for at least the outer silicate earth where Precambrian orogenic continental crust is the balance to the high-ε_{Hf} depleted mantle, and enriched lithospheric mantle is the balance to the low-ε_{Hf} depleted mantle.

Although the continental crust has been envisioned to have exceptionally high Re/Os ratios and very radiogenic Os isotope compositions, new data obtained on magnetite mineral separates suggest that some parts of the Precambrian continental crust are relatively Os-rich and non-radiogenic. It remains unclear how continental crust may obtain non-radiogenic Os isotope ratios, and these results have important implications for Re–Os isotope evolution models. In contrast, Phanerozoic batholiths and volcanic arcs that are built on young mafic lower crust may have exceptionally radiogenic Os isotope ratios. These results highlight the unique ability of Os isotopes to identify young mafic crustal components in orogenic magmas that are essentially undetectable using other isotope systems such as O, Sr, Nd and Pb.

KEY WORDS: Crustal evolution, granite, Lu–Hf isotope system, Re–Os isotope system.

Isotopic studies of granitic rocks have generated some of the first-order data required to trace the evolution of the continental crust, as well as constrain the sources of granitic batholiths. Detailed chemical and isotopic studies of modern island and continental volcanic arcs can determine the relative contributions of crust and mantle in young orogenic belts (e.g. Gill 1981; Thorpe 1982). However, the fact that large volumes of volcanic rocks are sporadically preserved in the geological record leads most workers who study Precambrian orogenic belts to focus on intrusive sequences. Granitic rocks therefore remain a key component to understanding crustal growth throughout most of the earth's history.

In this paper we highlight how the Lu–Hf and Re–Os isotope systems may be used in studies on the origin of granites and crustal evolution. These two isotope systems can provide information that is not obtainable from the commonly used isotope systems O, Rb–Sr, Sm–Nd and U–Th–Pb. The Lu–Hf isotope system will be shown to be very sensitive to the role of garnet during intracrustal melting or metamorphic processes, and the Re–Os isotope system to the presence of basalt composition precursors. Advances in analytical methods, as well as the identification of Os-rich phases in granitic rocks, will allow the Re–Os isotope system to be directly and broadly applied to continental crustal rocks for the first time. Although there is not yet a significant body of Hf and Os isotope data from the same samples, we hope to demonstrate that these

isotope systems have the potential to address some of the critical questions that remain regarding the origin of granites and the evolution of the continental crust.

1. Overview of isotopic studies of granitic rocks as applied to crustal evolution

For the purposes of the present discussion, the origin of granite and its bearing on crustal growth, it is useful to break down the applications of isotope geology to the origin of granitic rocks into three goals: (1) the determination of the relative contributions of crust and mantle in granitic rocks; (2) for granitic suites that are dominated by a crustal source component, determination of the age of the crustal source; and (3) for crust-dominated plutons, determination of the composition and depth of the crustal source. The applicability of the more common isotope systems (O, Sr, Nd and Pb) to these goals, according to the authors' biases, is illustrated in Table 1. Studies that have influenced Table 1 include those that evaluate the relative mass contributions of crust and mantle to batholiths (e.g. Kistler & Peterman 1973; DePaolo 1981a; Farmer & DePaolo 1983, 1984; Frost & O'Nions 1985; DePaolo et al. 1991; Johnson 1993), identification of the types of crustal components that may exist in granitic magmas (e.g. Taylor 1968; Zartman 1974; Zartman & Doe 1981; Zartman & Haines 1988) and the ages of crustal sources for crust-

Table 1 Applications of isotope geology to the origin of granite.

	Determination of percentage crust versus percentage mantle in granitoids		Crustal component: age of the crustal source		Crustal component: composition and depth of the crustal source	
Commonly used isotope systems						
O Isotopes	$+/-$	Possible only when crust is sedimentary in origin	$-$	No constraints	$+/-$	Only identify sedimentary sources; no depth constraints
Sr Isotopes	$+/-$	Not quantitative due to huge range in Sr isotope ratios of crust; most sensitive to upper crustal rocks. Insensitive to high-grade Rb-depleted crust	$-$	Rb/Sr ratios of the crust too variable to calculate ages of the crustal source	$+/-$	Most sensitive to upper crustal rocks; insensitive to middle and lower crustal rocks
Nd Isotopes	$+$	Relatively restricted range of Sm/Nd ratios in crust makes this system most suited to determining relative contribution of crust and mantle	$+$	Sm/Nd ratios of nearly all mafic to silicic rocks (except MORB) lie in a restricted range; Nd isotope ratios of the crustal source are well understood relative to age	$-$	Intracrustal processes do not fractionate Sm/Nd ratios in a manner that relates to depth or bulk composition
Pb Isotopes	$-$	The relatively high Pb contents of the crust makes mantle Pb contributions insignificant	$+/-$	Most sensitive to Early Archean components through anomalously high $^{207}Pb/^{204}Pb$ ratios	$+/-$	May distinguish radiogenic upper crust from cratonic (stable) middle and lower crust that may be generally non-radiogenic
'New' isotope systems as applied to genesis of granitic rocks						
Hf Isotopes	$+/-$	May be similar to Nd, although greater range in Lu/Hf ratios of the crust adds ambiguity; greater uncertainty in mantle isotope ratios is a further complication	$+/-$	Although generally similar to Nd, relatively large range in Lu/Hf ratios and mantle isotope compositions adds uncertainty	$+$	**Well positioned to detecting the presence or absence of garnet in deep mafic rocks or shallow silicic or pelitic rocks. Possible complications from accessory minerals**
Os Isotopes	$-$	Os content of fractionated mafic magmas very low and may be lower that those of silicic crust; potentially large ranges in crustal Os isotope ratios adds additional uncertainty	$+$	**Likely to be exceptional for detecting young, especially mafic crust that cannot be distinguished using other isotope systems; likely to be great uncertainty for old crustal sources**	$+/-$	Likely to be most sensitive to detecting young mafic crustal sources

dominated granitic rocks (e.g. Nelson & DePaolo 1985; Patchett & Arndt 1986; Bennett & DePaolo 1987), to name but a few.

However, two important questions remain concerning the origin of granitic magmas that cannot be adequately addressed with the O, Sr, Nd or Pb isotope systems: **can the lower and middle continental crust produce multiple generations of silicic melts over billion year periods; and are the 'mantle-like' isotope compositions of orogenic rocks reflective of major additions from the mantle or re-melting of mafic, possibly young, lower crust?**

Because intracrustal melting does not produce Rb/Sr, Sm/Nd and U/Pb fractionations that are sufficiently distinct from the intrinsic variations seen in most crustal rocks, these isotope systems are not sensitive to uniquely identifying intracrustal melting over the lifetime of a continent. In addition, the commonly used O, Sr, Nd and Pb isotope systems are essentially insensitive to the presence of young mafic lower crust. We divide this paper into two parts: the first is a discussion of the Lu–Hf isotope system and how it can be used to address the first question, and the second is a discussion of the Re–Os isotope system in light of the second question. How these relatively new isotope systems fit into studies of granitic rocks is summarised in Table 1.

2. Lu–Hf isotope evolution in the earth

In the broad sense, the Hf isotope evolution of the continental crust is similar to that of Nd, in that crustal rocks of increasing age have increasingly negative present day ε_{Hf} values and in that their age-corrected initial ratios generally follow the isotopic evolution of the chondritic to depleted mantle (Fig. 1). However, the Lu–Hf isotope evolution of the earth is distinct from that of the closest analogous system, Sm–Nd, because Lu/Hf fractionation in the mantle or crust is much more sensitive to the presence or absence of garnet than Sm/Nd fractionation. Recent Hf isotope work on mantle-derived basalts has highlighted the wide range in ε_{Hf} values for the depleted mantle, which is probably a result of the variable role for garnet in the source regions (e.g. Salters & Hart 1991; Johnson & Beard 1993). The fact that the present day ε_{Hf} values of the depleted mantle (MORB source) vary from near zero to $c. +20$ (Fig. 1), relative to a restricted range in ε_{Nd} values ($c. +8$ to $+12$), indicates that the Hf isotope evolution of the depleted mantle is best described graphically as defining a 'wedge' from 4·55 Ga to the present, encompassing both positive ε_{Hf} values and those near zero (Fig. 1). The large spread in initial ε_{Hf} values of Precambrian orogenic rocks is consistent with derivation from a depleted mantle of highly variable Hf isotope ratios (Fig. 1), compared with the relatively restricted range in ε_{Nd} values for such rocks (data sources cited in Fig. 1). This ε_{Hf}–time 'wedge' is a direct reflection of the unique and variable influence garnet has on Lu/Hf partitioning.

The strong influence of garnet in the continental crust is seen in measured Lu/Hf ratios, which are markedly lower than those of chondritic meteorites (Fig. 2). Regression of high-precision Lu/Hf ratios (determined by isotope dilution) that are available for granitic whole-rock samples from Precambrian orogenic terranes (3·8–1·7 Ga) produces an equation which describes the Lu/Hf ratio of silicic continental crust over time (T in Ga)

$$^{176}Lu/^{177}Hf_{CRUST}(T) = -0\cdot00157(T) + 0\cdot00857$$

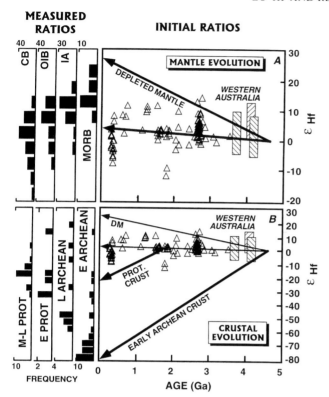

Figure 1 Hafnium isotope evolution in the mantle and crust. Initial compositions on ε_{Hf}–time plot (right-hand side) are from zircons, including the SHRIMP ion probe data from Early Archean detrital zircons from Western Australia (2σ errors shown as ruled bars; Kinny *et al.* 1991). The same initial Hf isotope compositions (Δ) plotted in parts (A) and (B). Histograms for present day Hf isotope compositions for mantle-derived lavas (part A) and continental crustal rocks (part B) are shown on the left-hand side. CB, continental basalts; OIB, ocean island basalts; IA, island arcs; MORB, mid-ocean ridge basalts; M-L PROT, mid- to late Proterozoic 'anorogenic' granitoids (SW U.S.A.); E PROT, early Proterozoic orogenic granitoids; L ARCHEAN, Late Archean (≈ 2.7 Ga) orogenic granitoids; and E Archean (*c.* >3 Ga) orogenic granitoids. Data from Patchett and Tatsumoto (1980a), Patchett *et al.* (1981), Patchett (1983a; 1983b), Stille *et al.* (1983, 1986, 1989), White and Patchett (1984), Hart *et al.* (1986), Smith *et al.* (1987), Woodhead (1989), Gruau *et al.* (1990), Barovich (1991), Salters and Hart (1991), Barovich and Patchett (1992), Chauvel *et al.* (1992), Corfu and Noble (1992), Schaltegger and Corfu (1992), Beard and Johnson (1993), Johnson and Beard (1993) and Johnson *et al.* (1995, unpublished data).

Two important observations can be made for Lu/Hf ratios from granitic rocks: (1) the exceptionally low Lu/Hf ratios strongly suggest derivation from sources that contained residual garnet (the effect is most striking for the Archean); and (2) few orogenic rocks, but most anorogenic granitic rocks, have significantly higher Lu/Hf ratios (although still less than chondrites) than those predicted by the crustal evolution model (Fig. 2). Melts derived from the mantle or crust that left garnet in the residue will have exceptionally low Lu/Hf ratios compared with melts that separated from sources without residual garnet (Fig. 2; Salters & Hart 1989, 1991; Beard & Johnson 1993; Johnson & Beard 1993). In contrast, Sm/Nd ratios are relatively insensitive to the presence or absence of residual garnet during melting of the crust or mantle, preventing the Sm–Nd isotope system from shedding light on the role of garnet during melting processes. Calculated Sm/Nd ratios for garnet-present or garnet-absent melting are essentially the same, particularly in view of the large range of Sm/Nd ratios measured for granitic rocks (Fig. 2).

3. Hf–Nd isotope balance of the outer earth

Hafnium–Nd isotope variations for ocean island basalts (OIB) have been known to be well correlated since the first Hf

isotope work (Patchett & Tatsumoto 1980a; Patchett 1983a, 1983b; Stille *et al.* 1983, 1986; Fig. 3). However, additional work on mid-ocean ridge basalts (MORB) (Salters & Hart 1989, 1991), continental basalts (Beard & Johnson 1993; Johnson & Beard 1993), as well as continental crustal rocks (references cited in Fig. 1), highlight the fact that Hf–Nd isotope variations in the earth have important subtleties (Fig. 3) that reflect distinct melting and metasomatic processes in different parts of the outer silicate earth. Recognising the need to emphasise these subtleties, Johnson and Beard (1993) defined the parameter $\Delta\varepsilon_{Hf}$, which is the deviation in ε_{Hf} value of a sample, at a given ε_{Nd} value, relative to the OIB Hf–Nd reference line

$$\Delta\varepsilon_{Hf} = \varepsilon_{Hf} - [1.36\varepsilon_{Nd} + 1.63]$$

Therefore, the $\Delta\varepsilon_{Hf}$ parameter serves as a useful means for highlighting differences in Hf isotope compositions for a given Nd isotope composition. Use of the OIB reference line in the $\Delta\varepsilon_{Hf}$ parameter is simply for convenience and does not imply a special role for the OIB source mantle in the genesis of a particular suite of rocks; in our discussion, it is the *relative* differences in $\Delta\varepsilon_{Hf}$ values for different rocks that is important, not their absolute $\Delta\varepsilon_{Hf}$ values.

We wish to stress an important aspect of ε_{Hf}–ε_{Nd} and $\Delta\varepsilon_{Hf}$–ε_{Nd} variations through time. Both Nd and Hf isotope evolution in the earth may be described using the linear equation

$$\varepsilon_{iT(1)} = \varepsilon_{iT(0)} + f_j * Q_i * [T(0) - T(1)]$$

where i = Nd or Hf and j = Sm/Nd or Lu/Hf (e.g. DePaolo 1988a). Because Q_i is similar for Nd and Hf ($Q_{Nd} = 25.1$ and $Q_{Hf} = 22.9$), the relative positions of reservoirs in the earth on ε_{Hf}–ε_{Nd} and $\Delta\varepsilon_{Hf}$–ε_{Nd} diagrams is essentially constant over the age of the earth. This fortuitous relationship (Q_i is a function of λ, parent/daughter ratio and bulk earth isotope composition) means that genetic interpretations based on where samples plot on these diagrams can be applied to rocks of all ages; this is an important aspect that is not generally true in correlations for most other radiogenic isotope systems.

Present day $\Delta\varepsilon_{Hf}$–ε_{Nd} variations for mantle-derived lavas and continental crust illustrate a striking Hf–Nd isotope balance for reservoirs that represent the outer (silicate) earth (Fig. 4). Archean and Proterozoic granitic crust from orogenic belts comprise the only reservoirs to exclusively plot at $\Delta\varepsilon_{Hf} < 0$ and $\varepsilon_{Nd} < 0$ (the 'SW quadrant'; Fig. 4). The present day Hf and Nd isotope compositions of Proterozoic and Archean *orogenic crust* are consistent with their moderately low Sm/Nd and exceptionally low Lu/Hf ratios (Fig. 2), which produces modern isotopic compositions that have $\Delta\varepsilon_{Hf} < 0$ and $\varepsilon_{Nd} < 0$. In addition, calculations using average crustal Sm/Nd and Lu/Hf ratios (e.g. Ben-Othman *et al.* 1994; compilation reported here), or models for Lu/Hf and Sm/Nd evolution of the crust (this study; DePaolo 1988b), always produce present day Hf and Nd isotope compositions for the crust that plot in the 'SW quadrant' ($\Delta\varepsilon_{Hf} < 0$, $\varepsilon_{Nd} < 0$). This conclusion holds regardless of the choice of initial isotopic composition (Fig. 4) and we consider it highly likely that most *orogenic crust* in the earth plots in the 'SW quadrant' today.

In contrast with orogenic continental crust, the mantle plots near $\Delta\varepsilon_{Hf} = 0$ (Fig. 4). By definition, OIB mantle has an average $\Delta\varepsilon_{Hf}$ value of zero and 90% of all analyses of OIBs plot within three $\Delta\varepsilon_{Hf}$ units of the OIB line (Figs 3, 4). Depleted mantle (MORB) has a wide range of $\Delta\varepsilon_{Hf}$ values over a restricted range of ε_{Nd}, which may reflect mixing between mantle that was depleted in the garnet peridotite stability field ('GM'; Fig. 4) and mantle that was depleted in the spinel peridotite stability field ('SM'; Fig. 4) (Johnson &

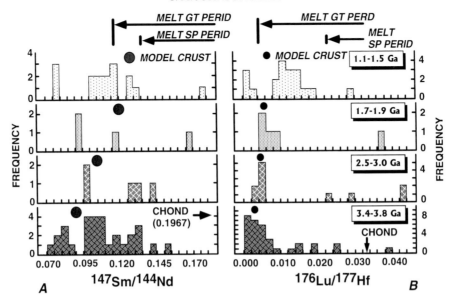

Figure 2 $^{147}Sm/^{144}Nd$ (A) and $^{176}Lu/^{177}Hf$ (B) and ratios (for same samples) for whole-rock granitic samples, separated by age. The lower three histogram sets represent data for orogenic rocks, whereas the top histogram (1·1–1·5 Ga rocks) represent data for 'anorogenic' granitic rocks. Lu/Hf ratios appear to vary systematically with age and model crust is shown as small closed circles (from this paper; see text). In contrast, Sm/Nd ratios do not appear to vary systematically with age and do not fit the model proposed by DePaolo (1988b) (shown as large circles). Arrows mark decreases in Sm/Nd and Lu/Hf ratios for 1% melt of garnet or spinel peridotite (as marked) of chondritic elemental ratios (melting parameters from Johnson & Beard 1993; Beard & Johnson 1993). The exceptionally low Lu/Hf ratios of orogenic granites strongly indicates that they were derived from sources that had residual garnet and that garnet was most important for the Archean. The significantly higher Lu/Hf ratios for Mid-Proterozoic 'anorogenic' granites reflects intracrustal processes, as discussed in text. Data sources as in Figure 1.

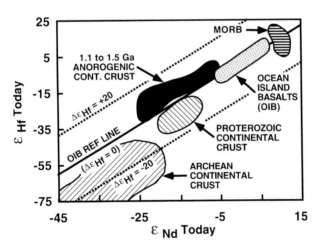

Figure 3   Present day ε_{Hf}–ε_{Nd} variations for whole-rock granitic samples, as well as ocean island basalts (OIB) and mid-ocean ridge basalts (MORB). OIB reference line from Johnson and Beard (1993), and equal to $\varepsilon_{Hf} = \varepsilon_{Nd}*1·36 + 1·63$. Lines of equal $\Delta\varepsilon_{Hf}$ values shown, which represent the deviation in ε_{Hf} value from the OIB reference line. Data sources for crustal fields are: *Archean* (Patchett *et al.* 1981; Smith *et al.* 1987; Barovich 1991); *Proterozoic* (Patchett *et al.* 1981); *anorogenic* (Barovich 1991; Barovich & Patchett 1992). Data for basalts from sources cited in Figure 1.

Beard 1993). The subcontinental mantle appears to form an array that lies between the 'NW quadrant' and the 'SE quadrant' (Fig. 4). This array has been interpreted to reflect mixing between an ancient mantle source that was metasomatised at shallow levels (in the spinel peridotite stability field), producing present day $\Delta\varepsilon_{Hf} > 0$ at $\varepsilon_{Nd} < 0$ (S1 and S2 components in Fig. 4), and the modern SM depleted mantle component (Fig. 4; Beard & Johnson 1993). Two mixing curves (lines with short dashes) are shown in Figure 4, both reflecting this mixing between melts that were derived from depleted spinel peridotite subcontinental mantle that was 'aged' isotopically for 1·8 billion years ('S1'; Fig. 4) and the

modern depleted mantle ('SM'; model depleted spinel peridotite, or an estimate for the depleted mantle beneath Colorado in the Early Proterozoic; Fig. 4). An alternative to producing the oblique subcontinental mantle array in this way would be mixing between ancient pelagic sediment, which contains relatively high Lu/Hf ratios compared with average continental crust (Patchett *et al.* 1984; White *et al.* 1986) and the SM mantle component (Fig. 4; Milling *et al.* 1994). The important point here is that processes which operate in the *shallow* mantle (spinel peridotite stability field) produce data arrays that lie between the 'NW quadrant' and 'SE quadrant' on a $\Delta\varepsilon_{Hf}$–ε_{Nd} diagram (Fig. 4).

Reservoirs in the earth that lie in the 'SW quadrant' and the 'NE quadrant' (Fig. 4) probably reflect melting and metasomatic processes that involve garnet. Until the recognition of Archean and Proterozoic orogenic crust in the 'SW quadrant' ($\Delta\varepsilon_{Hf} < 0$, $\varepsilon_{Nd} < 0$), there was no apparent isotope mass balance for the GM component (model depleted garnet peridotite) which sits in the 'NE quadrant' (Fig. 4). The present day $\Delta\varepsilon_{Hf}$ and ε_{Nd} values of ancient melts derived from batch or incremental melting of garnet peridotite (starting at 'Colorado depleted mantle') always plot in the 'SW quadrant' (Fig. 4). It is important to note that essentially the same 'SW quadrant' compositions are produced during ancient melting of garnet-bearing crustal rocks instead of garnet peridotite. It is a robust conclusion, therefore, that crust which plots in the 'SW quadrant' fundamentally owes its origin to melts which separated from sources that contained residual garnet.

4. The role of garnet in *intra*crustal processes and implications for the Lu–Hf isotope system

The present day, as well as initial, $\Delta\varepsilon_{Hf}$ and ε_{Nd} values of 'anorogenic' 1·1–1·5 Ga granitic rocks from the western U.S.A. plot in the 'NW quadrant' (Fig. 4), using data available to us at the time of writing. These unusual isotopic compositions

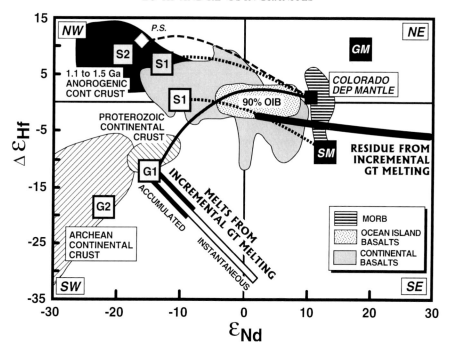

Figure 4 Present day $\Delta\varepsilon_{Hf}$–ε_{Nd} variations for the outer earth. Low-ε_{Nd} basalts from the subcontinental mantle are interpreted to reflect mantle metasomatism at shallow levels (fluids derived from spinel peridotite stability field, <80 km depth; Beard & Johnson 1993); 1% melts of depleted mantle source SM (model depleted spinel peridotite; Johnson & Beard 1993) at 1·8 and 2·7 billion years ago would have present day isotopic compositions of S1 (lower box) and S2, respectively. Also shown for comparison is a present day isotopic composition (upper S1 box) for a 1% melt generated 1·8 billion years ago from Colorado model depleted mantle in the spinel peridotite field (Beard & Johnson 1993). Mixing lines between the enriched metasomatic (melt) component (S1) and the modern depleted mantle shown by broken lines. P.S. is the present day isotopic composition of average pelagic sediment, assuming an age of 1·8 Ga. Mixing between ancient pelagic sediment or ancient enriched mantle and depleted mantle are the most likely explanations for the Hf–Nd isotope variations in the subcontinental mantle. Also shown are the present day isotopic compositions for 1% melts generated 1·8 and 2·7 billion years ago (G1 and G2, respectively) from Colorado model depleted mantle, assuming the phase mineralogy was that of garnet peridotite (>80 km depth). A mixing line between the melt component (G1) and original depleted mantle component is shown by solid curve. GM is a model depleted garnet peridotite (Johnson & Beard 1993). Also shown is the range of present day compositions of melt G1 assuming incremental melting of garnet peridotite (e.g. Milling *et al.* 1994). That present day Hf and Nd isotope compositions for Archean and Proterozoic orogenic crust (shown by diagonal-ruled fields) plot in the SW quadrant, suggesting that they were derived from melting a garnet-bearing source; the high $\Delta\varepsilon_{Hf}$ values of anorogenic granitic rocks (dark solid field) requires a different source (see text); data sources as in Figure 3.

require a more complex genesis than the single-stage melting of garnet-bearing sources that was proposed earlier for Archean and Proterozoic *orogenic* rocks. Although the tectonic setting of the mid-Proterozoic 'anorogenic' granites of the western U.S.A. is debated (e.g. Hoffman 1989; Windley 1993; Nyman *et al.* 1994), many workers consider these rocks to reflect the deep melting of crustal sources that are commonly aluminous (e.g. DePaolo 1981b; Anderson 1983; Anderson & Bender 1989). Direct melting of model depleted spinel peridotite ('SM'; Fig. 4), which produces the S1 and S2 components in Figure 4, is not a likely explanation for the unusual Hf and Nd isotope compositions of the 'anorogenic' granites, given the general consensus that these rocks represent nearly 100% crustal melts. Instead, their positive $\Delta\varepsilon_{Hf}$ values (at negative ε_{Nd} values) and anomalously high Lu/Hf ratios (Fig. 2) are best explained through the melting of a crustal source that had been *previously* melted with garnet as a residual phase. It will be shown later in this section that if such intracrustal melting occurred 300–600 Ma before the generation of the 1·1–1·5 Ga 'anorogenic' granites, very high $\Delta\varepsilon_{Hf}$ values for the residual crust may be produced; these high $\Delta\varepsilon_{Hf}$ values would be reflected in the younger ($\approx 1\cdot1$–1·5 Ga) melting episodes that produced the 'anorogenic' magmas.

Garnet is an important mineral that is present during intracrustal differentiation, not only as a primary metamorphic mineral in pelitic compositions, but also for dehydration

melting reactions involving most of the lower crust, particularly in areas of thick crust, as well as the middle crust for intermediate to silicic composition sources (Fig. 5 and references cited therein). The major dehydration melting reactions for pelitic rocks are

$$\text{Muscovite} + \text{quartz} + \text{plagioclase}$$
$$\Rightarrow \text{K-feldspar} + \text{biotite} \pm \text{cordierite} \pm \text{garnet} + \text{melt}$$

and

$$\text{Biotite} + \text{aluminosilicate} + \text{quartz} + \text{plagioclase}$$
$$\Rightarrow \text{garnet} + \text{K-feldspar} + \text{melt}$$

For amphibolite to tonalite protoliths, the major dehydration melting reaction is

$$\text{Amphibole} + \text{plagioclase}$$
$$\Rightarrow \text{garnet} + \text{clinopyroxene} + \text{quartz} + \text{melt}$$

A large number of recent fluid-absent (dehydration) melting experiments indicate that mafic composition sources will produce garnet during melting at ≈ 50–60 km depths; and intermediate to silicic composition rocks that did not originally contain garnet will produce garnet during melting at depths of less than 30 km (Fig. 5 and references cited therein). High-Al silicic ('S-type') rocks that did not originally contain garnet will produce garnet during melting at depths as shallow as

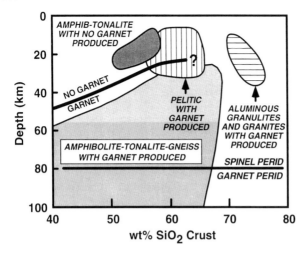

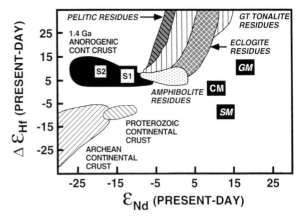

Figure 5 Depth–composition relations for dehydration melting experiments of various crustal lithologies. Garnet is produced in rocks that did not originally contain garnet during the dehydration-melting of intermediate to silicic composition rocks in the middle and lower crust; garnet is also produced during melting of pelitic rocks at intermediate crustal depths, demonstrating that garnet should be a common phase involved in magma production in the crust. Data from Green and Lambert (1965); Millhollen and Wyllie (1974), Merrill and Wyllie (1975), Stern and Wyllie (1978), Clemens and Wall (1981), Sekine *et al.* (1981), Thompson (1982), Le Breton and Thompson (1988), Rutter and Wyllie (1988), Vielzeuf and Holloway (1988), Beard and Lofgren (1989, 1991), Patiño Douce and Johnston (1991), Rapp *et al.* (1991), Rushmer (1991, 1993), Beard *et al.* (1993, 1994) and Skjerlei *et al.* (1993).

Figure 6 Present day $\Delta\varepsilon_{Hf}$–ε_{Nd} variations for residues generated by non-modal batch melting of continental crust of modal S1 initial isotopic and elemental abundances (Fig. 4) at 1·8 Ga, followed by isotopic evolution to the present day. S1 source is considered to be a likely continental source only for early orogenic crust in primitive arcs that are not strongly LREE enriched (see text and Beard & Johnson 1993). Four fields are shown, which encompass the range in initial and melting modes for four rock types noted in Table 2. Also shown are reference reservoirs and selected data fields from Figure 4. Present day isotopic compositions of residues calculated for 1–50% melt removal (or less if a specific model results in exhaustion of a phase). Initial and melting modes taken from dehydration melting experimental data and models of Stern and Wyllie (1978), Sekine *et al.* (1981), Le Breton and Thompson (1988), Waters (1988), Vielzeuf and Holloway (1988), Beard and Lofgren (1991), Rapp *et al.* (1991), Rushmer (1991). K_d values summarised in Johnson and Beard (1993), with the assumption that K_d values for clinopyroxene biotite, and amphibole are equal.

10–15 km (Fig. 5), overlapping the depths where pelitic rocks (which commonly have intermediate SiO_2 contents) will produce garnet during melting.

The $\Delta\varepsilon_{Hf}$ values that are produced over time in residual ('restite') crustal sources that experienced prior melt removal will reach very positive $\Delta\varepsilon_{Hf}$ values through time if ancient melt removal involved residual garnet. If later granitic magmas, such as the 1·1–1·5 Ga 'anorogenic' magmas, were generated by melting this residual crust (after some period of isotopic evolution), the younger anorogenic magmas would have high $\Delta\varepsilon_{Hf}$ values. The magnitude of the increase in $\Delta\varepsilon_{Hf}$ values is dependent on the source mineralogy and melting reactions, as well as the starting isotopic composition.

Assume, as a first example, that an original crust of S1 composition is melted 1·8 billion years ago and garnet is present in the residue. The relatively high original Lu/Hf ratios of this source (due to its origin from garnet-absent melting of the mantle, discussed in Section 3), and further increase in the Lu/Hf ratio of the residual crust because of the garnet that is present following melt extraction, produces exceptionally high $\Delta\varepsilon_{Hf}$ values through time (up to +150 for present day $\Delta\varepsilon_{Hf}$) in the residue (Fig. 6; Table 2). This example serves to illustrate the maximum increases in $\Delta\varepsilon_{Hf}$ values that may occur (using mantle K_d values). We consider an initial Hf isotopic composition as radiogenic as S1 to be rare in the

continental crust (see Section 3), although it may be an appropriate composition for primitive arc rocks that might be produced during the early stages of crust formation in an orogenic belt (e.g. Beard & Johnson 1993 and references cited therein). Note that no increase in $\Delta\varepsilon_{Hf}$ for the residue occurs from the melting of crustal sources that did not originally contain garnet, nor produced garnet during melting ('amphibolite residues' in Fig. 6).

An initial isotopic composition for the crustal source that lies in the 'SW quadrant' (G1 composition; Fig. 7) seems to be the most likely crustal composition that would be involved in intracrustal melting (Section 3). The relatively low initial Lu/Hf ratios of the G1 source (due to its separation from a garnet-bearing mantle source, discussed earlier), requires an even larger increase in Lu/Hf ratio in the crustal residue during an *intra*crustal melting event, if high $\Delta\varepsilon_{Hf}$ values are to be produced over time. For a G1 crustal source, the largest increases in $\Delta\varepsilon_{Hf}$ values for the residue always occurs for melting conditions that involve the *generation* of garnet during melting, rather than simply having garnet occur in the starting mode. This is well illustrated for melting eclogite, where the melting proportions were assumed to be generally equal to those of the modes (Fig. 7). The low initial Lu/Hf ratios of

Table 2 Starting and melting modes for Hf–Nd melting models.

	Pelite		Garnet tonalite		Amphibolite		Eclogite	
	Initial mode	Melting mode	Initial mode	Melting mode	Initial mode	Melting mode	Initial mode	Melting mode
Biotite	0·1–0·5	0·8	0·1	0·6				
Garnet	0·01–0·20	−0·6	0·05–0·20	−0·6			0·4–0·6	0·4–0·6
Feldspar	0·15–0·40	0·4	0·5	0·2	0·38–0·68	0·5		
Quartz	0·15–0·49	0·4	0·10–0·25	0·2	0·1	0·5		
Amphibole			0·1	0·6	0·2–0·5	0·8		
Clinopyroxene					0·01	−0·4	0·4–0·6	0·4–0·6
Orthopyroxene					0·01	−0·4		

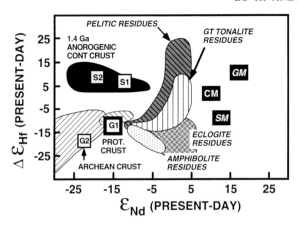

Figure 7 Present day $\Delta\varepsilon_{Hf}$–ε_{Nd} variations for residues generated by non-modal batch melting of continental crust of model G1 initial isotopic and elemental abundances (Fig. 4) at 1·8 Ga, followed by isotopic evolution to the present day. Melting fields and reference fields (Fig. 4) are the same as in Figure 6. The G1 source is considered to be the most likely composition for continental crust (see text).

eclogite of G1 composition prevents an increase in $\Delta\varepsilon_{Hf}$ in the residue after melting (although ε_{Hf} certainly increases), and in fact many model solutions indicate a decrease in $\Delta\varepsilon_{Hf}$ through time (this would not be the case for eclogite of S1 or chondritic initial Lu/Hf ratios; Fig. 6).

The production of positive $\Delta\varepsilon_{Hf}$ values in a residual crust after melt extraction requires melting reactions that produce garnet, if the initial isotopic and elemental composition of the crust before melting lies in the 'SW quadrant', the most likely quadrant for most continental crust. The requirement of garnet-producing melting reactions, rather than simply garnet in the mode, is consistent with experimental data on the dehydration melting reactions of pelitic and intermediate to silicic composition rocks in the lower and middle crust (Figs 5, 6, 7, and references cited therein). We envison, therefore, that the 1·1–1·5 Ga 'anorogenic' granites of the western U.S.A. owe their distinctive high $\Delta\varepsilon_{Hf}$ values to remelting of deep crustal sources that had been previously melted under conditions that produced garnet during dehydration melting reactions. It seems likely that the initial melting events occurred at the time of the major crust-forming orogenic events in this region at $\approx$ 1·7 Ga. This scenario also explains the relatively high Lu/Hf ratios of the 1·1–1·5 Ga anorogenic granites (Fig. 2), and the fact that *both* the measured and initial $\Delta\varepsilon_{Hf}$ values for the granites are markedly positive (Fig. 4).

Two problems remain: (1) an explanation of the low initial ε_{Nd} values of the 'anorogenic' granites, when the melting models involving garnet predict an increase in ε_{Nd} in the residual source (Figs 6, 7); and (2) prevention of the initial melting event ($\approx$ 1·7 Ga?) from creating a refractory assemblage that cannot be melted later. In terms of the first problem, the most likely explanation would seem to lie in either the accessory mineral assemblages that may be present, and/or the effects of REE metasomatism of the source rocks during dehydration metamorphism or interaction with the melt-inducing basaltic magmas. Accessory minerals such as apatite may commonly be residual during dehydration melting (Beard *et al.* 1993, 1994) and, in fact, the Lu/Hf ratios may be significantly changed from those modelled earlier, depending on the accessory minerals involved and whether they are consumed or produced during dehydration melting in the crust. The Lu/Hf ratios for most accessory minerals such as apatite and zircon will be much lower than those of garnet and would tend to counteract the effect of garnet on the production of high $\Delta\varepsilon_{Hf}$ values through time. In addition, the use of mineral/melt K_d values for mafic and ultramafic systems,

as are currently available, is not appropriate for the liquid compositions and lower temperatures of silicic melts that are produced in the continental crust. We might, in fact, expect Lu/Hf partitioning to be more extreme in crustal rocks, producing even more rapid increases in $\Delta\varepsilon_{Hf}$ values in garnet-bearing residues. Additional Lu–Hf and Sm–Nd data will be required from migmatitic and partially melted terranes to assess these factors. Lastly, it is clear that the interaction between basalt and crust during melting can produce significant changes in the major element compositions of the crustal melt (e.g. Patiño Douce 1995). It is possible that a LREE-enriched basalt may impart that enrichment to the crust during melting, preventing the increase in ε_{Nd} that is predicted to accompany the increase in $\Delta\varepsilon_{Hf}$ in Figures 6 and 7.

In terms of the second problem, we note that our view of crustal melting is changing rapidly with new experimental work and assessment of the dynamics of melting (e.g. Brown *et al.* 1995). Although we might guess that previously melted crust would commonly be too refractory to melt again, recent work has highlighted the importance of deformation (e.g. Rushmer 1995), basalt–crust interaction (e.g. Patiño Douce 1995), crustal heterogeneity (e.g. Skjerlie *et al.* 1993) and modal abundances and melt stoichiometry (e.g. Patiño Douce & Johnston 1991) in determining the melt fractions that can be produced in the crust and whether the crust can be melted multiple times. We envision, therefore, that large regions of the lower and middle crust may be partially melted during the intrusion of large volumes of magma during the orogenic activity that produces the initial crust formation events in a region. The result would be a heterogeneous crust that had experienced variable shifts in modal abundances and volatile fluxing, conditions which allow significantly greater melt production in a given section of crust than has been traditionally envisioned (Patiño Douce & Johnston 1991; Skjerlie *et al.* 1993).

5. Re–Os isotope evolution in the earth

The Re–Os isotope system stands in marked contrast with the Rb–Sr, Sm–Nd, Lu–Hf and U–Th–Pb systems because of its chalcophile/siderophile behaviour (e.g. Wolf & Anders 1980; Morgan *et al.* 1981; Morgan 1986; Birck & Allègre 1994) compared with the lithophile behaviour of the other systems. The large contrast in the partitioning behaviour of Re and Os in sulphide and silicate systems (e.g. Barnes *et al.* 1985; Fleet *et al.* 1991, 1993) has produced huge elemental abundance (seven orders of magnitude) and Re/Os ratio (six orders of magnitude) variations in the outer earth relative to the lithophile radiogenic isotope systems that have been used to explore crust–mantle evolution (Figs 8, 9, 10). Because of the analytical challenges of the Re–Os isotope system, early work focused on meteorites, ultramafic rocks and platinum group element (PGE) minerals and metal alloys in economic mineral deposits (e.g. Allègre & Luck 1980; Hart & Kinloch 1989; Martin 1989; Walker & Morgan 1989). Improvements in analytical methods have allowed extensive work on basalts (primarily oceanic) in the last half decade. Only recently, however, has it been possible to reliably work on old rocks, which demand reproducible spike equilibration and low-blank analytical methods (e.g. Shirey & Walker 1995).

Although the phases that host Re and Os in the mantle and crystallising basalts remain unclear, sulphides and PGE metal alloys seem to be the likely repositories for these elements (e.g. Barnes *et al.* 1985; Hattori & Hart 1991; Hattori *et al.* 1991; Reisberg *et al.* 1991; Fleet *et al.* 1993; Marcantonio *et al.* 1993). Therefore, the traditional approaches for modelling the evolution of lithophile radiogenic isotopes cannot be applied

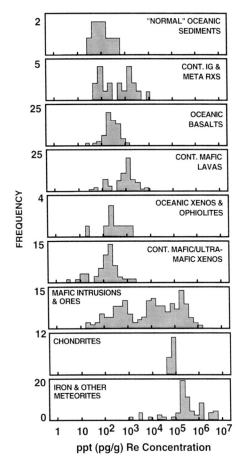

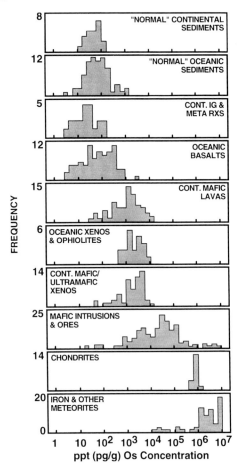

Figure 8 Histograms of log Re concentrations in meteorites and various mantle and crustal rocks in the earth. In the continental crust, ore deposits and related mafic intrusions contain the highest Re abundances, whereas oceanic and continental basalts, intermediate to silicic composition continental igneous and metamorphic rocks, and sedimentary rocks other than organic-rich shales ('normal' oceanic sediments) have much lower Re contents, generally varying from 10 to 1000 ppt. Data from Morgan and Lovering (1967), Morgan *et al.* (1981, 1992), Luck and Allègre (1983, 1984, 1991), Palmer and Turekian (1986), Esser and Turekian (1988, 1993), Walker *et al.* (1988), Lambert *et al.* (1989, 1994, 1995), Martin (1989), Ravizza and Turekian (1989, 1992), Walker and Morgan (1989), Walker *et al.* (1989a, 1989b, 1991a, 1991b, 1994), Reisberg *et al.* (1991, 1993), Ravizza *et al.* (1991), Dickin *et al.* (1992), Ellam *et al.* (1992), Horan *et al.* (1992, 1994), Pegram and Allègre (1992), Pegram *et al.* (1992), Hauri and Hart (1993), Koeberl and Shirey (1993), Marcantonio *et al.* (1993, 1994), Carlson and Irving (1994), Koeberl *et al.* (1994), Martin *et al.* (1994), Roy-Barman and Allègre (1994, 1995) and Peucker-Ehrenbrink *et al.* (1995).

Figure 9 Histograms of log Os concentrations in meteorites and various mantle and crustal rocks in the earth. As with Re, ore deposits and related mafic intrusions contain the highest Os abundances. Primitive (non-fractionated) basaltic magmas have Os contents ≈ 1000 ppt, although the Os contents for basaltic rocks extend to very low levels (10 ppt) with only modest amounts of fractionation. Osmium contents for the continental crust generally lie between 10 and 100 ppt, which presents considerable analytical challenges. Sediments that are termed 'normal' refer to non-organic-rich rocks. Data sources from Figure 8.

easily to the Re–Os system. Sulphide liquid segregation is clearly important in determining the Re/Os partitioning in some mafic/ultramafic magmas, and may also be a factor in Re/Os partitioning in intermediate to silicic composition magmas, depending on the degree to which such magmas can become saturated in sulphide and the sulphide composition (e.g. Carroll & Rutherford 1985). Similarly, although PGE minerals and metal alloys are important in influencing Re/Os partitioning in mafic magmas (e.g. Hattori & Hart 1991; Hattori *et al.* 1991), it is unknown how such phases may affect Re/Os behaviour in more evolved magmas. Overall, the bulk distribution coefficient for Os during melting of the mantle is high ($\geqslant 10$) and that of Re is low ($\leqslant 0.1$) (e.g. Hertogen *et al.* 1980; Walker *et al.* 1988; Hauri & Hart 1993; Martin *et al.* 1994), indicating that the oceanic and continental crust is likely to have Re/Os ratios that are far greater than those of the mantle (Fig. 10).

Re/Os and present day ^{187}Os/^{188}Os ratios of old continental crustal rocks are anticipated to be very high (e.g. Allègre &

Luck 1980) based on the large increase in Re/Os during even small amounts of basalt fractionation (Fig. 11) and the isotopic compositions measured for continentally derived sediments (Esser & Turekian 1993). The most primitive, unfractionated basaltic magmas may have a Re/Os ratio of ≈ 1, which over 2 billion years would evolve to a γ_{Os} value of ≈ 100 (Fig. 11). Fifty per cent crystallisation of basaltic magmas, which may produce basaltic andesite to andesite major element compositions, would be predicted to produce extremely high Re/Os ratios (≈ 1000), such that their γ_{Os} values would be $\approx 100\,000$ after 2 billion years (Fig. 11). The present day γ_{Os} values of the continents are estimated to be ≈ 1000, based on measurements of sediments (Esser & Turekian 1993), similar to the value calculated for moderate fractionation of basaltic magmas and isotopic evolution over ≈ 2 billion years (Fig. 11).

If the apparently high Os K_d value that is suggested from basalt studies is applicable for extensive fractionation of magmas to intermediate and silicic compositions, or during partial melting in the crust, the Os contents of granitic crust would be so low (Fig. 9) that it would not be possible to analyse such materials. Moreover, the Re/Os ratios would be expected to be so high, and the ancient crust so radiogenic (Figs 10 and 11), that meaningful initial Os isotope ratios would be impossible to obtain, although age information might be possible. It is understandable, therefore, that there

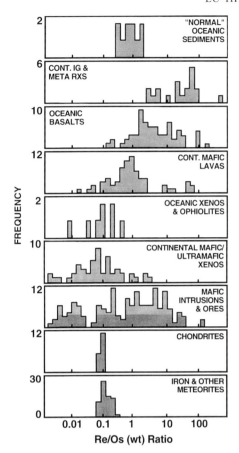

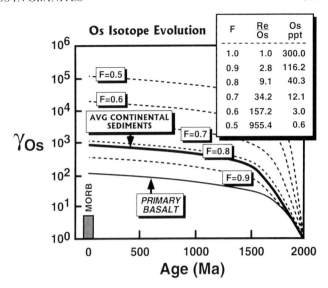

Figure 11 Predicted evolution of γ_{Os} values for 2 billion year old basaltic rocks that have undergone 10–50% fractional crystallisation ($F=0.9$–0.5), assuming Os $K_d=10$ and Re $K_d=0.1$. Estimates for primary basalt compositions from Martin *et al.* (1994) and the initial Os isotope composition is assumed to be chondritic at 2 Ga. We stress that most basalts will have significantly higher Re/Os ratios than our primary basalt values. The evolution of average continental sediments calculated assuming a present day γ_{Os} value of 1000 (Esser & Turekian 1993), an assumed age of 2 Ga and initial Os isotope composition equal to that of chondrites at 2 Ga. The strong curvature for isotopic evolution curves between 1.5 and 2.0 Ga is an artifact of fixing initial γ_{Os} to unity for the purposes of plotting γ_{Os} on a log scale. The calculations illustrate the large increase in Re/Os ratios that occurs during basalt fractionation, which quickly produces very radiogenic present day γ_{Os} values for mafic crust. Range of γ_{Os} for MORB also shown.

Figure 10 Histograms of log Re/Os (wt) ratios in meteorites and various mantle and crustal rocks in the earth. Meteorites have very restricted and low Re/Os ratios compared with those measured for mafic to silicic composition oceanic and continental rocks. The wide range in Re/Os ratios of ore deposits and related mafic intrusions reflects the diversity of host phases for Re and Os, which strongly control the Re/Os ratio (e.g. sulphides, which often have very high Re/Os ratios compared with PGE metal alloys, which may have very low Re/Os ratios). For basaltic rocks, the Re/Os ratios strongly increase during even slight amounts of fractionation, suggesting that the continental crust should contain the highest Re/Os ratios in the outer earth. Data sources from Figure 8.

have been few attempts at a systematic study of the Os isotope composition of continental crust through the direct measurement of such materials.

6. What can the Re–Os isotope system tell us about the origin of granites?

The very high Re/Os ratios for common basaltic rocks (Fig. 10) indicate that even young mafic crust will produce highly radiogenic Os isotope ratios in a short period of time (Fig. 11), showing that the Re–Os isotope system will be the most sensitive isotope system for detecting such crust in orogenic systems. Assuming that the difficulties of analysing crustal rocks, which contain very low Os contents (Fig. 9) can be overcome and that the system is resistant to open system behaviour during metamorphism, the Re–Os isotope system holds great promise for studying orogenic systems because it will detect remelted basaltic crust even over relatively short residence times.

Recent work indicates that oxide phases in granites may provide the answer to the difficulties involved in applying the Re–Os isotope system to ancient continental rocks, particularly granitic rocks. Stein *et al.* (1992) made the important discovery that magnetite in granites may contain Os contents that are markedly higher than those of whole rocks and that the Re/Os

ratios of some oxides may be sufficiently low, particularly for non-mineralised granites, that meaningful initial Os isotope ratios might be calculated. Our work (Johnson & Shirey 1995, unpublished data) on two distinct batholithic terranes indicates that magnetite may indeed be promising as a phase that contains relatively high Os contents and low Re/Os ratios, particularly for silicic samples (Fig. 12). Although magnetite mineral separates seem to contain very high Os contents relative to whole rocks (Figure 12), it is still unclear in what forms Re and Os occur in magnetite and it is likely that PGE metal alloy or sulphide inclusions are the true repository for Re and Os.

Our initial Re–Os isotope work has been targeted on two batholithic suites: (1) the Proterozoic (1.87–1.83 Ga) Penokean orogenic terrane in northern Wisconsin (Sims *et al.* 1989; Van Wyck 1995) and (2) the Mesozoic Sierra Nevada batholith, which forms an important component to the Cordilleran orogenic terrane of the western U.S.A. (e.g. Hamilton 1978). Granitic rocks analysed from both terranes have initial ε_{Nd} values that are significantly lower than those of the mantle, indicating an old crustal component in the magmas (Kistler *et al.* 1986; Van Wyck 1995; Kistler 1995, pers. comm.; Miller 1995, pers. comm.). Most of the Penokean rocks that have been analysed probably represent nearly 100% crustal melts based on their exceptionally low ε_{Nd} values, Pb isotope ratios that indicate an Archean source, the occurrence of inherited zircon populations as old as 3.2 Ga and the fact that most were intruded into Archean crust that is as old as 3.2 Ga (Van Wyck 1995).

It is striking that the Re and Os contents of magnetite from the crust-dominated Penokean suite are markedly higher than those of the Sierra Nevada suite (Fig. 12). The Os contents are highest and Re contents lowest in magnetite from the

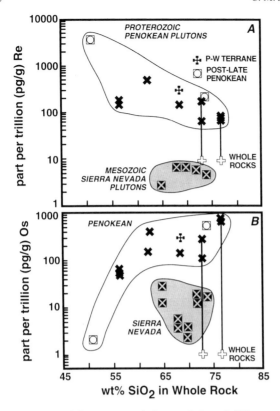

Figure 12 Re and Os contents, relative to whole-rock SiO_2 contents, for magnetite mineral separates from the Early Proterozoic Penokean Orogen in Wisconsin and the Mesozoic Sierra Nevada batholith in California. For the Penokean samples, a striking relation is observed of decreasing Re contents (and Re/Os ratios) and increasing Os contents with increasing whole-rock SiO_2 contents, suggesting that magnetite in granitic rocks may be a promising phase for Re–Os isotope studies of silicic continental crust. Corresponding whole rocks for two samples have Re and Os contents at near-blank levels, and they have been arbitrarily plotted at 1 ppt. The relatively low Re/Os ratios of the magnetite separates from the most silicic rocks are sufficiently low so that meaningful initial Os isotope ratios may be calculated for Precambrian rocks. The Re and Os contents for magnetite separates from the Sierra Nevada batholith are significantly lower than those for the Penokean rocks, although still sufficiently high for precise isotopic analysis. For comparison the average Re and Os contents of ocean island basalts (OIB) are shown. Data for magnetite from Johnson and Shirey (1995, unpublished data).

silicic Penokean rocks (Fig. 12). Re/Os (wt) ratios for magnetite vary from > 10 for mafic composition rocks to < 0·1 for silicic rocks. These concentration variations apparently make magnetite ideal for high-precision Os isotope analysis and the calculation of meaningful initial Os isotope ratios. This important discovery suggests that magnetite in some silicic rocks is well suited to studying the Re–Os isotope evolution of the continental crust. It is not possible to determine if the wide range in the Re and Os contents of magnetite relative to whole-rock SiO_2 contents is due to variations in apparent K_d values or variations in whole-rock or magmatic Re and Os contents because whole-rock contents are so low (Fig. 12).

Initial $^{187}Os/^{188}Os$ ratios, as expressed as γ_{Os} values (per cent deviation from chondrite $^{187}Os/^{188}Os$ ratios; Walker *et al.* 1989a, 1994), for the Penokean granitic rocks vary from those near chondrite ($\gamma_{Os}=0$) to c. +100, which overlaps the Os isotope compositions for mafic oceanic and continental basalts (Fig. 13). It is striking that a significant number of the Penokean granitic rocks have initial γ_{Os} values which overlap those of the mantle, considering the dominant ancient (Archean) crustal component that is in the rocks (above). The initial γ_{Os} values for the Penokean granites are significantly less radiogenic than estimates for average continental crust

based on continentally derived sediments (Fig. 13; Esser & Turekian 1993). These results, for the first time, provide robust initial γ_{Os} values for Precambrian continental rocks which suggest that some parts of the continents may have very non-radiogenic Os isotope ratios. It is unlikely that the Penokean rocks interacted to any degree with mafic crust, which would be expected to be very radiogenic (Fig. 11). The data presented by Stein *et al.* (1992) suggest that such non-radiogenic crust may also be recorded by Cenozoic granites. The abundance and distribution of continental crust that has non-radiogenic Os isotope ratios remains unknown, as well as the mechanisms by which non-radiogenic Os in such settings can be incorporated into granitic magmas. These initial results will have important implications for the Re–Os isotope evolution of the continental crust, as well as providing a framework for interpreting apparent temporal variations in Os isotope compositions of the continents based on oceanic sediments (Fig. 13; Pegram *et al.* 1992; Peucker-Ehrenbrink *et al.* 1995).

Initial γ_{Os} values for Sierra Nevada granitic rocks are markedly more radiogenic than those of the Penokean terrane and overlap those measured for oceanic and continental sediments (Fig. 13). The contrast in initial γ_{Os} values between the Penokean and Sierra Nevada granites is likely to reflect contrasts in the basement of these arcs; a large portion of the Sierra Nevada arc was probably built on young basaltic basement (Kistler & Peterman 1973), in contrast with the major part of the Penokean terrane studied, which was built on Archean crust (Sims *et al.* 1989; Van Wyck 1995). At present, we interpret the apparently low Os contents of the Sierran magmas (as inferred from the low Os contents of analysed magnetite), as well as their radiogenic Os isotope compositions, to reflect a major component in the magmas of young (Mesozoic) mafic crust that may underlie the batholith. Such crust may be very radiogenic in Os isotope composition (Fig. 11) and yet would have mantle-like O, Sr, Nd and Pb isotope ratios. We envision that an early to mid-Mesozoic mafic lower crust beneath the Sierra Nevada batholith (e.g. Domenick *et al.* 1983) would produce exceptionally radiogenic Os isotope ratios that may be later sampled during the major Late Cretaceous magmatic pulse in the batholith (Chen & Moore 1982). That many of the zoned plutonic complexes in the Sierra Nevada batholith have mafic cumulate components (e.g. Bateman & Chappell 1979) may be a likely explanation for the apparently low Os contents of the Sierra Nevada granites, compared with the higher inferred Os contents for the Penokean granites, which are interpreted to be nearly 100% crustal melts.

9. Summary and conclusions

We have discussed the application of two isotope systems that have been thus far little applied to the origin of granites, intracrustal igneous and metamorphic processes and the evolution of the continental crust. Promising recent data, however, indicate that the Lu–Hf and Re–Os isotope systems may provide new ways of looking at the origin of granites and the evolution of the crust. Application of these systems to granitic rocks presents formidable analytical challenges, including obtaining reproducible spike–sample isotope equilibration for Lu/Hf and Re/Os, both of which cannot be obtained through conventional approaches that are followed in Rb–Sr, Sm–Nd and U–Th–Pb isotope studies (Scherer *et al.* 1995; Shirey & Walker 1995). Routine Lu–Hf isotope analysis requires the standardisation of intricate chemical separation and mass spectrometry methods (Patchett & Tatsumoto 1980b; Barovich *et al.* 1995) and the development of low-abundance Re–Os isotope analysis of ancient rocks has only

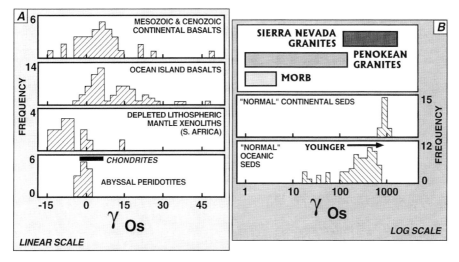

Figure 13 Histograms of present day γ_{Os} values for mantle (A) and crustal rocks (B). Note linear scale on left (A) and log scale on right (B). Although the mantle is significantly variable in γ_{Os} values relative to the spread observed for chondrite meteorites, the range is very small compared with the radiogenic γ_{Os} values inferred for the continental crust from sedimentary rocks that are not organic-rich ('normal' sediments). It is striking that the initial γ_{Os} values for Penokean rocks that contain a dominant crustal component (based on low ε_{Nd} values) range from near chondritic to +100, significantly less than that estimated for Precambrian continental crust based on sediment analyses. In contrast, the much younger Sierran batholith samples are more radiogenic, with γ_{Os} values that overlap those of average continental crust. Arrow for 'normal' oceanic sediments notes general trend of increasing γ_{Os} with decreasing age for Cenozoic sediments. Data for Sierra Nevada and Penokean samples from Johnson and Shirev (1995, unpublished data); other data sources from those cited in Figure 8.

become possible with new analytical methods (e.g. Shirey & Walker 1995).

The highly sensitive nature of Lu/Hf partitioning to the role of garnet during intracrustal differentiation suggests that the Lu–Hf isotope system may provide new insights into evaluating regions of the crust that have been repeatedly melted and at what depths; this information is not possible to obtain with the Rb–Sr, Sm–Nd or U–Th–Pb isotope systems. Garnet-bearing crust that has been partially depleted by melting is likely to be common in the crust, not only for pelitic compositions, but also for more typical metaluminous compositions because many vapour-absent melting reactions produce garnet, depending on the bulk composition and depth. Hafnium–Nd isotope variations for various reservoirs in the crust and mantle provide a striking mass balance for at least the outer earth that can be set into a context of deep and shallow level melting and metasomatic processes that has not been previously possible.

Although Re–Os isotope studies of fractionated basaltic rocks suggest that crustal rocks should have such high Re/Os ratios that it is impossible to calculate meaningful initial Os isotope ratios for Precambrian granites, recent data demonstrate that oxide minerals may provide a solution to these problems. Moreover, the data discussed here indicate that some parts of the continental crust may contain very non-radiogenic Os isotope ratios, similar to those found in the mantle, whereas others may be fairly radiogenic. If these differences can be ascribed to major differences in protolith types (e.g. basalt or granite), then the combined application of the Lu–Hf and Re–Os isotope systems should yield major new constraints on how continental crust is ultimately extracted from the mantle and differentiated.

10. Acknowledgements

We thank Brian Beard, Tom McCandless and Jeff Vervoort for reviews of an early version of the paper. This work was supported by NSF grants EAR-9105966 and EAR-9304455.

References

Allègre, C. J. & Luck, J.-M. 1980. Osmium isotopes as petrogenetic and geological tracers. EARTH PLANET SCI LETT **48**, 148–54.

Anderson, J. L. 1983. Proterozoic anorogenic granite plutonism of North America. GEOL SOC AM MEM **161**, 133–54.

Anderson, J. L. & Bender, E. E. 1989. Nature and origin of Proterozoic A-type granitic magmatism in the southwestern United States of America. LITHOS **23**, 19–52.

Barnes, S.-J., Naldrett, A. J. & Gorton, M. P. 1985. The origin of the fractionation of platinum-group elements in terrestrial magmas. CHEM GEOL **53**, 303–23.

Barovich, K. M. 1991. *Behavior of Lu–Hf, Sm–Nd and Rb–Sr isotopic systems during processes affecting continental crust.* Ph.D. Thesis, Univ. Arizona-Tucson.

Barovich, K. M. & Patchett, P. J. 1992. Behavior of isotopic systematics during deformation and metamorphism: a Hf, Nd and Sr isotopic study of mylonitized granite. CONTRIB MINERAL PETROL **109**, 386–93.

Barovich, K. M., Beard, B. L., Cappel, J. B., Johnson, C. M., Kyser, T. K. & Morgan, B. E. 1995. A chemical method for hafnium separations from high-Ti whole-rock and zircon samples. CHEM GEOL **121**, 303–8.

Bateman, P. C. & Chappell, B. W. 1979. Crystallization, fractionation, and solidification of the Tuolumne Intrusive Series, Yosemite National Park, California. GEOL SOC AM BULL **90**, 465–82.

Beard, B. L. & Johnson, C. M. 1993. Hf isotope composition of late Cenozoic basaltic rocks from NW Colorado, U.S.A.: new constraints on mantle enrichment processes. EARTH PLANET SCI LETT **119**, 95–409.

Beard, J. S. & Lofgren, G. E. 1989. Effects of waer on the composition of partial melts of greenstone and amphibolite. SCIENCE **244**, 195–7.

Beard, J. S. & Lofgren, G. E. 1991. Dehydration melting and water-saturated melting of basaltic and andesitic greenstone and amphibolites. J PETROL **32**, 365–401.

Beard, J. S., Abitz, R. J. & Lofgren, G. E. 1993. Experimental melting of crustal xenoliths from Kilbourne Hole, New Mexico and implications for the contamination and genesis of magmas. CONTRIB MINERAL PETROL **115**, 88–102.

Beard, J. S., Lofgren, G. E., Sinha, A. K. & Tollo, R. P. 1994. Partial melting of apatite-bearing charnockite, granulite, and diorite: melt compositions, restite mineralogy, and petrologic implications. J GEOPHYS RES **99**, 21 591–603.

Ben-Othman, D., Polvé, M. & Allègre, C. J. 1984. Nd–Sr isotopic composition of granulites and constraints on the evolution of the lower continental crust. NATURE **307**, 510–5.

Bennett, V. C. & DePaolo, D. J. 1987. Proterozoic crustal history of the western United States as determined by neodymium isotopic mapping. GEOL SOC AM BULL **99**, 674–85.

Birck, J. L. & Allègre, C. J. 1994. Contrasting Re/Os magmatic fractionation in planetary basalts. EARTH PLANET SCI LETT **124**, 139–48.

Brown, M., Rushmer, T. & Sawyer, E. W. 1995. Introduction to special issue: mechanisms and consequences of melt segregation from crustal protoliths. J GEOPHYS RES **100**, 15 551–63.

Carlson, R. W. & Irving, A. J. 1994. Depletion and enrichment history of subcontinental lithospheric mantle: an Os, Sr, Nd and Pb isotopic study of ultramafic xenoliths from the northwestern Wyoming Craton. EARTH PLANET SCI LETT **126**, 457–72.

Carroll, M. R. & Rutherford, M. J. 1985. Sulfide and sulfate saturation in hydrous silicate melts. J GEOPHYS RES **90**, 601–12.

Chauvel, C., Hofmann, A. W. & Vidal, P. 1992. HIMU-EM: the French Polynesian connection. EARTH PLANET SCI LETT **110**, 99–119.

Chen, J. H. & Moore, J. G. 1982. Uranium–lead isotopic ages from the Sierra Nevada batholith, California. J GEOPHYS RES **87**, 4761–84.

Clemens, J. D. & Wall, V. J. 1981. Origin and crystallization of some peraluminous (S-type) granitic magmas. CAN MINERAL **19**, 111–31.

Corfu, F. & Noble, S. R. 1992. Genesis of the southern Abitibi greenstone belt, Superior Province, Canada: evidence from zircon Hf isotope analyses using a single filament technique. GEOCHIM COSMOCHIM ACTA **56**, 2081–97.

DePaolo, D. J. 1981a. A neodymium and strontium isotopic study of Mesozoic calc-alkaline granitic batholiths of the Sierra Nevada and Peninsular Ranges, California. J GEOPHYS RES **86**, 10 470–88.

DePaolo, D. J. 1981b. Neodymium isotopes in the Colorado Front Range and crust–mantle evolution in the Proterozoic. NATURE **291**, 193–6.

DePaolo, D. J. 1988a. *Neodymium isotope geochemistry*. Berlin: Springer-Verlag.

DePaolo, D. J. 1988b. Age dependence of the composition of continental crust: evidence from Nd isotopic variations in granitic rocks. EARTH PLANET SCI LETT **90**, 263–71.

DePaolo, D. J., Linn, A. M. & Schubert, G. 1991. The continental crustal age distribution: methods of determining mantle separation ages from Sm–Nd isotopic data and applications to the southwestern United States. J GEOPHYS RES **96**, 2071–88.

Dickin, A. P., Richardson, J. M., Crocket, J. H., McNutt, R. H. & Peredery, W. V. 1992. Osmium isotope evidence for a crustal origin of platinum group elements in the Sudbury nickel ore, Ontario, Canada. GEOCHIM COSMOCHIM ACTA **56**, 3531–7.

Domenick, M. A., Kistler, R. W., Dodge, F. C. W. & Tatsumoto, M. 1983. Nd and Sr isotopic study of crustal and mantle inclusions from the Sierra Nevada and implications for batholith petrogenesis. GEOL SOC AM BULL **94**, 713–9.

Ellam, R. M., Carlson, R. W. & Shirey, S. B. 1992. Evidence from Re–Os isotopes for plume–lithosphere mixing in Karoo flood basalt genesis. NATURE **359**, 718–21.

Esser, B. K. & Turekian, K. K. 1988. Accretion rate of extraterrestrial particles determined from osmium isotope systematics of Pacific pelagic clay and manganese nodules. GEOCHIM COSMOCHIM ACTA **52**, 1383–8.

Esser, B. K. & Turekian, K. K. 1993. The osmium isotopic composition of the continental crust. GEOCHIM COSMOCHIM ACTA **57**, 3093–104.

Farmer, G. L. & DePaolo, D. J. 1983. Origin of Mesozoic and Tertiary granite in the western United States and implications for pre-Mesozoic crustal structure, 1, Nd and Sr isotopic studies in the geocline of the northern Great Basin. J GEOPHYS RES **88**, 3379–401.

Farmer, G. L. & DePaolo, D. J. 1984. Origin of Mesozoic and Tertiary granite in the western United States and implications for pre-Mesozoic crustal structure, 2, Nd and Sr isotopic studies of unmineralized and Cu- and Mo-mineralized granite in the Precambrian craton. J GEOPHYS RES **89**, 10 141–60.

Fleet, M. E., Tronnes, R. G. & Stone, W. E. 1991. Partitioning of platinum group elements in the Fe–O–S system to 11 GPa and their fractionation in the mantle and meteorites. J GEOPHYS RES **96**, 21 949–58.

Fleet, M. E., Chryssoulis, S. L., Stone, W. E. & Weisener, C. G. 1993. Partitioning of platinum-group elements and Au in the Fe–Ni–Cu–S system: experiments on the fractional crystallization of sulfide melt. CONTRIB MINERAL PETROL **115**, 36–44.

Frost, C. D. & O'Nions, R. K. 1985. Caledonian magma genesis and crustal recycling. J PETROL **26**, 515–44.

Gill, J. 1981. *Orogenic andesites and plate tectonics*. Berlin: Springer-Verlag.

Green, D. H. & Lambert, I. B. 1965. Experimental crystallization of anhydrous granite at high pressures and temperatures. J GEOPHYS RES **70**, 5259–69.

Gruau, G., Chavel, C., Arndt, N. T. & Cornichet, J. 1990. Aluminum depletion in komatiites and garnet fractionation in the early Archean mantle: hafnium isotopic constraints. GEOCHIM COSMOCHIM ACTA **54**, 3095–101.

Hamilton, W. 1978. Mesozoic tectonics of the western United States, *In* Howell, D. G., McDougall, K. A. (eds) *Mesozoic Paleogeography of the Western United States*. SOC ECON PALEONTOL MINERAL PACIFIC COAST PALEOGEOGR SYMP **2**, 33–70.

Hart, S. R. & Kinloch, E. D. 1989. Osmium isotope systematics in Witwatersrand and Bushveld ore deposits. ECON GEOL **84**, 1651–5.

Hart, S. R., Gerlach, D. C. & White, W. M. 1986. A possible new Sr–Nd–Pb mantle array and consequences for mantle mixing. GEOCHIM COSMOCHIM ACTA **50**, 1551–7.

Hattori, K. & Hart, S. R. 1991. Osmium-isotope ratios of platinum-group minerals associated with ultramafic intrusions: Os–isotopic evolution of the oceanic mantle. EARTH PLANET SCI LETT **107**, 499–514.

Hattori, K., Cabri, L. J. & Hart, S. R. 1991. Osmium isotope ratios of PGM grains associated with the Freetown Layered Complex, Sierra Leone, and their origin. CONTRIB MINERAL PETROL **109**, 10–8.

Hauri, E. H. & Hart, S. R. 1993. Re–Os isotope systematics of HIMU and EMII oceanic island basalts from the south Pacific Ocean. EARTH PLANET SCI LETT **114**, 353–71.

Hertogen, J., Janssens, M.-J. & Palme, H. 1980. Trace elements in ocean ridge basalt glasses: implications for fractionations during mantle evolution and petrogenesis. GEOCHIM COSMOCHIM ACTA **44**, 2125–43.

Hoffman, P. F. 1989. Speculations on Laurentia's first gigayear (2·0 to 1·0 Ga). GEOLOGY **17**, 135–8.

Horan, M. F., Morgan, J. W., Walker, R. J. & Grossman, J. N. 1992. Rhenium–osmium isotope constraints on the age of iron meteorites. SCIENCE **255**, 1118–21.

Horan, M. F., Morgan, J. W., Grauch, R. I., Coveney, R. M., Jr, Murowchick, J. B. & Hulbert, L. J. 1994. Rhenium and osmium isotopes in black shales and Ni–Mo–PGE-rich sulfide layers, Yukon Territory, Canada, and Hunan and Guizhou provinces, China. GEOCHIM COSMOCHIM ACTA **58**, 257–65.

Johnson, C. M. 1993. Mesozoic and Cenozoic contributions to crustal growth in the southwestern United States. EARTH PLANET SCI LETT **118**, 75–89.

Johnson, C. M. & Beard, B. L. 1993. Evidence from hafnium isotopes for ancient sub-oceanic mantle beneath the Rio Grande rift. NATURE **362**, 441–4.

Kinny, P., Compston, W. & Williams, I. S. 1991. A reconnaissance ion probe study of Hafnium isotopes in zircons. GEOCHIM COSMOCHIM ACTA **55**, 849–61.

Kistler, R. W. & Peterman, Z. E. 1973. Variations in Sr, Rb, K, Na and initial Sr^{87}/Sr^{86} in Mesozoic granitic rocks and intruded wall rocks in central California. GEOL SOC AM BULL **84**, 3489–512.

Kistler, R. W., Chappell, B. W., Peck, D. L. & Bateman, P. C. 1986. Isotopic variation in the Tuolumne Intrusive Suite, central Sierra Nevada, California. CONTRIB MINERAL PETROL **94**, 205–20.

Koeberl, C. & Shirey, S. B. 1993. Detection of a meteoritic component in Ivory Coast tektites with rhenium–osmium isotopes. SCIENCE **261**, 595–8.

Koeberl, C., Reimold, W. U. & Shirey, S. B. 1994. Saltpan impact crater, South Africa: geochemistry of target rocks, breccias, and impact glasses, and osmium isotope systematics. GEOCHIM COSMOCHIM ACTA **58**, 2893–910.

Lambert, D. D., Morgan, J. W., Walker, R. J., Shirey, S. B., Carlson, R. W., Zientek, M. L. & Koski, M. S. 1989. Rhenium–osmium and samarium–neodymium isotopic systematics of the Stillwater Complex. SCIENCE **244**, 1169–74.

Lambert, D. D., Walker, R. J., Morgan, J. W., Shirey, S. B., Carlson, R. W., Zientek, M. L., Lipin, B. R., Koski, M. S. & Cooper, R. L. 1994. Re–Os and Sm–Nd isotope geochemistry of the Stillwater Complex, Montana: implications for the petrogenesis of the J-M Reef. J PETROL **35**, 1717–53.

Lambert, D. D., Shirey, S. B. & Bergman, S. C. 1995. Proterozoic lithospheric mantle source for the Prairie Creek lamproites: Re–Os and Sm–Nd isotopic evidence. GEOLOGY 23, 273–6.

Le Breton, N. & Thompson, A. B. 1988. Fluid-absent (dehydration) melting of biotite in metapelites in the early stages of crustal anatexis. CONTRIB MINERAL PETROL 99, 226–37.

Luck, J.-M. & Allègre, C. J. 1983. ^{187}Re–^{187}Os systematics in meteorites and cosmochemical consequences. NATURE 302, 130–2.

Luck, J.-M. & Allègre, C. J. 1984. ^{187}Re–^{187}Os investigation in sulfide from Cape Smith komatiite. EARTH PLANET SCI LETT 68, 205–8.

Luck, J.-M. & Allègre, C. J. 1991. Osmium isotopes in ophiolites. EARTH PLANET SCI LETT 107, 406–15.

Luck, J.-M., Birck, J.-L. & Allègre, C. J. 1980. ^{187}Re–^{187}Os systematics in meteorites: early chronology of the Solar System and age of the galaxy. NATURE 283, 256–9.

Marcantonio, F., Zindler, A., Reisberg, L. & Mathez, E. A. 1993. Re–Os isotopic systematics in chromitites from the Stillwater Complex, Montana, U.S.A. GEOCHIM COSMOCHIM ACTA 57, 4029–37.

Marcantonio, F., Reisberg, L., Zindler, A., Wyman, D. & Hulbert, L. 1994. An isotopic study of the Ni–Cu–PGE-rich Wellgreen intrusion of the Wrangellia Terrane: evidence for hydrothermal mobilization of rhenium and osmium. GEOCHIM COSMOCHIM ACTA 58, 1007–17.

Martin, C. E. 1989. Re–Os isotopic investigation of the Stillwater Complex, Montana. EARTH PLANET SCI LETT 93, 336–44.

Martin, C. E., Carlson, R. W., Shirey, S. B., Frey, F. A. & Chen, C.-Y. 1994. Os isotopic variation in basalts from Haleakala Volcano, Maui, Hawaii: a record of magmatic processes in oceanic mantle and crust. EARTH PLANET SCI LETT 128, 287–301.

Merrill, R. B. & Wyllie, P. J. 1975. Kaersutite and Kaersutite Eclogite from Kakanui, New Zealand—water-excess and water-deficient melting to 30 kilobars. GEOL SOC AM BULL 86, 555–70.

Millhollen, G. L. & Wyllie, P. J. 1974. Melting relations of brown-hornblende mylonite from St. Paul's rocks under water-saturated and water-undersaturated conditions to 30 kilobars. J GEOL 82, 589–606.

Milling, M. E., Jr., Johnson, C. M. & Barovich, K. M. 1994. Hf isotope constraints on ancient depletion and enrichment events in the mantle beneath the SW U.S.A. ICOG-8 ABSTR, USGS CIR 1107, 219.

Morgan, J. W. 1986. Ultramafic xenoliths: clues to Earth's late accretionary history. J GEOPHYS RES 91, 12 375–87.

Morgan, J. W. & Lovering, J. F. 1967. Rhenium and osmium abundances in some igneous and metamorphic rocks. EARTH PLANET SCI LETT 3, 219–24.

Morgan, J. W., Wandless, G. A., Petrie, R. K. & Irving, A. J. 1981. Composition of the Earth's upper mantle—I. siderophile trace elements in ultramafic nodules. TECTONOPHYSICS 75, 47–67.

Morgan, J. W., Walker, R. J. & Grossman, J. N. 1992. Rhenium–osmium isotope systematics in meteorites I: magmatic iron meteorite groups IIAB and IIIAB. EARTH PLANET SCI LETT 108, 191–202.

Nelson, B. K. & DePaolo, D. J. 1985. Rapid production of continental crust 1·7 to 1·9 Ga ago: Nd isotopic evidence from the basement of the North American mid-continent. GEOL SOC AM BULL 96, 746–54.

Nyman, M. W., Karlstrom, K. E., Kirby, E. & Graubard, C. M. 1994. Mesoproterozoic contractional orogeny in western North America: evidence from ca. 1·4 Ga plutons. GEOLOGY 22, 901–4.

Palmer, M. R. & Turekian, K. K. 1986. ^{187}Os/^{186}Os in marine manganese nodules and the constraints on the crustal geochemistries of rhenium and osmium. NATURE 319, 216–20.

Patchett, P. J. 1983a. Importance of the Lu–Hf isotopic system in studies of planetary chronology and chemical evolution. GEOCHIM COSMOCHIM ACTA 47, 81–91.

Patchett, P. J. 1983b. Hafnium isotope results from mid-ocean ridges and Kerguelen. LITHOS 16, 47–51.

Patchett, P. J. & Arndt, N. T. 1986. Nd isotopes and tectonics of 1·9–1·7 Ga crustal genesis. EARTH PLANET SCI LETT 78, 328–38.

Patchett, P. J. & Tatsumoto, M. 1980a. Hafnium isotope variations in oceanic basalts. GEOPHYS RES LETT 7, 1077–80.

Patchett, P. J. & Tatsumoto, M. 1980b. A routine high-precision method for Lu–Hf isotope geochemistry and chronology. CONTRIB MINERAL PETROL 75, 263–9.

Patchett, P. J., Kouvo, O., Hedge, C. E. & Tatsumoto, M. 1981. Evolution of continental crust and mantle heterogeneity: evidence from Hf isotopes. CONTRIB MINERAL PETROL 78, 279–97.

Patchett, P. J., White, W. M., Feldmann, H., Kielinczuk, S. & Hoffmann, A. W. 1984. Hafnium/rare earth element fractionation

in the sedimentary system and crustal recycling into the Earth's mantle. EARTH PLANET SCI LETT 69, 365–78.

Patiño Douce, A. E. 1995. Experimental generation of hybrid silicic melts by reaction of high-Al basalt with metamorphic rocks. J GEOPHYS RES 100, 15 623–39.

Patiño Douce, A. E. & Johnson, A. D. 1991. Phase equilibria and melt productivity in the pelitic system: implications for the origin of peraluminous granitoids and aluminous granulites. CONTRIB MINERAL PETROL 107, 202–18.

Pegram, W. J. & Allègre, C. J. 1992. Osmium–isotopic compositions from oceanic basalts. EARTH PLANET SCI LETT 111, 59–68.

Pegram, W. J., Krishnaswami, S., Ravizza, G. E. & Turekian, K. K. 1992. The record of sea water ^{187}Os/^{186}Os variation through the Cenozoic. EARTH PLANET SCI LETT 113, 569–76.

Peucker-Ehrenbrink, B., Ravizza, G. & Hofmann, A. W. 1995. The marine ^{187}Os/^{186}Os record of the past 80 million years. EARTH PLANET SCI LETT 130, 155–67.

Rapp, R. P., Watson, E. B. & Miller, C. F. 1991. Partial melting of amphibolite/eclogite and the origin of Archean trondhjemites and tonalites. PRECAMBRIAN RES 51, 1–25.

Ravizza, G. & Turekian, K. K. 1989. Application of the ^{187}Re–^{187}Os system to black shale geochronometry. GEOCHIM COSMOCHIM ACTA 53, 3257–62.

Ravizza, G. & Turekian, K. K. 1992. The osmium isotopic composition of organic-rich marine sediments. EARTH PLANET SCI LETT 110, 1–6.

Ravizza, G., Turekian, K. K. & Hay, B. J. 1991. The geochemistry of rhenium and osmium in recent sediments from the Black Sea. GEOCHIM COSMOCHIM ACTA 55, 3741–52.

Reisberg, L. C., Allègre, C. J. & Luck, J.-M. 1991. The Re–Os systematics of the Ronda Ultramafic complex of southern Spain. EARTH PLANET SCI LETT 105, 196–213.

Reisberg, L., Zindler, A., Marcantonio, F., White, W., Wyman, D. & Weaver, B. 1993. Os isotope systematics in ocean island basalts. EARTH PLANET SCI LETT 120, 149–67.

Roy-Barman, M. & Allègre, C. J. 1994. ^{187}Os/^{186}Os ratios of mid-ocean ridge basalts and abyssal peridotites. GEOCHIM COSMOCHIM ACTA 58, 5043–54.

Roy-Barman, M. & Allègre, C. J. 1995. ^{187}Os/^{186}Os in oceanic island basalts: tracing oceanic crust recycling in the mantle. EARTH PLANET SCI LETT 129, 145–61.

Rushmer, T. 1991. Partial melting of two amphibolites: contrasting experimental results under fluid-absent conditions. CONTRIB MINERAL PETROL 107, 41–59.

Rushmer, T. 1993. Experimental high-pressure granulites: some applications to natural mafic xenolith suites and Archean granulite terranes. GEOLOGY 21, 411–4.

Rushmer, T. 1995. An experimental deformation study of partially molten amphibolite: application to low-melt fraction segregation. J GEOPHYS RES 100, 15 681–95.

Rutter, M. J. & Wyllie, P. J. 1988. Melting of vapour-absent tonalite at 10 kbar to simulate dehydration-melting in the deep crust. NATURE 311, 159–60.

Salters, V. J. M. & Hart, S. R. 1989. The hafnium paradox and the role of garnet in the source of mid-ocean-ridge basalts. NATURE 342, 420–2.

Salters, V. J. M. & Hart, S. R. 1991. The mantle sources of ocean ridges, islands and arcs: the Hf-isotope connection. EARTH PLANET SCI LETT 104, 364–80.

Schaltegger, U. & Corfu, F. 1992. The age and source of late Hercynian magmatism in the central Alps: evidence from precise U–Pb ages and initial Hf isotopes. CONTRIB MINERAL PETROL 111, 329–44.

Scherer, E. E., Beard, B. L., Barovich, K. M., Johnson, C. M. & Taylor, L. A. 1995. An improved method for determining the Hf isotopic composition of Lunar basalts. ABSTR 26TH LUNAR PLANET SCI CONF, 1235–6.

Sekine, T., Wyllie, P. J. & Baker, D. R. 1981. Phase relationships at 30 kbar for quartz eclogite composition in CaO–MgO–Al$_2$O$_3$–SiO$_2$–H$_2$O with implications for subduction zone magmas. AM MINERAL 66, 938–50.

Shirey, S. B. & Walker, R. J. 1995. Carius tube digestion for low-blank rhenium–osmium analysis. ANAL CHEM 67, 2136–41.

Sims, P. K., Van Schmus, W. R., Schulz, K. J. & Peterman, Z. E. 1989. Tectonostratigraphic evolution of the early Proterozoic Wisconsin magmatic terranes of the Penokean Orogen. CAN J EARTH SCI 26, 2145–58.

Skjerlie, K. P., Patiño Douce, A. E. & Johnston, A. D. 1993. Fluid absent melting of a layered crustal protolith: implications for the generation of anatectic granites. CONTRIB MINERAL PETROL 114, 365–78.

Smith, P. E., Tatsumoto, M. & Farquhar, R. M. 1987. Zircon Lu–Hf systematics and the evolution of the Archean crust in the southern Superior Province, Canada. CONTRIB MINERAL PETROL 97, 93–104.

Stein, H. J., Morgan, J. W., Walker, R. J. & Horan, M. F. 1992. Rhenium–osmium data for sulfides and oxides from climax-type granite-molybdenum systems: Mt. Emmons, Colorado. GEOL SOC AM ABSTR PROGRAM 24, A144.

Stern, C. R. & Wyllie, P. J. 1978. Phase compositions through crystallization intervals in basalt–andesite–H$_2$O at 30 kbar with implications for subduction zone magmas. AM MINERAL 634, 641–63.

Stille, P., Unruh, D. M. & Tatsumoto, M. 1983. Pb, Sr, Nd and Hf isotopic evidence of multiple sources for Oahu, Hawaii basalts. NATURE 304, 25–9.

Stille, P., Unruh, D. M. & Tatsumoto, M. 1986. Pb, Sr, Nd and Hf isotopic constraints on the origin of Hawaiian basalts and evidence for a unique mantle source. GEOCHIM COSMOCHIM ACTA 50, 2303–19.

Stille, P., Oberhänsli, R. & Wenger-Schenk, K. 1989. Hf–Nd isotopic and trace element constraints on the genesis of alkaline and calc-alkaline lamprophyres. EARTH PLANET SCI LETT 96, 209–19.

Taylor, H. P. Jr 1968. The oxygen isotope geochemistry of igneous rocks. CONTRIB MINERAL PETROL 19, 1–71.

Thorpe, R. S. 1982. *Andesites, orogenic andesites and related rocks.* New York: Wiley.

Thompson, A. B. 1982. Dehydration melting of pelitic rocks and the generation of H$_2$O–undersaturated granitic liquids. AM J SCI 282, 1567–95.

Van Wyck, N. 1995. *Oxygen and carbon isotopic constraints on the development of eclogites, Holsny, Norway and major and trace element, common Pb, Sm–Nd, and zircon geochronology constraints on petrogenesis and tectonic setting of pre- and Early Proterozoic rocks in Wisconsin.* Ph.D. Thesis, Univ Wisconsin-Madison.

Vielzeuf, D. & Holloway, J. R. 1988. Experimental determination of the fluid-absent melting relations in the pelitic system. CONTRIB MINERAL PETROL 98, 257–76.

Walker, R. J. & Morgan, J. W. 1989. Rhenium–osmium isotope systematics of carbonaceous chondrites. SCIENCE 243, 519–22.

Walker, R. J., Shirey, S. B. & Stecher, O. 1988. Comparative Re–Os, Sm–Nd and Rb–Sr isotope and trace element systematics for Archean komatiite flows from Munro Township, Abitibi Belt, Ontario. EARTH PLANET SCI LETT 87, 1–12.

Walker, R. J., Carlson, R. W., Shirey, S. B. & Boyd, F. R. 1989a. Os, Sr, Nd, and Pb isotope systematics of southern African peridotite xenoliths: implications for the chemical evolution of subcontinental mantle. GEOCHIM COSMOCHIM ACTA 53, 1583–95.

Walker, R. J., Shirey, S. B., Hanson, G. N., Rajamani, V. & Horan, M. F. 1989b. Re–Os, Rb–Sr, and O isotopic systematics of the Archean Kolar schist belt, Karnataka, India. GEOCHIM COSMOCHIM ACTA 53, 3005–13.

Walker, R. J., Echeverria, L. M., Shirey, S. B. & Horan, M. F. 1991a. Re–Os isotopic constraints on the origin of volcanic rocks, Gorgona Island, Colombia: Os isotopic evidence for ancient heterogeneities in the mantle. CONTRIB MINERAL PETROL 107, 150–62.

Walker, R. J., Morgan, J. W., Naldrett, A. J., Li, C. & Fassett, J. D. 1991b. Re–Os isotope systematics of Ni–Cu sulfide ores, Sudbury Igneous Complex, Ontario: evidence for a major crustal component. EARTH PLANET SCI LETT 105, 416–29.

Walker, R. J., Morgan, J. W., Horan, M. F., Czamanske, G. K., Krogstad, E. J., Fedorenko, V. A. & Kunilov, V. E. 1994. Re–Os isotopic evidence for an enriched-mantle source for the Noril'sk-type, ore-bearing intrusions, Siberia. GEOCHIM COSMOCHIM ACTA 58, 4179–97.

Waters, D. J. 1988. Partial melting and the formation of granulites facies assemblages in Namaqualand, South Africa. J METAMORP GEOL 6, 387–404.

White, W. M. & Patchett, P. J. 1984. Hf–Nd–Sr isotopes and incompatible element abundances in island arcs: implications for magma origins and crust-mantle evolution. EARTH PLANET SCI LETT 67, 167–85.

White, W. M., Patchett, P. J. & Ben-Othman, D. 1986. Hf isotope ratios of marine sediments and Mn nodules: evidence for a mantle source of Hf in seawater. EARTH PLANET SCI LETT 79, 46–54.

Windley, J. F. 1993. Proterozoic anorogenic magmatism and its orogenic connections. J GEOL SOC LONDON 150, 39–50.

Wolf, R. & Anders, E. 1980. Moon and Earth: compositional differences inferred from siderophiles, volatiles, and alkalis in basalts. GEOCHIM COSMOCHIM ACTA 44, 2111–24.

Woodhead, J. D. 1989. Geochemistry of the Mariana arc (western Pacific): source, composition, and processes. CHEM GEOL 76, 1–24.

Zartman, R. E. 1974. Lead isotopic provinces in the Cordillera of the western United States and their geologic significance. ECON GEOL 69, 792–805.

Zartman, R. E. & Doe, B. R. 1981. Plumbotectonics—the model. TECTONOPHYSICS 75, 135–63.

Zartman, R. E. & Haines, S. M. 1988. The plumbotectonic model for Pb isotopic systematics among major terrestrial reservoirs—a case for bi-directional transport. GEOCHIM COSMOCHIM ACTA 52, 1327–39.

CLARK M. JOHNSON, Department of Geology and Geophysics, University of Wisconsin, Madison, WI 53706, U.S.A.
STEVEN B. SHIREY, Department of Terrestrial Magnetism, Carnegie Institution of Washington, Washington, DC 20015, U.S.A.
KARIN M. BAROVICH, Department of Geology and Geophysics, University of Wisconsin, Madison, WI 53706, U.S.A. Current address: Department of Geology and Geophysics, University of Adelaide, Adelaide, South Australia 5005, Australia.

Transactions of the Royal Society of Edinburgh: Earth Sciences, **87**, 353–359, 1996

Two centuries after Hutton's 'Theory of the Earth': the status of granite science

D. Barrie Clarke

ABSTRACT: Granite is the final product of the high-temperature, magmatic, predominantly endogenic, chemical differentiation of the earth. Our understanding of the origin and evolution of granitoid rocks comes from a combination of direct observation, analogue experimentation and numerical modelling. A brief historical overview shows an exceptional level of such research activity over the last 50 years. The number and complexity of questions have resulted in both an absolute and a relative growth of the science since the plate tectonic revolution, largely consisting of refining the current magmatic paradigm within its overarching context. Current research activity involves large components of mineralogical–petrological–geochemical and structural–tectonic work, with much lower levels of experimental, geophysical and geochronological investigations. Many important questions concerning the thermal, physical and chemical aspects of the origin and evolution of granites remain. In keeping with the general progress of science, the complexity of the questions, the declining financial support and the revolution in information technology, directions of granite research in the foreseeable future will change from concrete and qualitative to abstract and quantitative, from expensive and active to cheap and armchair, from reductionist to holistic, and from periodic communication to continuous communication.

KEY WORDS: history, philosophy, Kuhn, paradigm, anomalies, future research.

I have been particularly anxious about this subject of granite.
James Hutton

At the end of its accretion from the solar nebula, the earth almost certainly contained no granite (*sensu lato*). Even now, 4·5 billion years later, the proportion of granitoid rocks is less than 0·001 of the bulk earth, but that small proportion concentrates at relatively high levels, mainly in the continental crust. As such, it is highly accessible and provides opportunities for both curiosity driven research (origin, emplacement and tectonic setting), as well as more practically driven investigations (building materials, china clays and mineral deposits). The non-trivial academic problem, simply stated, is one of determining the number, sequence, cause and effect of all the naturally occurring chemical and physical processes involved in converting solid carbonaceous chondrite, at one extreme of the compositional spectrum, to solid leucocratic alkali feldspar granite pegmatite at the other. No-one ever attempts to solve the entire problem; instead, following in the tradition of reductionist science, individual researchers work in great detail on one small piece of the problem and offer its findings as part of the larger solution.

How, then, do we find answers to the many questions concerning granitoid rocks? Our approach consists of only three types of activity:

1. *Direct observation.* Direct observations on granitoids involve such obvious activities as field and petrographic descriptions, but also include chemical and isotopic analyses of all types, geophysical measurements in the field and laboratory and geochronological measurements of all types. In all of these approaches we observe the products of 'experiments' conducted by Nature and attempt to deduce, from the physical and chemical clues left in these products, what the nature of the 'experiment' was (including the starting materials, the apparatus and the methods). In our attempts to solve a single specific problem, we normally have access to only one level of Nature's multi-storied experimental laboratory (with starting materials left mainly in the basement, methods filed throughout the edifice and run products normally stored on the upper floors). Thus it is not surprising that we derive only partial, and perhaps even equivocal, answers to the general problem.

2. *Analogue experiment.* Until the middle of this century, earth science was largely observational. Since then, we have tried to replicate physical processes (e.g. classical Ramberg-type experiments on the formation and ascent of magmas) and chemical processes (e.g. classical Winkler-type melting and crystallisation experiments to produce granite compositions) occurring within the earth, with some considerable degree of success. The experimentalists have provided us with valuable, but imperfect, information on suitable source rocks, melting reactions, crystallisation histories, ascent processes and partition coefficients.

3. *Numerical modelling.* In isolation, numerical modelling requires no direct physical contact with, or indeed even first-hand knowledge of, granite material; it only needs quantitative data from both direct observation and analogue experimentation. Examples of numerical modelling include finite element methods for the physical emplacement of granite bodies (e.g. Guglielmo 1993, 1994) and chemical mass balance equations for simulating assimilation–contamination processes of compositional evolution (e.g. DePaolo 1981).

The purpose of all three approaches is identical, namely, to acquire the basic facts about the granite system directly or to test the experimental and numerical models against the real world to discover the fundamental laws about the origin and evolution of granites.

1. Brief historical perspective

1.1. The distant past: Hutton (1795) to Holmes (1945)

Dean (1992:268) best records our debt to Hutton: 'Among Hutton's major contributions to the science of geology are his insistence on the basic orderliness of terrestrial processes; the centrality of heat; the duration of time; the significance of denudation, consolidation, and uplift; the efficacy of rivers; the intimate association between the earth's crust and subsurface forces; *the nature and significance of plutonic and metamorphic rocks*; and the ongoing nature of petrological creation. Other geological theorists have given us one or two good points, but none competes with Hutton in the number of sound principles established.' (my italics.) Space here does not allow even a brief review of the development of ideas in granite geology. Fortunately, Pitcher (1993) has donned the mantle of granite historian and has thoroughly documented the development of thought about granites, both through time and in terms of 'schools'.

1.2. The exceptional half-century: Holmes (1945) to the Third Hutton Conference (1995)

This period of time is 'exceptional' because of the large amounts of money to support pure science, the great number of researchers, the explosion of publications and the rate of scientific progress. Pitcher (1987) has also reviewed a significant part of this period of time and refined Read's important synthesis of 'granites and granites' in plate tectonic terms. Also, near the close of this period, Brown (1994) comprehensively summarised the case for granites originating in the continental crust.

Figure 1 illustrates our publication rate in this half century as a *proportion* of the total body of earth science publications. The trend shows three distinct parts: an erratic contribution rate averaging about 0·038 for the period 1945–60, a substantial decrease to 0·025 during the plate tectonic revolution when the attention of geoscience was focused on granite-poor ocean basins, and a reasonably steady recovery in the last 25 years. Given that not all disciplines can have increased their proportion of activity in that time, this most recent trend suggests that granite science still presents significant problems in the plate tectonic world.

1.3. The present

A compilation of current research activity in granite science should give some indication of what topics are deemed to be feasible at this time. In this respect, I have sampled three different sources, none of which is necessarily representative of the community as a whole. (Since writing the first draft of this paper, I have learned from Jean Louis Vigneresse of

Nancy that he has also been conducting a survey of publications in earth science involving the measurement of many more parameters in a selected number of journals.)

1. Third Hutton Symposium on Granites Abstracts. I have sorted 155 of the abstracts submitted to this conference (Brown & Piccoli 1995) into the following 10 categories: geochemical, mineralogical–petrological, mineral deposits, experimental (focus on mineralogy, petrology, geochemistry), experimental (focus on physical properties of granites and magmas), physical properties, geophysics (gravity, magnetics, remote sensing, etc.), structure, tectonics and geochronology. The first row of Figure 2 shows the relative levels of activity in these categories at this conference.

2. *Granite-research Internet Discussion Group*. In January 1995 a discussion group opened on the Internet with the address granite-research@ac.dal.ca. It offered an unmoderated discussion on matters of concern to those working on any aspect of granitoid rocks. The messages, most of which were exchanged over a period of three months, form a snapshot of the topics of interest to its approximately 300 subscribers. The second row of Figure 2 shows the relative levels of discussion in the same categories as above.

3. *GEOREF Search (1994–5)*. Although AGI cataloguing for this period of time is incomplete, the search expression ((syenogranit* or granit* or monzogranit* or granodiorit* or tonalit*) and (yr = 1994 or 1995)) yielded 740 references that could be classified into one of the 10 categories. The back row of Figure 2 shows the relative levels of activity in these categories.

Unsurprisingly, all three samples of current research work show strong components of geochemical, petrological and structural investigations. Clearly, the granite scientific community believes that answers to significant questions still lie in these disciplines. Noteworthy features of these histograms, however, include the low levels of activity in any kind of experimental work, the apparent under-representation in the Hutton conference and the Internet discussion of people working in geochronology and mineral deposits in granites, and the generally low level of activity in geophysical work of all kinds on granites. Apparently, we have largely failed to interest our geophysical colleagues in granite problems, despite our keen desire to know more about the shapes and emplacement mechanisms of batholiths and plutons.

2. In philosophical terms

From time to time, it is beneficial to try to think *about* our discipline, rather than always thinking *within* our discipline. (As an example of completely new directions in granite science, Nick Petford and John Clemens of Kingston are currently working on parallels between granite generation–ascent–emplacement and information theory.) In philosophical terms, what point has granite science reached?

The process of formulation of a 'theory of granites' is one of the induction of general laws from specific observations on granitic rocks. One aspect of such induced laws is that their conclusions contain more information than the combination of their premises and, unlike valid deductive arguments, the conclusions are not necessarily truth-preserving (Giere 1984). Thus they must be tested against further observations. Such testing involves only two kinds of outcome: either new observations support (confirm) the theory, or new observations do not support, or may even falsify, the theory (Popper 1992). In his landmark essay, entitled *The structure of scientific revolutions*, Kuhn (1970) advanced his own comprehensive view about the general progress of science. More recently, Hoyningen-Huene (1993) reviewed and refined Kuhn's syn-

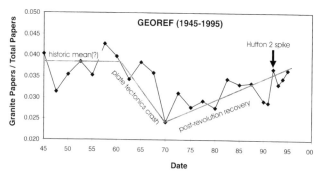

Figure 1 Publications in granite science as a proportion of all earth science publications for the period 1945–95. The data are the result of a GEOREF search expression ((syenogranit* or granit* or monzogranit* or granodiorit* or tonalit*) and (yr = 19xy)) every 2·5 years, except every year since 1990. The citations recovered were divided by the total number of citations in that year. The erratic trend between 1945 and about 1960 is difficult to explain. During the plate tectonics crash, attention of the earth science community focused on basalts of the ocean basins. Interest in granite problems has generally increased since 1970.

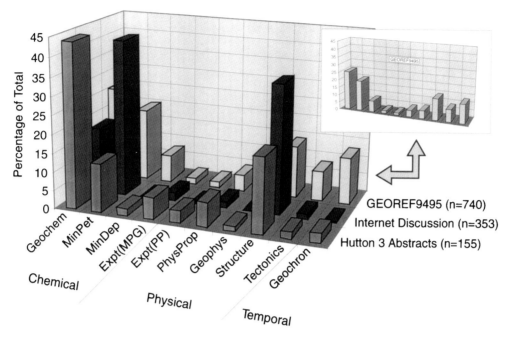

Figure 2 Current interest and activity in granite problems, by discipline. Sorting into categories is rather arbitrary, but is at least consistent for the three databases. Clearly, geochemistry, mineralogy–petrology and structure dominate all three measures of activity. See text for details. Abbreviations: Expt(MPG) = experimental (mineralogy, petrology, geochemistry); Expt(PP) = Experimental (physical properties).

thesis. What follows in this section is a summary of their position and an assessment of where we currently are in our discipline (all quotations are from Kuhn).

In short, any branch of science has three stages of development (Fig. 3).

1. *Pre-consensus, pre-paradigm, or pre-normal.* In the very early stages in the development of a branch of science, investigations proceed in the absence of any single theory that is capable of explaining the observations. In this type of world, each observation can have an explanation entirely independent of that for any other observation.

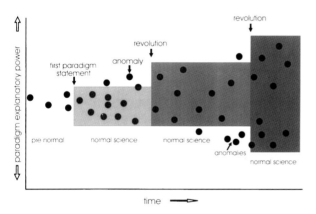

Figure 3 Graphic representation of the progress of science according to Kuhn. Each dot represents an observation. In the early 'pre-normal' stages, every observation is independent and not related by any encompassing theory. The first unifying and testable theory, or paradigm statement, begins the process of normal science (*sensu stricto*), during which, for a while at least, observations are explained in terms of the operating paradigm. With time, however, some observations fail to receive explanations in terms of the paradigm and are set aside as anomalies. The accumulating anomalies eventually force a revolutionary change in the unifying theory. Note that each new paradigm statement is broader than the anomalies it needs to explain, i.e. it has predictive power beyond the known set of observations (it can permit predictions and accommodate retrodictions). Note also, with time, most new work and new observations tend to occur near the margin of the paradigm, rather than in the old established core where few challenges remain.

2. *Normal, or paradigm-governed.* In this stage, a branch of science operates for relatively long periods of time within an accepted framework called a *paradigm* (a 'universally' recognised scientific achievement that for a time provides model problems and solutions to a community of practitioners). A good paradigm is simple, internally consistent, broad in scope and capable of successful applications and predictions. Kuhn uses terms such as 'puzzle solving', 'mopping-up operations' and 'cumulative' to describe the types of activity and products of normal science. However, such 'puzzle solving' can involve very difficult problems that require a high degree of technical, conceptual and mathematical sophistication to provide satisfactory answers. Any unsolved problem can be set aside, either because the solution exists but the researcher has not found it, or because the solution does not exist within the current paradigm and, therefore, it becomes an 'anomaly'. By definition, such anomalies can only ever occur against the backdrop of the operating paradigm and they accumulate over time.

3. *Revolution.* Over a relatively short period of crisis, the old paradigm is overthrown in favour of a new one which not only accounts for the anomalies under the old paradigm, but also has better explanatory and predictive power. 'The decision to reject one paradigm is always simultaneously the decision to accept another and the judgement leading to that decision involves comparison of both paradigms with nature *and* with each other.' Also, the new, post-revolution, paradigm incorporates much of the vocabulary and apparatus of the old one, but rarely in the same way. Interestingly, young scientists who have little or no investment in the old paradigm often lead such revolutions.

Pre-normal, catastrophic geology ended with the Hutton (1795)–Lyell (1830) paradigm of uniformitarianism. Their theory of the earth formed the first significant paradigm against which new observations could be tested. Then followed a very long period of normal science, during which various problems or 'puzzles' were tackled, most of which yielded satisfactory answers in terms of the operating paradigm. Other 'puzzles', however, did not, and they became part of the 'anomalies'. Those anomalies precipitated a crisis in the 1960s, which was resolved by acceptance of the current overarching paradigm in earth science, namely plate tectonics. (Ironically, Kuhn's two editions of *The structure of scientific revolutions* in 1962 and 1970 precisely bracket this most dramatic scientific revolution of the 20th century, yet he makes no mention of it in the second edition.) Today, no-one seriously questions the applicability of plate tectonic theory to solve a wide variety of problems in geoscience.

Within the two main earth science paradigms (first uniformitarianism and then plate tectonics), each subdiscipline developed its own accepted framework. In granite geology specifically, Hutton (1795) also brought about the first major revolution by offering a new and more attractive plutonist paradigm in place of the old neptunist one. That paradigm lasted for about 150 years, until again the accumulating anomalies, this time over the room problem, forced a major change in thinking from plutonism, in general, to a choice between magmatism or transformationism, with the magmatists prevailing. Since then, we have operated satisfactorily within the magmatic paradigm.

3. A statement of the current granite paradigm

Kuhn (1970) states that 'Despite occasional ambiguities, the paradigms of a mature science can be determined with relative ease' and 'Some of the generalizations...to describe the shared beliefs will present no problems. Others, however...will seem a shade too strong. Phrased in just that way, or in any other way...they would almost certainly have been rejected by some members of the group...' The following paradigm statement, necessarily made lengthy by the complexity of the subject, attempts to distill the essence of our shared beliefs concerning the physical, chemical, spatial and temporal origin and evolution of granitoid rocks. (This statement began its development on the granite-research discussion group on the Internet. Two drafts of the statement were scrutinised by about 300 members of the granite community and a third was subsequently discussed by another subset of the community who attended the Third Hutton Symposium in Maryland. This version of the statement owes its present form, in part, to the unselfish contributions of those cited in the Acknowledgements.)

Members of the granite family (alkali feldspar granite, syenogranite, granite, monzogranite, granodiorite, tonalite) are magmatic rocks that:

- form indirectly by partial melting of pre-existing rocks such as mantle peridotite (thereby requiring subsequent compositional modification) or directly by partial melting of mantle-derived mafic igneous rocks, felsic igneous rocks and crustal siliciclastic metasedimentary rocks with high normative quartz and feldspar contents, or combinations of these types of source materials;
- originate under a wide range of pressure conditions (mantle to crust), temperature conditions (thermal anomalies resulting from some combination of subduction, delamination, overthrusting, radioactive decay, friction, mafic underplating/intraplating) and chemical conditions (including both $a_{H_2O} < 1$ and $a_{H_2O} = 1$);
- in which the partial melt, or mobilisate, in response to its low density and low viscosity relative to the surrounding rocks, as well as to the prevailing stress regime in the host rocks, undergoes mechanical segregation by a combination of porous flow, transport along growing or pre-existing fractures, flow during crystal–plastic deformation and/or wholesale melting and buoyancy-driven accumulation in the

region of partial melting (which, at least in the continental crust, may resemble a migmatite terrane) and thereby attains the status of an independent magma;

- that magma may ascend through the mechanically heterogeneous crust by dyking, diapirism, transport in shear zones or a combination of these processes, and displaces country rocks to form magma conduits or chambers by a combination of fracturing and elastic contraction, folding and/or crystal–plastic flow, faulting, stoping and roof uplift;
- during which, at all stages from source to emplacement, changes occur in the direction of thermal, mechanical and chemical equilibrium without necessarily attaining the equilibrium state;
- the magma may entrain, and partially or completely react and mix with, enclaves of all materials (restite, country rock, autoliths, coeval mingled mafic magma) which it encounters, during all stages of its movement;
- during some stage of its crystallisation, depending on the pressure and dissolved volatile content ($H_2O \gg CO_2$, Li, B, F, P), the magma normally reaches water saturation ($a_{H_2O} = 1$) and evolves a separate water-rich fluid phase;
- the magma, throughout its crystallisation history, also undergoes a variety of open–system and closed–system differentiation processes involving fractionation of solid + melt ± aqueous fluid phases, which in the closing, lowest temperature magmatic stages may produce pegmatites;
- the aqueous fluid has a principal role in the alteration of primary igneous minerals (although such metasomatic processes are restricted to modifying granite compositions, not producing them), hydrothermal veins and polymetallic mineralisation (although externally derived fluids may also be involved in these processes);
- the resulting igneous rock has a peraluminous, metaluminous or peralkaline chemical composition that reflects some unique combination of sources (as indicated largely by isotope geochemistry) and processes (as indicated largely by mineralogy, textures and major and trace element geochemistry), and for which the granite 'end-point' represents a convergence of evolution for both fractional crystallisation of more primitive magmas and partial melting of metasedimentary and metaigneous rocks;
- the granite body ultimately varies significantly in style of emplacement and chemical composition as a function of depth of emplacement, time of emplacement relative to crustal tectonics and plate tectonic setting (e.g. predominantly metaluminous granites in subduction-related settings and oceanic ridge settings, predominantly peraluminous granites in continental collision zones including late orogenic collapse and transcurrent movements along shear zones, and characteristically peralkaline granites in anorogenic settings).

This statement inherits only its first few words from Hutton (1795), otherwise it is a position that neither Holmes (1945) nor Read (1948) could have formulated in their time. Like any acceptable paradigm statement, it permits without necessarily being explicit, e.g. it makes no mention of restite unmixing, but it allows such a process within the more general statement 'fractionation of solid, melt and aqueous fluid phases'. Most of the above paradigm statement reflects *our* accomplishments over the past 50 years. An even more expanded statement of the 1995 paradigm is the thoughtful collection of papers contained in this issue. The fully expanded statement of the current paradigm is embedded in the entire body of publications on granitoid rocks.

4. Divining the future of granite research

Most earth scientists are comfortable with the concept and practice of retrodiction, but not of prediction. For us, determining the past physical and chemical history of a pluton is much easier than determining the future direction of granite research. Any such predictions about the future of the science must be based on comparisons with what happens in other branches of science, as well as on projections of any trends from the recent past. In preparing to make such predictions, I have considered four components that seem to be important.

1. *Comparing granite science with more mature sciences generally.* The progress from visual-observational and classificatory, to indirect-observational and comparative, to experimental and fully quantitative, is the trend in any maturing science. Recognition of this progression is not exactly novel; Roger Bacon asserted the primacy of mathematics more than 700 years ago when he stated that 'the things of this world cannot be made known without a knowledge of mathematics'. Mathematical models, however, form only part of the solution and they are inherently recursive; their successes may reveal new kinds of observations that could be made in the field, and their failures to explain the real world may show the need for better field observations, more appropriate experimental data or simply better mathematical representations. Only when observational, experimental and mathematical aspects are in complete harmony is the problem (at least temporarily) solved. Also, an elegant mathematical representation should not be considered any more valuable in advancing the science than a completely qualitative, but brilliant, conceptual insight.

2. *Reliance on technology.* In a sense, we in the earth sciences have been rather passive in developing technology, but very active in applying it (e.g. the wide variety of chemical analytical techniques that we have used to determine the compositions of minerals and rocks). So, to some extent, we wait for opportunities to transfer new technology to our discipline to provide new kinds of data that will shed light on the origin of granites. The newest advance available to us is that of information technology, which largely removes the old barriers to communication.

3. *Projecting scientific trends from the recent past, including compiling the anomalies that currently exist.* Some of today's unanswered questions, our accumulating anomalies, include the following.

(a) *Thermal–mechanical–structural problems*
- Thermal energy—what is the source of heat for crustal melting? What are the proportions of mantle heat (subduction, delamination, underplating, intraplating) and crustal heat (radioactive decay, burial, friction) in each tectonic setting for granite? Can we determine the thermal histories of granite bodies? What do geothermometers in granites really record?
- Region of melting—in the zone of partial melting, what are the melt production rates, extraction rates, extraction dynamics (wet melting of large volumes in a lithostatic stress field, dry melting of small volumes in a differential stress field or some other combination of these variables) and for both physical and chemical reasons, do those melts pond, homogenise and differentiate before ascent? How do the rheological properties of such magmas vary as a function of temperature, pressure, composition?
- Physical state of the crust—what are the thermal, spatial, temporal and genetic relationships between mechanisms of segregation, ascent and emplacement

and the physical state of the crust at all levels? Is the movement of magma by fracture flow or diapirism (the evidence for which is so easily destroyed in the ductile crust)? What are the important physical properties and appropriate values for them? When is emplacement distinct from ascent, and when does emplacement just equal end of ascent?

(b) *Mineralogical–petrological–geochemical problems*
- Equilibrium generally—every physical change and every chemical reaction in granite systems is in the direction of mechanical–thermal–chemical equilibrium, but how close to the equilibrium state do these granite processes operate?
- Geochemical modelling—at what level of confidence are geochemical models valid? Are the systems linear or non-linear (even chaotic)? Is any geochemical modelling possible given the uncertainties in the compositions of sources, refractory accessory minerals, partition coefficients and uncertainties about equilibrium–disequilibrium? Are diffusion rates in crystals a limiting factor in a truly dynamic 'equilibrium' melting environment in which some or all of the crystals may melt and reform many times over the period of the melting event? Do the fine points of kinetics even matter, given the heterogeneity of the source materials? How do we uniquely measure the 'involvement coefficients' of the various processes such as fractional crystallisation, restite unmixing, assimilation, recharge, mixing, unmixing, fluid interaction, etc. in producing the chemical variation observed in granites (Clarke 1992: 91–3)?
- Tectono-magmatic discriminators—even if the major element compositions of granites are related to their tectonic environment, what is the reliability of discriminators based largely on trace elements?
- Role of metasomatism—because granites inevitably stew in late-stage fluids of their own origin, or in a variety of fluids that can circulate though them from the outside, how can we quantify the degree of metasomatic overprint?
- The mafic–felsic connection—can we determine, in every tectonic setting for granite, how to assess the physical and chemical input of the mantle as represented by physical evidence such as associated mafic plutons, dykes and enclaves, or by chemical evidence such as trace element concentrations and isotopic ratios?
- Enclaves—although they can only be restites, country rock xenoliths, quenched mafic or felsic magmas, cumulates, autoliths or quenched compositional equivalents of the main magma, questions about them never seem to go away. What are the indisputable mineralogical, textural and chemical characteristics of each type? How do they react with the granite magma (i.e. change in size, shape and composition as a function of T–P–X–t)? What is their importance in granite petrology?
- Granite textures—textures store information about the interplay among temperature (especially cooling rate), pressure (static or dynamic) and composition, but can that textural information ever be reliably decoded back into those parameters?
- Migmatites and granites—are they related or not and, if so, in what way? What are migmatite domes? What is the meaning of fertile granulites and undepleted restites in terms of classical ideas on crustal origins for granites? Are there any scaled-up

analogue experiments we can do, or do we have to rely exclusively on numerical modelling?

- Fluids—what are the conditions necessary for the appearance of a separate fluid phase, what is the composition of that fluid and what are the physical and chemical consequences of the appearance of the fluid?

The variety and complexity of these problems in themselves accounts for much of the recovery from the depressed days of 1970. Even in the face of an unprecedented onslaught of careful observation, sophisticated conceptualisation, quasi-religious dogmatism and even occasional bombast these problems have not yet yielded wholly acceptable solutions.

4. *Projecting funding trends from the recent past.* The phenomenal scientific progress of the 'exceptional half-century' was fuelled by an unprecedented level of public support for science. Financial excesses, however, have caught up with governments in the highly developed world and their current preoccupation with debt reduction will mean less support for science generally, and considerably less support for basic research than for applied research, for at least as long as it takes to retire those debts and probably longer. In blunt terms, society wants more accountability for its financial expenditures, and it wants answers to its problems, not ours.

Thus, as seen against the dual backdrop of the general progress of science and the current financial constraints, it seems safe to predict that granite science, at least in the highly developed world, will change in the following ways.

- *From concrete-qualitative to abstract-quantitative,* in response to the general progress of science, and in response to the microcomputer revolution [e.g. from analogue experimental petrology to chemical thermodynamics; from heat flow measurements to thermal modelling (e.g. Zen 1992); from traditional metamorphic and structural geology to three-dimensional, thermally coupled, deformation-enhanced fluid-flow finite-element modelling!)].

- *From expensive to cheap, and from active to armchair,* in response to declining financial support, and manifest as a general shift from work based in the field→laboratory→office→home.

- *From reductionist to holistic,* in response to the complexity of questions we are addressing, and in recognition that reductionism only works well for a period of time, depending on the state of understanding of the problem and then new opportunities for progress may lie on the boundaries between the subdisciplines (e.g. from working on one-dimensional problems like chemical zoning in single mineral grains to integrating information from various vertical tiers in Nature's laboratory to assemble the big picture through space, time and composition—for example, a research team involving a structural geologist, a geochemist, an experimentalist, a mechanical engineer, a fluid mechanics engineer, a materials scientist and a computer scientist would be well positioned to bring some new thinking to the segregation and ascent debate).

- *From periodic communication to continuous communication,* in response to information technology (e.g. our Internet discussion)—although the novelty of an *unstructured* discussion on the Internet may have worn off, the network may still serve several important functions. Firstly, the network can serve as a large consulting 'firm' to deal with problems of any type that arise within the community of researchers. Secondly, as more researchers have access to World Wide Web browsers and the ability to make 'home pages', we should be able to hold, say, the poster presentations of Hutton Conferences electronically (a 'virtual' poster session) and in advance of face to face meetings, thereby saving those meetings exclusively for rock walk and shop talk. A logical extension of this idea is to use video-conferencing, also on the Internet. Thirdly, and most importantly, I suggest we use subsets of the granite-discussion network to collectively write some rules for such problems as geochemical modelling, ascent of granite magmas and the relation between granites and migmatites. We could have ongoing *structured* subcommissions on every major problem that concerns us. Those subcommissions could report their progress to the community periodically and informally on the Internet, and formally at future Hutton conferences. Also, as I have suggested elsewhere (Clarke 1992), granite science would benefit from a globally accessible, on-line, computer-based 'expert system' that would allow the integration of all types of observational and experimental data on granites. Such a system could become the repository of the granite database and the major tool for integration in the discipline. Its creation and maintenance would require an unprecedented level of co-operation within the discipline, as well as between the discipline and computing science.

In summary, then, we are now entering a period of relatively inexpensive, abstract, synthetic, numerical, theoretical, focused, integrative, collaborative, co-ordinated, reflective inquiry until we can afford to enter another phase of expensive data gathering in the field and in the laboratory, at least at a rate comparable with that of the last 50 years. (The alternative, of course, is always the unstructured, unco-ordinated, unfocused, etc. approach.) Groups of researchers interested in specific aspects of a problem could then define the problem, list all the variables, systematically consider all imaginable end-member outcomes, list the physical and chemical evidence and suggest ways to find solutions. The rest of the granite research community could then be educated and enlisted in a quest for the answers.

5. Conclusions

Granite is the final product of the high-temperature, magmatic, predominantly endogenic, chemical differentiation of the earth. The principal lesson we have to keep re-learning in granite science, however, is that many combinations of mechanical and chemical processes can have granite as their product, thus there is no *single* right answer to problems concerning their origin, ascent and emplacement. As individuals, we are prisoners of our own limited experience and we must be careful when generalising from that experience. We must also respect each other to make accurate observations and reasonable inferences from those observations. And, as we are beginning to focus on problems we cannot easily observe directly (such as segregation and ascent), we must become more experimental and numerical in our approach.

Hutton said, in reference to earth processes generally, 'The result, therefore, of our present inquiry is, that we find no vestige of a beginning, and no prospect of an end'. In reference to our understanding of granites, however, the time *will* come when the granite problem will be 'solved', or, more precisely, when enough answers are satisfactory and too few important anomalies remain. Then, according to the law of diminishing returns, we will make less progress and get less satisfaction for the same investment of time, effort and money. (We can already see this phenomenon to some extent in *one* aspect of experimental petrology—what new experiment could we design involving melting of source rocks or crystallisation of granitic melts that would significantly change the way we already think about granites? True, not every *possible* melting experiment has been done, but every known *type* of melting experiment has been done, and will one more experiment on melting a sandstone from Saskatchewan really change the way we think about the role of partial melting of sedimentary rocks in the formation of granitic magmas?) When that time arrives, as surely it must, researchers will see few opportunities for significant advance and will simply direct their energies elsewhere.

I conclude with two different perspectives on granite. First, Doris Lessing (1993), in her book *African laughter—four visits to Zimbabwe* describes the feelings of one of her characters in this passage: 'This man says that when he is away from Zimbabwe he feels exiled from granite. It is the oldest rock in the world, says he: it came bubbling up from the world's secret interior, slowly rising through layers of other rock to surface here. He can't live without granite.' Surely his sentiment is one that most of us share. Secondly, although this quotation is slightly out of context, I leave the last words to Hutton (1785: 30): 'In this manner, there is opened to our view a subject interesting to man (sic) who thinks; a subject on which to reason with relation the system of nature; and one which may afford the human mind both information and entertainment.' More than two centuries later, few of *us* at this Hutton conference would disagree!

6. Acknowledgements

I thank two of my colleagues at Dalhousie University, Richmond Campbell and Eric Mills, for significantly influencing the way I now think about science. I also want to acknowledge those whose input clarified and expanded the paradigm statement originally posted on the granite-research discussion network: Michael Brown, Phil Candela, Eric Christiansen, John Clemens, Ken Currie, David Greene, Jean-Paul Liegeois, David London, Jonathan Miller, Peter Nabelek, Scott Paterson, Bob Pankhurst, Nick Petford, Jim Roddick, Virginia Sisson, Paul Sylvester, Bob Trumbull, Ron Vernon and Roberto Weinberg. For their insightful formal reviews of this paper, I thank John Clemens and David London; for his supportive informal review, I am indebted to Wallace Pitcher. And finally, for creating the open environment in which unconventional ideas, such as those contained in this paper, are acceptable, I wish to thank the principal organisers of the Third Hutton Symposium: Michael Brown, Phil Candela and E-an Zen.

7. References

Brown, M. 1994. The generation, segregation ascent and emplacement of granite magma: the migmatite-to-crustally-derived granite connection in thickened orogens. EARTH SCI REV **36**, 83–130.

Brown, M. & Piccoli, P. M. 1995. The origin of granites and related rocks: Third Hutton Symposium abstracts. U.S. GEOL SURV CIRC **1129**, 170 pp.

Clarke, D. B. 1992. *Granitoid rocks.* London: Chapman & Hall.

Dean, D. R. 1992. *James Hutton and the history of geology.* Ithaca: Cornell University Press.

DePaolo, D. J. 1981. Trace element and isotopic effects of combined wallrock assimilation and fractional crystallization. EARTH PLANET SCI LETT **53**, 189–202.

Giere, R. N. 1984. *Understanding scientific reasoning,* 2nd edn. New York: Holt, Rinehart and Winston.

Guglielmo, G. 1993. Interference between pluton expansion and non-coaxial tectonic deformation: three-dimensional computer model and field implications. J STRUCT GEOL **15**, 593–608.

Guglielmo, G. 1994. Interference between pluton expansion and coaxial tectonic deformation: three-dimensional computer model and field implications. J STRUCT GEOL **16**, 237–52.

Holmes, A. 1945. Natural history of granite. NATURE **155**, 412–5.

Hoyningen-Huene, P. 1993. *Reconstructing scientific revolutions: Thomas Kuhn's philosophy of science.* Chicago: University of Chicago Press.

Hutton, J. 1785. System of the earth, its duration, and stability. TRANS R SOC EDINBURGH, abstract.

Hutton, J. 1795. *Theory of the earth with proof and illustrations, I and II.* Edinburgh: Creech.

Kuhn, T. 1970. *The structure of scientific revolutions,* 2nd edn. INT ENCYCL UNIFIED SCIENCE **2**, No. 2, 210 pp.

Lessing, D. 1993. *African laughter—four visits to Zimbabwe.* Harper Perennial.

Lyell, C. 1830–3. *Principles of geology: being an attempt to explain the former changes of the earth's surface, by reference to causes now in operation,* Vols 1–3. London: J. Murray.

Pitcher, W. S. 1987. Granites and yet more granites, forty years on. GEOL RUNDSCH **76**, 51–79.

Pitcher, W. S. 1993. *The nature and origin of granite.* Glasgow: Blackie Academic & Professional.

Popper, K. R. 1992. *The logic of scientific discovery.* New York: Routledge.

Read, H. H. 1948. Granites and granites. GEOL SOC AM MEM **28**, 1–19.

Zen, E-an 1992. Thermal modelling of stepwise anatexis in a thrust-thickened sialic crust. TRANS R SOC EDINBURGH EARTH SCI **79**, 223–35.

D. BARRIE CLARKE, Department of Earth Sciences, Dalhousie University, Halifax, Nova Scotia, Canada B3H 3J5. E-mail: clarke@is.dal.ca

INDEX – Authors of papers